2nd International Conference on Advanced Materials for Better Future (ICAMBF 2017)

IOP Conference Series: Materials Science and Engineering
Volume 333

Surakarta, Indonesia
4 – 5 September 2017

ISBN: 978-1-5108-6113-8
ISSN: 1757-8981

Printed from e-media with permission by:

Curran Associates, Inc.
57 Morehouse Lane
Red Hook, NY 12571

Some format issues inherent in the e-media version may also appear in this print version.

Printed by Curran Associates, Inc. (2018)

For permission requests, please contact the Institute of Physics
at the address below.

Institute of Physics
Dirac House, Temple Back
Bristol BS1 6BE UK

Phone: 44 1 17 929 7481
Fax: 44 1 17 920 0979

techtracking@iop.org

Additional copies of this publication are available from:

Curran Associates, Inc.
57 Morehouse Lane
Red Hook, NY 12571 USA
Phone: 845-758-0400
Fax: 845-758-2633
Email: curran@proceedings.com
Web: www.proceedings.com

Table of contents

Volume 333

International Conference on Advanced Materials for Better Future 2017

4–5 September 2017, Surakarta, Indonesia

Accepted papers received: 9 March 2018
Published online: 6 April 2018

Preface

Papers

Material Chemistry and Physics (MCP)

Metal & Alloy Material/Metallurgy (MA)

Functional Materials (FM)

Polymer & Composite (PC)

Civil & Construction Materials (CM)

Miscellaneous (MS)

IOP Conf. Series: Materials Science and Engineering **333** (2018) 011001 doi:10.1088/1757-899X/333/1/011001

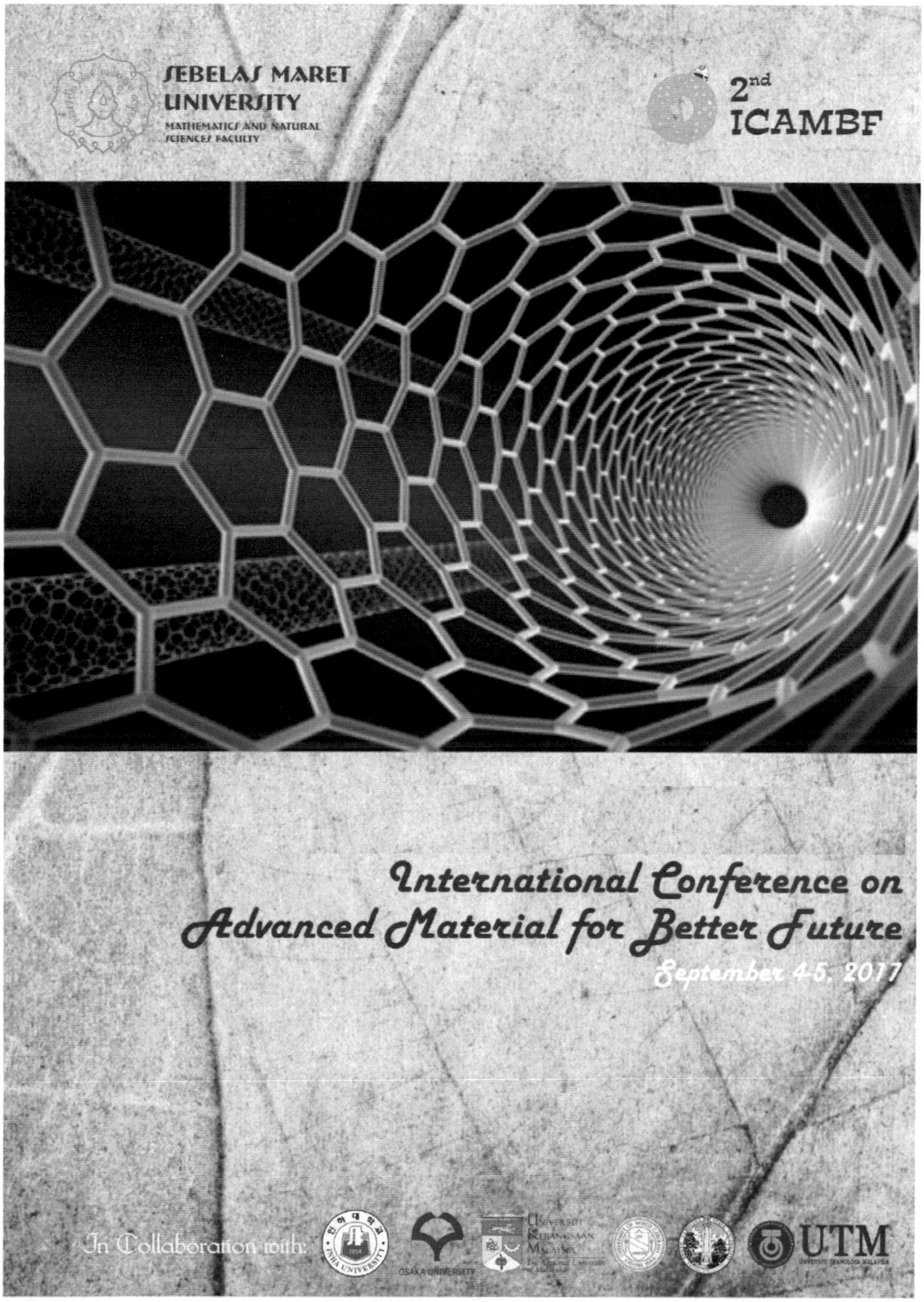

IOP Conf. Series: Materials Science and Engineering **333** (2018) 011001 doi:10.1088/1757-899X/333/1/011001

PREFACE

2nd International Conference on Advanced Material for Better Future 2017 (ICAMBF2017)

The 2nd International Conference on Advanced Material for Better Future 2017 (ICAMBF2017) is an international scientific forum that was initiated to overcome material problems as well as factors affecting them that have emerged as the main issue and challenge attracting more attention from various scientists, government, and regulatory bodies worldwide. The ICAMBF 2017 was attended by leading scientists from countries near and abroad who work in the field of material.

The ICAMBF 2017 aims to provide a forum for scientists, researchers, and engineers within the nation and from around the world to present, discuss, share, and exchange their knowledge, experiences, and research finding on all aspects of material. Thus, this conference can give an excellent overview in all scope of fields including Metal & Alloy Material/ Metallurgy, Functional Materials, Biomedical Materials/Biomaterial Technology, Material Chemistry and Physics, Polymer & Composites, Civil & Construction Materials, and Energy Materials/Nanomaterials.

We are pleased to our outstanding keynote speakers Prof. Roger Narayan, Prof. Dr. Santiago Gómez-Ruiz, Prof. Takuji Ogawa, Prof. Seung Bok Choi, Prof. Saiful Amri Mazlan, Prof. M. Mat Salleh and other invited speakers for sharing the best knowledge in Advanced Material.

We also thanks to all researchers and student who participate in conference by submitting their papers to convey the current science discussed at ICAMBF 2017, all of the organizing committee wish you get awesome conference experience, and reviewers for their valuable revision of these papers.

Prof. Ir. Ari Handono Ramelan, M.Sc (Hons), Ph.D

Chairman of ICAMBF 2017

Dean of Faculty Mathematics & Natural Sciences

IOP Conf. Series: Materials Science and Engineering **333** (2018) 011002 doi:10.1088/1757-899X/333/1/011002

Peer review statement

All papers published in this volume of *IOP Conference Series: Materials Science and Engineering* have been peer reviewed through processes administered by the proceedings Editors. Reviews were conducted by expert referees to the professional and scientific standards expected of a proceedings journal published by IOP Publishing.

IOP Conf. Series: Materials Science and Engineering **333** (2018) 012001 doi:10.1088/1757-899X/333/1/012001

Rheological properties of plate-like shape carbonyl iron particles compositions based magnetorheological grease in oscillatory mode

N Mohamad[1], S A Mazlan[1], Ubaidillah[2] and F Imaduddin[3]
[1] VSE Research Laboratory, Malaysia-Japan International Institute of Technology, Universiti Teknologi Malaysia, Jalan Sultan Yahya Petra (Jalan Semarak), 54000 Kuala Lumpur, Malaysia
[2] Mechanical Engineering Department, Faculty of Engineering, Universitas Sebelas Maret, Jl. Ir. Sutami 36A, Kentingan, Surakarta, 57126, Central Java, Surakarta, Indonesia
[3] Faculty of Engineering and Technology, Multimedia University, Jalan Ayer Keroh Lama, Bukit Beruang, 75460 Melaka, Malaysia

E-mail: amri.kl@utm.my

Abstract. This paper investigates dynamic viscoelastic properties of magnetorheological (MR) grease with different fractions of plate-like Carbonyl Iron (CI) particles (MRGP). The plate-like CI particles were prepared through milling process. Subsequently, three types of MRGP was developed by dispersing 30 to 70 wt% of plate-like CI particles into grease using mechanical stirrer. The rheological test was performed using oscillatory shear rheometer under flux density influence of 0 to 0.7T and strain amplitude ranging from 0.001 to 10% at constant temperature of 25°C. The experimental results show that the storage and loss moduli of the MRGP is dependent on the fractions of the plate-like CI particles. The enhancement in the MR effect in the plate-like CI particles based MR grease is observed due to the anisotropy shape which leads to the strong inter-particle configuration and chain formations.

1 Introduction

Magnetorheological (MR) fluid consists of dispersion of magnetic particles in a carrier fluid [1]. It can be considered as a smart material that have controllable rheological properties on the variation of external magnetic field [2]. MR fluid has unique characteristics that are reversible and tunable from liquid-like to solid-like state [3]. MR fluid is widely investigated in many industries for the application of shock absorbers, brakes, dampers, valves and torque transfer devices [4–6]. Therefore, many researchers have studied the MR fluid due to its potential to be applied in many fields. The rheological properties of the MR fluid are mainly influenced by the features of the magnetic particles such as size, shape and saturation magnetization. There are some researchers investigated about the plate-like of magnetic particles in MR fluid. For example, Shah et al. [7] had studied the effect of different size of plate-like magnetic particles in MR fluid. The result showed that the rheological properties of MR fluid was affected by the size of magnetic particles. The larger diameter of plate-like magnetic particles with 19 μm exhibited high yield stress compared to the smaller diameter of plate-like Carbonyl Iron (CI) particles which is 2 μm. In addition, study done by Upadhay et al. [8] stated that the plate-like MR fluid has a higher storage modulus of 6 MPa compared to sphere-based MR fluid. This is due to the field-dependent solid friction between the CI particles. The plate-like magnetic particles can be easily magnetize even though at low magnetic field due to the long axis aligned with the external field that contributed for the formation of strong structures [9]. However, the sedimentation rate in MR fluid can only be reduced up to 20% by using plate-like CI particles [10]. Therefore, the researchers have come up with MR grease that has ability to eliminate the sedimentation problem which occur in MR fluid.

MR greases are identical to the MR fluid that it can be stimulated by the magnetic field [11]. The difference between the MR fluid and MR grease is the viscosity of the carrier fluids [11]. The MR grease is able to suspend the magnetic particles in equilibrium in the grease medium [12]. Recently, MR grease had received a great attention from researchers by investigating the rheological and characteristics of MR grease under various magnetic field strengths. Sahin et al [11] studied the rheological behavior of MR grease and found that the shear stress of MR grease increases with the increment of magnetic field, up to 50 kPa at 0.53T. Another study conducted by Park et al. [13] reported that the maximum dynamic yield stress of the MR grease under influence of magnetic field was 10 kPa. The study was continued by investigating the viscoelastic properties of MR grease under different weight percentages of magnetic particles and magnetic field strengths by Mohamad et al. [14]. The result showed that the MR grease with 70 wt% of magnetic particles exhibited the highest viscosity which was up to 120 kPa with the increment of applied current, yet decreases as the shear rate increases. The viscoplastic properties of MR grease can be described by referring to its yield stress [15]. This MR grease exhibited a maximum yield stress for a similar amount of applied current of 5A which was 52.7 kPa. Besides that, it was found that MR grease has a certain value of yield stress although in the absence of magnetic field due to the intrinsic grease property [13].

Another researcher [16] found that the hexagon shape of magnetic particles in MR grease has a higher shear stress compared to spherical shape of magnetic particles through simulation study. The shear stress of the hexagonal shape particles increased from 7.4 to 21.7 kPa whereas for the spherical shape particles, the shear stress increased from 6.2 to 16.6 kPa. This phenomenon was due to the contact area between the particles. Although many researchers have investigated the synthesis and characterization of MR grease, so far, there are no investigation and experimental work have been done related to MR grease with plate-like CI particles. Therefore, in this paper, the investigation about the effect of different fractions of plate-like shape CI particles on the dynamic viscoelastic properties of MR grease is presented.

2 Experimental

2.1 Materials

Soft magnetic CI particles in spherical shape with OM series purchased from BASF Germany were used to produce plate-like CI particles. The average size and density of the spherical CI particles were 5 μm and 7.874 g/cm^3, respectively. The plate-like CI particles were attained from milling process by utilizing rotary ball mill (QM-5 model from Tencan Company). The details of procedure is according to the method described by Shilan et al. [10]. In this work, the duration of milling is 40 hours. Three types of MR grease (will refer as MRGP after this) with different ratios (30, 50 and 70 wt%) were prepared between plate-like CI particles and commercial grease (NPC Highrex HD-3 Grease, Nippon Koyu Ltd, Japan). Firstly, commercial grease was stirred for five minutes using mechanical stirrer. Subsequently, the plate-like CI particles were added and stirred another two hours at room temperature until the suspension is homogeneous.

2.2 Characterization

MRG-P samples were characterized using commercial rheometer (Anton Paar, Physica, MCR 302) equipped with electromagnetic apparatus (MRD 70/1T) under rotational and oscillatory mode. A 20-mm diameter parallel plate measuring system at a gap of 1-mm was used. The storage modulus and loss modulus of MRGP samples were measured in dynamic oscillation test with strain amplitude ranging from 0.001 to 10%. The applied magnetic field is varied from 0 to 0.7T by adjusting the coil applied current from 0 to 5A at room temperature.

3 Results and Discussions

Figure 1 shows storage modulus (G') and loss modulus (G") as function of strain amplitude under influence of different fractions of plate-like CI particles at applied current 3A. It was found that storage and loss moduli of MRGP is dependent on the fractions of the plate-like CI particles. The storage and loss moduli of MRGP increased by increment of the plate-like CI particles fractions at constant applied magnetic field strength. This outcome is expected to be occurred in MRGP due to the larger surface contact areas which leads to the strong attraction between plate-like CI particles [10,17]. In addition, strong attraction between CI particles is rearranged in line to form chain structures; parallel to the applied magnetic field. The chain formations become strong and thick as the applied magnetic field increases. As results, the separation gap between particles has decreased which resulted in hard chain structures formation and indirectly increase the suspension storage modulus [18].

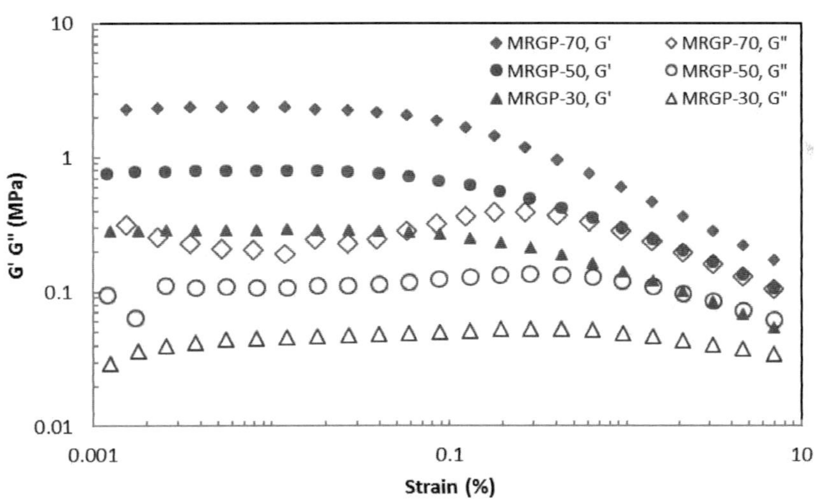

Figure 1 Storage and loss modulus as a function of strain amplitude under influence of fractions of plate-like CI particles at constant (3A) applied current

The effect of strain amplitude towards storage modulus of MRGP under variation of applied current is depicted in Figure 2. The storage modulus of MRGP-50 increase by increasing current applied. This pattern is also observed for samples MRGP-30 and MRGP-70. The increment of storage modulus is an indication of strong chain structures formation in the suspension [19]. By increasing the strain amplitude, the chain structures initiate to break and slowly starts to flow when exceed the linear viscoelastic (LVE) region [8]. From Figure 2, the LVE region of MRGP-50 is determined at strain amplitude 0.1%. In addition, it has been noticed that the increment of applied magnetic field strength has almost insignificant effect on the LVE range of the MRGP, while the storage modulus is promptly changed with the presence of magnetic field strength. Besides, this phenomenon is closely related to the dipole-dipole interactions which occurs in the shape anisotropy due to the stimulated magnetic field in which irregular shape particles can be only attained. The shape anisotropy can produce one or more axis along the long axis of plate-like shape particles which depends on the favorable direction of magnetization. Otherwise, the demagnetization field for shape anisotropy is more complex due to the three-dimensional inhomogeneity of magnetization [20]. Therefore, a greater value of retentivity is stimulated for the plate-like shape particles due to their easy magnetization axis and smaller demagnetization factor in their long direction [20,21].

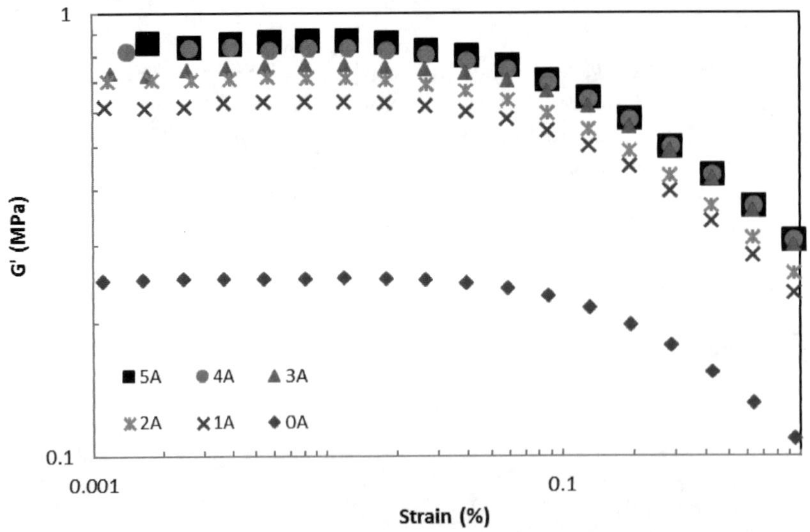

Figure 2 Storage modulus as a function of strain amplitude under various applied current for MRGP-50

Consequently, the relative MR effect of MRGP for all samples were calculated from storage modulus were shown in Table 1. As mentioned earlier, the shape anisotropy that are in the plate-like CI particles responsible to the inter-particles configuration of chain formation and indirectly promotes the higher relative MR effect [21].

Table 1 The zero-field modulus and MR effect of MRGP with different plate-like CI particles fractions

Samples	Zero-field modulus, MPa	Absolute MR effect, MPa	Relative MR effect, %
MRGP-70	0.149	1.05	708.03
MRGP-50	0.160	0.232	144.59
MRGP-30	0.129	0.064	49.54

4 Conclusion

Three types of MRGP with 30 to 70 wt% of plate-like CI particles have been synthesized via mixing process using mechanical stirrer. The plate-like particles itself are prepared from the spherical particles which have undergone milling process. The rheometer test under oscillatory mode has been conducted to measure the rheological properties of the samples. The experimental results have shown a significant increment of the storage modulus which is parallel to the increment of plate-like CI particles fractions and magnetic field strength. Meanwhile, the LVE of the MRGP which is determined at 0.1% strain has insignificant effect towards increment of magnetic field strength. In addition, the MR effect of 708.03% is achieved by utilizing 70 wt% of plate-like CI particles which proves that the anisotropy shape, which obtained in the plate-like CI particles, has improved the interactions between inter-particles. The anisotropy shape also indirectly improves the particles configurations during on-state condition. Finally, it can be concluded that the rheological characteristics of MRGP is showing advantage for practical applications which require high MR effect.

Acknowledgement
This research was financially supported by the Ministry of Education Malaysia and Universiti Teknologi Malaysia under PRGS Grant (Vote No: 4L667) and Tier 1 Grant (Vote No: 13H55). Authors also acknowledge Hibah Kolaborasi Internasional 2017 by Universitas Sebelas Maret.

References
[1] Bossis G, Lacis S, Meunier A and Volkova O 2002 *J. Magn. Magn. Mater.* **252** 224–8
[2] Ashtiani M, Hashemabadi S H and Ghaffari A 2015 *J. Magn. Magn. Mater.* **374** 716–30
[3] de Vicente J, Klingenberg D J and Hidalgo-Alvarez R 2011 *Soft Matter* **7** 3701
[4] Ichwan B, Mazlan S A, Imaduddin F, Ubaidillah, Koga T and Idris M H 2016 *Smart Mater. Struct.* **25** 37001
[5] Ubaidillah, Imaduddin F, Nizam M and Mazlan S A 2015 *Int. J. Electr. Eng. Informatics* **7** 308–22
[6] Pokaad A Z, Hudha K, Nasir M Z B M and Ubaidillah 2011 *Int. J. Struct. Eng.* **2** 164
[7] Shah K, Oh J S, Choi S B and Upadhyay R V. 2013 *J. Appl. Phys.* **114** 213904
[8] Upadhyay R V, Laherisheth Z and Shah K 2014 *Smart Mater. Struct.* **23** 15002
[9] Shah K, Phu D X, Seong M S, Upadhyay R V and Choi S B 2014 *Smart Mater. Struct.* **23** 27001
[10] Shilan S T, Mazlan S A, Ido Y, Hajalilou A, Jeyadevan B, Choi S and Yunus N A 2016 *Smart Mater. Struct.* **25** 95025
[11] Sahin H, Gordaninejad F, Wang X and Fuchs A 2007 *Proceedings of SPIE* vol **6525**, ed Y Matsuzaki, M Ahmadian and D J Leo p 65250D
[12] Kintz K and Carlson J 2003 *US Pat. 6,547,986* 1
[13] Park B O, Park B J, Hato M J and Choi H J 2011 *Colloid Polym. Sci.* **289** 381–6
[14] Mohamad N, Mazlan S A, Ubaidillah, Choi S and Nordin M F M 2016 *Smart Mater. Struct.* **25** 95043
[15] Sahin H, Wang X and Gordaninejad F 2009 *Proceedings of SPIE* vol **7288**, ed M Ahmadian and M N Ghasemi-Nejhad p 72881E–72881E–9
[16] Wei J, He G T, Song L, Yan M, Li M, Xu Z Y, Ran Y C and Zhang D S 2012 *Appl. Mech. Mater.* **246–247**
[17] de Vicente J, Segovia-Gutírrez J P, Andablo-Reyes E, Vereda F and Hidalgo-Lvarez R 2009 *J. Chem. Phys.* **131**
[18] Siebert E, Laherisheth Z and Upadhyay R V 2015 *Smart Mater. Struct.* **24** 75011
[19] Yang P, Yu M, Fu J and Luo H 2017 *J. Intell. Mater. Syst. Struct.* 1045389X1769205
[20] Chekanova L A, Denisova E A, Yaroslavtsev R N, Komogortsev S V, Velikanov D A, Zhizhaev A M and Iskhakov R S 2015 *Solid State Phenom.* **233–234**
[21] de Vicente J, Vereda F, Segovia-Gutiérrez J P, del Puerto Morales M and Hidalgo-Álvarez R 2010 *J. Rheol. (N. Y. N. Y).* **54** 1337

IOP Conf. Series: Materials Science and Engineering **333** (2018) 012002 doi:10.1088/1757-899X/333/1/012002

Copper-Graphene Composite: Electrochemical Synthesis and Structural Characterization

D Ni'maturrohmah[1], D Maharani[1,4], O Ruzicka[2,4], U H Gitasari[2], E Adhitama[3,4], T E Saraswati[1*]

[1]Chemistry Department, Faculty of Mathematic and Natural Science, Sebelas Maret University, Surakarta 57126, Indonesia.

[2]Chemical Engineering Department, Faculty of Engineering, Sebelas Maret University, Surakarta 57126, Indonesia.

[3]Physic Education Department ,Faculty of Teacher Training and Education, Sebelas Maret University, Surakarta 57126, Indonesia.

[4]Research and Technology Department, Studi Ilmiah Mahasiswa, Sebelas Maret University, Surakarta 57126, Indonesia.

*E-mail: teguh@mipa.uns.ac.id

Abstract. Graphene utilization as multifunctional conductive material still has been obstacled in the scalling up process. Some methods, such as mechanical exfoliation only results few products, otherwise Chemical Vapor Deposition (CVD) needs high cost and high temperature. Synthesis of copper modified graphene by electrochemical with low cost has been studied in this research. Cu-Graphene composite was characterized using X-ray diffraction (XRD), Scanning Electron Microscopy (SEM). XRD analysis indicates conformity with simulated pattern peaks 2θ at 26.73°; 31.9° and 45.8°, that shows presence of graphene and Cu. In addition, SEM analysis shows that the Cu-Graphene morphology was in exfoliated flake structure. The electrochemical method performed here is expected can be used for scaling up the Cu-graphene production.

1. Introduction

Recently graphene material has attracted due to it's high charge mobility carry, up to 200,000 cm^2 Vs. In Indonesia, graphene synthesis has not been developed. The most common method of graphene synthesis are mechanical exfoliation by peeling mechanical on graphite and Chemical Vapor Deposition (CVD). However, the mechanical exfoliation method only produces a small amount of graphene [1]. In addition, the CVD method requires high temperatures, inert systems and relatively high cost [2]. An alternative method of graphite exfoliation to form graphene under development is the chemical process [3]. Chemically reduction of graphene oxide can produce large single-layered graphene with relatively low cost [4]. To improve it's conductivity, graphene can be combined with transition metals such as Ti, Cu, Zn, Sc, and Fe. Several methods of combining metals with graphene over reported are hydrothermal [5], microwave-assisted [6], self-assembly [7] and electrolysis [8]. Cu metal is selected as particles embedded in graphene because of it's considerable abundance in nature, low cost and it has better electrical and thermal conductivity. Cu-Graphene composites would be suitable for electrofriction materials with low coefficient of friction and can be used on electric motors and generators [9]. In this research, graphene synthesis using sonication method continued electrolysis to produce copper-graphene composite. Copper-graphene composite have higher electrical and thermal conductivity than copper for various applications [2].

2. Experimental

2.1 Materials and equipment

Graphite as raw material, certain amounts of aquades, NaCl 98% (Sigma Aldrich), $KMnO_4$ 98% (Sigma Aldrich), H_2SO_4 98% (Sigma Aldrich), and H_2O_2 30% (Sigma Aldrich), carbon rod and copper plate 99% as electrodes. The equipments are pH meter, power supply, ultrasonic cleaner, oven MEMMERT (type 67 UN30), *Hot plate* Termo Scientific CIMAREX.

2.2 Graphene oxide synthesis

Graphene oxide synthesis was prepared using Hummer and sonication method based on the reported procedures [5]. 2 g of graphite was dissolved in 46 mL H_2SO_4 98% then was stirred for 30 minutes in the ice bath. 6 g $KMnO_4$ was added to the mixture then was stirred for 1 hour. The mixture was continuously stirred at 35 °C for 20 hours. 92 mL of deionized H_2O was slowly added to the mixture and strried 2 hour then 280 mL of deionized H_2O was diluted for 15 minutes. 20 mL H_2O_2 was added to the dilute mixture then stirred for 15 minutes. The precipitate was separated and washed by deionized H_2O until it's pH was neutral then it was dried at 60 °C for 12 hours to obtaine graphite oxide. 1 g of graphite oxide was diluted into 100 mL deionized H_2O then it was sonicated for 1 hour to obtaine graphene oxide.

2.3 Cu-Graphene composite synthesis

Due to the lack of previous study [10] that reduction of graphene oxide using hydrogen with inert system was not simple and highly cost, we provide a simple electrochemical method of graphene reduction. 0.5 g of graphene oxide and 0.1 M of NaCl were diluted into 25 mL of deionized H_2O, stirred to get homogenous mixture. Electrolysis was carried out in a electrochemical cell under 3 and 5 Volt for 30 minutes at room temperature using carbon rod as cathode and copper plate as anode. Finally the yellow-black precipitate was dried at 100 °C for 1 hour.

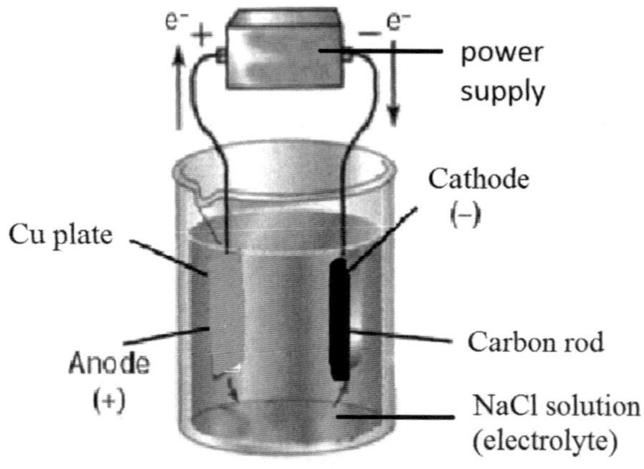

Figure 1. The schematic of electrolysis method for Cu-graphene synthesis

IOP Conf. Series: Materials Science and Engineering **333** (2018) 012002 doi:10.1088/1757-899X/333/1/012002

2.4 Characterization

The X-ray powder diffraction data were collected on Rigaku Miniflex 600 Benchop. The morphology of the final materials were recorded using scanning electron microscopy (SEM) type: inspect S50-FEI.

3. Results and Discussion

Graphene was prepared through graphite oxidation to form graphite oxide. $KMnO_4$ and H_2SO_4 were used as oxidation agent. In order to disperse multilayer structure of graphite, graphite oxide was sonicated to form graphene oxide. After that, graphene oxide was reduced through electrochemical synthesis to form graphene. Conterminous with Cu plate used as anode was oxidized and deposited at graphene surface forming Cu-graphene. Illustration of graphite change to be Cu-graphene was shown in Figure 2.

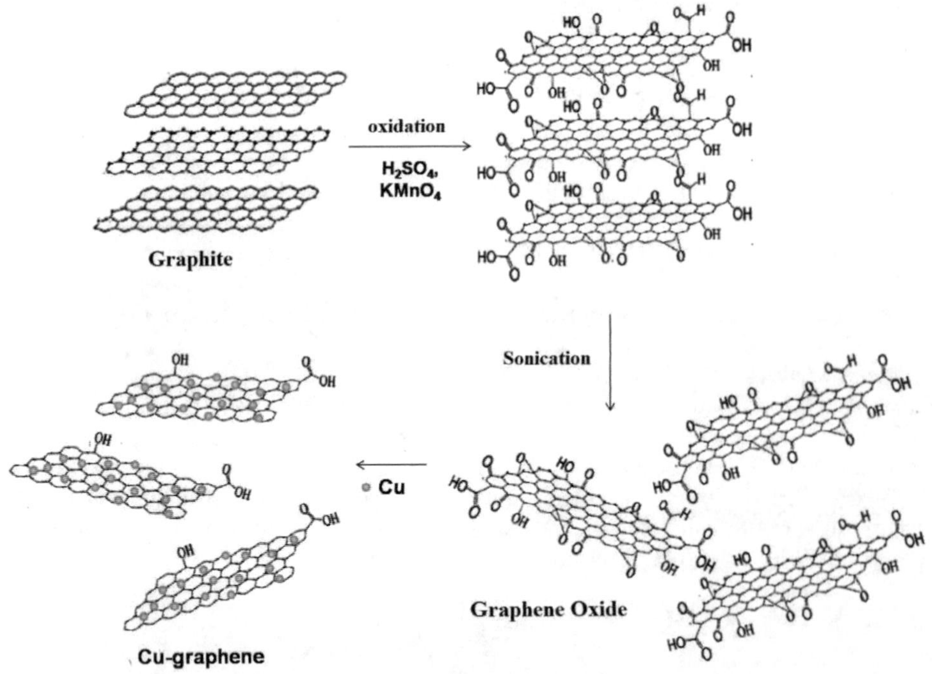

Figure 2. Illustration of Synthesis route of Cu-graphene

The diffraction pattern of the obtained materials is confirmed through similarities with the standart pattern of graphene and Cu which is generated from ICSD Number. 040290. Diffragtogram of Cu-Graphene is shown at Figure 3. The peak detected at 2θ: 28° indicating the presence of graphene and two peaks at 2θ 31.8° dan 45.83° may be assigned that Cu metal was successfully deposited into active sites to form Cu-Graphene composite

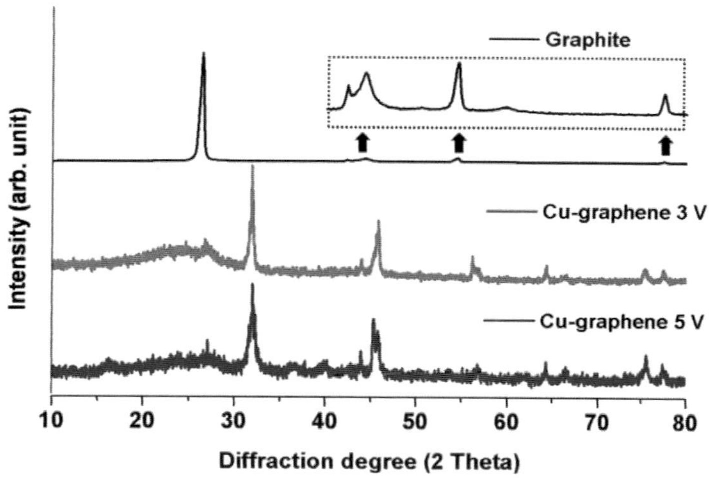

Figure 3. Diffractogram of graphite, Cu-Graphene synthesized by electrolysis method using 3 and 5 volt

According to the SEM analysis (Figure 3), morphology of graphite oxide is a flake structure. After being reduced to form graphene, the shape was changed to be exfoliated layer of Cu-graphene with increase pore size distribution.

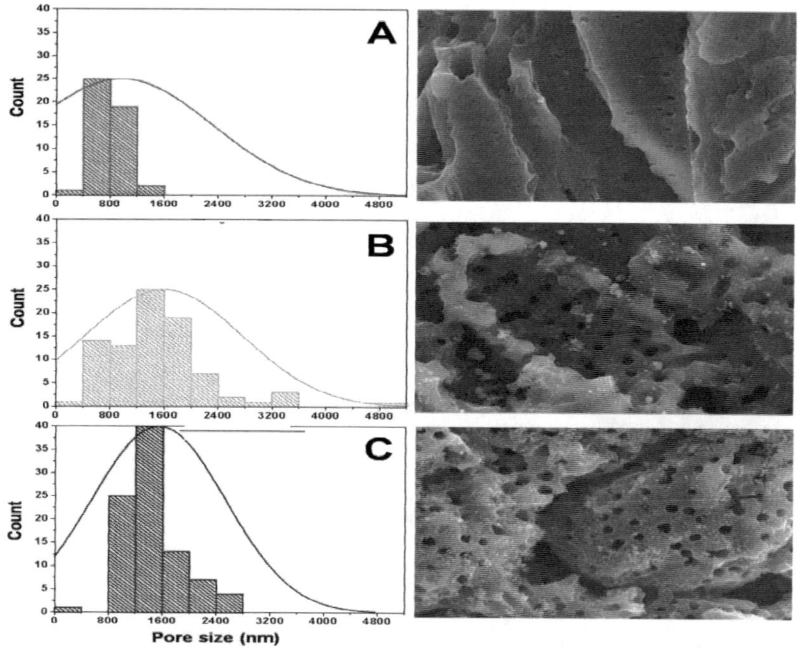

Figure 4. Pore size histogram of graphite (A) and Cu-graphene under electrolysis using 3 volt (B) and 5 volt (C) and their SEM images.

4. Conclusion

In this present study, we synthesized Cu-Graphene via sonication and electrolysis method. The result indicates that the presence of graphene and copper on the obtained materials. Moreover, the graphene morphology is exfoliated structure with increase pore size distribution. Due to low cost and simplicity, electrolysis method can be applided for upscale graphene.

References

[1]. Artur C and Paolo S 2014 *Chem. Soc. Rev.* 43 381-398
[2]. Stephan H, Felicitas S, Rolf C, Christof W and Roland A F 2005. *J. Am. Chem. Soc.* **127** 13744-13745
[3]. Novoselov, K.S., Fal V I, Colombo L, Gellert P R, Schwab M G and Kim K 2012 *Nature* **490** 192-200.
[4]. Ji C, Bowen Y, Chun L and Gaoquan S 2013 *Carbon* **64** 225-229
[5]. Lita N A and Diah S 2014 *Jurnal Teknik ITS 3* 209-214
[6]. Na L, Ying X, Changwen H and Minhua C 2013 *Chemistry–An Asian Journal* **8** 1960-1965
[7]. Chao C, Weimin C, Mingce L, Baoxue Z, Yahui W, Deyong W and Yujie F 2010 *Acs Nano* **4** 6425-6432
[8]. Kasichainula K 2012 *Journal of Vacuum Science & Technology B, Nanotechnology and Microelectronics: Materials, Processing, Measurement, and Phenomena* **30**
[9]. Ferry I and Yahdi B R 2015 *Advanced Materials Research* **1112** 290-293 Trans Tech Publications.
[10]. Jagannadham, K 2012 *Journal of Vacuum Science & Technology B, Nanotechnology and Microelectronics: Materials, Processing, Measurement, and Phenomena* **30**(3)

IOP Conf. Series: Materials Science and Engineering **333** (2018) 012003 doi:10.1088/1757-899X/333/1/012003

Synthesis and Characterization of A Coordination Complex of *Tetrakis(*diphenylamine)copper(II) Sulfate Hexahydrate

H Syaima[1], S B Rahardjo[2], and E Suciningrum[2]
[1]Chemistry Graduate Program, Graduate School, Universitas Sebelas Maret, Surakarta, 57126, Indonesia
[2]Department of Chemistry, Faculty of Mathematics and Natural Sciences, Universitas Sebelas Maret, Surakarta, 57126, Indonesia

E-mail: husna_syaima@yahoo.com

Abstract. $CuSO_4{\cdot}5H_2O$ with diphenylamine formed a complex compound in 1:4 mole ratio of metal to the ligand in methanol. The forming of the complex was indicated by shifting of UV-Vis spectra of $CuSO_4{\cdot}5H_2O$ and the complex from 819 nm to 593 nm. The result of analysis Cu(II) in the complex showed the copper content in the complex was 6.43 % therefore the empirical formula of the complex was Cu(diphenylamine)$_4$SO$_4$(H$_2$O)$_6$. The electrical conductivity of complex showed the charge ratio of cation and anion = 1:1. Therefore, the proposed formula of the complex was [Cu(diphenylamine)$_4$]SO$_4{\cdot}$6H$_2$O. Based on infrared spectra, it was determined that the functional group of N-H of diphenylamine was coordinated to the center ion Cu^{2+}.The electronic spectral study of the complex showed a transition peak on λ = 593 nm (υ = 16863 cm^{-1}) corresponding to the $^2B_{1g} \rightarrow {}^2A_{1g}$ transition. The complex was paramagnetic with effective magnetic moment 1.72 B.M. It was indicated square planar geometry around Cu(II).

1. Introduction

The coordination chemistry of copper plays an important role in many fields. Synthesized copper complex has been reported to have a potentiality in catalytic system [1-2], solar cell as dye sensitizer [3-4], antibacterial [5-6], antifungal [7-9], anticancer [10-11], and etc. Copper complexes are able to form many geometries with different types and numbers of ligands. It is reported that copper complex forming square planar [12-14], octahedral [15-16], tetrahedron [17-18], trigonal planar [18-19], and square pyramidal [13],[20] geometry. Although a large number of complexes have been obtained, the synthesis and design of novel copper complexes by varying the nature of the reactants and synthetic conditions are still under investigation.

Diphenylamine is an organic compound with the formula $(C_6H_5)_2NH$, colorless solid but often yellow due to oxidized impurities [21]. Diphenylamine and its derivatives are commonly used as stabilizers in nitrocellulose-containing explosives and propellants, in the perfumery, and as antioxidants in the rubber and elastomer industry, also widely used for the production of dyes, pharmaceuticals, photography chemicals and further smallscale application [22]. Diphenylamine is a monodentate ligand as well, having one donating site with a lone pair of electron on the nitrogen atom [21]. Besides, our attention is focused on ligands containing N-H group during last years . The ligands have been interesting compounds because of their coordination chemistry and ability to be an antibacterial agent.

In this work, we report the new complex of copper with diphenylamine ligand. The Cu(II) complex has been characterized through Atomic Absorption Spectroscopy (AAS), molar conductivity, UV-Visible, IR, and magnetic susceptibility measurements.

2. Experimental

2.1. Materials

All chemicals and solvents were of analytical grade and used as received without further purification. The metal salts ($CuCl_2{\cdot}2H_2O$, $CuSO_4{\cdot}5H_2O$, $NiCl_2{\cdot}6H_2O$, $NiSO_4{\cdot}6H_2O$, $AlCl_3{\cdot}6H_2O$) and diphenylamine were purchased from Merck.

2.2. Physical Measurements

The copper content was determined by Atomic Absorption Spectrometer (AAS) Shimadzu AA-6650. Infrared spectra were recorded on Prestige-21 Shimadzu spectrophotometers as KBr pellets in the frequency range of 4000–450 cm^{-1}. Molar conductivity (Λ*m) of 1 mM solution in methanol was measured on Jenway CE 4071 conductivity meter at 25 °C. The magnetic moment was measured using Auto Sherwood Scientific 10169 Magnetic Susceptibility Balance. Spectra UV-VIS of metal complex and ligand was recorded in methanol solvent using Shimadzu UV-3601 spectrophotometer.

2.3. Synthesis of Cu(II) complex

The complex has been prepared by reaction of $CuSO_4{\cdot}5H_2O$ and diphenylamine in methanol. Methanolic solution (10 mL) of $CuSO_4{\cdot}5H_2O$ (0.748 g; 3 mmol) was added dropwise to methanolic solution (20 mL) of diphenylamine (2.031 g; 12 mmol) with constant stirring for 6 hours. The solution was then concentrated to 15 mL of volume and left to stand for 48 hours. The precipitate was filtered with filter paper, then dried in a desiccator for 24 hours.

3. Result and Discussion

3.1. Electronic spectra

Figure 1 shows the shift of $CuSO_4{\cdot}5H_2O$ maximum absorption wavelength (819 nm) towards smaller wavelengths in the complex Cu(II)-diphenylamine (593 nm). Shifting λ_{max} indicates the formation of the complex due to the substitutions of H_2O molecules with diphenylamine. It was estimated that diphenylamine acting as a stronger ligand than H_2O. Electronic spectra of the complex Cu(II)-diphenylamine show a wide absorption at λ_{max} 593 nm (16863 cm^{-1}) assigned to the $^2B_{1g} \rightarrow {}^2A_{1g}$ transition. It was reported that the maximum wavelength of copper complexes in 613 [23], 586 [24], or 579 [24] nm attributed to square planar geometry.

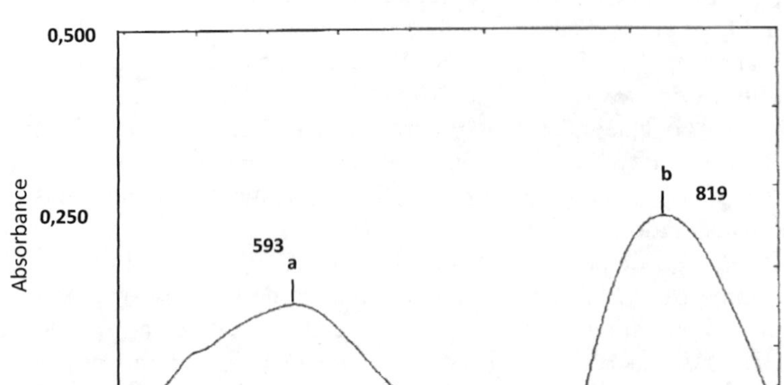

Figure 1. Electronic spectra of $CuSO_4{\cdot}5H_2O$ solution (a) and the complex solution (b)

3.2. Analysis of the Amount of Copper with Atomic Absorption Spectroscopy

The results of measurements of the copper content in complex $CuSO_4{\cdot}5H_2O$ with diphenylamine was 6.43 ± 0.16%. If the results of these measurements as compared with the levels of copper are theoretically on the various possibilities of complex formulas, then the formula can be approximated complex $CuSO_4{\cdot}5H_2O$ with diphenylamine which most likely is Cu(diphenylamine)$_4$$SO_4$$^{\cdot}$($H_2O$)$_6$. It needs more methods to confirm the number of presence of water molecules, such as thermogravimetry and differential thermal analysis.

3.3. Analysis of Electrical Conductivity

The molar conductivity of the complex was determined in methanol (10^{-3} M) at room temperature. The molar conductance values of the complex lie in 13.39 ± 0.10 S.cm^2.mol^{-1}, thus indicating that is an electrolyte. By comparing the conductivity values of the complex sample with a molar conductivity of the standard solution, it is seen that the charge ratio of cationic: anions= 1:1. This shows that the ion SO_4^{2-} is not coordinated to the central metal ion Cu(II) and only acts as an anion [25]. The complex dissociates in solution to be [Cu(diphenylamine)$_4$]$^{2+}$ and SO_4^{2-}. Thus, it strengthens the proposed formula of the complex is probably [Cu(diphenylamine)$_4$]$SO_4{\cdot}6H_2O$ namely *Tetrakis*diphenylaminecopper(II) sulfate hexahydrate.

3.4. Infrared Analysis

In the infrared spectra, shown in Figure 2 and Figure 3, the secondary N-H vibration of the ligand (3406 and 3383 cm^{-1}) was shifted to lower frequencies (3404 dan 3381 cm^{-1}) upon complex formation revealing coordination of the amine nitrogen atom. The aryl-NH vibration of the ligand (1319 cm^{-1}) is shifted to lower frequencies as well (1317 cm^{-1}). No absorption band of -OH was noticed due to overlapping of the N-H and –OH absorption band.

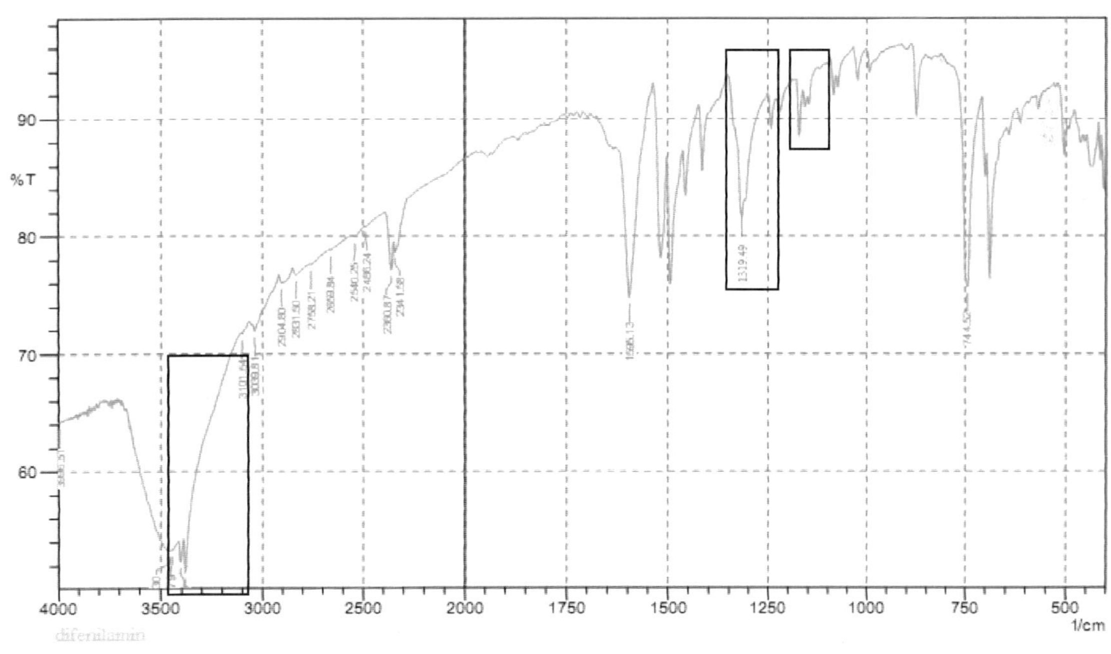

Figure 2. Infared spectra of diphenylamine

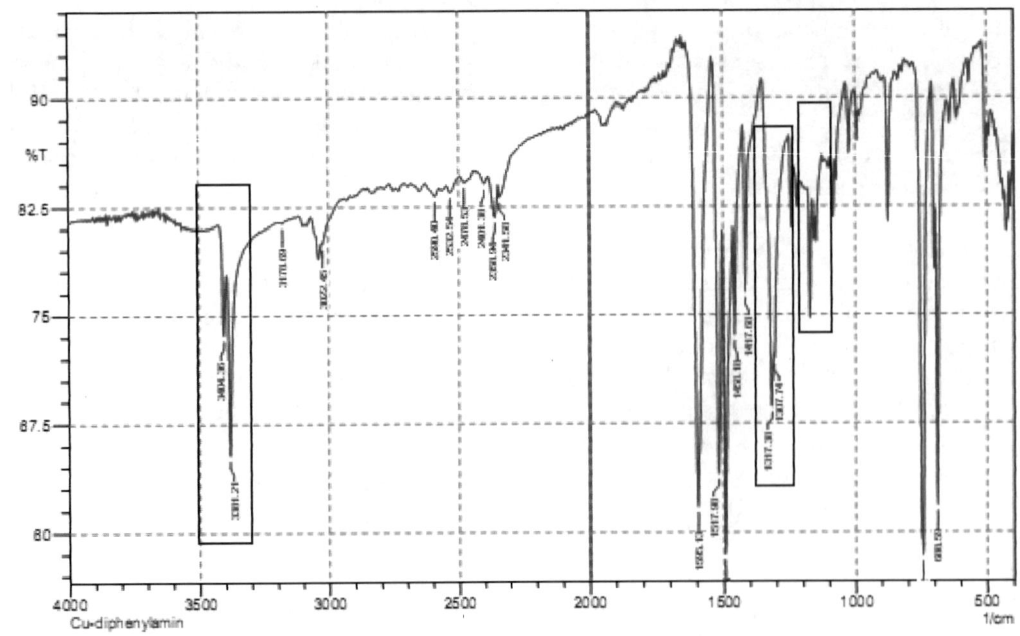

Figure 3. Infared spectra of the Cu(II) complex

3.5. Magnetic Properties (μ_{eff})

The μ_{eff} value measured for the present Cu(II) complex is 1.72 BM which has one unpaired electron B.M. [26] and discrete magnetically non-coupled spin-only value for copper(II) ion [27]. The magnetic moment value of Cu(II) complex indicates that [Cu(diphenylamine)$_4$]$SO_4 \cdot 6H_2O$ is present in square planar geometry [26],[28]. The copper complex was formed by coordination of four N-H groups of diphenylamine. Each atom will produce the same strength of repulsion. The distance and position of each atom will be the same to achieve the least possible repercussion [29]. Based on the above results, the structure shown in Figure 4 is suggested for the complex.

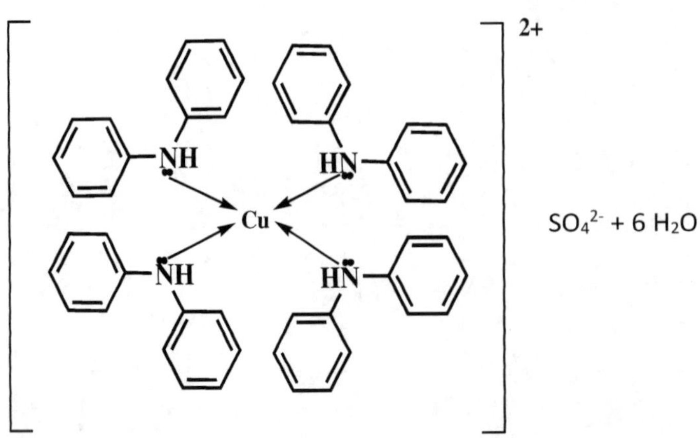

Figure 4. Suggested structure of [Cu(diphenylamine)$_4$]$SO_4 \cdot 6H_2O$

4. Conclusion

The complex [Cu(diphenylamine)$_4$]$SO_4 \cdot 6H_2O$ was able to be synthesized by mixing a methanolic solution of Cu(II) in and a methanolic solution of diphenylamine solution with a mole ratio of 1:4 and stirring for 6 hours. The complex was paramagnetic. The functional group coordinated to the Cu(II) was possibly N-H, forming square planar geometry.

Acknowledgement
The authors gratefully express acknowledgments to Indonesian Directorate General of Higher Education (DIKTI) for financial support.

References

[1] Abdolmaleki S and Ghadermazi M 2017 *Inorg. Chim. Acta.* **461** 221-232
[2] Zhou Y, Tao J, Lv Q, Jia W, Yun R and Cheng Y 2015 *Inorg. Chim. Acta.* **426** 211-220
[3] Wills K, Mandujano-Ramírez H, Merino G, Oskam G, Cowper P, Jones M, Cameron P and Lewis S 2016 *Dyes Pigm.* **134** 419-426
[4] Chen X, Liao Y, Liu Y, Zhu C, Chen T and Zhong C 2017 *Dyes Pigm.* **139** 420-430
[5] Joseph J, Suman A, Nagashri K, Joseyphus R and Balakrishnan N 2017 *J. Mol. Struct.* **1137** 17-26
[6] Sabounchei S, Ahmadianpoor M, Hashemi A, Mohsenzadeh F and Gable R 2017 *Inorg. Chim. Acta.* **458** 77-83
[7] Kutyreva M, Gataulina A, Kutyrev G, Ulakhovich N, Newman T, Khasanova E, Bondar O, Yurtaeva S, Ziganshina S and Khaldeeva E 2016 *Inorg. Chim. Acta.* **450** 101-111
[8] Liu W, Qin Y, Liu S, Xing R, Yu H, Chen X, Li K and Li P 2017 *Carbohydr. Polym.* **160** 97-105
[9] Lobana T, Indoria S, Sood H, Arora D, Randhawa B, Garcia-Santos I, Smolinski V and Jasinski J 2017 *Inorg. Chim. Acta* **461** 248-260
[10] Li X, Fang C, Zong Z, Cui L, Bi C and Fan Y 2015 *Inorg. Chim. Acta* **432** 198-207
[11] Mokhtaruddin N, Yusof E, Ravoof T, Tiekink E, Veerakumarasivam A and Tahir M 2017 *J. Mol. Struct.* **1139** 1-9
[12] Palmer J, Wu J and Upmacis R 2015 *J. Mol. Struct.* **1091** 177-182
[13] Gaur A, Klysubun W, Nitin Nair N, Shrivastava B, Prasad J and Srivastava K 2016 *J. Mol. Struct.* **1118** 212-218
[14] Kose M, Hepokur C, Karakas D, McKee V and Kurtoglu M 2016 *Polyhedron* **117** 652-660
[15] Sharma R, Saini A, Kumar J, Kumar S, Venugopalan P and Ferretti V 2017 *Inorg. Chim. Acta* **457** 59-68
[16] Olalekan T, Ogunlaja A, VanBrecht B and Watkins G 2016 *J. Mol. Struct.* **1122** 72-79
[17] Nanjundan N, Narayanasamy R, Geib S, Velmurugan K, Nandhakumar R, Balakumaran M and Kalaichelvan P 2016 *Polyhedron* **110** 203-220
[18] Huang M, Guo Y, Shi Y, Zhao L, Niu Y, Shi Y and Li X 2017 *Inorg. Chim. Acta* **457** 107-115
[19] Khalaji A, Peyghoun S, Akbari A, Feizi N, Dusek M and Eigner V 2017 *J. Mol. Struct.* **1127** 511-514
[20] Chun M, Jeong A, Min K and Jeong J 2017 *Inorg. Chem. Commun.* **78** 82-84
[21] Ejelonu B C and Olagboye S A 2015 RJPBCS **6** 2 184-190
[22] Srinivasan K, Vaheethabanu J, Manisankar P and Stalin T 2011 *J. Mol. Struct.* **987** 214-224
[23] Regupathy S and Sivasankaran Nair M 2014 *Arab. J. Chem.* **7** 1003-1012
[24] Mautner F, Fischer R, Spell M, Acevedo A, Tran D and Massoud S 2016 *Crystals* **6** 91
[25] Ababei L V, Kriza C, Andronescu A M, Musuc, 2011 *J. Serb. Chem. Soc.* **2** 1-24
[26] Turan N, Körkoca H, Adigüzel R, Çolak N and Buldurun K 2015 *Molecules* **20** 9309-9325
[27] Bhaumik P, Bauzá A, Drew M, Frontera A and Chattopadhyay S 2015 *Cryst. Eng. Comm.* **17** 5664-5671
[28] Shakir M, Abbasi A, Khan A and Khan S 2011 *Spectrochim. Acta A Mol. Biomol. Spectrosc.* **78** 29-35
[29] Rahardjo SB, Lestari WW, and Syaima H 2017 *IOP Conf. Series: Earth and Environmental Science* **75** 012025 do i :10.1088/1755-1315/75/1/012025

IOP Conf. Series: Materials Science and Engineering **333** (2018) 012004 doi:10.1088/1757-899X/333/1/012004

Synthesis and Characterization of *Tetrakis*-aqua-*bis*-isonicotin-amide(itmd)nickel(II) Sulfate

S B Rahardjo[1], S Hastuti[1], N Amanati[1], H Syaima[2]

[1]Department of Chemistry, Faculty of Mathematics and Natural Sciences, Universitas Sebelas Maret, Surakarta, 57126, Indonesia
[2]Chemistry Graduate Program, Graduate School, Universitas Sebelas Maret, Surakarta, 57126, Indonesia

E-mail: sentotbr@yahoo.com

Abstract. The complex of *Tetrakis*-aqua-*bis*-(isonicotinamide)nickel(II) sulfate has been synthesized in 1:2 mole ratio of metal to ligands in methanol. The formula of the complex predicted from analysis nickel content in the complex by Atomic Absorption Spectroscopy (AAS) was $Ni(itmd)_2SO_4(H_2O)_4$. The conductivity of the complex in methanol was measured by conductivity meter correspond to 1:1 electrolyte. Thermal analysis of the complex was determined by Differential Thermal Analyzer (DTA) indicating that the complex contains four H_2O molecules as ligands. The magnetic susceptibility measurement showed that the complex was paramagnetic with μ_{eff} = 3.02 BM. The electronic spectra of the complex appear due to two transition peak on λ = 398 nm and 664 nm. The Infrared spectra showed a shift of NH_2 stretching vibration of $Ni(itmd)_2SO_4(H_2O)_4$. These facts indicated that these functional groups were coordinated to the center ion of the complexes. The proposed structure of the complex was octahedral therefore the possibility formula of this complex was $[Ni(itmd)_2(H_2O)_4]SO_4$.

1. Introduction

Isonicotinamide (itmd), shown in Figure 1, has function as strong anti-tubercular, fibrinolytic, anti-pyretic, and anti-bacterial [1-3]. That pyridine derivative with an amido group(–$CONH_2$) is often used in the manufacture of drugs due to its strong pharmacological effects [1]. Therefore, it is so easily to obtained. Metal complexes of biologically important ligands sometimes have more effective activity than the free ligands [4-6]. Isonicotinamide can also be used as ligand because it has electrons donor atoms. It is reported that thiocyanate anion and one neutral isonicotinamide ligand forming complex with zinc as central metal ion shown in Figure 2 [7].

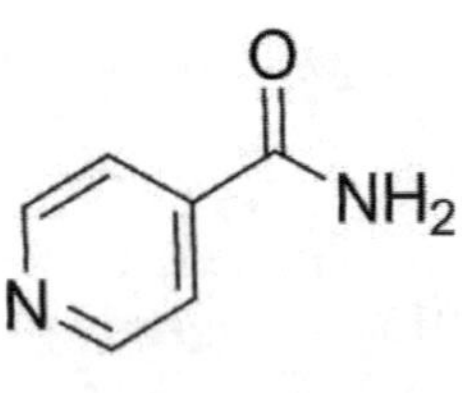

Figure 1. Stucture of isonicotinamide

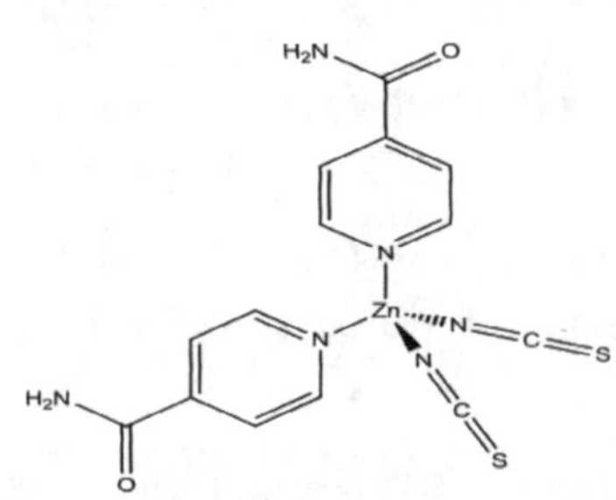

Figure 2. Stucture of $[Zn(SCN)_2(C_6H_6N_2O)_2]$

Nickel is a metal transition element and abundant the most as Nickel(II) [8]. Ni(II) is considerable to learn due to their stability and abundance in nature [9]. The synthesized nickel(II) complex may form several geometries such as square-planar [10] or tetrahedral [11], square-pyramidal [12], trigonal-bipyramidal [13], and octahedral [14-15]. The most common geometry of nickel(II) complex geometry is octahedral and square-planar [8]. In this work, we learn synthesis and the formation of the nickel complex with isonicotinamide. Thus, the complex is characterized by several instrumental analysis to know the formula, electronic spectra, and magnetic property of the complex. We also determine the probability of the functional group in isonicotinamide that bind to nickel ion.

2. Experimental

2.1. Materials

The chemicals and solvents were of reagent grade and used without further purification. All chemicals such as isonicotinamide, $NiSO_4 \cdot 6H_2O$, $FeSO_4 \cdot 7H_2O$, $CuCl_2 \cdot 2H_2O$, $NiCl_2 \cdot 6H_2O$, $CuSO_4 \cdot 6H_2O$, $AlCl_3 \cdot 6H_2O$, and nitric acid were purchased from E. Merck.

2.2. Physical measurements

Spectra UV-VIS of metal complex and $NiSO_4 \cdot 6H_2O$ were recorded in methanol solvent using UV-Vis Double Beam Shimadzu PC 1601 spectrophotometer. The nickel content was determined by Atomic Absorption Spectrometer (AAS) Shimadzu AA-6650. The presence or absence of H_2O molecules in the complex was estimated from the results of thermal analysis using Differential Thermal Analyzer Shimadzu 50. Molar conductivity (Λ*m) of 1 mM complex and standard solution in water were measured on Jenway CE 4071 conductivity meter at 25 °C. The magnetic moment was measured using Auto Sherwood Scientific 10169 Magnetic Susceptibility Balance. Infrared spectra were recorded on Prestige-21 Shimadzu spectrophotometers as KBr pellets in the frequency range of 4000–450 cm^{-1}.

2.3. Synthesis of Ni(II) complex

$NiSO_4.6H_2O$ (0.263 g; 1 mmol) in methanol (5 mL) was added dropwise on isonicotinamide (0.244 g; 2 mmol) in methanol (20 mL) while continuously stirring for ± 2 hours. The solution is allowed to stand for 2 days until a precipitate was formed. The precipitate formed was filtered, washed with cold methanol then dried in a desiccator.

3. Result and Discussion

3.1. Formation of the complex

The electronic spectra of the formed complex in methanol are shown in Figure 3.

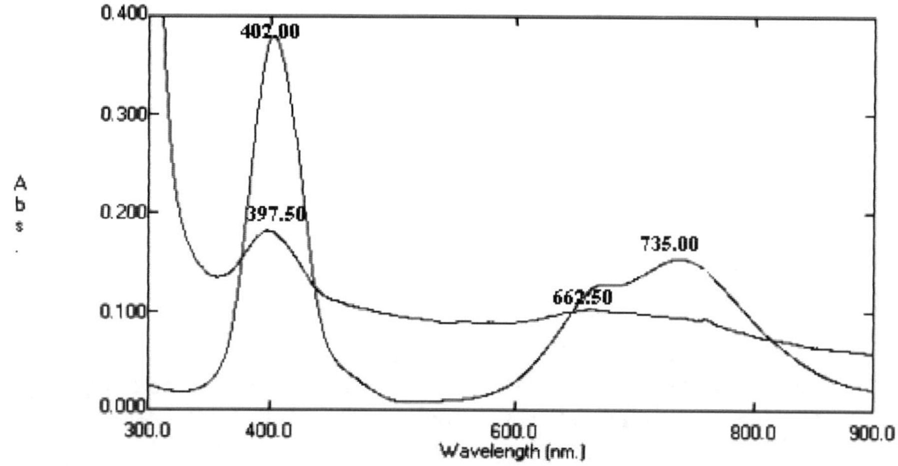

Figure 3. Electronic spectra the complex (a) and $NiSO_4 \cdot 6H_2O$ (b) in methanol

Figure 3 shows the UV-Vis spectra of the complex and $NiSO_4 \cdot 6H_2O$ in methanol. The absorption spectra of the nickel and the produced compound were shown in two peaks. It is similar with the electronic spectra of the nickel and nickel complex reported by Rahardjo et al. [16]. The maximum wavelength shifted from 735 and 402 nm ($NiSO_4 \cdot 6H_2O$) to 664 and 398 nm (Ni-itmd). The shifting of complex wavelength toward smaller than metal ion showed the formation of the complex. It indicated that itmd changes molecules of water and bind to nickel ion.

3.2. Analysis of the amount of nickel with Atomic Absorption Spectroscopy (AAS)

The result of nickel content analyzed by AAS in the Ni(II)-itmd complex was 12.60%. If the result of the measurement was compared to the theoretical analysis shown in Table 1, it could be predicted that the formula of Ni(II)-itmd complex. The different value of theoretical calculation and experiment by AAS was sometimes happen in the research [17-18]. Therefore, the formula of the formed complex was close to $Ni(itmd)_2SO_4(H_2O)_4$. It needed thermal analysis methods to confirm the number of presence of water molecules as well as the state whether as hydrates or ligands.

Table 1. Percentage of nickel in the complex theoretically.

Empirical Formula	**Mr**	**% Nickel**
$Ni(itmd)_2SO_4(H_2O)_2$	435.04	13.50
$Ni(itmd)_2SO_4(H_2O)_3$	453.06	12.96
$Ni(itmd)_2SO_4(H_2O)_4$	**471.07**	**12.46**
$Ni(itmd)_2SO_4(H_2O)_5$	489.09	12.00
$Ni(itmd)_2SO_4(H_2O)_6$	507.10	11.58

3.3. Analysis of electrical conductivity

The results of molar conductivity measurement of the standard solution and complex in water are shown in Table 2. Molar conductivity of the complex was 158 ± 2 S cm^2 mol^{-1}. If the result was compared to the standard solution, the conductivity of the Ni(II)-itmd complex was close to the electrical conductivity of $FeSO_4 \cdot 7H_2O$, $CuSO_4 \cdot 5H_2O$, and $NiSO_4 \cdot 6H_2O$. The value of the measurement was consistent with the 1:1 electrolyte, indicating that the SO_4^{2-} ions were not coordinated to Ni^{2+}.

Table 2. Molar conductivity of metal salts and Ni(II)-itmd in water

Solution	Λ_m (S cm^2 mol^{-1})	Cation:anion Charge
Water	0	-
$FeSO_4 \cdot 7H_2O$	103 ± 1	1:1
$CuSO_4 \cdot 5H_2O$	174 ± 2	1:1
$NiSO_4 \cdot 6H_2O$	147 ± 2	1:1
$CuCl_2 \cdot 2H_2O$	216 ± 1	2:1
$NiCl_2 \cdot 6H_2O$	252 ± 1	2:1
$AlCl_3 \cdot 6H_2O$	315 ± 1	3:1
Ni(II)-itmd	**158 ± 2**	**1:1**

3.4. Thermal analysis

The result of Differential Thermal Analysis (DTA) was shown at Figure 4. There was no endothermic peak at about 100 ^{0}C, but there was an endothermic peak at 212.42 ^{0}C. Lattice water molecules would release at about 100 ^{o}C as water boiling point. This peak indicated the release of coordinated H_2O from the Ni(II)-isonicotinamide complex. Four molecules of water in the complex acted as ligands.

Thus, at 401.08 ^{0}C there was an exothermic peak which indicating the occurrence of sulfate ion decomposition [19]. Therefore, the formula of the complex was $[Ni(itmd)_2(H_2O)_4]SO_4$, namely *tetrakis*-aqua-*bis*-(isonicotinamide)nickel(II) sulfate.

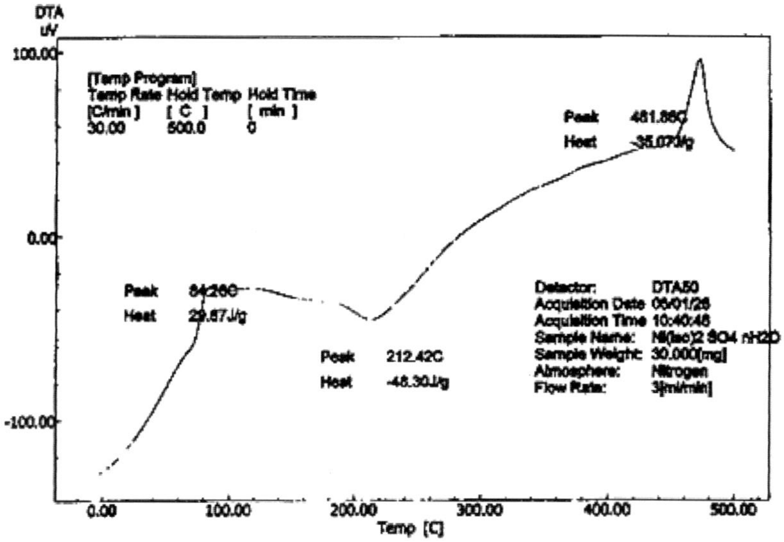

Figure 4. Thermogravimetric/differential thermal analysis spectra

3.5. Infrared analysis

Table 3 shows the infrared analysis of isonicotinamide and Ni(II)-itmd complex. This indicated that the coordinated $-NH_2$ group of isonicotinamide at the Ni(II) central atom. The smaller wavenumber than isonicotinamide was possible due to the delocalized electrons in the sixth circumference therefore a large enough force was required to push the electron density towards $-NH_2$. The N-H bond became long and the vibrational energy was small. The shift towards smaller wave numbers of the N-H groups also occured in the $[NiL_2](ClO_4)_2$, L= uramido benzoic acid, from 3224 cm^{-1} to 3125 cm^{-1} [20].

Table 3. Infrared spectra of functional groups isonicotinamide and the complex

Compounds	υ N-H bending	υ C=O cyclic	υ C=N aromatic
Isonicotinamide	3371	1670	1597
	3302		1555
			991
Ni(II)-itmd	3267	1682	1609
			1555
			976

3.6. Magnetic properties (μ_{eff})

The effective magnetic moment value (μ_{eff}) of the Ni(II)-itmd complex was 3.02 ± 0.09 B.M. The value showed the paramagnetic complex with two unpaired electrons. It was reported that nickel complexes with a value of magnetic moment 3.0, 3.1, 3.2, and 3.3 B.M. forming octahedral geometry [20-21]. The result of the magnetic moment value 2.8-3.5 in the complex of nickel showed octahedral structure [22]. Therefore, the Ni(II)-itmd complex was predicted to form octahedral geometry.

3.7. Electronic spectra

The solution UV–Vis spectrum of the nickel complex showed two peaks at 664 and 398 nm. Basically, the octahedral complex Ni(II) has three electronic transitions $^3A_{2g}\rightarrow^3T_{2g}$ (F), $^3A_{2g}\rightarrow^3T_{1g}$ (F) and

$^3A_{2g} \rightarrow {}^3T_{1g}$ (P) in the region 10,000 nm, 600 nm and 400 nm which appear as three peaks in the UV-Vis spectrum [8]. Sometimes the transitions are not visible and just appear two transitions. Tetrakis(2-amino-3-methylpyridine)di(aqua)nickel(II) also showed two transitions at 620 nm (16116 cm^{-1}) and 375 nm (26630 cm^{-1}). The complex formed octahedral geometry [21]. It was assumed that Ni(II)-itmd complex forming octahedral geometry as well.

3.8. Proposed structure
From all of the characterization analysis above, the proposed structure of the Ni-itmd complex is shown in Figure 5.

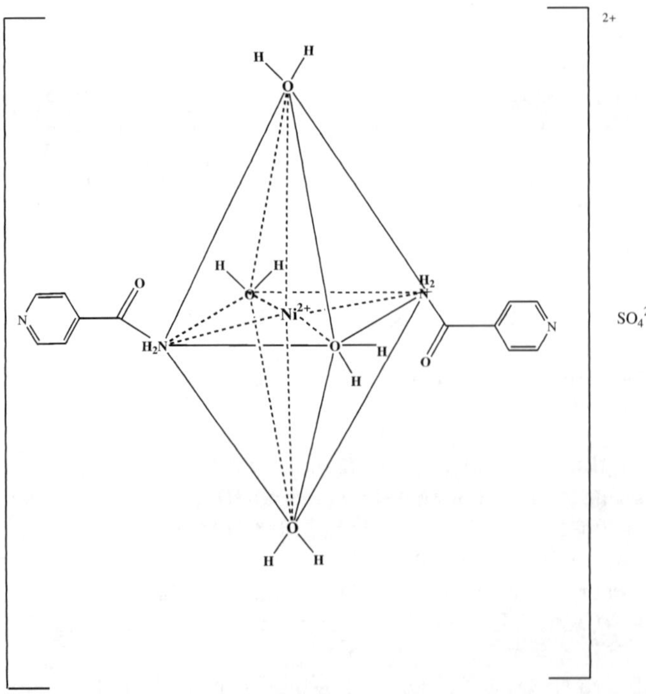

Figure 5. Suggested structure of $[Ni(itmd)_2(H_2O)_4]SO_4$.

4. Conclusion
The Ni-itmd complex, namely *tetrakis*-aqua-*bis*-(isonicotinamide)nickel(II) sulfate, could be synthesized from $NiSO_4{\cdot}6H_2O$ and isonicotinamide in refluxing methanol with a mole ratio of metal to ligand = 1:2. The formula of the complex was $[Ni(itmd)_2(H_2O)_4]SO_4$. The ligands were coordinated to central metal ion through $-NH_2$ group in the isonicotinamide and O-H group of water molecules. The complex was paramagnetic. It had two peaks of UV-Vis absorption at 664 and 398 nm. It was indicated that the complex forming octahedral geometry.

Acknowledgement
The authors gratefully express acknowledgments to Ministry of Research, Technology and Higher Education (RISTEKDIKTI) of the Republic of Indonesia for financial support.

References
[1] Atac A, Yurdakul S, Ide S 2006 *J. Mol. Struct.* **783** 79–87
[2] Wang R, Yu P, Zhou C, Lin F, Chen C, Pan Y, Mei *2015 Bioorganic Med. Chem. Lett.* **25** 1036-1039
[3] Li B, Wu Y, Zhu J, Chen K, Wu B, Ji L 2016 *Thermochimica Acta* **627–629** 55–60
[4] Ahuja L S, Prasad I 1976 *Inorg. Nucl. Chem. Lett.* **12** 777

[5] Xie J, Li Y, Song L, Pan Z, Ye S, and Hou Z 2017 *Drug Deliv.* **24(1**) 707-719
[6] Loboda D, Rowińska-Żyrek M 2017 *J. Inorg. Biochem.* **174** 150-155
[7] Neumann T, Jess I, Nather C 2016 *Acta Cryst.* **E72** 922–925
[8] Lee J D 1994 *Concise Inorganic Chemistry Fourth Edition* (London: Chapman and Hall)
[9] Hazari N, Melvin P R, Beromi MM 2017 *Nature Reviews* **1** 0025 1-16
[10] Saha P, Samanta D, Ghosh P 2017 *New J. Chem.* **41** 1149-1159
[11] Powers X B, Ghiassi K B, Greenfield J T, Olmstead M M, and Balch AL 2017 *Cryst. Eng. Comm.* **19** 3244-3253
[12] Allali M, Ichou Y, Allali N, Zarrouk A, Habbadi N 2016 *J. Mater. Environ. Sci.* **7(6)** 2042-2050
[13] Chastain B B, Rick E A, Pruett R L, Gray H B 1968 *J. Am. Chem. Soc*. **90(15**) 3994–4000
[14] Nikiforov A A, Eremin A V, Gurzhii V V, Misharev A D, Medvedskii N L, Ponyaev A I, Belyaev A N 2017 *Russ. J. Coord. Chem.* **43(5)** 269–277
[15] Sohtome Y, Nakamura G, Muranaka A, Hashizume D, Lectard S, Tsuchimoto T, Uchiyama M, Sodeoka M 2017 *Nat. Commun.* **8** 14875
[16] Rahardjo S B, Aditya H P 2017 *IOP Conf. Series: Materials Science and Engineering* **17** 012048
[17] Mahdi A S, Awad A A, Hasson M M *2017 Acta. Chim. Pharm. Indica* **7 (1)** 1-7
[18] Chaudhary N K, Mishra P 2017 *Bioinorg. Chem. Appl.* 6927675
[19] Cotton F A, Willkinson G, Gauss G P 1995 *Inorganic Chemistry 3rd Edition* (New York: John Wiley and sons)
[20] Prema S, Pasupathy A, Bheeter S R 2016 *IJSRP* **6(5)** 721-723
[21] Mukherjee P, Biswas C, Drew M G B, Ghosh A 2007 *Polyhedron* **26** 3121–3128
[22] Sonmez M 2003 *Inorg. Chem.* **77** 397-402

Synthesis and characterization of mesoporous NaY zeolite from natural Blitar's kaolin

S N Khalifah[1*], Z N aini[1], E K Hayati[1], N Aini[1], A Prasetyo[1]
Department of Chemistry, Faculty Science and Technology, Universitas Islam Negeri Maulana Malik Ibrahim, Jalan Gajayana 50, 65144, Malang, Indonesia.

[*]E-mail: susikhalifah@gmail.com

Abstract. Mesoporous NaY Zeolite has been synthesized from calcined natural Blitar's kaolin with the addition of NaOH and CTABr surfactant as mesoporous template by hydrothermal method. Natural kaolin was calcinated with different time and temperature to change kaolin to metakaolin. X-ray diffraction data showed that mesoporous NaY zeolite was formed with impurities compound of sodalite, kaolin and quartz phases. The BET analysis resulted that the pore of NaY Zeolite belongs to mesoporous type with pore size 9,421 nm. Characterization from FTIR confirmed about the functional group of zeolites (988, 776, 663, 464 cm^{-1}). Scanning electron microscopy characterization showed that the morphological of mesoporous NaY zeolites have uniform and crystalline particles formed.

1. Introduction

NaY zeolite is a material with regular architecture and micropores consist of cavities and channels [1]. Mesoporous NaY with larger specific surface area than micropores NaY zeolite has a wider application as catalyst [2,3], adsorption [4,5] and drug delivery system [6]. Mesoporous NaY zeolites can be synthesized using high-purity chemical reagent, but its relatively high cost. Natural Kaolin is an alternative raw material for low cost synthesis of mesoporous zeolite. This raw material is very cheap and exist abundantly in many areas in Indonesia, such as in Blitar. Kaolin is often used as the raw material for synthesis of low-silica zeolite such as zeolite X [7], zeolite P [8] and zeolite A [9,10], because the contents of SiO_2 and Al_2O_3 are relatively similar with the product. Zeolite with a higher Si/Al ratios, such as zeolit Y, can also be prepared using kaolin with additional sodium silicate solution [11,12] to increase the silica content. However, Blitar's kaolin has a high silica content, so the synthesis of zeolite NaY can be carried out from Blitar's kaolin without the addition of silica thus can lowering the zeolite production cost.

The synthesis of mesoporous NaY zeolites from natural kaolin involves two basics principle steps: (a) the conversion of kaolin to metakaolin with calcinations at high temperature to increase the reactivity of kaolin in the metastable phase [10, 13, 14, 15] and (b) the reaction of calcined kaolin with sodium hydroxide and surfactants in conventional hydrothermal method. The addition of surfactants serving as mesoporous templates. In this study, mesoporous NaY zeolite synthesized from Blitar's kaolin without the addition of silica by hydrothermal method using cationic surfactants of Cethyl Trimethyl Ammonium Bromide (CTABr) as mesoporous template.

IOP Conf. Series: Materials Science and Engineering **333** (2018) 012005 doi:10.1088/1757-899X/333/1/012005

2. Experimental

Kaolin was prepared by washing HCl 1M and calcination at 630 °C, 700 °C and 800 °C. The calcined kaolin reacted with aluminum oxide and sodium hydroxide at molar ratio $9Na_2O$: 2 Al_2O_3: $6SiO_2$: 249 H_2O. The mixture was stirred for 24 h and added by 18.73 g of CTABr. Stirring process continued until gel formed and aged at room temperature for 24 hours, then transferred into a hydrothermal reactor at 100 ° C for 48 hours. The obtained powder filtered, washed, dried and calcined at 550 °C for 1 h.

Characterization of synthesized material were determined for the functional group, elemental analysis, structure, morphologycal characteristics and surface area using Fourier Transform Infra-Red (FT-IR), X-ray fluorescence (XRF), X-ray powder diffraction (XRD), Scanning Electron Microscopy (SEM) and BET analysis.

3. Result And Discussion

The element contents of Blitar's kaolin are given in Tabel 1. The Si/Al rasio found to be 5.45 and after washing HCl increased to 8.55. The other metals percentage decreased after washing with HCl.

Table 1. The element content of Blitar's kaolin

Element	before washing HCl (%)	after washing HCl(%)
Al	12	8.8
Si	65.4	75.2
K	8	5.83
Ca	2.68	2.38
Ti	3.68	3.18
V	0.08	0.06
Mn	0.22	0.19
Fe	8.61	2.50
Ni	0.19	1.11
Cu	0.12	0.18
Eu	0.11	0.05

Figure 1 shows the XRD pattern of Blitar's kaolin and calcined kaolin. The Blitar's kaolin in Figure 1 (c) contains kaolin structure and high impurities of quartz phase. Further calcinations until 800 °C for 24 hours did not show any significant differences in quartz content. It was previously reported that the major impurities in natural kaolin is the high content of quartz which can be removed by further alkali fusion treatment [16,17]. In this research, natural kaolin was only calcinated as physical treatment without additional treatment by alkali fusion, so the content of quartz was still high. The stucture of synthesized zeolite material show in Figure 2 indicated that crystalline stucture of NaY zeolite has been successfully synthesized according to IZA literature with some impurities were detected for sodalite, kaolin phase and quartz phase. The sodalite [16] and quartz phase [17] has reported as known impurities in zeolite which was synthesized from natural kaolin. The presence of impurities may caused by incompletely transformation of natural kaolin into metakaolin phase during physical treatment.

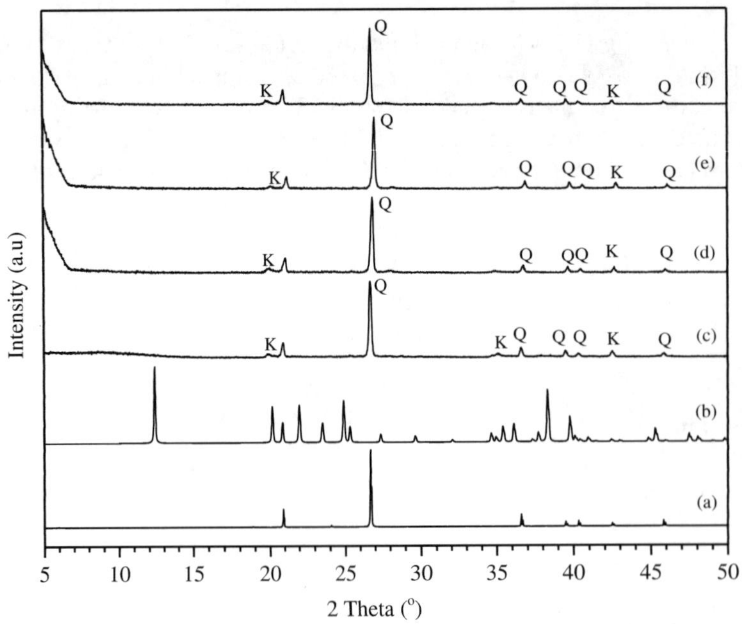

Figure 1. XRD pattern of quartz standard (ICPDS) (a), kaolin standard (ICPDS) (b), Blitar's kaolin (c), calcined kaolin 630 °C 3 h (d), calcined kaolin 700 °C 3 h (e), calcined kaolin 800 °C, 24 h (f). Peaks marked with letter are: Q- quartz, K-kaolin.

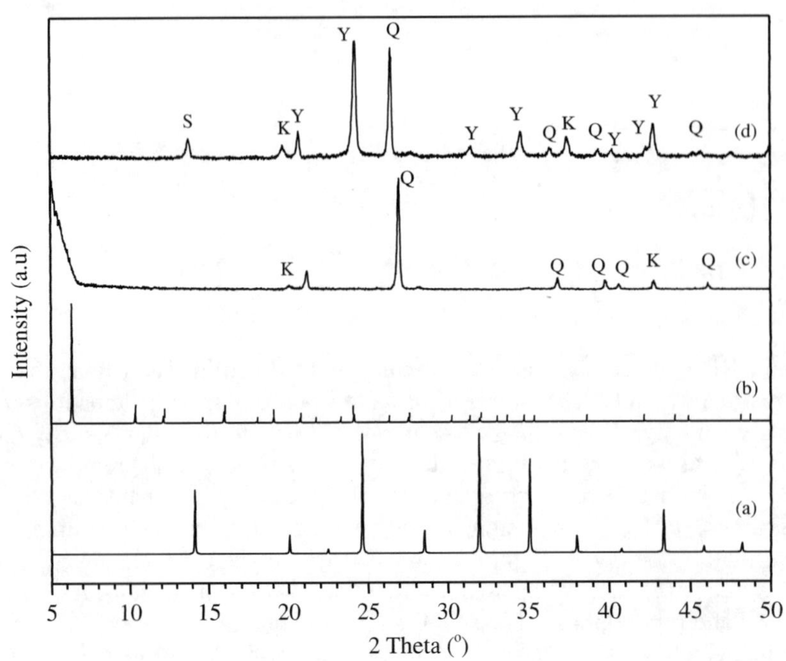

Figure 2. XRD pattern of sodalit standard (IZA) (a), zeolit Y standard (IZA) (b), calcined kaolin 700 °C 3 h (c), Mesoporous NaY zeolite by conventional hydrothermal (d). Peaks marked with letters are: Q- quartz, K-kaolin, S-sodalit, Y-NaY zeolite

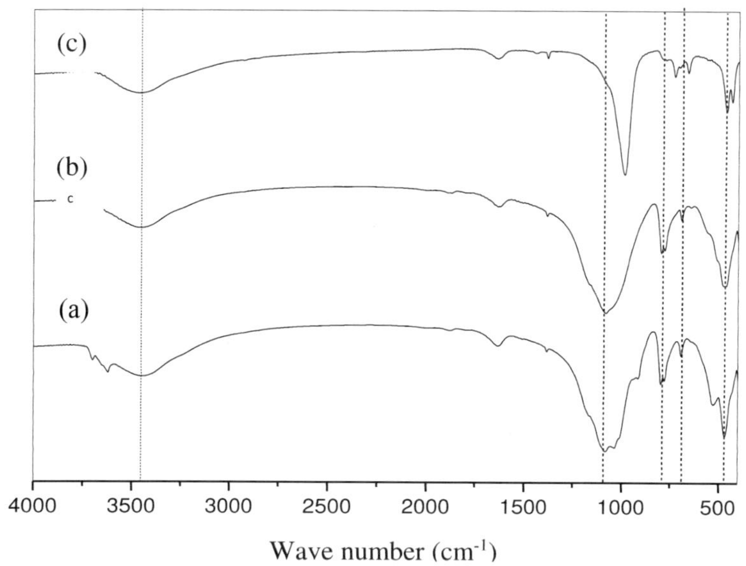

Figure 3. FTIR spectra of Blitar's kaolin (a), Calcined kaolin 700 °C for 3 h (b) Mesoporous NaY zeolite by conventional hydrothermal (c)

The transformation from kaolin to metakaolin (calcined kaolin) and NaY zeolite can be analyzed from its functional group by infrared spectroscopy in Figure 3. Blitar's kaolin showed characteristic bands at 3623 cm^{-1} which is assigned as stetching vibration of –OH and at 529 assigned as vibrations Al-O in octahedral alumina structure. The bands at 1037.811 cm^{-1} is assigned as stretched vibrations of Si-O, while the bands at 796 cm^{-1} is assigned as vibrations of O-Si-O, and the bands at 468 cm^{-1} is assigned as bending vibrations Si-O. The bands at 3623 cm^{-1} and 529 cm^{-1} were dissapeared for the calcined kaolin spectra in Figure 3(b). Its indicated that octahedral alumina structure of kaolin was destroyed during calcination step. The spectra of NaY zeolite in Figure 3(c) contains new vibration bands at 988 cm^{-1} assigned to the characteristics vibration of SiO_4 or AlO_4 tetrahedral units in zeolite structure. Thus can be concluded that synthesis product was zeolite material.

Figure 4. SEM morphologies of Blitar's kaolin (a), Calcined kaolin 700 °C for 3 h (b), Mesoporous NaY zeolite by conventional hydrothermal (c).

Figure 4 showed the morphological transformation of kaolin, calcined kaolin and NaY zeolite. Figure 4 (a) showed that the kaolin morphology is a coated sheet while the calcined kaolin in Figure 4(b) showed a damaged layered sheets (began to form amorphous). NaY zeolites in Figure 6 (c) shows the uniform morphology which is indicated that crystalline particles of zeolite has formed from natural kaolin as raw material.

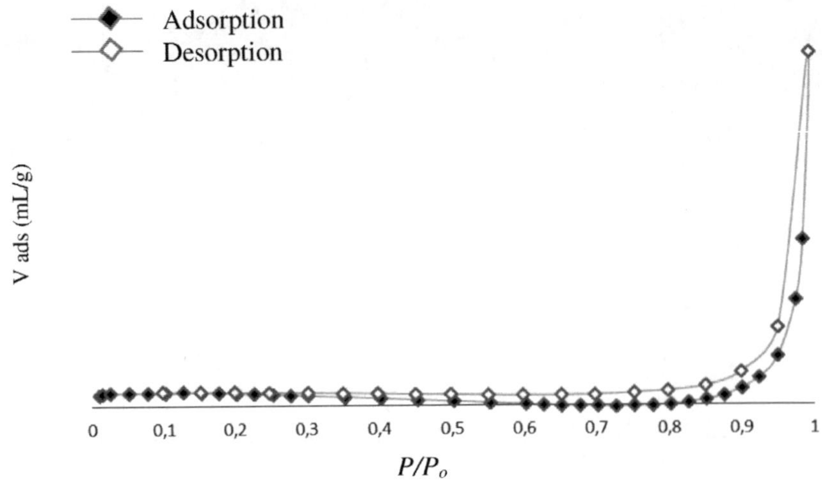

Figure 5. N_2 adsorption-desorption isotherm of mesoporous NaY zeolite

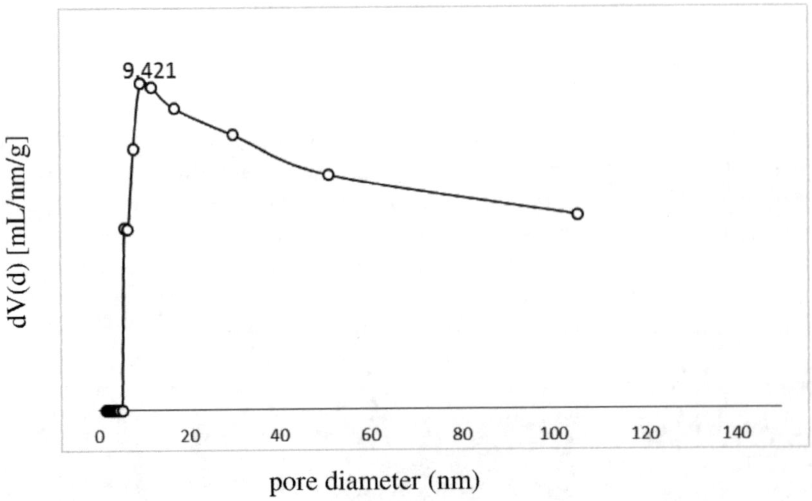

Figure 6. BJH pore size of mesoporous NaY zeolite

Figure 5 showed the N_2 adsorption-desorption isotherm of NaY zeolite. The NaY zeolite exhibited the type IV curve with H3-type hysteresis loop which indicated as mesoporous zeolite. The Brunauer-Emmett-Teller (S_{BET}) of the synthesized mesoporous NaY zeolite was measured to be 434.2 m^2/g. This result showed significant enhancement of specific surface area than the Blitar's kaolin which only has S_{BET} 16,16 m^2/g. However this result is still lower than commercial NaY zeolite (S_{BET} 626 m^2/g) [18] which may contributed by incompletely transformation of the raw material. The mesopore size according to Barrett-Joyner-Halenda (BJH) algorithm in Figure 6 showed the pore diameter of mesoporous NaY zeolite is 9.421 nm. This pore diameter was larger than the pore of raw material for Blitar's kaolin which is 1,72 nm.

4. Conclusion

Mesoporous NaY zeolite has been successfully synthesized from Blitar's kaolin with sodalite, quartz and kaolin impurities. The NaY zeolite has uniform morphology and crystalline particle with the vibration of SiO_4 or AlO_4 tetrahedral units detected at 988 cm^{-1}. The synthesized zeolite was the mesoporous material with type IV of isotherm curve, 434.2 m^2/g of surface area and 9.421 nm of pore diameter.

References

[1] Baerlocher C, Lynne B M, and David H 2007 *Atlas of Zeolite Framework Types (Sixth Edition)* ISBN: 978-0-444-53064-6 Elsevier B.V.
[2] Doyle A M, Albayati T M, Abbas A S, and Alismaeel Z T 2016 *Renewable Energy* **97**:19-23.
[3] Zhao J, Wang G, LihongQin, HaiyanLi, YuChen, and Liu B 2016 *Catalysis Communications* **73**:98–102.
[4] Kevin X Lee and Valla J A 2017 *Applied Catalysis B: Environmental* **201**:359-369.
[5] Zhou W, Zhou Y, Wei Q, Ding S, Jiang S, Zhang Q, and Liu M 2017 *Chemical Engineering Journal* **330**:605-615.
[6] Egodawatte S, Dominguez S, Larsen S C 2017 *Microporous and Mesoporous Materials* **237**:108-116 .
[7] Chandrasekhar S and Pramada P N 1999 *Journal of Porous Material* **6**:283-297.
[8] Bessa R D A, Costa L D S, Oliveira C P, Nascimento R F D, Sasaki J M, Bohn F, and Loiola A R, 2017 *Microporous and Mesoporous Materials* **245**:64-72.
[9] Prokof'ev V Y, Gordina N E, Zhidkova A B, and Efremov A M 2012 *J Mater Sci* **47**:5385-5392.
[10] Ayele L, Pariente J P, Chebude Y, and Díaz I 2015 *Elsevier Microporous and Macroporous Materials* **215**:29-36.
[11] Qiang L, Ying Z, Zhijun C, Wei G, and Lishan C 2010 *Pet.Sci* **7**: 403-409.
[12] Ahmedzeki N S, Rashid H A, Alnaama A A, Alhasani M H, and Abduhulhussain Z 2013 *Korean J. Chem. Eng.* **30**:2213-2218.
[13] Alkan M, Hopa C, Yilmas Z and Guler H 2005 *Elsevier Microporous and Macroporous Materials* **86**:176-184.
[14] Feng H, Li C, Shan H 2009 *Appl. Clay Sci.* **42**(3-4):439-445.
[15] Kovo A S, Hernandez O, and Holmes S M 2009 *J. Mater Chem* **19**(34):6207-6212.
[16] Ríos C A, Williams C D, and Fullen M A 2009 *Applied Clay Science* **42**:446-454.
[17] Ayele L, Pariente J P, Chebude Y, and Díaz I 2016 *Applied Clay Science* **132-133**:485-490.
[18] Keselj D, Lazic D, Scudric B, Skudric J P, Perusic M 2015 *IJSR*. **4**(2):37-41.

X-RAY Diffraction and Fourier Transform Infrared Study of Ca-Mg-Al Hydrotalcite from Artificial Brine Water with Synthesis Hydrothermal Treatments

Heriyanto[1], E Heraldy[2,3] and K D Nugrahaningtyas[2,3]
[1)] Department of Chemistry, Master Program Graduate School, Sebelas Maret University
[2)] Department of Chemistry, Faculty of Mathematics and Natural Sciences, Sebelas Maret University
[3)] Solid State Chemistry & Catalysis Research Group, Sebelas Maret University
Jl. Ir. A. Sutami 36A, Kentingan, Surakarta 57126, Indonesia

E-mail: eheraldy@mipa.uns.ac.id

Abstract. Synthesis of various hydrothermal treatments for synthesis Ca-Mg-Al hydrotalcite from artificial brine water had been performed. The hydrothermal treatment has four variations with a temperature at 180 °C into Teflon lined stainless steel for 18 hours. The XRD result of the synthesis materials after various hydrothermal treatments has hydrotalcite phase and $CaCO_3$. Therefore a preparation method might not discover a mixture of Ca and Al cations uniformly, and then a number of Ca cation agglomerated and enabled evidence that the radius of Ca^{2+} was larger than Mg^{2+}. Le Bail refinement of the value of reliability index parameter especially X^2 values has concern limitation value of reliability index. Lattice parameter a and c have a value of around 22 and 3 Å, respectively. Ca-Mg-Al hydrotalcite possesses crystallite size lower than Mg-Al hydrotalcite commercial. Infrared spectra has found hydroxyl band at region 3474.91 and 3072.74 cm^{-1}, carbonate band at region 1365.66; 1491.04; 857.40; and 685.72 cm^{-1}, and metal band at region 447.50; 553.59; 772.52 cm^{-1}; and 975.06 cm^{-1}.

1. Introduction

Hydrotalcite, layered double hydroxide, is material unique and interested in latest years due to their potential application in several fields as adsorbent [1], catalyst [2], ion exchange [1], pharmaceutical [3] and photocatalysis [4]. Structurally, hydrotalcite can explain as containing brucite $Mg(OH)_2$ like layers [5], where each Mg^{2+} cation is octahedral side share edge to form infinite sheets. According to [6], in the hydrotalcite, substitution isomorphous of Mg^{2+}/Al^{3+} is in the octahedral site of the hydroxide sheet cause a net positive charge, which is to be neutralized by anionic species in the interlayer. Hydrotalcite can be stated with $[M^{2+}_{1-x}M^{3+}_{x}(OH)_2]^{x+}(A^{n-})_{x/n}.\,yH_2O$ [7], where M^{2+} is a divalent cation, M^{3+} is a trivalent metal cation, and A^{n-} is the replaceable anion (usually carbonate, nitrate or chloride) and $0.2 < x < 0.33$; y is the amount of interlayer H_2O.

In the previous study, synthesis hydrotalcite from brine water with co-precipitation method especially Ca-Mg-Al hydrotalcite has another hydrotalcite phase such as $Al(OH)_3$, $Mg(OH)_2$, $Ca(OH)_2$ and $CaCO_3$ [8]. Rives (2001) reported that the parameter of an experiment may be affect the hydrotalcite crystallized, such as pH, concentration metallic salts, an alkaline solution, the flow rate of reactants, temperature in the reactor, and aging time of the precipitate. In many cases, an optimization condition to make improvement contain hydrotalcite phase by hydrothermal treatment is in the presence of water vapor at high temperature [10]. Synthesis with various hydrothermal treatment has been studied by Hickey et al. (2000), which is used mother liquor and washed slurry hydrotalcite. For temperature hydrothermal treatment has been reported by Kovanda et al. (2005) and Sharma et al. (2007). The hydrothermal treatment for aging time has been reported Oh et al. (2002) and Sharma et al. (2007).

The objective of the current work is to study the impact of various hydrothermal treatments from artificial brine water. The synthetic hydrotalcite with Ca-Mg/Al molar ratio of 2 was synthesis under hydrothermal conditions at various treatment washed and centrifuged. An influence of hydrothermal treatment was assessed by powder X-ray diffraction (XRD) and Fourier transform infrared (FTIR).

2. Experimental

2.1 Materials

Magnesium chloride ($MgCl_2.2H_2O$); calcium chloride ($CaCl_2.2H_2O$); sodium carbonate (Na_2CO_3); aluminum chloride ($AlCl_3.6H_2O$); aquadest was used for the synthesis of Ca-Mg-Al hydrotalcite in p.a. grade from Merck, Germany or equivalent and without further treatment.

2.2 Preparation of Ca-Mg-Al hydrotalcite

Ca-Mg-Al hydrotalcite compounds with the molar ratio (Ca+Mg)/Al is 2 was prepared by co-precipitation method [14]. Artificial brine water contained a magnesium and calcium source, added $AlCl_3.6H_2O$ solution and the alkaline solution Na_2CO_3 0.1 M until pH was ≈ 10. The suspension was stirred for 1 h at 65 °C. The slurry was divided into three portions to undergo different hydrothermal treatment. One portion (called A sample) was hydrothermal treatment at 180 °C into Teflon lined stainless steel for 18 h [6], and then the sample was washed with aquadest until free ion Cl^- ($AgNO_3$ test). The other portion (called B sample) was washed with aquadest until free ion Cl^- and then hydrothermal treatment at 180 °C for 18 h [12]. The last portion (called C sample) was centrifuged at 2000 rpm for 15 min, and then the slurry was divided into two portions (called C1 and C2 sample). A sample of C1 was hydrothermal treatment like as A sample method, and a sample of C2 was hydrothermal treatment like as B sample method. After all hydrothermal treatment, the slurry was separated by centrifugation at 2000 rpm for 15 min and then dried overnight at 105 °C for the subsequent characterizations.

2.3 Characterization

The samples were recorded by a Bruker D8 Advance X-ray diffractometer using monochromatized Cu-Kα radiation. The samples were measured in the diffraction angle range 2θ from 10° to 70° with steps of 0.02°. A Jade software package (Material Data, Inc.) was used to analyze the data, identification of the existing crystalline phase being approached standard JCPDS diffraction data files. Data manipulation and interpretation were performed using OriginPro 8.0 software. The XRD pattern was analyzed by Le Bail refinement method using RIETICA program [15]. Peak profiles were modeled using a pseudo-Voigt peak shape. Experimental parameters refined were the instrument zero, scale factor, the lattice parameters and the peak shape parameters u, v, w, γ_0 and γ_1 [16]. Each refinement was conducted at 30 cycles [17].

The IR spectrum of samples was recorded at room temperature with a Fourier transform infrared (FTIR) spectroscopy Shimadzu IR-Prestige21. FTIR spectra measurements were conducted to identify the functional groups contained in the hydrotalcite crystal. Sample were contained in a KBr pellet in the range 4000-400 cm^{-1}, typically, 45 scans were performed at a resolution of 2 cm^{-1}. Data manipulation, such as baseline adjustment and fitting multi peak interpretation, was performed using OriginPro 8.0 software.

3. Result and Discussion

3.1 XRD

The XRD patterns of the synthesis materials after various hydrothermal treatments are very similar (Figure 1). The sharp and symmetric basal reflection of the 003 and 006 planes at 2θ value 11-23° and broad, asymmetric reflections at higher 2θ value 34-65° were observed in XRD patterns. The hydrotalcite phase (JCPDS #890460) was identified as the main crystallite phase and $CaCO_3$ (JCPDS #712396) as the secondary crystallite phase. Hydrotalcite has diffraction specific identify at 2θ value 11,5° (d_{003}); 23,45° (d_{006}) and 34,57° (d_{009}), although 2θ value 61,9° (d_{110}) was detected diffraction specific for anion interlayer CO_3^{2-} [1]. Gao et al. (2010) reported crystal of $CaCO_3$ can be formed because a preparation method might not discover a mixture of Ca and Al cations uniformly, then a number of Ca cation agglomerated. Another explained from the evidence that the radius of Ca^{2+} was larger than Mg^{2+} [18]. After Mg^{2+} in layers was partially replaced with Ca^{2+}, the interaction between interlayer anions and cations in layers was decreased, causing that the interaction between anion in interlayer and Ca^{2+} in layers is weaker than between anions in interlayer and Mg^{2+} in layers [9].

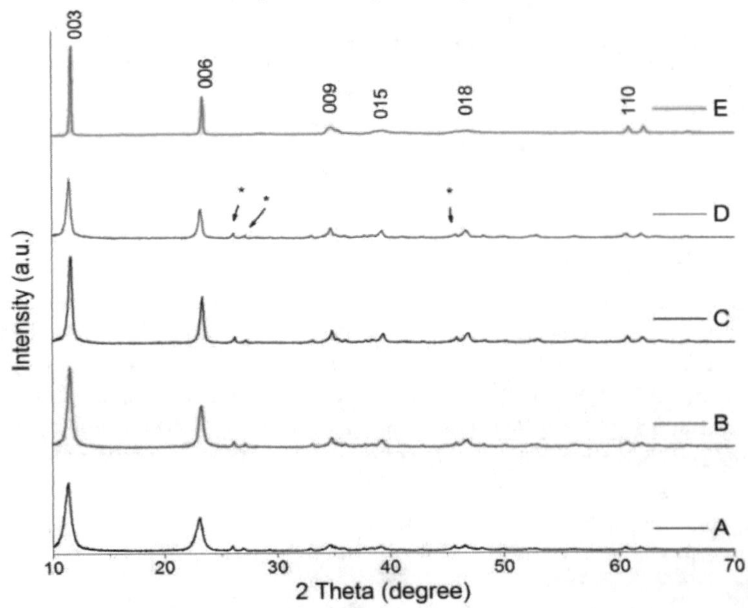

Figure 1. Diffractogram XRD Ca-Mg-Al hydrotalcite with variation treatment hydrothermal (A) sample A, (B) sample B, (C) sample C1, (D) sample C2 and Mg-Al hydrotalcite commercial (E). * = $CaCO_3$.

IOP Conf. Series: Materials Science and Engineering **333** (2018) 012006 doi:10.1088/1757-899X/333/1/012006

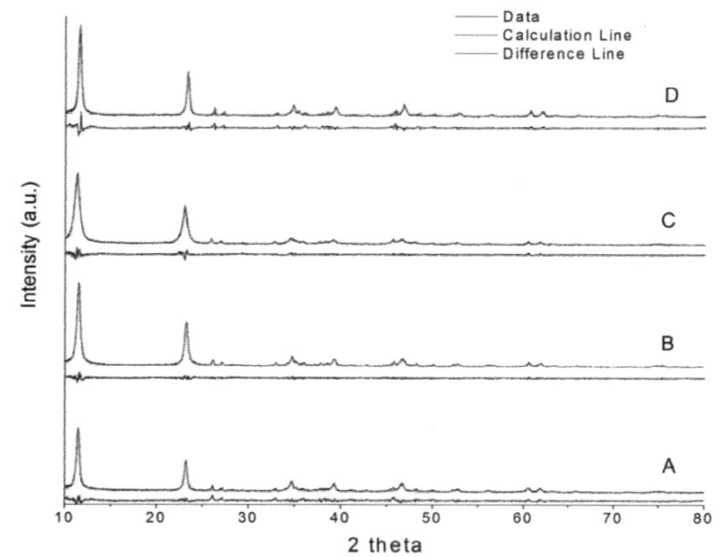

Figure 2. Observed (black), calculated (red) and difference (blue) curves of the Le Bail refine patterns of Ca-Mg-Al hydrotalcite sample that were produced hydrothermal treatment (A) sample A, (B) sample B, (C) sample C1 and (D) sample C2.

The Le Bail refinement results were shown in Figure 2. It could be seen that there was peak corresponding to the calculation results. There was peaks similarity both sample and JCPDS standard, but *Rp* and *Rwp* value result is the difference between Ca-Mg-Al hydrotalcite and standard at 2θ 11° and 23°. It is implied on the value of reliability index parameters i.e. *Rp*, *Rwp*, *Rexp*, and X^2. The limitation of value of reliability index parameters for *Rp* and *Rwp* value are 8.0 [19], while X^2 value is 4.0 [20,21]. In Table 1, all samples Ca-Mg-Al hydrotalcite has low reliability index in *Rp* and *Rwp* value, but X^2 value has concern by value of reliability index parameter.

Table 1. Le Bail reliability factor of sample Ca-Mg-Al hydrotalcite.

Sample	Le Bail reliable factors			
	Rp	*Rwp*	*Rexp*	X^2
A	10.88	14.57	11.89	1.501
B	9.16	12.74	11.03	1.333
C	10.48	14.26	11.15	1.634
D	9.82	13.35	11.16	1.433

Another sense observed was a shift toward lower and higher 2θ values of the reflections. According to the Bragg's law:

$$n\lambda = 2d\ sin\theta \quad (1)$$

where n is 1, λ is wavenumber Cu Kα, 1,54063 Å, d is basal spacing (reflectance) and θ is Bragg angle [7,9]. In the first time, hydrotalcite has a brucite-like $Mg(OH)_2$ network, and then Al^{3+} isomorphous substitution Mg^{2+} in octahedral site. In a case, three element metals like as Mg^{2+}, Al^{3+} and Ca^{2+} in hydrotalcite-like which is Ca^{2+} ions can be replaced Mg^{2+} ion in octahedral site, caused by ionic radius of Ca^{2+} (0,98 Å) was larger than that of Mg^{2+}(0,65 Å) and Al^{3+} (0,50 Å) [7]. Consequently, basal reflectance of d_{003} has become large than Mg-Al hydrotalcite commercial. In Figure 3, sample A has higher basal reflection (7.7968 Å) than Mg-Al hydrotalcite commercial (7.5632 Å). Millange et al. (2000a) reported Ca-Al hydrotalcite (hydrocalumite) has basal reflectance d_{002} at 8.7 Å. Lv et al. (2012) reported that the crystal structure of hydrotalcite could be partly damaged with the increase of amount Ca^{2+} ion to replaced Mg^{2+} in octahedral site. This clarifies calculate Le Bail refinement has higher than limitation of value of reliability index parameters.

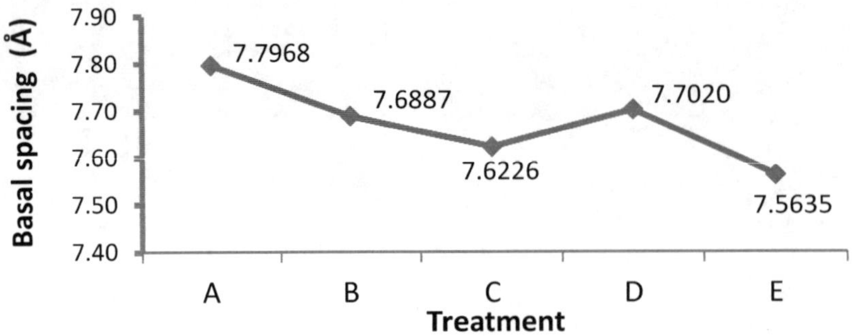

Figure 3. Basal reflection 003 Ca-Mg-Al hydrotalcite with variation treatment hydrothermal (A) sample A, (B) sample B, (C) sample C1, (D) sample C2 and (E) Mg-Al hydrotalcite commercial.

The lattice parameters *c* and *a* were determined from the basal reflectance positions of d_{003}; d_{006}; d_{009} and d_{110}, with formula $c = 3d_{003}$ and $a = 2d_{110}$ [7,10]. The *c* parameter provides data on the character, charge and size of the anion placed between the brucite-like layers. In the same way, parameter *a* can be investigated by a simple way of analyzing the isomorphic substitution of Mg^{2+} by the Al^{3+} cations. [10]. The increase in lattice parameter *c* from 22.6896 Å (sample Mg-Al commercial) to 23.3889 Å (sample C1), and each hydroxide layer increases by in 0.6993 Å thickness layer brucite-like. The increase in lattice parameter *c* (Table 2) is caused by the decrease in the electrostatic attraction between the positive brucite-like sheets and the interlayer, with modification of the OH- - -O bond strength [7]. For lattice parameter *a* has the value of around 3 Å. This is good conformity that informed in the literature [8,24]. The replacement of magnesium with aluminum or calcium leads to increase in lattice parameter *a* from 3.0446 Å (Mg-Al hydrotalcite commercial sample) to 3.5078 Å (C1 sample), reflecting the evidence that the ionic radii of Mg^{2+}, Ca^{2+} and Al^{3+} are 0.66, 0.9 and 0.56 Å, respectively [25].

The crystallite size of Ca-Mg-Al hydrotalcite was computed from the main reflection (11°) using the Scherer formula [26]:

$$\varepsilon = \frac{K\, x\, \lambda}{\beta\, x \sin\theta} \tag{2}$$

Where ε is the crystallite size (nm); K is the shape factor (0.89); λ is the wavelength of the Cu Kα radiation (0.1540 nm), β is the full width at half maximum (FWHM) and θ is the Bragg diffraction angle. Mg-Al hydrotalcite commercial has crystallite size wider than sample hydrothermal treatment. Which is Mg-Al hydrotalcite commercial more crystalline in XRD patterns. In Figure 1, that shown peak in d_{003} and d_{110} has wider than Mg-Al hydrotalcite commercial, so FWHM Mg-Al hydrotalcite commercial is very low than other.

Table 2. Crystallite size Debye-Scherer and lattice parameter (a and c).

Sample	Crystallite size (nm)		Lattice parameter (Å)	
	003	110	a	c
A	2.5995	10.9154	3.0611	23.1779
B	3.6290	6.0817	3.0555	23.0971
C1	3.8499	7.0155	3.0500	22.9673
C2	4.1107	4.7634	3.0573	23.1161
Mg-Al Commercial	8.1507	5.8883	3.0446	22.6896

3.2 FTIR

The FTIR spectra for sample synthesis at hydrothermal treatment are shown in Figure 4. The region between 2600-4000 cm^{-1}, also known as the hydroxyl stretching region was characterized by broad bands at 3474.91 and 3072.74 cm^{-1}. The 3474.91 cm^{-1} band is attributed to the H-bonded interlayer H_2O surrounding the interlayer region and the metal-OH stretching mode. In this one, band cannot distinguish between the various metal (Mg, Ca and Al) bonded to the hydroxyl groups in the infrared spectrum. The 3072.74 cm^{-1} band is in general figure out as the CO_3-H_2O bridging mode [27]. The broad band region at 1635,71 cm^{-1} was attributed the –OH deformation mode of the H_2O molecules in interlayer hydrotalcite [28]. The sharper band region at 1365.66 cm^{-1} is ascribed to the stretching mode of antisymmetric of the carbonate species existing in the interlayer, and the band region 1491.04 cm^{-1} was observed for the free CO_3^{2-} ion in the interlayer region [11].

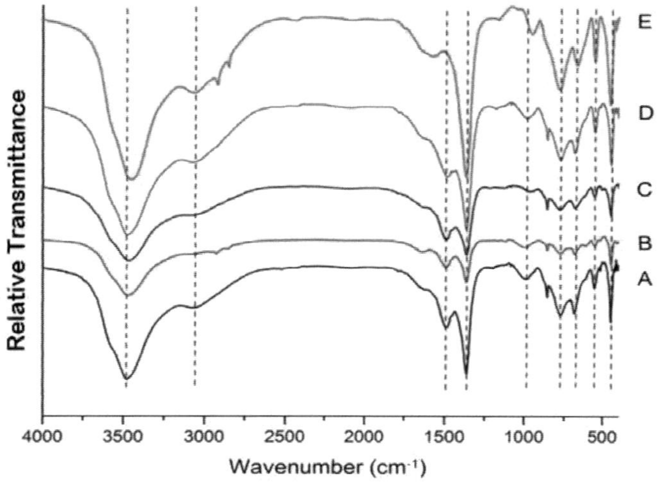

Figure 4. FTIR with variation treatment hydrothermal variation treatment (A) sample A, (B) sample B, (C) sample C1, (D) sample C2 and Mg-Al hydrotalcite commercial (E).

In Figure 4 the band region around 1400 cm^{-1} shown at Mg-Al hydrotalcite commercial has not peak of free carbonate ion. Which is due to the carbonate ion in the interlayer has bonded hydroxyl (HO-M) in layer brucite-like. The band at 685.72 and 857.40 cm^{-1} are characteristic of vibration carbonate symmetric stretching mode [29]. The band region around at 1000-400 cm^{-1} ascribed specific bonding metal (M-O) such as Mg-O, Ca-O and Al-O vibrational modes in the brucite-type layer. The band at 447.50 cm^{-1} is assigned to M^{2+}-O and band at 772.52 cm^{-1} is ascribed vibration M^{2+}-O deformation mode. Kagunya et al. (1998) reported for Ca-O and Mg-O in infrared assignments show that hydrotalcite translation modes are significantly higher frequencies than those of Mg-O. This investigation is reliable with a shorter O–H bond in hydrotalcite than in Mg-O as a result of the increased electrostatic attraction within a hydrotalcite cationic lattice due to the existing of Al cations in site of some of the Mg cations. The band Al-O is assigned to the band at 553.59 cm^{-1} and band at 975.06 cm^{-1} as vibration Al-O deformation mode [28].

4. Conclusion

In this work, Ca-Mg-Al hydrotalcite synthesized with various hydrothermal treatments was successfully prepared. Hence, artificial brine water has contained magnesium and calcium, which is can be a source of hydrotalcite synthesis. The result from this study confirmed that various hydrothermal treatment found hydrotalcite and $CaCO_3$ phase in XRD patterns. Le Bail refinement of the value of reliability index parameter especially X^2 values has concern limitation value of reliability index. A basal reflection, d, has shifted to become higher than basal reflection Mg-Al hydrotalcite commercial (7.5632 Å). The increase in lattice parameter *c* and *a* improved thickness layers brucite-like. The crystallite size has higher than Mg-Al hydrotalcite commercial. Infrared spectra have found hydroxyl band at region 3474.91 and 3072.74 cm^{-1}, carbonate band at region 1365.66; 1491.04; 857.40; and 685.72 cm^{-1}, and metal band at region 447.50; 553.59; 772.52; and 975.06 cm^{-1}.

Reference

[1] Heraldy E, Santosa S J, Triyono and Wijaya K 2015 *Indones. J. Chem.* **15** 234–41
[2] Gao L, Teng G, Xiao G and Wei R 2010 *Biomass Bioenergy* **34** 1283–8
[3] Heraldy E, Nurcahyo I F and Ainurofiq A 2012 *Prosiding InSINas 2012* pp 37–44
[4] Sarkarat M, Komarneni S, Rezvani Z, Wu X, Yin S and Yan Z 2013 *Appl. Clay Sci.* **80–81** 390–7
[5] Millange F, Walton R I and O'Hare D 2000 *J. Mater. Chem.* **10** 1713–20
[6] Kovanda F, Koloušek D, Cílová Z and Hulínský V 2005 *Appl. Clay Sci.* **28** 101–9
[7] Cavani F, Trifirò F and Vaccari A 1991 *Catal. Today* **11** 173–301
[8] Heraldy E, Nugrahaningtyas K D and Heriyanto 2017 *IOP Conf. Series: Materials Science and Engineering* vol 176
[9] Rives V 2001 *Layered Double Hydroxides Present and Future* (New York: Nova Science Publishers, Inc.)
[10] Sánchez-cantú M, Pérez-díaz L M, Rubio-rosas E, Abril-sandoval V H, Merino-aguirre J G, Reyes-cruz F M and Orea L 2014 *Chem. Pap.* **68** 638–49
[11] Hickey L, Kloprogge J T and Frost R L 2000 *J. Mater. Sci.* **35** 4347–55
[12] Sharma S K, Kushwaha P K, Srivastava V K, Bhatt S D and Jasra R V 2007 *Ind. Eng. Chem. Res.* **3** 4856–65
[13] Oh J M, Hwang S H and Choy J H 2002 *Solid State Ionics* **151** 285–91
[14] Heraldy E, Prasasti D, Wijaya K, Santosa S J and Triyono 2010 Studi Pendahuluan Pemanfaatan Limbah Desalinasi Untuk Pembuatan Mg/Al Hydrotalcite-Like *J. Bumi Lestari* **12** 16–23
[15] Sanches E A, Carolino A D S, Santos A L Dos, Fernandes E G R, Triches D M and Mascarenhas Y P 2015 *Adv. Mater. Sci. Eng.* **2015**
[16] Peterson V K 2005 *Powder Diffr.* **20** 14–7
[17] Heraldy E, Rahmawati F, Heriyanto and Putra D P 2017 *J. Environ. Chem. Eng* **5** ***1666-1675***
[18] Zhang M L, Gao Y and Li L F 2011 *Adv. Mater. Res.* **287–290** 569–72
[19] Fei L, Yanhuai L, Zhongxiao S, Kewei X, Zhi Z and Hong C 2015 *Rare Met. Mater. Eng.* **44** 2716–20
[20] Lu X, Shih K, Li X, Liu G, Zeng E Y and Wang F 2016 *Water Res.* **90** 9–14
[21] Yan H and Shih K 2016 *Water Res.* **95** 310–8
[22] Millange F, Walton R, Lei L and O'Hare D 2000 *Chem. Mater.* 1990–4
[23] Lv T, Ma W, Xin G, Wang R, Xu J, Liu D, Liu F and Pan D 2012 *J. Hazard. Mater.* **237–238** 121–32
[24] Prikhod'ko R V, Sychev M V, Astrelin I M, Erdmann K, Mangel A and van Santen R A 2001 *Russ. J. Appl. Chem.* **74** 1621–6
[25] Li F, Jiang X, Evans D G and Duan X 2005 *J. Porous Mater.* **12** 55–63
[26] Langford J I and Wilson A J C 1978 *J. Appl. Crystallogr.* **11** 102–13
[27] Kloprogge J T, Hickey L and Frost R L 2004 *J. Solid State Chem.* **177** 4047
[28] Kloprogge J T, Hickey L and Frost R L 2005 *Mater. Chem. Phys.* **89** 99–109
[29] Kloprogge J T, Hickey L and Frost R L 2004 *J. Raman Spectrosc.* **35** 967–74
[30] Kagunya W, Baddour-Hadjean R, Kooli F and Jones W 1998 *Chem. Phys.* **236** 225–34

Performance prediction of serpentine magnetorheological valves under various gap size

M H Idris[1], Ubaidillah[2],S A Mazlan[1], F Imaduddin[3], A Abd Fatah[4] and D D D P Tjahjana[2]

[1] VSE Research Laboratory, Malaysia-Japan International Institute of Technology, Universiti Teknologi Malaysia, Jalan Sultan Yahya Petra (Jalan Semarak), 54100 Kuala Lumpur, Malaysia
[2] Mechanical Engineering Department, Faculty of Engineering, Universitas Sebelas Maret, Jl. Ir. Sutami 36A, Kentingan, Surakarta, 57126, Central Java, Surakarta, Indonesia
[3] Faculty of Engineering and Technology, Multimedia University, Jalan Ayer Keroh Lama, Bukit Beruang, 75460 Melaka, Malaysia
[4] UTM Razak School of Engineering and Advanced Technology, Universiti Teknologi Malaysia, Jalan Sultan Yahya Petra (Jalan Semarak), 54100 Kuala Lumpur, Malaysia

E-mail: ubaidillah_ft@uns.ac.id

Abstract. Magnetorheological (MR) valve is one of the key components in regulating the magnetorheological (MR) fluid flow in MR devices. Due to MR properties enabling magnetic modifications, many MR valve designs have been introduced and are widely utilized in MR devices. Serpentine MR valve has a potential in improving the MR devices performance, whereby using this method it can help to increase the effective region of the valve. Thus, this article is comparing the pressure drops generated under various gap sizes in the effective regions of MR valve. Three different gap sizes are considered for this study. To provide a fair comparison, several parameters are fixed; the size of the MR valves, the number of coil turns and the power consumption. In order to demonstrate the MR valve performance, the pressure drops are calculated based on the mathematical equations derived from MR valve models and the magnetic fields derived from Finite Element Method Magnetics (FEMM) software. The results of the simulations are compared to various gap sizes and, discussed.

1 Introduction

Recently, applications and devices utilizing magnetorheological (MR) fluids are experiencing a rise due to its property uniqueness of magnetic controllable and fast-response rheological change with the magnetic field. MR fluid is a type of smart material where the viscosity of the fluid increases when the magnetic field increases [1]. Currently, MR fluid devices use several modes of operations, and the most commonly used is the valve mode, which is also known as MR valve [2–7]. A conventional MR valve usually includes an electromagnetic coil to generate the magnetic field intensity. With the use of electromagnetic circuit, the pressure drop of the MR valve can be increased by raising the current in the electromagnetic circuit, which in turn increases the magnetic field strength.

The valve dimensions and configurations had been observed to affect the performances of the MR valve [3]. Researchers have developed several studies on the geometrical arrangements, which utilizes the gaps for the fluid flow path known as annular and radial gaps [8]. An annular gap is a flow channel that allows the MRF to flow axially in the MR valve and the placement of it is parallel to the inlet and outlet of the valve. Meanwhile, for the radial gaps, the fluid flows transversely in a radial manner, forcing the fluid to flow perpendicular to the inlet and outlet of the valve. The applications of these two gaps have been established in the MR devices [9–11]. The progress of the MR valve studies also continues with a new type of configuration which is the serpentine configuration [5,12]. The magnetic flux in the design of serpentine MR valve is weaved into the annular gap by alternating the non-magnetic and magnetic materials. Previous study only investigate two gap sizes for the serpentine configuration therefore to wider the variations of the gap size, three gap sizes are considered [5].

One study reported a methodology in configuring and comparing different MR valve configurations using analytical models and Finite Element Analysis [FEA] [13]. This type of analysis is remarkably time consuming with various processes involved in the evaluation to determine the most suitable design for a particular application. In order to shorten the process, this study aims to compare the pressure drop generated by each types of MR valves involved (no previous mention on how many valves are involved). The operating range (of what?) and the pressure magnitude (pressure drop?) are also discussed. Thus, the analysis serves as a basis in selecting or designing tools for MR valves.

2 MR valve modelling

The performance of an MR valve is analyzed based on its capability and flexibility in producing the pressure drop between upstream and downstream sides. Valve ratings are mainly determined by the pressure drop hence the mathematical modelling of pressure drop is used for each configuration in each section. The basic expression of the pressure drop in the MR valve is observed as [14]:

$$\Delta P = \Delta P_{\text{Viscous}} + \Delta P_{\text{Yield}} \tag{1}$$

where, $\Delta P_{\text{Viscous}}$ and ΔP_{Yield} are the viscous and field-dependent yield stress of pressure drop. The mathematical expressions of viscous and field dependent pressure drop for annular and radial flow path are observed as [3,15] :

Table 1 Equation for annular and radial gap

Gaps	Viscous	Yield
Annular	$\Delta P_{\text{Viscous}} = \frac{6 \quad QL}{\pi d^3 R}$	$\Delta P_{\text{Yield}} = \frac{c\tau\,(B)L}{d}$
Radial	$\Delta P_{\text{Viscous}} = \frac{6 \quad Q}{\pi d^3}\text{In}\left(\frac{R_0}{R_i}\right)$	$\Delta P_{\text{Yield}} = \frac{c\tau\,(B)}{d}(R_0 - R_i)$

where d is the valve gaps for both annular and radial gaps, is the fluid base viscosity, Q is the flow rate, L is the annular channel length of the valve, R is the channel radius, $\tau(B)$ is the field dependent yield stress value and c is the flow velocity profile coefficient. Also the R_0 and R_i referring to the inner and outer radius of radial gaps respectively.

The coefficient of c can be obtained through the calculation of ratio between viscous pressure drop and the total pressure drop as seen following[3]:

$$c = 2.07 + \frac{12Q}{12Q \quad + 0.8\pi R d^2 \tau\ (B)}. \tag{2}$$

Meanwhile, for the equation on the orifice gaps, the expression is slightly different since there is no field dependent yield stress on the MR fluid. The pressure drop equation of orifice gaps solely expressed in viscous resistance as shown [2]:

$$\Delta P = 2\frac{8 \quad QL}{\pi R^4}. \tag{3}$$

3 MR valve configuration

In this research paper, the strength of magnetic flux densities was evaluated to predict the performance of MR valve. The magnetic field strength of MR valves are difficult to measure through experimental method thus the numerical approach was chosen to aid the prediction of MR effect. FEMM (Finite Element Method Magnetics) software was chosen to analyze magnetic simulations in this study.

Parameters are required when conducting the simulation to obtain the magnetic flux distributions. The coils of each MR valve were designed and fixed with 450 turns of 24 AWG copper wires and a total resistance of 5.5 Ω. Due to the maximum current applied to the coil being 1 A, the power consumption is only 5.5 watt. The simulation was conducted in 2D axis-symmetric with different total elements and node according to the type of MR valves. The permeability of MR fluid was obtained from the manufacturer. Meanwhile, the permeability of magnetic materials of MR valve is assumed to be similar to the B-H curve of AISI 1020 while the permeability of non-magnetic materials are following the B-H curve of AISI 6061-T6. The results are discussed in each section for each type of MR valve.

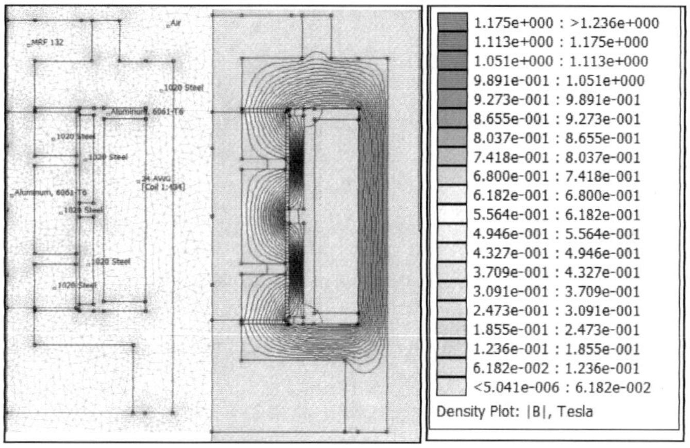

Figure 1 Configuration of Serpentine type MR valve

4 Pressure drop generated by serpentine MR valve

Figure 2 depicts the pressure drop versus current for the serpentine MR valve that was proposed in this study with three gap sizes and flow rate of 10 ml/s. The figure shows the MR valve could experience various pressure drop under various gap sizes due to the MR effect was generated by the coils and also the fluid path. During the off-state, the MR valve would work based on the viscosity of the MR fluid. It was identified that the serpentine MR valve with a 0.5 mm gap size can produce a pressure drop of 0.13 MPa in the absence of the input current and could be increased up to 7.3 MPa by applying the input current of 1.0 A with flowrate of 10 ml/s. Meanwhile serpentine MR valve with a 0.75 mm gap size can produce a pressure drop of 0.06 MPa in the absence of the input current and could be increased up to 3.9 MPa by applying the input current of 1.0 A with flowrate of 10 ml/s. The lowest pressure drop generated is from 1.0 mm gap size, with a pressure drop of 0.03MPa in the absence of the input current and could increase up to 2.9 MPa by applying the input current of 1.0 A in the flowrate of 10 ml/s. Specifically, the smallest gap size generates the highest pressure drop.

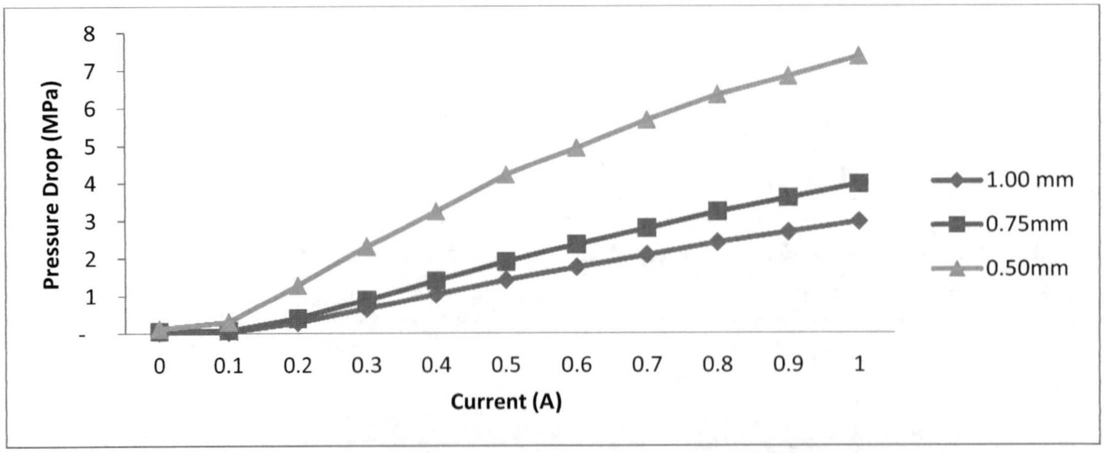

Figure 2 Pressure drop comparison

Figure 3 depicts the off-state pressure drop generated by each gap size for various flowrate. The figure shows 0.5mm gap size having the highest force and 1.0mm gap size is the lowest. Meanwhile, Figure 4 shows the on-state pressure drop generated for each gap size with various flow rate. Results shows that smallest gap size is the highest pressure drop generated follows by the 0.75 mm and the 1.0 mm gap size.

Figure 5 illustrates the dynamic range of the pressure drop generated from each gap size of serpentine MR valves. Valve with 0.50mm gap size produced the biggest value compared to the other gap size. Meanwhile, valve with 0.75 mm is ranked the second highest in pressure drop generated the lowest falls to 1.0 mm gap size.

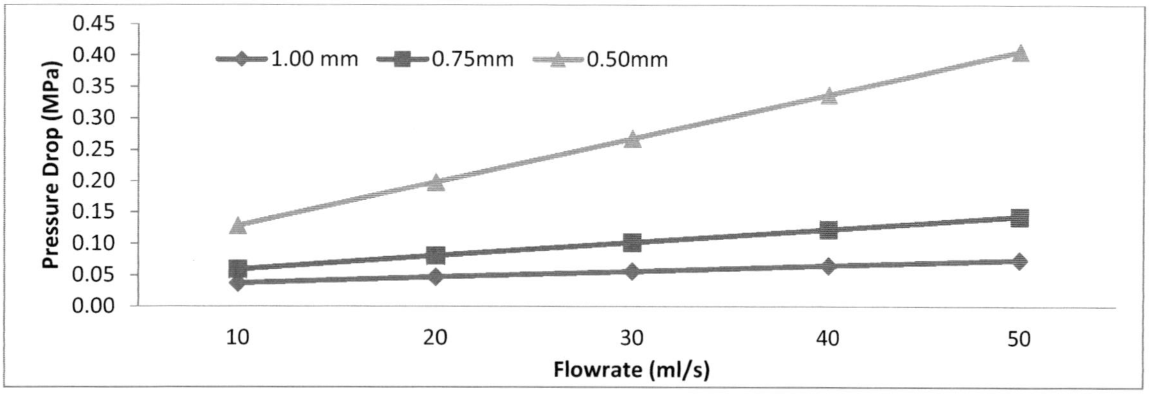

Figure 3 Off-state pressure drop of serpentine MR valve

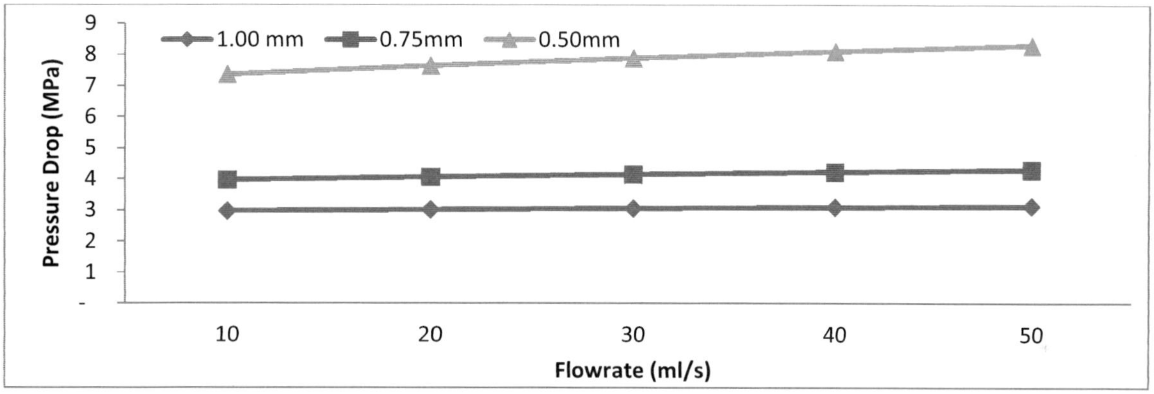

Figure 4 On-state pressure drop of MR valve

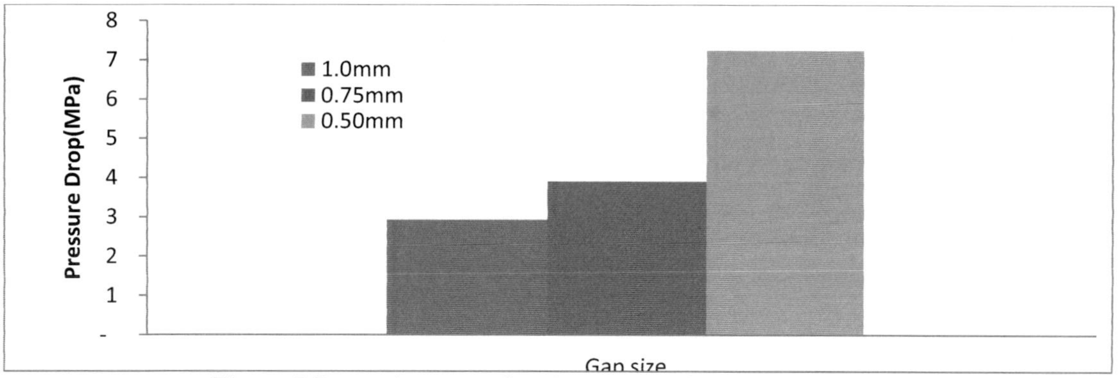

Figure 5 Dynamic range of each gap size of serpentine MR valve

5 Conclusion

The presented study has discussed the comparison of three gap sizes of serpentine MR valve: 0.50, 0.75, and 1.0 mm. For each type of gap size, the same mathematical modelling of pressure drop for serpentine MR valves were developed and the results were compared. The proposed approach aims to study the performance of serpentine MR valve based on different gap sizes. The analysis also considers constant size of MR valves, number of turns and power consumption. The simulation results have shown that the smaller gap size, which is gap 0.50 mm can generate a higher pressure drop and higher dynamic range. Finally, the proposed studies have reported useful insight of MR valve types and new information for the selection of gap size of serpentine MR valves with different application requirement and valves performances.

Acknowledgement

This research was financially supported by Universitas Sebelas Maret under Hibah Mandatory 2017 as well as the Ministry of Education Malaysia and Universiti Teknologi Malaysia under PRGS Grant (Vote No: 4L667) and Tier 1 Grant (Vote No: 13H55).

References

[1] P P Phulé 2001 *Smart Mater. Bull.* 7–10
[2] A Grunwald, A G Olabi 2008 *Sensors Actuators, A Phys.* **148** 211–223
[3] Q -H Nguyen, S -B Choi, N M Wereley 2008 *Smart Mater. Struct.* **17** 25024
[4] F Imaduddin, S Amri Mazlan, M Azizi Abdul Rahman, H Zamzuri, Ubaidillah, B Ichwan 2014 *Smart Mater. Struct.* **23** 65017
[5] A Y Abd Fatah, S A Mazlan, T Koga, H Zamzuri, F Imaduddin 2016 *Int. J. Appl. Electromagn. Mech.* **50** 29–44
[6] M Mao, W Hu, Y -T Choi, N M Wereley 2007 *J. Intell. Mater. Syst. Struct.* **18** 1227–1232
[7] X Bai, N M Wereley, Y Choi 2016 *J. Intell. Mater. Syst. Struct.* **27** 944–958
[8] H Sahin, F Gordaninejad, X Wang, Y Liu 2012 *J. Intell. Mater. Syst. Struct.* **23** 949–957
[9] F Gordaninejad, X Wang, G Hitchcock, K Bangrakulur, S Ruan, M Siino 2010 *J. Struct. Eng.* **136** 135–143
[10] H Hu, X Jiang, J Wang, Y Li 2012 *J. Intell. Mater. Syst. Struct.* **23** 635–645
[11] A Z Bin Pokaad, K Hudha, M Z B M Nasir, N A Ubaidillah 2011 *Int. J. Struct. Eng.* **2** 164
[12] D Senkal, H Gurocak, *Proc.-3rd Jt. EuroHaptics Conf. Symp. Haptic Interfaces Virtual Environ. Teleoperator Syst. World Haptics* 91–96
[13] T Xu, M Liang, C Li, S Yang 2015 *J. Sound Vib.* **355** 66–85
[14] H X Ai 2006 *J. Intell. Mater. Syst. Struct.* **17** 327–334
[15] D H Wang, H X Ai, W H Liao 2009 *Smart Mater. Struct.* **18** 115001.
[16] A Z B Pokaad, K Hudha, M Z B M Nasir, Ubaidillah 2011 *Int. J. of Structural Engineering* **2** 164-187

IOP Conf. Series: Materials Science and Engineering **333** (2018) 012008 doi:10.1088/1757-899X/333/1/012008

Design of magnetic Circuit Simulation for Curing Device of Anisotropic MRE

N Hapipi[1], **Ubaidillah**[2], **S A Mazlan**[1], and **P J Widodo**[2]

[1] VSE Research Laboratory, Malaysia-Japan International Institute of Technology, Universiti Teknologi Malaysia, Jalan Sultan Yahya Petra (Jalan Semarak), 54000 Kuala Lumpur, Malaysia

[2] Mechanical Engineering Department, Faculty of Engineering, Universitas Sebelas Maret, Jl. Ir. Sutami 36A, Kentingan, Surakarta, 57126, Central Java, Surakarta, Indonesia

E-mail: amri.kl@utm.my

Abstract. The strength of magnetic field during fabrication of magnetorheological elastomer (MRE) plays a crucial role in order to form a pre-structured MRE. So far, gaussmeter were used to determine the magnetic intensity subjected to the MRE during curing. However, the magnetic flux reading through that measurement considered less accurate. Therefore, a simulation should be done to figure out the magnetic flux concentration around the sample. This paper investigates the simulation of magnetic field distribution in a curing device used during curing stage of anisotropic magnetorheological elastomer (MRE). The target in designing the magnetic circuit is to ensure a sufficient and uniform magnetic field to all the MRE surfaces during the curing process. The magnetic circuit design for the curing device was performed using Finite Element Method Magnetic (FEMM) to examine the magnetic flux density distribution in the device. The material selection was first done instantaneously during a magnetic simulation process. Then, the experimental validation of simulation was performed by measuring and comparing the actual flux generated within the specimen type and the one from the FEMM simulation. It apparent that the data from FEMM simulation shows an agreement with the actual measurement. Furthermore, the FEMM results showed that the magnetic design is able to provide sufficient and uniform magnetic field all over the surfaces of the MRE.

1 Introduction

Magnetorheological elastomer (MRE) is one of the smart materials in which their mechanical and rheological properties can be quick, persistently and reversibly changed by controlling the external magnetic field. MRE comprises carbonyl iron particles (CIP) implanted in a rubber-like matrix. Generally, MRE can be classified into two types which isotropic and anisotropic MRE fabricated with and without external magnetic field [1]. Curing the MRE with an external magnetic field can prompt the formation of a chain-like structure that trapped inside the rubber matrix [2]. These chain-like structures are beneficial to improve the MR effect of MRE [3]. Because of these special properties, anisotropic MRE received increasing attention and gained wide application prospects mainly in applications that employ variable stiffness such as vibration absorbers [4,5].

The formation of chain-like structures exceptionally reliant on the strength of magnetic field during curing process [6]. The formation of chains can be fluctuated by the strength of magnetic field. It can be a single chain-like, three-dimensional structures or even more complex structure [7]. The variability of magnetic field strength during curing will result in the formation of anisotropic MRE with different magnetorheological effect. For instance, Li et al [6] have fabricated MRE with different magnetic field strengths from 0 to 350 mT and the results revealed the differences of MR effect for each of the sample. Wu et al. [3] cured polyurethane MRE under a magnetic field strength of 900 mT, the relative MR effect achieved was only 21%. Ju et al. [8] fabricated anisotropic polyurethane based MREs with a constant magnetic field strength of 700 mT. The result revealed that the MR effect of the anisotropic MRE was 33.7 %.

In order to fabricate the MRE with anisotropic configuration, curing device with an attachment of the external magnetic field device or a couple of permanent magnets could be considered to be utilized. Previous researchers have built up their own curing device to fabricate the anisotropic MRE. Lu et al [9] developed the curing system with the attachment of magnetic field and temperature control to fabricate the styrene-b-ethylene-co-butylene-b-styrene (SEBS) based MRE. The MRE was cured using the homemade magnet-heat coupled device and the actual intensity was measured with a Tesla meter. In the mean time, Kaleta et al [1] had used a couple of permanent magnets to cure the MRE samples. The strength of the magnetic field applied during curing was tested in between the gap of the mold and the value is equal to 0.5 T. However, a noteworthy issue with this sort of technique is the correct estimation of the magnetic field intensity subjected to the MRE sample can't be resolved. Furthermore, most of the researchers have focused on studying the rheological properties of anisotropic MRE itself, such as its dynamic stiffness, damping factor and MR effect. A little report has been noticed discussing on the magnetic intensity distribution inside the curing device to ensure the uniform distribution to the MRE samples. Therefore, this paper focuses on the study of magnetic field distribution in curing device using FEMM. This FEMM analysis is used to illustrate the efficiency of the design. Finally, an actual measurement was measured in the chamber of the device and the result was compared with the simulation result. It was done to verify the distribution of the magnetic flux within the MRE samples as the real magnetic intensity within the MRE sample cannot be measured with the gauss meter probe

2 Experimental

2.1 Design Consideration

The important consideration in designing the curing device is to make sure that all the magnetic lines must pass through the area inside the magnetic coil. Therefore, the strongest and uniform magnetic field distribution can be guaranteed at the MRE sample areas. The magnetic field is uniformly concentrated in the center of the coil, while at the outside region of the coil, the strength of the field is weak and divergent. Figure 1 shows the proposed idea of the mold and curing device that would be developed. The curing device mainly consists of a magnetic coil, bobbin, steel cover customized mold. The details of the components include the raw materials are listed in Table 1.

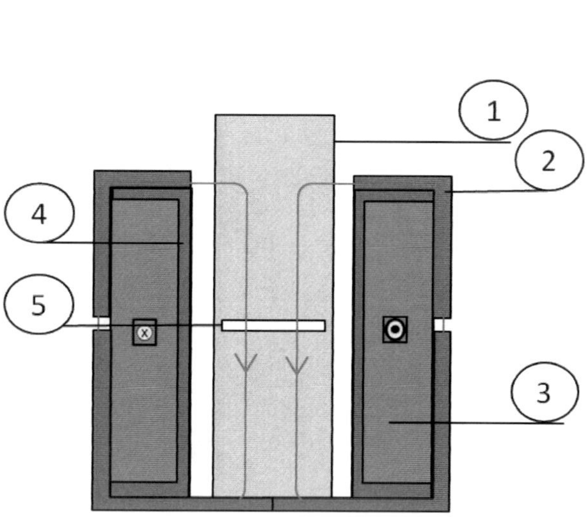

Figure 1. (a) Cross-section view for curing device.

Table 1. Detail of material selections for each part.

No.	Part list	Material
1	Mold	Low carbon steel, AISI 1117
2	Steel cover	Low carbon steel, AISI 1117
3	Coil	Copper wire (0.9mm)
4	Bobin	Plastic
5	MRE sample	-

The space between the coil and the bobbin are separated by the ambient air to allow heat release generated during magnetization [10]. Due to the magnetic principle, all the magnetic lines that enter any region must also leave that region without any remanence [11]. The customized mold was made from low carbon steel and will be placed in the center of the coil. The region of the MRE sample is about 1 x 50 mm in thickness and diameter respectively. The steel from the same types of mold was used to cover the bobbin of the coil. It is necessary to create an enclosed path for magnetic flux so that the flux is passed and went to the inside core through the mold. The magnetic coil then connected to an external direct current (DC) power source to generate the magnetic field throughout the coil. When the current is "ON", thus the charges will move around the coil and produce the magnetic field which is called as electromagnetism. This phenomenon can be explained by Faraday's Law where it explained that a changing magnetic field produces an electric field. For example, if a loop of wire is placed in a magnetic field so that the field passes through the loop, a change in the magnetic field will induce a current in the loop of wire.

2.2 Simulation

The magnetic flux distribution around the curing device was simulated first using FEMM 4.2 software. FEMM is one of the useful software to analyze the magnetic behavior around the magnetization area. In order to validate the result from the FEMM software, the actual measurement from the curing device was measured using Hirst Magnetic Instrument GM08 Gaussmeter. Figure 2 shows a step by step of the magnetic circuit simulation using FEMM software.

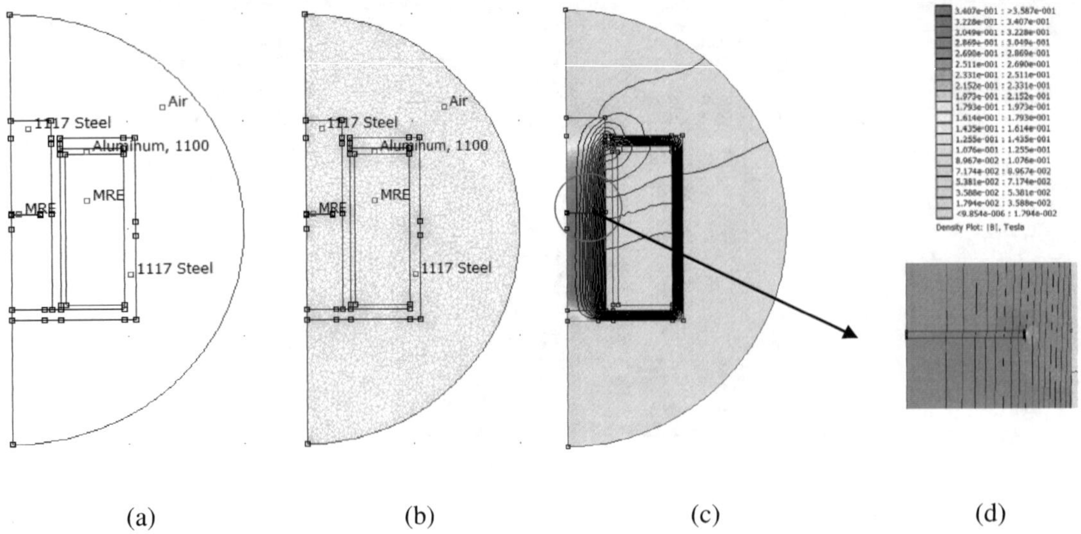

Figure 2. Model of curing device for anisotropic MRE in FEMM simulation (a) designing stage, (b) meshing area of the model, (c) magnetic flux density distribution, and (d) close up at MRE samples space.

The simulation was performed in a 2D axis-symmetric plane. The sketch map of the curing device includes the mold are illustrated in Fig. 2 (a). The simulation was conducted in 2D axis symmetric were the cross section of the desired design was drawn to represent the magnetic flux distribution around the design. Then, the type of materials was chosen at the FEMM built in library of block property definitions. The selection of material was based on the availability of the material on the market. Thus, low carbon steel, AISI 117 was chosen as the main material for the mold and the steel cover due to the abundant source in the market. The coil was firmly wounded on the plastic bobbin and the total winding number of a coil is 1500 turns with the diameter of 0.9 mm. After all the block were labeled by the selected materials, the mesh was created by 5545 nodes as shown in Fig. 2 (b) before the flux line plotting as illustrated in Fig. 2 (c). The magnetostatic simulation plotted in Fig. 2 (c) has resulted from 0.5 A of applied current in each coil. From the figure, the flux lines pass through the steel cover right into the mold by forming a closed loop magnetic circuit. Fig. 2 (d) shows the close up of the MRE region at the center of the mold. From the magnetic simulation, the region for MRE samples shows the highest strength and uniform magnetic flux concentration.

3 Results and Discussion

The verification of the FEMM simulation then was performed by comparing the result from the simulation results with the actual measurement at the test chamber of curing device. The measurement was made in the region between the top of the steel cover and the mold. The currents were controlled by external DC power supply and varied from 0 to 1 Tesla by an interval of 0.1 A. Fig. 3 presents the comparison between the simulation and the actual measurement. The differences between these two are still considered acceptable.

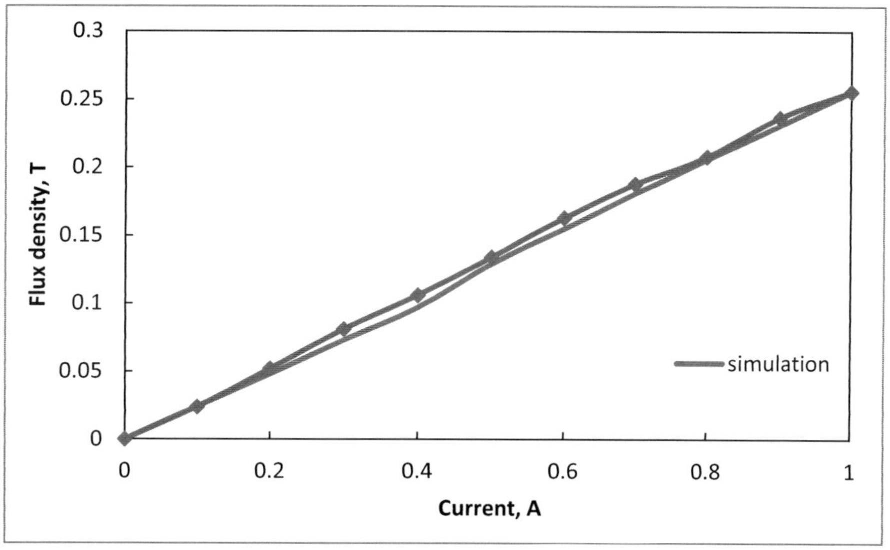

Figure 3. Validation result between actual measurement and FEMM simulation.

Next, the simulation was made in the MRE region to evaluate the flux density that passes through the MRE sample. This is because, in real measurement, the gaussmeter cannot be placed nearby the sample. Therefore, by doing this simulation, the magnetic flux density passes through the MRE sample can be predicted. Table 2 presents the simulation measurement for the MRE region. As shown in Table 2, the value of the magnetic field strength can be achieved when the regulated current supply is 1T was about 681 mT approximately.

Table 2. Measurement of magnetic flux density at MRE region.

Current, A	Magnetic field strength, mT (MRE region)
0.1	70
0.3	209
0.5	345
0.7	482
1.0	681

4 Conclusions

Through the finite element analysis, this paper presents a thorough investigation of the magnetic flux intensity distribution inside the curing device. The curing device for anisotropic MRE was developed based on the rough simulation through FEMM software. The magnetic flux density distribution has been successfully stimulated by the software in order to identify the uniform distribution to all over the MRE sample. Experimental testing was conducted to obtain the actual magnetic field passes through the MRE region. The result of the actual measurement was then compared with the simulation analysis and it is found that the result shows a correlation with the measurement result. Furthermore, through the simulation, it is observed that the middle of the mold possessing the strongest and uniform magnetic field distribution. Therefore, it is suggested the MRE samples should be placed in the middle of the mold.

Acknowledgement

This research was financially supported by the Ministry of Education Malaysia and Universiti Teknologi Malaysia under PRGS Grant (Vote No: 4L667) and Tier 1 Grant (Vote No: 13H55).

References

[1] Kaleta J, Królewicz M and Lewandowski D 2011 *Smart Mater. Struct.* **20** 085006
[2] Stepanov G, Borin D and Odenbach S 2009 *J. Phys. Conf. Ser.* **149** 012098
[3] Wu J, Gong X, Fan Y and Xia H 2010 *Smart Mater. Struct.* **19** 105007
[4] Sun S, Deng H, Yang J, Li W and Du H 2015 *Smart Mater. Struct.* **24** 45045
[5] Kim H K, Kim H S and Kim Y 2017 *Smart Mater. Struct.* **26** 015016
[6] Li J, Gong X, Xu Z and Jiang W 2008 *Int. J. Mater. Res.* **99** 1358–64
[7] Zhang W, Gong X, Xuan S and Jiang W 2011 *Ind. Eng. Chem. Res.* **50** 6704–12
[8] Ju B, Tang R, Zhang D, Yang B, Yu M, Liao C, Yuan X, Zhang L and Liu J 2016 *Polym. Compos.* **37** 1587–95
[9] Lu X, Qiao X, Watanabe H, Gong X, Yang T, Li W, Sun K, Li M, Yang K, Xie H, Yin Q, Wang D and Chen X 2012 *Rheol. Acta* **51** 37–50
[10] Ubaidillah, Mazlan S A, Yahya I and Harjana 2015 *Asian Des. Eng. Work.*
[11] Li Y and Li J 2015 *J. Intell. Mater. Syst. Struct.* **26** 1861–70

IOP Conf. Series: Materials Science and Engineering **333** (2018) 012009 doi:10.1088/1757-899X/333/1/012009

Magnetic field dependence of the current flowing in the spin-coated chlorophyll thin films

J R P Aji, Kusumandari, B Purnama[1]

Physics Department Sebelas Maret University, Jl. Ir. Sutami 36A Kentingan Jebres Surakarta 57126, INDONESIA

E-mail: bpurnama@mipa.uns.ac.id

Abstract. The magnetic dependence of the current flowing in the spin coated chlorophyll films on a patterned Cu PCB substrate has been presented. Chlorophyll was isolated from *Spirulina* sp and deposited by spin coated methods. The reducing of current by the change of magnetic field (magneto conductance effect) was performed by inducing the magnetic field parallel to the inplane of film at room temp. The magnetoconductance ratio decreases as the increase of voltage. It was indicated that the origin of carrier charge in chlorophyll films should be different with the carrier charge injection (electron).

1. Introduction

Organic material has attracted many researchers because its weak molecules bond in the solid state can make organic materials as insulators and semiconductors. Semiconductors based organic has various advantages compared to semiconductors based inorganic in applications of the optics and electronics devices. Furthermore, the manufacture of semiconductors based organic is simple and cheap [1,2].

One of the most common organic materials studied in electronic application is chlorophyll. The chlorophyll forms a nanometer-discrete layer as a result of a surface substrate interaction [3-5]. So that, recently the chlorophyll thin films is the one of the important part in the electronics based nano hybrid system [6-8].

Chlorophyll is commonly isolated using chromatography method from natural materials such as spinach or micro-algae [9]. The products of isolation become the source materials of thin layer that can be prepared by various methods such as evaporation, vapor deposition, spin coating, and Langmuir Blodgett deposition technique [10].

In this paper, the magnetic dependence of current flowing in chlorophyll films is discussed. The isolated solution of chlorophyll was confirm by UV-Vis spectroscopy and the electrical property was evaluated by I-V measurement.

2. Experimental

2.1 Preparation of natural chlorophyll.

Fabrication of natural chlorophyll coating was done through isolation process including extraction, chromatography, and evaporation stage as experimental procedures that had been done before [11]. The material used was *Spirulina* sp fresh water production of PT Neo-algae Sukoharjo, Central Java, Indonesia.

2.2 Growth and characterization of thin layers

The *Spirulina* sp solution was deposited on the PCB substrate. The PCB Cu Substrate Pattern is shown in Fig. 1. Before using the PCB, its surface must be cleaned. PCB was conducted by washing it with 99% alcohol and acetone via ultrasonic cleaner for 15 minutes. Then, PCB was allowed to dry at room temperature conditions.

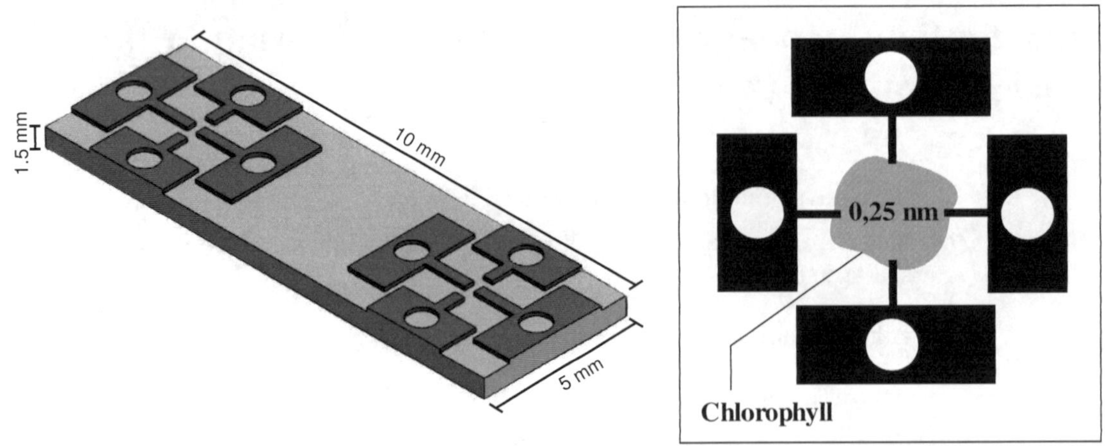

Figure 1. The modified substrates of Cu PCB used in the experiment.

The layer growth of chlorophyll was carried out by Chemat Technology KW-4A spin coating. Deposition process was done at using 2000 rpm rotational speed for 30 s, then followed by heating process at temperature of 70°C for 90 s.

The characterization of this research included observing the curve of absorbance performed UV-Vis Spectrometer. Finally, the IV meter was carried out to observe the magnetic field effect of current flowing samples (magneto conductance effect) shown as in Fig. 2.

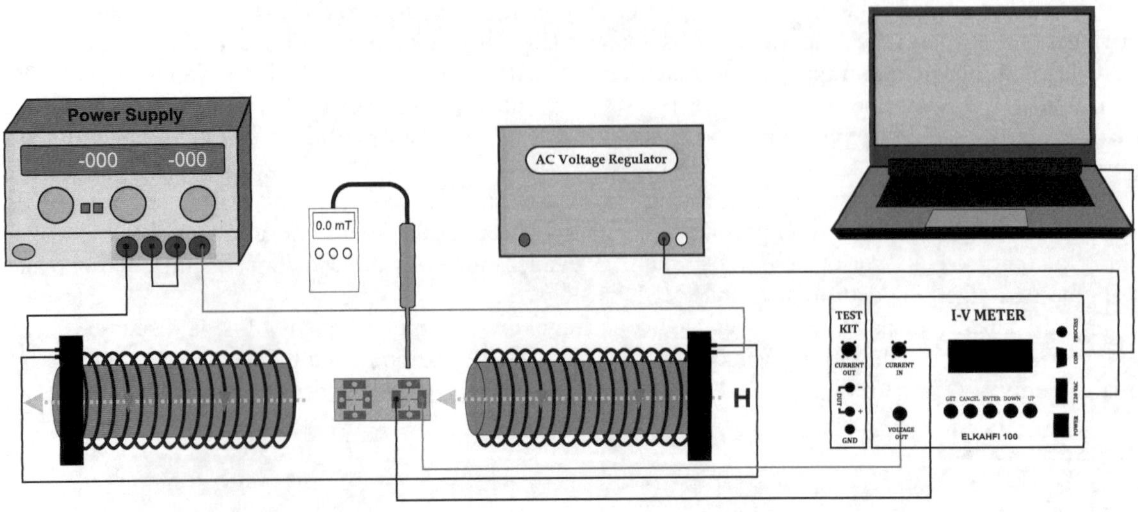

Figure 2. The experimental set up of magnetoconductance measurement.

3. Results and Discussion

UV-Vis characterization was performed to confirm the existence of chlorophyll in *Spirulina* sp solution. Fig. 3 shows the typical of absorbance profile for chlorophyll solution. As the results, two peaks of absorbance curve at λ = 430.50 and 662.50 nm are confirmed as chlorophyll [12].

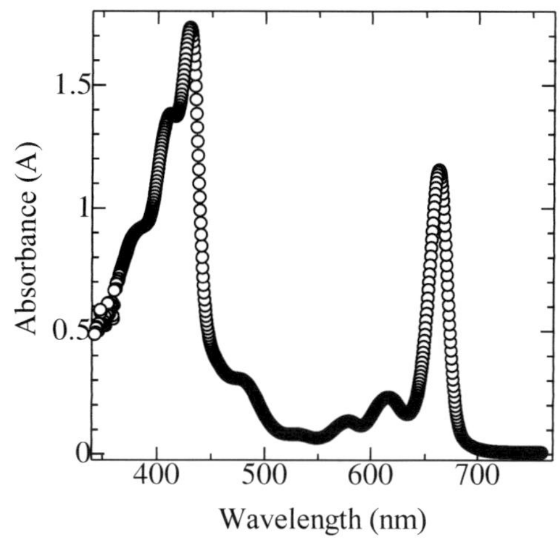

Figure 3. The absorbance of isolated chlorophyll from microalgae *Spirulina* sp.

The reducing current *I* within the increasing magnetic field *H* is displayed by Fig. 4. The current *I* is maximum at zero magnetic field and then gradually decrease with the increase of the magnetic field. After exceeding a given magnetic field, the current tends to be constant. It clearly observed by Fig. 4 that the typical reduce of the current change for different voltage used for supply current. In case of V = 0.4 volt, at *H* = 0 mT, the current of 0.478 nA is obtained. The current is reduced to 0.317 nA when the field of 20 mT is applied. At V = 0.8 volt and *H* = 0 mT, the current obtained is 0666 nA and it tends to be constant at 0.435 nA at magnetic field *H* of 15 mT.

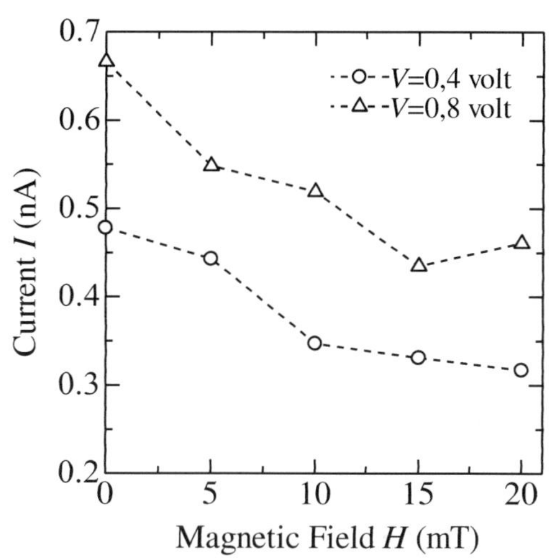

Figure 4. Graph of the current as a function of magnetic field evaluated at V= 0.4 and 0.8 volt.

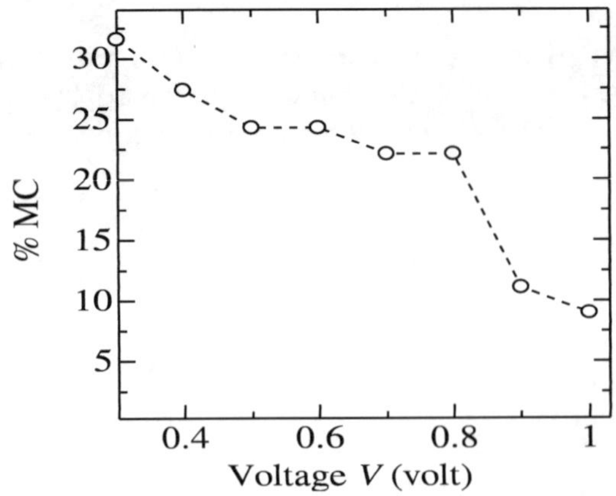

Figure 5. The MC ratio with the increase of magnetic field evaluated for various voltage of chlorophyll film with N = 9.

Fig. 5 shows the reduced current as a function of magnetic field evaluated for various voltages. The reduced current in applied magnetic field is defined as magnetoconductance (MC), so the MC ratio may express as MC = $[I(H)–I(H=0)]/I(H=0)$ in which $I(H)$ and $I(H=0)$ are the current when the magnetic field is applied and zero magnetic field, respectively. Fig. 5 presents the MC ratio decrease as the increase of voltage. The increase of voltage means the increase of supplying of carrier charge into the samples. It should be noted that the supplying of the carrier charge caused by the increase of the voltage is electron with negative carrier charge. The reduction of the MC ratio with the increase voltage should indicate that the dominant carrier charge of the chlorophyll films is different to carrier charge supplied by the increase voltage. So, the response of the available carrier charge pair to the induction of the magnetic field is the recombination, with in result the reduction of the MC ratio.

4. Conclusion

The magnetic dependence of the current flowing in the spin coated chlorophyll films on a patterned Cu PCB substrate has been presented. The reducing of current by the change of magnetic field (magneto conductance effect) was performed by inducing the magnetic field parallel to the inplane of film at room temp. The magnetoconductance ratio decreases as the increase of voltage. It is indicated that the origin of carrier charge in chlorophyll films should be different with the carrier charge injection (electron).

References

[1] Bloom F L, Wagemans W, Kemerink M, and Koopman B 2007 *Phys. Rev. Let.* **99** 257201
[2] Punke M, Mozer S, Stroisch, Gerken M, Bastian G, Rabus D, and Henzi P 2006 *SPIE* 6185 618505-1
[3] S Boussad, A Tazi, and R M Leblanc 1999 *J. Colloid Interface Sci.* **209** 341-346
[4] S Boussad, J A DeRose, and R M Leblanc 1995 *Chem. Phys. Let* **246** 107-113
[5] M Crevecoeur *et al* 2000 *Protoplasma* **212** 46-55
[6] P Bombelli *et al* 2012 *Phys. Chem. Chem. Phys* **12** 12221-12229
[7] S Y Chen *et al* 2013 *Carbon* **63** 23-29
[8] Patrascu *et al* 2014 *Beilstein J. Nanotech.* **5** 2316-1325
[9] Jubert C and Bailey G 2007 *J. Cromatography* **1140** 95-100
[10] El- Nahass M M, Farag A A M, El-Metwally M, Samaha F.S.H.A, and Elesh E 2014 *Synthetic Metals* **195** 110-116
[11] Elvira F F 2014 *Jurnal Fisika dan Aplikasinya* **2** 99-102
[12] Li Y, Scales N, Blankenship R E, Willows R D, and Chen M 2012 *Bio et Biophy Acta* **1817** 1292–1298

IOP Conf. Series: Materials Science and Engineering **333** (2018) 012010 doi:10.1088/1757-899X/333/1/012010

Derivation and constants determination of the Freundlich and (fractal) Langmuir adsorption isotherms from kinetics

Patiha[1], M Firdaus, S Wahyuningsih, K D Nugrahaningtyas and Y Hidayat
Department of Chemistry, Sebelas Maret University, Surakarta 57126, Indonesia
Tel./Fax. +62 271 669376

E-mail : patiha31@yahoo.co.id

Abstract. The Freundlich adsorption isotherm is mostly presented as an empirical equation and the (fractal) Langmuir adsorption isotherm is derived from kinetics. However, their constants are determined by thermodynamic approach. As yet, both are used as an independent tool for determining the type of adsorption, just physical or chemical adsorption. This study aims to introduce an alternative way of driving Freundlich adsorption isotherm from kinetics, to elaborate its relation with the (fractal) Langmuir adsorption isotherm, and to determine their constants simultaneously by kinetics approach. An alternative way of determining the constants in both isotherm equations by kinetics approach was introduced. The results were then compared with that obtained by the conventional method. The study is theoretical but the validity of determination technique is based on the statistical tests on data adapted from literature. The Freundlich adsorption isotherm is introduced; the isotherm is the (fractal) Langmuir adsorption isotherm at relatively low concentration. It is irrelevant to use both adsorption isotherm separately. The new method is more reliable and its represent the reaction.

1. Introduction

One of the first isotherm equations introduced to describe the adsorption phenomenon is the Freundlich Adsorption Isotherm (*FAI*):

$$S = KC^b \tag{1}$$

where S is the mass adsorbed per adsorbent mass; C is the concentration in solution; K and b are fitting constants [1]. The value of K is obtained from the intercept and b from the slope of the double-logarithmic plot of log S versus log C of equation (Eq.):

$$\log S = \log K + b \log C \tag{2}$$

The other is the Langmuir Adsorption Isotherm (*LAI*):

$$S = \frac{K_{ads} C}{k + C} \tag{3}$$

where K_{ads}, and k are fitting constants [2,3]. The curve pattern of S versus C of (Eq.) (3) is a rectangular hyperbola. Thus, it is difficult to determine the values of K_{ads} and k directly from Eq. (3). By taking the reciprocal of both sides of Eq. (3), however, is transformed into its linear form

$$1/S = 1/K_{ads} + (k/K_{ads})1/C \tag{4}$$

The K_{ads} value is obtained from the intercept and k from the slope of the curve ($1/S$) versus. ($1/C$). However, for some reason, Eq. (4) cannot be used to get the exact values of K_{ads} dan k [4].

In many textbooks, the *FAI* is presented as an empirical equation with nearly no, or limited in its usefulness for its ability to fit data or to interpret the physical significance of the coefficient [5]. The kinetics approach has also been introduced [6]. It is based on equation:

$$-dC/dt = k_1C^{n_1} - k_2S^{n_2} \tag{5}$$

At equilibrium the derivative is equal to zero. Then, after some steps Eq. (5) becomes:

$$S = (k_1/k_2)^{1/n_2} C^{n_1/n_2} \tag{6}$$

where n_1 and k_1 are the order and fractal rate constant for adsorption and n_2 and k_2 for desorption. Eq. (6) is identical with Eq. (1) if $(k_1/k_2)^{1/n_2} = K$ and $(n_1/n_2) = b$.

The modified or fractal *LAI* has also been introduced by the same author. It is based on equation:

$$-dC/dt = k_1(S_o - S)C^{n_1} - k_2S \tag{7}$$

At equilibrium the derivative is equal to zero. Then, after some steps Eq. (7) becomes:

$$S = \frac{S_oC^{n_1}}{k_2/k_1 + C^{n_1}} \tag{8}$$

where S_0 is the maximum adsorption capacity and k_1 and k_2 are the fractal rate constants. Or, if $S_o = K_{ads}$; $n_1=b$, and $k_2/k_1 = k$ becomes

$$S = \frac{K_{ads}C^b}{k + C^b} \tag{9}$$

Eq. (9) is the fractal *LAI* and identical with Eq. (3) if $b = 1$. Another form of equation has also been reported earlier [7].

The reciprocal of both sides of Eq. (9) is

$$1/S = 1/K_{ads} + (k/K_{ads})1/C^b \tag{10}$$

Eq. (10) is identical with Eq. (4) if $b = 1$. The value of K_{ads} is obtained from the intercept of the linear plot of 1/S versus $1/C^b$ and then used to determine k from the slope of the line.

It is interesting to compare Eq. (10) with the equation obtained from Eq. (9) on the condition of C^b which is relatively higher and lower than k.

If C^b is higher than k, then $C^b+k \approx C^b$, Eq. (9) becomes

$$S = K_{ads} \tag{11}$$

The reciprocal of Eq. [11] is

IOP Conf. Series: Materials Science and Engineering **333** (2018) 012010 doi:10.1088/1757-899X/333/1/012010

$$1/S = 1/K_{ads} \tag{12}$$

which is differ with Eq. (10).

Inserting Eq. (12) into Eq. (10) gives

$$1/K_{ads} = 1/K_{ads} + (k/K_{ads})1/C^b \qquad or$$

$$k = 0 \tag{13}$$

That is to say, Eq. (10) does not apply if C^b is relatively higher than k.

If C^b is lower than k, and so $C^b + k \approx k$, Eq. (9) becomes

$$S = (K_{ads}/k)C^b \tag{14}$$

The reciprocal of Eq. (14) is

$$1/S = (k/K_{ads})1/C^b \tag{15}$$

which is also differ with Eq. (10).

By assumption, $S_o = K_{ads}$ and therefore, if all surfaces occupied (or $S = S_o$), then $S = K_{ads}$ and Eq. (15) becomes

$$k = C^b \tag{16}$$

That is to say, Eq. (10) also does not apply if C^b is relatively lower than k. Thus, it can be concluded that, theoretically, Eq. (10) is ambiguous and therefore it cannot be used to determine both K_{ads} and k.

There are some problems arises. First, Eq. (6) does not always apply to every reaction. The fitting constant b has a value of $0 < b < 1$ [8]. But, the reaction orders are usually integers (1, 2, ..,) or half-integers (1/2, 3/2, ...), positive or negative [9]. If, n_1 is higher than n_2, then $b > 1$. Furthermore, unlike Eq. (7), in Eq. (5), the rate of adsorption is assumed to be independent on the vacant sites, S_1 ($= S_o - S$). This goes against theory. Normally, the amount of vacant sites and the types of adsorbent always affect the rate. Thus, the rate of adsorption in Eq. (5), must depend on S_1. Moreover, the order with respect to S is n_2 but in Eq. (7) is the *first*-order.

Second, usually the *FAI* and *LAI* equations are treated separately. For example in determining the type of bond formed in the adsorption process; chemical or just physical bond [10]. However, the *FAI* can be derived from *LAI* [11]. Allegedly, it is suspected that this statement can be confirmed by using Eq. (9) and Eq. (1). If so, then it is possible to determine the constants of both *FAI* and *LAI* simultaneously.

Third, as yet, despite being derived from kinetics, the constants in both isotherms are still determined by the thermodynamic approach. The data used are always obtained from equilibrium condition at the same contact time. In practice, by using a certain concentration, the first step is to find the time needed to achieve the equilibrium condition. Then by using the time obtained to find the S change from some other concentrations. For some reasons, this treatment is not recommended. It is possible that the data are just taken from a certain condition and only valid to Eq. (11) *i.e* C^b is higher than k or Eq. (14) *i.e* C^b is lower than k. This is not representing the sample as a whole. Basically, the S concept is resemble to the rate concept; the changes of the amount of reactant or product at a given time interval. To be more specific, it is similar to the differential method in kinetics. The order obtained by differential method is fixed but the value of the constant must be checked with that obtained by integral method [12]. In addition, the only criterion used to determine the validity of the calculation result is based on the regression coefficient. The result is always claimed to be valid if the regression coefficient is close to ±1.000. It is never questioned whether the data used have met the adsorption isotherm conditions for *FAI* and / or *LAI* or not.

Based on the above discussion, the objectives of this study are to derive *FAI* by kinetics approach, to elaborate the relation of *FAI* and *LAI*, and to determine the *FAI* and *LAI* constants simultaneously by kinetics approach.

2. Experimental

2.1. Theory

This research is a literature study. The first goal is to drive Eq. (1) from Eq. (7). The second goal is to drive Eq. (1) from Eq. (9). The third goal is to find an easier and reliable technique for the simultaneous determination of b, K, K_{ads}, and k by kinetics approach and will be achieved by the following steps.

i. By Eq. (14), the order (b) is obtained from the slope of the curve of S versus C^b for $b = 1$ and 0.5. This because, by kinetics, to get $b < 1$, the possible orders are 1 or 0.5. The curve should cross the y-axis at y = 0 and x-axis at x= 0 with regression coefficient closer to ±1.00000 and intercept closer to y = 0.00. The k value is determined using the slope and K_{ads} obtained by Eq. (2).

The more exact b value (at the conditions of C^b higher and lower than k) will be obtained by using integral equations:

$$[A_o] - [A] = akt \qquad \text{for } n = 0 \qquad (17)$$

$$[A_o]^{0.5} - [A]^{0.5} = \tfrac{1}{2}akt \qquad \text{for } n = 0.5 \qquad (18)$$

$$\ln[A]/[A_0] = -akt \qquad \text{for } n = 1 \qquad (19)$$

The equation is selected with regression coefficient closer to ±1.000.

ii. The value of K_{ads} is determined simply, by making the curve of S versus C at the conditions needed for Eq. (10) is valid to be used. This is possible because by Eq. (10) the reaction is *zeroth*-order with respect to C. The curve should cross the x-axis at x = 0. The intercept is the value of $K_{ads.}$ The curve chosen is that with slope closer to x = 0.00 and regression coefficient closer to ±1.000.

2.2. Model

This study uses data, not from adsorption experiment, but taken from the bromination of acetone: $(CH_3)_2CO + Br_2 \rightarrow CH_3COCH_2Br + Br^- + H^+$. Br_2 is assumed to be the adsorbate and acetone, A, (which is in excess) is the adsorbent, H^+ is the catalyst (which is also in excess) to activate acetone.

The mechanism of the reaction has been given [13] and from that, the rate law of

$$-d[P]/dt = \frac{(k_1k_2)/k_{-1}[A][H^+][Br_2]}{(k_{-2}/k_3)[H^+] + [Br_2]} \qquad (20)$$

If in the beginning, $[Br_2]_o$, is higher than $(k_{-2}/k_3)[H^+]$ (actually higher than $[H^+]$ released from acetone), at the early-stage, the reaction is *zeroth*-order with respect to $[Br_2]$ but at the near-end is *first*-order. During the reaction process, $[Br_2]$ is decreased while $[H^+]$ is almost constant (in this reaction H^+ acts as a catalyst and is added excessively). At this condition, this reaction is resemble to adsorption reaction. But, if in the beginning, $[Br_2]_o$ is lower than $(k_{-2}/k_3)[H^+]$, for the same reason, the reaction is always *first*-order [14]; [15].

As seen in Table 1, the data are not presented in concentration unit but in absorbance of Br_2 in the solution, measured at λ=400 nm (absorbance index 160 M^{-1}) and room temperature. The data were obtained from the experiment where, originally, $[Br_2]_o$ is higher than $[H^+]$ released from acetone.

Table 1. Absorbance, A, of Br_2 in Bromination of Acetone: 10 mL Acetone 4.0 M, 5.0 mL HCl 1.0 M, 25 mL H_2O, 10 mL Br_2 0.05 M [18]

	Early stage		**Near end**	
No.	***t / s***	**A**	***t / s***	**A**
1	10	0.483	590	0.073
2	20	0.473	600	0.066
3	30	0.463	610	0.059
4	40	0.454	620	0.053
5	50	0.444	630	0.048
6	60	0.434	640	0.043

In order to make it more similar to the experimental technique on adsorption, each observation which was taken at the same time interval and the observation resulting at the end of one time interval being considered as the initial value for a new time interval. Each direct observation is taken as C_o and the following as C. The difference between each pair is taken as S (thus the amount of S is always taken at the same time–interval). The setting of the data is presented in Table 2.

The value of b is determined by integral method using Eq. (17), Eq. (18), and Eq. (19). It was based on the value of regression coefficient (closest to ± 0.00000) and intercept (closer to 0.00). This step is of special important. This is because the reaction proceeds in different order at the early-stage and near-end.

Table 2. Setting the data of Table 1. for $n_1 = 1$

		Early-stage				**Near-end**		
No.	*t/s*	C_o	C	$C_o - C = S$	*t/s*	C_o	C	$C_o - C = S$
1.	10	0.483	0.473	0.010	590	0.073	0.066	0.007
2.	20	0.473	0.463	0.010	600	0.066	0.059	0.007
3.	30	0.463	0.454	0.009	610	0.059	0.053	0.006
4.	40	0.454	0.444	0.010	620	0.053	0.048	0.005
5.	50	0.444	0.434	0.010	630	0.048	0.043	0.005
6.	60	0.434			640	0.043		

3. Result and Discussion

3.1. Derivation of the FAI

Derivation of the Freundlich isotherm is based on the equation:

$$-dC/dt = k_1(S_o - S)C^{n_1} - k_2S$$

In Eq. (5), the order with respect to S_1 (= $S_o - S$) and S are assumed to be *first*-order because there is no change in the structure of the adsorbent and the adsorption is assumed to be monolayer. It is just a process of bending and releasing of an adsorbate. It is differing with C. Here, the order is n_1 (which is can be an integer or non-integer). The mechanism of the bending process of C may be differing. A change in the structure of an adsorbate could happen. As an example, for a given adsorbent, an acid HA will ionize and being adsorbed as its ion. By ionization, $HA \leftrightarrow H^+ + A^-$, then $A^- = K_a^{0.5}C^{0.5}$.

If, $S << S_O$, Eq. (5) becomes

$$-dC/dt = k_1 S_o C^{n_1} - k_2 S \tag{21}$$

At equilibrium the derivative is equal to zero. The equation can then be solved for S:

$$S = (k_1/k_2) S_o C^{n_1} \tag{22}$$

or, if $k_1/k_2 = 1/k$ and $S_o = K_{ads}$, then Eq. (22) becomes

$$S = (K_{ads}/k) C^{n_1} \tag{23}$$

Eq. (23) is identical to the standard Freundlich isotherm, Eq. (1), if $K_{ads}/k = K$ and $n_1 = b$.

3. 2. Relation of FAI and LAI

At high concentration or C^b is relatively higher than k, Eq. (9) becomes:

$$S = K_{ads} \tag{11}$$

If S approaches the limiting value or $S = S_o$, all surfaces are covered and $S_o = K_{ads}$. From kinetics point of view, the reaction is of the *zeroth*-order with respect to C.

At low concentration or C^b is relatively lower than k, Eq.(9) becomes

$$S = (K_{ads}/k) C^b \tag{14}$$

Eq. (14) is identical with Eq. (1) if $K_{ads}/k = K$. From kinetics point of view, the reaction is of the *b-th*-order with respect to C. This means, the *FAI* can be derived from the fractal Langmuir equation; the same as discussed above. The *FAI* is the *LAI* at low concentration C.

3. 3. Simultaneous determination of LAI and FAI from kinetics

Based on above discussion, it is important to determine the order of the reaction at both conditions; at the early-stage (to determine K_{ads}) and at the near-end (to obtain K_{ads}/k, and then used to determine k). This treatment is then called the New Method (*NM*). For easier calculation, the A_o for the early- stage is 0.483 and for the near-end is assumed to be 0.073 for $n = 1$ and for the early-stage is $0.483^{0.5}$ and for the near-end is assumed to be $0.073^{0.5}$ for $n = 0.5$. The results of the calculation by integral method for $n = b = 0$, 0.5, and 1 are presented in Table 3.

Table 3. Order, $n = b$, of the Reaction at the Early-stage and Near-end

Position	Early-stage				Near-end			
$n = b$	A	B	r		A	B	R	
0	0.00030	0.00097	0.99984	*	0.00210	0.00057	0.99709	
0.5	-0.00005	0.00072	0.99976		-0.00202	0.00123	0.99897	
1	0.00094	-0.00214	0.99965		0.00249	-0.01063	-0.99981	*

Based on Table 3, the new method confirms that at the early-stage, the process is *zeroth*-order (r = 0.99984) and *first*-order at the near-end (r = 0.99984). Thus, by the *NM*, the value of K_{ads} is determined from the intercept of the curve of S versus C of the data at the early-stage (Eq. 11) and k from the slope of the curve of S versus C of the data at the near-end (Eq. 14). The results are then compared with the values of K_{ads} and k obtained by conventional method using Eq. (10) and K_{ads} / k (obtained from Eq. (14) with K obtained from Eq. (2). The results of the determination of *FAI* by conventional (Eq. 2) are presented in Table 4.

Table 4. The Results of Statistical Analysis for the Determination of *FAI* Constants

Data Source	*A*	*B*	*r*
Early-stage	-2.01973	-0.03079	-0.02212
Near-end	-1.02415	0.94386	0.94269

The slope ***B*** in Table 4, is used to determine b (of Eq. (2)). In this case its value is positive. In Table 4, the slope ***B*** at the early-stage is negative. This means that the Freundlich constants b of the reaction, cannot be determined from the data at the early-stage. Then, the Freundlich constants of the reaction, can only be determined from the data at the near-end.

The slope ***B*** in Table 5. is used to determine k / K_{ads}. Both intercepts have positive value. But, the slope ***B*** at the early-stage is negative. The slope can not be used to determine k / K_{ads} of Eq. (4) or Eq.

Table 5. The Results of Statistical Analysis for the Determination of *LAI* Constants

Data Source	*A*	*B*	*R*
Early-stage	106.5711	-1.9708	-0.0297
Near-end	9.39179	8.47298	0.94110

(10) for = 1. This means that the Langmuir constants of the reaction, can only be determined from the data at the near-end.

Table 6. The Results of Statistical Analysis for the Determination of K_{ads} and k by the *NM*

Data Source	*A*	*B*	*R*
Early-stage	0.00999	-0.00043	-0.01458
Near-end	0.00040	0.10404	0.94039

The slope ***B*** at the early-stage in Table 6. is close to to zero: the curve is parallel to x-axis and so, the intercept can be used to determine K_{ads}(of Eq. (11)). The intercept ***A*** at the near-end is close to zero. Then, its slope can be used to determine K_{ads} / k (of Eq. (14)).

Table 7. The values of K_{ads} and k by the Conventional and the NM

Method	Freundlich			Langmuir			New Method	
	B	K	r	K_{ads}	k	r	K_{ads}	K
Early-stage	-	-	-	-	-	-	0.00999	-
Near-end	0.94386≈1	0.09459	0.94269	0.10648	0.90217	0.94110	-	0.09602

There are five interesting facts reveal in Table 7. that need to be discussed in their relation with the aims of this study.

1. The *FAI* and *LAI* constants could not be determined using the conventional method from the data at the early-stage (in the condition of $C > k$) . This is consistent with Eq. (11). This means, the constants of *FAI* and *LAI* counted by conventional method do not represent the reaction as a whole.
2. By the *NM*, K_{ads} is determined from the data at the early-stage and k from the data at the near-end.These mean, the values of K_{ads} and k determined by the *NM* represent the reaction as a whole.
3. The value of $b \approx 1$, which is the same as that by integral method, indicates that the *FAI* is the fractal *LAI* at low concentration *C*. And, that, relatively, the *K* value (of $FAI = 0.09459$) $\approx K_{ads} / k$ (of $LAI = 0.00999 / 0.09602 = 0.10404$). This again confirms the fact that the *FAI* is a part of fractal *LAI*. (This is because, *FAI* is determine by differential while *NM* by integral method)
4. The fact that, at low concentration *C*, *FAI* has regression coefficient ($r = 0.94269$) higher than *LAI* ($r = 0.94110$) not just mean, *FAI* works better than *LAI* but, simultaneously, confirms that *FAI* is the fractal *LAI* at that condition. So, it is irrelevant to distinguish both isotherms.
5. At first sight, the value of K_{ads} and k obtained from the reciprocal of Langmuir equation, Eq. (4), are the real values. Both are higher (than that obtained by *NM*.) However, the data used for the calculation is the data at the near-end (fact 1.) At this condition, *C* is relatively lower than k and the reaction is *first*-order; the curve of *S* versus *C* cross x-axis at x = 0 and y-axis at y = 0. There is no maximum value. K_{ads} can not be determined. Thus, the *NM* is best used to determine K_{ads} and k. The values given are exact and the process is simple; directly using observation data.

It is interesting to note that the K_{ads} and k values derived from the reciprocal of *LAI* (Eq. (4)) is approximately ten times higher than that from *NM*. Is it because, unlike *NM*, in Eq. (4), the time factor (in this case 10 seconds) is not included in the calculation? It is unlikely; the value of *K* (of $FAI = 0.09459$) obtained from Eq. (2) is relatively the same with K_{ads} / k (of $LAI = 0.00999 / 0.09602 = 0.10404$) obtained by using *NM*.

4. Conclusion

In summary, the *FAI* equation has been derived from kinetics. The equation is the fractal *LAI* at low concentration *C*. By *NM*, K_{ads} and k of the fractal *LAI* should be determined separately: K_{ads} from the intercept of the linear plot of Eq. (17): (relatively high *C*) and k from the calculated-value of K_{ads} and the slope of the linear plot of Eq. (19): (relatively low *C*.)

References

[1] Freundlich HMF 1906 *Z. Phys. Chem* **57** 385–470
[2] Langmuir I 1916 Part I *The Research Laboratory of The General Electric Company* 2221
[3] Langmuir I 1918 Part II *The Research Laboratory of The General Electric Company* 1848
[4] Patiha, Heraldy E, Hidayat Y, Firdaus M 2016 *IOP Conf. Ser. : Mat. Sci Eng.* **107 (1)**
[5] Spark D L 2003 *Environmental Soil Chemistry* 2nd (San Diego : Academic Press) pp. 151
[6] Skopp J 2009 *J. Chem. Ed.* **86** 1341-1342
[7] Sposito G 1980 *Soil Sci. Soc. Am. J.* **44** 652-54
[8] Gregory T, Karns C L, Shimizu K D 2005 *Anal. Chim. Acta.* **528** 107-113
[9] Levine I N 2003 *Physical Chemistry 6th Edition* (New York : McGraw-Hill Companies, Inc) pp. 517
[10] Castellan G W 1983 *Physical Chemistry 3rd Edition* (Massachusetts : Addison-Wesly Publishing Company, Inc) pp. 428
[11] Sawyer C N, McCarty P L, Parking G N 2003 *Chemistry for Environmental Engineering and Science* 5th *Edition* (New York : McGraw-Hill Companies, Inc) pp.100-105
[12] Laidler K J 1987 *Chemical Kinetics* 3rd *Edition* (New York : Harper Collins Publisher, Inc) pp. 28

[13] Daniels F, Mathews J H, Williams J W 1970 *Experimental Physical Chemistry, 7th Edition* (Tokyo : Kogakusha Company, Ltd) pp. 152-155
[14] Patiha 2013 *Kajian Kritis terhadap Persamaan-persamaan dan Teknik untuk Penentuan Tetapan Michaelis-Menten* Penelitian Fundamental FMIPA UNS (Surakarta)
[15] Patiha, Heraldy E, Hidayat Y, Firdaus M. 2016 IOP Conf. Ser.: *Mat. Sci. Eng.* **107 (1)**

IOP Conf. Series: Materials Science and Engineering **333** (2018) 012011 doi:10.1088/1757-899X/333/1/012011

Synthesis and Characterization of Chitosan-p-t-Butylcalix[4]arene acid

D S Handayani, W Frimadasi, T Kusumaningsih, Pranoto
Department of Chemistry, Faculty of Mathematics and Natural Sciences, Sebelas Maret University

E-mail: desisuci@staff.uns.ac.id

Abstract. The synthesis of chitosan-p-t-butylcalix[4]arene acid was done with DIC (N, N'-diisopropylcarbodiimide) as the coupling agent. The structural analysis of the chitosan-p-t-butylcalix[4]arene acid was conducted by spectrophotometer Fourier Transform Infra Red (FTIR) and X- Ray Diffraction (XRD).Meanwhile, the surface area was investigated by Surface Area Analysis, the Scanning Electrone Microscope (SEM) analysed the surface morphology, and also the melting point temperature was determined. FTIR analysis on Chitosan-p-t-butylcalix[4]arene provides an overlapped absorption of -OH and -NH groups at 3438.26 cm-1. Meanwhile, a C = C aromatic bond present at 1480.43 cm-1. XRD analysis shows some broaden peaks due to the amorphous phase of the prepared material. The prepared material is a brownish yellow solid, odorless and porous. The melting point, surface area, and the average pore radius are above 300 °C, 9.42 m^2 / g, and 52.5938 Å, respectively.

1. Introduction

Chitosan has been widely used as an adsorbent because of its ability to bind metal ions and the possibility of relatively easy retrieval of metal ions bound to chitosan by using a particular solvent. The advantage of chitosan is its ability for multi handling absorption [1]. However, Chitosan is highly soluble in a particular solvent such as acid. It could be a weakness due to its low stability. Therefore, to improve the quality of chitosan by modifying chitosan is required in order to reduce solubility in a liquid waste. Because a liquid waste may have various pH condition, therefore chitosan must be modified to become a material with low soluble property even under acidic condition, and the modified chitosan will has ability to be used as adsorbent. A study conducted by Meriatna in 2008 [2] found that the metal wastewater of metallic coating industry has a pH of 5 that still does not meet the quality standards permitted by the Government pursuant to under KEP-51/MENLH/10/1995. In addition, Research on the utilization of chitosan has always improved both in pure chitosan and chitosan modified, primarily in its application as heavy metal adsorbents. Types of chitosan form as adsorbents in flakes [3-4] and powder [5-6], while modified chitosan forms of crosslinked chitosan [6-7].

IOP Conf. Series: Materials Science and Engineering **333** (2018) 012011 doi:10.1088/1757-899X/333/1/012011

Increasing the ability of chitosan as adsorbent can be done through cross-linking method, binding crown ether compounds as cross-linker, or cross-connecting of chitosan with calix[4]arena to form a chelate polymers [8-9]. Chitosan modification can also be carried out by binding the cross-linking of macromolecular crown ether compounds to chitosan as has been done by Tabakci and Yilmaz (2008) [10] by cross-connecting between chitosan and calix[4]arene to form chelate polymers. Other modification of chitosan was performed with calix [4] crown by Xiaoting [11] and with calix[4]resorsinarene by Jumina [12].

Modification of calixarene such as by add carboxyl groups to all four hydroxyl group in the p-t-butylcalix[4]arene compounds. This modification will produce p-t-butyl calix[4]arene acid that can be used as a phosphate ion adsorbent [13]. Our previous research on chitosan modification was to attach a macromolecule compound of p-t-butylcalix[4]arene to the chitosan molecules [14]. The produced material was used as adsorbent for Remazol Yellow FG [15], Procion Red MX 8B [16], and also Cd ions [17].

This study aims to synthesize chitosan-p-t-butylcalix[4]arene acid by binding p-t-butylcalix[4]arene acid to chitosan with DIC as a coupling agent, to analyze the structure of the prepare material and also to characterize its properties. The presence of acidic groups in the produced compound is expected to increase the adsorption capacity even larger than our previous research[14].

2. Experimental

2.1. Ingredients:

All chemical used in this reaseach were procured from E. Merck except special ones, Chitosan (PT Biotech Surindo) and , DIC, Dichloromethane, HCl 37%, Methanol, Ethanol, Chloroform, NaOH Gas N_2 (PT Samator Gas), p-t-butylcalix[4]arene [14].

2.2. Instruments:

X-Ray Diffraction (XRD) type Bruker D8, Fourier Transform Infra Red (FT-IR, Shimadzu type FT-IR-8201 PC), Surface Area Analyzer (SAA) Quantachrome Nova 1200e, Scanning Electron Microscopy (SEM, Inspect-S50 EDAX)

2.3. Procedure:

2.3.1. *Characterization of chitosan.* Characterization of chitosan was done by determining ash content (ASTM Standard E, 1755) and deacetylation degree by baseline method formulated by Baxter et al. [18].

2.3.2. *Synthesis of p-t-calix[4]arene acid.* A total of 50 mL of dry acetone was added to a round bottom flask containing 0.16 g (0.25 mmol) p-t-butylcalix[4]arene, 0.29 g (2.35 mmol) of chloroacetic acid, 0.36 g (2, 35 mmol) NaI and 0.43 g (3.25 mmol) of K_2CO_3. The mixture was refluxed for 24 hours at a temperature of 60 °C and after that the mixture was filtered. The filtrate was then evaporated at room temperature and obtained crystals.

2.3.3. *Synthesis of chitosan-p-t-calix[4]arene acid.* The p-t-calix[4]arene acid (0.60 mmol; 0.38 gram), chitosan (0.55 gram), DIC coupling agent (1.30 mmol; 0.20 mL) and dichloromethane (20 mL) were pour into a flask and stirred at room temperature for 3 x 24 hours under atmospheric nitrogen conditions. Then, the mixture was cooled, filtered and washed successively with dichloromethane and methanol three times. The product was dried at 45 °C for 3 hours and stored in the desiccator. Structural analysis was performed by FTIR Shimadzu type FT-IR-8201 PC) and XRD (brand and specification). Those both analysis is important to investigate whether the functional groups and the diffraction pattern are matched well with the targeted compound. Meanwhile, the material characterization was conducted to investigate the specific surface by SAA analysis, the surface morphology through SEM (SEM, Inspect-S50 EDAX) analysis, and also the solubility of compound by conducting a solubility test. The solubility test was carried out with a solution at various pH of 2, 4, 6, 8 and 10.

3. Results And Discussion

3.1. Characterization of chitosan

Characterization of chitosan was done by determining the degree of deacetylation and determining chitosan ash content. The chitosan characterization results are shown in Table 1.

Table 1. Results of Chitosan characterization

Description	Result
Form	Powder
Color	Brownish white
Smell	Odourless
Ash Content	1.07+0.27%
Degree of Deacetylation	89.93 /

3.2 Synthesis of p-t-calix[4]arene acid

The carboxyl groups addition to p-t-calix[4]arene produced a white and odorless powder. The reaction occurred during synthesis was etherification reaction, in which the addition of the carboxyl group causes the formation of ether compound. The formation of ether lead to release some Cl atoms from chloroacetic acid and atom I from NaI. The Na^{+} atom is a very electropositive ion that can be binded to a highly electronegative Cl atom to produce NaCl. Chlor (Cl) atom of chloroacetic acid is replaced by atom I of NaI. The addition of a K_2CO_3 base caused the proton to escape from p-t-calix[4]arene, and also caused I atom to be detached and reacted with K to form KI. The residual chloro acetic acid having a positive partial C atom then attacked the negative partial O at p-t-calix[4]arene to enter the carboxyl group. The description on reaction mechanism is depicted in Figure 1.

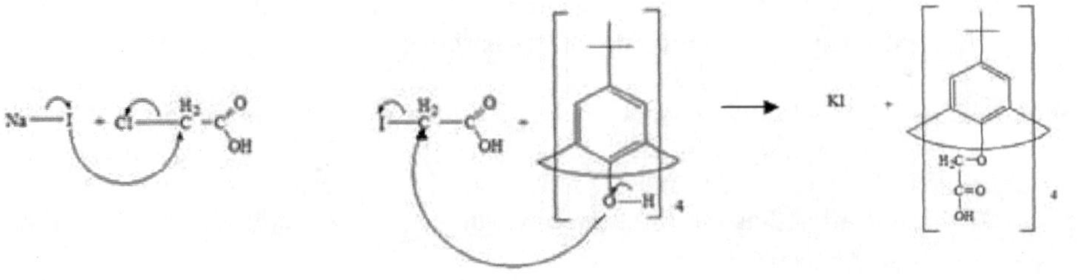

Figure 1. Mechanism of the synthesis reaction of p-t-calix[4]arene acid

The success of synthesis can be known by analyzing the structure of p-t-calix[4]arene acid using FTIR as shown in Figure 2. in the spectrum shows a sharp absorbance at 3210.65 cm^{-1} refers to -OH adsorption of carboxylic acids. The absorption indicates that the -OH group was in considerable amount. A peak at 1087.28 cm^{-1} is a typical C-O-C-ether absorbance. The adsorption of the -COOH group peak at 1743.72

cm^{-1} indicates the presence of carboxylic acid group in the chloro acetic acid compound. Meanwhile, the presence of the aromatics C = C groups is shown by absorbance at 1487.18 cm^{-1}.

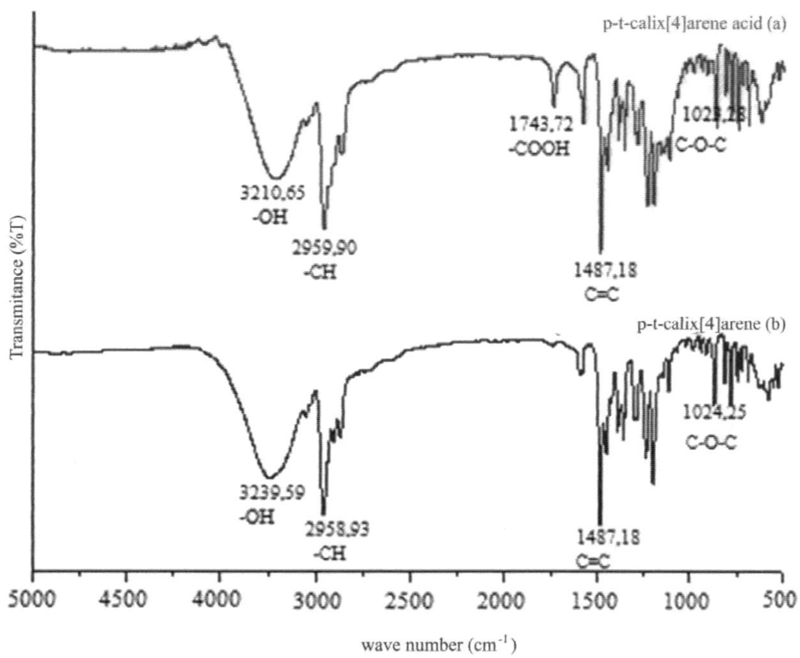

Figure 2. FTIR spectra of p-t-calix[4]arene acid (a) and p-t-calix[4]arene (b)

3.3 Synthesis of chitosan-p-t-calix[4]arene acid.

Synthesis of chitosan-p-t-calix[4]arene acid was using DIC as a coupling agent. According to Han and Young-Ah Kim (2004) N, N'-diisopropylcarbodiimide (DIC) is a linear carbodiimide compound that will easily bind two organic compounds [19]. The reaction was carried out at room temperature under atmospheric nitrogen to avoid side reactions with oxygen. The binding process of chitosan with p-t-calix[4]arene acid occurred through a substitution reaction with DIC as an intermediate agent and will re-form at the end of the reaction. The resulting compound of the synthesis was a brownish yellow solid, odorless and has a yield of 67.29%. Possible reactions are shown in Figure 3.

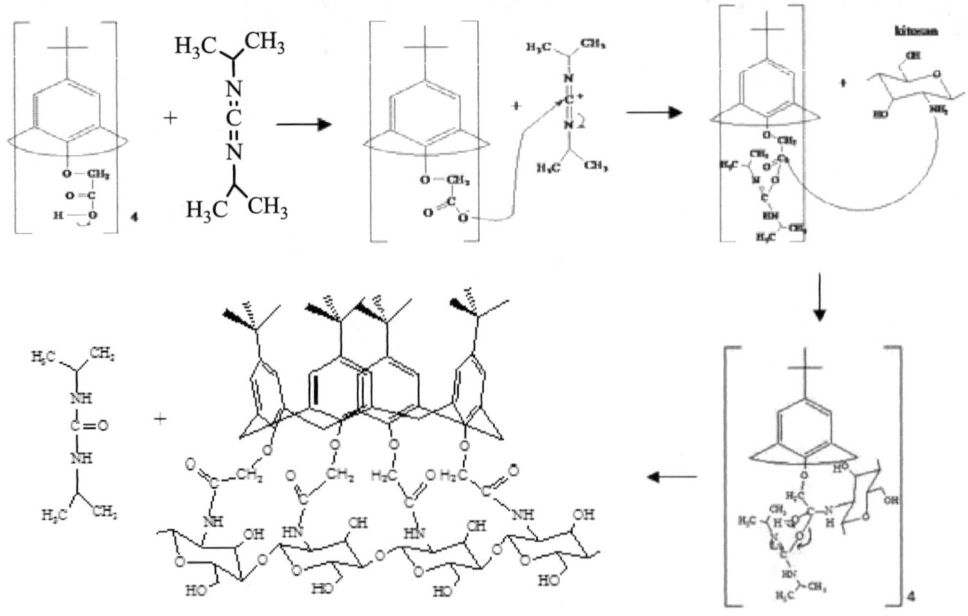

chitosan with p-t-calix[4arene acid

Figure 3. Possible reaction of binding chitosan with p-t-calix[4]arene acid

The reaction mechanism as described in Figure 3 is as follows at first, there is a proton release on the hydroxyl group present in the p p-t-calix[4]arene acid, irrespective of H+ thus causing a partially negative charged O atom. One of C = N on the DIC coupling agent is cut off into a single bond resulting in a carbocation. Atom O of p-t-calix[4]arene acid a partially negative arena attacks the carbocation of the DIC coupling agent. The bonding results between p-t-calix[4]arene acid with DIC will then react with the NH_2 group of chitosan through nucleophilic substitution. Atom N on chitosan has a lone pair electron that is nucleophile and will attack the carbocation. The result of a nucleophile attack to a carbocation produces a secondary amine by releasing a DIC coupling compound.

Analysis structure of the synthesis of p-t-butylcalix[4]arena acid using FTIR obtained spectra shown in Figure 4, while functional group analysis is shown in Table 2.

IOP Conf. Series: Materials Science and Engineering **333** (2018) 012011 doi:10.1088/1757-899X/333/1/012011

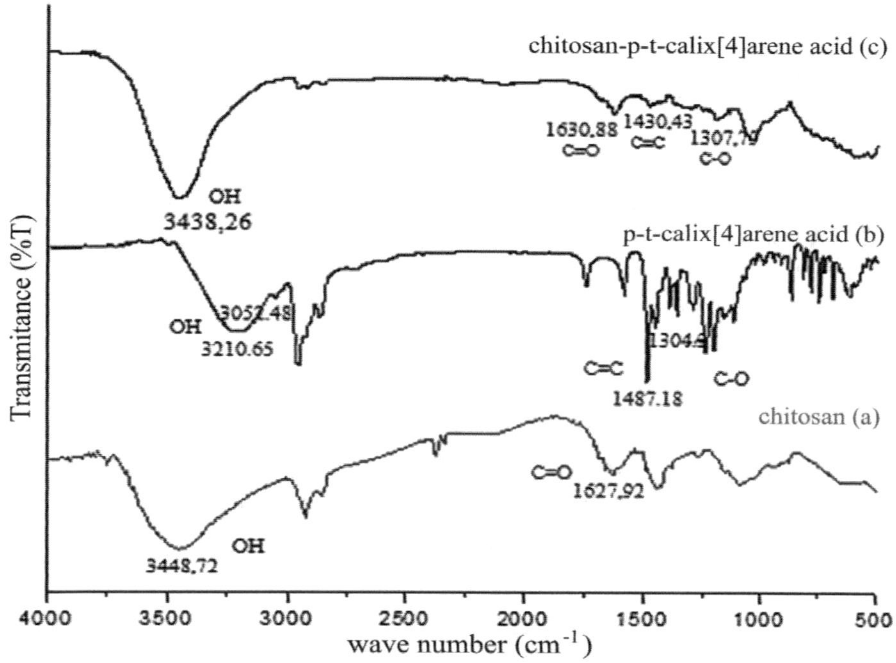

Figure 4. Spectra of chitosan (a), p-t-calix[4]arene acid (b) and chitosan-p-t-calix[4]arene acid (c)

Table 2. Analysis functional group of chitosan-p-t-calix[4]arene acid synthesis

Wave number (cm^{-1})			Functional group
Chitosan	p-t-calix[4]arene acid	Chitosan-p-t-calix[4]arene acid	
3448.72	-	3438.26	υ –NH
-	3210.65	-	υ Hydroxy –OH
-	3052.48	-	υ -CH stretching aromatic (sp2)
2924.09	2959.90	2926.14; 2957.93	υ -CH stretching asymmetric (sp3)
2854.65	2867.31	2855.73	υ -CH stretching symmetric (sp3)
1627.92	-	1630.88; 1697.43	υ C=O, N Acetyl (-NHCOCH3)
-	1487.18	1480.43	υ C=C aromatic
1442.75; 1381.03	1452.46; 1361.8	1381.09; 1437.99	υ -CH Bending symmetrically (sp3)
-	1304.9	1307.79	υ -CH stretching
1026.13	1023.28	1038.71	υ C-N

The XRD analysis was conducted at 2θ = 5°-50°. The result of diffractogram obtained was shown in Figure 5. Based on the diffraction pattern obtained, no new diffraction pattern appeared in the Chitosan-p-t-calix[4]arene acid. Change only occurs in intensity. The change of intensity occurs both chitosan peaks of 10° and 20° where at 2θ = 10° there was a change of intensity from 1040 to 667 and at 2θ = 20° there is a change of intensity from 1733 to 898. It can be assumed a considerable decrease in intensity indicates that there is an increasing nature amorphous chitosan by the addition of p-t-calix[4]arene acid.

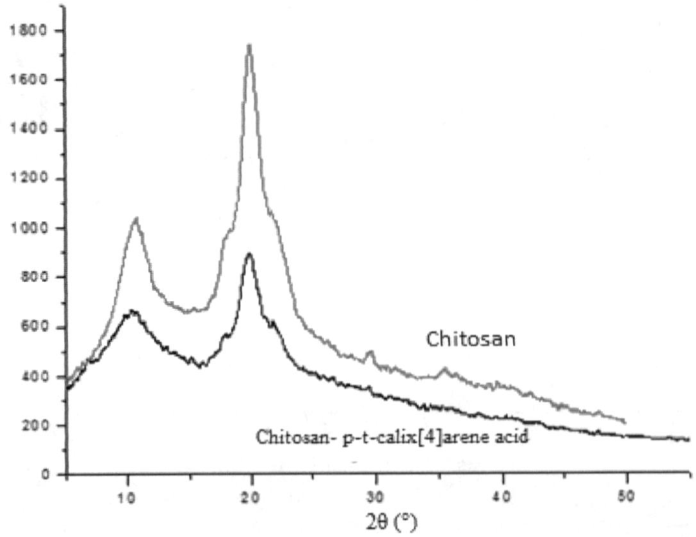

Figure 5. Diffractogram of chitosan and Chitosan-p-t-calix[4]arene acid

The p-t-calix[4]arene had a melting point of 342-344 °C [20] whereas its derivatives in the form of p-t-calix[4]arene acid had a melting point of 292-293 °C [13]. The result of the measurements shows a chitosan- p-t-calix[4]arene acid of more than 300 °C.

Surface analysis to determine the specific area area as well as the pore distribution of chitosan-p-t-butylcalix4]arene acid with SAA (Surface Area Analyzer). The results of SAA testing are shown in Table 3. The increased surface area of the modified chitosan is due to the presence of an active group of p-t-butylcalix4]arene acid binding to chitosan.

Table 3. Analysis SAA of Chitosan and chitosan-p-t-butylcalix4]arene acid

Parameter	Chitosan	chitosan-p-t-butylcalix4]arene acid
Surface area	6.42 m^2/g	9.42 m^2/g
Total pore volume	1.423x10-2 cc/g	2.478x10-2 cc/g
Average pore radius	44.3001 Å	52.5931 Å

Analysis of surface morphology was done with SEM. The results of analysis using SEM shown chitosan-p-t-butylcalix[4]arene had porous surface morphology as shown in figure 6.

a b

Figure 6. Morphology of (a) Chitosan and (b) Chitosan-p-t-butylcalix[4]arene

4. Conclusion

The synthesis of chitosan-p-t-butylcalix[4]arene acid can be conducted with DIC as a coupling agent. The product obtained is a powder with melting point above 300 °C, surface area of 9.42 m^2 / g and an average pore radius of 52.5938 Å.

Acknowledgment

The financial support from Indonesian Directorate General of Higher Education through the research project of Penelitian Unggulan Perguruan Tinggi (PUPT) 2015 is greatly appreciated.

References

[1] Darjito D, Purwonugroho D, and Ningsih R 2014 *J. Pure App. Chem. Res.* **3**(2) 53–61
[2] Meriatna 2008 *Use of Chitosan Making to Reduce Chrom Metal (Cr) and Nickel (Ni) in Waste Liquid Industry Coating Metal* (Medan : Universitas Sumatra Utara Indonesia)
[3] Jansson-Charier J, Guibal M, Roussy E 1996 *Water Res.* **30**(2) 6285–6290
[4] Lima C. I. S and Airoldi 2000 *Thermochim. Acta* **421**(1–2) 133–139
[5] Vakili M 2014 *Carbohydr. Polym.* **113** 115–130
[6] S Cao, Z Ge, H Lai 2002 *Eur. Polym. J.* **37**(10) 115–130
[7] X. jiang Hu 2011 *J. Hazard. Mater* 185(1) 306–314
[8] Wan R, Endud N W S, Mayanar C S 2002 *React. Funct. Polym.* **50**(2) 181–190
[9] Anitha A, Rejinold S N, Bumgardner J D, Nair S V, and Jayakumar R 2012 *Chitosan-Based Syst. Biopharm. Deliv. Target. Polym. Ther.* 107–124
[10] Mustafa T and Mustafa Y 2008 *Bioresour. Technol.* **99**(14) 6642–6645
[11] Fang X, Guo H, Yang F, and Bai X 2015 *Chem. Res. Chinese Univ.* **31**(6) 1051–1055
[12] Jumina, Siswanta D, Anggraeni M, Mardjan M I D, Mulyono P, and Ohto K *Asian J. Chem.* 2015 **27**(6) 2273–2276
[13] Maming, Jumina, Siswanta D, and H. Sastrohamidjojo 2007 *Indo. J. Chem.* **7**(1) 172–179
[14] Handayani D S 2015 *ICOPIA* 21–25
[15] Handayani D S, Purnawan C, Pranoto, Hastuti S, and Hilmiyana D 2016 *IOP Conf. Ser. Mater. Sci. Eng.* **107** (12011)
[16] Handayani D S, Kusumaningsih T and Hak L A 2017 *IOP Conf. Ser. Mater. Sci. Eng.* **176** (12016)

[17] Handayani D S, Kusumaningsih T, Pranoto, and Ihsaniati N 2017 *IOP Conf. Ser. Mater. Sci. Eng.* **172** (12012)
[18] Baxter A, Dillon M, and Taylor K D A 1992 *Int. J. Biol. Macromol.* **14**(3) 166–169
[19] K. Han S, Young-Ah J. 2004 *Tetrahedron* **60**(11) 2447–2467
[20] Gutsche C D 1998 *Calixarene Revisited, Monograph in Supramolecular Chemistry* (Cambridge: Royal Society of Chemistry)

IOP Conf. Series: Materials Science and Engineering **333** (2018) 012012 doi:10.1088/1757-899X/333/1/012012

Graphene Oxide from Carbon Rod Waste

F Rahmawati[1*], B L W Prasasti[1], M Mudjijono[1]
Research Group of Solid State Chemistry & Catalysis, Chemistry Department, Sebelas Maret University, Jl.Ir. Sutami 36 A Kentingan Surakarta INDONESIA 57126

E-mail: fitria@mipa.uns.ac.id

Abstract. Carbon rods extracted from Zn-C primary battery waste was used as raw material for graphene oxide (GO) synthesis. The synthesis used a modified Hummers method by providing potassium permanganate-sulfuric acid as the oxidizing agent. XRD analysis confirms a significant change between the graphite waste pattern and the produced graphene oxide pattern. A major peak at 2θ 27 ° which present in the graphite waste pattern is disappeared after it converts to the product, as well as a broad peak under 25 ° referring the presence of amorphous carbon. A broad peak at low angle of 12.02 ° dominantly present in the prepared GO pattern as a characteristic peak of GO. Meanwhile, some small peaks at 2θ of 17.76 °, 28.58 °, and 37.28 ° confirming the presence of manganese oxide which was used as oxidizing agent. A sharp peak at 1700 – 1500 cm^{-1} in the FT-IR spectrum indicates the presence of –C=O group, and at 1600 cm^{-1} refers to –C=C group. It confirms that this research has produced the targeted GO. Even though, the purity is need to be enhanced by removing the rest of oxidizing agent that still exist in the material.

1. Introduction

Graphene has the excellent mechanical, electrical, thermal, and optical properties. It can be produced by micro-mechanical exfoliation of highly ordered pyrolytic graphite, epitaxial growth, chemical vapor deposition, and the reduction of graphene oxide (GO). GO itself can be produced from graphite as raw material which is inexpensive. GO also can easily produce a stable aqueous colloid for large scale graphene production or reduced GO (rGO) by a simple solution process [1]. Graphene oxide contains a range of reactive oxygen functional groups allowing this GO to be a good candidate for some applications through chemical functionalizations [2]. GO is formed by oxidizing graphite by a strong oxidation agent to exfoliate the flakes of graphitic stack into mono-and few layer sheets [3].

Hummer's method is known to successfully produce GO from graphite by mixing it with potassium permanganate (KMnO4) and concentrated sulfuric acid (H_2SO_4)[4]. However, the products may show strong variance, depending not only on the particular oxidants used, but also on the graphite source and reaction conditions [2]. Therefore, this research used a graphite extracted from Zn-C primary battery waste. It is known that waste batteries increase year by year due to high demand on the portable electronic devices. More than 80 % of batteries sold in Europe are a primary battery, in which the large majority of primary batteries are alkaline and zinc carbon batteries [5]. Many of the components of these batteries could be recycled, avoiding the release of hazardous substances to the environment. These will contribute to the improvement of environment quality. Zinc carbon battery is

[1] fitria@mipa.uns.ac.id

IOP Conf. Series: Materials Science and Engineering **333** (2018) 012012 doi:10.1088/1757-899X/333/1/012012

a primary or disposable battery in which a carbon rod is inserted into electrolyte mixture and serve as a cathode. The carbon rod is a good electronic conductor with inert properties. An effort to reuse those carbon rod waste became another reason to use it as a raw material for graphene oxide synthesis or other well synthesized carbonaceous materials.

2. Experimental

The graphene oxide was synthesized by pouring a 69 mL of high concentrate sulfuric acid in a three necks bottle and the bottle was dipped into an ice bath and adjusted the temperature at around 0 – 3°C. The 1.5 g of graphite powder was poured into the sulfuric acid solution, and followed by 6 g of Kalium permanganate powder under stirred condition. The temperature must be kept under 15 °C. After the mixture became homogeneous, the stirring then to be continued for 1 hour under 35 °C. a 100 mL of distillied water was then poured into the mixture. The reaction released heat due to the exothermic process. Therefore, the temperature must be kept under 100 °C for 30 minutes under stirred condition. After 30 minutes, the mixture was poured into 300 mL of distilled water. The oxidation reaction was terminated by add 10 mL of hydrogen peroxide 10% until the mixture turn into yellow suspension.

The suspension was leached with HCl 5% to remove the sulphate ions, and then continued by centrifugation and water leaching. The precipitate as resulted by centrifugation was dried at 60 °C in the oven. After dry, the powder was collected to be characterized. X-ray diffraction (Bruker D8 X-ray diffraction, Cu Kα source at λ of 1.5406 Å) analyzed the diffraction pattern of the prepared materials at 2θ = 10-80°. The diffraction pattern then compared with the diffraction pattern of graphene oxide. X-ray Fluorescent analysis (Bruker S2 Ranger) was conducted for elemental analysis. Meanwhile, the change in the functional group at before and after the reaction was tracked by doing FTIR analysis (Prestige-21 Shimadzu, 4 cm^{-1} of resolution) at the wave number range of 4000- 400 cm^{-1}.

3. Result and Discussion

The XRD pattern of the waste graphite as raw material is described in Figure 1, in comparison with the diffraction of the standard graphite ICSD#76767, and also with the diffraction pattern of the product of reaction.

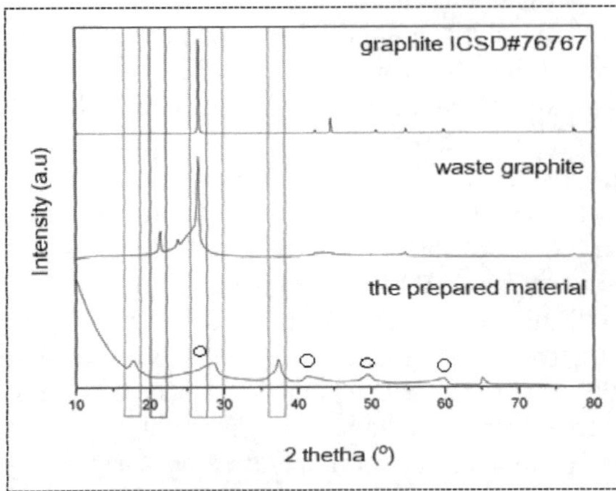

Figure 1. The diffraction pattern of waste graphite as raw material in comparison with the diffraction pattern of the prepared material and the graphite ICSD #76767

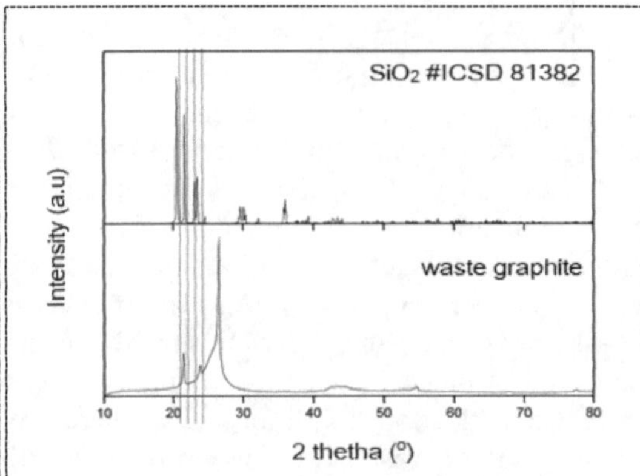

Figure 2. The diffraction pattern of waste graphite in comparison to SiO_2 diffraction ICSD#81382

IOP Conf. Series: Materials Science and Engineering **333** (2018) 012012 doi:10.1088/1757-899X/333/1/012012

Figure 1 shows a significant molecular change between the graphite waste pattern and the product of synthesize. The major peak in graphite waste is at 2θ 27°. Meanwhile, when the graphite waste pattern is compared to the standard graphite diffraction ICSD #76767, the major peak is similar, at 2θ 27°, with some small peaks at 2θ of 42°, 44°, 50°, 55°, 60°, and 77°. Meanwhile, a broad peak at 2θ below 25 ° indicates the presence of amorphous carbon [6]. A peak at 2θ 21.45° also present in the graphite waste diffraction which is not present in the standard graphite diffraction. This peak indicates the presence of impurities. The indication is silica, which was present when the graphite waste powder crushed with other reagent in a porcelain mortar. The peaks itself are in agreement with Si standard diffraction ICSD#81382 (Figure 2). The diffraction pattern of the prepared material is compared to the diffraction pattern of graphene oxide found by other researchers in Figure 2[7],[8]. The graphene oxide by Chen [7] shows a characteristic peak at 2θ of 14.6 °. Meanwhile, the graphene oxide pattern by other researcher shows a peak at 2θ 12.02 °. Those patterns are also in agreement with a standard diffraction found in a graphite when an oxygen functionality is introduced into the graphite, i.e., a peak present at around 2θ10.3° [9].

Some peaks at larger 2thetha might be some impurities as it was confirmed by Figure 6, by comparing the XRD pattern with standard diffraction of $MnO2$, $SiO2$. Peaks at 2θ 17.76°, 28.58° and 37.28° are match well with $MnO2$ standard diffraction ICSD#73363. Manganese dioxide might be produced from the remaining oxidation reaction. Meanwhile, silica is the oxide contains in the graphite waste.

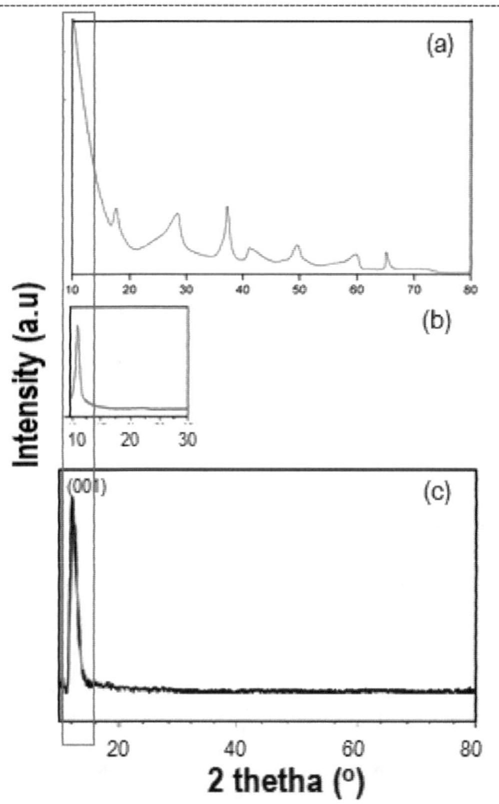

Figure 3. XRD patterns of the prepared GO (a) in comparison with GO XRD pattern (b)(c)

IOP Conf. Series: Materials Science and Engineering **333** (2018) 012012 doi:10.1088/1757-899X/333/1/012012

Some peaks at larger 2thetha might be some impurities as it was confirmed by Figure 6, by comparing the XRD pattern with standard diffraction of MnO_2, SiO_2. Peaks at 2θ 17.76°, 28.58° dan 37.28° are match well with MnO_2 standard diffraction ICSD#73363. Manganese dioxide might be produced from the remaining oxidation reaction. Meanwhile, silica is the oxide contains in the graphite waste. XRF analysis also confirms the presence of MnO_2 in the prepared GO, among other minerals that may also exist in the material. Even though, their characteristic peaks do not appear in the diffraction pattern, probably due to their amorphous phase. The XRF result is depicted in Table 2. Due to XRF instrument measure the wavelength of individual fluorescence emission after X-ray radiation, therefore only a specific weight element that able to emit X-ray radiation as a fluorescence light. Carbon, C, is a light element that does not emit such fluorescence light.

Table 1. Elemental content in the prepared GO as analyzed by XRF

Formula	Z	Content (%)
MnO	25	62.08
Na_2O	11	14.34
SO_3	16	5.36
K_2O	19	4.27
SiO_2	14	3.63
MgO	12	2.69
CaO	20	2.09
Al_2O_3	13	1.58
Trace element		3.96

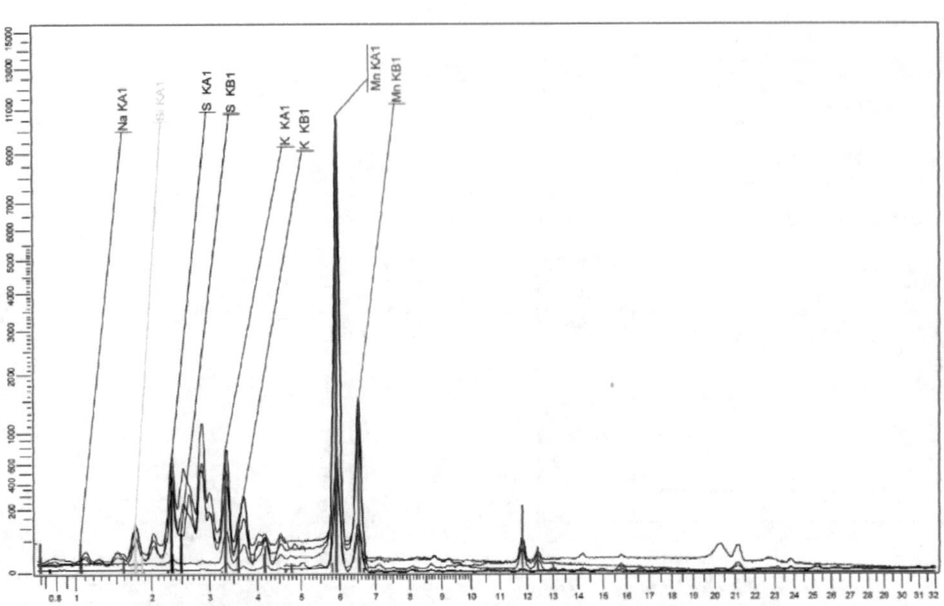

Figure 4. The XRF spectrum of the prepared material

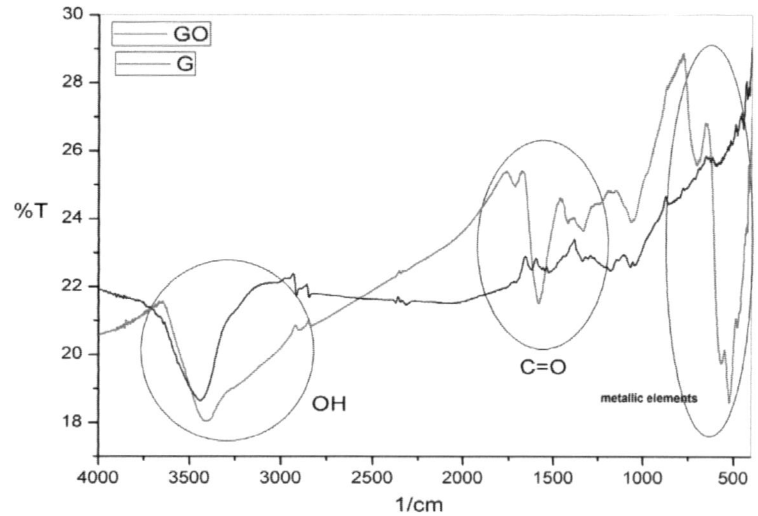

Figure 5. FTIR spectrum of the prepared GO in comparison with the previous graphite waste.

FTIR analysis spectrum as described in Figure 8, shows some vibrational changes. The graphite waste shows a sharp peak at 3500 – 3250 cm^{-1} indicating the presence of OH vibration. The OH vibration might coming from the adsorbed H_2O molecules. FTIR spectrum of the prepared GO shows a broad peak at 3000 cm^{-1} indicating that the OH vibration is not only from the adsorbed H_2O molecules, but also from the specific OH groups contained in the GO, as it can be seen in Figure 10 the presence of OH group in a GO compound. Graphite waste does not provide a peak at 1700 – 1500 cm^{-1} as it is provided by GO compound. The peak confirms the presence of carbonil, C=O vibration as a functional group in a GO. The C=C peak is also detected at around 1600 cm^{-1}. Meanwhile, the presence of Mn is confirmed by a sharp peak at around 500 cm^{-1}, as the characteristic peak also appear in the FTIR spectrum of dispersed Mn in acetone [10].

4. Conclusion

Analysis of the prepared material indicates the production of graphene oxide (GO), even though the material still consist some impurities dominated by Mn that possibly remains from the oxidation agent. Therefore, the effort to purify the graphite waste as raw material is required for this GO synthesis, as well as the requirement to purify the prepared GO after production.

Aknowledmengement

This research is a part of Hibah Mandatory 2017 funded by Universitas Sebelas Maret. Authors acknowledge for the support.

References

[1] Pei S, and Cheng H.-M 2012 *Carbon N. Y.* **50(9)** 3210
[2] Daniel R. S. R, Dreyer R, Park S, Bielawski C. W 2010 *Chem. Soc. Rev.* **39** 228
[3] Dave S. H, Gong C, Robertson A. W, Warner J. H, and Grossman J. C 2016 *ACS Nano* **10(8)**

7515
[4] william R. E. O, Hummers S 1958 *J. Am. Chem. Soc.* **80(6)**1339
[5] Rahmawati F, Yuliati L, Alaih I. S, and Putri F. R 2017 *J. Environ. Chem. Eng.* **5 (3)** 2251
[6] Walkery Jr P. L, Rakszawski J. F, and Amington F 1954 *Chemistry* **46(7)** 1620
[7] Chen J, Yao B, Li C, and Shi G 2013 *Carbon N. Y.* **64(1)** 225
[8] Shahriary L and Athawale A 2014 *Int. J. Renew. Energy Environ. Eng.* **2(1)** 58
[9] Crystallographic Open Database, http://www.crystallography.net/cod, accessed at October 21, 2017
[10] Shengxue Yang Y. D., Yang H, Ma H, Guo S, Cao F, Gong J 2011 *Chem. Commun.* **47(9)** 2619

IOP Conf. Series: Materials Science and Engineering **333** (2018) 012013 doi:10.1088/1757-899X/333/1/012013

A feasibility work on the applications of MRE to automotive components

S H Kim[1], Y J Park[1], A R Cha[1], G W Kim[1], J H Bang[2], C S Lim[2] and S B Choi [3]

[1]Department of Mechanical Engineering, 100 Inha-ro, Nam-gu Incheon, Korea
[2]NVH Competence Team, R&D Dev. Div. Hyundai Motor Group, 150, Hyundaiyeonguso-ro, Namyang-eup, Hwaseong-si, Gyonggi-do, Korea
[3]Department of Mechanical Engineering, 100 Inha-ro, Nam-gu, Incheon, Korea
Fellow Professor, Inha Unverisity,

E-mail: seungbok@inha.ac.kr

Abstract. A feasibility work on the application of magneto-rheological elastomers (MREs) to automotive components, such as engine mounts is presented. While vehicle components require the high resonance frequency in terms of ride quality and handling, it is required to have the low resonance frequency to isolate the incoming vibration. With the conventional automotive technologies, it is challenging to combine these two conflicting performance trade-offs, ride quality including handling, and NVH (noise, vibration and harshness). Over the last decades, MREs, one of the new emerging smart materials, have been widely used to resolve this technical limitation. For example, an advanced engine mount was developed by using MRE to isolate the vibration transmitting from engines. In this paper, we will focus on rear cross member bushes, which is a key component for isolating the vibration from the road, and demonstrate their improved performance by utilizing MRE. The resonance frequency shift induced by the stiffness change of MRE will be presented through the frequency response functions estimated by simulation result.

1. Introduction

Magnetorheological (MR) fluid, foam, elastomers are group of smart materials whose rheological properties can be controlled by magnetic field. One of smart material, MR Fluid is composed of silicone oil and magnetic particles, which allow controlling fluid flow in accordance with external magnetic field, thereby forming a very agile response to the flow of fluid and forming yield stress [1, 2 and 3]. Over the past decade, industrial and academic and commercialize research have been utilized in many vibration reduction systems using the MR fluid characteristics. For example, the landing gear system, vibration control of the bridge and washing machine, building damping system, vibration isolation system of the semi-conductor equipment as well as vehicle suspension, seat damper and so forth. In particular, MR Fluid dampers and mount as a semi-active system have been effectively reduced vibration by adjusting damping force [1, 7]. In spite of MR fluid advantages, it is impossible to derive significantly changes of vibration transmissibility on the vehicle by means of damping forces.

MR elastomer, which belongs to the MR material family, is a composite material with magnetic-sensitivity particles suspended or arranged within a non-magnetic elastomer matrix. With a magnetic field, the particles exhibit an MR effect, providing a field-dependent material property to the material, i.e. a controllable modulus and damping. Compared with MR fluids, MR elastomers overcome the problems that accompany the application of MR fluid, such as deposition, sealing issues and environmental pollution etc., which makes MR elastomers a favourable candidate for various and controllable vibration control application. MR elastomer consist of three basic components, e.g. polarized magnetic particles and elastomer/rubber matrix and additives materials (usually used silicon oil). Conventional matrix of MR elastomers is usually made of natural rubber or silicone rubber and are adjusted according to need as shown figure 1. In addition, MR elastomer classified into two groups, i.e. isotropic and anisotropic (or sometimes it is used aligned) MR elastomers, which attribute to different ways of curing. Firstly, three basic components, namely silicon rubber, silicone oil and iron particles, are mixed thoroughly into a homogenous mixture. The isotropic MR elastomer is cured without the presence of a magnetic field, while the anisotropic MR elastomer is cured with the action of a magnetic field [4, 12].

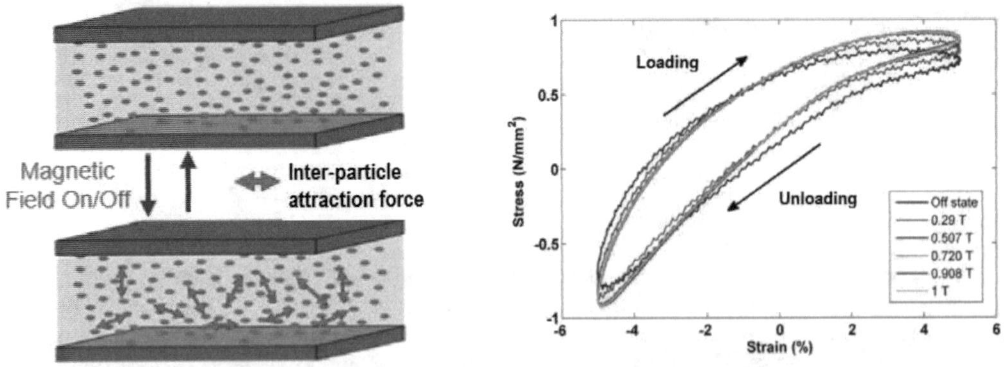

Figure 1. Principle and material characteristics of MR Elastomer.

In case of the passenger vehicle development, the passive type bush and mount have limited to satisfying both NVH (Noise, Vibration and Harshness) and ride, handling performance simultaneously. MR elastomer used as an alternative to overcome the limitation of conventional passive bush and mount, MR fluid mounts [2]. Based on the principles of MR elastomer, if the magnetic field switched off, the stiffness of the system reduced for NVH characteristics, on the contrary, if the magnetic field is applied, dynamic stiffness can be significantly increased then ride quality and handling performance will be improved. By utilizing the controllable characteristics (storage modulus and loss modulus) of MR elastomer, it improved and derived from the trade-off performance both NVH performance and ride quality, handling performance of the vehicle system. In this research, we will focus on rear cross member bushes, which are a key component for isolating the vibration from the road, and demonstrate their improved performance by utilizing the MR elastomer. The resonance frequency shift induced by the stiffness change of as shown in figure. 7. The aims to design and assess the stiffness change of MR elastomer will be presented through the frequency response estimated by simulation result.[4,5 and 11]

2. Fundamentals of MR Elastomer

Figure 2 illustrates three basic operation mode of the MR elastomer, in the compression modes, the direction of motion or vibration is parallel to line of magnetic induction while in the shear mode, the direction of motion is perpendicular to the line. Compared with the one in the compression mode, the MR elastomer in the shear mode accomplishes a smaller storage modulus and greater than MR effect and thus is widely utilized. Moreover, two types of particle arrangement both isotropic and anisotropic MR elastomers classified for the three basic operation modes.[4, 8] Therefore, different configurations found for device with isotropic and anisotropic MR elastomers, e.g. combination of field directions and aligned chain directions. Also sometimes MR elastomers that work in field active mode used to design various actuators. In shear mode devices, include vibration absorbers or vibration isolators and base isolators. Examples of squeeze mode devices are vibration absorbers, engineering mounts and compressive spring elements [5].

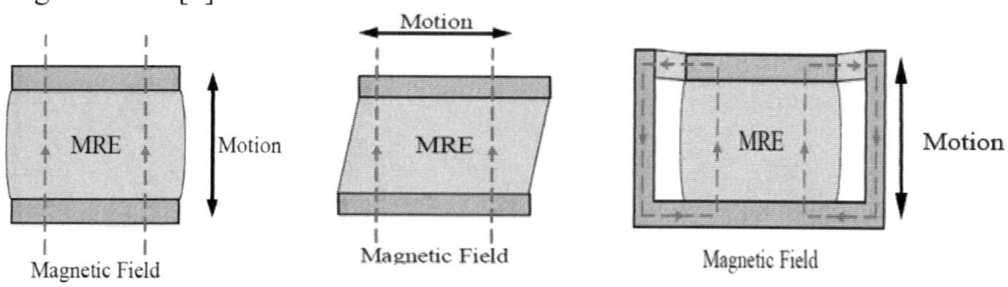

Figure 2. Three basic operation modes of MR elastomers [4,5,8, and 9]

The mechanical properties of MR elastomers are a combination of the matrix properties and the properties of the magnetic particles embedded in the matrix. It is recommend-using particles with a high saturation magnetization, because the maximum possible field induced MR effect occurs when the particles become magnetically saturated. According to rubber material, particle arrangement and iron particle content i.e., Carbonyl Iron Power, curing condition and mixing time and so on. MR effect could define as MR elastomer performance calculated from the sweep current. It is also experimental formula, which used widely as follows and the higher of MR effect, the higher dynamic rate. Figure 3 illustrates three moduli, i.e., shear modulus, storage modulus and loss modulus of MR effect for viscoelastic material including MR elastomer.

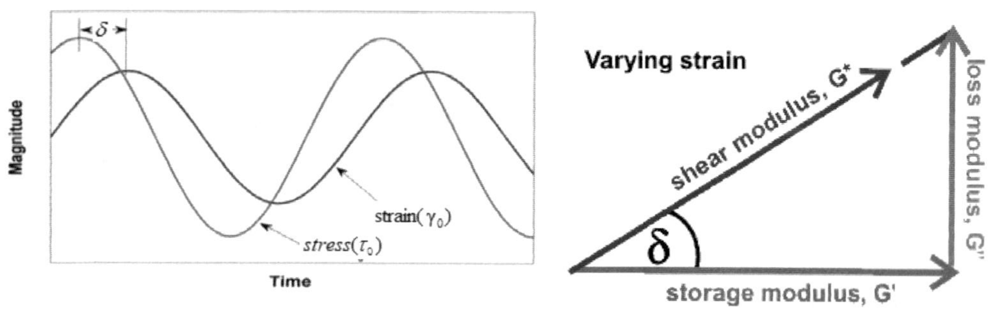

Figure 3. Illustrated of the shear modulus, storage modulus, loss modulus, of MR elastomer.

$$G^{*} = G' + i\,G'' \tag{1}$$

$$G' = G^{*}\cos\delta \text{ and } G'' = G^{*}\sin\delta\ ,\ \tan\delta = \frac{G''}{G'} \tag{2}$$

where G^{*} is shear modulus, G' is storage modulus(real part of shear modulus), G'' is loss modulus(imagery part of shear modulus), δ is defined phase angle respectively, as shown in figure 3. Thus, MR effect formula expressed as following:

$$\frac{G'_{max} - G'_{0}}{G'_{0}} \times 100\ \ (\%) \tag{3}$$

Where G'_{max} is maximum on status of magnetic field, G'_{0} is off state, i.e. not applied magnetic field.

IOP Conf. Series: Materials Science and Engineering **333** (2018) 012013 doi:10.1088/1757-899X/333/1/012013

3. Design of MRE Bush and Simulation Result.

3.1 Design and Magnetic Analysis of MRE bush for rear cross member

Based on the passive type bush product analysis, we have designed MRE bush for rear cross member of 3D math models as shown in figure 8. These 3D models show isotropic view and quarter section view. In order to find optimal design and performance for magnetic flux, intensity, and magnetic force, we have to take into consider not only effective number of turns of coils but also MRE shape, arrangement and so forth. Therefore, we have conduct the optimal shape and MRE arrangement through a magnetic field analysis in five cases of 3D design model.

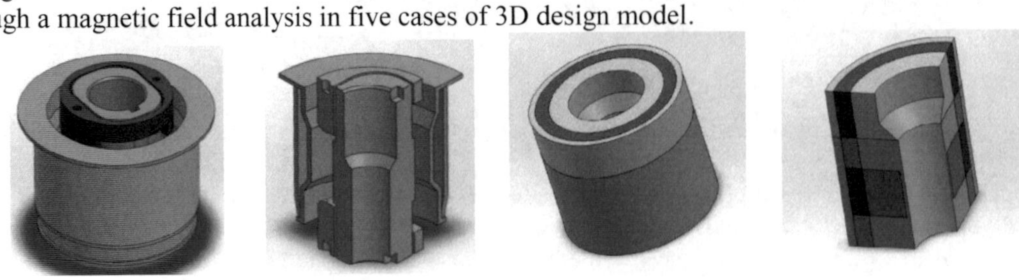

Figure 5. 3D model of cross member front and rear bush quarter section and assembly

With a magnetic field is the key role to the MRE bush working properly, it is worthwhile to study its magnetic field. According to Kirchhoff's magnetic law can be used to analyse the magnetic circuit design.[5] To verify the above theoretical calculation, a finite element analysis conduct with commercial software in ANSY 14.5 version. In order to find optimization feature of MR elastomer bush design, we have conduct magnetic analysis of five cases with 3D cad model design. Moreover, to maximize magnetic flux density and magnetic field intensity, we adapt two coils parallel structure model, which named proto #5 specification for optimization shape. From proto #5 design case, it increased to coil arrange to improve magnetic flux and field intensity. Thus, the magnetic flux path of MR elastomer was uniformly applied the magnetic field on three points, which located on MR elastomer. In addition, this model has a tendency to array of magnetic flux density and magnetic field formation compare with other four design. Therefore, we conclude fifth design for shape optimization design as shown in figure 6. [5, 6]

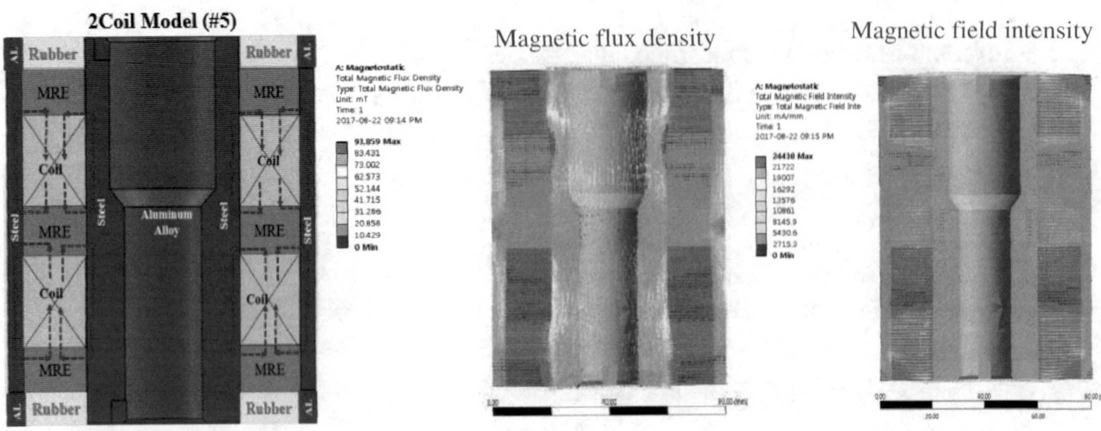

Figure 6. The final MRE bush design model with magnetic analysis

3.2 Dynamic model of MR elastomer and frequency shifting

The MR elastomer vibration isolator and absorbers may work in shear and squeeze mode or combined mode of those two. As variable stiffness devices, their effects alter resonance frequencies as following equation. [5, 6 and 11].

$$f_n = \frac{1}{2\pi}\sqrt{\frac{k}{m}} \quad \text{and} \quad f_d = f_n\sqrt{1-\xi^2} \tag{4}$$

where, k and m are the stiffness and lumped mass, respectively. In addition to the damping in the system, such as by using MR fluid damper is less efficient in changing the natural frequency as equation (4). Where, f_d is the damped natural frequency after adding damping into the system, and ξ is the damping ratio of the system. The performance of the isolators achieved by evaluating their fundamental frequency, which relates to the stiffness of isolator. For a shear mode device with a cross-section area of an MR elastomer layer as rectangular square form expressed as equation (5).

$$K_s = K\frac{GA}{h} \quad \text{or} \quad K_s = \frac{EA}{nh} \tag{5}$$

where, G is the shear modulus of MR elastomer, A and h are the cross-section area and the thickness of the rubber layer, respectively, n is the number of MR elastomer layer in isolator. For squeeze mode of MR elastomer, the compressive stiffness K_s expressed as equation (5). E is the compression modulus of the material. For single degree of freedom based and principle of vibration system shown in figure 7 (a) and (b), the vibration transmissibility expressed as equation (6).

$$T = \frac{F_T}{F_0} = \sqrt{\frac{1+(2\zeta\eta)^2}{(1-\eta^2)^2+(2\zeta\eta^2)}} \tag{6}$$

where, ζ is the damping ratio and η is frequency ratio, $\eta = f/f_n$, F_T is the transmitted force, F_0 is the excited force, respectively. The resonance frequency shift induced by the stiffness change of MRE will be presented through the frequency response functions estimated by simulation result for transmissibility as well as phase angle changed as shown in figure 7. After verify not only to evaluate for the MR elastomer performance of our proto design, but also the resonance frequency shift induced by the stiffness change of MRE will be presented through the frequency response functions estimated by simulation work and experimental measured result.

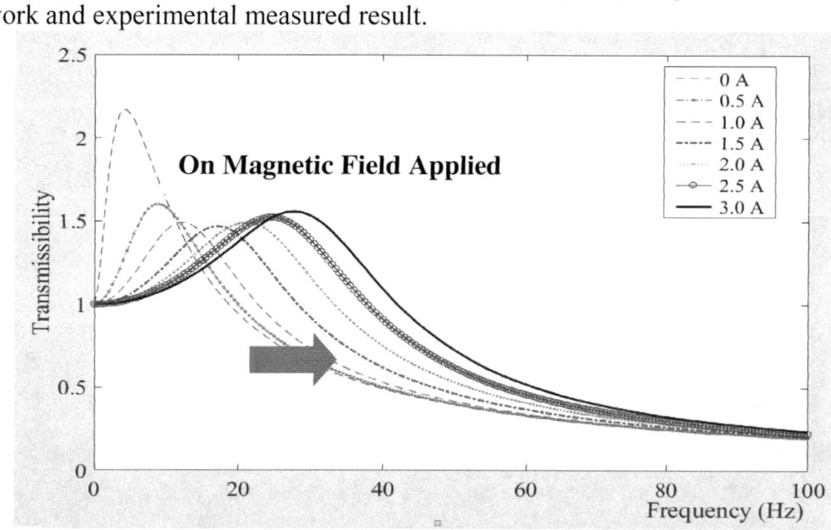

Figure 7. Simulation result of transmissibility with different currents

4. Conclusion and Future work

This paper describes as a preliminary research and aims to confirm the feasibility of MR elastomer applicable to rear wheel drive vehicles, which through analysis of conventional passive type bushes on the rear cross member product. In addition, we have conduct literature review for MR elastomer basic theory, which is a key component for isolating the vibration from the excited vibration. Thus, we will ongoing to research not only MRE bush design but also to find optimization shape through case study of magnetic analysis. From our initial research, we will verify MR elastomer bush design and shape optimization. Furthermore, we will fabricate a proto design of the rear cross member bush by utilizing the MR elastomer. Then after will verify the MR elastomer performance of our proto design, as well as the resonance frequency shift induced by the stiffness change will be presented through the frequency response functions estimated by simulation work and experimental measured result.

Acknowledgments
Hyundai Motor Company supports this research.

References

[1] J. D. Carlson and Jolly M R 2000 *Mechatronics* **10** 555-569
[2] J.M. Ginder, M.E. Nicols, L.D. Elie and J.L. Tardiff 1999 *Part of the SPIE Conference on Smart Materials Technologies SPIE Vol.* 3675 0277-786X
[3] J. M. Ginder, W.F Schlotter and M.E Nicols 2001 *Smart Structures and Materials 2001:Damping and Isolation Proceedings for SPIE Vol.* **4331** pp.103-110
[4] S S Sun, Y Chen, J Yang, T F Tian, H X Deng, W H Li, H Du and G Alici *Smart Material and Structures* **23** 075009 pp.1-8
[5] Yancheon Li, Jianchun Li, Weihua Li and Haiping Du 2014 *Smart Material and Struc.s* **23**123001 pp.1-11
[6] Jie Fu, Peidong Li, Miao Yu, Guanyao Liao, Junjie Lai, Zhiwei Xing 2017 *IEE Transaction on Mechatronics.* **11**
[7] Xiaojie Wang and Faramarz Gordaninejad 2009 *Smart Material and Structures* **18** 095045 pp.1-9
[8] M Kallio, T Lindroos, S Aalto, E Jäervinen, T Kärnä and T Meinander 2007 *Smart Material and Structures* **16** 020506 pp.506-514
[9] Hossein Vatandoost, Mahmood Norouzi, Seyed Masound Sajjadi Alehashem and Stoyan K Smoukov 2017 *Smart Material and Structures* **26** 065011 pp.1-26
[10] N Hoang, N Zhang and H Du 2013 *Smart Material and Structures* **18** 074009, pp.1-10
[11] Hua-xia Deng, Xing-long Gong 2008 *Communications in Nonlinear Science and Numerical Simulation* **13** 1938-1947
[12] Lujie Zhao Miao Yu, Jie Fu, Mi Zhu and Binshang Li 2017 *Smart Material and Structures* **26** 047001(16pp)

IOP Conf. Series: Materials Science and Engineering **333** (2018) 012014 doi:10.1088/1757-899X/333/1/012014

Optical Sensor based Chemical Modification as a Porous Cellulose Acetate Film and Its Application for Ethanol Sensor

S Mulijani[1], D Iswantini[1], R. Wicaksono[2], D Notriawan[1]
[1]Department of Chemistry Bogor Agriculture University Bogor, Indonesia
[2]Department of Agro Industrial Technology, Faculty of Halal Food Science, University of Djuanda, Bogor, Indonesia

E-mail: srimulijani@ipb.ac.id

Abstract. A new approach to design and construction of an optical ethanol sensor has been developed by immobilizing a direct dye at a *porous* cellulosic polymer fllm. This sensor was fabricated by binding Nile Red to a cellulose acetate membrane that had previously been subjected to an exhaustive base hydrolysis. The prepared optical ethanol sensor was enhanced by adding pluronic as a porogen in the membrane. The addition of pluronic surfactant into cellulose acetate membrane increased the hydrophilic and porous properties of membrane. Advantageous features of the design include simple and easy of fabrication. Variable affecting sensor performance of dye concentration have been fully evaluated and optimized. The rapid response results from the porous structure of the polymeric support, which minimizes barriers to mass transport. Signal of optical sensor based on reaction of dye nile red over the membrane with ethanol and will produce the purple colored product. Result was obtained that maximum intensity of dye nile red reacted with alcohol is at 630-640 nm. Linear regression equation (r^2), limit of detection, and limit of quantitation of membrane with 2% dye was 0.9625, 0.29%, and 0.97%. Performance of optical sensor was also evaluated through methanol, ethanol and propanol. This study was purposed to measure the polarity and selectivity of optic sensor toward the alcohol derivatives. Fluorescence intensity of optic sensor membrane for methanol 5%, ethanol 5% and propanol 5% was 15113.56, 16573.75 and 18495.97 respectively.

1. Introduction

In everyday life, brewers supply several liquor (alcoholic drink) such as beer, brandy, wine, and whisky and people are fond of them. However, in brewery, it takes many times to check alcohol concentration in product liquors, by using distillation method and gas chromatography or liquid chromatography. Therefore, development of a simple and quick detection system of alcohol concentration in liquors is strongly required. Alcohol contained in several liquors is ethanol. Ethanol also used in food, and medicinal products [2]. In recent year, several methods has been used in detecting ethanol such as time FTIR, gas chromatography, and raman spectrophotometer [3]–[5]. However, these traditional methods have some disadvantage, such as more time consuming and require large sample, complicated operations, and well trained operation [6]. Hence, a rapid and accurate method for the on-site detection of ethanol is an emerging need from the perspective for safety to monitor alcohol consumers.

Recently, ultraviolet-visible absorption spectroscopy based on optical fiber sensors have received much attention in relation to ethanol sensing because of their high sensitivity, real-time and onsite detection capability, and rapid response and recovery time. Another option is biosensor. Biosensor is a sensor that uses biological materials such as an enzyme, antibody and DNA to identify the molecules. Biosensor for ethanol commonly used alcohol oxydase enzyme (AOx) and alcohol dehydrogenase. Biosensor technique have the ability to detecting ethanol quickly, low limit detection and accurate. The use of enzymes in biosensor technique need high cost, susceptible to temperature, pH and pressure [7]. Optical sensor can be one of the method that able to detect ethanol quickly and easy fabrication.

Optical sensor based on the reaction of analyt with a sensitive layer that caused the change of absorbance, light is reflected of objects. Optical sensor development very interesting application in biology, biotechnology and ecology. Optical sensor advantage because easy to fabrication, low cost, high sensitivity and selectifity. Optical sensor consisting the solution of dye reagents that moved to the organic or inorganic matrix [8]. Optical sensors capable to detecting ethanol in the form of a liquid or gas. Solvatochromic dye used to detect the analyt. Commonly solvatochromic dye used in optical sensor are nile red, nile blue, and Reichardt's dye. Immobilization of ionofor into optical sensors could be by entrapping, multilayer film, sol-gel and by covalent bond [9]–[11]. Cobalt optical sensor using red pyrogallol dye immobilized into cellulose acetate membrane. Ensafi et al. 2015 reported that cellulose acetate as immobilize dye matrix good accuracy and precision.

Cellulose acetate membrane is a synthetic biodegradable polymer that have the properties high permaebility, nontoxic and could be composited with various materials. Membrane modificated by mixing with supporting material to support the nature of the membrane. The material commonly used in membrane modification such as pluronic and polyethylene glycol [12], [13]. Kalathimekkad *et al.* [14] had making the gas sensor based on alcohol fluorescence using nile res as dye and immobilized on Polydimethylsiloxane (PDMS) polymers. The sensor have the ability to detect the presence of ethanol vapor based on color changing and optical properties changing of the dye. Nile red dye has potential to detect the ethanol on ethanol sensor. The develop of red nile on optical sensor which the analite is liquid still on progress. The aims of this study to make and evaluate the etanol optic sensors based on cellulose acetate/pluronic/red nile.

2. Experimental

2.1 Fabrication of Membrane

Cellulose acetate (CA) membrane (Sigma Aldrich) were fabricated by inversion phase process. The membrane were prepared by flooding the surface of glass plate with polymer solution. Polymer solution contained CA and pluronic (80:20) in 100 ml acetone. After drying, the membrane were hydrolyzed in 0.1 M KOH for 24 h. Nile red as a fluorescent indicator was obtained from Sigma Aldrich. The Nile red was immobilized at the porous polymeric membrane with a classical dye bath recipe. This consisted of immersing a hydrolyzed membrane in an electrolytic dye bath for 20 min. The temperature of the dye bath was held 60°C. After removal from the dye bath, excess reagent was washed from the substrate with deionized water. This terminology derives from the ability of these dyes to adsorb strongly to cellulose-based polymers simply by immersing the polymer into a hot dye solution with a high ionic strength. Nile red is almost nonfluorescent in water and other polar solvents but undergoes fluorescence enhancement and large absorption and emission blue shifts in nonpolar environments (excitation/emission maxima ~552/636 nm in methanol). The factors that govern the formation and adsorption strength of such dyes are a complex mixture of chemical, physical, and structural effects. For Nile Red, it is apparent the chemical interactions of tertiary amine with the hydroxyl groups of the cellulosic support play an important role in the formation of the immobilized structure.The relative importance of the molecular planarity (flatness)and solubility, both of which have been found to contribute at differing degrees to the binding strength.

2.2 Measurement of Optical Sensor

An operational diagram of the flow cell for the sensor in a fiber-optic transmission configuration is shown in figure 1. The incoming light entered the solution cavity, which was defined by the front window, the sensor, and the thickness of a silicone rubber gasket, through an array of quartz optical fiber (Laboratory Scale at Physic Department Bogor Agriculture University). The beam then passed through the thin-film sensor and was collected and transmitted to a monochromator by a second array of fibers. The arrays, which consisted of five fibers arranged in a vertical orientation, were sealed into slotted Plexiglas cylinders and mounted to the cell with set screws. Performance of optical sensor can be measured withsize of the membran was 1x2.5 cm then mixed with ethanol 1-40% v/v with range 5%, and the membrane rinses for 5 minutes then put in cuvette holder to measured the fluorecent intensity. Measurment the intensity of fluorescents by spectrophotometer Optic Ocean.

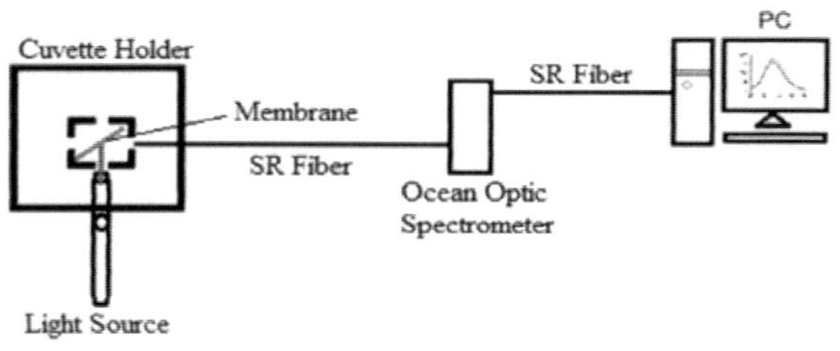

Figure 1. Experimental setup of the flow cell of sensor

2.3 Membrane characterization
Performance of morphology membrane surface before and after immobilize the nile red dye were characterized by Scanning Electron Microscopy (SEM) (Carl Zeis EVO MA 10). Characterization the functional group of the membrane and compared with cellulose acetate, pluronic and nile red dye was conducted by FTIR (Perkin Elmer).

3. Result and Discussion

3.1 Optical Sensor Ethanol
Fluorescence intensity of optical sensor ethanol is shown in figure 2, the result showed optimum intensity of optic sensor is at λ range 600-640nm. This result was similar to Kalithimekad *et al.* [14] that maximum emmision of the ethanol gas sensor membrane is at λ range 590-610 nm. Flourescence intensity depend on ethanol concentration, the value of fluorescence intensity proportional with ethanol concentration.

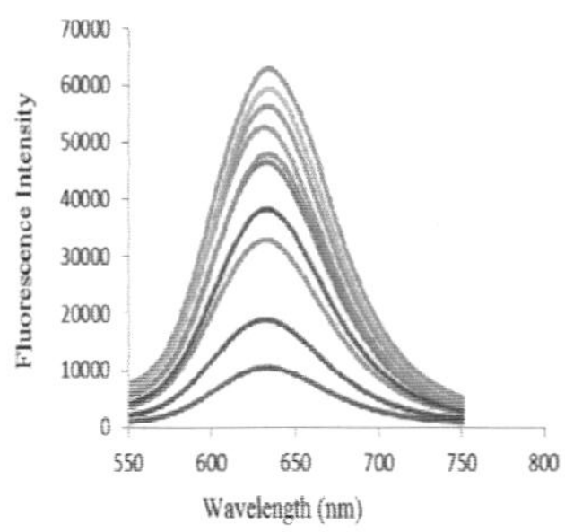

Figure 2. Flourescence intensity of ethanol membrane sensor.

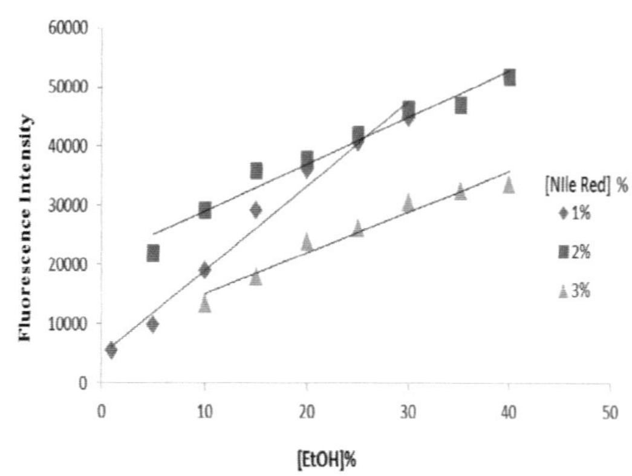

Figure 3. Linearity of ethanol sensor membrane with various nile red dye concentration.

Figure 4 showed the difference of flourescence intensity of three alcohol derivates, and this is due to the difference of polarity and chain structure of the alcohol derivates [15]. However, this can be concluded optical sensor membrane poses ability to detect the different of alcohol derivates by fluorescence intensity.

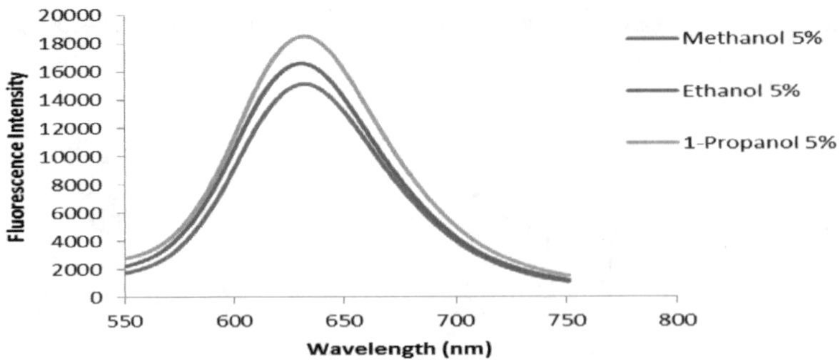

Figure 4. Fluorescence Intensity of optical sensor membrane selectivity of ethanol derivatives

3.2 Ethanol sensor membrane performance

Evaluation and confirmation the interaction among cellulose acetate, nile red, and Pluronic was conducted by FTIR. Figure 5b showed the cellulose acetate spectrum that representing the cellulose aceteate fingerprint. The wavenumber 3430 cm^{-1} for OH, 1754 cm^{-1} for C=O , 1235 $cm^{-1}$for C-O ester and 1052 cm^{-1} for C-O ester. Figure 5c showed the fingerprint of Pluronic$^{@}$, the wavenumber 3564 cm^{-1} for OH, 2905 cm^{-1} for C-H alkane, and 1107 cm^{-1} for C-O ether. Our result have the similarity with Shao D *et al* [16], indicated that the of peak of wavenumber has the spectrum pluronic. Streching overlay spectrum was for OH, C-H, C=O and C-O, this caused high range intensity. The results showed that functional group of ethanol sensor membran did not change among the analysis.

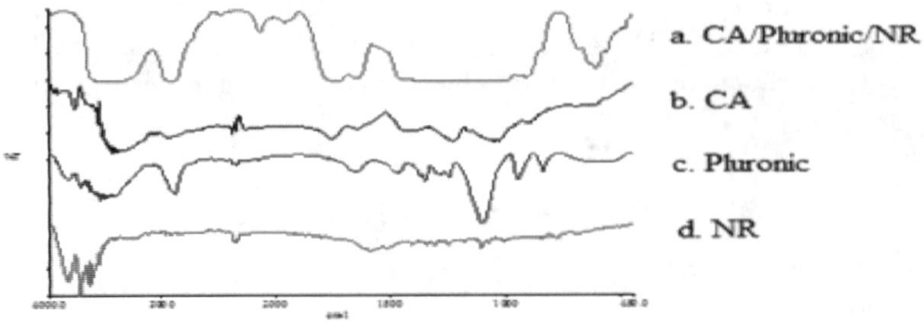

Figure 5. FT-IR characterization of etanol sensor membran: (a) CA/Pluronic/Nile Red, (b) Celluloce acetate, (c) Pluronic, (d) Nile red dye

3.3 Morphology of optic sensor membrane

Morphology of optic sensor membrane is shown in figure 6. Optic sensor membrane based cellulose acetate is occupied by nile red (figure 6a),this indicated that immobilization process of membrane was successful and Nile Red dye which mixed with cellulose acetate and pluronic was immobilized into the membrane pores. In contrast, pluronic as a porogen in membrane was not successful to form pores, and this is appropriate with the figure 6b.

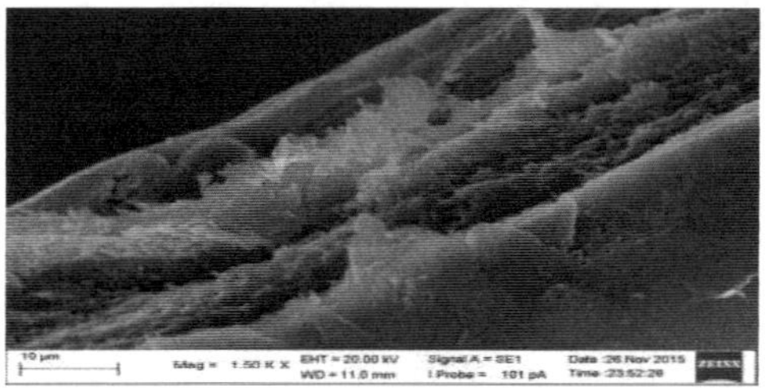

Figure 6a. Morphology of cross section optic sensor membrane CA/Pluronic/Nile Red

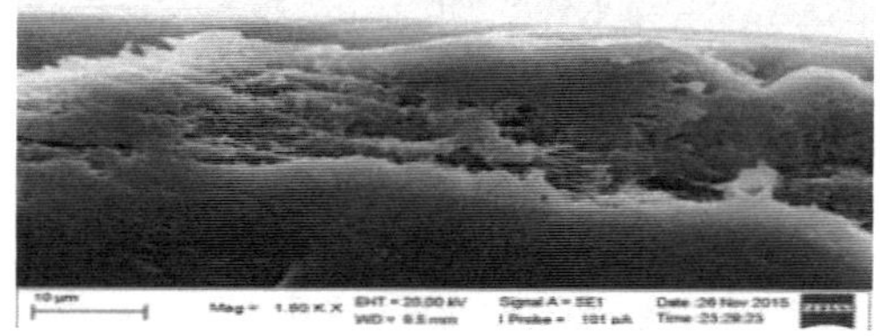

Figure 6b. Morphology of cross section optic sensor membrane cellulose acetate/pluronic membrane/pluronic

4. Conclusion

We concluded, that cellulose acetate/Pluronic@/ nile red possed potential as optical sensor ethanol membrane. Maximum intensity of optical sensor membrane at λ range 630-640 nm and flourescence intensity increased with increase ethanol concentration. Optic sensor membrane able specific detect the ethanol with concentration range 1-30% and distinguish the different of alcohol derivates. .Based on membrane characteristic with FTIR and SEM, Nile red was better immobilized to membrane pores.

References

[1] C. Staub and A. Staub Spörri, 2014 *Toxicol. Asp. Drug-Facilitated Crimes* 93–119

[2] E. Akyilmaz and E. Din, 2005 *Biosens. Bioelectron.*, **20** **7** 1263–1269

[3] T. E. Coldea, C. Socaciu, F. Fetea, F. Ranga, R. M. Pop, and M. Florea, 2013 *Not. Bot. Horti Agrobot. Cluj-Napoca* **41** **1** 143–149,.

[4] M. Wang, Y. Choong, N. Su, and M. Lee, 2003 *J. Food Drug Anal.* **11** **2** 133–140

[5] L. S. Mendes, F. C. C. Oliveira, P. A. Z. Suarez, and J. C. Rubim, 2003 *Anal. Chim. Acta* **493** **2** 219–231

[6] A. Aziz, H. N. Lim, S. H. Girei, M. H. Yaacob, M. A. Mahdi, N. M. Huang, and A. Pandikumar 2015 *Sensors Actuators, B Chem.* **206** 119–125.

[7] N. Chandrasekharan, S. Ibrahim, Y. Kostov, and G. Rao 2007 *Bioautomation* 31–39

[8] N. Sarlak and M. Anizadeh 2011 *Sensors Actuators, B Chem.***156** **1** 176–180

[9] M. Plaschke, R. Czolk, and H. J. Ache 1995*Anal. Chim. Acta* **304** **1** 107–113

[10] M. Hosseini, R. Heydari, and M. Alimoradi 2014 *Spectrochim. Acta - Part A Mol. Biomol. Spectrosc.* **128** 864–867

[11] S. T. Dubas, C. Iamsamai, and P. Potiyaraj 2006 *Sensors Actuators, B Chem.* **113** **1** 370–375

[12] H. Bai, Y. Zhou, X. Wang, and L. Zhang 2012 *Procedia Environ. Sci.* **16** 346–351

[13] C. Lv, Y. Su, Y. Wang, X. Ma, Q. Sun, and Z. Jiang 2007 *J. Memb. Sci.***294** **1–2** 68–74

[14] I. Levitsky, S. G. Krivoshlykov, and J. W. Grate 2001 *Anal. Chem.* **73** **14** 3441–3448,

Physical and Optical Studies of Bi^{3+}-Modified Erbium Doped Tellurite Glasses

Ahmad Marzuki[a*], Devara Ega Fausta[a]

[a]Department of Physics, Sebelas Maret University, Surakarta 57126
Jl. Ir. Sutami 36A Surakarta 57126 Indonesia
*Email: amarzuki@mipa.uns.ac.id

Abstract. Er^{3+}-doped tellurite glasses with various compositions (in mol%): $54TeO_2$-(41-x)ZnO-xBi_2O_3-$2Na_2O$-$3Er_2O_3$ (x = 1, 2, 3, 4, and 5) were prepared with melt quenching method. Studies was aimed at investigating the effect of Bi^{3+} ion content on the physical and optical properties of the glasses. The density, refractive index, optical absorption, and optical energy band gap measurements were carried out at room temperature using pycnometer, Brewster angle method, and UV-VIS-NIR spectrophotometer, respectively. From the experiment, it was shown that the density and refractive index of the glasses increased with the increase of Bi^{3+} ions concentration. The absorption band intensity of electronic transition from $^4I_{15/2}$ to $^4H_{11/2}$ exhibited an increase as the Bi^{3+} ions concentration increase suggesting that incorporating Bi^{3+} ions into this glasses might improve the pumping efficiency.

Keywords: Tellurite glasses, Erbium doped glasses, Glass density, glass refractive index, glass band gap.

1. Introduction

In the last view decades, tellurite glasses have been the subject of several spectroscopic investigations mainly due to their potential for laser glass host [1-5]. These glasses are characterized by their high index of refraction [6], low phonon energy [7, 8], and thermally stable against devitrivication [8, 9]. Large numbers of tellurite glasss composisions have been synthesized. Among them are doped with rare-earth ions Y^{3+}, Sm^{3+}, Eu^{3+}, Gd^{3+}, Tb^{3+}, Dy^{3+}, Lu^{3+}, Er^{3+}. Nd^{3+}, Ho^{3+}, Tm^{3+} and Yb^{3+} [8, 10-12].

Erbium doped glasses are one those intensively investigated mainly for the optical amplifier application used in the third] window telecommunication corresponding to the electronic transition from $^4I_{13/2} \rightarrow {}^4I_{15/2}$ emitting 1.5 μm. A lot of works on the spectroscopic properties of Er-doped tellurite glasses have been done by different researchers [13-15].

In order that the optical amplifiers can be connected to the installed optical fiber network with better confinement and high quantum yield, optical amplifiers are formed in a defined optical path either in the form of planar waveguide or optical fiber. In either form, slighly different composition of Er-doped tellurite glasses have to be taylored to from a thin film of Er-doped tellurite glass with higher refractive index. Similarly, a fiber core containing Er^{3+} ions having slighly higher refractive index than that for the cladding is required. Adding small amount of heavy ions with higher ionic polarizability into the chosen glass substrates is a common way done by most researchers to increase glass refractive index. This index change, however, can lead to change to any other spectroscopic properties of the glasses.

This paper aims at investigating the spectroscopic properties change of tellurite glasses as small amount of Zn^{2+} ions concentrations were substituted by Bi^{3+} ions. Studies were focused on the effect of the incorporation of Bi^{3+} ions to the physical and optical properties glasses: density, refractive index, absorption spectra, and optical energy band gap.

2. Experiment

The composisition of Er-doped tellurite glasses used in this experiment is (in mol%): $54TeO_2$-(41-x)ZnO-xBi_2O_3-2Na_2O-3Er_2O_3 (x = 1, 2, 3, 4, and 5). The purity of all chemicals were more than 99.9%. Mixture of these chemicals were prepared in a glove box. After thoroughly mixed, mixture was contained in a platinum crucible and heated in an electric furnace. Na_2O used in these glass compositions were obtained from Na_2CO_3 that upon heating released CO_2 and leaft Na_2O. In order that this reaction to occur, mixtures were initially heated at 300^{0}C for 30 minutes and then brought to 900^{0}C. Molten was hold at this temperature for about 60 minutes and periodically stirred using platinum rod. Casting was performed by pouring the melt in a preheated square brass mold and let it to cool to room temperature. To remove any inhomogenity in the glass occurs during natural rapid cooling, glass was then annealed at 375^{0}C for 6 hours. Cooling at 1^{0}C/minute was performed to ensure that all part of the glass at any time during cooling is at the same temperature.

For optical characterization, the obtained glasses were cut and polished to optically polished. Glass density was measured at room temperature by applying Archimedes law using pycnometer. Glass refractive index was determined from Brewster's angle measured using spectrometer. Spectroscopic properties of all glasses were studied from the glasses absorption spectra measured within UV-VIS-NIR region. All of these measurement were carried out at room temperature.

3. Results and Discussion

3.1. Density and refractive index

Figure 1 shows densities and molar volume of tellurite glasses containing different concentration of Bi_2O_3. As shown, glass density increases with the increase of Bi_2O_3 content. This result is mainly attributed by the high value of Bi_2O_3 molecular weight which is 465.96 g/mol. This value is higher than those for other oxides composing glass structure which are 159.60 g/mole for TeO_2 and 81,379 g/mole for ZnO. Linear increase of sample molecular weight (Figure 1b) and linear decrease of ionic packing ratio (Figure 1c) as Bi_2O_3 content increases do not make the glass density change in the same way (linear increase).

Glass molar volume is another factor affecting glass density. Figure 1d is molar volume of the glass calculated using equation [15]

$$V_m = \frac{\sum x_i M_i}{\rho} \quad (1)$$

Where ρ is glass density, x_i and M_i are molar fraction and molecular weight of an oxide composing the the glass. As molar volume can be associated with the spatial dispersion of oxygen in the glass matrix characterized by oxygen packing density (OPD) expressed by [15]

$$OPD = \frac{1000 * N_O}{V_m} \quad (2)$$

where N_O is number of oxygen atom present in the glasses, the abrupt change in molar volume within the range of the increasing Bi_2O_3 content from 2 mol% to 4 mol% can then be interpreted as the abrupt change in the number of the non-bridging oxygen. The way the molar volume change as the Bi_2O_3 content content change which is not similar to both the change in molecular weight and ionic packing ratio can be linked to the appearance of a very little change in glass density within the range of Bi_2O_3 content from 2 mol% to 4 mole% (Figure 1d).

Figure 2 shows refractive index of tellurite glasses which changes as function of Bi_2O_3 content. As shown, refractive index of glasses increase with the increase of Bi_2O_3 content. The cationic molar polarizability of Bi^{3+} (1.508 Å^3), which is higher than that of Zn^{2+} (0.283 Å^3), is the main caused of this increase. In addition, creation of non-bridging oxygen in the glass network resulted from the substitution of ZnO by Bi_2O_3 can also be considered as an additional factor contributing to the refractive index increase.

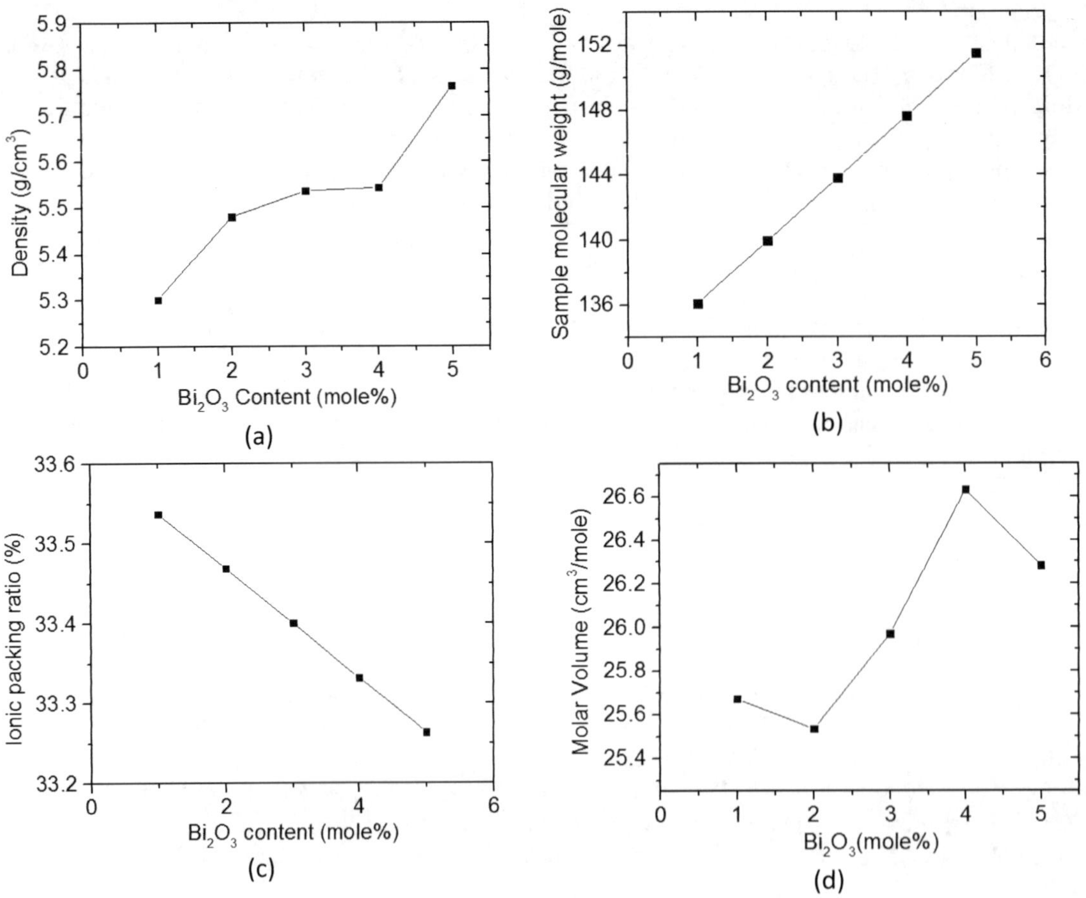

Figure 1. Variation of density (a), sample molecular weight (b), ionic packing ratio (c) and molar volume (d) of tellurite glasses as function of Bi_2O_3 content

3.2. Absorption spectra

UV-VIS-NIR absorption spectra of Er-doped tellurite glasses containing different concentration of Bi^{3+} are given in Figure 3. All the observed absorption bands; which have peaks at 414 nm, 451 nm, 460 nm, 498 nm, 532 nm, 555 nm, and 665 nm; correspond to electronic transitions from ground state ($^4I_{15/2}$) to the exited states of $^2H_{9/2}$, $^4F_{3/2}$, $^4F_{5/2}$, $^4F_{7/2}$, $^4H_{11/2}$, $^4S_{3/2}$, and $^4F_{9/2}$, respectively. Among them, absorption corresponding to electronic transition from $^4I_{15/2}$ to $^4H_{11/2}$ is the strongest one. The intensity of this band increases with the increase of Bi^{3+} content. This result might be attributed by the glass density which leads to the increase of Er^{3+} density and consequently increase the absorption coefficient. Moreover, it is seen that the fundamental absorption of all glasses raise rapidly just below the absorption band at 414 nm and therefore absorption band below this band cannot be observed.

Figure 3b can roughly illustrate the fundamental absorption change as the Bi^{3+} content increases. Although, incorporating Bi^{3+} ions into the Er-doped tellurite glasses might benefit in term of pumping efficiency, it is seen that the UV absorption edge is blue-shifted as the Bi^{3+} ions content is increased.

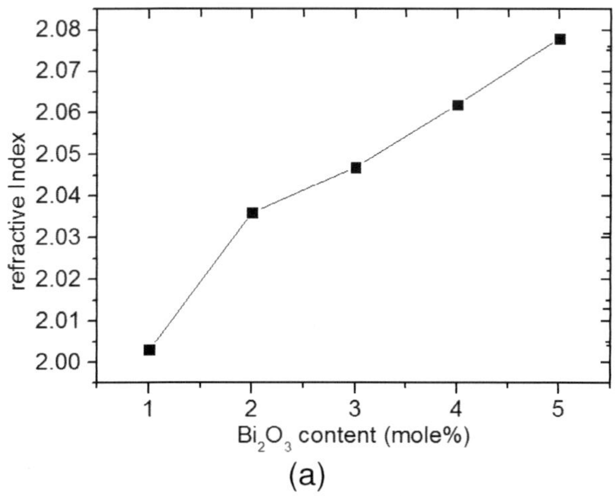

Figure 2. Refractive index of tellurite glasses as function of Bi_2O_3 content

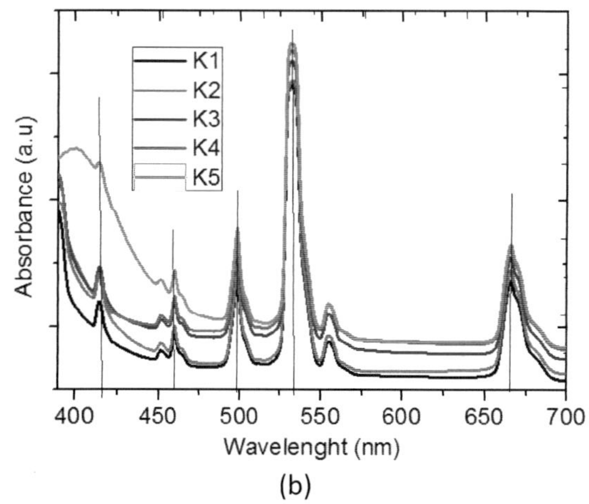

Figure 3. Absorption spectra of tellurite glasses as function of Bi_2O_3 content.

Figure 4a is UV absorption edge from which optical energy band gap was calculated. In this paper, the optical energy band gap was calculated using expression

$$(ahv)^{1/2} = C(hv - E_{opt}) \quad (3)$$

where α, hv, E_{opt}, and C are absorption coefficient, photon energy, optical band gap, a Constance which is independent to the photon energy, respectively. In term of the optical energy band gap, incorporating Bi^{3+} ions into the glasses can lead to decreasing energy band gap. Figure 4b indicates how the optical energy band gap changes as the Bi^{3+} ions content in the glass changes. The energy band gap decreases as the Bi^{3+} ions content in the glass increases. This decrease of energy band gap indicates that incorporating Bi^{3+} ions into the glass causes the glass becomes more ionic which is in a good agreement with has been mention above that the increase of Bi^{3+} ions content increases the number of non-bridging oxygen, and therefore increases the number of free electrons in the glass matrix.

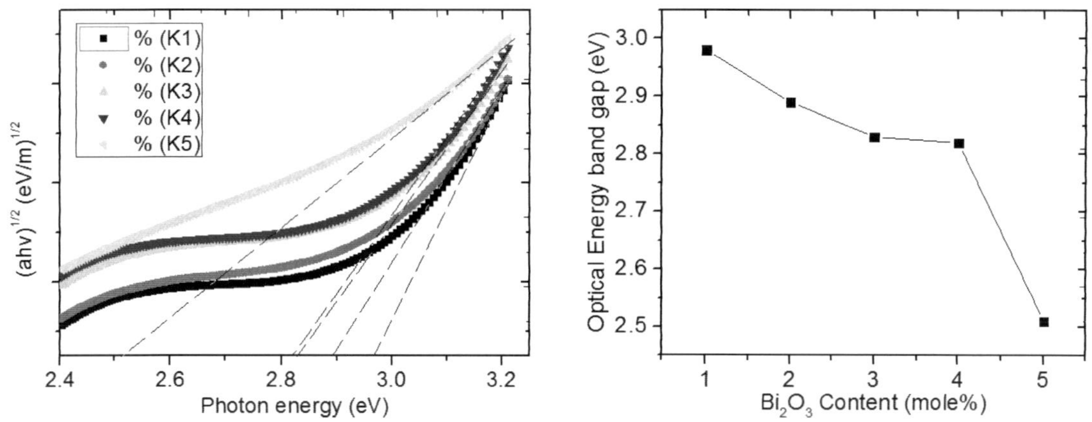

Figure 4. (a). UV-absorption edge of tellurite glasses with different concentration of Bi^{3+} ions. (b). Optical energy band gap of glasses as function of Bi^{3+} ions content.

4. Conclusions

Series of Erbium doped tellurite glasses containing different concentration of Bi_2O_3 were fabricated using conventional melt quenching method. The changes in density, refractive index, absorption spectra, and optical energy band gap for the same concentration of Er^{3+} ions concentration were analyzed in order to determine the effect of Bi^{3+} ions concentration on the glass properties. The obtained density and refractive index of the glasses were increase with the increase of Bi^{3+} ions concentration.

The absorption band intensity of electronic transition from $^4I_{15/2}$ to $^4H_{11/2}$ exhibited an increase as the Bi^{3+} ions concentration increase suggesting that incorporating Bi^{3+} ions into these glasses might improve the pumping efficiency.

Acknowledgment
The authors wish to thank to Sebelas Maret University for financial support through the scheme of Hibah KI 2017.

References

[1] Rivera V A G and Barbosa L C 2014 *J. Lumin.* **156** 116-123.
[2] Dousti M R, Sahar M R, Ghoshal S K, Amjad R J and A.R. Samavati A R 2013 *J. Mol. Struct.* **1035** 6-12.
[3] Marzuki A, Wahyudi W, Pramuda A and Nurmalasari I 2016 *AIP Conf. Proc.* **1725** 020044.
[4] Marzuki A, Wahyudi W and Pramuda A 2017 *MSF* **904** 98-101.
[5] Kundu R S, Dhankhar S, Punia S R, Nanda K and Kishore N 2014 *J. Alloys Compd.* **587** 66-73.
[6] El-Mallawany R 2014 *Optik* **125** (20) 6344-6346.
[7] Sajna M S, Gopi S, Prakashan V P, Sanu M S, Joseph C, Biju P R and Unnikrishnan N V 2017 *Opt Mater* **70** 31-40.
[8] Kaur A, Khanna A and Lyubomir I. Aleksandrov L I 2017 *J. Non-Cryst. Solids* **476** 67-74.
[9] Elkhoshkhany N, Abbas R, El-Mallawany R and Hathot S F 2017 *J. Non-Cryst. Solids* **476** 15-24.
[10] Wang J S, Vogel E M and Snitzer E 1994 *Opt. Mater* **3** (3) 187-203.
[11] Wang C, Tian Y, Gao X, Liu Q, Huang F, Li B, Zhang and Shiqing Xu S 2018 *J. Lumin* **194** 791-796.
[12] Conti G N, Berneschi S, Bettinelli M, Brenci M, Chen B, Pelli S, Speghini A and Righini G C 2004 *J. Non-Cryst. Solids* **345-346** p 343-348.
[13] Savikin A P, Grishin I A, Sharkov V V and Budruev A V 2013 *J. Solid State Chem* **207** 80-86.
[14] Damodaraiah S, Prasad V R and Ratnakaram Y C 2018 *J. Alloys Compd.* Paper in Press.
[15] Yusof N N, Ghoshal S K and Azlan M N 2017 J. Alloys Compd **724** 1083-1092.
[16] Kashif I, Ratep A and El-Mahy S K 2017 *Mater Res Bull* **89** 273-279.

Evanescent Wave Absorption Based Fiber Sensor for Measuring Glucose Solution Concentration

Ahmad Marzuki[1a], Arni Candra Pratiwi[1b], Venty Suryanti[2]
[1]Department of Physics, Sebelas Maret University, Surakarta 57126 Indonesia
[2]Department of Chemistry, Sebelas Maret University, Surakarta 57126 Indonesia

[*]amarzuki@mipa.uns.ac.id

Abstract. An optical fiber sensor based on evanescent wave absorption designed for measuring glucose solution consentration was proposed. The sensor was made to detect absorbance of various wavelength in the glucose solution. The sensing element was fabricated by side polishing of multimode polymer optical fiber to form a D-shape. The sensing element was immersed in different concentration of glucoce solution. As light propagated through the optical fiber, the evanescent wave interacted with the glucose solution. Light was absorbed by the glucose solution. The larger concentration the glucose solution has, the more the evanescent wave was absorbed in particular wavelenght. Here in this paper, light absorbtion as function of glucose concentration was measured as function of wavelength (the color of LED). We have shown that the proposed sensor can demonstrated an increase of light absorption as function of glucose concentration.

Keywords: Evanescent wave, fiber sensor, glucose solution concentration measurement.

1. Introduction

Glucose is a simple types of sugar that can be found in many aspect of life. This substance has molecular formula of $C_6H_{12}O_6$ and can be dissolved in the water because of its polarity. In human and animal body, glucose circulates trough the blood vessel as a blood sugar and the body keeps it up to a certain level for metabolic homeostatis. In food industries, many types of sugar were used to give flavor. We can say that measurement of glucose concentration will accelerate the technology development in industrial aspect and medical research.

For many decades, glucose concentration measurement has been an interesting subject of research. Many methods have been developed utilizing different glucose properties (e.g., those based on chemical reaction and biochemical-sensors). Two examples of the chemical methods used to detect glucose concentration are copper-iodometric glucose and enzyme method. The first method uses copper as a titration for glucose reduction. In the other hand, enzymatic methods were usually combined with biochemical sensors, utilize the change of electrical properties which happens because of the enzyme reacted with the sample. The electrical properties of a compound change whenever a reaction that produce or consumes electron occur. This kind of reaction were often catalyzed by redox enzymes [1].

Other glucose detection method that uses enzyme is ampere-metric glucose sensor based on dehydrogenase enzymes. In this case, glucose is reacted with the dehydrogenase enzymes to produce an electro-active substrate. The produced substance is then detected by the sensor in the form of a current change [2]. Other chemical reaction used to detect glucose is Maillard reaction which detect the product using the change in Fluorescence intensity [3]. In this work, glucose concentration

IOP Conf. Series: Materials Science and Engineering **333** (2018) 012016 doi:10.1088/1757-899X/333/1/012016

measurement using evanescence field based optical fiber sensor is proposed. The field will interact with the glucose obeying Beer-Lambert's law [4]. The leak field out of the core can be imagined as tunneling effect on quantum mechanics [5]. In fact, the optical fiber sensor utilizing evanescent wave phenomenon [6] have been widely used for chemical detections with many different modifications [6-12].

2. Material and Methods

Glucose ($C_6H_{12}O_6$) solutions were made with different concentrations (0.1 – 0.6 M). This maximum value of the glucose concentration (0.6 M) was taken by considering that this value corresponded to the refractive index of the polymer optical fiber (POF) used in this experiment. The refractive index of the fiber core is 1.49 and its cladding is 1.42. In order to create an evanescent field based fiber sensor, POF with 0.5 mm in diameter and the length of 8.0 cm was partially polished to the depth of 0.10 mm to form a D-shape.

Figure 1 is the experimental setup used in this experiment. Absorption of light passing through this sensing area was carried out by launching the light (i.e., LED with different color) from one end the fiber and the output light was then detected by a photo detector. Experiments were carried out by covering the sensing area with different glucose concentrations at room temperature.

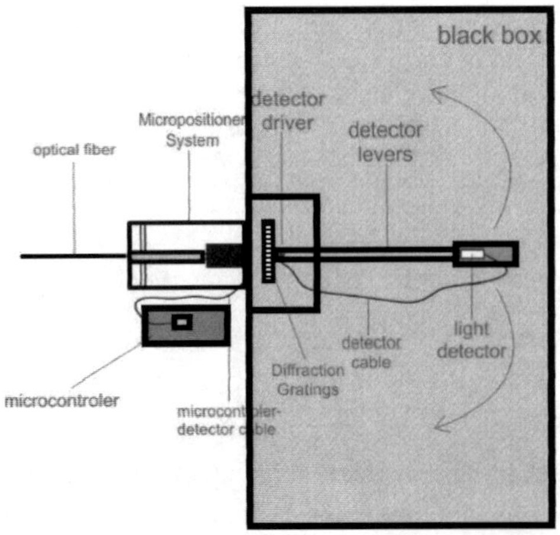

Figure 1. Experimental setup of optical fiber based spectrophotometer used in this experiment

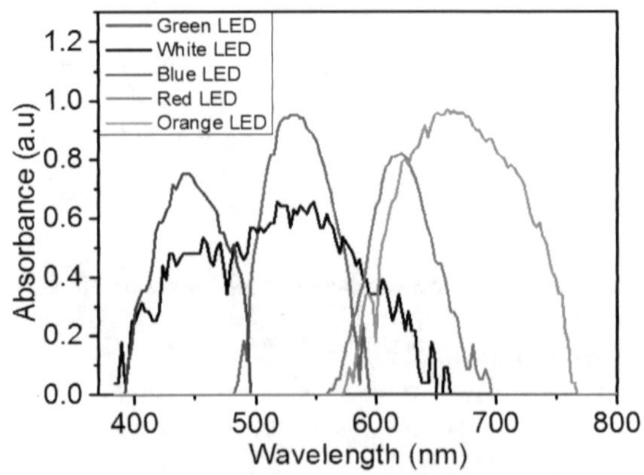

Figure 2: Spectral absorbance from various Light source (for 0.1 M solution).

3. Results and Discussion

Figure 2 shows the output light detected for different color of LED at which the sensing area was covered with 0.1 M glucose. All light spectra as shown were obtained by dividing the output light of glucose covered fiber sensing area by that before being covered. Strong absorption at 626 nm is typical absorption of glucose as has been recorded using UV-VIS-NIR spectrophotometer.

Figure 3 shows the output light from red LED detected by photo-detector as the sensing area was covered by different concentration of glucose. From figure 4, the trend of increase was also observed for other light sources (especially for red and orange LED). Repetition of the experiment showed almost the same results (Figure 5). It is clearly seen that absorption increases (the output voltage decreases) as the glucose concentration increases. This result can be understood from the frame of Maxwell-Botlzmann particles distribution (N_i) over various micro-states or energy levels (ε_i) at a temperature T as expressed by equation:

$$N_i = \frac{N.exp\left(-\frac{\varepsilon_i}{kT}\right)}{\sum_j exp\left(-\frac{\varepsilon_j}{kT}\right)} \quad (1)$$

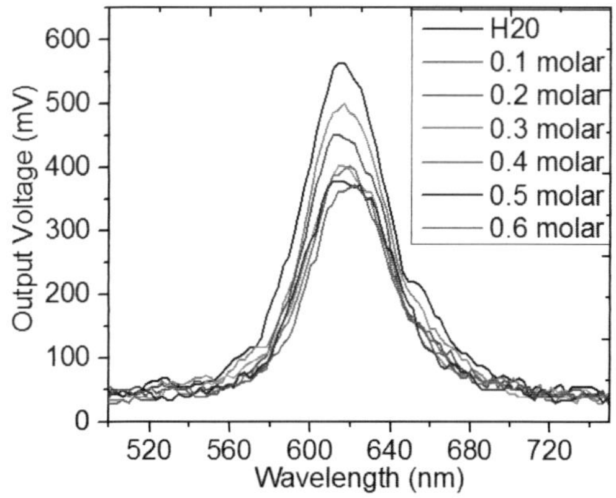

Figure 3: Intensity of light in the wavelength peak of Red LED in various glucose concentration

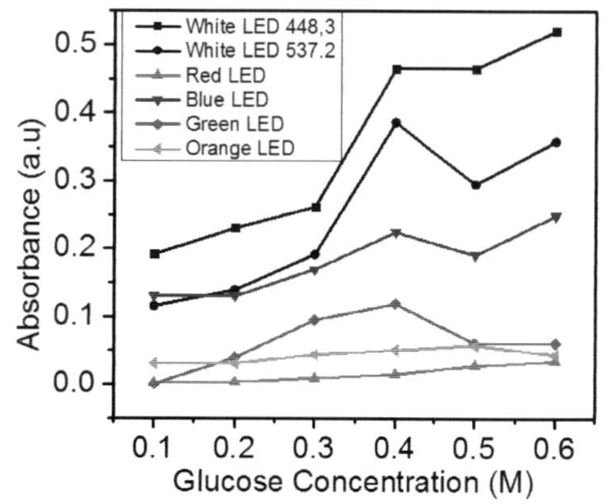

Figure 4: Absorbance as a function of glucose concentration.

where i and j are indices of a single particle micro-states while N and k are the total number of particles in the system and Boltzmann constant, respectively. As the number of particles in the system (N) increases, i.e., glucose concentration increases, the possible number of particles that can absorb a photon energy ε_i in order to be in an energy level ε_i is also increased.

Figure 4 indicates that an increase of glucose concentration results in a linear increase of their refractive index. Taking the relationship between the cladding (glucose covering the D-shape fiber sensor) refractive index (n_{Gl}) and the penetration depth (d_λ) of evanescent field at a wavelength λ as

$$d_\lambda = \frac{\lambda}{2\pi\sqrt{n_C{}^2 sin^2\theta - n_{Gl}{}^2}} \qquad (2)$$

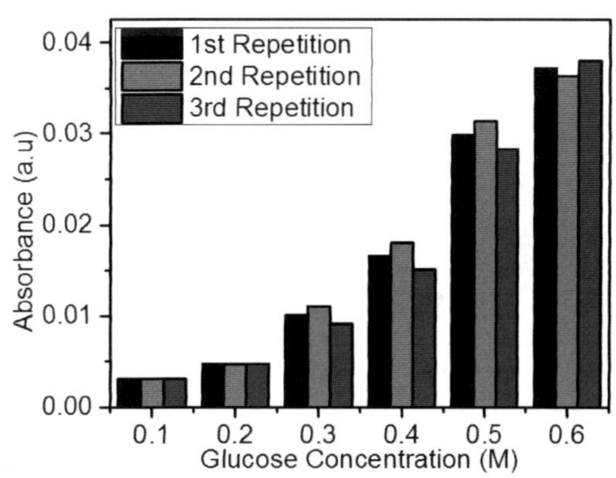

Figure 5: Repeatability of absorbance measurement of light coming out of the fiber as function of glucose concentrations.

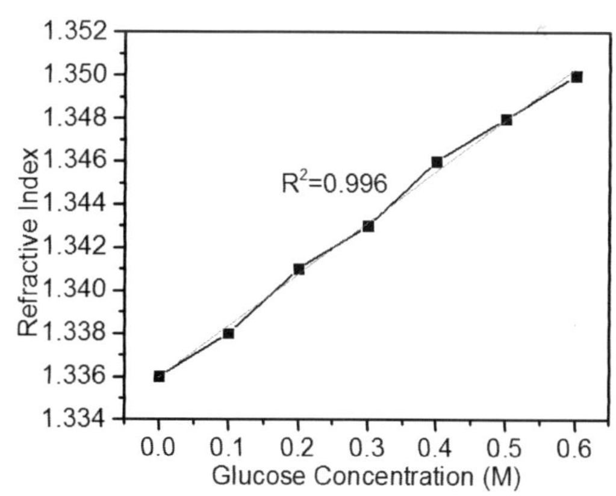

Figure 6. Refractive index of glucose with various concentrations

where θ is incident angle and n_C is refractive index of the core, It can be seen that the increase of the cladding refractive index (glucose concentration) will result in an increase in penetration depth or portion of light travelling in glucose. As absorbance (A) and glucose concentration (c_{Gl}) is related by

$$A = \alpha c \epsilon \qquad (2)$$

where α and ϵ are absorption coefficient and molar absorptivity, respectively; it can be seen that the increasing glucose concentration as well as the absorption coefficient will lead to an increase of absorbance. Stated in Beer-Lambert law relationship,

$$I_{t-Gl} = I_{0-Gl}. exp(-\alpha x) \qquad (3)$$

Increasing the absorption coefficient will finally reduce the out-coming light passing through the cladding detected by the photo-detector from $I_{0\text{-}Gl}$ to $I_{t\text{-}Gl}$.

4. Conclusion

A fiber sensor based on evanescent field absorption made by side polishing that used to measure glucose concentration have been fabricated and their performance have been demonstrated. The sensor was made by side polishing in such a way that cladding at certain portion of fiber was removed. We have shown that light absorption increased with the increase of glucose concentration measured with different color of LED (wavelength). as function of glucose concentration was measured as function of wavelength (the color of LED). We have shown that the proposed sensor can demonstrated an increase of light absorption as function of glucose concentration.

Reference

[1] Galant A, Kaufman R. and Wilson J 2015 *Food. Chem.* **188** 149-160.
[2] Cardosi M and Liu Z, 2012. *Intech.* **48491** 319-349.
[3] Jiang W, He X, Yang H, Xiang X, Hu S and Li S 2017 *Food Control* **82** 136-144.
[4] Hussain I, Das M, Ahmad K U and Nath P 2016 *Sensor Actuat. B-Chem.* **239** 1042-1050.
[5] Iizuka K, 2002. Elements of Photonics **1,** New York: Wiley–Interscience.
[6] Qiu, HW, Xu, SC, Jiang SZ, Li Z. Chen PX, Gao SS, Zhang C and Feng DJ 2014 *Appl. Surf. Sci.* **329** 390-395.
[7] Dachao L, Songlin Y, Changyue S, Chongwei Z, Haixia Y and Kexin X 2015 *Acs. Sym. Ser.* **72** 370-375.
[8] Parukhar VG, Tamgadgeb YS, Gambhirec AB and Muleya GG 2014 *Measurement* **61** 9-15.
[9] Klonoff D C 2012 *J Diabetes. Sci. Technol.* **6** 1242-1250.
[10] Fang Yu L, Wang Chen T, and Chiang Chia C 2016 *Sensors* **16**1460.
[11] Marzuki A, Sari N W and Riyatun, 2016, *IOP Conf. Ser. Mater. Sci. Eng.* **107** 012032.
[12] Wang C, Kaya M, Wang C, 2012 *J Biomed. Opt.* **17(3)** 037004.

IOP Conf. Series: Materials Science and Engineering **333** (2018) 012017 doi:10.1088/1757-899X/333/1/012017

Identification of magnetic minerals in the fine-grain sediment on the Bengawan Solo River

B Purnama[1], R Kusuma[1], B Legowo[1], Suharyana[1], A T Wijayanta[2]
[1] Physics Department Sebelas Maret University, Jl. Ir. Sutami 36A Kentingan Jebres Surakarta 57126, INDONESIA
[2] Department of Mechanical Engineering Sebelas Maret University, Jl. Ir. Sutami 36A Kentingan Jebres Surakarta 57126, INDONESIA

E-mail: bpurnama@mipa.uns.ac.id

Abstract. The magnetic mineral content in the fine sediment of Bengawan Solo River is discussed. The fine sediment is obtained on the upper part of the tributary of Bengawan Solo River. Magnetic minerals are separated using permanent magnets. Furthermore the magnetic minerals are overnight dried using oven at 100°C. FTIR characterization indicate that the magnetic minerals in the fine sediment of Bengawan Solo River have the same characteristics as minerals in Cilacap and Purwokerto areas. Magnetic minerals are estimated to form at wave number k = 569.03 cm^{-1}. This hematite content increase to 70.67% after anealling treatment at 600°C for 4 hours in atmosphere condition. This result is interesting because the heating process increases the hematite content. Within results, the magnetic properties of the sample will change.

1. Introduction

Analysis of magnetic rocks from deposited sediments is a technique that has been established in the field of earth science. This is primarily with environmental magnetism. This method has the advantage of relatively fast, non-destruction, relatively cheap cost [1-4]. Characteristics of magnetic rocks in sediments are useful information for various field such as paleoceanography, paleoclimatology, sedimentology and soil science [4, 5].

In general, magnetic minerals include magnetite (Fe_3O_4), maghemite (γ-Fe_2O_3), hematite (α-Fe_2O_3), and goethite (α-FeOOH) formed well drained soils in response to external factors i.e. physical, chemical and biological factors [3]. Hematite and goethite are antiferromagnetic materials. Both of these minerals are stable and have a small magnetization saturation M_S (0.4 Am^2 kg^{-1} and 0.05~0.3 Am^2 kg^{-1} for pure hematite and goetite respectively). However the both minerals indicate as hard magnetic material owing a relatively large coercive field i.e. H_c > 300 mT [6].

In this paper, preliminary study of a magnetic mineral identification on the fine sediment of Bengawan Solo River is presented. The formed of magnetic minerals evaluate by FTIR spectroscopy. Finally, the magnetic property measure by vibrating sample magnetometer (VSM) at room temperature.

2. Experimental

Preparation and evaluation of the natural mineral in a fine sediment of Bengawan Solo Rivers are conducted by the following steps. First, separation of the fine sediments by using permanent magnet. Thus the sample is then washed with water and alcohol. This procedure is repeated until clean. Subsequently the sample was heated overnight at a temperature of 100°C to remove the water content.

The second step is annealing process. Here the sample is annealed at a temperature of 600°C for 4 hours under atmospheric conditions. Then the sample is homogenized using mortar. The next step is characterization including the determination of metal mineral oxides using FTIR and x-ray fluorescence. Finally, the magnetic property of the fine sediment evaluate by using vibrating sample magnetometer at room temperature.

3. Results and Discussion

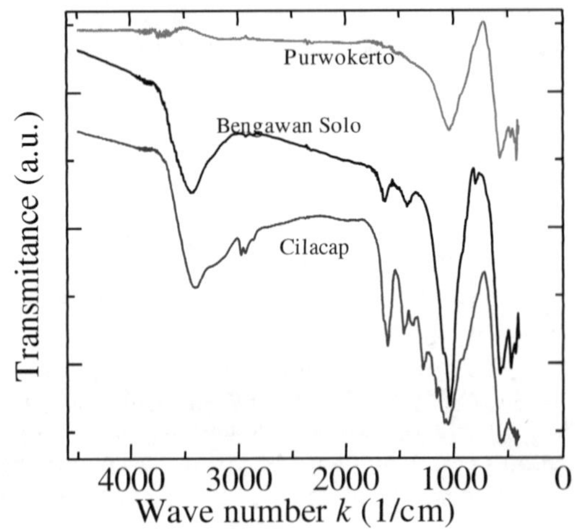

Figure 1. FTIR spectrum of a fine sediment sample of Bengawan Solo River compare with Cilacap and Purwokerto sediments.

Figure 1 shows a Fourier Transform Infra-Red (FTIR) spectrum characteristic for the fine sediments of Bengawan Solo Rivers compare to the fine-magnetic sediment of Cilacap and Purwokerto. The FTIR characterization indicates the formation of the oxide bond resonance of the fine sediment of Bengawan Solo including metal oxide bonds such kind Fe–O, Si–O and others. FTIR in a fine sediments of the Bengawan Solo River typically have the same spectrum as a magnetic fine sediment of Cilacap. Absorbance peak of FTIR spectrum at wave number 1447 cm^{-1} to 3000 cm^{-1} is associated to a vibration stretching and bending of H–O–H bond. Whereas a peak absorbance for wave number $k = 1023$ cm^{-1} is the anti symmetric stretch of the Si-O-Si bond attributed to the oxygen movement. It is known that the Fe-O bond observe at the wave number k of about 600–700 cm^{-1}.

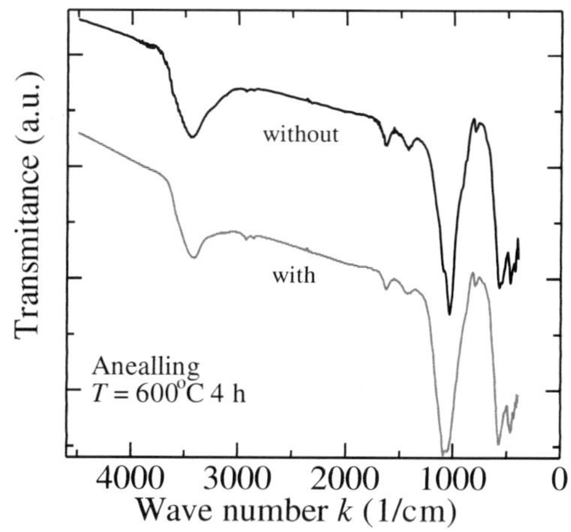

Figure 2. The FTIR spectrum evaluate for the fine sediment of Bengawan Solo River under with and without annealing at 600°C for 4 hours.

Table 1. The contents of the fine sediment sample Bengawan Solo River with and without annealing @600°C for 4 h

No	Compounds	Concentration (%)	
		Without annealing	With annealing @600°C for 4 h
1	Fe_2O_3	68.85	70.67
2	SiO_2	12.34	12.59
3	TiO_2	6.78	6.31
4	Al_2O_3	4.71	3.57
5	CaO	2.8	2.82
6	P_2O_5	1.12	0.84
7	Others (consist 12 compounds)	3.35	3.14

The annealing effect on the FTIR spectrum the fine sediment of Bengawan Solo River under with and without annealing temperature at 600°C for 4 h is depicted at Fig 2. It is clearly observed that there is no significantly change or shift in the absorption peak from the original condition. This indicates that the formed metal oxide is stable with high temperature treatment. The Fe–O bond realize at k = 569 cm^{-1} for case without annealing and a few change become k = 571.9 cm^{-1} for case with annealing.

In order to know more details about the oxide composition especially Fe–O in the sample, an XRF analysis is performed as shown at Table 1. From the result table XRF analysis confirmed that the Fe-O compound formed is hematite (Fe_2O_3). The hematite composition dominates up to 68.85% and increases to 70.67% by annealing treatment. The increase in concentrations of other compounds that increase in the sediment is only silicon dioxide i.e. from 12.34% to 12.59%. While concentration of the other compounds show decrease with annealing treatment. It can be said that only Fe_2O_3 and SiO_2 are stable with heat treatment.

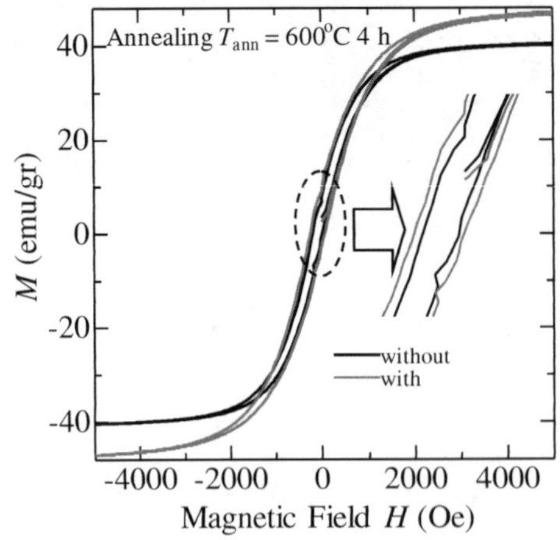

Figure 3. The hysteresis curve of the fine sediment on Bengawan Solo River without and with annealing temperature of 600°C for 4 h.

Finally, the magnetic property of the fine sediment evaluate by using vibrating sample magnetometer (VSM) at room temperature. Figure 3 show the VSM results of the fine sediment on Bengawan Solo River under without and with annealing temperature. It was clearly observed that the coercive field did not significantly change with the annealing treatment at temperature of 600°C for 4 h (see zoom figure). Coercive field changes slightly from 100.63 Oe to 157.46 Oe after annealing. This can be attributed the occurrence of the domain wall depining. While the magnetization of saturation increased from 32.97 emu/g to 39.98 emu/g due to the increase of hematite concentration in the samples sediments of Bengawan Solo River.

4. Conclusion

The magnetic mineral content in the fine sediment of Bengawan Solo River is discussed. FTIR characterization indicate that the magnetic minerals in the fine sediment of Bengawan Solo River have the same characteristics as minerals in Cilacap and Purwokerto areas. Magnetic minerals are estimated to form at wave number k = 569.03 cm^{-1}. This hematite content increase to 70.67% after annealing treatment at 600°C for 4 hours in atmosphere condition. Within results, the coercive field change from 100.63 Oe to 157.46 Oe. Finally the saturated magnetization increase from 100.63 emu/g to 157.46 emu/g.

Acknowledgements

This study was financially supported by Direktorat Riset dan Pengabdian Masyarakat Direktorat Jenderal Penguatan Riset dan Pengembangan Kementerian Riset, Teknologi, dan Pendidikan Tinggi sesuai dengan Kontrak Penelitian Nomor: 098/SP2H/LT/DRPM/IV/2017.

References

[1] Jiang Z, Liu Q, Zhao X, Roberts A P, Heslop D, Barrón V & Torrent J 2016 *Journal of Geophysical Research: Solid Earth* **112(6)** 4195-4210.
[2] Oldfield F 1991 *Quaternary Science Reviews* **10(1)** 73-85.
[3] Maxbauer D P, Feinberg J M & Fox D L 2016 *Earth-Science Reviews* **155** 28-48.
[4] Scheidt S, Egli R, Frederichs T, Hambach U & Rolf C 2017 *Geophysical Journal International* **210(2)** 743-764.
[5] Brachfeld S A & Banerjee S K 2000 *Earth and Planetary Science Letters* **176(3)** 443-455.
[6] Schwertmann U & Cornell R M 2007 *The iron oxides. Iron Oxides in the Laboratory: Preparation and Characterization* 5-18.

IOP Conf. Series: Materials Science and Engineering **333** (2018) 012018 doi:10.1088/1757-899X/333/1/012018

Optical properties of anthocyanin dyes on TiO_2 as photosensitizers for application of dye-sensitized solar cell (DSSC)

A H Ahliha[1], **F Nurosyid**[1,2], **A Supriyanto**[1,2] **and T Kusumaningsih**[2]
[1]Physics Department of Graduate Program, Universitas Sebelas Maret, Indonesia
[2]Department of Physics, Faculty of Mathematics and Natural Science, Universitas Sebelas Maret, Indonesia
[3]Department of Chemistry, Faculty of Mathematics and Natural Science, Universitas Sebelas Maret, Indonesia

E-mail: nurosyid@yahoo.com

Abstract. Dye-sensitized solar cell (DSSC) is one of the alternative energy that can convert light energy into electrical energy. The component of DSSC consists of FTO substrates, TiO_2, electrolyte, dye sensitizer, and counter electrode. This study aim was to determine the effect of optical properties of anthocyanin dyes on efficiency of DSSC. The dye sensitizer used can be extracted from anthocyanin pigments such as dragon fruit, black rice, and red cabbage. The red cabbage sensitizer shows lower absorbance value in the visible range (450-580 nm), than dragon fruit and black rice. The chemical structure of each dye molecules has an R group (carbonyl and hydroxyl) that forms a bond with the oxide layer. Red cabbage dye cell has the highest efficiency, 0.06% then dragon fruit and black rice, 0.02% and 0.03%.

1. Introduction

Dye-sensitized solar cell is a dye-based solar cell that converts photons energy into electrical energy by using semiconductor materials that have the wide band gap energy$[1-4]$. A DSSC is composed of a transparent conductive oxide substrate, a wide semiconductor, dye sensitizers, a counter electrode, and a electrolyte$[5, 6]$. The dye sensitizer plays a key role in developing the high performance of DSSC. It has been the subject of more recent studies as the efficiency of DSSC depends mainly on its properties such as the absorption spectrum, dye-to-TiO_2 charge transfer and the anchorage of the dye to the surface of $TiO_2$$[7-10]$.

One of the natural dyes used as a dye sensitizer is anthocyanin. Anthocyanin absorbance spectrum in the visible ranges (450-580 nm). Anthocyanin is a pigment that this is red, blue, and purple on the fruits, flowers, and leaves. Pratiwi *et al.* using dragon fruits, red cabbages, and mangosteen peels as the dye sensitizer. Dragon fruit extract gave the lowest efficiency, despite the highest absorbance value$[1]$. This is due to the chemical structure of electrons reaction between the dye and TiO_2, which the difference is not well understood.

In this study, the dye sensitizer used can be extracted from anthocyanin pigments such as dragon fruit, black rice, and red cabbage. This study is expected to find the composition of the dye compounds in order to absorb light better and generate more efficient electrical energy.

2. Experimental

The materials used are natural dyes from dragon fruit, red cabbage, black rice, and spinach leaves. The anthocyanin dyes were dissolved in acetic acid: quads: methanol solvents. The substrates used Fluorine-doped Tin Oxide (FTO) glass with a size of 2.0 cm x 2.0 cm and active cell area of 1.0 cm x 1.0 cm. The working electrodes was coated using TiO_2 nanopowder by the spin coating method. The counter electrode was prepared from platinum on the FTO conductive glass.

Characterization by using UV-Visible Spectrophotometer 1601 PC to determine the wavelength of dye absorbance, Fourier Transform Infrared Spectrophotometer Shimadzu Prestige 21 to know the chemical bond of dye. The working electrode and counter electrode were assembled into a sandwich type arrangement. The electrolyte solution has filled the space between the two electrodes. *I-V* meter *Keithley* 2602A to determine the efficiency of the DSSC prototypes.

3. Results and Discussion

3.1. Optical Characteristics of Anthocyanin Dyes

The absorption characteristics of each sample using a spectrophotometer UV-Vis 1601 PC. Figure 1 show the plots of absorbance spectra from the anthocyanin dyes. Figure 1 is that the red cabbage shows lower absorbance in the visible range (450-580 nm), than the other dye. The absorption characteristics is a very important properties in DSSC as it directly reflects the optical transitions probability[11]. Anthocyanin dyes show absorption in the visible region, 450-580 nm. Anthocyanin can absorb light and transfer that the light energy by resonance energy transfers to the anthocyanin pair in the reaction center of the photosystems. Anthocyanin absorbs strongly in the red and violet regions of the absorption spectrum. The absorption peaks for anthocyanin dye solution from red cabbage, black rice, and dragon fruit are 533.5 530.5 nm; 526.5 nm; and 533.5 nm with the absorbance is 0.3091 a.u; 0.6259 a.u; and 0.3857 a.u, respectively. The absorbance value is influenced by the anthocyanin compound absorbed on the surface of TiO_2, where the anthocyanin compound is proportional to the absorbed light.

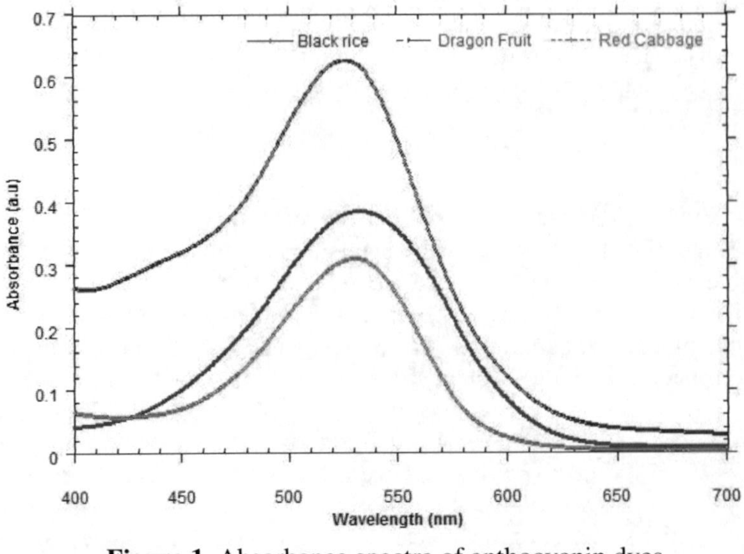

Figure 1. Absorbance spectra of anthocyanin dyes

3.2. Fourier Transform Infrared (FTIR) Characterization of Anthocyanin Dyes

Figure 2 shows the FTIR analysis to determine the functional groups on TiO_2 and TiO_2-dye. The spectra of TiO_2-dye red cabbage adsorption widened at wave number 3200-3600 cm^{-1} is the stretching vibration of OH groups on the surface of TiO_2. The intensity dye of red cabbage had 92.3% than the other TiO_2-dye. This is due to the presence of hydroxyl groups forming intermolecular and intramolecular hydrogen bonds. The dye of anthocyanin adsorbed on the surface of TiO_2. The spectrum of FTIR shifted absorption wave numbers of TiO_2 and TiO_2-dye are presented in Table 1. The intensity of TiO_2 nanoparticle is 57% thus the vibration of -OH stretching. The intensity of dye adsorbed on the surface of TiO_2 is 92.3% of red cabbage, 84.4% of black rice, and 76.1% of dragon fruit. The intensity of TiO_2-dye occurred in the group of Ti-O-Ti are 67.8%, 38.89%, 35.8%, 38.4%, and 29%. This is due to the dye adsorbed on the surface of TiO_2.

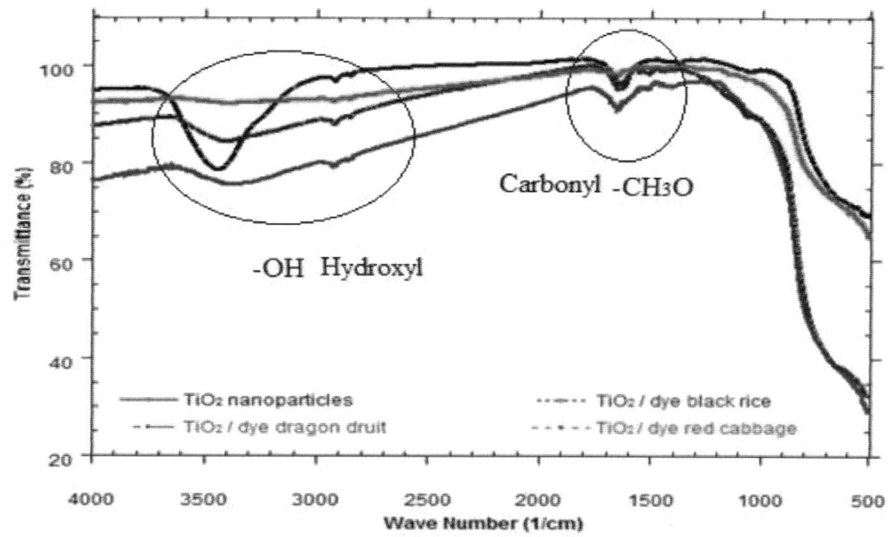

Figure 2. FTIR spectra of the dye from (a) TiO_2 nanoparticle, (b) TiO_2-dye red cabbage, (c) TiO_2-dye dragon fruit, and (d) TiO_2-dye black rice

Table 1. Characterization of FTIR from anthocyanin dye and TiO_2

Absorption frequency (cm^{-1})								Assigment and remarks
TiO_2		Red Cabbage		Black Rice		Dragon Fruit		
Peak	T (%)	Peak	T (%)	Peak	T (%)	Peak	T (%)	
3444	57	3400	92.3	3415	84.4	3281	76.1	Ti-O-H
		2927	92.7	2924	88	2924	79.2	C-H
		2856	93.1	2855	89.4	2855	80.5	
		1655	97.9	1654	95.17	1654	90.9	C=O
				1560	98.7			C=O
1640	89.2							C=C
				1511	98.4	1512	95.2	C=C
						1410	95.7	C-H
		1290	99.3					C-O
		1152	98.3	1065	89.6	1017	89.1	
		709	74.1	656	38.5	584	35.4	Ti-O-H
672	53.7	546	67.8	566	35.8	502	29	

The result of spectra showed the different functional groups between the TiO_2 and TiO_2-dye. The new absorption in TiO_2-dye indicated the absorption from the dye adsorbed on the surface of TiO_2. In the wave number at 3000-2700 cm^{-1} is a vibration of C-H stretching, which has two bonds and the wave number at 1655 cm^{-1} and 1654 cm^{-1} is the absorption of C=O stretching. The highest in spectra of FTIR is TiO_2-dye red cabbage because this is due to the dye adsorbed on the surface of TiO_2. The chemical structure of the dye of anthocyanin consisting of 7 hydroxyl flavyllium cations, these molecules function in the absorption of light and form the natural. Anthocyanin color formed in the hydroxyl group (-HO) at position R_3 and always connected to the glucose molecule required for thermal equilibrium and the position of R_5 is one or more hydroxyl group or methoxyl ($-CH_3O$) in ring of B. The variety of colors shown by anthocyanin depends on the number and position of the group[12]. The widening of absorption occurs due to the reduction between the already-formed spectra. The reduced spectrum causes the breaking of the bonds of the reduced chains which then join the other bonds. Based on the result of a UV-Vis spectrum and FTIR spectrum it is concluded that the compounds are extracted of anthocyanin pigments.

3.3. I-V Characterization of DSSC

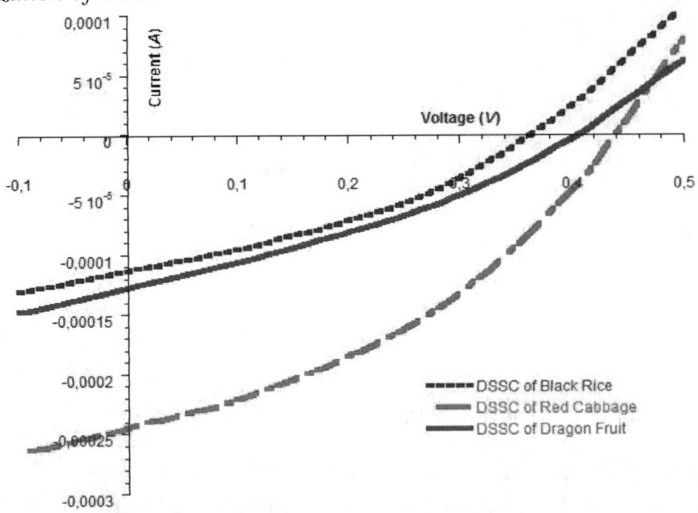

Figure 3. *I-V* characteristics of DSSCs with the anthocyanin dye

Table 2 and Figure 3 shows it can be seen that DSSC sensitized with black rice the lowest measured $\boldsymbol{V_{oc}}$ of 0.57 V. The black rice gave current densities value of 0.14 mA, this is the lowest value than the other dyes. The difference in the current densities of the cells may be attributed to the difference in the structure of the dyes, and how fast the charge is injected into the conduction band of the TiO_2 film[**10**]. The photo-energy conversion efficiency of red cabbage, black rice, and dragon fruit are 0.06%; 0.03%, and 0.02%, respectively. Red cabbage extract gave the highest efficiency, despite the lowest absorbance value. This may be the analysis of the chemical properties of the dye and electrons reactions between the dye and TiO_2. High efficiencies may be attributed to rich adsorption of dye molecules onto the TiO_2 particle. A nanoparticle of TiO_2 provided paths for the electrons while maintaining a low surface area for dye absorption. Therefore, availability of bonds between the dye and TiO_2 molecules through which electrons can transport from the excited dye molecules to the TiO_2 film has been reported to result in high efficiencies [9]. Table 2 show the *I-V* curve of a DSSC based on TiO_2-dye under irradiation of 1000 W/m^2, these include I_{SC}, V_{OC}, FF, dan η.

Table 2. Parameters of DSSCs with different anthocyanin natural dyes

Dyes	V_{oc} (V)	I_{sc} (mA)	FF	η (%)
Red Cabbage	0.6	0.21	0.46	0.06
Black Rice	0.57	0.14	0.3	0.03
Dragon Fruit	0.6	0.15	0.3	0.02

IOP Conf. Series: Materials Science and Engineering **333** (2018) 012018 doi:10.1088/1757-899X/333/1/012018

4. Conclusion

The dye of anthocyanin was used to make dye-sensitized solar cells. The absorption of dyes and attached to TiO_2 thin films in the visible region, 450-580 nm. The red cabbage extract gave the highest efficiency, despite the lowest absorbance value. The R groups in spectra of FTIR indicate that the presence of carbonyl and hydroxyl groups of substances. High efficiencies may be attributed to rich adsorption of dye molecules onto the TiO_2 particle.

Acknowledgment

This work was supported by Institute for Research and Community services, Universitas Sebelas Maret by Penelitian Unggulan UNS (Contract Number. 1075/UN27.21/PP/2017).

References

[1] Pratiwi D D, Nurosyid F, Supriyanto A, and Suryana R 2016 *J. Phys.: Conf. Ser.* **776** 012007
[2] Dey A, Moyez Sk A, Mandal M K, and Roy S 2016 *Material Today: Proceedings* **3** 3498-3504
[3] Cakar S, Guy M, Ozacar M, and Findik F 2016 *Electrochimica Acta* **209** 407-422
[4] Gomez T, Jaramillo F, Schott E, Arratia-Perez R, and Zarate X 2017 *Solar Energy* 215-223
[5] Suyitno S, Zainal A, Ahmad A S, Argatya T S, and Ubaidillah 2014 *Applied Mechanics and Materials* **575** 689-695
[6] Sengupta D, Mondal B, and Mukherjee K 2015 *Spectrochim. Acta. A. Mol. Biomol. Spectrosc.* **148** 85-92
[7] Chang K, Kao M, Chen T, Chen C, Cho K, and Lai X 2013 *J. Photoenergy* **8**
[8] Maabong K, Muiva C M, Monowe P, Sathiaraj T S, Hopkins M, Nguyen L, Malungwa K, and Thobega M 2015 *IJRER* **5** 54-59
[9] Nwanya A C, Ugwuoke P E, Ejikerne P M, Oparaku O U, and Ezerna F I 2012 *Int. J. Electrochem. Sci.,* **7** 11219-11235
[10] Moustafa K F, Rekaby M, El Shenawy E T and Khattab N M 2012 *JASR* **8** 4393-4404
[11] Lai W H, Sub Y H, Teoh L G and Hon M H 2008 *J. Photochem. Photobiol. A Chem.* **195** 307-313
[12] Fernando J M R C and Senadeera G K R 2008 *Current Science* **95** (5) 663-666

IOP Conf. Series: Materials Science and Engineering **333** (2018) 012019 doi:10.1088/1757-899X/333/1/012019

The Synthesis of Photocatalyst Material ZnO using the Simple Sonication Method

R Faradis[1], E N Azizah[1], S D Marella[1], N Aini[1], A Prasetyo[1*]
[1]Department of Chemistry, Faculty of Science and Technology, Universitas Islam Negeri Maulana Malik Ibrahim Malang, Jl. Gajayana 50, Malang, 65144, Indonesia

[*]E-mail:antoniaprasetyo@gmail.com

Abstract. ZnO is well known as photocatalyst material therefore potentially to applied in many purposes. The particle size of photocatalyst material influenced the catalytic activities. In this research, ZnO was synthesized using the simple sonication method to obtain the the smaller particle with sonication time variation respectively: 30, 60, 160, 360 minute. X-ray diffraction data showed that the synthesized material have wurtzite structure with space group $P6_3mc$. The synthesized ZnO with 30 minutes sonication time produced the smallest particle size and have the lowest band gap energy (2.79 eV). The photocatalytic test at methylene blue also showed that the optimum activity was gained from ZnO which synthesized at 30 minute sonication time (degradation percentage of metylene blue is 77.93%).

1. Introduction

The dye pollutant from textile industries have become environmentally problem. Photocatalysis is one of technology have potentially to degradate the dye pollutant so it can applied to solve the dye pollutant problem.Photocatalysis is a process of chemical reaction which involving photocatalyst material and light as initial reaction [1]. ZnO is one of material that used for photocatalyst material and have some advantages such as high photocatalyst activity with a band gap 3.20-3.37 eV, cheap, more stable structure, and non-toxic [2].

The particle size of photocatalyst material influenced to the photocatalytic activities [3]. So it is important to develop the synthesis method of ZnO to obtain the smaller particle as results enhancing the photocatalytic activities of ZnO. There are many method to produce ZnO.Sonochemical synthesis is more effective as well as simple method to produce the sample with smaller particle size [3,4]. The mechanism of sonochemical method involved: (a) the formation, (b) growth, and (d) collapse of bubbles in liquid and then can produce high temperature and pressure. This condition can use to chemical reaction [3-5]. Therefore, in this research we synthesized ZnO using sonochemical method to obtain the smaller particle and in this method, we use the simple ultrasound reactor withlow frequency to synthesized ZnO.

2. Experimental

Zinc acetate dihydrate was dissolved of double-distilled water and then aqueous solution of 1 M NaOH was poured drop wise in this solution while stirring it continuously at room temperature until pH 12. The result solution was transferred into ultrasonic reactor for 30 (ZnO 30 min), 60 (ZnO 60 min), 180 (ZnO 180 min), and 360 minutes (ZnO 360 min)at room temperature.The white precipitate was aged for 24 h and filtered, washed with water and ethanol for several times and then dried. The frequency of sonicator is 42 kHz with power output 100 W.

The sample results were characterization using X-ray diffraction (XRD), scanning electron microscopy (SEM), and diffuse reflectance spectroscopy (DRS). The test of photocatalytic activities applied to methylene blue (MB) using ultra violet (UV) light source with wavelength 365 nm from 0 to 60 minutes.

3. Result and Discussion

Figure 1 shows XRD patterns of synthesized ZnO at time variation of sonication.The all of XRD pattern sample matched with standart ZnO wurtzite (JCPDS No. 36-1451). All the reflection can be indexed to the pure hexagonal phase of ZnO with space group *P63mc* with the crystal lattice a = 3.2495 (Å), c = 5.2069 (Å). The XRD pattern also showed that at 30 minutes sonication, ZnO is successfully formed and there are no change when the sonication time was longer.

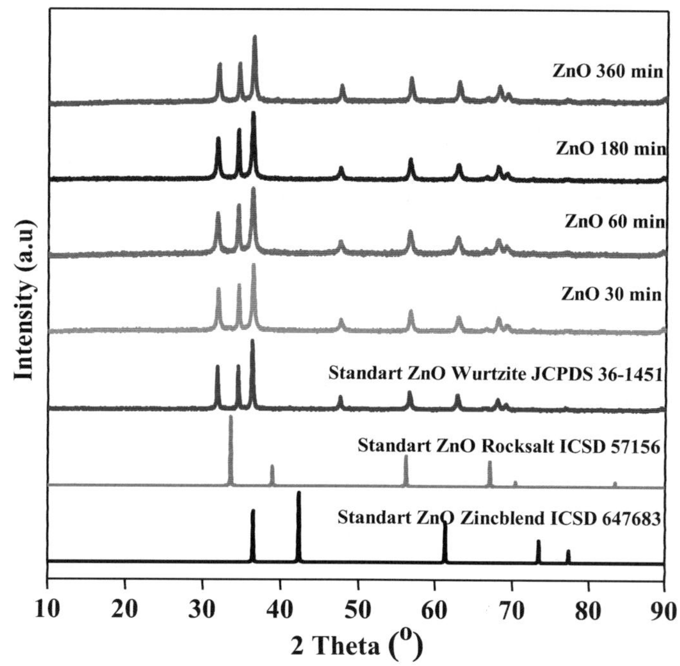

Figure 1. The X-ray powder diffraction pattern of ZnO

The micrograph of SEM are shows in Figure. 2. ZnO morphology which synthesized using sonication method has better distribution than without sonication method. It indicated that the time of sonication influenced to the morphology of particle [6]. Optimum distribution achieved at 30 minutes of ultrasonic irradiation because the morphology of other time variation and without sonication method is not uniform

IOP Conf. Series: Materials Science and Engineering **333** (2018) 012019 doi:10.1088/1757-899X/333/1/012019

and produce agglomeration on ZnO morphology. It indicated at more 30 minutes of ultrasonic irradiation, the smaller particle attracted each other and produces the agglomerates.

Figure 2. SEM image of synthesized ZnO at time variation of sonication:(a) 30(b) 60 (c) 180, and (d) 360 minutes.

Figure. 3 shows the DRS data of synthesized ZnO. As can be seen that the reflectance value was decreased in visible area for ZnO at sonication time 30 minutes if compared with others. The increasing of absorbance in visible range gave opportunity to work in visible light range, so it is expected to have more effective photocatalic activity. The value of bandgap energy was obtained from % reflectance and calculated using Kubelka-Munk equation and was shown in Table 1.The lowest bandgap energy which obtained at ZnO 30 minutes sonication is 2.79 eV which equivalent with wavelength 444.44 nm. It indicates that ZnO 30 min has more effective potential to utilized as photocatalys material.

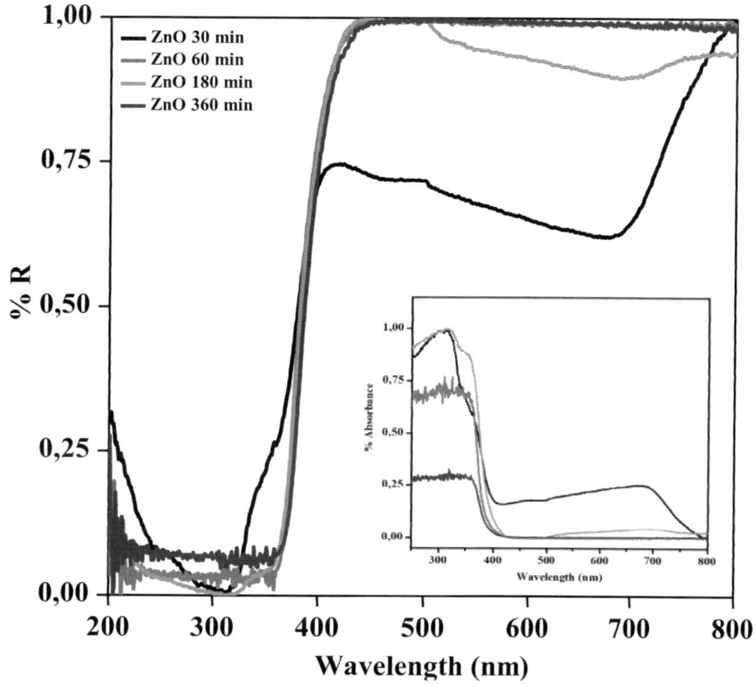

Figure 3. DRS spectra of samples (inset: % absorbance spectra data)

Table 1. The bandgap energy of ZnO

Material	Bandgap Energy (eV)	Wavelength (nm)
ZnO 30 min	2.79	444.44
ZnO 60 min	3.18	389.93
ZnO 180 min	3.10	400.00
ZnO 360 min	3.16	392.40

Figure.4 shown that degradation MB without UV and ZnO catalyst less than 4%, while degradation MB without ZnO and using UV until 5%. The optimum degradation MB using ZnO catalyst UV was obtained from ZnO 30 min at value 77.94%. It relates to ZnO 30 min is (a) smallest particle size(b) lowest bandgap energy, and (c) light absorption at visible light. And also it indicated that the size of particle influenced to the photocatalytic activity [7].

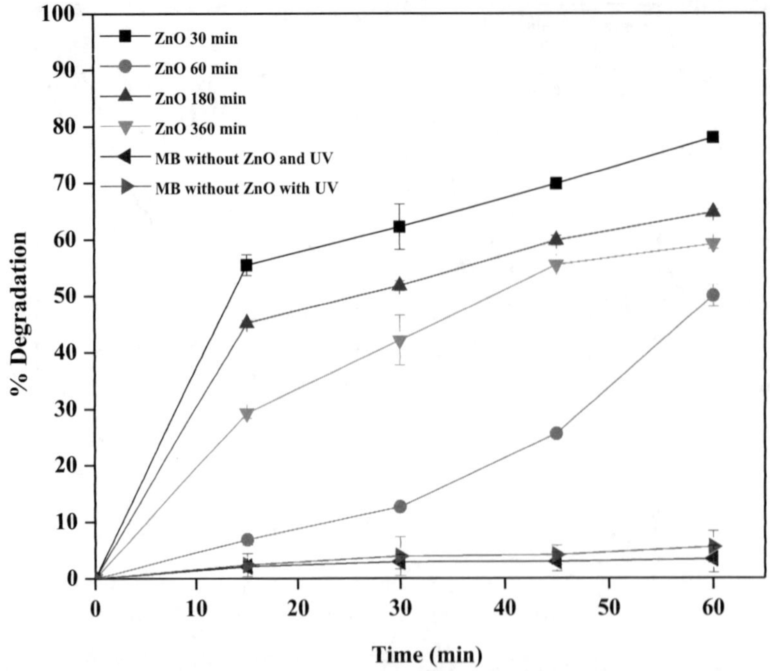

Figure.4. Degradation test of metylene blue using ZnO catalyst

4. Conclusion

Photocatalytic material ZnO was synthesized using sonochemical method have wurzite structure with space group *P6₃mc*. The morphology particle of ZnO which synthesized using sonication method has better distribution than without sonication method, while the optimum distribution achieved at 30 minutes of ultrasonic irradiation.

Reference

[1]. Chakrabarti S and Dutta B. K. 2004 *J. Hazard. Mater.* **112** 269-278
[2]. Young J Y, Simer C, Ohm T 2006 *Mater. Res. Bull.* **41** 67-66
[3]. Yadav R S, Mishra P, Pandey A C 2008 *Ultrason. Sonochem.* **15** 863-868
[4]. Hou X, Zhou F, Sun Y, Liu W 2007 *Mater. Lett.* **61** 1789-1792
[5]. Xiao Q, Huang S, Zhang J, Xiao C, Tan X 2008 *J. Alloy Comp.* **459** 18-22
[6]. Mahdavi R, Siamak S, Talesh A 2017 *Ultrason. Sonochem.* **39** 504-510
[7] Yang J, Kong X, Jiang W, Cao J, Zou P, Luan H, Yang L 2015 *Mat. Sci. Semicon. Proc.* **40** 713-719.

IOP Conf. Series: Materials Science and Engineering **333** (2018) 012020 doi:10.1088/1757-899X/333/1/012020

Visible light driven photocatalyst of vanadium (V^{3+}) doped TiO_2 synthesized using sonochemical method

N Aini[1*], R Ningsih[1], D Maulina[1], F F Lami'[1], S N Chasanah[1]
[1]Department of Chemistry, Faculty of Science and Technology, Universitas Islam Negeri (UIN) Maulana Malik Ibrahim Malang, Jl. Gajayana 50, Malang 65144, Indonesia

[*]E-mail: nuraini.kkfamaliki@gmail.com

Abstract. TiO_2 has been widely investigated due to its superior photocatalytic activity under ultraviolet irradiation among the photocatalyst materials. In this research, vanadium (V^{3+}) was doped into TiO_2 to enhance its light response under visible irradiation for wider application. Vanadium was introduced into TiO_2 lattice at various concentration respectively 0.3, 0.5, 0.7 and 0.9% using simple and fast sonochemical method. X-Ray Diffraction data show that vanadium doped TiO_2 crystallized in anatase phase with $I4_1amd$ space group. X-Ray Diffraction pattern shifted to lower value of 2θ due to vanadium dopant. It indicated that V^{3+} was incorporated into anatase lattice. UV-Vis Diffuse Reflectance Spectra was revealed that the doped TiO_2 has lowered reflectance and enhanced absorption coefficient in visible region than undoped TiO_2 and commercial anatase TiO_2. Band gap energy for undoped and doped TiO_2 were respectively 3.22, 3.05, 2.93, 3.03 and 2.40 eV. Therefore vanadium doped TiO_2 had potential to be applied under visible light.

1. Introduction

Modification anatase TiO_2 with various metal dopants have been reported as one of strategy to enhance its optical absorption and photocatalytic activity under visible irradiation. Vanadium dopant has produced a higher optical absorption and lower band gap energy than Ce, Cr, Fe, and Ni dopant [1-3] thus vanadium is an attractive dopant to be studied. Vanadium in anatase lattice shown the urbach tail phenomenon [4] therefore resulted in decreasing the band gap energy and enhancing the absorption of anatase in visible light. Vanadium also affected on the phase transformation of anatase TiO_2 since vanadium has the low solubility in anatase phase (about 5 at %) [5]. Low level of vanadium (0.3%) does not induced the phase transformation and only leads a small shift in absorption edge through solid state synthesis [6]. However, 0.3% of vanadium was reported to induce the phase transformation from anatase into rutile [3] while sol gel method was employed. It can be concluded that the structural transformation and the band gap reduction can be influenced by concentration of vanadium and the synthesis method employed in photocatalyst's production.

Sonochemical method is fast and simple method which nowdays developed in synthesis of semiconductor materials. Ultrasound wave can assisted the chemical reaction and resulted in smaller particle size, homogenous particle distribution, high surface area and high purity of material. Irradiation of ultrasound wave into solution given an acoustic cavitation which is produced through bubble formation, bubble growth and collapsing the bubbles in liquids [7]. Acoustic cavitation provide an extremely high temperature and pressure in short duration to facilitate the chemical reaction on synthesis process. Sonochemical method has been applied in synthesis of TiO_2 nanoparticle [8], Pt, Ru doped TiO_2 [9] and V doped TiO_2 1-5 at % [10]. No sufficient work reported to produce low content of vanadium doped TiO_2 using sonochemical method, therefore it became important to explore this efficient method on synthesis of vanadium doped TiO_2. Sonochemical method at low frequency was applied in this research to study the effect of vanadium concentration (0.3-0.9 at %) on structure and optical charasteristics of anatase TiO_2.

2. Experimental

Titanium isopropoxide (TTIP, sigma Aldrich) was dissolved into 40 mL ethanol:deionize water (3:1v/v). Stoichiometric amount of vanadium acetylacetonate ($V(acac)_3$, sigma aldrich) in 20 mL ethanol: deionize water (3:1v/v) added into titanium solution at 0.3, 0.5, 0.7 and 0.9% atomic ratio of vanadium III (V^{3+}). The homogenous mixture transferred into ultrasonic bath reactor and irradiated with low frequency ultrasound wave 40 KHz for 40 minutes. The obtained precipitates aged, filtered and washed with ethanol then dried at 105°C for 2h. The dry powders subsequently pressed into pellet and calcined for 2h at 500°C. Undoped TiO_2 was also synthesized using the same method with the absence of vanadium. The white powder produced for undoped TiO_2 and pale brown to brown for vanadium doped TiO_2 corresponding with increasing the vanadium concentration. The structure of synthesized materials were studied by Powder of X-Ray Diffracton (XRD). The optical properties were characterized by UV-Vis Diffuse Refflectance Spectroscopy (UV-Vis DRS). For further information, the synthesized materials also compared with commercially anatase TiO_2 purchased from Sigma Aldrich, ≥99% purity (TiO_2 SA No.248576) without further purification.

3. Result and Discussion

XRD pattern in figure 1 shows that all synthesized materials have agreement with commercially anatase TiO_2 Sigma Aldrich (TiO_2 SA). Introducing vanadium (III) at 0.3 to 0.9% into TiO_2 lattice using sonochemical method resulted in the pure phase of anatase and does not induced the phase transformation from anatase into rutile or brukite. Further insight on XRD pattern at highest intensity peak shows that V^{3+} doped TiO_2 shifted to lower value of 2θ compared to undoped TiO_2. The shift of 2θ attributed to alteration of d spacing material caused by specific vanadium content. It was indicated that vanadium was doped in TiO_2 lattice.

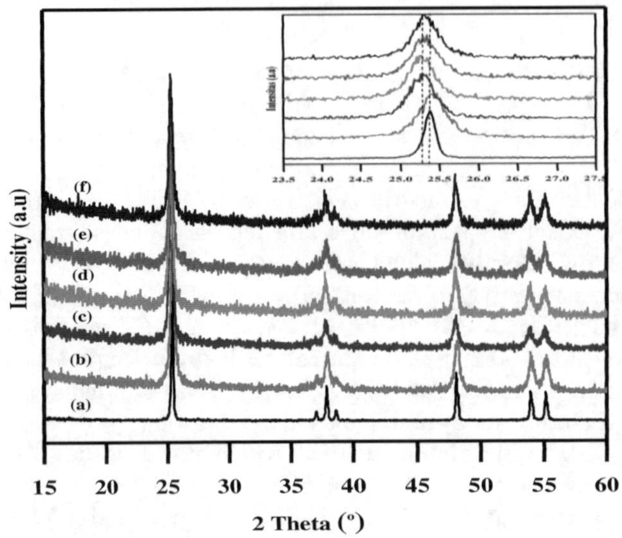

Figure 1. XRD Pattern for (a) TiO_2 SA (b) undoped TiO_2 and V^{3+} doped TiO_2 at (c) 0.3% (d) 0.5% (e) 0.7% and (f) 0.9%.

Compared to commercially TiO_2 SA, the XRD pattern of undoped and V^{3+} doped TiO_2 have boardened peaks which indicated smaller particle size. Table 1 shows the average crystallite size were calculated by Scherrer's equation at the highest intensity peak (101 plane). Average crystallite size of undoped and doped TiO_2 (0.3-0.7%) were smaller than TiO_2 SA (53.19 nm) while 0.9% of V^{3+} was larger than TiO_2 SA. Further XRD data analysis using Le Bail refinement method with anatase standart (JCPDS No. 21-1272) revealed that all materials crystallized in tetragonal lattice with $I4_1amd$ space group and 4 asymmetric unit (Z). Unit cell parameter (a, c) and unit cell volume of V^{3+} doped TiO_2 observed to decrease than undoped TiO_2. It can be affected by different size ion between Ti^{4+} and V^{3+}. Vanadium was known as multivalence ion with different ionic radii as different oxidation state. At six coordination number, V^{3+} (0.78Å) is larger than Ti^{4+} (0.74Å) while V^{4+} (0.72Å) and V^{5+} (0.68Å) is smaller ionic radii than Ti^{4+}[4]. Substitution Ti^{4+} ion with larger V^{3+} ion can result in increasing the unit cell and crystal volume. However, decreasing the unit cell and crystal volume in this research indicating that vanadium might be present at higher valence or mixed valence in TiO_2 lattice. The larger decrease was obtained at 0.7% of Vanadium than another concentration.

Table 1. Unit Cell, Cell Volume and Average Crystallite size of undoped and doped TiO_2.

	Undoped TiO_2	TiO_2-V 0.3%	TiO_2-V 0.5%	TiO_2-V 0.7%	TiO_2-V 0.9%
a,b (Å)	3.803	3.786	3.792	3.784	3.786
c (Å)	9.559	9.515	9.526	9.509	9.514
V (Å^3)	138.217	136.414	136.990	136.195	136.419
Cryst. (nm)	34.28	34.18	26.42	26.43	79.40

UV-Vis DRS data in figure 2 shows the reduced reflectance for all vanadium doped TiO_2 materials compared to undoped TiO_2 and commercial TiO_2 SA at visible wavelength (400-800 nm). Reflectance tend to continuous decrease as increasing V^{3+} concentration and therefore the enhancement of absorption coefficient (F(R)) in inset figure was obtained. The reduced reflectance and enhanced absorption coefficient in visible wavelength indicated that vanadium doped TiO_2 which synthesized by sonochemical method had potential to be applied under visible irradiation.

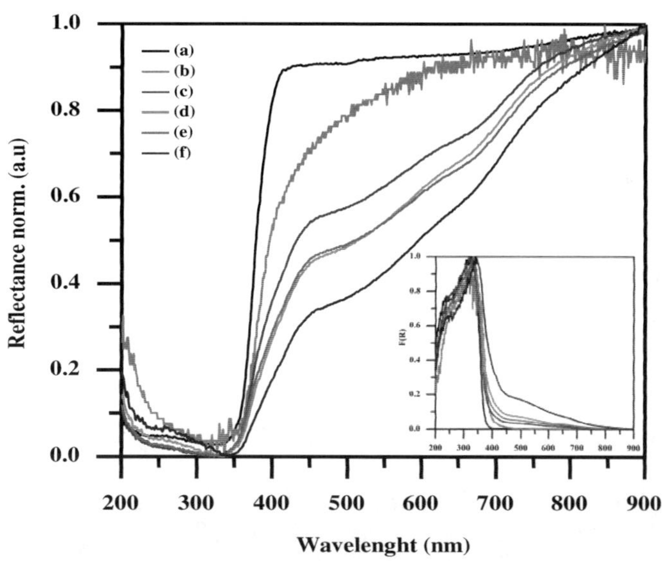

Figure 2. UV-Vis DRS data of (a) TiO_2 SA (b) undoped TiO_2 and V^{3+} doped TiO_2 at (c) 0.3% (d) 0.5% (e) 0.7% and (f) 0.9%.

Determination of bandgap energy by Kubelka-Munk transformation shown in figure 3. The bandgap energy for TiO_2 SA, undoped TiO_2 and V doped TiO_2 0.3%, 0.5%, 0.7% , 0.9% were respectively 3.25, 3.22, 3.05, 2.93, 3.03 and 2.40 eV. The energy tend to decrease as increasing vanadium content with the highest reduction achieved at 0.9% vanadium. 0.7% of V doped TiO_2 shows slightly alteration from the trend, respecting to the refinement result which showed the largest srinkage in unit cell parameter. This evidence may contributed by the nature of vanadium state in TiO_2 lattice at 0.7%. Furthermore, substitution of Ti^{4+} by vanadium at different oxidation states and different concentration may result in various spin states, oxigen vacancy, and reduction level of band gap energy as the effect of compensation method in substitutional [11] or interstitial doping.

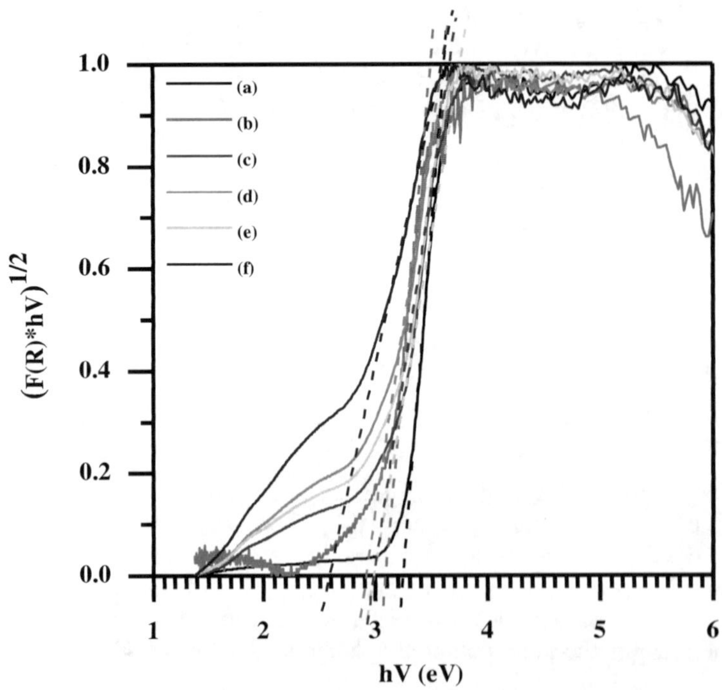

Figure 3. Bandgap energy of (a) TiO_2 SA (b) undoped TiO_2 and V^{3+} doped TiO_2 at (c) 0.3% (d) 0.5% (e) 0.7% and (f) 0.9%.

4. Conclusion

Vanadium doped TiO_2 was successfully synthesized using sonochemical method. Photocatalyst materials crystallized in pure anatase structure with tetragonal lattice and *I*41*amd* space group. Lower shifted of *2θ* and decreased unit cell parameter indicated that anatase structure distorted due to vanadium dopant. Vanadium doped TiO_2 shows decreased bandgap energy, lowered reflectance and enhanced absorption in visible light. As a result, it had potential photocatalytic activity under visible irradiation (400-800 nm)

References

[1] Shao G N, Jeon S J, Haider M S, Abbas N and Kim H T 2016 *J. Colloid Interface Sci* **474** 179-189
[2] Padmini E and Miranda L R 2013 *Chem. Eng. J* **232** 249-258
[3] Choi J, Park H, and Hoffman M R 2009 *J. Mater. Res* **25** 149-158
[4] Khatun N, Rini E G, Shirage P, Rajput P, Jha S N and Sen S 2016 *Mater. Sci. Semicond. Process* **50** 7-13

[5] Kubacka A, Garcia M F, and Colon G 2008 *J. Catal.* **254** 272–284
[6] Aini N and Ningsih R 2016 *ALCHEMY: Journal of Chemistry* **5(1)** 24-30
[7] Bang J H and Suslick K S 2010 *Adv.Mater* **22** 1039-1059
[8] Arami H, Mazloumi M, Khalifehzadeh R and Sadrnezhaad 2007 *Mater. Lett.* **61** 4559-4561
[9] Singh S A and Madras G 2016 *Appl. Catal., A* 518 102-114
[10] Kahatta C, Wongpisutpaisan N, Vittayakorn N, and Pecharapa W 2013 *Ceram. Int.* **39** S389-S393
[11] Islam M M, Bredow T, and Gerson A 2011 *Chem. Phys. Chem.* **12** 3467 – 3473

IOP Conf. Series: Materials Science and Engineering **333** (2018) 012021 doi:10.1088/1757-899X/333/1/012021

Synthesis of Optode Thin Layer using Sol Gel Hybrid of Trietoxysiloxane monomer and 3-(Trimethoxysilyl) Propilamine with Ionophore 4-(2-Pyridilazo)-1,3-Benzenadiol (PAR)

S Wahyuningsih[1*], F Rahmawati, S Kamal, S Slamet, M Yunianto, P Rahmawati, F N Aini

[1]Inorganic Material Research Group, Faculty of Mathematics and Natural Sciences, Sebelas Maret University, Ir. Sutami Street No. 36 A, Surakarta, Central Java, 57125, Indonesia.

*E-mail: sayekti@mipa.uns.ac.id

Abstract. Optode (*Optical sensors*) is one of the modern chemical sensors in the field of analytical chemistry that has utilized of inorganic polymers. The optode based on MLCT (*Metal to Ligand Charge Transfer*) (or MMLL'CT, *Mixing Metal-Ligand to Ligand Charge Transfer*) or LMCT (*Ligand to Metal Charge Transfer*) phenomenons have beed generated from oktyltrietxysilane, aminopropyltrimethoxysilane and 4-(2-pyrydilazo) resorcinol (abbreviated as OTES-APTS-PAR) for Cu(II), Cr(III), Ni(II), Fe(III), Cd(II), and Zn(II) ions target. The syntheses of thin layer optode were performed by sol gel method followed by evaporation in glass substrat. The formation of 4-(2-pyrydilazo) resorcinol complexes with ions target have gained strong absorption spectras in visible region because of charge transfer phenomenons. The optical sensor of OTES-APTS-PAR was analysed thermal properties using Differential Thermal Analysis (DTA). DTA thermogram showed a glass transition peaks at a temperature of 315.5 °C. Fourier transform Infrared (FTIR) spectras have showed that the optode materials consisted NH aryl groups indicated IR absorption at 1577.7 cm^{-1} and also –CH aromatic at 1469.0 cm^{-1}. Synthesized optode materials have strong broad visible absorption with the maximum wavelengths (λ_{max}) = 405 nm and 508.5 nm, respectively. This material have excellent optical responds to several metal ions such as Cu(II), Cr(III), Ni(II), Fe(III), Cd(II), and Zn(II) that was showed from huge $\Delta\lambda_{max}$ and the increase of K_{total}

1. Introduction

Optode (optical sensor) is one of chemical sensor that brings unique properties. This optical sensor doesn't need a comparative material, so that it becomes easier maintenance, reproducible and portable. Optode can be used without magnetic field's interference thereby it will decrease mistaken measurement and increase its accuracy [1]. Optode technology is based on MLCT (*Metal to Ligand Charge Transfer*) (or MMLL'CT, *Mixing Metal-Ligand to Ligand Charge Transfer*) or LMCT (*Ligand to Metal Charge Transfer*) phenomenons. Those phenomenons have huge benefit as a recorder of gas optical sensor luminescent, PH sensor, standpoint technology, sensitizer on dye-sensitized photovoltaic cells and photocatalytic material [2].

IOP Conf. Series: Materials Science and Engineering **333** (2018) 012021 doi:10.1088/1757-899X/333/1/012021

pH optical sensor has been used based photoluminescent on europium and terbium complex [3]. Other researches have been introduced a ruthenium complex which has long excitation MLCT period and strong absorption for O_2 sensor [4] and the use of 4-desiloxy-2-(2-pyridilazo-1-naphtol) (PAN) ligand mobilized on polyvinylchloride (PVC) matrix as optode material for determining Cu(II) ion [5]. PAN is first line transition metal strong bonded ligand because charge transfer that is occuring along with building of bond produces sharp complex's color intensity. Color changing concept resulted in maximum absorption alteration of visible light absorption is the basic of choosing optical sensor.

Some of the previous researchers argue that the use of plasticizer in optical sensor material is always necessary to produce lipophilization and prevent exfoliation and rigidity. Plasticizer can be obtained from organic compounds with long hydrocarbon chain. PVC membrane plasticizer based optical sensor has several problems such as bad adhesive property and active surface which easy to collapse. But, the problem of plasticizer can be handled with the use of long hydrocarbon chain organic compound ligand or polymer matrix which has hydrocarbon group. Bond strength between organic compound and supported material, an inorganic material can be engineered through chemical merging using '*silylating agent*' compound.

Organic compound can be used as ionophore dyes, plasticizer or additive material (selectophore). Ionophore dyes can be used as plasticizer because of hydrocarbon chain existence. PAN is one of ionophore dye. Ionophore which can give a color when attached to ion target is chromo ionophore. Chromoionophore has big molar absorptivity ($\varepsilon > 10^5$ L mol^{-1} cm^{-1}), maximum wavelength and stolen color change. In addition, inorganic material as supported material which is usually used is silica. Porous silica has high surface area and volume ratio so that it becomes very sensitive to chemical species. tetramethoxysilane and tetraethoxysilane with their derivatives are usually used.

In this paper, thin layer material is synthesized from triethoxyoxylsilane (n-octyltriethoxysilane, OTES), 3-(trimethylsilyl)propylamine (aminopropyltrimethoxysilane, APTS) and 4-(-pyridylazo)-recorcinol (PAR) using sol-gel and chemical deposition methods. Material performances for optical sensor and metal ion target-ionophore study are also being analyzed.

2. Experimental

2.1 Materials

The materials were used for synthesis are octyltrimethoxysilane, aminopropyltrimethoxysilane, 4-(-2 pyridilazo) resorcinol (PAR), methanol, and NH_4OH. The instruments were used for analysis are UV-Vis spectrophotometer Lambda 25 (Pelkin Elmer), Fourier Transform Infra Red (FTIR) Prestige 21 (Shimadzu), and *Differential Thermal Analysis* (DTA).

2.2 Preparation

Optode thin layer material was synthesized by sol gel method followed by solvent evaporation using borosilicate glass as substrat. Synthesis of another optode material with the same method also was conducted without glass substrat. The synthesis formula as shown in Table 1.

Table 1. Thin-layer material synthesis conditions of OTES-APTS-PAR optode

Optode preparation formula				
OTMS	APTS	Methanol	NH_4OH 0.05M	PAR 10^{-4} M
8.1 mL	3 mL	20 mL	6 mL	100 mL

IOP Conf. Series: Materials Science and Engineering **333** (2018) 012021 doi:10.1088/1757-899X/333/1/012021

3. Results and Discussion

3.1 Thermal Analysis of the Optode Materials

Synthesis of OTES-APTS material was done through sol-gel methods using NH_4OH as a catalyst. Polymerization process of octyltriethoxysilane and monomer of aminopropyltrimethoxysilane was built and through their branch chain results three dimension structures. DTA thermogram (Figure 1) shown excellent thermal stability of OTES-APTS polymer where there was not structural transformation until temperature of 206.32 ^{0}C. Exothermic peak at relatively huge heat released (2472.24 J) at 315.53 ^{0}C was estimated as glass transition (Tg) of the polymer. Over this temperature, polymer structure was flexible.

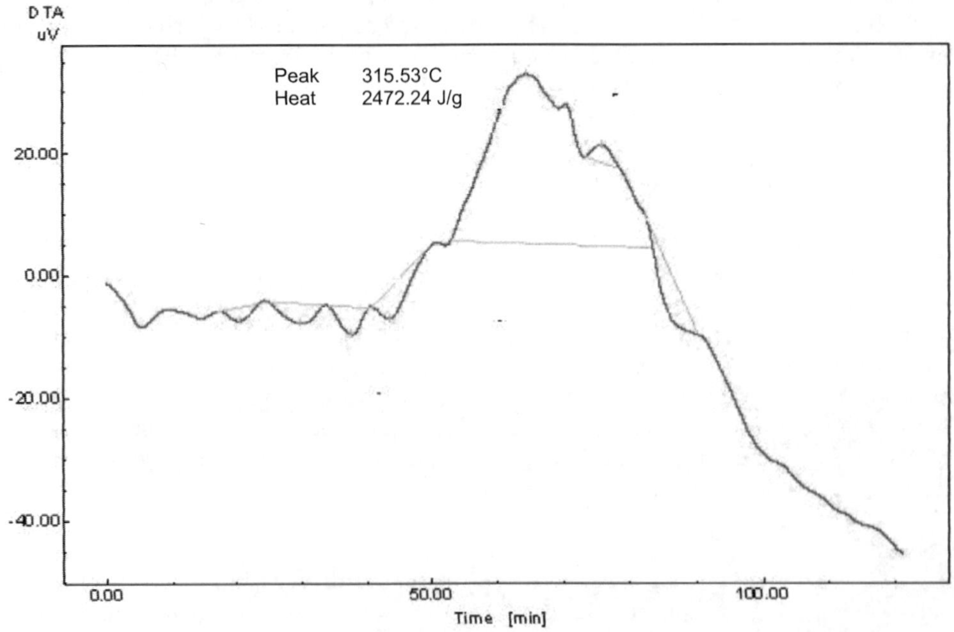

Figure 1. DTA Thermogram of OTES-APTS Polymer

3.2 Identification of Functional Group using FTIR spectroscopy

OTES shows Si-O-C group represented at 1188.1 cm^{-1}, while APTS shows two adjacent absorptions of Si-O-C at 1103.2 cm^{-1} and 1088.1 cm^{-1}. Weak absorption of –NH aryl appears at 1627 cm^{-1} and 1593 cm^{-1}. OTES-APTS polymer shows sharp absorption at 1134.1 and 1033.8 cm^{-1} as an asymmetric stretching of Si-O-Si and Si-O-C respectively. Other functioal groups such as –CH aliphatic show absorption region at 2927.7 cm^{-1} and 2854.5 cm^{-1}. The presence of PAR in the optical sensor materials is proven by addition absorptions at 1577.7 cm^{-1} and 1469 cm^{-1} as –NH aryl and –CH aromatic from C=C bond respectively.

3.3 Study respon optical material

The presence of PAR on OTES-APTS polymer improved λ_{max} respond to materials and probably absorb visible light as showed at Figure 2.

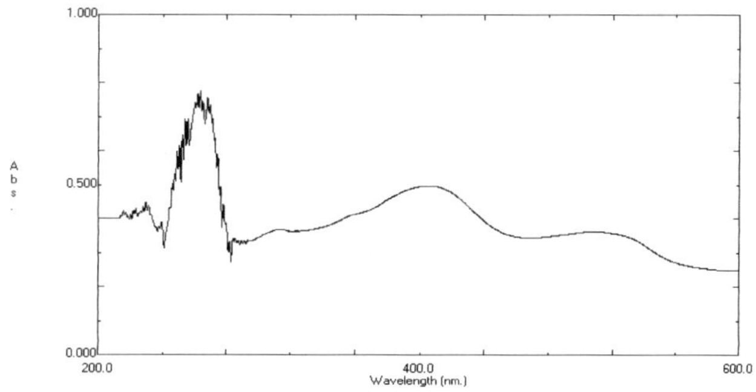

Figure 2. Electronic spectrum of OTES-APTS-PAR

The formation of complexes compound between ions of Cu(II), Cr(III), Zn(II), Cd(II), Ni(II) and Fe(III) with PAR created optical shift respond of electronic spectrums (Figure 3). PAR and metal complexes have high molar existing coefficient (ε) about 10^4 L mol^{-1} cm^{-1}. Maximum wavelength shifts occurred was based on optical responding. PAR spectrum has two peaks as a deprotonated and protonated stage. Interaction between metal ions as ions target and PAR throught complexes formation mechanism. It shows that there was a maximum peak at protonated state from several complexes spectrums.

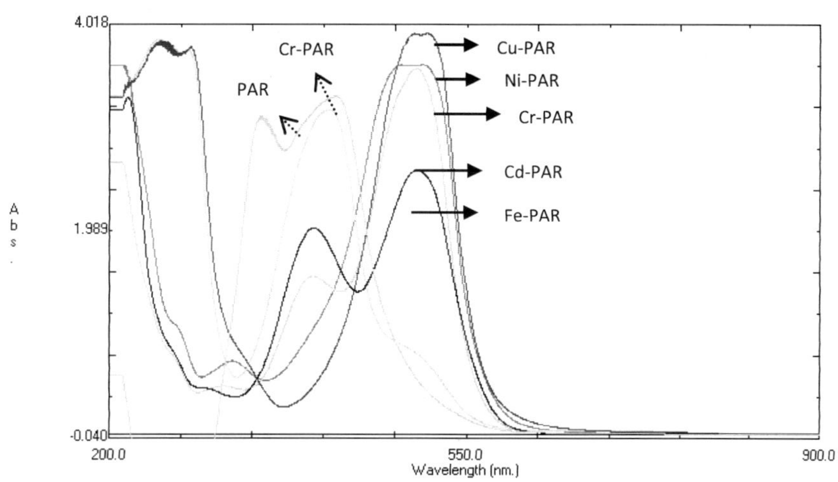

Figure 3. Electronic spectrum of PAR after being complexes with Cd(II), Cu(II), Zn(II), Cr(III), Ni(II) and Fe(II)

3.4 Calculation of optical respond of maximum wavelength shift ($\Delta\lambda_{max}$)

The interaction of optode materials with metal ions of Cd(II), Zn(II), Cr(III), Ni(II), Fe(III) and Cu(II) produced $\Delta\lambda_{max}$ as summarized in Table 2. The result obtained show that $\Delta\lambda_{max}$ which was determined from the difference of $\Delta\lambda_{max}$. $\Delta\lambda_{max}$ of PAR was increased in order : Cr(III), Ni(II), Zn(II), Fe(III), Cd(II) and Cu(II).

Table 2. Determination of maximum absorbance values and $\Delta\lambda_{max}$ after the interaction between target metal ions and optode materials at a contact time10 seconds (λ_{max} of optode materials = 405 nm)

Target metal ion	$\lambda\lambda_{max}$ (nm)	$\Delta\lambda_{max}$ (nm)
Cd(II)	513	108
Zn(II)	488	83
Cr(III)	298	7.0
Ni(II)	480	75
Fe(III)	508	103
Cu(II)	520.5	114.5

3.5 Relatively respons of optodes with metal ions target to within Cu^{2+} ion

Relativelly respon of optodes with metal ions compared within Cu^{2+} were valued as $A/A_{Cu(II)}$ at their λ_{max} (Table 3).

Table 3. Relative respon value of absorbance ($A/A_{Cu(II)}$)

Metal ions target (10^{-4} M)	$\lambda\lambda_{max}$	Abs	$A/A_{Cu(II)}$
Cu(II)	520.5	0.0169	1.00
Cr(III)	398.0	0.0427	2.52
Ni(II)	480.0	0.0101	0.60
Fe(III)	508.5	0.0126	0.75
Cd(II)	513.5	0.0236	1.40
Zn(II)	488.0	0.0210	1.24

Relativelly respon of optodes with metal ions compared within Cu^{2+} gained various values. Cr(III) has the highest $A/A_{Cu(II)}$ value but only has small $\Delta\lambda_{max}$. Nevertherless, this condition was unfavorable for optode based measurement.

3.6 Separation facor (1-α)

Based on the λ_{maks} of M-Optode complex can be calculated of α:

$$\alpha = \lambda_{maksM1} / \lambda_{maksM2} \quad (1)$$

If $\lambda_{maxM1} = \lambda_{maxM2}$, α value = 1, but (1-α) value = 0, in this situation the separation factor does not exist, it can't be distinguished M1 and M2 analytic response.

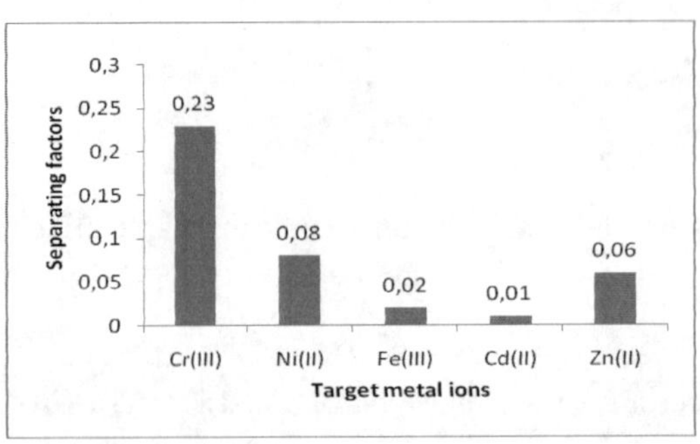

Figure 4. The separation factor (1-α) of Cr(III), Ni(II), Fe(III), Cd(II) and Zn(II) metal ions to Cu(II) metal ions (contact time 1 seconds, $\alpha = \lambda_M/\lambda_{Cu(II)}$)

The separation factors (1-α) of Cr (III), Ni (II), Fe (III), Cd (II) and Zn (II) metal ions are calculated on the Cu (II) metal ion increased in the order of Cd (II), Fe (III), Zn (II), Ni (II), and Cr (III) (separation factor for zero Cu ions) (Figure 4). Qualitative measurements of Cu (II) ions with optode material are interference in magnitude inversely proportional to the separation factor values. The presence of metal ions Cd(II) and Fe(III) most interferes with the qualitative measurement of Cu(II) ions by this optode.

3.6 Determination of total displacement constant (K_{total})

K_{total} value is a comparison number between the analyte species present in the sensor material and the analytical species present in the external water phase

$$K_{total} = \frac{[M^{n+}]_o}{[Mn+]_a} \quad (2)$$

$[Mn^+]_s$ is the concentration of target metal ions in the sensor material whereas $[Mn^+]_a$ is the concentration of target metal ions remained in the water phase. Measurment of each the metal ions is carried out with an atomic absorption spectrophotometer (AAS).
Total displacement constant value (K_{total}) was determined by comparing target metal ion concentration on optode with target metal ion as residual in water phase that is showed at Figure 5. Measurement of Cu (II) ion has great interference with Fe(III) and Cd(II) ions.

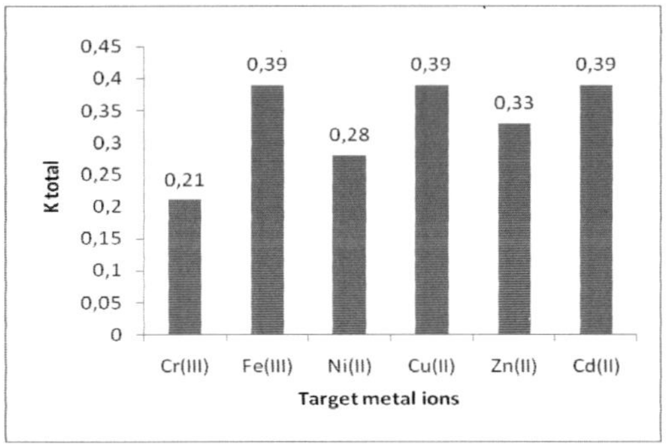

Figure 5. Total transfer constant (K_{total}) of metal ions target to the optode, 10 ppm concentration of metal ions target along 10 seconds contact time

4. Conclusion
Thin layer optode built from OTES-APTS-PAR was succesfully prepared by sol gel method and evaporation. Bond formation of the polymer matrix and PAR on optode materials was identified by the appearance of –NH aryl at 1577.7 cm^{-1} and –CH aromatic in 1469.0 cm^{-1} , While the electronic transitions was absorb visible ligh at 404 nm and 508.5 nm. The glass transition (Tg) optode was present at a temperature of 315.5 ^oC. The use of Cr(III), Ni(II), Zn(II), Fe(III), Cd(II) and Cu(II) by arrange as metal ion target decreases $\Delta\lambda_{max}$, nevertheless K_{total} increase by the following arrangement Cr(III), Ni(II), Zn(II), Fe(III) Cu(II) and Cd(II). Measurement of Cu (II) ions with this optode has great interference with Fe(III) and Cd(II) ions, respectivelly.

References

[1] Dybko A 2001 *Sensors* **1** 29-37
[2] Whittle C Ed, Weinstein J A, George M W, Schanze K S 2001 *Inorg Chem* **40** 4053-4062
[3] Blair S, Lowe M P, Mathieu C E, Parker D, Senanayake P K, Kataky R 2001 *Inorg Chem* **40** 5860-5867
[4] Arvidsson J, Stean Hulth 2000 Götheborg University
[5] Amiet 2001 *J. Chem* **54** 27-30

IOP Conf. Series: Materials Science and Engineering **333** (2018) 012022 doi:10.1088/1757-899X/333/1/012022

Variation of MEH-PPV layers with/without PEDOT:PSS layer in organic solar cells

R Suryana[1,2], L K Tyas[1] and F Nurosyid[1,2]

[1]Physics Department of Graduate Program, Universitas Sebelas Maret, Indonesia
[2]Department of Physics, Faculty of Mathematics and Natural Sciences, Universitas Sebelas Maret, Indonesia

Email: rsuryana@staff.uns.ac.id

Abstract. Organic solar cell in structures of Glass/FTO/TiO_2/MEH-PPV and Glass/FTO/TiO_2/PEDOT:PSS/MEH-PPV were fabricated using spin coating method. The MEH-PPV layer was variated at 1 layer, 2 layer, 3 layer and 4 layer. All samples without/with PEDOT:PSS for each MEH-PPV layer were characterized using the UV-Vis spectroscopy to evaluate the absorbance, the cross-section SEM to determine the thickness of MEH-PPV layer and the I-V meter to calculate the performance of organic solar cell. The thickness for each MEH-PPV layer was 0.88 μm (1 layer), 1.14 μm (2 layer), 2.35 μm (3 layer) and 5.81 μm (4 layer). The absorbance of samples without/with PEDOT:PSS increase with increasing the thickness of MEH-PPV. Adding PEDOT:PSS can improve the performance of samples. It is considered that PEDOT:PSS can enhance hole collection, exciton diffusion and photocarrier transport.

1. Introduction

Organic solar cells have great potential to be developed as a renewable energy source and environmentally friendly. However, the power efficiency is still low and cannot be accepted commercially. Therefore, it is needed a comprehensive study to improve the efficiency of organic solar cells [1]. Poly[2-methoxy-5-(2-ethylhexyloxy)-1.4-phenylenevinylene] (MEH-PPV) is still attractive polymers for organic solar cells which has the absorption of light at a wavelength of 450-550 nm [2]. In addition, MEH-PPV has properties which compatible with organic solar cell applications such as absorption of the highest energy of sunlight at a wavelength of 500 nm [3], as electron donor (p-type semiconductor) material [4], has orbital energy at 5.3 eV for highest occupied molecular orbital (HOMO) and lower unoccupied molecular orbital (LUMO) at 3.2 eV [5].

Due to its stability and easily controlled size and shape, the TiO_2 nanoparticle semiconductor material is added as an acceptor in a MEH-PPV-based hybrid solar cell [6]. TiO_2 as an acceptor (n-type semiconductor) has a valence band 7.6 eV and conduction band 3.7 eV [7].

One of important materials in an organic solar cell is poly (3,4-ethylenedioxythiophene) polystyrene sulfonate (PEDOT:PSS) which can improve the roughness of active layer [8]. In addition, PEDOT:PSS can improve the conversion performance by enhancement of hole collection, exciton diffusion and photocarrier transport [9]. In this work, we present the influence of PEDOT:PSS on the performance of hybrid solar cell based on MEH-PPV/TiO_2 nanoparticle with variation of MEH-PPV layer.

2. Experimental

TiO_2 nanoparticle (21 nm, anatase phase) Sigma Aldrich, poly[2-methoxy-5-(2-ethylhexyloxy)-1.4-phenylenevinylene] (MEH-PPV) Sigma Aldrich, and poly[3,4-ethylenedioxythiophenepolystyrene sulfonate] (PEDOT:PSS) Sigma Aldrich was used as main material in hybrid solar cell while ethanol and chlorobenzene was used as solvent. Aluminum layer was coated on MEH-PPV layer using evaporator as metal contact.

The FTO was cut in size of 2 cm × 1.5 cm. The FTO surface was patterned using zinc powder and HCl. The width of conductive part was 2 mm while the width of non-conductive (glass) part was 4 mm, 3 mm and 4 mm as shown in Fig. 1. Conductive part was covered by the scotch tape that is resistant to acid. Zinc powder was coated onto entire the FTO surface then HCl solution was dropped onto non-covered part and waiting for 3 min. The remaining of zinc powder and HCl solution was cleaned using the cotton bud and was rinsed into aquades with ultra-sonic cleaner for 3 min. The non-conductive (glass) was checked using a multi-meter to confirm that the FTO surface has been removed.

0.5 g TiO_2 was dissolved in 2 ml ethanol then stirred for 30 min at 300 rpm using a vortex stirrer. 0.05 g MEH-PPV was dissolved in 5 ml chlorobenzene then distilled at 300 rpm for 18-24 h until seen homogeneous solution.

The hybrid solar cell structure was conducted in two types i.e. FTO/TiO_2/MEH-PPV/Al and FTO/TiO_2/PEDOT:PSS/MEH-PPV/Al. Later, it was named type I and type II, respectively. The FTO surface was coated by TiO_2, PEDOT:PSS, and MEH-PPV using the spin coating method. For type I, the TiO_2 solution was dropped on center of the FTO conductive glass and was spun by spin coater at 1000 rpm for 30 s. Then it was annealed at 450°C for 30 min. While MEH-PPV solution was coated on the TiO_2 layer at 1000 rpm for 60 s then annealed at 80°C until dried. MEH-PPV layer was varied i.e. 1 layer, 2 layer, 3 layer, and 4 layer. For type II, PEDOT:PSS was coated on the TiO_2 layer at 800 rpm for 30 s before coating the MEH-PPV. Then, it was annealed at 110°C for 15 min.

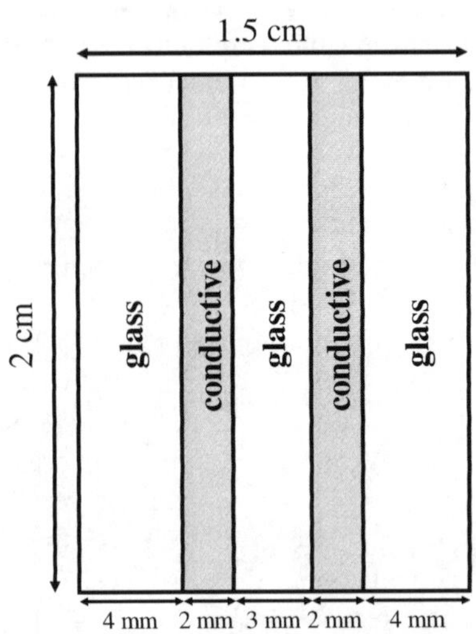

Figure 1. The pattern of the FTO surface after the dry etching

Finally, aluminum was coated on the top of MEH-PPV layer using evaporator. A shadow mask in column size of 10 cm × 3 mm was put onto the sample at perpendicular to Fig. 1 pattern. Thus, an active area was obtained in size of 4 mm^2. When the pressure was reached at about 7×10^{-6} Torr, the DC current was increased slowly until aluminum evaporate at range of 16–20 A. The sample was placed on the PCB holder, then the active area was contacted using the silver paste to measure I-V characteristic.

Absorbance of the sample at wavelength of 200 – 800 nm with/without PEDOT:PSS was measured by ultraviolet visible (UV-Vis) spectroscopy. The thickness of each MEH-PPV layer was determined by scanning electronic microscopy (SEM). The performance of the sample was analyzed by I-V measurement using the 2062A system source Keithley with irradiation using Xenon lamp at 1000 W/m^2.

3. Results and Discussion

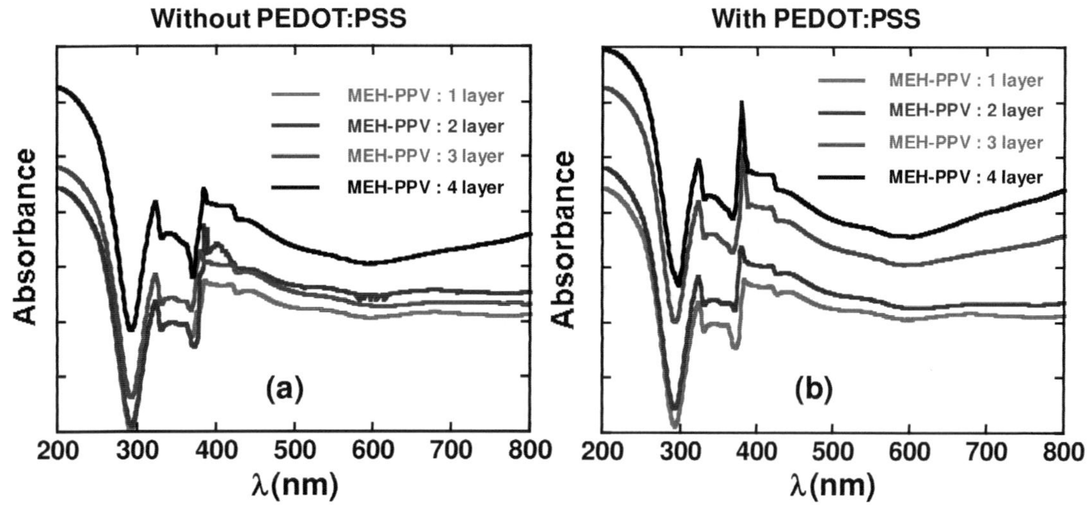

Figure 2. Absorbance of FTO/TiO_2/MEH-PPV (a) and FTO/TiO_2/PEDOT:PSS/MEH-PPV (b) samples with variation of MEH-PPV at 1 layer, 2 layer, 3 layer and 4 layer

Absorbance of FTO/TiO_2/MEH-PPV (type I) and FTO/TiO_2/PEDOT:PSS/MEH-PPV (type II) samples are shown in Fig. 2 (a) and (b), respectively. Absorbance patterns are almost similar between type I (without PEDOT:PSS) and type II (with PEDOT:PSS). Absorbance increases with increasing the MEH-PPV layer. There are three peaks which can be observed by UV-Vis spectroscopy i.e. about 200 nm, 325 nm and 400 nm. These peaks correlate to TiO_2 at 200 nm [10] and MEH-PPV+TiO_2 at 325 and 400 nm [11] while PEDOT:PSS curve increase from 400 nm to 1200 nm [12]. We cannot observe PEDOT:PSS absorption especially in peak of about 400 nm in Fig. 2 (b). However, the adding PEDOT:PSS can increase absorption in type II.

Fig. 3 shows the cross-section SEM images of MEH-PPV layer on FTO substrate. The thickness of MEH-PPV increase with increasing the amount of layer i.e. 0.88 μm (1 layer), 1.14 μm (2 layer), 2.35 μm (3 layer) and 5.81 μm (4 layer). Based on Fig. 2, type I and type II without Al layer absorb ultraviolet (UV) or visible (Vis) light, the values of the absorbance bands depend on the MEH-PPV thickness according to Beer's law [13]. The smaller thickness for 1 layer which also has the lowest absorbance.

Effect of PEDOT:PSS layer on the performance of organic solar cell with variation of MEH-PPV layer were determined through I-V characterization. I-V curves for each MEH-PPV layer under irradiation of Xenon light are shown in Fig. 4(a) without PEDOT:PSS and 4(b) with PEDOT:PSS. The procedure to obtain the efficiency of organic solar cell is similar to dye-sensitized solar cell (DSSC) case which has been demonstrated in the previous work [14].

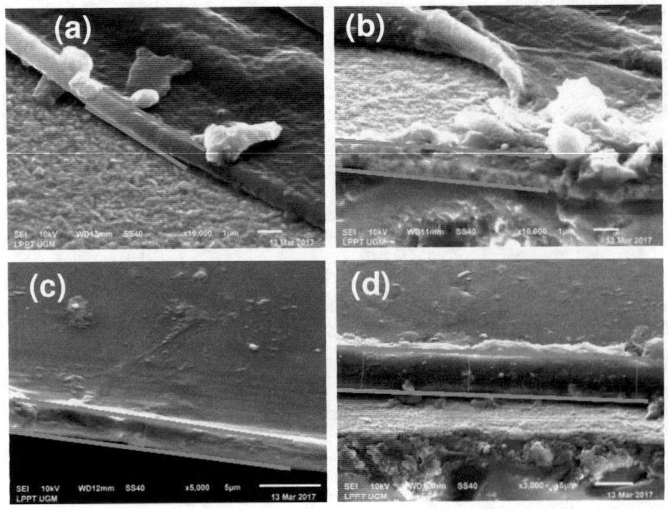

Figure 3. Cross-section SEM images of MEH-PPV layer on FTO with two orange lines indicate thickness for each layer on (a) 1 layer, (b) 2 layer, (c) 3 layer and (d) 4 layers

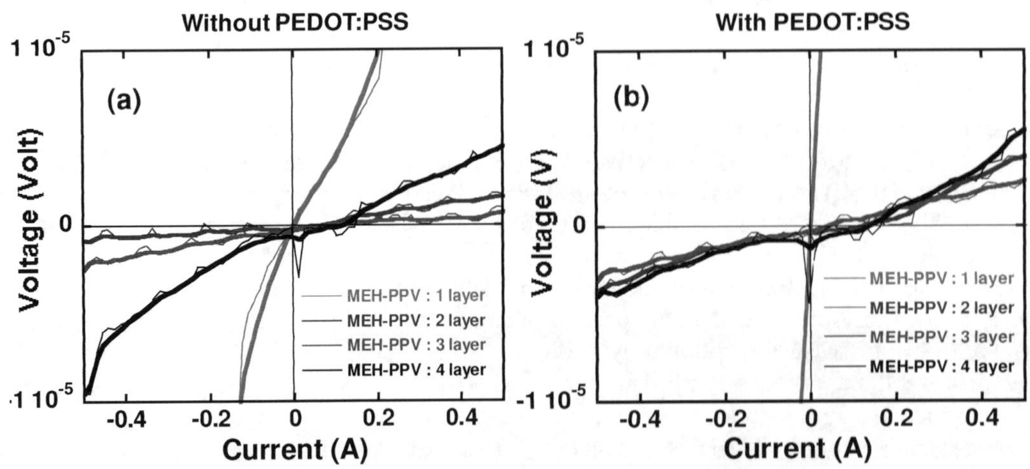

Figure 4. I-V characteristics for each MEH-PPV layer on the organic solar cell without PEDOT:PSS (a) and with PEDOT:PSS (b) under Xenon irradiation

Table 1 summarize the I-V measurement consisting the open circuit voltage (*Voc*), the current short circuit (*Isc*), the voltage maximum (*Vm*), the current maximum (*Im*) to calculate the fill factor (*FF*) and the efficiency (η). In the case both without and with PEDOT:PSS, the performance on samples for MEH-PPV 1 layer and 2 layer are almost similar. For MEH-PPV 3 layer, the performance decrease while for MEH-PPV 4 layer, the performance increase. Main processes involved in the organic solar cell consists of photon absorption by the MEH-PPV to generate the exciton, exciton diffusion toward the donor/acceptor interface, exciton dissociation at the donor/acceptor through an ultrafast electron transfer [15]. Then electron/hole will be transported toward correct electrodes. Therefore, many excitons are generated when MEH-PPV layer increase and diffused toward the donor/acceptor interface to produce the electron and hole. It is considered that the performance of sample increase with increasing the MEH-PPV layer. However, until this work is reported, we have not understood why the performance on the organic solar cell with MEH-PPV at 3 layer decrease.

The performance of organic solar cell with adding the PEDOT:PSS are higher than the samples without adding the PEDOT:PSS. This is an evidence that PEDOT:PSS can improve the conversion performance by enhancement of hole collection, exciton diffusion and photocarrier transport [9].

Table 1. Summary of the efficiency of organic solar cell without/with PEDOT:PSS for each MEH-PPV layer

MEH-PPV	*Voc* (Volt)	*Isc* (Ampere)	*Pm* (Watt)	*Vm* (Volt)	*Im* (Ampere)	*FF*	η ($\times 10^{-2}$ %)
GLASS/ FTO/ TiO_2/ MEH-PPV							
1 layer	0.009914	8.64×10^{-8}	8.57×10^{-9}	0.009914	8.64×10^{-7}	0.234	0.0214
2 layer	0.009872	2.9×10^{-8}	9.50×10^{-9}	0.054932	1.73×10^{-7}	0.322	0.0238
3 layer	0.009898	6.89×10^{-9}	2.44×10^{-9}	0.039805	6.12×10^{-8}	0.265	0.0061
4 layer	0.009872	1.38×10^{-7}	6.52×10^{-8}	0.114997	5.67×10^{-7}	0.138	0.1630
GLASS/ FTO/ TiO_2/ PEDOT:PSS/ MEH-PPV							
1 layer	0.009933	3.67×10^{-7}	3.65×10^{-8}	0.009933	3.67×10^{-7}	0.221	0.0912
2 layer	0.018084	8.57×10^{-8}	2.58×10^{-8}	0.055066	4.68×10^{-7}	0.162	0.0644
3 layer	0.009919	7.26×10^{-9}	3.64×10^{-9}	0.009919	3.67×10^{-8}	0. 219	0.0091
4 layer	0.009991	5.23×10^{-7}	9.50×10^{-8}	0.114994	8.26×10^{-7}	0.393	0.2375

4. Conclusion

The highest efficiency of FTO/TiO_2/PEDOT:PSS/MEH-PPV/Al with MEH-PPV 4 layer is 0.2375×10^{-2}%. The performance of organic solar cell increase with adding the PEDOT:PSS layer. It is considered that PEDOT:PSS can improve the conversion performance by enhancement of hole collection, exciton diffusion and photocarrier transport.

Acknowledgments

Authors wish to thank PNBP Universitas Sebelas Maret 2015 for financial support through Hibah Mandatory

References

[1] Supriyanto A, Kusminarto, Triyana K, Roto, Salleh M M and Umar A A 2010 *J. Mater. Sci. Eng.* **04** 40
[2] Omer B M 2012 *J. Nano Electron. Phys.* **4** 04006
[3] Kim M S 2009 *Understanding Organic Photovoltaic Cells: Electrode, Nanostructure, Reliability, and Performance* (Dissertation: The University of Michigan, US)
[4] Aernouts T 2006 *Organic Bulk Heterojunction Solar Cells From Single Cell Towards Fully Flexible Photovoltaic Module* (Dissertation: Khatolike University Leuven Belgia)
[5] Oku T, Nagaoka S, Suzuki A, Kikuchi K, Hayashi Y, Sakuragi H and Soga T 2008 *J. Ceram. Process. Res.* **9(6)** 549
[6] Kim S S, Jo J, Chun C, Hong J C and Kim D Y 2007 *J. Photochem. Photobiol. A Chem.* **188(2-3)** 364
[7] Tiwana P, Docampo P, Johnston M, Snaith H and Herz L 2011 *ACS Nano* **5** 5158
[8] Iftimie S, Radu A, Radu M, Besleaga C, Pana I, Craciun S, Girtan M, Ion L and Antohe S 2011 *Dig. J. Nanomater. Biostruct.* **6(4)** 1631
[9] Brincoveanu O A, Ioanid A, Iftimie S and Antohe S 2016 *Romanian Reports in Phys.* **68(3)** 1097
[10] Zhou K, Zhu Y, Yang X, Jiang X and Li C 2011 *New J. Chem.* **35** 353
[11] Habelhames F, Lamiri L, Zerguine W and Nessark B 2013 *Mater. Sci. Semicond. Process.* **16(3)** 727
[12] Luo J, Billep D, Blaudeck T, Sheremet E, Rodriguez R D, Zahn D R T, Toader M, Hietschold M, Otto T and Gessner T 2014 *J. Appl. Phys.* **115** 054908
[13] Seoudi R, El-Bahy G S, El Sayed Z A 2006 *Opt. Mater.* **29** 304

[14] Suryana R, Khoiruddin, Supriyanto A 2013 *Material Science Forum* **737** 15
[15] Chen C, Wu F, Geng H, Shen W and Wang M 2011 *Nanoscale Res. Lett.* **6** 350

IOP Conf. Series: Materials Science and Engineering **333** (2018) 012023 doi:10.1088/1757-899X/333/1/012023

The Influence of Cr^{3+} on TiO_2 Crystal Growth and Photoactivity Properties

S Wahyuningsih[1*], W N Hidayatika[1], P L Sari[1], P P Sari[1], R Hidayat[1], H Munawaroh[1], A H Ramelan[1]

[1]Inorganic Material Research Group, Faculty of Mathematics and Natural Sciences, Sebelas Maret University, Ir. Sutami Street No. 36 A, Surakarta, Central Java, 57125, Indonesia.

*E-mail: sayekti@mipa.uns.ac.id

Abstract: The photocatalyst technology is an integrated combination of photochemical processes and catalysis in order to carry out a chemical transformation reaction. One of the semiconductor materials that have good photocatalytic activity is TiO_2 anatase. This study aim to determine the effect of the Cr^{3+} addition on the growth of TiO_2 rutile crystal and the increasing of TiO_2 photoactivity. Diffractogram X-Ray of the samples showed that the synthesized TiO_2 at 400 °C has been produced 100% TiO_2 anatase. Synthesis of TiO_2 doped Cr^{3+}composite was using wet impregnation method. The TiO_2 doped Cr^{3+}composites have beed grown by annealed at a temperature of 300, 400, 500, 600 and 700 °C, respectively Annealing process have capabled to gain to the TiO_2 doped Cr^{3+}nanocomposite. The result product annealed at 500 °C only appear anatase phase due to the Cr^{3+} addition influence that was able to suppress the growth of rutile. Identification of TiO_2 doped Cr^{3+}composite using Fourier Transform Infra-Red (FT-IR) showed O-Cr vibration at 2283.72 cm^{-1}. The TiO_2 doped Cr^{3+} photoactivity was studied to degrade Rhodamin B. The best result on photodegradation of Rhodamin B was performed by using TiO_2 doped Cr^{3+}composite which was annealed at 700 °C i.e. 74.71%.

1. Introduction

The photocatalyst technology is a combination of integrated photochemical methods and catalysts in order to carry out a chemical transformation reaction. Titanium dioxide (TiO_2) is widely reported as an active semiconductor material as a photocatalyst. How to improve the photocatalytic activity of TiO_2 in the visible region is the main focus of the recent TiO_2 photocatalysis research. Many efforts have been made to achieve the utilization of visible light for TiO_2 material, such as transitional metal ion doping [1–3], nonmetal element doping [4,5] and dye sensitization [6,7]. Nanomaterial TiO_2 anatase phase has photocatalytic activity higher than rutile phase [2,4,5].

The TiO_2 is one of the widely use photocatalyst because it has high photocatalytic efficiency, low cost, inert biologically and chemically [2,4,5]. Degradation photocatalytic is a technique used to water and air pollutants treatment. The degradation process is known to have two of compounds are added to accelerate the process of organic compounds degradation, namely chemical oxidants and photocatalyst. Photocatalyst can be utilized in the processing of dye waste [8-9]. The color quantity caused by Rhodamin B is very sharp, this is due to the presence of two auxochrome groups. Rhodamin

B is a synthetic dye waste used for textile, paint and paper industries. This dye can cause pollution, irritation, and carcinogen [4 10].

In previous study, TiO_2 doped Cr^{3+}electrode have driven visible light response. and confirmed that the impurity band is formed near the π*conduction band of the TiO2 [10]]. In this study, the addition of Cr^{3+} to TiO_2 semiconductors was conducted to suppress the growth of TiO_2 phase crystal rutile thus yield more anatase phase. This controlled crystal phase has an effect in increasing photocatalytic activity on the TiO_2 degradation of Rhodamine B.

2. Experimental

2.1. The Preparation of TiO_2

10 mL of TTIP (Titanium Tetra Iso Propoxide) was hydrolyzed with 100 mL acetic acid. After that, the temperature of the solution was maintained to 10-15 °C. The mixture was heated to a temperature of 90 °C until a TiO_2 sol gel was formed. Then the TiO_2 gel was heated to 150 °C for 24 h to assemble a white TiO_2 xerogel. Xerogel TiO_2 was calcinated for 4 h at 150, 300, 400, 500, 600, and 700 °C. The resulted TiO_2 was characterized with XRD and FTIR.

2.2. The Preparation of TiO_2 doped Cr^{3+} Composites

0.8 gram of TiO_2 was added with 4 gram of $Cr(NO_3)_3.9H_2O$ and 3 gram of urea dissolved in 100 mL of aquadest. The mixture was then heated at 110 °C for 2 days. After the powders were obtained, it was calcined at 300°C, 400°C, 500°C, 600°C, and 700°C for 4 hours, respectively. Furthermore, The resulted TiO_2 doped Cr^{3+} was characterized using XRD and FTIR.

2.3. Photodegradation of Rhodamine B

0.01 grams of TiO_2 and TiO_2 doped Cr^{3+} samples were added to 25 mL of 5 ppm Rhodamin B then stirred. Samples were irradiated with visible light for 0, 30, 60, 120, and 180 min. Subsequently, the solution containing Rhodamine B were analyzed with UV-Vis spectrophotometer at a wavelength region of 400-600 nm.

3. Result and Discussion

The synthesis of TiO_2 material was carried out by sol-gel process. Figure 1 shows that pure TiO_2 diffractogram from annealing results. TiO_2 400 crystals show as anatase phase, in accordance with JCPDS No. 782-486 ie peak at $2\theta = 25.35^o$ (d_{101} = 3,5091 Å), $2\theta = 37.95^o$ (d_{004} = 2.3679 Å), and $2\theta = 48.15^o$ (d_{200} = 1.8874 Å), $2\theta = 54.86^o$ (d_{211} = 1.6716 Å) and $2\theta = 55.01^o$ (d_{211} = 1.6674). While on TiO_2 500, rutile TiO_2 was found with a peak at $2\theta = 27.50^o$ (d_{110} = 3.2394 Å) in accordance with JCPDS No. 870-710, this rutile peak at TiO_2 600 and on TiO_2 700 found more. At an annealing temperature of 700 °C, the peak intensity of anatase TiO_2 was decreased. Some of the anatase peaks of TiO_2 disappear while the peaks represented as a rutile TiO_2 was increased.

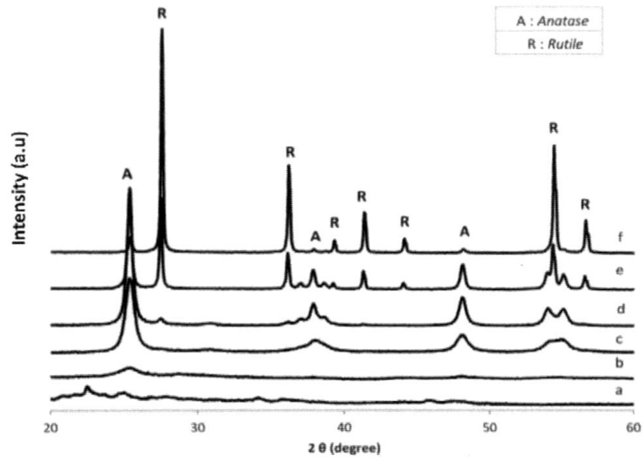

Figure 1. X-Ray Diffractogram of TiO_2 annealed at temperature of (a) 150 °C (b) 300 °C (c) 400 °C (d) 500 °C (e) 600 °C (f) 700 °C

The synthesis of TiO_2 doped Cr^{3+}composite was conducted with wet impregnation method. The growth mechanism can be explained by the reaction as follows:

$$Cr^{3+} + 3H_2O \longrightarrow Cr(OH)_3 + 3H^+$$
$$(NH_2)_2CO + 3H_2O \longrightarrow 2NH_4^+ + 2OH^- + CO_2$$
$$Cr(OH)_3 \longrightarrow CrOOH \longrightarrow Cr_2O_3$$
$$Cr_2O_3 + \text{Xerogel } TiO_2 \longrightarrow \text{Xerogel composite } TiO_2\text{-}Cr_2O_3$$
$$\text{Xerogel composite } TiO_2\text{-}Cr_2O_3 \longrightarrow \text{Powder composite } TiO2\text{-}Cr_2O_3$$

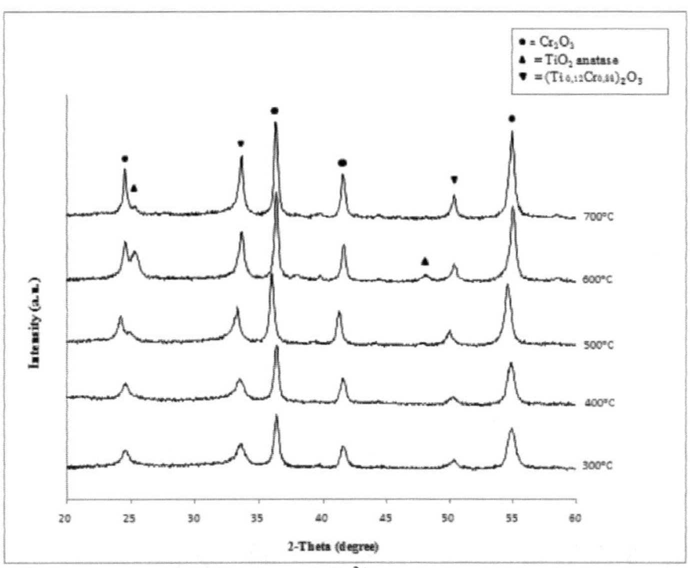

Figure 2. Diffractogram of TiO_2 doped Cr^{3+}composite annealed at temperature of (a) 150 °C (b) 300 °C (c) 400 °C (d) 500 °C (e) 600 °C (f) 700 °C

IOP Conf. Series: Materials Science and Engineering **333** (2018) 012023 doi:10.1088/1757-899X/333/1/012023

The XRD diffractogram of TiO_2 doped Cr^{3+}composite with various annealling can be seen in Figure 2. The presence of new peaks in Figure 2 represented Cr_2O_3 peak at 2θ = 24.60° (d_{102} = 3.6144 Å), 2θ = 36.30° (d_{110} =2.4717 Å), 2θ = 41.55° (d_{113} = 2.1707 Å), 2θ = 54.80° (d_{116} = 1.6730 Å) according to JCPDS No. 381-479. That is also produced $TiCrO_3$ which appears at 2θ = 33.50° (d_{104} = 2.6716 Å), 2θ = 50.15° (d_{024} = 1.8167 Å) in according with JCPDS No. 820 -211. From the diffractogram in Figure 2 (d), a new peak indicates the presence of TiO_2 anatase shown with a peak at 2θ = 25.30° (d_{101} = 3.5159Å) in according with JCPDS No.782-486. The result on diffractogram in Figure 2 (e) was indicated two peaks of TiO_2 anatase at 2θ = 25.35° (d_{101} = 3,5091 Å) dan 2θ = 48.05° (d_{200} = 1,8911Å) according to JCPDS No. 782-486. Annealing at 700 °C the peak of TiO_2 anatase begins to decrease in intensity.

When the difractograms were compared to the difractogram of pure TiO_2, we conclude that TiO_2 doped Cr^{3+} have increasing in crystallinity, because TiO_2 annealed at 300 °C exhibit more pointed peaks whereas pure TiO_2 only shows a broad peak shown in Fig. 1 and Fig. 2. TiO_2 doped Cr^{3+} be able to inhibit the rutile growth at high temperatures, supported by the absence of rutile peaks at 500 °C in XRD data. At temperatures of 600 °C until 700 °C, peaks intensity of TiO_2 rutile was increased and the peaks of TiO_2 anatase was decreased in intensity. Therefore, TiO_2 doped Cr^{3+} can built thermal resistance at high temperatures compared with pure TiO_2 by mean no rutile peaks present in the composite. The result of FT-IR analyzes TiO_2 and TiO_2-Cr_2O_3 composite shown in Fig. 3 and Table 1.

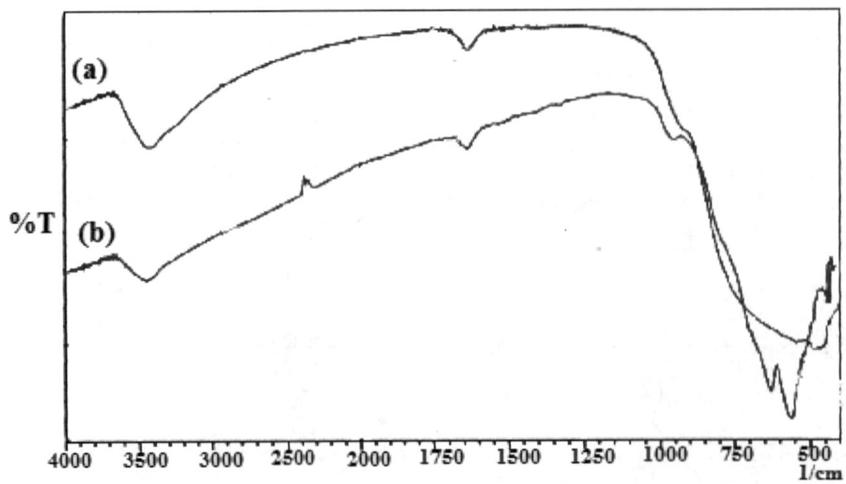

Figure 3. FT-IR spectra at 400 °C (a) pure TiO_2 (b) TiO_2- doped Cr^{3+} composite

Table 1. Result of FTIR absorption in pure TiO_2 and TiO_2 doped Cr^{3+} composite

	Wavelength	Function Group
TiO_2	459.06	Ti-O-Ti
	536.21	Ti-O
	1531.78	-OH
	3406.72	Ti-OH
TiO_2-doped Cr^{3+} composite	549.71	Ti-O
	1625.99	-OH
	3404.36	Ti-OH
	2283.72	O-Cr

The photoactivity test of TiO_2-doped Cr^{3+} composite for the degradation of Rhodamine B was performed in several temperatures (400 °C-700 °C), and was measured at certain times. The result of Rhodamine B degradation is shown in Fig. 4. The longer of irradiation time the greater decreasing of Rhodamine B absorbance. Degradation percentage of Rhodamine B is shown in Table 2.

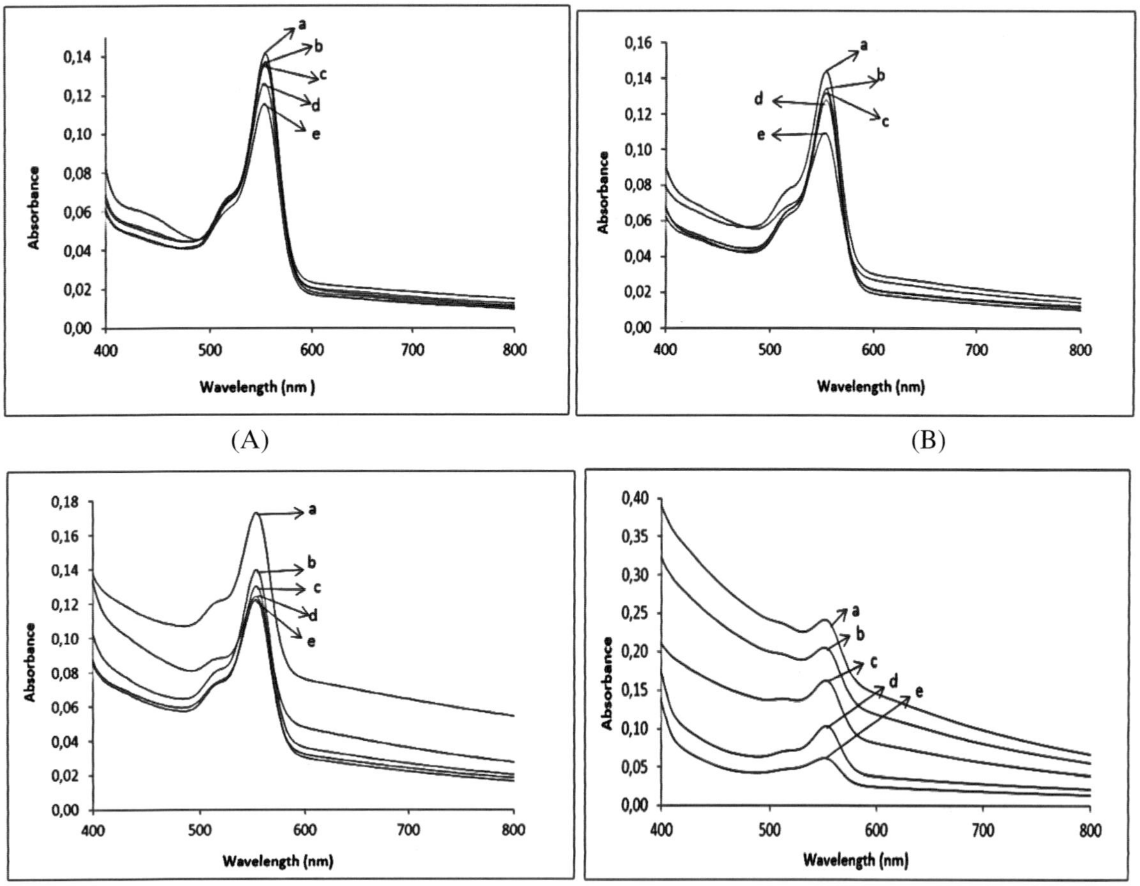

Figure 4. The spectras reprensented of decreasing Rhodamin B concentration by expose the visible light along (a) 0 minute (b) 30 minute, (c) 60 minute (d) 120 minute and (e)180 minute. The TiO_2-doped Cr^{3+} were annealed at (A) 400 ° (B) 500 °C (C) 600 °C (D) 700 °C

Table 2. Degradation result of Rhodamine B using TiO_2 doped Cr^{3+} composite in radiation time various.

Annealing	Percentage of degradation (%)*			
	30 min	60 min	120 min	180 min
400 °C	2.99	3.80	10.59	18.13
500 °C	6.73	8.41	11.16	24.02
600 °C	19.34	25.01	28.31	29.49
700 °C	15.00	32.92	57.34	74.71

* percentage of degradation was calculate by A_0-A_t/A_0 x100%

The Cr^{3+} in TiO2 photocatalys act as photosensitizer so it can be used to overcome the limitations of the spectral sensitivity of the TiO_2 semiconductor with a high energy gap of 3.2 eV. The TiO_2 surface have been modified using color sensitizer molecules which lead to improvement of TiO_2 properties. The exposure of visible light in a certain time will have an effect on the increasing composite photocatalytic activity (Table 5). This because of the longer of visible light irradiation the more

IOP Conf. Series: Materials Science and Engineering **333** (2018) 012023 doi:10.1088/1757-899X/333/1/012023

electrons are continuously excited. Electron generated in semiconductor material is importance step in the photocatalytic degradation process of Rhodamine B. The best result on photodegradation of Rhodamine B was achieved when TiO_2 doped Cr^{3+} composite was annealed at 700 °C amount 74.71%.

4. Conclusion

The TiO_2-Cr_2O_3 composite can be synthesized using wet impregnation method. The addition of Cr^{3+} prevents the growth of the TiO_2 crystal rutile phase. Influence of Cr^{3+} increases the TiO_2 photoactivity on photocatalytic degradation process of Rhodamine B.

References

[1] Choi W, Termin A, Hoffmann M R 1994 *J. Phys. Chem.* **98** 13669
[2] Zhu J, Deng Z, Chen F, Zhang J, Chen H, Anpo H, Huang J, Zhang L 2006 *Appl. Catal. B* **62** 329
[3] Yuan S, Sheng Q, Zhang J, Chen F, Anpo M, Zhang Q 2005 *Micropor. Mesopor. Mater* 79-93
[4] Asahi R, Morikawa T, Ohwaki T, Aoki K, Taga Y 2001 *Science* 293-269
[5] Cong Y, Zhang J, Chen F, Anpo M 2007 *J. Phys. Chem. C* **111** 69-76
[6] Bae E, Choi W, Park J, Shin H S, Kim S B, Lee J S 2004 *J. Phys. Chem. B* **108** 14093
[7] Bae E, Choi W 2003 *Environ. Sci. Technol.* **37** 147
[8] Wahyuningsih S, Purnawan C, Kartikasari P, Praistia N 2014 *Chemical Papers* **68** 1-9
[9] Wahyuningsih S, Ramelan A H, Hidayat R, Fadillah G, Munawaroh H, . Saputri L N M Z, . Hanif Q A 2015 *Proc. Biophotonics Japan 97921E*
[10] Matsumoto Y, Kurimoto J, Shimizu T, Sata E 1981 *J. Electrochem. Soc.* 1040-1044

Plasma Discharge with Different Electrode Diameters for Reducing Methylene Blue Concentration

H Rasyidah[1], **Kusumandari**[1], **T E Saraswati**[2], **M Anwar**[3]

[1] Department of Physics, Faculty of Mathematics and Natural Sciences, Sebelas Maret University, Jl. Ir. Sutami 36A Kentingan Surakarta 57126, Indonesia
[2] Department of Chemistry, Faculty of Mathematics and Natural Sciences, Sebelas Maret University, Jl. Ir. Sutami 36A Kentingan Surakarta 57126, Indonesia
[3] Department of Electrical Engineering, Faculty of Engineering, Sebelas Maret University, Jl. Ir. Sutami 36A Kentingan Surakarta 57126, Indonesia

E-mail: kusumandari@staff.uns.ac.id; teguh@mipa.uns.ac.id

Abstract. Recently, plasma technology has gained attention since it overcomes the shortcomings of water treatment. This research studies the effect of electrode diameter of plasma discharge reactors on the concentration reduction of methylene blue as an organic solution. The plasma discharge reactor was built from a pair of stainless needle electrodes connected with high-AC voltage. The electrodes were placed approximately 2 mm above the solution and stirred at 5.5 rpm. The diameters of the electrodes were 2, 3.2 and 4 mm. The times for plasma treatment were set at 2, 4, 6, 8 and 10 min. Absorbance, temperature and pH of the solution were measured to know the effects of electrode diameter of the plasma reactor. Absorbance and pH significantly decreased after plasma treatment. The best of the absorbance reduction were obtained when the sample was treated under plasma discharge using the smallest diameter electrodes for 8-10 min.

1. Introduction

Water is one of the most important natural resources for living creatures, and is used in hydroelectric power plants to produce electricity and for various other purposes. Increasing the number of living things causes the need for water to increase. However, the increasing number of living creatures, especially humans, increases water pollution, which makes it difficult to obtain clean water. Therefore, a solution is needed to overcome water pollution by water treatment. Current water treatment methods use physics, chemistry and biology. However, such methods, for example adsorption method, are inefficient and may cause secondary pollutants [1].

Plasma technology began to be used as a water treatment process because it does not cause secondary pollutants or require a large area or span of time [2]. Plasma technology oxidizes or decomposes organic pollutants in water using active species, such as negative and positive ions, radical, etc., by decomposing the organic compounds in water [1].

The oxidation process of organic compounds can occur in plasma in water [3]. Electrical discharge is formed on the water (and or in the air) and then diffuses into a thin layer of contaminated water. Ions and electrons generated in the plasma process have very high energy, which causes the water to decompose and generate active species such as OH, oxygen, hydrogen and H_2O_2 [4]. Active species are strong oxidants that can oxidize various organic compounds, such as carbon, hydrogen, nitrogen and sulphur, that are contained in liquid waste. In this study, this is done using a solution of methylene blue ($C_{16}H_{18}ClN_3S$).

The great benefits of plasma technology in water treatment make it an interesting topic of research. This study examines the effects of stainless steel electrode diameter used in plasma discharge generation on the reduction of methylene blue concentration.

2. Experimental

The plasma discharge reactor was built with a pair of stainless steel needle electrodes connected with high-AC voltage at 46.5 kV. The diameters of the electrodes were 2, 3.2 and 4 mm. The distance between the two electrodes mounted parallel to the plasma reactor was set at 2 cm. Materials used in this study were 0.006 grams of methylene blue powder dissolved in 1,000 mL of distilled water.

Initial measurements of absorbance, pH and temperature in methylene blue were performed before plasma treatment generation. A sample without the plasma treatment was also prepared as a reference. A 30 ml sample was stirred in a glass beaker with a magnetic stirrer at 5.5 rpm. Plasma treatments lasted 2, 4, 6, 8 and 10 mins for each sample. Absorbance, pH and temperature measurements were performed using UV-Vis, a pH meter and a thermometer, respectively, for each solvent that had been treated by plasma. The same procedure was applied to all electrodes. A detailed schematic of the experimental setup is presented in Figure 1.

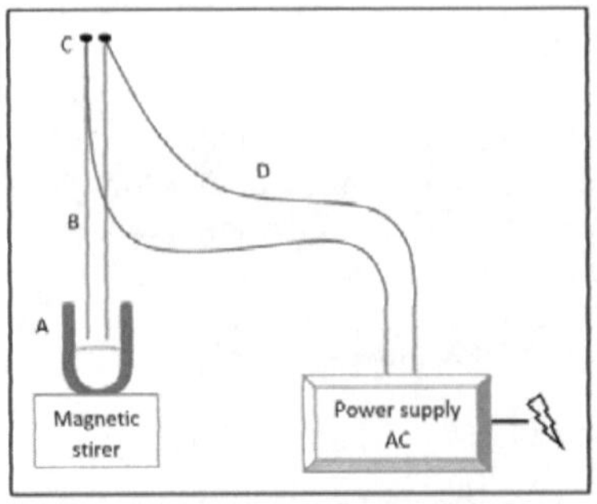

Figure 1. Experimental setup scheme. Beaker glass (A); Stainless steel electrodes (B); Clamp (C); necting cable (D)

3. Results and Discussion

Figure 2 shows the decreasing methylene blue absorbance with time on various electrode diameters. With plasma treatment, the methylene blue absorbance decreased with the increase of contact time. Contact times in the plasma process were 2, 4, 6, 8 and 10 mins. As a reference, we used a control sample, the methylene blue samples without plasma discharge.

Figure 2 (a) shows the experiment using a 2-mm-diameter electrode. The percentages of absorbance reduction at contact times 2, 4, 6, 8 and 10 mins were 30.4%, 82.5%, 94.8%, 99.5% and 100%, respectively. Experimental results using a 3.2-mm-diameter electrode are shown in Figure 2 (b). The percentages of absorbance reduction at the same contact time variations as for the 2-mm-diameter electrode were 20%,

76.9%, 93.8%, 97.5% and 99.7%, respectively. Figure 2 (c) shows the experiment using a 4-mm-diameter electrode. The percentages of absorbance reduction at the same contact time variations as in previous trials were 14.1%, 58.8%, 91.8%, 93.6% and 97.6%, respectively.

The highest peak of each graph indicates the initial absorbance of the methylene blue before plasma treatment. Experiments with plasma treatment at the given time variations show the decrease of absorbance in each graph. The decrease of absorbance occurred because the methylene blue decomposed as a result of the active species produced during the plasma process. Longer contact times meant that more methylene blue decomposed. The highest percentage of absorbance decrease was obtained by using the 2-mm-diameter electrode.

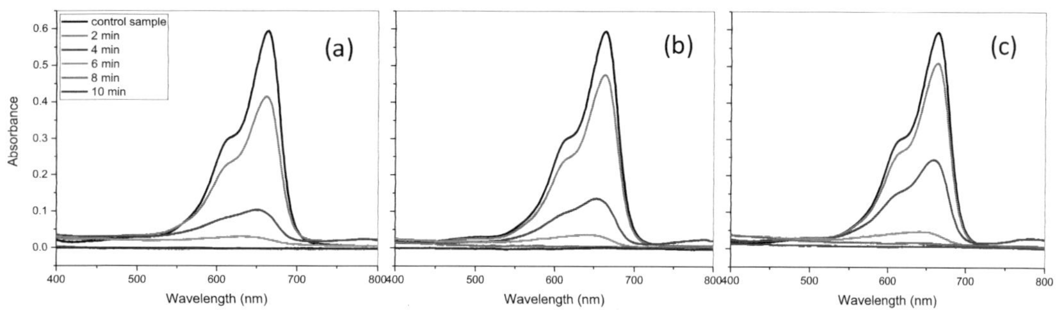

Figure 2. Decrease of absorbance at the given time variations of plasma treatment for electrode diameter of (a) 2 mm (b) 3.2 mm (c) 4 mm.

The effects of electrode diameter variation on the absorbance at given times is shown in Figure 3. The results show that the smaller the diameter of the electrode, the lower the absorbance. This happens because the electrodes with short diameters produce high-intensity plasma so that more active species form. As more active species form, more organic compounds decompose [5]. The graphic in Figure 3 is drawn from data with a contact time of 2 mins to show the most significant absorbance reduction.

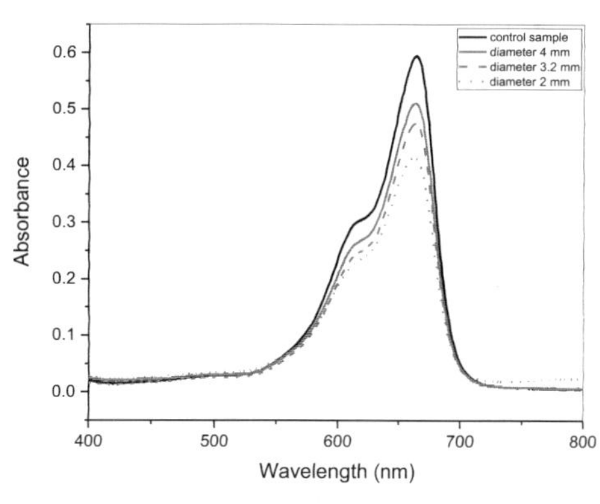

Figure 3. The effects of electrode diameter variation on the absorbance of methylene blue at a contact time of 2 mins.

The effects of electrode diameter and contact times on pH and temperature are shown in Figures 4 (a) and (b), respectively. Figure 4 (a) shows that smaller electrode diameters result in lower pH. This phenomenon happens because the oxidation process in the plasma reactor breaks the complex bonds in the methylene blue solution into mild acids. Longer contact times mean that more mild acids form so that the solution becomes more acidic. The formation of mild acids during the process decreases the solution's pH with the addition of contact time [6].

The lowest pH in this study was obtained with the 2-mm-diameter electrode. Figure 4 (b) shows the effects of electrode diameter on temperature. The results show that smaller electrode diameters are correlated with higher resulting temperatures. This occurs because the electric discharges generated from short diameter electrodes have a high velocity that causes strong collisions between water molecules, increasing heat. Along with increased contact time, the collisions lead to the increasing of solution temperature.

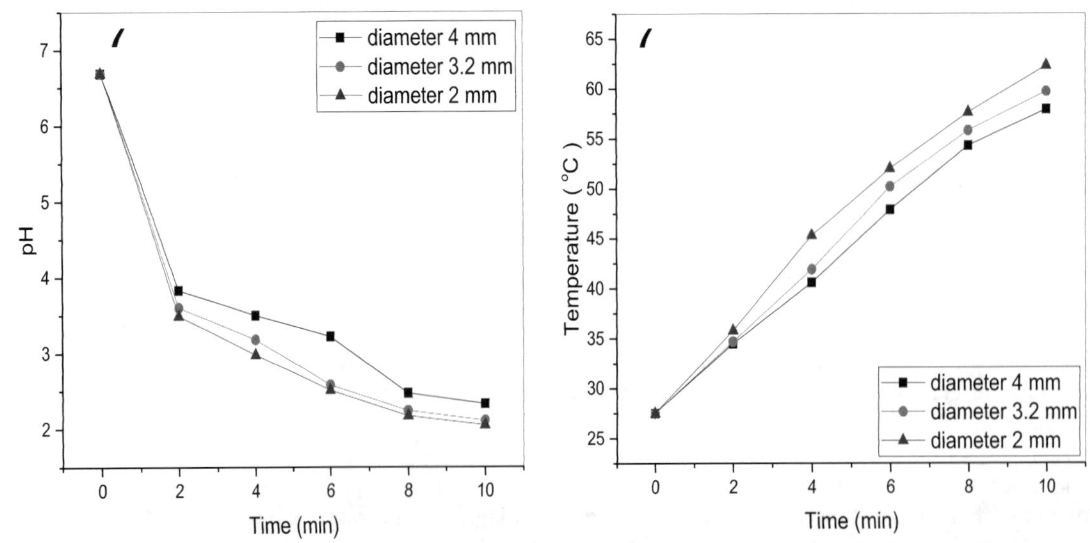

Figure 4. The relationship between (a) pH and (b) the temperature with contact times and electrode diameter variations

4. Conclusion

The plasma discharging performed in this study successfully decreased the absorbance of methylene blue. The results show that the methylene blue absorbance after plasma treatment significantly decreased compared with the concentration before the plasma process. Smaller electrode diameters are correlated with a greater reduction of methylene blue absorbance. Moreover, smaller electrode diameters are also linked with higher temperatures and lower pH. The lowest methylene blue absorbance and best results obtained in this study were achieved under plasma treatment for 10 mins using 2-mm-diameter electrodes.

IOP Conf. Series: Materials Science and Engineering **333** (2018) 012024 doi:10.1088/1757-899X/333/1/012024

References

[1] Chang J S 2009 *Thermal Plasma Solid Waste and Water Treatments* (Canada: McMaster University) 67-84
[2] Chen J & Davidson J H 2002 *J. Appl. Phys.* **22** 199-224
[3] Clements J S, Sato M and Davis R H 1987 *J. IEEE Trans.* **IA-23** 224-235
[4] Conrads H & Schmidt M 2000 *J. Sci. Technol.* **9** 441-454
[5] Lunak S, Sedlak P 1992 *J. Photochem. Photobiol. A: Chem* **68** (**1**) 1-33
[6] Muradia S and Nagatsu M 2013 *J. Appl. Phys.* **102** 144105

Water Treatment Using Plasma Discharge with Variation of Electrode Materials

N Chanan[1], Kusumandari[1], T E Saraswati[2]
[1] Department of Physics, Faculty of Mathematics and Natural Sciences, Sebelas Maret University, Jl. Ir. Sutami 36A Kentingan Surakarta 57126, Indonesia
[2] Department of Chemistry, Faculty of Mathematics and Natural Sciences, Sebelas Maret University, Jl. Ir. Sutami 36A Kentingan Surakarta 57126, Indonesia

E-mail: kusumandari@staff.uns.ac.id ; teguh@mipa.uns.ac.id

Abstract. This research studied water treatment using plasma discharge. Plasma generated in this study produced active species that played a role in organic compound decomposition. The plasma reactor consisted of two needle electrodes made from stainless steel, tungsten, aluminium and grafit. It placed approximately 2 mm above the solution and connected with high-AC voltage. A solution of methylene blue used as an organic solution model. Plasma treatment times were 2, 4, 6, 8 and 10 min. The absorbance, temperature and pH of the solution were measured before and after treatment using various electrodes. The best electrode used in plasma discharging for methylene blue absorbance reduction was the graphite electrode, which provided the highest degradation efficiency of 98% at 6 min of treatment time.

1. Introduction

Water is the most important source of life in an ecological system. The problems related to water resources has been increasing due to industry, agriculture, and rising population [1]. Therefore, it is necessary to perform water purification so that water-soluble pollutants can be separated, improving water quality.

Plasma is the fourth state of matter, beside gas, liquid and solid material [2]. Plasma is defined as an ionized gas. A gas will turn into plasma if the temperature is raised, causing the atoms to release some oral of their electrons. During this process, the remaining atom is positively charged, while the negative electrons move freely. Plasma technology has now begun to be used for water purification. Plasma can be generated in gas and water, forming some radicals, ions, ozone, or namely active species, which have high potential oxidation rates, break down organic compounds and kill microorganisms by decomposing pollutants without leaving secondary pollutants [3].

There are several plasma technologies that are commonly used for water purification, such as ozone, electron beam and electrical discharge. In this study, we used the electrical discharge method due its simplicity. The plasma discharge generates ions and electrons with very high energy as active species, causing water to decompose. The active species also kill the bacteria and further oxidize the compounds contained in water [4].

Due to their physical and chemical properties, the material used for the electrodes affects the chemical process of plasma in water, necessitating a study of variation in electrode material in plasma discharge applied in organic solution removal. The electrode materials used in this study were stainless steel, tungsten, aluminium and grafit. The effect of electrode material variation on plasma discharging

IOP Conf. Series: Materials Science and Engineering **333** (2018) 012025 doi:10.1088/1757-899X/333/1/012025

before and after treatment was studied not only by absorbance reduction but also in the changes of pH solution and temperature.

2. Experimental

The initial solution was prepared by dissolving 6 mg of methylene blue powder in 1 litre of distilled water. Before treatment, the temperature and pH of untreated solution was measured as a control sample. The absorbance of untreated solution was also analysed by using a UV-Vis spectrometer with a scanning method at a wavelength range of 400-800 nm.

The plasma reactor used two parallel electrodes installed 2 cm apart. An AC power supply with a voltage of 46.3 kV was connected to the electrodes to start the discharge. The electrodes were made from grafit, stainless steel, tungsten or aluminium. A sample of 30 ml methylene blue solution placed in a beaker was treated under the electrodes while it was stirred at 6 rpm. The distance between electrode and sample was set at 2 mm. Plasma discharging was applied for 2, 4, 6, 8 and 10 min. The pH, temperature and absorbance after treatment were measured and compared with pH, temperature and absorbance results in the untreated sample before treatment.

3. Results and Discussion

Figure 1 shows the absorbance of methylene blue after treatment from 0 to 10 minutes for (a) aluminium, (b) grafit, (c) stainless steel and (d) tungsten electrode, respectively, compare to the absorbance of control sample. It can be seen that the methylene blue absorbance more decreased with longer time of plasma treatment. Reaction between active species of plasma will decompose the methylene blue [5]. The longer treatment time will resulting much more decomposition and reduce more absorbance.

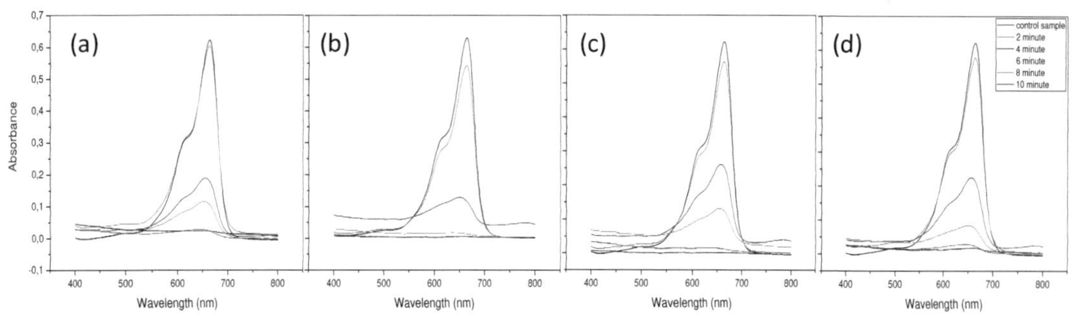

Figure 1. The absorbance of methylene blue for control sample and (a) aluminium, (b) grafit, (c) stainless steel and (d) tungsten elctrode at treatment time of 2, 4, 6, 8 and 10 minutes.

Moreover, Fig. 2 resume percentage degradation of methylene blue after plasma treatment for a given time compare to control sample. At treatment time of 2 to 4 min, plasma treatment using grafit electrode show the highest degradation. During plasma generation with high voltage, grafit which is consist of C-C bond might be broken and react to decompose the methylene blue [6]. Furthemore, for all electrode materials, after conducting the plasma treatment for 8 min, the absorbance reduction reached 99%. This indicates that the methylene blue was no longer present in the sample. It also indicated in the colour differences at each time as shown in Figure 3.

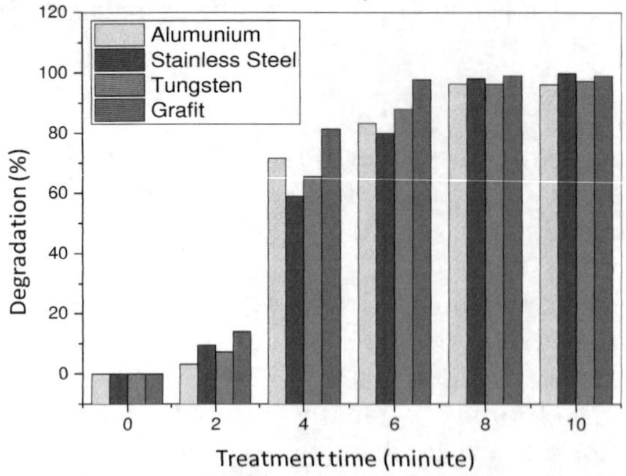

Figure 2. Degradation of methylene blue after plasma treatment for a given treatment time.

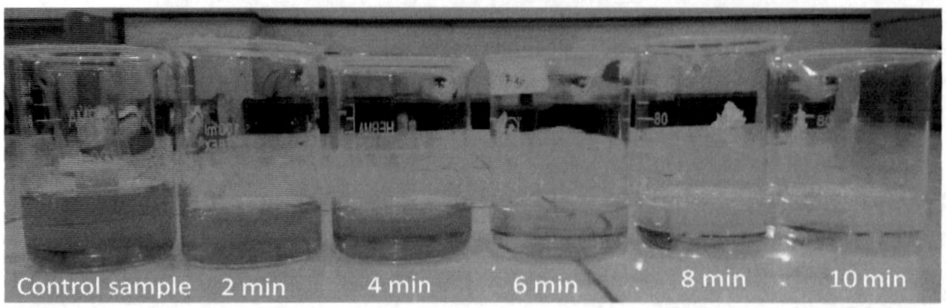

Figure 3. The colour difference of methylene blue after plasma treatment for a given treatment time.

Figure 4(a) shows the relationship between temperature and treatment time. Longer treatment times resulted in higher solution temperature. It is due to the electric discharges generated have a high velocity that causes strong collisions between water molecules, increasing heat. Along with increased treatment time, the collisions lead to the increasing of solution temperature. Figure 4(b) shows the relationship between pH and treatment time where the pH decrease with increasing treatment time. It occurs because the oxidation process in the plasma reactor breaks the complex bonds in the methylene blue solution into mild acids [7]. Longer treatment times mean that more mild acids form and make the solution becomes more acidic. The formation of mild acids during the process decreases the solution's pH with the addition of treatment time [8].

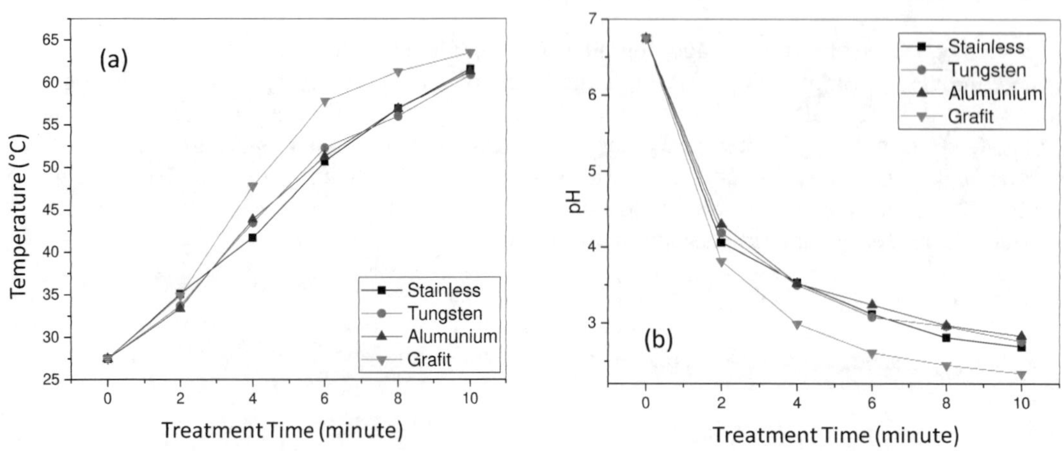

Figure 4. The relationship of (a) temperature and (b) pH with plasma treatment time

4. Conclusion

The study of variation of electrode materials used in plasma discharging for water treatment with methylene blue as solution model provided different degradation efficiencies. Grafit showed the best results in plasma treatment because it provided higher absorbance reduction efficiency in a shorter time compared to the other materials.

References

[1] Xu M, Bai X, Pei L & Pan H 2016 *Int. J. of Hydrogen Energy*.**41** 15930-15937
[2] Joshi A A, Locke B R, Arce P, & Finney W C. *J. Hazard Materials*. **44**
[3] Hazmi A, Desmiarti R, Waldi E P, Hadiwibowo A, & Darwison 2012 *J. Rekayasa Genetika*. **101**
[4] Agung T and Winata H S 2012 *J. Ilmiah Teknik Lingkungan*.**22**
[5] Huang F, Chen L, Wang H, Yan Z 2010 *J. Chem Eng* **162250–256**
[6] Parkansky N, Vegerhof A, Boris A, Alterkop, Berkh O, Raymond L, Boxman 2012 *Plasma Chemistry and Plasma Processing*. **32 933–947**
[7] Muradia S 2013 Study of Low-Voltage Pulsed Plasma Discharges Inside Water Using a Bubble Generating Porous Ceramic Electrode for Wastewater Treatment *Doctoral Thesis* **27**
[8] Grabowski L R 2006 Pulsed Corona in Air for Water Treatment *Thesis*. (Technische Universiteit Eindhoven)

Preparation of amine-modified Fe_3O_4/carbon nanoparticles by submerged arc discharge in ethylenediamine/ethanol

N Rismana[1], A R Astuti[1], Y H Suselo[2], M Anwar[3] and T E Saraswati[1*]
[1] Department of Chemistry, Faculty of Mathematics and Natural Science
[2] Department of Medicine, Faculty of Medicine
[3] Department of Electrical Engineering, Faculty of Engineering
Sebelas Maret University, Ir. Sutami 36A, Kentingan, Surakarta 57126 Indonesia

E-mail: teguh@mipa.uns.ac.id

Abstract. The preparation of amine-modified iron oxide/carbon (Fe_3O_4/C) nanoparticles was performed by the submerged arc discharge method using graphite electrodes and a liquid medium consisting of 50% ethanol and added ethylenediamine. The arc discharge was conducted by passing a voltage of 20–40 V. Two graphite electrodes were used, where one of them was filled with Fe_3O_4, carbon graphite and binder in a mass ratio of 1:3:1. The properties of the fabricated Fe_3O_4/C nanoparticles were analysed using X-Ray diffraction (XRD), scanning electron microscopy (SEM) and transmission electron microscopy (TEM). The XRD diffractogram pattern of the prepared Fe_3O_4/C showed definitive peaks at 26.5238° and 35.6374°, which are the main characteristic peaks of C graphite and Fe_3O_4, respectively. SEM analysis showed that the nanocomposite had a spherical morphology. TEM analysis found that the structure of the nanocomposite was Fe_3O_4-coated carbon.

1. Introduction

Magnetic nanoparticles (MNPs) are of great interest for researchers from a wide range of fields, including those working in magnetic fluids, catalysis, biotechnology/biomedicine, magnetic resonance imaging, data storage and environmental remediation [1]. Recently, iron oxide (Fe_3O_4) MNPs have been intensively investigated because of their superparamagnetism, high coercivity and low Curie temperature [2–5].

In addition to these characters, Fe_3O_4 MNPs are non-toxic. These simple nanoparticles are vulnerable to rapid degradation in the environment due to the high surface area to volume and reactivity ratios. One way to protect nanoparticles from degradation is by coating them with a stable chemical species, such as graphite. Carbon is one of the best solutions for encapsulation because it is light, inexpensive and very stable under extreme chemical or physical environments. The carbon layer protects the nanoparticles from environmental degradation so as to retain their intrinsic nanocrystalline properties [6].

Magnetic carbon nanoparticles are synthesized using several methods, such as arc discharge [7], microwave heating [8], thermal plasma [9], explosion [10], chemical vapour condensation [11] and hydrothermal reaction [12]. In this paper, the arc discharge method was used to synthesize Fe_3O_4/C because it is a simple, convenient and economical method [13]. According to Sano (2002), the arc-discharge method in liquid media, as opposed to a vacuum system, will produce nanomaterial products with a high level of purity [14].

Carbon-based MNPs are hydrophobic, whereas in biological applications they must be hydrophilic. Therefore, surface modification is needed to make the nanoparticles hydrophilic, which will allow them to be coated with useful molecules, such as antibodies [15]. Amino-group functionalization is suitable for graphite because of its ability to enhance the reactivity and hydrophilic nature of the nanoparticles [16]. These magnetic nanoparticles has superparamagnetic property, which means they will only be magnetic when present in a magnetic field. The amine-group functionalization allows the nanoparticles to have hydrophilic surfaces allowing for an easy dispersion, and preventing clumping without losing its magnetic property [17–18].

In this work, the surface of Fe_3O_4/C nanoparticles were synthesized and modified in one step via the arc discharge method in a liquid medium. The amine-modified Fe_3O_4/C, prepared by adding ethylenediamine into the ethanol medium, provided an attachment point for an amine functional group on the nanocomposite surface. Additionally, ethylenediamine is expected to improve the hydrophilic properties of the Fe_3O_4/C nanoparticles. This should make the nanoparticles more biocompatible, allowing them to be used for bio-applications, such as coating them with useful molecules like antibodies.

2. Experimental details

2.1. Materials

The materials that were used for fabrication process include carbon electrodes, carbon powder, iron oxide (Fe_3O_4), 99% ethylenediamine (Merck), 70% ethanol (technical grade), fructose binder (technical grade), and deionized water.

2.2. Fabrication of Fe_3O_4/C nanoparticles

A modified graphite electrode as the anode was filled with a mixture of Fe_3O_4, carbon powder and fructose binder in a ratio of 1:3:1 (w/w/w). This modified electrode was heated at 180°C for six hours. The cathode was made from a graphite electrode with a sharp tip.

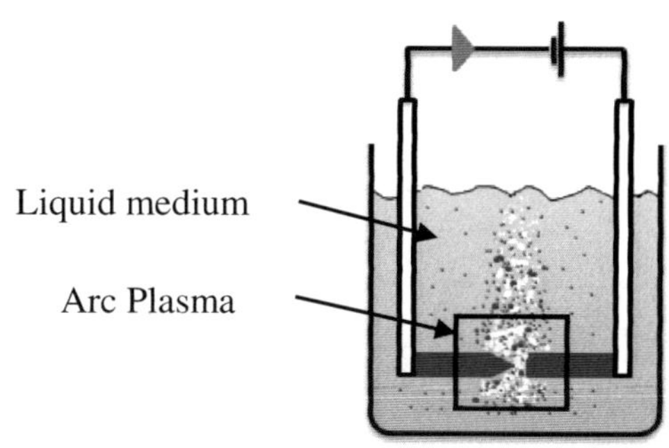

Figure 1. Experimental setup for the fabrication of amine-modified Fe_3O_4/C nanoparticles via the arc discharge method.

The cathode and anode were placed a certain distance apart in 2 mm in order to achieve the arcing. A liquid medium consisting of 50% ethylenediamine and 50% ethanol in 300 mL total volume was used.

The fabrication process was carried out using 20 V with 10 A of current. The experimental set-up is presented in Figure 1.

The Fe_3O_4/C nanoparticles produced were collected and further characterized by X-ray diffraction (XRD) (Cu; 40 kV; 35 mA) to determine their crystallinity and crystal structure. Scanning electron microscopy (SEM) and transmission electron microscopy (TEM) were used to identify the surface structure morphology, particle size and to determine the structure of the nanoparticles. Dispersion tests were also performed to determine the quality of the nanoparticle dispersion prepared from the fabrication process.

3. Results and discussion

Preparation and amino-group surface modification of the Fe_3O_4/C nanoparticles were successfully achieved using the arc discharge method in a liquid medium. When an electrical current was passed from the cathode to anode, the electrical ions produced an arc that leaped between both electrodes. During this process, the Fe_3O_4 and carbon present in the prepared electrodes were evaporated in the hot arc zone and then condensed to a nanocomposite in the cooler surrounding zone. This interaction produced amine-modified Fe_3O_4/C nanoparticles.

Following the fabrication process, the Fe_3O_4/C nanoparticles were characterized using XRD to determine the crystallinity changes after fabrication and surface modification. The XRD patterns for the Fe_3O_4/C and amine-modified Fe_3O_4/C nanoparticles are shown in Figure 2.

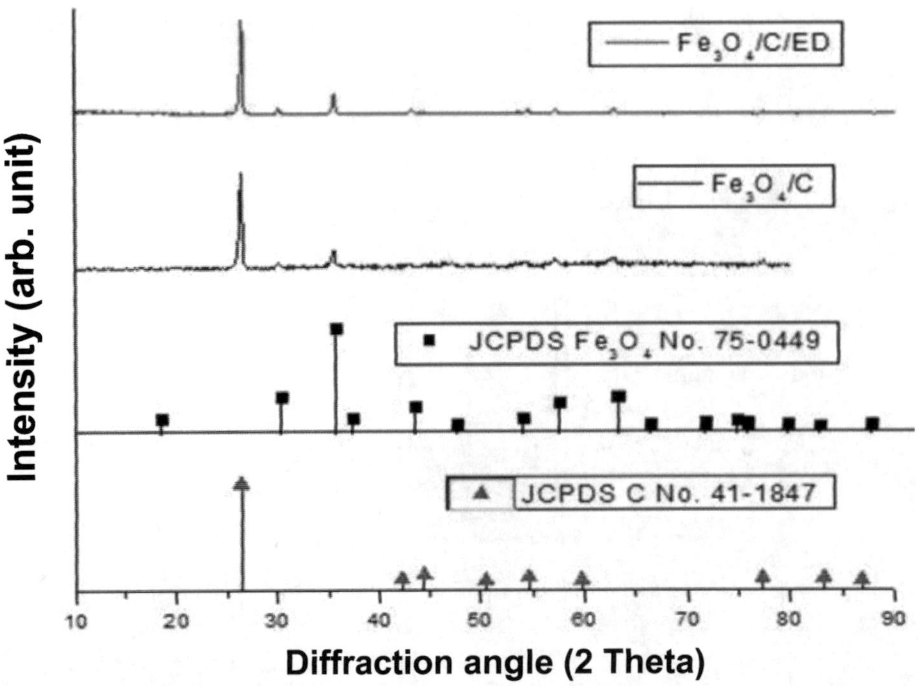

Figure 2. X-ray diffraction (XRD) patterns of Fe_3O_4/C and amine-modified Fe_3O_4/C (Fe_3O_4/C-ED).

The diffractogram of the amine-modified Fe_3O_4/C shows peaks comparable to those in the Joint Committee on Powder Diffraction Standards (JCPDS) pattern. The peak at 2θ of 26.38 (002) is the main peak characteristic of carbon (JCPDS No. 41-1487), and the one at 35.76 (311) is the main peak characteristic of Fe_3O_4 (JCPDS No. 75-0449). The main peaks for both Fe_3O_4 and carbon existed in the XRD pattern of the amine-modified Fe_3O_4/C, as well as in the Fe_3O_4/C, indicating that the both of nanoparticles were composed of crystalline Fe_3O_4 and elemental carbon.

The images of the amine-modified Fe_3O_4/C nanoparticle were analysed using SEM and TEM, and the results are shown in Figure 3. Figure 3A shows the SEM image of an amine-modified Fe_3O_4/C demonstrating a spherical morphology. The structure of these nanoparticles was further observed in detail by TEM, as shown in Figure 3B. The Fe_3O_4/C nanoparticles were observed to be nanoparticles coated by carbon layers. The imaging results shown in Figure 3 have a good agreement with the XRD pattern, which shows that the produced nanoparticles consisted of Fe_3O_4 and carbon.

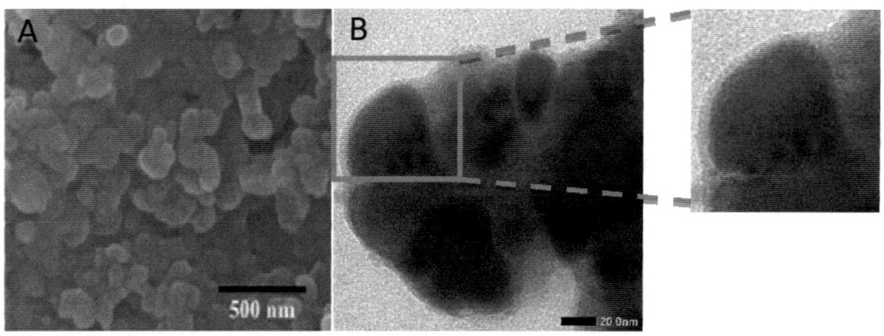

Figure 3. (A) SEM and (B) TEM images of amine-modified Fe_3O_4/C.

Figure 4 shows the histogram of the particle cluster size of the amine-modified Fe_3O_4/C, as measured from the SEM images in a spot observation (the images are not shown here). The diameter of the particle cluster size of the amine-modified Fe_3O_4/C produced using arc discharge has a dominant size range of 100–150 nm as an agglomerated formation, indicating that the particles produced are nanoparticles because each particle diameter is probably less than 100 nm in size.

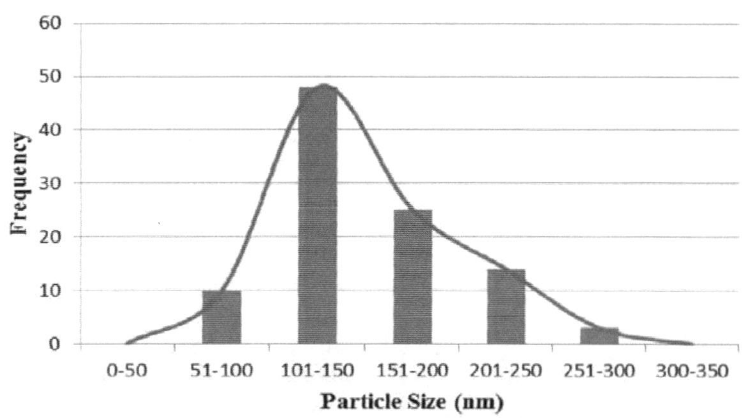

Figure 4. Histogram of particle cluster size of amine-modified Fe_3O_4/C.

In order to visually observe the successful surface modification, the produced nanoparticle was dispersed in distilled water. The purpose of this dispersity test, shown in Figure 5, was to determine the quality of the dispersion of the material produced using this fabrication process. Figure 5 compared the dispersion of Fe_3O_4/C produced using submerged arc discharge in ethanol and Fe_3O_4/C produced using submerged arc discharge in ethanol with addition of ethylenediamine.

The dispersion test pictured in Figure 5A and 5B showed that the Fe_3O_4/C produced in ethanol/ethylenediamine has a better dispersity in the aqueous liquid medium than the Fe_3O_4/C produced in ethanol only. These latter particles remained on the top of the water surface, while the amine-modified particles were dispersed completely, as indicated by the darker liquid.

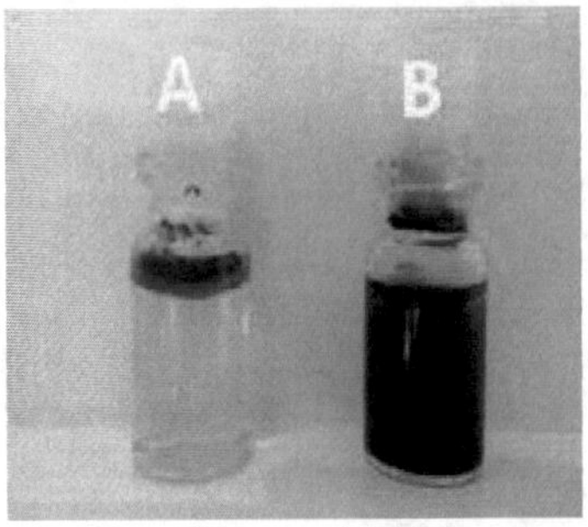

Figure 5. Dispersion test of (A) unmodified Fe_3O_4/C produced using submerged arc discharge in ethanol only (B) and amine-modified Fe_3O_4/C produced using submerged arc discharge in ethanol/ethylenediamine.

This clearly demonstrates that the nanoparticles that had a better dispersion have a hydrophilic surface character provided by amino groups attached to the nanoparticles' surfaces, called the amine-modified Fe_3O_4/C. The amino groups contained in the ethylenediamine/ethanol liquid medium were successfully attached to the nanoparticle surface during the arc discharge process. On the other hand, the unmodified Fe_3O_4/C nanoparticles with a worse dispersion have hydrophobic surface character due to the lack of amine groups on their surface.

4. Conclusion

The XRD diffractogram pattern of the prepared Fe_3O_4/C showed main peaks characteristic peaks of C graphite and Fe_3O_4 at 2θ 26.5238° and 35.6374°, respectively. SEM and TEM analysis showed that the nanoparticle had a spherical morphology and that the structure of the nanoparticle was Fe_3O_4 coated with carbon layers. The amine-modified Fe_3O_4/C had a better dispersion in the aqueous liquid medium than Fe_3O_4/C, indicating the ethylenediamine addition into the liquid medium during submerged arc discharge successfully modified the particles, producing hydrophilic surfaces.

Acknowledgment

The author would like to thank the Minister of Research, Technology and Higher Education, Republic of Indonesia, for providing the Research Grants under project No. 873/UN27.21/PP/2017.

References

[1] Lu A H, Salabas E E and Schüth F 2007 *Angewandte Chemie Int. Ed.* **46(8)** 1222-1244
[2] Kim Y S, Kim Y H 2003 *J. Magn. Magn. Mater* **267** 105-110
[3] Raj K, Moskowitz R 2002 *Transactions on Magnetics* **16** 358- 363
[4] Beydoun D, Amal R, Low G K C and McEvoy S 2000 *J. Phys. Chem.* B **104** 4387-4396
[5] McMichael R D, Shull R D, Swartzendruber L J, Bennett L H and Watson R E 1992 *J. Magn. Magn. Mate.* **111** 29-33
[6] Wu A, Yang X and Yang H 2012 *J. Alloys Compd.***513** 193-201

[7] Scott J H and Majetich S A 1995 *Phys. Rev.* B **52** 12564-12571
[8] Jacob D S, Genish I Klein L and Gedanken A 2006 *J. Phys. Chem.* B **110** 17711–17714
[9] Bystrzejewski M, Huczko A, Lange H, Baranowski P, Cota-Sanchez G, Soucy G, Szczytko J, and Twardowski A 2007 *Nanotechnology* **18** 145608
[10] Wu W Z, Zhu Z P, Liu Z Y, Xie Y N, Zhang J and Hu T D 2003 *Carbon* **41** 317–321
[11] Wang Z H, Zhang Z D, Choi C J and Kim B K 2003 *J. Alloy. Compd.* **361** 289–293
[12] Wang Z F, Xiao P F and He N Y 2006 *Carbon* **44** 3277–3284
[13] Miranzadeh M, and Kassaee M Z 2014 *Chem. Eng. J.* **257** 105–111
[14] Sano N, Wang H, Alexandrou I, Chhowalla M, and Teo K B K 2002 *J. App. Phys.* **92** 2783-2788
[15] Poplawska M, Bystrzejewski M, Grudziński I P, Cywińska M A, Ostapko J and Cieszanowski A 2014 *Carbon* **74** 180-194
[16] Saraswati T E, Ogino A, and Nagatsu M 2012 *Carbon* **50 (3)** 1253-1261
[17] Mikhaylova M, Kim D K, Bobrysheva N, Osmolowsky M., Semenov V, Tsakalakos T, and Muhammed M 2004 *Langmuir* **20(6)** 2472-2477
[18] Chiang C L, Sung C S, Wu T F, Chen C Y, and Hsu C Y 2005 *J. Chromatogr.* B ***822*(1)** 54-60

IOP Conf. Series: Materials Science and Engineering **333** (2018) 012027 doi:10.1088/1757-899X/333/1/012027

Quantitative analysis by UV-Vis absorption spectroscopy of amino groups attached to the surface of carbon-based nanoparticles

T E Saraswati[1*], A R Astuti, and N Rismana

Department of Chemistry, Faculty of Mathematics and Natural Sciences, Sebelas Maret University, Jl. Ir. Sutami 36 A, Kentingan, Surakarta 57126 Indonesia

Email: teguh@mipa.uns.ac.id

Abstract. Carbon-based nanoparticles must be modified due to their wide array of applications, especially when they are used as biomaterials. After modifying, quantitative analysis of the functional group is essential to evaluate a number of the available functional groups applied for further functionalization. In this study, we modified the carbon-based nanoparticles by amino group using submerged arc discharge in different liquids. The attached amino groups were then characterised and quantified by UV-Vis spectroscopy. This amino group functionalization was also confirmed by Fourier transform infrared (FTIR) spectra. The FTIR spectra of amine-modified nanoparticles show the definitive absorption peaks of N—H amine, C—H, C═O, C—N and Fe—O at 3418.97; 3000–2850; 1700–1600; 1400–1100; and 480-550 cm^{-1}, respectively. The amine groups have different performance signals between the amine-modified and unmodified nanoparticles. The FTIR spectra results were correlated with the UV-Vis absorption spectroscopy method using acidic methyl orange. The UV-Vis absorption spectroscopy shows that the absorbance of methyl orange represented to amino groups number was 1.3 times higher when the pH of the solution was increased. The absorbance intensity was then used to estimate the quantity of amine groups attached.

1. Introduction

Fe_3O_4 materials have been studied in multiple reports due to their several phases such as ferromagnetic magnetite (Fe_3O_4), maghemite (γ-Fe_2O_3) and hematite (α-Fe_2O_3) [1,2]. Furthermore, iron magnetic nanoparticles can be synthesised by various methods, such as electron beam irradiation methods, coprecipitation methods and emulsion precipitation methods [3-5]. The iron nanoparticles are also coated using other carbon materials to increase the absorption ability of the nanoparticles. Moreover, carbon coating can be used as an adhesion point for functional groups.

The magnetic-based nanocarbon material can also be synthesised by the arc discharge method. In the conventional arc discharge method, the vacuum pump and gas flow can be replaced using a liquid medium become a simple, easy and economical arc discharge method [6]. According to Sano et al. (2010), the arc discharge methods in the liquid produce nanomaterial products with high yield and lower impurities [7,8].

We previously reported the changing of the hydrophobic property of the carbon surface of TiO_2/C photocatalysts being hydrophilic during submerged arc discharge by using ethanol liquid with ethylenediamine or urea added [9,10]. One of the successful surface modifications of TiO_2/C with NH_2 groups was performed in the medium of ethanol:ethylenediamine (1:1 v/v) and qualitatively analysed by N—H and C—N peaks shown in its Fourier transform infra-red (FTIR) spectrum [10]. However, a quantitative study has not yet been performed. Therefore, this study merged FTIR characterisation and chemical derivatization analysis using UV spectroscopy techniques to qualitatively and quantitatively examine the successful surface modification of amine groups attached to the Fe_3O_4/carbon nanoparticles prepared by submerged arc discharge using the ethanol/ethylenediamine medium.

2. Experimental details

2.1. Fabrication of amine-modified nanoparticles

Amine-modified nanoparticles were fabricated by the arc discharge method using carbon electrodes. The anode used was a carbon electrode filled with a mixture of carbon, Fe_3O_4 and fructose binder with a weight ratio C:Fe_3O_4 of 3:1. Before filling, the mixture was sonicated for 480 s. After filling, the filled carbon electrodes were heated in a furnace at a temperature of 180°C for 6 h. The cathode was a sharp pencil-like graphite electrode, and the cathode and anode were positioned at a very close distance (less than 2 mm separation) to produce stepping electrons. Both the cathode and anode were submerged in a 600 mL glass beaker containing 300 mL ethanol 50%:ethylenediamine 50% (1:1 v/v). A current of 10 A (~30 V) then was passed through the electrodes, which causes arcing to occur. During this process, the nanomaterial was formed as a black powder dispersed in the liquid. For comparison, the arc discharge also was performed in a solution of only ethanol.

2.2. Characterizations of amine-modified nanoparticles

Black powder obtained from the previous section was characterised using FTIR instruments to qualitatively analyse the amine groups attached to the surface of the nanoparticles. The other analysis was performed by UV-Vis spectroscopy that followed the procedure in [11]. The amine-modified nanoparticles (0.005 g) were treated with 10 mL acidic methyl orange solution (0.05% methyl orange in a 0.1 m solution of sodium dihydrogen phosphate in water) for 2 minutes in a vortex mixer and then centrifuged at 8000 rpm for 5 minutes. After removing the excess dye by rinsing, the attached methyl orange was resolved by 5 mL of a 0.1 M potassium carbonate solution. The filtrate separated by centrifugation was then tested using UV-Vis absorption spectroscopy to quantitatively analyse the amine groups attached on the nanoparticle surfaces at a wavelength of 464 nm. The molar extinction coefficient of the methyl orange was $\varepsilon = 21600$ L/(mol cm).

3. Results and discussion

The modified Fe_3O_4 nanocomposite with carbon was successfully fabricated by using the arc discharge method in a liquid media of ethanol/ethylenediamine. Thus, the material product comprised amine-modified nanoparticles.

During arc discharging, a section of the tip of the filled carbon electrodes submerged in liquid was destructed and then evaporated in the arc area. The carbon vapour that moved away from the arc area then condensed on the Fe_3O_4 particle, resulting in carbon encapsulated iron oxide (Fe_3O_4/C) nanoparticles, which are illustrated in Figure 1. During the carbon layer condensation, the amine group from the ethylenediamine liquid was attached, covalently linked to sp^3 carbon atoms. The ethyl group possibly helped the hexagonal based structure carbon to further form as graphite layers. The amine functional groups attached on the surface were responsible for the dispersion quality of Fe_3O_4/C in water, and the modified nanoparticle surface with the amine group is illustrated in Figure 2.

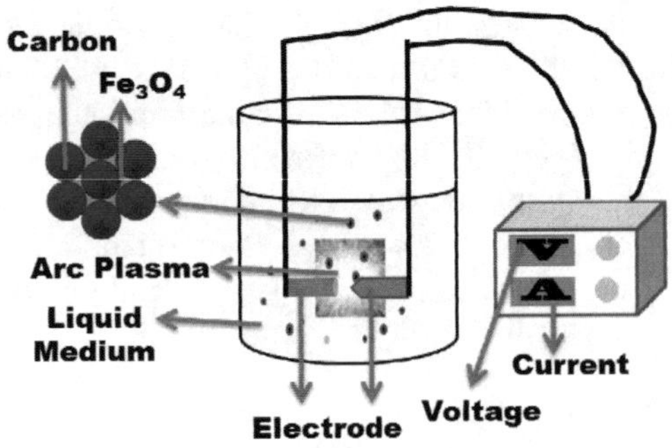

Figure 1. Illustration of arc discharging in liquid to form Fe_3O_4/C nanoparticles

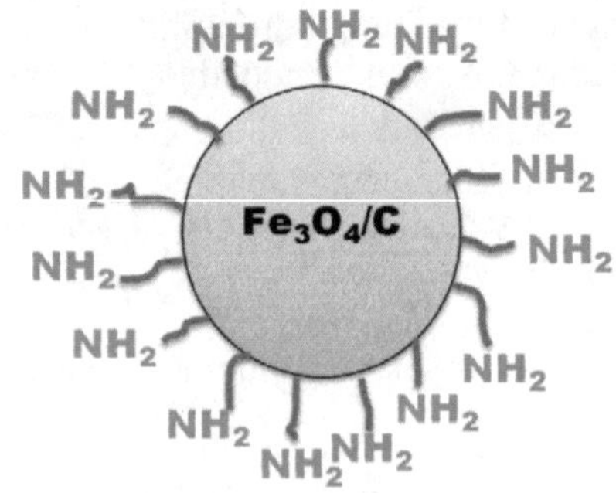

Figure 2. Illustration of amine group modified Fe_3O_4/C nanoparticles

To characterise the presence of nitrogen- containing functional groups, FTIR characterisation was performed to identify the successful attachment of amine groups on the nanoparticle surface. The FTIR characterisation shown in the spectra in Figure 3 confirmed that the studied method synthesised Fe_3O_4/C functionalized amine groups. The figure revealed the transmittance of N—H amine, C—H, C=O, C—N and Fe—O at 3418.97; 3000–2850; 1700–1600; 1400–1100; and 480-550 cm^{-1}, respectively.

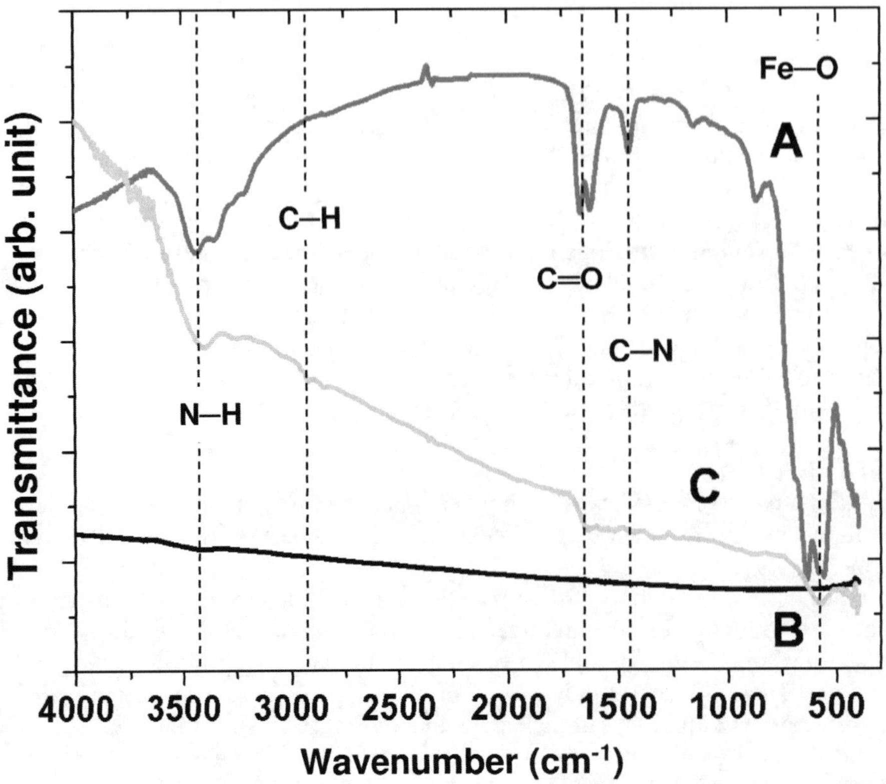

Figure 3. FTIR spectra: iron oxide (Fe_3O_4) (A), graphite powder (B), and amine-functionalized Fe_3O_4-C (C)

The presence of the N–H amine indicates the successful surface modification of Fe_3O_4/C nanoparticles by amine groups provided from the breakdown of the liquid medium of ethanol and ethylenediamine during the arc discharge process. The changes of surface characteristics of the particle due to the addition of a functional group derived from the liquid medium when synthesised in plasma arc discharge is illustrated in Figure 2. The amine groups were attached on the carbon layer via covalent bonds, and the covalent bonds were linked with carbon atoms in the hexagonal structure of graphite with the nitrogen atom of NH_2.

To quantitatively analyse the amine groups on the nanoparticle surfaces, the chemical derivatization of amine group was performed by a UV-Vis spectrometer using an acidic methyl orange (MO) solution. MO covalently bonds to the amine groups in the lower pH, and the bound MOs were released when the solution was in higher pH controlled by the potassium carbonate addition. The released MOs were easily detected by UV-Vis spectroscopy.

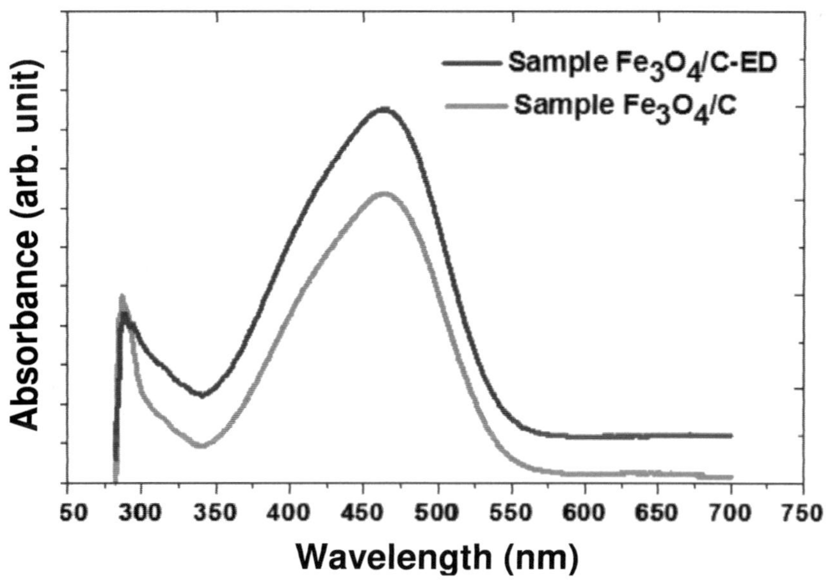

Figure 4. The UV-Vis absorption spectroscopy of acidic methyl orange released from Fe_3O_4/C and Fe_3O_4/C-ED

Figure 4 shows the spectra of the MO released from the modified and unmodified nanoparticles. The spectra in the figure show that the absorption peak of MO was significantly different between two of those samples. The MO absorbance of amine-modified nanoparticles was 1.3 times higher than that of unmodified nanoparticles. Consequently, the amine group concentration presented on Fe_3O_4/C-ED was equal to 0.022×10^{-3} mol NH_2 per one gram nanoparticles or 0.13×10^{20} NH_2 groups per one gram nanoparticles, if the initial value is multiplied by Avogadro's number.

4. Conclusion

The amine group being attached on the surface of the nanoparticles was successfully confirmed by FTIR analysis and UV-Vis spectroscopy techniques. The FTIR spectra of Fe_3O_4/C synthesised in the liquid medium ethanol/ethylenediamine revealed transmission peaks representing N—H amine, C—H, C═O, C—N and Fe—O at 3418.97; 3000–2850; 1700–1600; 1400–1100; and 480-550 cm^{-1}, respectively. The UV-Vis absorption spectroscopy technique shows that the peak of Fe_3O_4/C produced in the arc discharge in ethanol/ethylenediamine was 1.3 times higher than the peak of Fe_3O_4/C produced in only ethanol.

Acknowledgment
The author would like to thank the Minister of Research, Technology and Higher Education, Republic of Indonesia, for providing the Research Grants under project No. 873/UN27.21/PP/2017.

References
[1] Aji M P, Yulianto A and Bijaksana S 2007 *Jurnal Sains Materi Indonesia* October 2007 106-108
[2] Chirita M and Grozescu I 2009 *Chemistry Bulletin "Politehnica" Univ. Timisoara* **5** 1-8
[3] Zhao B, Wang Y, Guo H, Wang J, He Y, Jiao Z and Wu M 2007*Materials Science-Poland* **25** 1143-1148
[4] Sahoo, Agarwal K, Singh A K, Polke B G and Raha K C 2010 *International Journal of Engineering, Science and Technology* **2**118-126
[5] Berkmans A, Joseph M, Jagannatham S, Priyanka and Prathap Haridoss 2014 *Waste Management. (Oxford)* **3** 2139–2145
[6] Miranzadeh M and Kassaee M Z 2014 *Chem. Eng. J.* **257** 105–111
[7] Sano N, Wang H, Chhowalla M, Alexandrou I and Amaratunga G A J 2001 *Nature* **414** 506-507
[8] Sano N, Wang H, Alexandrou I, Chhowalla M and Teo K B K 2002 *Journal of Applied Physics* **92(5)** 2783-2788
[9] Arikawati E, Pranoto P and Saraswati T E 2017 *IOP Conference Series: Materials Science and Engineering* **176** 012045
[10] Arikawati E 2016 *Bachelor Thesis,* Sebelas Maret University 43-44
[11] Hartwig A and Mulder M 1994 *Advances in Colloid and interface Science* **52** 65-78

IOP Conf. Series: Materials Science and Engineering **333** (2018) 012028 doi:10.1088/1757-899X/333/1/012028

Effect of sintering on transparent TiO_2 18NR-T type thin films as the working electrode for transparent solar cells

A Supriyanto[1], Nandani[1], S Wahyuningsih[2], A H Ramelan[1]
[1]Physics Department of Mathematics and Natural Science Universitas Sebelas Maret, Jl. Ir. Sutami No. 36 A Kentingan, Surakarta 57126
[2]Chemistry Department of Mathematics and Natural Science Universitas Sebelas Maret, Jl. Ir. Sutami No. 36 A Kentingan, Surakarta 57126

E-mail: agusf22@yahoo.com

Abstract. The working electrode based on semiconductor transparent TiO_2 type 18NR-T for transparent solar cells have been grown by screen printing method. This study aim is to determine the effect of sintering on TiO_2 thin films transparent as the working electrode of transparent solar cells. TiO_2 films will be sintered at temperature 450°C, 500°C, 550°C and 600°C. TiO_2 films optical properties were characterized using UV-Vis spectrophotometer, electrical properties were characterized using 4 point probemethods and the crystallization was characterized by X-Ray Diffraction (XRD). The lowest transmittance due to the treatment of annealing temperature variations is 550°C because the 550°C TiO_2 layer is more absorbing. The peaks resulted from the annealing temperature treatment show that the high temperature the more anatase peaks.Characterization using four-point probe showed that the highest conductivity of TiO_2 18NR-T thin film was $2.42 \times 10^2\ \Omega^{-1}m^{-1}$at annealing temperature 550°C.

1. Introduction

Solar cells as a device to convert sunlight into electrical energy is always growing, in principle solar cells is an alternative technology that produces electrical energy by using free electrons produced by sunlight. This development is evidenced by the innovations that are divided into three generations. The first generation of silicon solar cells [1], the second thin layer solar cell [2] and the third generation of Dye-Sensitized Solar Cell (DSSC) [3-5]. One application of DSSC is a solar window, the main requirement is that windows are transparent and usually use transparent semiconductor material TiO_2 transparent 18NR-T. The transparent window used is a transparent conductor of transparent condcting oxide (TCO). This material has been widely used in optoelectronic applications and photovoltaics.

Titanium dioxide (TiO_2) is a versatile material in several applications whether in light absorption or transparency for the addition of electrical conductivity. Titanium dioxide has three distinct crystalline structures: rutile (tetragonal), anatase (tetragonal) [6,7] and brookite (orthorhombic) [8-11]. The type of crystal phase of TiO_2 frequently used in DSSC is the anatase phase, this is because the anatase phase has a high degree of photoactivity and large surface area. Factors affecting the photocatalytic activity of TiO_2 are phase structure, surface area, crystal size, particle shape and surface hydroxyl number [12-18]. With the addition of annealing temperatures, TiO_2 can transform into anatase and rutile forms [19]. Therefore, in this study focused on TiO_2 layer due to the treatment of variation of annealing temperature given. The heat treatment of the TiO_2 coating results can produce anatase phase which is well used in DSSC, low temperature to produce anatase phase on TiO_2 ranging from 400°C - 600°C [20]. In this observation, the TiO_2 layer was deposited by screen printing method. TiO_2 layer is the result of fabrication of transparent TiO_2 paste type 18NR-T.

2. Experimental

The substrate used in the study is a glass preparation having a resistance specification of thickness 1,1 mm.The first substrate was washed using an ultrasonic cleaner with 4 stages that are using soap water, quads, acetone, and alcohol which each step done for 10 minutes. TiO_2 used is brand dyesol with type 18NR-T shaped paste yellow. Pasta TiO_2 18NR-T weighed 1 gram and measured 1 ml of ethanol with a micropipette. The sample is stirred using a magnetic stirrer for 1 hour with a rotation speed of 300 rpm at room temperature.

The $TiO_2$18NR-T solution is deposited on a glass substrate using screen printing method, the screen used is the T-49 type which has 125 holes per inch. The type of screen affects the volume deposited on the glass substrate. Deposition of TiO_2 18NR-T solution above the substrate measuring 2.5 cm x 2 cm, the deposition area of 1.5 cm x 1 cm. TiO_2 18NR-T thin layer than in drying using the oven at 130°C for 10 minutes.The subsequently finished TiO_2 18NR-T thin layer was heat treated with 4 temperature variations of 450°C, 500°C, 550°C and 600°C for 60 minutes. The heat treatment is carried out for 1 hour with atemperature rise of 15° per minute. Sintering is intended for crystallinity of TiO_2 18NR-T thin films to form.

Characterization is performed to determine the result of a thin layer formed after sintering. A given sintering temperature difference may affect the formation of a rutile crystal or anatase TiO_2, characterization using XRD will show the anatase and rutile peaks formed.TiO_2 18NR-T thin layer will also be directed by UV-VIS to determine the value of absorbance, transmittance and bandgap energy.TiO_2 18NR-T thin electrical propertieswere characterized by Four Point Probe.

3. Result and Discussion

TiO_2 18NR-T thin layer has been deposited by screen printing method. Figure 1 shows the results of X-Ray Diffraction TiO_2 18NR-T thin film with annealing temperature treatment of 450°C, 500°C, 550°C and 600°C.

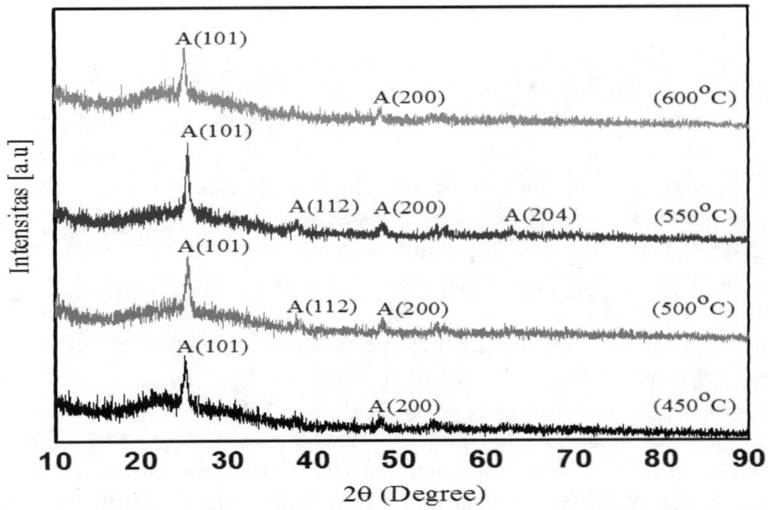

Figure 1. X-Ray Diffraction Pattern TiO_2 18NR-T with annealing temperature variation.

The variation of annealing temperature provided affects the formation of the number of diffraction peaks. Based on matching with ICDD 21-1272 database, the peaks formed are TiO_2 anatase phases.

Higher annealing temperatures result in the formation of more anatase phase peaks, proving that the formation of the best anatase phases at 550°C. TiO_2 semiconductors in photovoltaic applications, the most stable phase used is anatase phase. This is because the anatase surface area is wider so that in the absorbing process the anatase phase is larger absorbs

In Figure 2 showed the transmittance value of the deposited TiO_2 18NR-T layer with annealing temperature variation. The resulting value supports the absorbance value obtained, where the transmittance results with increasing temperature increase the value of transmittance is decreased. This value proves that low transmittance indicates more absorbing material.

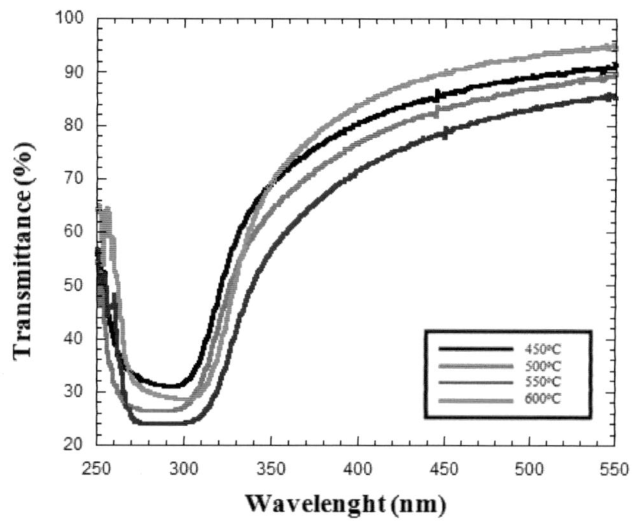

Figure 2. TiO2 18NR-T transmittance curve with annealing temperature variation

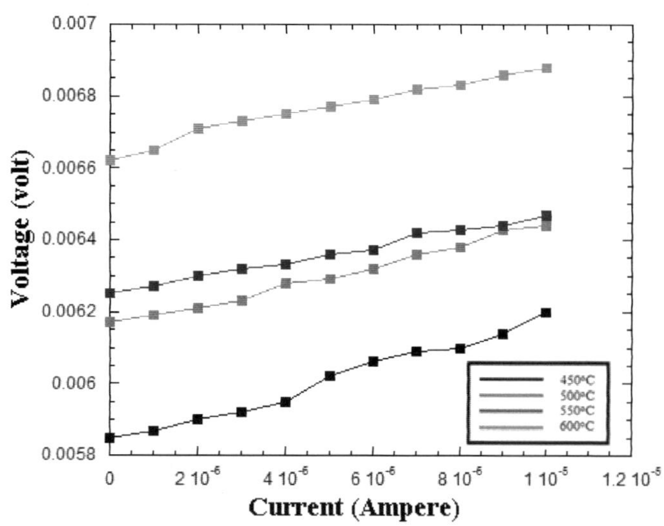

Figure 3. The current-voltage (I-V) characteristics of TiO_2 18NR-T with variation of annealing temperature

Figure 3 shows the current-voltage (I-V) characteristics of TiO_2 18NR-T thin film using a four-point probe methods. Characterization is done by varying the current value so as to obtain the voltage value of TiO_2 18NR-T thin layer. Annealing temperature variations result in a higher resistance value, which means that the TiO_2 18NR-T thin film conductivity value is greater. The conductivity can be obtained by first calculating the value of TiO_2 18NR-T thin film resistivity with equation 1:

$$\rho = 2\pi s \frac{V}{I} \tag{1}$$

Where s is the distance between the probes used in the measurement. Equation of conductivity value:

$$\sigma = \frac{1}{\rho} \tag{2}$$

The conductivity value of TiO_2 18NR-T thin film due to annealing effect can be seen in Table 1.

Table 1. Conductivity of TiO_2 18NR-T Thin Film with Variation of Annealing Temperature.

Annealing Temperature (°C)	**Conductivity x 10^2 (Ωm)$^{-1}$**
450	1.51
500	1.87
550	2.42
600	2.15

The $TiO_2$18NR-T thin film with an annealing temperature of 450°C has a conductivity value that is (1.51 x 10^2 $\Omega^{-1}m^{-1}$). The Conductivity value at annealing temperature 450°C-550°C increased. While at annealing temperature 600°C the conductivity value of TiO_2 18NR-T thin film decreased by (2.15x10^2 $\Omega^{-1}m^{-1}$). It can be concluded that TiO_2 18NR-T thin film with an annealing temperature of 550°C has good electrical properties.

4. Conclusion

The structure and optical properties of the TiO_2 18NR-T layer have been exposed to the substrate using screen printing method with annealing temperature variation treatment. Characterization using XRD showed anatase phase growth in the presence of peak intensity at each of the annealing temperature treatments. Anatase phase grows more at 550°C. UV-Vis characterization results show different transmitance values due to the temperature difference, transmittance results with increasing temperature increase the value of transmittance is decreased. this may be caused by crystallinity due to annealing temperature. This is supported by the conductivity of TiO_2 18NR-T thin film characterized by Four Point Probe, TiO_2 18NR-T thin film with an annealing temperature of 550°C has good electrical properties.

Acknowledgments

The authors would like to acknowledge the grant of MRG -Universitas Sebelas Maret (PUT- MRG UNS), 2017.

References

[1] Kim DS 2003 *Proc. 3rd World. Conf. Photovoltaic. Energy. Conv.* **2** 1293-1296 ISBN 4-9901816-0-3
[2] Edoff, Marika 2012 *AMBIO* **41**(2) 112-118
[3] O'Regan B, Grätzel M 1991 *Nature* **353** 737-739
[4] Zhou H, Wu L, Gao Y 2011 *J. Photocem. Photobiol.* A. **219** 188-194
[5] Gratzel M 2003 *J. Photochem. Photobiol. C. Photochem. Rev.* **4** 145-153
[6] Joseph A R, Viswanathan B 2009 *Indian. J. Chem* **48** 1378-1382
[7] Gang L, Xuewen W, Zhigang C, Hui-Ming C, Gao Qing Max L 2009 *J. Colloid. Interface. Sci.* **329** 331–338
[8] Bin Z, FengC, Qiwei Hang, Jinlong Z 2009 *Chem. Commun.* **34** 5115-5117

[9] Haifeng L, Liping L, Minglei Z, Xinsong H, Xiaomei, Guangshe L, Richeng Y 2012*J. Am. Chem. Soc.***134** (20) 8328-8331

[10] Yukiaki O, Koji T, Yukihiro K, Takaaki T, Ken-ichi K, Nobuhiro M, Toshihiro K, Kiyoshi O 2011 *Cryst. Growth. Des.***11** (11) 4831-4836

[11] Bernardini C, Cappelletti G, Dozzi M V, Selli EJ 2010 *Photochem. Photobiol. A.* **211** 185

[12] Chin S M, Park E S, Kim M S, Jurng J S 2010 *Powder. Technol.* **201** 171–176

[13] Kim D S, Han S J, Kwak S Y 2007 *J Colloid. Interface Sci.* **316** 85–91

[14] Cassiers K, Linssen T, Mathieu M, Bai Y Q, Zhu H Y, Cool P, Vansant E F 2004 *J. Phys. Chem. B.* **108** 3713–3721

[15] Sakatani Y, Grosso D, Nicole L, Boissiere C, de G C, Soler-IlliaA A, Sanchez C 2006 *J. Mater. Chem.***16** 77–82

[16] Akurati K K, Vital A, Fortunato G, Hany R, Nueesch F, Graule T 2007 *Solid. State. Sci.* **9** 247–257

[17] Christensen A, Dilks A, Egerton T A, Temperley J 2000 *J. Mater. Sci.* **35** 5353–5358

[18] Corrales T, Peinado C, Allen N S, Edge M, Sandoval G, Catalina F 2003 *J. Photochem Photobiol. A Chem.***156** 151–160

[19] Swati S Pradhan, Pradhan S K, Bhavanasi V, SambitaSahoo, Sarangi S N, Anwar P K, Barhai 2012 *Thin. Solid. Films* **520** 1809

[20] Ahmadi M, Ghasemi M R, Rafsanjani H H 2011 *J. Mat. Eng.* **5**87-93

[21] Dongsun Yoo, Ilgon Kim, Sangsoo Kim, Chang Hie Hahn, Changyu Lee, Angseongjin Cho 2007 *Appl. Surf. Sci.* **253** 3888–3892

IOP Conf. Series: Materials Science and Engineering **333** (2018) 012029 doi:10.1088/1757-899X/333/1/012029

Modification of CuI based Hole Transport Material for Solid State DSSC Application

Q A Hanif[1], A H Ramelan[2], L N M Z Saputri[1], and S Wahyuningsih[1*]
[1]Inorganic Material Research Group, Faculty of Mathematics and Natural Sciences, Universitas Sebelas Maret, Ir. Sutami street 36A, Surakarta Indonesia
[2]Electronic Materials and Energy, Faculty of Mathematics and Natural Sciences, Universitas Sebelas Maret, Ir. Sutami street 36A, Surakarta Indonesia

Email: sayekti@mipa.uns.ac.id

Abstract. Modification of Hole Transport Material (HTM) with addition of tetramethylethylenediamine (TMED) and ammonium thiocyanate (NH_4SCN) have been conducted. Copper Iodide (CuI) were used as the main component of HTM. Several volume variations of TMED (0.1; 0.2; and 0.4 mL) was added into 0.05 M CuI solutions. While TMED: NH_4SCN ratio were 1:1,1:2,2:1 also introduced to the dissolved CuI. Optical properties of these materials showed the band gap energy value ranging from 2.38 to 3.79 eV. The conductivity of HTM has been measured and showed the maximum value in CuI added with 0.4 mL TMED = 0.29 S m^{-1}, and ratio of TMED: NH_4SCN=2:1 added into CuI = 0.39 S m^{-1}, these value were increased compared to the CuI conductivity itself (0.26 S m^{-1}). The effect of this modification towards SS DSSC efficiency also has been monitored. The SS DSSC construction consists of TiO_2 nanorods sensitized by N3, HTM, and platinum as a counter electrode. The performance of SS DSSC showed rising efficiency as follows TiO_2|N3|CuI<CuI-TMED<CuI-TMED: NH_4SCN|Pt. The highest conversion was increased 3.28 times when CuI-TMED: NH_4SCN used as a HTM. Therefore, both TMED and NH_4SCN incorporated with CuI be able to improve the electrical properties in SS DSSC system.

1. Introduction

Global energy demands is increased in the last past years. Its increase predicted will occur to a couple years ahead because the dependence of many sectors towards energy. Transportation, industrial, residential, and commercial sectors contribute to the high energy consumption [1]. Until now, the most utilized energy, comes from unrenewable energy fossil fuel. Therefore, the study of renewable energy is the interesting field topic to be concern about. One of massive renewable energy source is sun/ solar. Silicon solar cell as a device to convert solar energy to electrical energy was commercially producing. Despite the relatively high efficiency, the synthesis of this material requires a high purity of Si that lead to high cost fabrication [2]. Another type of solar cell device pioneered by O'Regan B and Grätzel is Dye Sensitized Solar Cell (DSSC). Compared to the silicon solar cell, DSSC is relatively low cost, less toxic, easy in fabrication, good performance, doesn't require very high purity of material, and environmentally friendly [3,4]. Typically, DSSC consists of 1) photoanode, TiO_2 is commonly used, 2) complex compound as dye, usually ruthenium based complex, 3) liquid electrolyte I^-/ I_3^-, and 4) cathode or counter electrode [4]. Photovoltaic performance of DSSC containing protonated N3 complex as

* To whom any correspondence should be addressed.

IOP Conf. Series: Materials Science and Engineering **333** (2018) 012029 doi:10.1088/1757-899X/333/1/012029

sensitizer yielding efficiency 11.18% [5]. Unfortunately, DSSC shows several drawbacks such as corrosion, photoreactive, dye desorption, evaporation, and leakage of solvent while using liquid I^-/I_3^- as electrolyte, thus lead to changing to quasi solid and solid state DSSC [6,7]. Replacement of liquid electrolyte with inorganic hole transport material was investigated by Tennakone and coworkers with copper iodide (CuI) [8]. CuI more promising for HTM compared to CuSCN [9], since it difficult to be dissolved [10] and CuBr [11]. However, CuI remains unstable and to obtain fulfilled area with CuI is not easy because its rapid crystallization leaves porous surface thus giving only few contact interface [13]. Chelating agent compounds such as 1-methyl-3-ethyl-imidazolium thiocyanate (MEIT) [14] and triethylamine hydrothiocyanate (THT) could improve the efficiency of SS DSSC [15]. Another research shown that interaction CuI with functional group NCS from dye also supports for the better performance [16]. Because of the fairly expensive of MEIT, Zainun *et al.* add tetramethylethylenediamine (TMED) into CuI. TMED addition prevents CuI from fast crystallization, filling pores, yielding well covered layer [17]. Nevertheless, the effect of TMED and NH_4SCN addition to performance of SS DSSC haven't been studied.

2. Experimental

2.1. Synthesis of CuI and its Modification TMED

The synthesis of CuI was conducted according to reference [9], with minor addition procedure. Copper iodide 0.19 gram was then dissolved in 20 mL acetonitrile. The TMED (Sigma Aldrich, 96%) was added into this 0.05 M CuI. The addition of TMED was varied in three different volumes of 0.1, 0.2, and 0.4 mL. The mixture was then stirred in 3 hours, yield CuI-TMED.

2.2. Modification of CuI-TMED with NH_4SCN

The first step of this modification was conducted similar with modification with TMED, but we only used same volume addition (0.4 mL). NH_4SCN (Merck) 1.5 gram was dissolved in 20 mL acetonitrile. Then the CuI-TMED was added by NH_4SCN in ratio of 1:2,1:1, and 2:1 (TMED: NH_4SCN) (v/v). All solution after modification were characterized with uv vis spectrophotometer (Perkin Elmer Lambda 25), Fourier Transform Infra-Red (FTIR, Prestige-21 Shimadzu type 8201PC, KBr pellet), I V measurements (El Kahfi 100, Cu electrode) and Scanning Electron Microscopy (SEM, Quanta 250).

2.3. Application in SS DSSC

Each CuI, CuI-TMED, and CuI-TMED: NH_4SCN compound was utilized in SS DSSC system as a HTM. SS DSSC in this test consists of TiO_2 nanorods as working electrode [18], platinum (Brataco) as counter electrode (Sigma Aldrich) each coated in separate Fluorine doped Tin Oxide (FTO) (dyesol), dye N3 complex (N3=cis-bis(isothiocyanato) bis(2,2'-bipyridyl-4,4'dicarboxylato ruthenium (II)) (Sigma Aldrich), and HTM. FTO containing sensitized TiO_2 was added with HTM via drop coating method in 80 °C. I V measurement of SS DSSC was conducted with I V Keithley 2602 System source.

3. Results and Discussion

3.1. Optical Properties of CuI and Modified CuI

The spectra of CuI with and without TMED shows different profile. CuI itself has no absorption peak in visible area, similar with previous result [19]. The shifting peak in UV ranged (200-300 nm) to the lower wavenumber or higher energy may the cause of the complexation between inorganic compound (Cu^+) and bidentate ligand TMED [20]. TMED as chelating agent for CuI, contributes to the broad absorption in visible area (550-800 nm), and its intensity increase while increasing amount of TMED was added. Not only TMED, but NH_4SCN also shown higher absorption at visible range, depicted in Figure 1.

The energy band gap was calculated with Kubelka-Munk equation [21]. According to Beranek and Kisch [22], CuI likely to have direct band gap, so the n value is 0.5. Energy band gap of CuI, CuI-TMED

0.4 mL addition, and CuI-TMED: NH_4SCN 2:1 are 3.79, 3.63, 3.55 eV, respectively. The utilization of smaller band gap, relates to the better conductivity and electrical property [23].

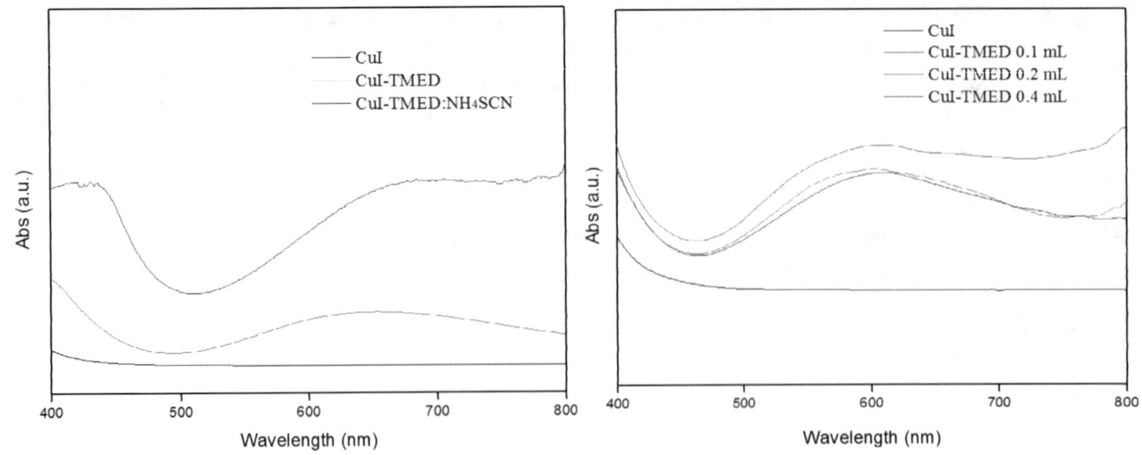

Figure 1. UV Vis absorption spectra of CuI and modified CuI

3.2. Infra-Red Spectra

FTIR spectrum with specific atomic bond signal of CuI, TMED, and NH_4SCN can be seen in Figure 2. The broad peak between 3400-3100 cm^{-1} in NH_4SCN represent N-H or amine bond [24]. This peak was changed into hydrogen bonding characteristic after mixed with TMED, shown from a broaden peak in the same area. The transformation into hydrogen bonding may occur due to the effect of interaction between hydrogen atom from NH_4SCN with lone pair electron in nitrogen from TMED. The interaction causes extending distance between hydrogen and nitrogen in NH_4SCN. Besides that, hydrogen atom attracted by nitrogen yielding closer distance, thus hydrogen bond rather dominates than amine bond.

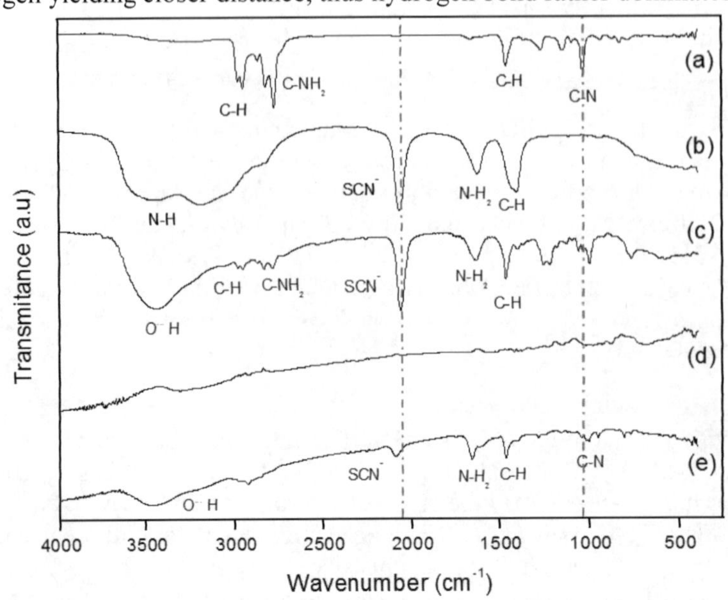

Figure 2. IR Spectra of a) TMED, b) NH_4SCN, c) TMED: NH_4SCN, d) CuI, e) CuI-TMED: NH_4SCN

The intensity of SCN^- group in frequency of 2050 cm^{-1} decreased after it mixed together with TMED and CuI. It indicates that other functional group from each compounds and metal Cu influence the SCN^-

nature. Furthermore, it is also possible for SCN^- act as counter ion for Cu^+ beside I^-. Because there are two peaks reveal in 461.97 and 428.22 cm^{-1} in e) spectrum, represent interaction of C and I. The decrease intensity also found in 1036.78 cm^{-1} as C-N signal [24]. The interaction of Cu^+ with nitrogen from both TMED and NH_4SCN may the cause of its decrease.

From the FTIR characterization, we assumed that the interaction between CuI and TMED is dominates in the coordination bonding between Cu^+ and lone pair electron from nitrogen in TMED then form complex compound. When NH_4SCN introduced to CuI-TMED, it has many probabilities of interaction. Ion SCN^- can play role whether as counter ion of Cu^+ or another ligand source beside TMED.

3.3. Surface Morphology

Figure 3 is the morphology of HTM observed with SEM instrument. In the absence of TMED, CuI apparently cannot fulfill all of the surface because the existence of porous in some area. After addition of TMED, there is no porous have been found. The reaction between CuI and bidentate ligand TMED yield complex compound that lead to formation of agglomerates as can be seen in the Fig. 2b. Crystallization of CuI, happened spontaneously in a short time. The chain in crystallite binding leaves empty porous which not good for SS DSSC application. Addition of TMED could prevent CuI from rapid crystallization and facilitate pores filling [17]. The incorporation of NH_4SCN into CuI-TMED prevents the large agglomeration of CuI-TMED. Therefore, relatively more distributed materials in the surface layer, smaller size aggregates, and no porosity was obtained.

a) b) c)

Figure 3. Morphology of CuI a) pretreatment 1000x magnification, b) after addition of TMED, and c) TMED: NH_4SCN 100x magnification

3.4. The Conductivity of HTM

The result of conductivity of all HTM measured with I V el kahfi depicted in Figure 4. It may conclude that the better conductivity achieved when 0.4 mL TMED added (0.29 Sm^{-1}), also the incorporation of NH_4SCN in 2:1 ratio (TMED: NH_4SCN) perform maximum conductivity (0.39 Sm^{-1}). This result is an agreement with the previous research, stated that both conductivity and photoenergy conversion were greatly improved with the addition of thiocyanate salt [25].

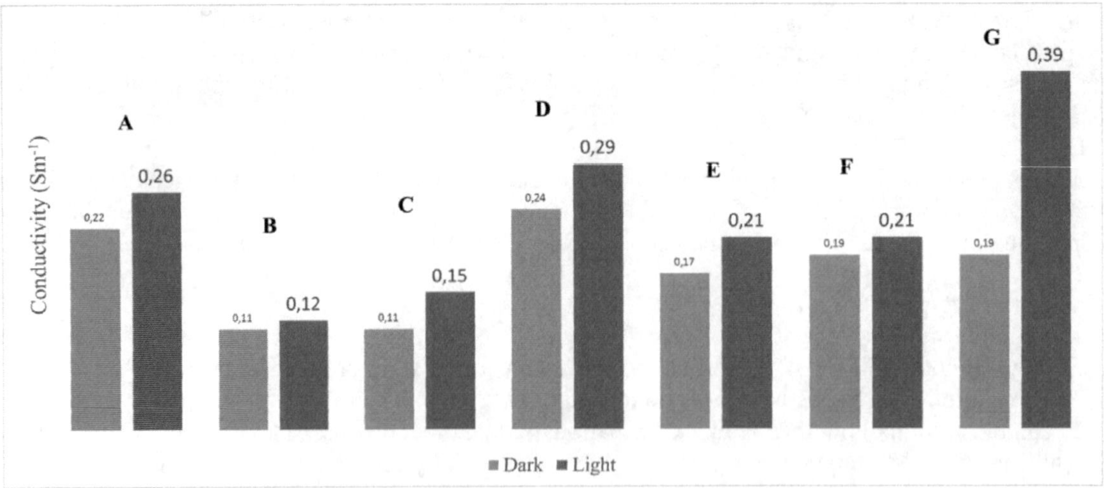

Figure 4. The conductivity value of A) CuI, B) CuI-TMED 0.1 mL, C) CuI-TMED 0.2 mL, D) CuI-TMED 0.4 mL, E) CuI-TMED: NH_4SCN (1:1), F) CuI-TMED: NH_4SCN (1:), G) CuI-TMED: NH_4SCN (2:1)

3.5. The Electrical Properties of SS DSSC

The result of I V measurement of SS DSSC conducted with Keithley is present in Table 1. The decrease of open circuit voltage as well as short circuit current is followed by the increase of conversion efficiency. The highest absorption capability, smallest band gap energy, highest conductivity, are contribute to the highest photoenergy conversion of CuI-TMED: NH_4SCN compared to other HTM. Both TMED and NH_4SCN provide intense interactions, that prevent CuI from fast crystallization, have an impact in pore filling, thus be able to improved conductivity, and photoenergy efficiency [13, 17].

Table 1. Electrical properties of SS DSSC

HTM	Voc (V)	Isc (10^{-3} mA)	η (10^{-3} %)
CuI	0,70	8,37	0,46
CuI-TMED	0,67	8,35	0,90
CuI-TMED: NH_4SCN	0,52	5,33	1,52

4. Conclusion

TMED compound introduced to the CuI solution, influence the optical characteristic which is shown from broaden absorption spectrum into visible area. The conductivity of the material was increased after addition of TMED due to the interaction that occur between these two compounds and the change of its optical characteristic. The addition of NH_4SCN into mixture CuI-TMED also be able to improve both optical characteristics and conductivity. The electrical properties of all HTM together in SS DSSC system with the use of TiO_2 as photoanode, N3 as dye, and platinum as counter electrode was monitored. The highest efficiency was achieved when HTM CuI-TMED: NH_4SCN utilized. Its efficiency value was increased 3.28 higher compared to the control CuI. The addition of organic compound (TMED) and thiocyanate salt (NH_4SCN) not only lead to better electrical properties, but also prevent the CuI deterioration which affects to the stabilization of SS DSSC system.

Acknowledgment
Authors express gratitude for The Ministry of Research Technology and Directorate General of Higher Education (KEMENRISTEK DIKTI), Indonesia for support this research in financial, through Hibah Mandatory Project.

References

[1] Diefenderfer J R, Conti J J, Paul D, James R, Napolitano S A, Michael A, Turnure J T, and Westfall L D 2015 *The Annual Energy Outlook 2015*
[2] Tob´ıas I, Canizo C, and Alonso J 2003 *Handbook of Photovoltaic Science and Engineering: Crystalline Silicon Solar Cells and Modules* New York John Wiley & Sons
[3] O'Regan B and Grätzel M 1991 *Nature* **353** 737-740
[4] Grätzel M 2006 *C. R. Chim.* **9** 578-583
[5] Yum J H, Chen P, Grätzel M, and Nazeeruddin M K 2008 *Chem. Sus. Chem.* **1** 699-707
[6] Yanagida S, Yu Y H, and Manseki K, 2009 *Acc.Chem. Res.* **42** 1827-1838
[7] Olson C, Veldman D, Bakker K, and Lenzmann F 2011 *Int. J. Photoenergy* Article ID 513089
[8] Tennakone K, Kumara G R R A, Kumarasinghe A R, Wijayantha K G U, and Sirimanne P M 1995 *Semicond. Sci. Technol.* **10** 1689-1693
[9] Tennakone K, Kumara G R R A, Kottegoda I R M, Perera V P S, Aponsu G M L P, and Wijayantha K G U 1998 *Sol. Energ. Mat. Sol. Cells* **55** 283-289
[10] Zhenzhen Y, Katherine C P, Jia L D, Yang R, and Tao X 2011 *J. Renew. Sustain. Ener.* 3 Article ID 063101
[11] Li B, Wang L, Kang B, Wang P, and Qiu Y 2006 *Sol. Energ. Mat. Sol. Cells* **90** 549-573
[12] Kumara G R R A, Konno A, Senadeera G K R, Jayaweera P V V, Silva D B R A D, and Tennakone K 2001 *Sol. Energ. Mat. Sol. Cells* **69** 195-199
[13] Sakamoto H, Igarashi S, Niume K, and Nagai M 2011 *Org. Electron.* **12** 1247-1252
[14] Kumara G R R A, Konno A, Shiratsuchi K, Tsukahara J, and Tennakone K 2002a *Chem. Mater.* **14**(3) 954-955
[15] Kumara G R R A, Kaneko S, Okuya M, and Tennakone, K 2002b *Langmuir* **18** 10493-10495
[16] Sakamoto H, Igarashi S, Uchida M, Niume K, and Nagai M 2012 *Org. Electron.* **13** 514-518
[17] Zainun R, Mamat M H, Noor U M, and Rusop M 2013 *Adv. Mater. Res.* **667** 447-451
[18] S Wahyuningsih, Ramelan A H, and Saputri L N M Z 2017 *J. Phys. Conf. Ser.* **75** 012002 1-6
[19] Johan M R, Siwen K, Hawari N, Azri N, and Aznan K 2012 *Int. J. Electrochem. Sci.* **7** 4942-4950
[20] Zainun R, Rusop M, and Noor U M 2010 *Electrical and Optical Properties of Nanostructured Copper(I) Iodide (CuI) Incorporate with Ligand Agent for DSSC* New Jersey IEEE Publisher
[21] Gaya U I 2014 *Heterogeneous Photocatalysis Using Inorganic Semiconductor Solids* New York Springer Dordrecht Heidelberg
[22] Beranek R and Kisch H 2008 *Angew. Chem. Int. Ed.* **47** 1320-1322
[23] Amalina M N, Rasheid N A, and Rusop M 2012 *J. Nanomater.* 1-6
[24] Stuart B 2004 *Infrared Spectroscopy: Fundamentals and Applications* Chichester John Wiley & Sons
[25] Konno A, Kitagawa T, and Kida H 2005 *Curr. Appl. Phys.* **5** 149-151

IOP Conf. Series: Materials Science and Engineering **333** (2018) 012030 doi:10.1088/1757-899X/333/1/012030

Effect of composition of chlorophyll and ruthenium dyes mixture (hybrid) on the dye-sensitized solar cell performance

D D Pratiwi[1], F Nurosyid[1,2], Kusumandari[1,2], A Supriyanto[1,2] and R Suryana[1,2]
[1]Physics Department of Graduate Program, Universitas Sebelas Maret, Indonesia
[2]Department of Physics, Faculty of Mathematics and Natural Sciences, Universitas Sebelas Maret, Indonesia

E-mail: nurosyid@yahoo.com

Abstract. The fabrication of dye-sensitized solar cell (DSSC) has been conducted by varying the composition of natural dye from moss chlorophyll (Bryophyte) and synthesis dye from ruthenium complex N719. The sandwich structure of DSSC consists of the working electrode using TiO_2, dye, electrolyte, and counter electrode using carbon. The composition of chlorophyll and synthesis dyes mixture were 100% and 0%, 80% and 20%, 60% and 40%, 40% and 60%, and 20% and 80%. The UV-Vis absorption spectra of moss chlorophyll showed the first peak in the wavelength range of 450-500 nm and the second peak at wavelength of 650-700 nm. The peak value of absorbance at wavelengths of 450-500 nm was 6.1004 and at wavelengths of 650-700 nm was 3.5835. The IPCE characteristic curves showed the absorption peak of photon for DSSCs occurred at wavelength of 550-650 nm. It considered that photon in this wavelength can contribute dominantly to produce the optimum electrons. The *I-V* characteristics of DSSCs with composition of chlorophyll and synthesis dyes mixture of 100% and 0%, 80% and 20%, 60% and 40%, 40% and 60%, and 20% and 80% resulted the efficiency of 0.0022; 0.0194; 0.0239; 0.0342; and 0.0414, respectively. It suggested that the addition of a little composition of the ruthenium complex dye into moss chlorophyll dye can increase the efficiency significantly.

1. Introduction

Organic solar cell is a type of photovoltaic that uses light energy (photons) from the sun. It also called renewable energy or alternative energy. It converts solar energy into eco-friendly electrical energy. One of the solar cell development is dye-sensitized solar cell (DSSC) by Grätzel. DSSC is solar cell based on a dye that can absorb the incident sunlight and uses the light energy to produce electrons transfer reaction. The advantages of DSSC are the simple and low cost to fabricate, the materials also easily obtained, and good performance for the future [1-3].

DSSC components include semiconductor layer on the Transparent Conductive Oxide (TCO) substrate as the anode, a catalyst on the counter electrode as the cathode, the dye as a photosensitizer, and redox electrolyte. One of the components that affect the performance of DSSC is the dye. The dye such as synthesis from complex compounds or natural pigments from the parts of a plant [2,4].

The researchers have been developed the synthetic and natural dyes in the DSSCs application. The type of synthesis dye is generally uses the ruthenium complexes. It can result the high efficiency, but it has high cost production, difficult to synthesize, and contains heavy metals that are not eco-friendly.

Therefore, the natural dye is alternatively as eco-friendly dye. Natural dye can be extracted from the parts of plant such as leaves, flowers, or fruit [5-11].

The research has been conducted in Japan used plant pigments as photosensitizer in the dye sensitized solar cell application. The research proved that natural dye can provide the photovoltaic effect. The pigments of plant parts consist of chlorophyll [12], anthocyanin [13-15], carotenes [16] or betalains [17,18].

This research used natural dye of chlorophyll extract from moss (Bryophyte) mixed with synthesis dye of ruthenium complex N719. The dyes mixture (hybrid) aim to improving the efficiency of DSSC. Chlorophyll is the main pigment in the photosynthesis of green plant which absorbs the maximum wavelength at 670 nm. Therefore, chlorophyll is an attractive natural pigment as photosensitizer in the visible spectrum [12].

The moss was used in this experiment because it easy to obtain in the land, water, or damp places. The moss plant can live with the minimum sunlight, but it can carry out own photosynthesis. The moss plant certainly has chlorophyll because it performs photosynthesis, so that it contains the many chloroplasts. This article will discuss the composition of dyes mixture on the *I-V* and Incident Photon to Current Conversion Efficiency (IPCE) characteristics of DSSC. The composition of dyes mixture plays vital role in the DSSC application. The function of dye absorbs the photons to produce the electrons excitation.

2. Experimental

Natural dye was extracted from moss chlorophyll (Bryophyte), while synthesis dye was obtained from ruthenium complex N719. The dyes mixture (hybrid) was obtained from both chlorophyll and synthesis. Concentration between chlorophyll and synthesis dyes were 100% and 0%, 80% and 20%, 60% and 40%, 40% and 60%, and 20% and 80%. The substrates used Fluorine-doped Tin Oxide (FTO) glass from Aldrich with a size of 2.0 cm × 2.5 cm and active cell area of 1.0 cm × 2.0 cm. The working electrodes were coated using TiO_2 nanopowder 21 nm from Aldrich by the spin coating method, the counter electrodes using carbon produced from pencil and candle soot. Electrolyte solution was made by mixing the KI and I_2 in the *Polyethylene glycol* (PEG) solution.

Characterization included UV-Visible Spectrophotometer Lambda 25 to determine the wavelength of dye absorbance, Keithley *I-V* meter to determine the efficiency of the DSSC prototypes, and Measurement System QEX7 Serial #150 to determine the Incident Photon to Current Efficiency (IPCE) of the DSSC prototypes.

3. Results and Discussion

3.1. Optical Characterization of Chlorophyll Dye Solution

The absorption spectra of chlorophyll dye solution use UV-Vis spectrophotometer. The result of UV-Vis characterization is shown in Fig.1. The absorption spectra of moss chlorophyll show the first peak in the wavelength range of 450-500 nm and the second peak at wavelength of 650-700 nm. The peak value of absorbance at wavelengths of 400-500 nm is 6.1004 and at wavelengths of 550-700 nm is 3.5835. These correlates with the pigment in the moss chlorophyll, it absorbs color spectrum of greenish blue (400-500 nm) and red (610-800 nm). It indicates that chlorophyll dye can be used as a photosensitizer of DSSC due to its absorbance in the visible spectrum.

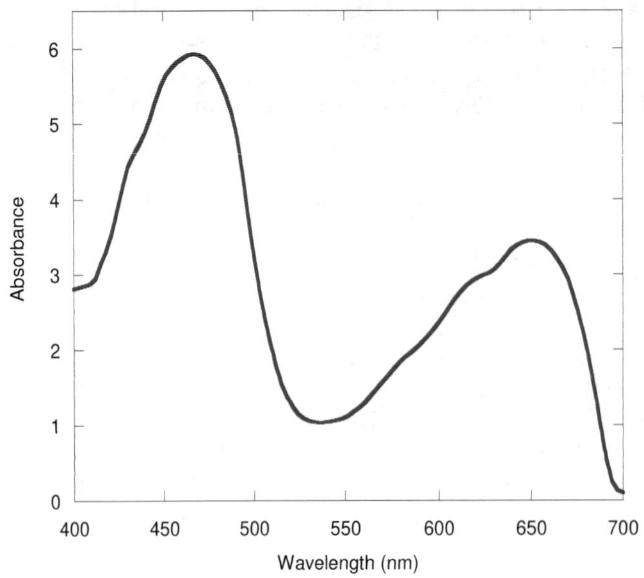

Figure 1. Absorption peak of moss chlorophyll dye

3.2. Incident Photon-to-Current Efficiency (IPCE) Characterization in the DSSC Application

IPCE characterization is used to determine the absorption of dye that produces the current in the DSSC devices. The IPCE characteristic shows a graph with quantum efficiency (QE) percentage versus wavelength (λ). QE is the comparison of electrons (currents) and photons which is given in the sample of certain wavelength. Figure 2 is the graph of absorbance and QE versus wavelength of chlorophyll dye.

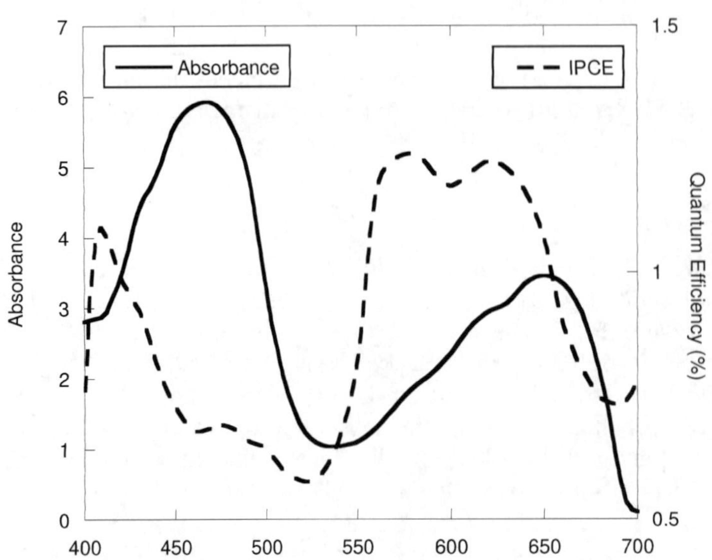

Figure 2. The graph of absorbance and quantum efficiency versus wavelength of moss chlorophyll dye

Figure 2 shows that the quantum efficiency (QE) is higher at the wavelength of 550-650 nm than the other wavelengths. It means that at the wavelength range of 550-650 nm, the absorption produces more current than another wavelength. It happens because the absorption produces the exciton (hole-electron pairs) and current. In the wavelength range of 450-500 nm, QE is lower than in the wavelength range of 550-650 nm in spite of its high absorbance of chlorophyll dye solution. It is considered when chlorophyll dye is exposed the light, it's not overall wavelengths can excite the electrons of the dye in DSSC.

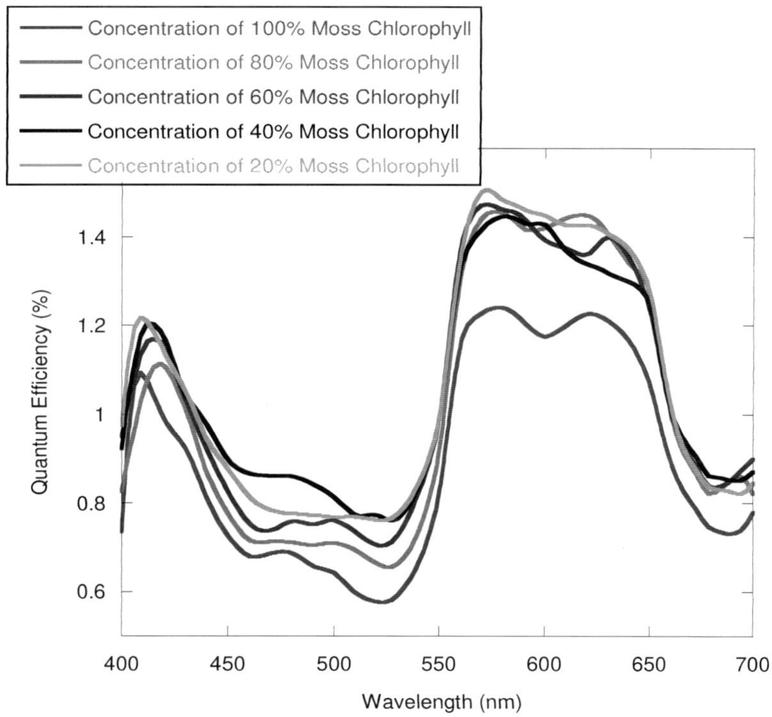

Figure 3. IPCE characteristics of DSSCs with the composition of moss chlorophyll and ruthenium dyes mixture

Figure 3 shows that DSSC with 100% moss chlorophyll dye has the lowest quantum efficiency. Peak absorbance of all DSSC prototypes occur at the wavelength of 550-650 nm. It exhibits that photon is absorbed by the dye in this wavelength produces high energy and photocurrent. The DSSC with more concentration of ruthenium dye increase the photons contribution, accordingly the efficiency is bigger. The efficiency will increase when the chlorophyll dye is added ruthenium dye. It can be seen in the green curve with sample of 100% chlorophyll dye has lowest QE, while the red curve shows that the efficiency increases at sample of 80% chlorophyll dye (20% ruthenium dye). The QE is higher as comparable to the addition of ruthenium dye into chlorophyll dye.

3.3. *I-V Characterization*

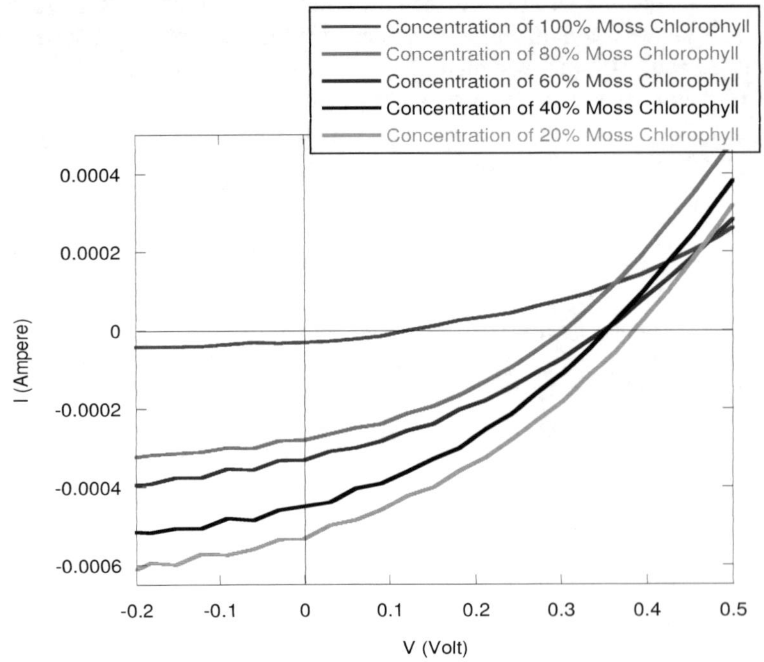

Figure 4. *I-V* characteristics of DSSCs with the composition of moss chlorophyll and ruthenium dyes mixture

Figure 4 is the *I-V* characteristics of DSSC prototypes with composition of moss chlorophyll and ruthenium dyes mixture. All curves show the diode characteristics correlating the p-n junction. The p-n junction consists of electron donor (n-type) and electron acceptor (p-type). Electron donor is dyes mixture, while electron acceptor is TiO_2 semiconductor. It is shown that ruthenium concentration affects the performance of DSSC. Concentration of 100% moss chlorophyll (0% ruthenium) has lowest current, while chlorophyll and ruthenium dyes mixture have higher currents. It indicates that chlorophyll and ruthenium dyes mixture absorb the photons dominantly as compared to moss chlorophyll dye. In addition, concentration of ruthenium dye will increase the DSSC efficiency.

Previous study [19] we have demonstrated how to obtain the open circuit voltage (V_{oc}), the current short circuit (I_{sc}),the voltage maximum (V_{max}), the current maximum (I_{max}), the fill factor (*FF*), the power maximum(P_{max}) and efficiency (η) from the *I-V* curves in Fig. 4. Table 1 represents those values for composition of moss chlorophyll and ruthenium dyes mixture.

Table 1. Photovoltaic performance of DSSCs with the composition of moss chlorophyll and ruthenium dyes mixture.

Characteristics	**Ratio of moss chlorophyll dye : ruthenium dye (%)**				
	100 : 0	**80 : 20**	**60 : 40**	**40 : 60**	**20 : 80**
V_{oc} **(Volt)**	0.1200	0.3202	0.3603	0.3602	0.3802
I_{sc} **(mA)**	0.0458	0.3120	0.3410	0.5200	0.6120
V_{max} **(Volt)**	0.1000	0.1604	0.2203	0.2403	0.2202
I_{max} **(mA)**	0.0400	0.2410	0.2170	0.2850	0.3760
P_{max} **(mW)**	0.0045	0.0387	0.0478	0.0684	0.0829
FF	0.7278	0.3877	0.3892	0.3656	0.3560
η **(%)**	0.0022	0.0194	0.0239	0.0342	0.0414

Table 1 shows that DSSC prototypes with increasing composition of ruthenium dye will improve the efficiency value. Addition of 20% ruthenium dye into chlorophyll dye, the efficiency of DSSC is increasing significantly. The DSSC with 100% chlorophyll dye has the lowest efficiency of 0.0022%. Then, it is added by 20% ruthenium dye can increases its efficiency of 0.0194 or up to 780%. Furthermore, the addition of composition of ruthenium dye (40%, 60%, and 80%) does not significantly increase the efficiency. It is only 20-25%. It proves the addition of a little ruthenium dye can improve the efficiency of moss chlorophyll dye. As another important aspect, it provides low production cost for DSSC application due to substitute dominants of ruthenium dye with concentration of natural dye (like moss chlorophyll).

4. Conclusion

In this research, the composition of chlorophyll and ruthenium dyes mixture (hybrid) has successfully increased the efficiency of DSSCs. Addition of a little ruthenium complex N719 dye of 20% into natural dye (moss chlorophyll) increased the efficiency significantly. As a result, the DSSC with 100% chlorophyll dye resulted the efficiency of 0.0022%, while addition of 20% ruthenium dye into chlorophyll dye, the efficiency increased 0.0194% or up to 780%.

Acknowledgment

This work was supported by Higher Education and Institute for Research and Community services, Universitas Sebelas Maret by MRG Grant Program (Contacts Number: 623/UN27.21/PP/2017)

References

[1] O'regan B and Grätzel M 1991 *Nature.* **353** 737
[2] Grätzel M 2003 *J. Photochem. Photobiol.* **4** 145
[3] Grätzel M 2005 *Inorg.Chem.* **44** 6841
[4] Chiba Y, Islam A, Watanab Y, Komiya R, Koide N and Han L 2006 **45** 1638
[5] Calogero G and Marco G 1999 *Inorg. Chem.* **38** 6298
[6] Wongcharee K, Meeyoo V and Chavadej S 2007 *Solar. Energy. Materials. Solar Cells.* **91** 566
[7] Roy M, Balraju P, Kumar M and Sharma G 2008 *Solar. Energy. Materials. Solar Cells.* **92** 909
[8] Ito S, Saitou T, Imahori H, Ueharad H and Hasegawad N 2010 *Energy. Environmental. Sci.* **3** 905
[9] Zhou H, Wu L, Gao Y and Ma T 2011 *J. Photochem. Photobiol.* **219** 188
[10] El-Agez T, El Tayyan A, Al-Kahlout A, Taya S A and Abdel-Latif M 2012 *Int. J. Materials. Chem.* **2** 105
[11] Taya S, El-Agez T, El-Ghamri H and Abdel-Latif M 2013 *Int. J. Materials. Sci. Appl.* **2** 37
[12] Amao Y, Yamada Y and Aoki K 2004 *J. Photochem. Photobiol. A. Chem.* **164** 47
[13] Wongcharee K, Meeyoo V and Chavadej S 2007 *Solar. Energy. Materials. Solar Cells.* **91** 566
[14] Calogero G, Yum H J, Sinopoli A, Di Marco G, Grätzel M and Nazeeruddin M K 2012 *Sol. Energy.* **86** 1563
[15] Garcia C G, Polo A S and Iha N Y M 2003 *J. Photochem. Photobiol. A. Chem.* **160** 87
[16] Yamazaki E, Murayama M, Nishikawa N, Hashimoto N, Shoyama M and Kurita O 2007 *Sol. Energy.* **81** 512
[17] Calogero G, Di Marco G, Cazzanti S, Caramori S, Argazzi R, Carlo A D and Bignozzi C T 2010 *Int. J. Mol. Sci.* **11** 254
[18] Martinez A R H, Esteves M, Vargas S, Quintanilla F and Rodriguez R 2012 *Int. J. Mol. Sci* **12** 5565
[19] Suryana R, Khoiruddin and Supriyanto A 2013 *Material. Sci. Forum.* **737** 15

Optimization Study of Hydrogen Gas Adsorption on Zig-zag Single-walled Carbon Nanotubes: *The Artificial Neural Network Analysis*

Nasruddin[1*], M Lestari[1], Supriyadi[2], Sholahudin[1]
[1] Department of Mechanical Engineering, Universitas Indonesia, Depok, Jawa Barat 16424, Indonesia
[2] Department of Mechanical Engineering, Universitas Trisakti, Jakarta Barat 11440, Indonesia

E-mail: nasruddin@eng.ui.ac.id

Abstract. The use of hydrogen gas in fuel cell technology has a huge opportunity to be applied in upcoming vehicle technology. One of the most important problems in fuel cell technology is the hydrogen storage. The adsorption of hydrogen in carbon-based materials attracts a lot of attention because of its reliability. This study investigated the adsorption of hydrogen gas in Single-walled Carbon Nano Tubes (SWCNT) with chilarity of (0, 12), (0, 15), and (0, 18) to find the optimum chilarity. Artificial Neural Networks (ANN) can be used to predict the hydrogen storage capacity at different pressure and temperature conditions appropriately, using simulated series of data. The Artificial Neural Network is modeled as a predictor of the hydrogen adsorption capacity which provides solutions to some deficiencies in molecular dynamics (MD) simulations. In a previous study, ANN configurations have been developed for 77k, 233k, and 298k temperatures in hydrogen gas storage. To prepare this prediction, ANN is modeled to find out the configurations that exist in the set of training and validation of specified data selection, the distance between data, and the number of neurons that produce the smallest error. This configuration is needed to make an accurate artificial neural network. The configuration of neural network was then applied to this research. The neural network analysis results show that the best configuration of artificial neural network in hydrogen storage is at 233K temperature i.e. on SWCNT with chilarity of (0.12).

1. Introduction

The most potential energy for future that environmental friendly with massive of quantity in this universe is hydrogen [1]. However, the problem is about how to keep hydrogen gas on storage cover up by single-walled carbon nano tube (SWCNT) as adsorbent [2, 3]. SWCNT is a structure of nano tube with one carbon graphene builder, which is optimum in adsorbed hydrogen with inside and outside of the nano tube as illustrated in Figure 1.

Figure 1. The hydrogen adsorption in SWCNT with inside and outside confirguration.

SWCNT rolled by three type rolling method as showed in Figure 2 [4]. It is result a different tube. SWCNT that rolled by zig zag model was selected for this research because more effective to adsorb hydrogen with it peaks form.

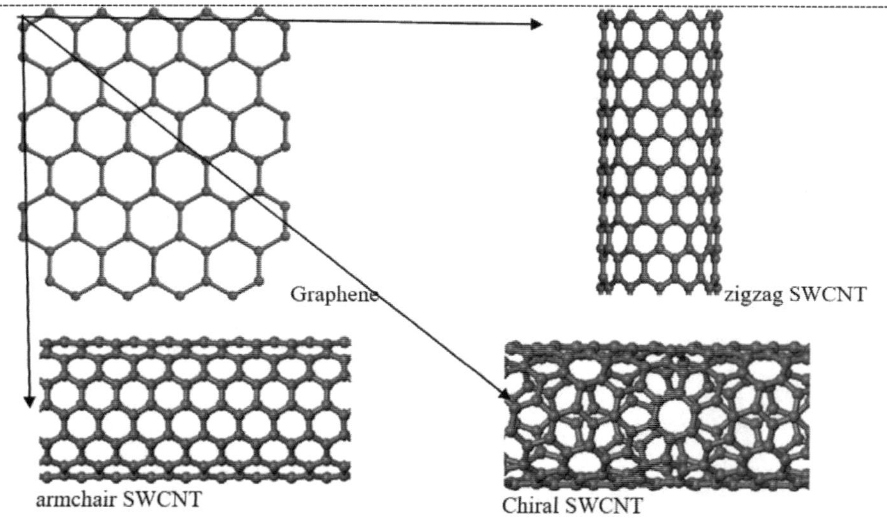

Figure 2. Step of SWCNT graphene sheet rolled into a tube [4]

With rolling zigzag method, SWCNT, as the adsorbent of hydrogen gas with various chilarity or diameter of tube giving difference result on the adsorption capacity [5]. In the figure 1, the molecular dynamic simulation of hydrogen adsorption by SWCNT is showed. Molecular dynamic simulation (MDS) is a simulation method that used for simulating the adsorbed hydrogen by CNT on storage [6, 7]. Actually, the experiment and simulation has conducted an observation of the adsorption hydrogen by SWCNT [3, 8, 9]. And there has shortcoming from both research methods. Then it is cover up by artificial neural network (ANN). Neural network is a tool on MatLab software data apps that one of its function is used for predict hydrogen storage capacity on different pressure condition precisely without conducting simulation [10].

In the previous research, ANN was modeled, as an effort to preparing ANN to be a smart prediction tool that resulting good output. There is a configuration of ANN is needed [11]. So in the previous research neural network configuration was observed [12]. The result is an ANN configuration to predict hydrogen storage capacity called weight percent (wtp) without simulation was found. This configuration is for three variances of hydrogen temperature storage which is 77k, 233k, and 298k resulting difference configuration but with same optimum neuron that is neuron 10 [13]. The configuration of ANN formed with 3 scale range variance for wtp data clarity, 2 type of training set and validation set election variance, and 4 neuron election [14]. The best configuration that is chosen from each temperature is the configuration that resulting wtp value closer to both the experiment and simulation wtp resulted. The configuration that formed shown in Table 1 – Table 3.

Table 1. The error result with scale of 100-1000 and first election variant at temperature 77K.

	(a). Error result of training output			(b). Error result of validation output		
	RMS	CV	R2	RMS	CV	R2
N3	0.140146	2.414148	0.999436	0.338206	6.149633	0.996284
N5	0.140146	2.414148	0.999432	0.318123	5.784464	0.9967209
N7	0.112371	1.935687	0.999635	0.343534	6.246516	0.996686
N10	0.111936	1.928196	0.99964	0.332033	6.037397	0.996374

The focus of attention from Table 1 is the value of coefficient of determination parameter (R2) and the best regression neuron that is neuron 10 i.e 0.999641. Then the graphs of target and validation data are presented in Figure 3.

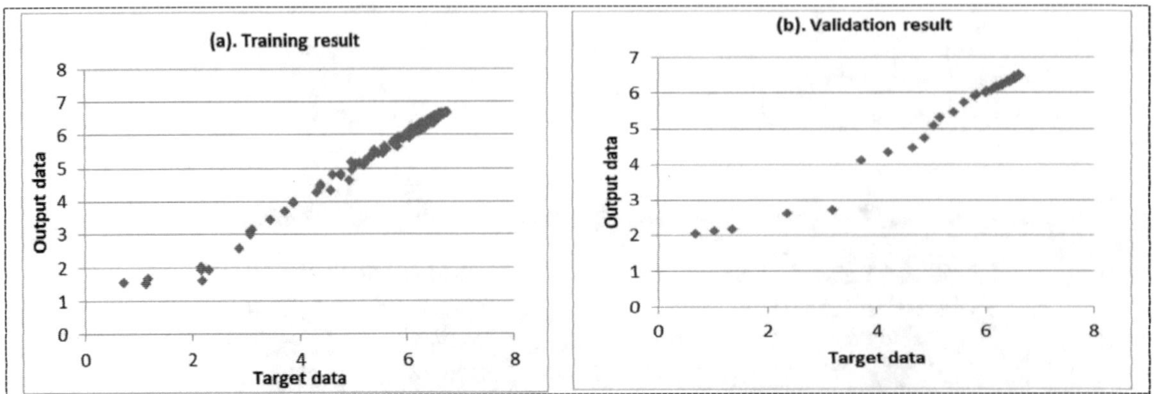

Figure 3. Regression result of training set and validation set at 77K[m3], (a). Training result with R2 value Test 0.9948, (b). Validation result with R2 value Test 0.97903.

The configuration for temperature of 233 K is shown in Table 2.

Table 2. The error result with scale of 100-10000 and first election variant at temperature 233K

	(a). Error result of training output			(b). Error result of validation output		
	RMS	CV	R2	RMS	CV	R2
N5	0.000957422	0.086474	0.998832	0.092853	9.040775	0.994739
N3	0.00375281	0.338953	0.995352	0.157781	15.37459	0.98192
N7	0.000830143	0.074978	0.998985	0.091785	9.814819	0.9949889
N10	0.000534122	0.048242	0.99935	0.190719	18.07104	0.974673

It can be shown in Table 2 the same result as 77 K temperature, that the best regression neuron is neuron 10 configuration. The graphs of target and validation data at temperature 233 K are presented in Figure 4.

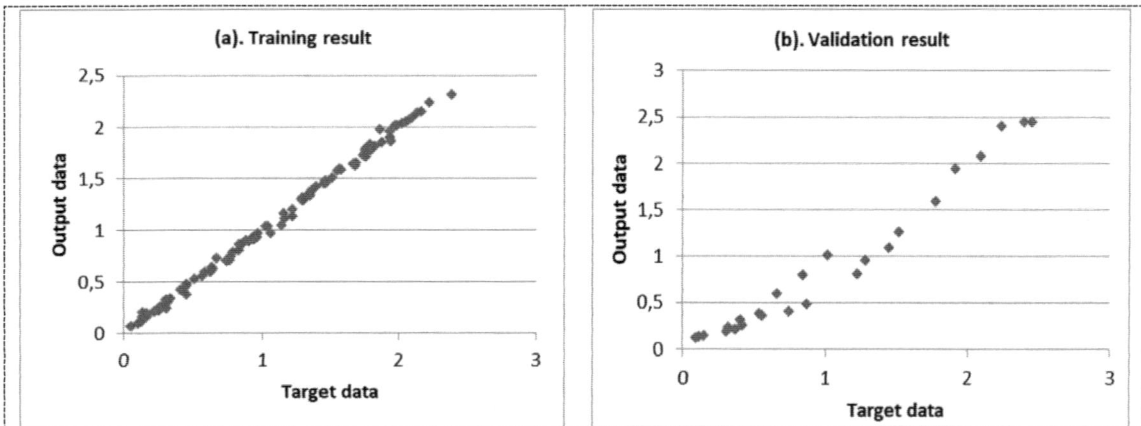

Figure 4. Regression result of training set and validation set at 233 K[m3], (a). Training result with R2 value Test 0.99808, (b). Validation result with R2 value Test 0.99819.

The result for temperature 298 K is presented in Table 3 and Figure 5. The result of the best configuration is also neuron 10.

Table 3. The error result with scale of 0-100 and second election variant at temperature 298K.

	(a). Error result of training output			(b). Error result of validation output		
	RMS	CV	R2	RMS	CV	R2
N3	0.055917	6.441731	0.996896	0.077435	8.337538	0.995263
N5	0.036758	4.230856	0.998657	0.066016	6.81065	0.996264
N7	0.030488	3.512508	0.999078	0.036559	3.904285	0.998822
N10	0.026574	3.060571	0.9993	0.120471	13.18335	0.984872

As the previous result from other temperature, on this temperature, neuron 10 showing it best regression where the R square closer to 1 than the other neuron.

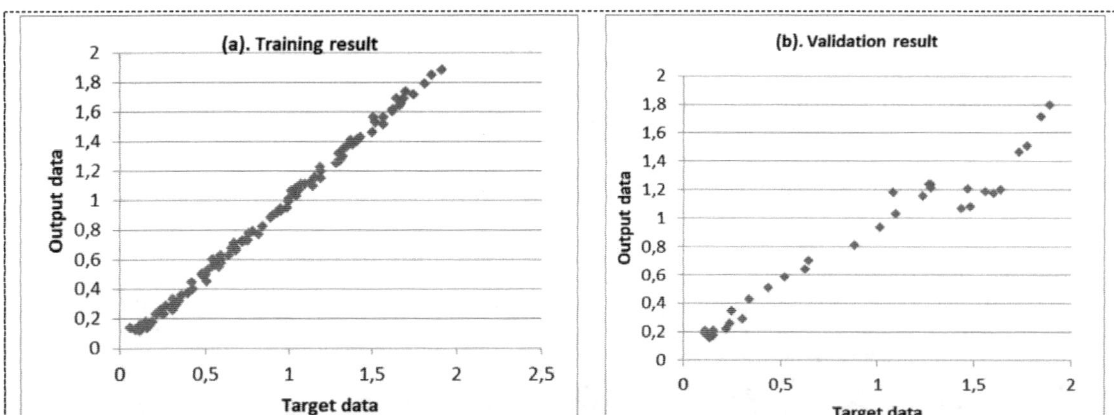

Figure 5. Regression result of training and validation at 298 K[m3], (a). Training result with R2 value of 0.99792, (b). Validation result with R2 value Test 0.99672.

IOP Conf. Series: Materials Science and Engineering **333** (2018) 012031 doi:10.1088/1757-899X/333/1/012031

In this research the configuration has been tested to predict wtp for ideal pressure. Hydrogen storage ideal pressure is the pressure that used in hydrogen storage in gas form. The pressure that conditioned to keeping hydrogen gas on storage has range 0-120 bar according to the previous research on hydrogen storage. More than 120 bar is limit closer to compressed tank pressure. It is unsuitable to implement the compressed tank on this purpose vehicle that is light weight vehicle, such as vehicle. Less than 0 bar or closer to -100 bar is the pressure on liquid gas storage. Hydrogen liquid storage absolutely no effective implemented for land transportation like vehicle.

(0,12), (0,15), and (0,18) chirality of SWCNTs were observed to find the optimum chirality to adsorb the hydrogen gas [15]. This chirality has used on previous research finding neural network configuration. And now this configuration is implemented to find the optimum chirality on optimum temperature in hydrogen adsorption.

2. Experimental

2.1. Computation method

With each best neural network configuration for each temperature, the testing method is used. On the three variances of temperature and chirality that used in the previous research that is temperature 77K, 233K, and 298K with (0,12), (0,15), and (0,18) chirality of SWCNT[16]. The required pressure is inputted with range 0-120 bar, the approachment of wtp resulted is analogically [15]. Data is grouped according with neural network characteristic.

2.2. Training ANN on MatLab

After the data has grouped, the procees is training on MatLab with neural network tools. It is applied for each chilarity on each temperature. For the final step is analyzing the optimum chirality on the optimum temperature consider from wtp resulted from the pressures inputted equally for dependent variables.

3. Result and Discussion

On chirality (0,12) that showing by Figure 6 (a), conditioning storage with temperature 77K wtp value going to be constant after pressure given more than 50 bar and increasing on the pressure more than 140 bar. Meanwhile, on temperature of 233K, wtp value going steadily increases according to the increasing of pressure value. And on 298K, the wtp is going to be constant after given pressure more than 10 bar.

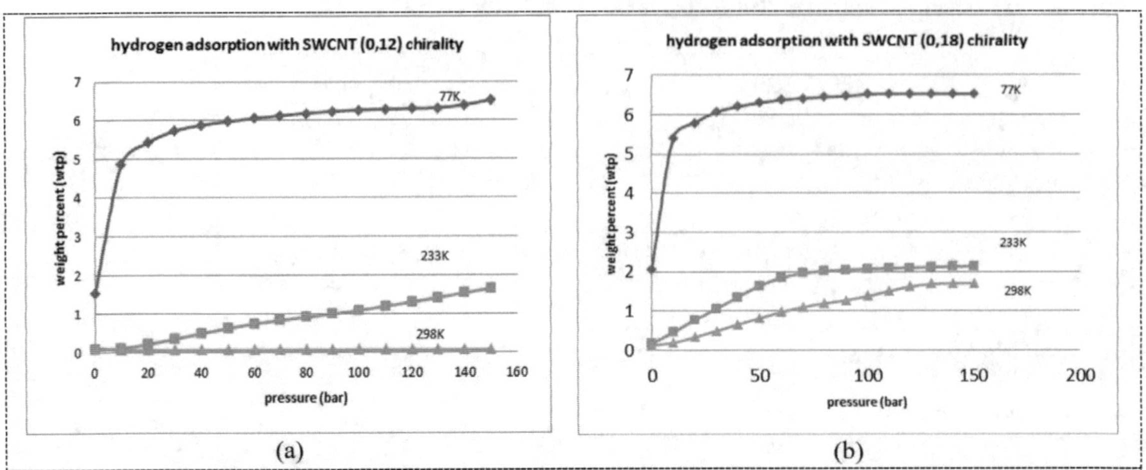

Figure 6. Weight percent (wtp) value with SWCNT; (a). (0,12) chirality, (b). (0, 18) chirality.

Figure 6 (b) shows graphic formed by chirality (0,15), conditioning storage with temperature 77K giving the wtp value highest than given by chirality (0,12). The weight percent (wtp) value chirality (0,15) on the pressure more than 30 bar has getting wtp more than 6, while the wtp on chirality (0,12) getting wtp value more than 6 on the pressure more than 70 bar. Then on temperature 233K wtp value is going constant on the pressure more than 100 bar. And on 298K, the wtp is going increase with a bit difference.

The last is graphic resulted by chirality (0,18) on Figure 8 showing that conditioning storage with temperature 77K, wtp 6 has got on pressure just more than 20 bar, it is easier to got then on chirality (0,15). Then on temperature 233K wtp value is going constant on the pressure more than 60 bar. And on 298K, same graphic on chirality (0,15) is resulted on chirality (0,18) too, the wtp is going to increase with a bit difference.

The results shown in Figure 6-8 are obtained from molecular dynamics simulations using the LAMMPS program developed by the Department of Energy (DoE) United States [7, 17, 18]. For setting the molecule in initial condition used packmol program [19].

4. Conclusion

Conditioning storage on 77K resulting wtp value that higher than other temperature. From Figure 6, Figure 7, and Figure 8 showing that conditioning storage with temperature 77K as increasing the chirality so the wtp value is easy to get with low pressure. In conclusion the recommended temperature and chirality that optimum to adsorb hydrogen from this observation is use SWCNT chirality 18 with storage temperature on 77K. Regression result of training set and validation set at 298 K[N3], (a). Training result with R^2 value Test 0.99792, (b). Validation result with R^2 value Test 0.99672.

References

[1] Srinivasa M S and Kumar E A 2014 *J. App Thermal Eng.* **72(2)** 176-189
[2] Zhao L, Cheng J R and Huang D C 2004 *Chinese J of Chem. Phys.* **17(5)** 572-576
[3] Barghi S H, Tsotsis T T, and Sahimi M 2014 *Int. J. Hyd. En.* **39(3)** 1390-1397
[4] Dai H 2002 *J. Surf. Sci.* **500(1)** 218-241
[5] Nasruddin, Kosasih E A, Budhi K, supriyadi, Zulkarnain I A 2016. *IJTech* **7(2)** 264-273
[6] Peng L 2011 *Atomistic Modeling of Hydrogen Storage in Nanostructured Carbons* (PhD Diss-University of Tennessee)
[7] Plimpton, S 1995 *J. comp. phys.* **117(1)** 1-19
[8] Tarasov BP, Maehlen JP, Lototsky MV, 2003 *J. alloys comp.* **356** 510-514
[9] Klechikov AG, Mercier G, Merino P, Blanco S, Cesar M and Alexandr VT 2015 *J. Micr. Mesop. Mat.* **210** 46-51
[10] Sholahudin S and Han H 2016 *Energy.* **115** 1672-1678
[11] Malinova T and Guo Z 2004 *Mat. Sci. Engi.* **365(1)** 219-227
[12] Yao H M, Vuthaluru H B, Tade M O and Djukanovic D 2005 *Fuel.* **84(12)** 1535-1542
[13] Yıldız, Z. and H. Uzun 2015 *J. Micr. Mesop. Mat.* **208** 50-54
[14] Tian q, Zhang Y, Wu Y, Tan Z 2009 *Int. J. Hyd. En.* **34(4)** 1931-1936
[15] Supriyadi 2017 *Optimization of Doping Elements on Carbon-based Nanostructures as Hydrogen Storage Sedia: Molecular Thermodynamics Analysis*, (PhD Diss-Universitas Indonesia)
[16] Nasruddin, Kosasih E A, Supriyadi, Zulkarnain I A 2016 *J. Mech. Eng. Mechatron.* **1(01)**
[17] Banerjee S and Puri I K 2008 *Nanotechnology.* **19(15)** 155702
[18] Pei Q, Zhang Y, and Shenoy V 2010 *Carbon.* **48(3)** 898-904
[19] Martínez L, Andrade R, Birgin E G AND Martínez J M. 2009 *J. comp. chem.* **30(13)** 2157-2164

IOP Conf. Series: Materials Science and Engineering **333** (2018) 012032 doi:10.1088/1757-899X/333/1/012032

The Influence of NiO Addition in TiO_2 Structure and Its Photoactivity

S Wahyuningsih[1*], A H Ramelan[1], P D Purwanti[1], H Munawaroh[1], S Ichsan[1], Y R Kristiawan[1]
[1]Inorganic Material Research Group, Faculty of Mathematic and Natural Sciences, Universitas Sebelas Maret, Surakarta, Indonesia

E-mail: sayekti@mipa.uns.ac.id

Abstract. The synthesis of TiO_2 together with the TiO_2-NiO composite using various annealing temperatures has been studied. The synthesis of TiO_2 was performed by sol gel method using Titanium Tetra Isopropoxide (TTIP) precursor, whereas the synthesis of TiO_2-NiO composite was done by wet impregnation method using $NiNO_3.6H_2O$ precursor. This study aims to determine the influence of NiO addition in its structure and photoactivity. The diffraction of synthesized TiO_2 at 400 °C temperature shows anatase TiO_2 peak at $2\theta = 25.35$ °. The addition of NiO dopant to the synthesis of TiO_2 process is carried out by annealing at 300 °C, 400 °C, 500 °C, 600 °C, and 700 °C, respectively. The TiO_2-NiO composite has been prepared and shows the diffraction peak of NiO at 2θ=43° about 33.08 to 36.68%. The optimum result of Rhodamine B photodegradation with TiO_2 was 43.15%, while the optimum result of Rhodamine B degradation with TiO_2-NiO composite was 92.85%.

1. Introduction

The utilization of inorganic materials as catalyst has grown quite rapidly. The catalyst that has been used in photochemical processes, or better known as photocatalysts. Photocatalyst is a mixture of photochemical processes and a catalyst to carry out a reaction chemistry that takes place on the surface of the charged semiconductor catalyst light. Titanium dioxide (TiO_2) is one of the catalysts that have been applied as a photocatalyst semiconductor that has photoactivity and high chemical stability is commercially available. The advantages of TiO_2 include: inert, capable of absorbing ultraviolet light well, high chemical stability, large pH intervals (0 to 4), resistant to photodegradation, insoluble in reaction either biologically or chemically, non-toxic, has redox properties, has a wide band gap of 3.2 eV and relative cheap [1, 9, 12]. The structure of TiO_2 exists in three forms that are rutile, anatase, and brookite. Rutile is a stable form of TiO_2, whereas the form of anatase and brookite is a stable meta structure [2]. Anatase has been shown to have good photocatalytic activity for degradation of various organic compounds, including dyes [3,8]. Anatase was formed at temperature 120-500 °C while after 700 °C it changed into rutile formed with a highly decreasing of photocatalytic activity [4].

In order to improve the TiO_2 crystal structure a doping modification was conducted that inhibiting the electron-hole recombination process on the surface of TiO_2 during the catalytic process. According to Ezema et al. (2008) [5] microstructure, surface area, particle size distribution, and dopant distribution may affect the photocatalytic activity of TiO_2. The transition metal oxide has a character can be used as a catalyst, since the transition metal oxide has *d* orbitals on its partially filled metal ion. It can capture electrons from reactants and forming a strong bond, to activate the reacting species.

Metal oxides often used as catalysts include oxides V, Cr, Mn, Fe, Co, Ni and Cu. Ni^{2+} is transition metal ion that has high crystallinity, low valence and can enhance photocatalytic activity of photocatalyst semiconductors [6, 11].

Nickel oxide (NiO) is a semiconductor material formed from transition metal ions Ni^{2+} in the form of metal oxide. NiO has been used as a sensor material with an energy band gap of 2.1-3.2 eV and showing the photoelectrochemical response at visible region. This material was combined with TiO_2 will produce a composite TiO_2-NiO which can degrade organic compounds when subjected to light. The photocatalytic composites were irradiated for photocatalytic activity by visible light sources to activate the catalyst. Several studies have used photocatalysts on dye waste treatment. Rhodamine B is one of synthetic dye which is used for textile, paint, and paper industries. This dye can causing irritation and is a carcinogenic substance [7].

In this study have been composite TiO_2 semiconductors and NiO metal oxides to improving activity of TiO_2 at visible light and maintain the structure of TiO_2 anatase to transform a rutile when annealing in order to degrade the Rhodamine B dye effectively.

2. Experimental

2.1. Preparation of TiO_2

TiO_2 was synthesis by sol gel method from titanium tetra-iso-propoxide (TTIP) precursor. Firstly TTIP solution was hydrolyzed with excess acetic acid to maintain suspension of particles against agglomeration. The TTIP was mixed with glacial acetic acid at pH 2 in low temperature at least 30 minutes to form TiO_2 transparent at 90 °C to formed TiO_2 sol. TiO_2 sol was heated at 150 °C for 24 hours (overnight) to formed xerogel TiO_2 colored white as a standard TiO_2.

2.2. Preparation of TiO_2-NiO composite

1.145 grams of xerogel TiO_2 was added to 0.81 gram of $NiNO_3.6H_2O$ which has been dissolved in 25 ml of distilled water and mixed by stirrer about an hour to form a homogeneous solution. The mixture was heated at 110 °C for 3 hours in oven. Then, the precipitated was annealed at a temperature of 150 °C, 300 °C, 400 °C, 500 °C, 600 °C, and 700 °C, respectively.

2.3. Characterization of TiO_2 and TiO_2-NiO composite

Sample of TiO_2 (150, 300, 400, 500, 600, and 700 °C) and TiO_2-NiO (150, 300, 400, 500, 600, and 700 °C) composite were analyzed by using X-ray Diffraction (Shimadzu XRD-6000) to determine the crystal phase at 2θ = 20-60 degree.

2.4. Photodegradation of Rhodamine B

Samples of TiO_2 and TiO_2-NiO composite were placed to an erlenmeyer, each 0.01 gram of TiO_2 and TiO_2-NiO composite was added 25 ml of Rhodamine B 5 ppm. Further, the solution was irradiated using visible light source lamp in the reactor for 0, 30, 120, 240, and 360 minutes. After irradiation process, each sample was decanted to obtain clear solutions. The remaining Rhodamine B solutions were analyzed with a UV-Vis spectrophotometer at wavelength between 500-600 nm.

3. Results and Discussions

TiO_2 is relatively fine grain or white powder, with good crystallinity which has three phases: rutile, anatase and brookite. The synthesized TiO_2 have been prepared using TTIP as a precursor into excess glacial acetic acid solution (pH=2) under below room temperature. Addition of TTIP into 98.5% of glacial acetic acid was performed at low temperatures of about 10 -15 °C, it was proposed to restrain the fast agglomeration of TTIP produced a white gel. Sol gel formation was conducted by evaporate the glacial acetic acid at a temperature of 90 °C. In order to find the annealing temperature optimum to form TiO2 anatase phase, annealing was conducted in a temperature of 150 °C, 300 °C, 400 °C, 500 °C, 600 °C, and 700°C, respecctively. The diffraction pattern in Figure 1 have been determined the TiO_2 anatase at various annealing treatment.

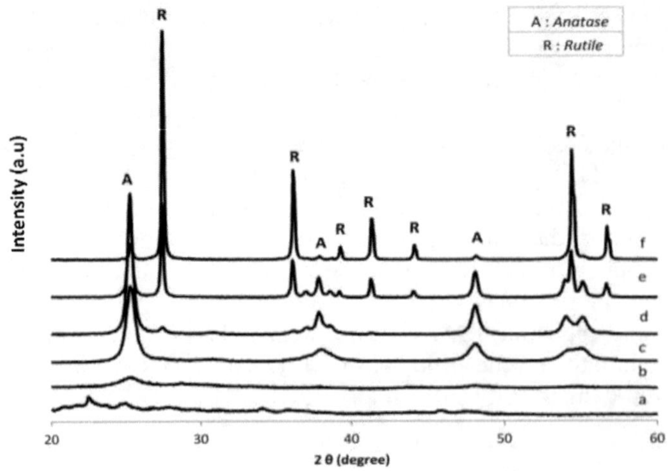

Figure 1. X-Ray Diffractogram of TiO_2 annealed at temperature of (a) 150 °C (b) 300 °C (c) 400 °C (d) 500 °C (e) 600 °C (f) 700 °C

The X-Ray diffraction in Figure 1 (a) and Figure 1 (b) shows the peak of TiO_2 at an annealing temperature of 150 °C and 300 °C, which is still in the amorphous phase, only small crystals have been formed. Figure 1 (c) further shows the TiO_2 diffractogram at a temperature of 400 °C so that the annealing temperature produces the highest anatase phase and no rutile phase occurs. This conclusion is supported by the diffraction peak at 2θ = 25,35° (d_{101} = 3,5091Å), 2θ = 37,95° (d_{004} = 2,3679 Å), 2θ = 48,15° (d_{200} = 1,8874 Å), 2θ = 54.11° (d_{105} = 1,6931 Å) And 2θ = 55.01° (d_{211} = 1,6674 Å) which is a characteristic of TiO2 in accordance with JCPDS 782-486.

Based on XRD results, TiO_2 anatase 100% will be produced by annealing treatment at 400 °C. A/R ratio calculation have been resumed in Table 1.

Table 1. Anatase rutile ratio (A/R) of TiO_2 crystal synthesized in various annealing temperature

Ratio	Annealing of TiO_2 (°C)			
	400	500	600	700
A:R	1:0.0	1:0.064	1:0.716	1:9.929

These ratios show that at temperature 400 °C was formed TiO_2 pure anatase phase while increasing the annealing temperature caused the transformation of anatase to rutile.

TiO_2-NiO composites with various annealing can be seen in Figure 2. In Figure 2 (a) and Figure 2 (b), the overall spectra of XRD did not show the existence of TiO_2 due to the amorphous phase. The presence of new peaks in Figure 2 (c) - Figure 2(f) represented NiO at 2θ = 43° (d_{200} = 2.08 Å) according to JCPDS No. 471-049. That is also produced $NiTiO_3$ which appears at 600 and 700 °C along with NiO formation in high temperature. The peaks of $NiTiO_3$ represented at 2θ = 24° (d_{012} = 3.6584 Å), 2θ = 33° (d_{110} = 2.6911 Å), 2θ = 49° (d_{024} = 1.8356 Å), and 2θ = 57° (d_{018} = 1.5969 Å) in according with JCPDS number 753-757.

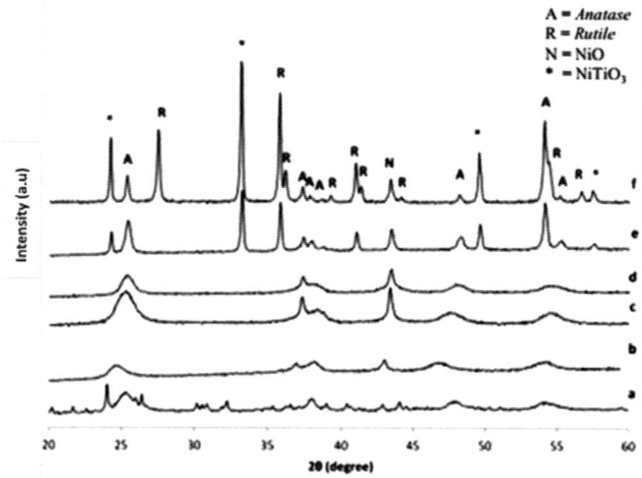

Figure 2. The X-ray Diffractogram of the composite TiO_2-NiO (a) 150 °C (b) 300 °C (c) 400 °C (d) 500 °C (e) 600 °C (f) 700 °C

The results of various annealing temperatures obtained that the ratio of anatase has been smaller due to increasing temperatures and it is mean the ratio of rutile has been greater (Table 2). While the additions of NiO dopant were inhibited anatase transform into rutile. It can explain by comparing the results without the addition of dopants (Table 1). The occurrence of such structural changes influenced the photoactivity of TiO_2, further information be explained in the activation of TiO_2-NiO composite degrade the Rhodamine B.

Table 2. Anatase ratios: rutile (A/R) on TiO_2-NiO composite result with various annealing.

Ratio	Annealing of TiO_2-NiO (°C)			
	400	500	600	700
A:R	1:0.20	1:0.21	1:0.41	1:1.75

The activity of TiO_2 and TiO_2-NiO composites for Rhodamine B photodegradation was performed by experiments at various temperatures and annealing times. Rhodamine B degradation results have shown in Figure 3, Figure 4, Figure 5 and Figure 6 for TiO_2 and Figure 7, Figure 8, Figure 9 and Figure 10 for TiO_2-NiO composites. Each of samples was irradiated by Halogen lamps for 30, 120, 240, and 360 minutes, respectively. Rhodamine B has been degraded by TiO2 at different time points and shows that the increase in irradiation time has further decreased Rhodamine B due to increasing electron excitation as an initiator in photodegradation process.

Table 3. Degradation result of Rhodamine B using TiO_2 in various irradiation times.

TiO_2 sample	Percentage of degradation (%)			
	30 min	120 min	240 min	360 min
TiO_2 400	1.88	9.72	31.91	43.15
TiO_2 500	2.46	6.73	14.53	37.97
TiO_2 600	12.03	14.87	24.95	30.73
TiO_2 700	6.20	7.40	12.65	14.04

IOP Conf. Series: Materials Science and Engineering **333** (2018) 012032 doi:10.1088/1757-899X/333/1/012032

Table 4. Degradation result of Rhodamine B using TiO_2-NiO in various irradiation times.

TiO_2-NiO sample	Percentage of degradation (%)			
	30 min	120 min	240 min	360 min
TiO_2-NiO 400	2.32	4.95	5.45	11.32
TiO_2-NiO 500	2.82	4.86	5.32	8.60
TiO_2-NiO 600	0.80	6.64	8.94	28.99
TiO_2-NiO 700	90.63	90.86	91.58	92.85

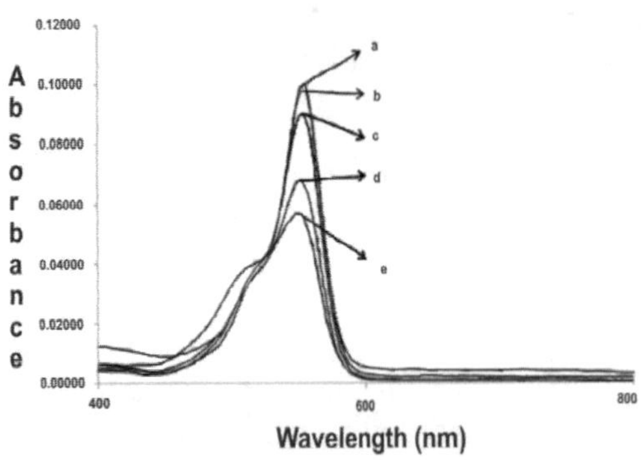

Figure 3. The UV Vis absorption spectrum of Rhodamine B degradated by TiO_2 400 at irradiation time of (a) 0 (b) 30 (c) 120 (d) 240 (f) 360 minutes

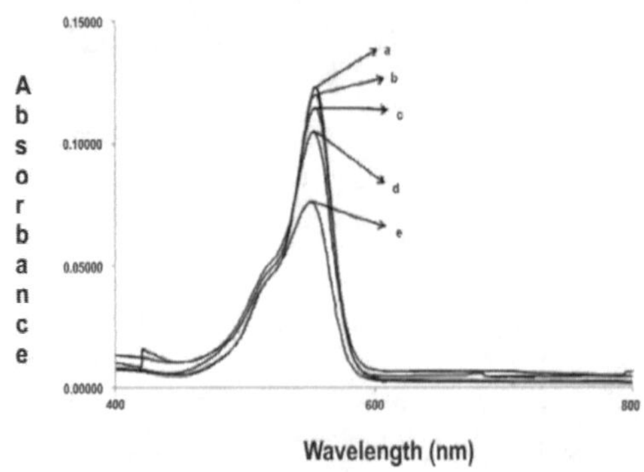

Figure 4. The UV Vis absorption spectrum of Rhodamine B degradated by TiO_2 500 at irradiation time of (a) 0 (b) 30 (c) 120 (d) 240 (f) 360 minutes

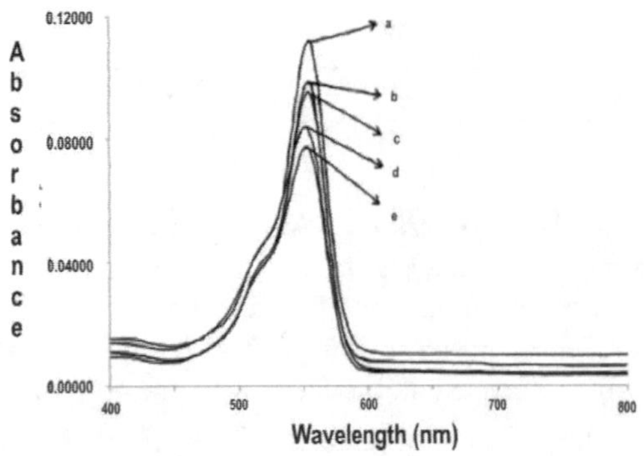

Figure 5. The UV Vis absorption spectrum of Rhodamine B degradated by TiO_2 600 at irradiation time of (a) 0 (b) 30 (c) 120 (d) 240 (f) 360 minutes

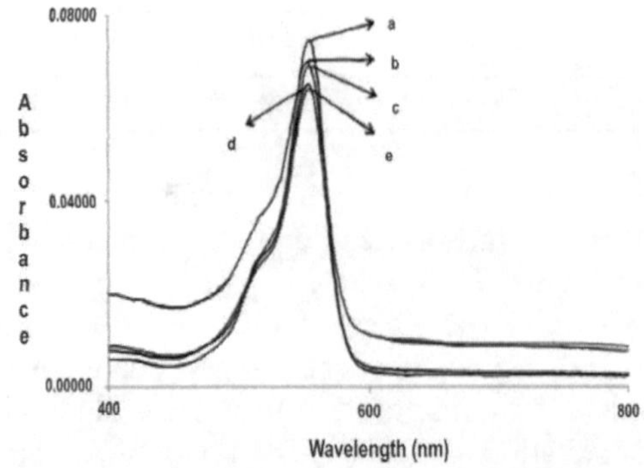

Figure 6. The UV Vis absorption spectrum of Rhodamine B degradated by TiO_2 700 at irradiation time of (a) 0 (b) 30 (c) 120 (d) 240 (f) 360 minutes

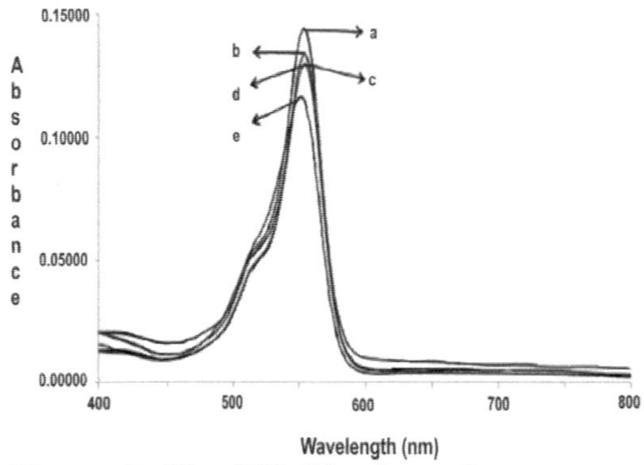

Figure 7. The UV Vis absorption spectrum of Rhodamine B degraded by TiO_2-NiO 400 at irradiation time of (a) 0 (b) 30 (c) 120 (d) 240 (f) 360 minutes

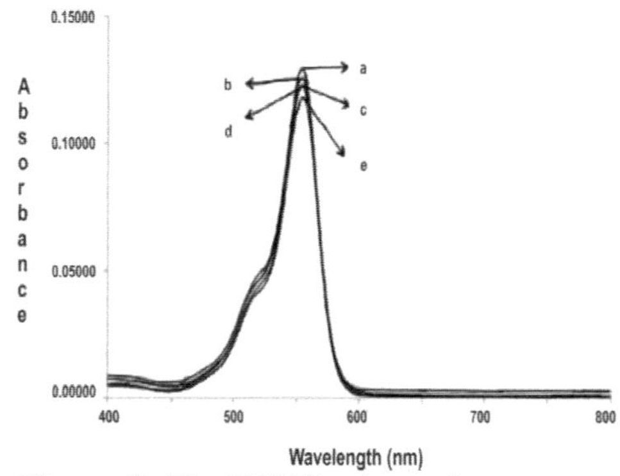

Figure 8. The UV Vis absorption spectrum of Rhodamine B degraded by TiO_2-NiO 5 00 at irradiation time of (a) 0 (b) 30 (c) 120 (d) 240 (f) 360 minutes

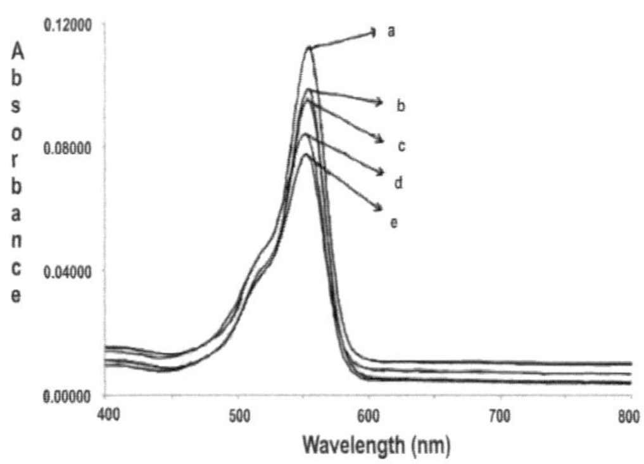

Figure 9. The UV Vis absorption spectrum of Rhodamine B degraded by TiO_2-NiO 600 at irradiation time of (a) 0 (b) 30 (c) 120 (d) 240 (f) 360 minutes

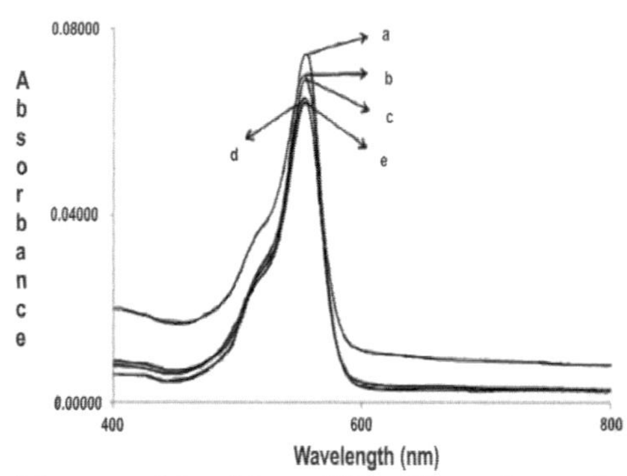

Figure 10. The UV Vis absorption spectrum of Rhodamine B degraded by TiO_2-NiO 700 at irradiation time of (a) 0 (b) 30 (c) 120 (d) 240 (f) 360 minutes

From this evidences, it can be assumed that the presence of NiO in TiO_2-NiO composite influenced the Rhodamine B degradation as shown in Table 4. Degradation increases with increasing time of irradiation.

4. Conclusion

In conclusion, the presence of NiO in the TiO2-NiO composite was able to improve the highest photodatal photodegradation of Rhodamol B. Material photocatalyst was a TiO_2-NiO 600 composite with the result of Rhodamine B degradation reaching 92.85% at 360 minutes irradiation time.

References

[1] Riyas S, Yasir V A, and Das P N M 2002 *Bull. Mater. Sci.* **25** 267-273
[2] Riyas S, Krishnan G, and Das P N M 2008 *J. Braz. Chem. Soc.* **19** 1023-1032
[3] Linsebigler A L, Lu G, and Yates J T Jr 1995 *Chemical Reviews.* **95** 735-758
[4] Begum N S, Ahmed H M F, and Gunashekar K R 2008 *Bull. Mater. Sci.* **31** 747–751
[5] Ezema F I, Ekwealor A B C, and Osuji R U 2008 *Superficies y vacto* **21** 6-10
[6] Karami A 2010 *J. Iran. Chem. Soc.* **7** 154-160
[7] Di Li, Hanedaa H, Hishita S, Ohashi N, and Labhsetwar N K 2005 *J. Fluor. Chem.* **126** 69–77

[8] Ruslimie C A, Razali H, and Khairul W M 2011 *Sains Malaysia* **40** 897-902
[9] Traistaru G A, Covaliu C I, Matei V, Cursaru D, and Jitaru I 2011 *Digest J. Nanomat. Biostruct.* **6** 1257-1263
[10] Wang P, Wang D, Li H, Xie T, Wang H, and Du Z 2007 *J. Coll. Interface Sci.* **314** 337-340
[11] Wei Y L and Chen K W 2009 *J. Vac. Sci. Techno* **23** 1385-1387
[12] Reddy K, Reddy C, and Manorama S 2001 *J. Sol. State Chem.* **158** 180-186

IOP Conf. Series: Materials Science and Engineering **333** (2018) 012033 doi:10.1088/1757-899X/333/1/012033

The Influence of Fe_2O_3 Addition on the Tio_2 Structure and Photoactivity Properties

S Wahyuningsih[1], A H Ramelan[1], L Prasetyawati[1], L N M Z Saputri[1], S Ichsan[1], Y R Kristiawan[1]

[1]Inorganic Material Research Group, Faculty of Mathematic and Natural Sciences, Sebelas Maret University, Surakarta, Indonesia

E-mail: sayekti@mipa.uns.ac.id

Abstract. The influence of Fe_2O_3 addition on the TiO_2 structure and photoactivity properties have been studied. The addition of Fe_2O_3 on the TiO_2 done by TiO_2-Fe_2O_3 synthesized with variation of annealing temperature. The result showed that peak of anatase TiO_2 at $2\theta = 25.35°$ and Fe_2O_3 at $2\theta= 54.20°$. The XRD of TiO_2 show annealing temperature at 400°C is anatase phase and the composite with annealing at temperature 150°C, 300°C, 400°C and 500°C is crystalline anatase phase, due to the addition of Fe_2O_3. Photodegradation of Rhodamin B with TiO_2 at 400°C annealing temperature showed optimum degradation 36.2 %, and the composite with annealing at 400°C showed optimum degradation 44.3% for 300 minutes irradiation.

1. Introduction

Titanium dioxide (TiO_2) is a semiconductors that have photocatalytic activity. TiO_2 has three distinct crystalline structures i.e. anatase, rutile, and brookite. The synthesis of TiO_2 has been widely studied, Yan (2005) have prepared TiO_2 using $TiCl_4$ [1], Libanori (2009) performed TiO_2 using Titanium Tetraisopropoxide (TTIP) precursor by sol-gel technique [2], Lu (2010) synthesized TiO_2 using TTIP precursors [3], Wahyuningsih (2007) synthesized TiO_2 with TTIP sol gel method [4]. Among the frequently used methods for the synthesis of TiO_2, the sol-gel technique is an easy preparation than others. TiO_2 anatase has a good photocatalytic activity for the degradation of various organic compounds, including dyes. However, TiO_2 anatase can be irreversibly transformed into rutile in high temperatures. Therefore, the modified TiO_2 in order to form stable TiO_2 is required.

The modifications to form stable TiO_2 is addition of metal oxide into the molecule TiO_2 known as doping. The addition of dopant aims to inhibit the formation of rutile at high temperatures. According to Rao (1959) the presence of impurities (dopant) inhibits anatase-rutile transformation [5]. The presence of the dopant can increase the activity and catalytic efficiency of TiO_2. The performance of dopant can increase the uptake of TiO_2 in the visible region, thus enhancing the photocatalytic. Some types of metal oxides such as Fe_2O_3, CuO, and MnO_2 can be used as dopants. Riyas (2002) has study of comparing two different methods of co-precipitation and wet-impregnation in the synthesis of TiO_2 with Fe_2O_3 addition. The result indicated that Fe_2O_3 can affect the transformation of TiO_2 and obtained rutile crystal structure by co-precipitation method, and anatase crystalline structure with wet-impregnation method at temperature 700°C and 900°C [6]. Music (2004) explained the transformation of β-FeOH to α- Fe_2O_3 was begin at temperature of 300°C and formed α- Fe_2O_3 crystalline phase at temperature of 400°C [7].

In this research, the synthesize of TiO_2 and addition of Fe_2O_3 to know the change of crystal structure of TiO_2. The addition of Fe_2O_3 in variation of temperature can maintain the stability of TiO_2 crystal structure in anatase phase to increase the photocatalytic activity, which is used to degradation of Rhodamin B.

2. Experimental

2.1. Synthesis of TiO_2

CH_3COOH 98.5% (pH=2) was added 10 mL TTIP solution in water bath at ±10°-15°C till formed white solution. The mixture was stirred for 30 minutes formed transparent solution. The solution was heated at ±90°C to reduce CH_3COOH until sol gel form. The sol gel was heated at 150°C overnight and formed white TiO_2 powder. Then the TiO_2 was annealed at various temperature i.e. 150 °C, 300 °C, 400 °C, 500 °C, 600 °C dan 700 °C for 4 hours [4].

2.2. Synthesis of TiO_2-Fe_2O_3 composites

$Fe(NO_3)_3.9H_2O$ 5.05 gram was added 25 ml aquadest and stirred until homogeneous. The mixture was added 11.45 g TiO_2 and stirred for an hours to orange colour. Then the sample was heated at 110°C overnight and obtained reddish-brown TiO_2-Fe_2O_3 composites. The composites of TiO_2-Fe_2O_3 was annealed at 150°C, 300°C, 400°C, 500°C, 600°C, 700°C and characterized by X-Ray diffraction (XRD), Fourier Transform Infrared and UV-Vis spectrophotometer.

2.3. Photodegradation of Rhodamin B with composites of TiO_2-Fe_2O_3 at various of annealing and time irradiation

The composites of TiO_2-Fe_2O_3 at temperature annealing of 150°C, 300°C, 400°C, 500°C, 600°C and 700°C were added 25 mL Rhodamin B (5 ppm), respectively. The solution was irradiated into black box for 0, 15, 30, 120, and 300 minutes. Then, the solution was filtered and the filtrate was characterized by UV-Vis spectrophotometer.

3. Results and Discussion

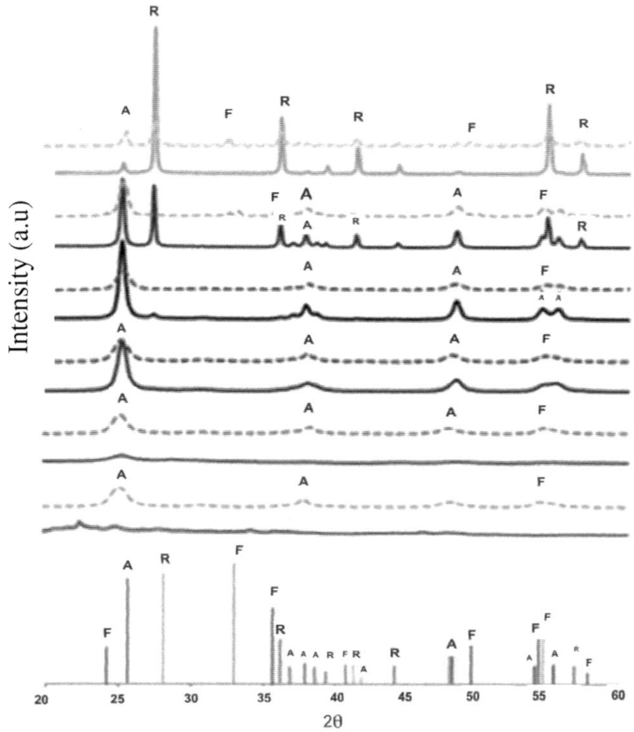

Figure 1. X-Ray Diffractogram of TiO_2 and TiO_2-Fe_2O_3 composites at annealing of 150°C (—), 300°C (—), 400°C (—), 500°C (—), 600°C (—), and 700°C (—).

X-Ray diffractogram of TiO_2 are shown in Figure 1. The synthesize of TiO_2 annealing at 150 °C and 300 °C were obtained the amorphous peaks which indicated the crystalline of TiO_2 was formed yet, while TiO_2 at 400°C, 500°C, 600°C, and 700°C have formed peaks of TiO_2 anatase at 2θ=25.35(d_{101}=3,509Å) according to JCPDS No. 78-2486 and new peaks appear at 2θ=27.50(d_{110}=3,239Å) that indicated the rutile phase according to JCPDS No. 87-0920. The ratio between anatase and rutile phase of TiO_2 (A:R) at various annealing are shown in Figure 2 and Table 1.

Table 1. Ratio of Anatase/Rutile at various annealing

Ratio	Annealing of TiO_2			
	400°C	500°C	600°C	700°C
A:R	1:0.0	01:0.1	01:0.7	01:9.9

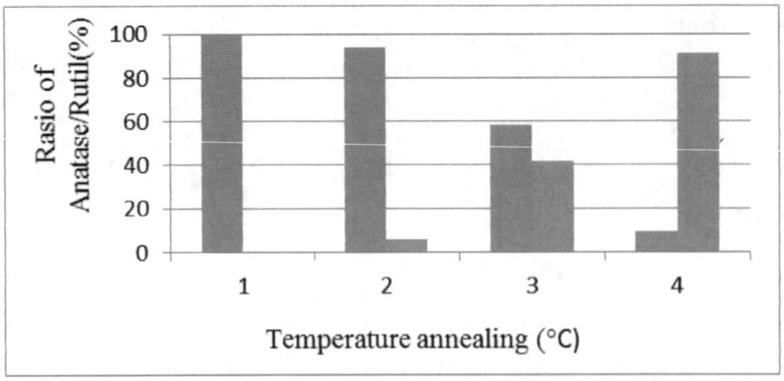

Figure 2. Ratio of Anatase/Rutile at various annealing

X-Ray diffractogram of TiO_2-Fe_2O_3 composites are shown in Figure 1 (strip line). TiO_2-Fe_2O_3 composites that annealed at 150°C and 300°C was obtained peaks of TiO_2 anatase at 2θ = 25° (d_{101}=3.537Å) according to JCPDS No.78-2486, Fe_2O_3 at 2θ =53.90 (d_{116}=1.699Å) according to JCPDS No.87-1165, and appear peak of FeOH at 2θ= 47.65 (d_{041}=1.906Å) according to JCPDS No.81-0462.

The synthesis of TiO_2 that annealed at 150°C and 300°C were formed amorphous phase, whereas after TiO_2-Fe_2O_3 composites formed a sharp peak indicating crystalline TiO_2. The presence of Fe_2O_3 dopant capable accelerates the formation of TiO_2 at low temperature. The TiO_2-Fe_2O_3 composites with annealing at 500 °C indicated the dominant peak of the anatase phase at 2θ = 25.35 (d_{101} = 3.509Å), the dominant peak of Fe_2O_3 at 2θ = 48.05 (d_{230} = 1.891Å), 2θ = 54.95 (d_{060} = 1,669Å) and not formed yet the rutile phase in TiO_2 due to Fe_2O_3 dopant at 500°C this can resist the transformation TiO_2 anatase to rutile.

Table 2. Degradation percentage (%) of Rhodamin B at various annealing temperatures and time irradiation

	Degradation percentage (%) of Rhodamin B at various time irradiation				
Variation annealing of TiO_2-Fe_2O_3 composites	0 minutes	15 minutes	30 minutes	120 minutes	300 minutes
150	0	3.2	5.0	10.9	17.6
300	0	1.3	3.1	7.2	24.8
400	0	0.0	5.4	16.2	44.3
500	0	0.0	3.0	7.5	23.5
600	0	0.0	5.1	12.5	13.9
700	0	5.2	8.1	8.2	9.8

The degradation curve (A/Ao vs the variation of irradiation times) indicated that the longer time of irradiation can increase of degradation activity. The TiO_2-Fe_2O_3 composite that annealed at 400°C has high degradation (44.3%) compared with other temperature at 300 minutes irradiation due to the structure of TiO_2-Fe_2O_3 composite has a high percentage of anatase and hematite structure. TiO_2-

Fe_2O_3 composite that annealed at 500°C, 600 °C, and 700 °C with 300 minutes irradiation, the percentage of degradation decrease due to the presence of anatase-hematite structure decreases while rutile phase has been established. Thus the presence of Fe_2O_3 dopant into TiO_2 increases the photocatalytic activity to high irradiance time and inhibits the formation of rutile phases at high temperatures.

4. Conclusion

The influence of Fe_2O_3 addition on the TiO_2 structure and photoactivity properties has been studied. The TiO_2-Fe_2O_3 composite with annealed at 400°C has high degradation (44.3%) compared with other temperature at 300 minutes irradiation. And the TiO_2-Fe_2O_3 composite with annealed at 500°C, 600 °C, and 700 °C showed decrease of the percentage of degradation because decreases of the presence of anatase-hematite structure and rutile phase has been formed.

References

[1] Yan M F, Cheng J, Zhang J, and Anpo M 2005 *J. Phys. Chem. B* **109** 8673-8678.
[2] Libanori R, Tania R G, Elson L, Edson R L, and Caue R 2009 *J. Sol Gel Sci. Tech.* **49** 95-100.
[3] Luu C L, Quoc T N, and Si Thoang Ho 2010 *Adv. Nat. Sci: Nanosci. Nanotech* **1** 1-5.
[4] Wahyuningsih S, Narsito, and Kartini L 2007 *International Conference on Chemical Science* Yogyakarta.
[5] Rao C and Turner 1959 *J. Phys. Chem. Solids* **11** 173-175
[6] Riyas S, Yasir A, and Mohan D 2002 *J. Bull. Mater. Sci.* **25** 267-273
[7] Music S, Krehula S, Propovic S 2004 *Material Lettters* **58** 444-448

The morphological study of porous silicon formed by electrochemical anodization method

R Suryana[1], D K Sandi[2] and O Nakatsuka[3]
[1]Department of Physics, Faculty of Mathematics and Sciences, Sebelas Maret University, Indonesia
[2]College of Engineering, Graduate Institute of Energy Engineering, National Central University, Taiwan
[3]Department of Materials Physics, Graduate School of Engineering, Nagoya University, Japan

E-mail: rsuryana@staff.uns.ac.id

Abstract. Due to its good physical and chemical properties, porous silicon (PSi) is very attractive to study. In this research, PSi has been fabricated on n-type Si (100) by the electrochemical anodization method. The electrolyte solution used was a mixture of HF (40%), ethanol (99%) and aquadest with volume ratio of 1:1:2, respectively. It was anodized on Si(100) surface at different current densities of 10 mA/cm^2 and 20 mA/cm^2 with the anodization time at each current density for 10 min, 20 min, and 30 min. The Scanning Electron Microscope (SEM) images showed that the PSi surfaces have inhomogeneous sized pores in the range of 95.00 nm–1.46 µm. The PSi layers with current density and anodization time of 10 mA/cm^2 (10 min), 10mA/cm^2 (20 min), and 20mA/cm^2 (10 min) have spherical shaped pores while the others have some uncommon (cross sectional) shaped pores on surfaces. It is considered that the cross sectional shaped maybe caused by unstable the current during the electrochemical anodization process.

1. Introduction

Porous silicon (PSi) is silicon crystal consisting of voids in the size of nanometer–micrometer on surface [1-2]. PSi remains very attractive to study because it has good physical, chemical, and optical properties that can be applied in technologically electronic devices such as light emitting diodes, waveguides, gas testing devices, chemical and biological sensors, photonic crystal, photovoltaic, etc [1-7].

There are several methods to produce PSi such as electrochemical anodization, focused ion beam, electron beam lithography, metal assisted etching, and rapid thermal annealing [8, 9]. Electrochemical anodization is an easiest way to form porous. Besides, the experimental setup of this method is simply replicable and easy to modify [7]. This method involves two electrodes dipped in hydrofluoric acid (HF) as electrolyte solution while it is supplied by constant current density [3, 10]. In the PSi manufacture, the two electrodes are silicon substrate as the anode and platinum or the other conductive materials resistant to HF as the cathode. The pores formation is affected by some parameters during the electrochemical process such as current density, anodizing time, electrolyte solution and its concentration, and silicon type [7, 11, 12]. Therefore, the electrochemical anodization method remains challenge in formation of PSi.

In this research, the morphological study of porous silicon fabrication has been conducted by electrochemical anodization method. The parameters of the method such as current density and anodization time were varied to investigate their effects on the pore formation. The Scanning Electron Microscopy (SEM) instrument was performed on PSi surfaces to observe the pore shapes.

2. Experimental

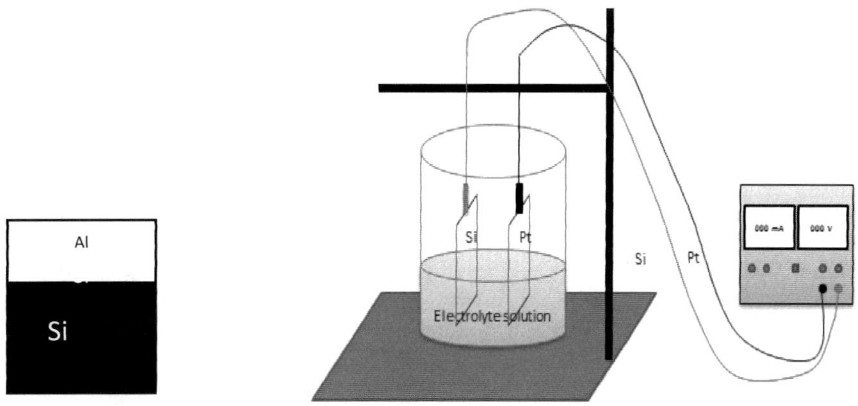

(a) (b)

Figure 1. (a) Si substrate with aluminum plaster (b) the experimental setup of electrochemical anodization method

This experiment used n-type Si (100) substrate as an anodized material. It was cut into rectangular shape with size of 1.0×1.5 cm^2. Firstly, the Si samples were cleaned through the RCA cleaning process method. A Si sample was rinsed in NH_4OH: H_2O_2: aquadest solution with volume ratio of 1:1:2 for 5 min at 70°C-80°C. Then, it was soaked in aquadest with overflow for 3 min and immersed in 40% HF:aquadest solution with volume ratio of 1:30 for 15 s. Finally, it was dried with Ar gas. The aim of the RCA cleaning process is to remove organic and metallic contaminant, native oxide contaminant and ionic contaminant on layer of commercial silicon wafer [13]. The Si sample was covered by aluminum plaster on 0.5 cm^2 area as shown in Figure 1(a) to make a conductive contact.

The experimental setup as demonstrated by Figure 1(b) was carried out to anodize the Si surface. The Si was placed as anode while the Pt was placed as cathode. They were dipped into electrolyte solution that was a mixture of 40% HF: 99% ethanol: aquadest with volume ratio of 1:1:2. It was then applied in different current densities of 10 mA/cm^2 and 20 mA/cm^2 with the anodizing times of each current density of 10 min, 20 min, and 30 min using DC power supply (SUNSHINE P-3005D). After the anodization process, samples were cleaned in distilled water via ultrasonic cleaner and then dried with Ar gas. The obtained PSi samples were observed by SEM instrument (FEI Quanta 200 FEG).

3. Results and discussion

Figure 2 shows the SEM images of Si surfaces after the electrochemical anodization process at different currents and anodization times. The PSi has been successfully formed on each silicon surfaces. The current density and the anodizing time affect the PSi formation. The pores have inhomogeneous size in the range of 95 nm–1450 nm. Nevertheless, the pores size increases as increasing the current density and the anodizing time. In addition, the PSi layers with current density and anodizing time of 10 mA/cm^2 (10 min), 10mA/cm^2(20 min), and 20mA/cm^2(30 min) have spherical shaped pores, while the others have some uncommon (cross sectional) shaped pores on their

IOP Conf. Series: Materials Science and Engineering **333** (2018) 012034 doi:10.1088/1757-899X/333/1/012034

surfaces. It is considered that those inhomogeneous size pores and irregular shaped pores are caused by the unstable current density during the anodization process which is a problem in this experiment.

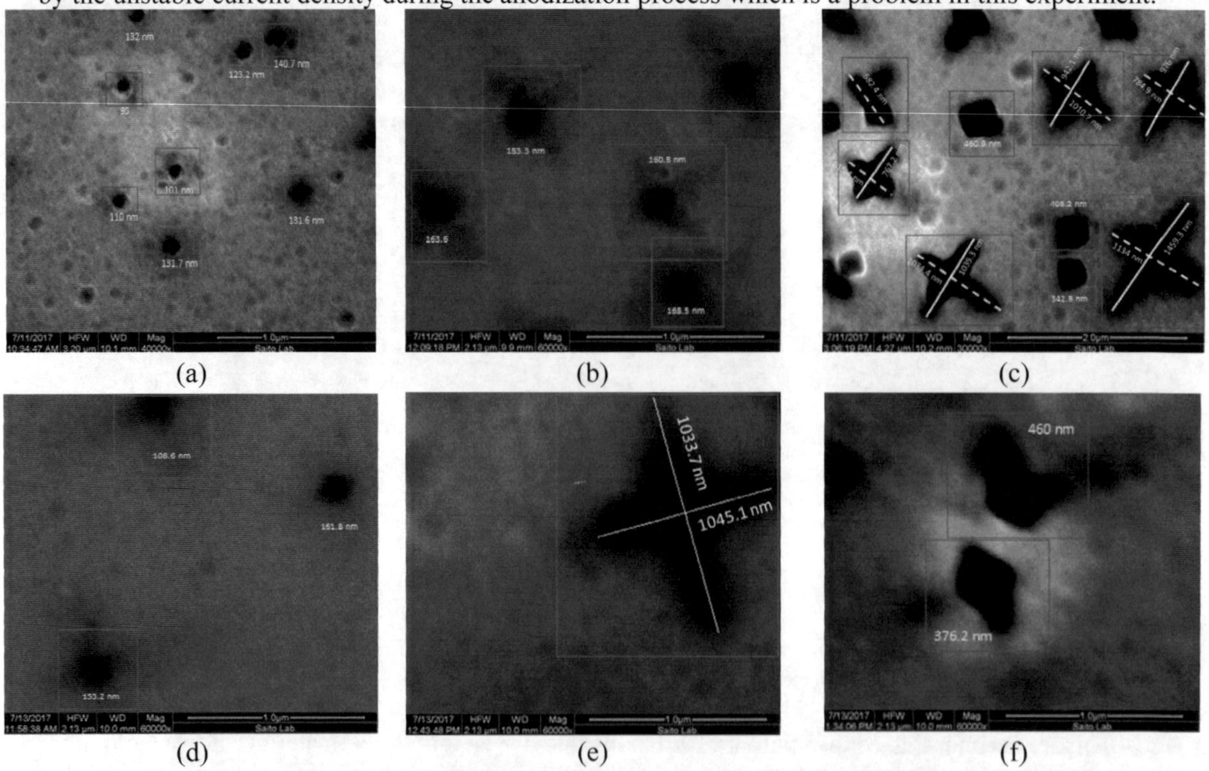

Figure 2. The SEM images of PSi after electrochemical process at various currents and etching times (a) 10 mA/cm^2-10 min (b) 10 mA/cm^2-20 min (c) 10 mA/cm^2-30 min (d) 20 mA/cm^2-10 min (e) 20 mA/cm^2-20 min (f) 20 mA/cm^2-30 min

Furthermore, it is agreed with Burham *et al.* [7] which declares that the unstable current influence on the inhomogeneous size pores and pore structures. Therefore, in the next experiment, we have to make a new DC power supply to control the current density in stable condition during the electrochemical anodization process.

The mechanism of the pores formation has been explained in refs 7, 10, and 13. When the current is in unstable condition, the current increases or decreases at certain current during the electrochemical anodization process, the amount of the holes generated are different. That hole will be attacked by F^- ions thus a pore formed on Si surface. When the currents increase, more holes are attacked by F^- ions to form the pores in larger sizes. While when the currents decrease, less holes are attacked by F^- ions to form the pores in smaller sizes. Therefore, unstable currents will produce inhomogeneous size pores including cross shapes on Si surface

4. Conclusion

PSi has been successfully formed by the electrochemical anodization method on Si(100) surfaces. The pores have inhomogeneous pores size in the range of 95 nm–1450 nm. The PSi structures with current density and anodizing time of 10 mA/cm^2 (10 min), 10mA/cm^2(20 min), and 20mA/cm^2(30 min) have spherical shaped pores, while the others have some irregular shaped pores. It is considered that both inhomogeneous sized and uncommon shaped pore are caused by the unstable current during the electrochemical anodization process. The increase current density and the longer anodizing time tend to make the bigger pore size and the narrower pores distribution. Therefore, it can be concluded that the parameters of the electrochemical, the current density and the anodizing time, affect the pores formation.

Acknowledgement

Authors wish to thank PNBP Universitas Sebelas Maret 2017 for financial support through Hibah Peningkatan Kapasitas Laboratorium Penelitian (PKLP). RS would like to thank Saito Laboratory, Department of Applied Physics, Graduate School of Engineering, Nagoya University, Japan for supporting SEM instrument.

References

[1] Nayef U M 2013 *Int. J. Basic & Appl. Sci. (IJBAS-IJENS)* **13(02)** 61
[2] Singh P, Sharma S N and Ravindra N M 2010 *JOM* **62** 15
[3] Kadhim R G, Ismail R A and Abdulridha W A 2015 *Int. J. Fil. Sci. Tec.* **4(3)** 199
[4] Mohammed I M and Shnieshil A H 2013 *Int. J. Application in Engineering and Management* **2(9)** 77
[5] Kumar P, Lemmens P, Ghosh M, Ludwi F and Schilling M 2009 *J. Nanomater.* **18** 7
[6] Russo L, Colangelo F, Cioffi R, Rea I and Stefano L D 2011 *Materials* **4(6)** 1023
[7] Burham N, Hamzah A A and Majlis B Y 2017 Self-Adjusting Electrochemical Technique for Producing Nanoporous Silicon *New Research on Silicon-Structure, Properties, Technology* (Materials Science) ed V I Talanin (Croatia: INTECH) chapter 6 pp 125-154
[8] Striemer C C, Gaborski T R, McGrath J L, and Fauchet P M 2007 *Nature* **445(7129)** 749
[9] Hamzah A, Abidin H Z, Majlis B Y, Nor M M, Ismardi A, Sugandi G, Tiong T Y, Dee C F and Yunas J 2013 *J. Micromech. Microeng.* **23** 074007
[10] Sailor M J 2011 *Porous Silicon in Practice: Preparation, Characterization and Applications* (Wiley-VCH Verlag GmbH & Co. KGaA)
[11] Zhang G X 2003 *J. Electroch. Soci.* **151** 65
[12] Vashpanov Y, Jung J I and Kwack K D 2011 *Sensors* **11(2)** 1321
[13] Santos A and Kumeria T 2015 Electrochemical Etching Methods for Producing Porous Silicon *Electrochemically Engineered Nanoporous Materials Methods, Properties and Applications* (Springer Series in Materials Science vol 220) ed D Losic and A Santos (Switzerland: Springer) chapter 1 pp 1-36

Optical Properties and Microstructure of Barium Titanate Thin Film ($BaTiO_3$) for Solar Cell Applications

A U L S Setyadi[1], Y Iriani[2*], F Nurosyid[2]

[1]Department of Physics, Graduate Program, Sebelas Maret University Surakarta, Indonesia
[2]Department of Physics, Faculty of Mathematics and Natural Sciences, Sebelas Maret University, Surakarta, Indonesia

*E-mail: yopen_2005@yahoo.com

Abstract. Barium Titanate thin films were prepared with variations in the number of layers and variation of the solution on a Quartz substrate using the sol-gel method with spin coating technique, at rotation speed 3000 rpm for 30 seconds. The first solution was made with heated and the second with stirred and heated. In this experiment, $BaTiO_3$were heated at 900°C for 2 hours. The characterization of optical properties was performed by UV-Vis spectrometer and microstructural characterization was performed by X-Ray Diffraction (XRD).Variation of layers number affects the intensity of the diffraction peaks. The more layers of the intensity are also greater. The variation of solution making process affects the intensity of diffraction peak. The process of making the solution with stirred and heated has greater intensity than the process of solution by simply heating it. When stirred at the same time heated to produce atoms diffuses more easily with other atoms so the bonds between atoms are more orderly and strong. The process of making the solution in the heated is larger in the crystallite size of than preparation of solution by stirred and heated. The stirred which the solution is produced influences the appearance of the size of the crystal. Variation number of layers influences the absorbance value of layer. The absorbance increases with increasing number of layers. The absorbance of the sample was made with heated the higher than with stirred and heated.

1. Introduction

The thin film gave new hope in the development of solar cell devices due to its low cost and good material stability [1][2]. Thin Film can be made with several methods there are vacuum method and nonvacuum method. The vacuum method consists of Pulsed Laser Deposition (PVD), Ablation Laser, ion planting and Chemical Vapour Deposition CVD[1][3]. For nonvacuum methods consisting of electrodeposited, Dip Coating, Spin Coating, Electrophoresis, Screen Printing, Pyrolysis [1], and Chemical Solution Deposition or sol-gel method[3]. The sol-gel method is a way of making the thin film by the sol-gel method above the substrate, then prepared by spin coating at a certain rotational speed. This method is superior because the stoichiometric easily controlled with either [4].

Ferroelectric has material that great potential as a new photovoltaic material, as a driving light and an optical sensor [5]. Material features have included in the ferroelectric material spontaneous polarized regions, the formation of the hysteresis curve (curve relationship between polarization and electric field)[2]. Ferroelectric is a material has a spontaneous polarization in the absence of electric field interference from outside material with Curie Temperature (Tc) [6]. Polarization is the application of the field resulting in the existence of a ferroelectric material. The connection between electrical polarization and electric field in the sample is indicated by the formation of the hysteresis curve [7].

One of the materials can be used in producing the thin film is Barium Titanate ($BaTiO_3$). $BaTiO_3$ has an interesting properties that have been used for 60 years that is chemically and mechanically very stable, exhibits ferroelectric properties above room temperature, has Curie temperature 120°C [7], high dielectric constant [8] at room temperature ≥1500, low dielectric loss [2] , enormous band gap energy [8]. Barium titanate is used in applications such as capacitors [2], high-density multilayer ceramic capacitors [9], Ferroelectric Random Access Memory (FRAM), Dynamic Random Access Memory (DRAM), characteristic of piezoelectric can be used for microactuator and sensor, characteristic of polarizability can be used Nonvolatile Ferroelectric Random Access Memories (NvFeRAMs) [10]. The $BaTiO_3$ ferroelectric material has used as a solar cell material since it generally has a gap energy ± 3 eV and a conductivity of 105 S/cm so that a small band gap can enhance the photovoltaic effect of ferroelectrics [5].

In this study, $BaTiO_3$ was made using Chemical Solution Deposition (sol-gel method) with spin coating process at rotation speed 3000 rpm. Annealing at 400°C was held for 30 minutes, 900°C was held for 2 hours with a heating rate of 5°C/min. The variations have made in this study are a variation of the number of layers and variation of solution making process. The $BaTiO_3$ thin layer is characterized using XRD to obtain crystal structure information. Characterization of BaTiO3 optical properties uses UV-Vis Spectrophotometer.

2. Experimental

2.1 Preparation of Solutions

The materials used in this study were Acetic Acid [(CH_3COOH)], Barium Acetate [Ba (CH_3COO)$_2$], Titanium Isopropoxide [Ti ($C_{12}O_4H_{28}$)] and Etylen Glycol ($HOCH_2CH_2OH$). Preparation of $BaTiO_3$ solution using the sol-gel method with molarity 0.4 M. In the preparation of this solution has two different treatments i.e. first treatment, the solution was made by heated and stirred; second treatment, the solution was made just heated. Step early, Acetic Acetic and Barium Acetate powder are mixed. The solution was heated at 60°C for 30 minutes to obtain a homogeneous solution. Both solutions were added with Titanium Isopropoxide and heated at 60°C for 30 minutes. The third Ethylen Glycol was added to the solution and heated to 60°C for 30 minutes. The solution is heated again at 90°C for 30 minutes.After the $BaTiO_3$ solution was stored for 24 hours, then deposition of the solution above the Quartz substrate by spin coating technique at a rotation speed of 3000 rpm for 30 minutes. To get some layers could be repeated by dripping the solution above the Quartz substrate and then rotated with a predetermined rotation speed. The next step was the sample heated at temperature 120°C for 5 minutes. Then the substrate has been coated, annealing with a temperature of 400°C with a resistance time of 30 minutes, 900°C with a hold time of 2 hours with a heating rate of 5°C/minute.

2.2 Characterization of thin film

The Barium Titanate thin film was characterized using XRD equipment, Bruker D8 Advance to obtain crystal structure information with wavelength 1.54184 Å, the increment of 0,1105 and θ of 10°-60°. The result of the characterized were data by showing the relationship between 2θ with intensity. Data were plotted in the origin software so that a graph showing the diffraction peaks of the sample appears. The peaks diffraction was compared to the ICDD database so that it could be known that the peaks were the diffraction peaks of the sample. From the diffraction peaks can be known the value of Full Width Of Half Maximum (FWHM), could be known the level of crystallinity, crystallite size, and crystal structure of the sample. The crystallinity level of a material could be determined by following equation (1).

$$\text{Crystallinity} = \frac{I_{peak\,max} - I_{peak\,min}}{I_{peak\,max}} \quad (1)$$

The crystallite size of $BaTiO_3$ thin layer can be calculated using the Scherrer equation in the equation (2)

$$D = \frac{k\lambda}{\beta \cos\theta} \quad (2)$$

The value of β is the FWHM value, θ is the diffraction angle, k is the Scherrer constant of 0.9, and λ is the wavelength of X-ray.

Barium Titanate's optical characterization used UV-Vis spectrophotometer with wavelengths between 200-800 nm. The UV-Vis test result data was plotted in the origin software so that a graph showing the value of the relationship between absorbance and wavelength occurs. The absorbance test was performed to see the thin film absorption spectra of $BaTiO_3$. Spectroscopy used to detect the properties of absorbance, transmittance, and reflectance of thin film BST. The data obtained from these measurements are further processed to obtain the absorbance characteristics of $BaTiO_3$. Absorption in the film takes place through the excitation of electrons from the filled or valence states into an empty state (conduction).

3. Result and Discussion

3.1 Microstructure

Characterization of microstructure has done using XRD equipment. From the test results obtained in the form of the diffraction pattern was a graph showing the relationship between the intensity (I) with the receding diffraction (2θ). The formation of crystal fields on characterization using XRD is laid with the appearance of peaks of diffraction patterns that have been matched with ICDD databases. The diffraction data of $BaTiO_3$ has been identified with database # 812205.

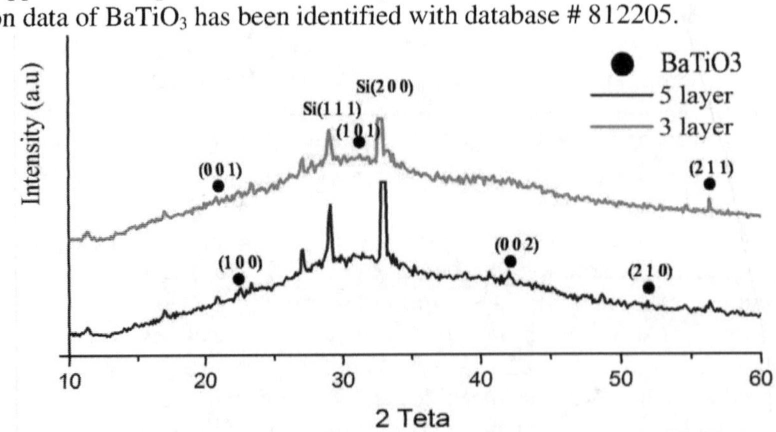

(a)

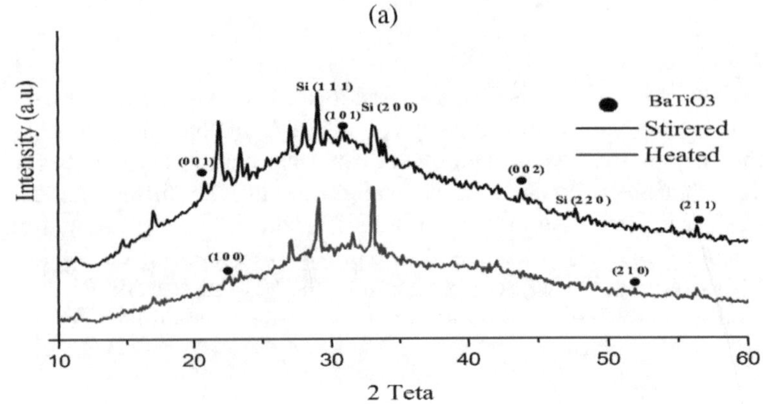

(b)

Figure 1. Diffraction pattern $BaTiO_3$ variation of (a) layers number (b) solution making process

Figure 1 is pattern diffraction $BaTiO_3$ that has matched the database Figure 1 (a) The XRD result for variation of layers number. It is known that there is a peak of $BaTiO_3$ in 3 layers and 5 layers sample marked with a sign (•). The presence of the $BaTiO_3$ peak indicates that the $BaTiO_3$ solution has not fully reacted (not yet homogeneous). Sample 5 layers have peaks of $BaTiO_3$ appeared more peaks of $BaTiO_3$ 3 layers. The number of layers resulted in the more $BaTiO_3$ peaks. The highest intensity in the 3layersand 5layerssample was at the field of $BaTiO_3$ crystals (1 0 1). Quantitatively, the greatest peak intensity based on the variation of the number of layers in each diffraction pattern is shown in table 1. The thinner the number of layers so the higher the intensity of the diffraction peak.

Figure 1(b) is diffraction pattern for variation of solution making process. It is known that there is a peak of $BaTiO_3$ in the sample with a sign (•).In stirred samples, more $BaTiO_3$ peaks appear than in heating samples. The temperature at which the solution was produced influences the appearance of the $BaTiO_3$ peak. The intensity of the sample was stirred higher than the heated sample. Quantitatively, the greatest peak intensity in each diffraction pattern based on the variation of the solution was to saw in table 2. When given the effect of stirrer on the sample make it can increase the intensity of diffraction peak. When it was heated and stirred the solute atoms can diffuse from one to the other so that the reaction can occur perfectly and the atoms were arranged regularly [12].

Table 1. Intensity field (1 0 1) for variation of layers number

Number of Layers	Intensity (Count)
3 Layers	2053
5 Layers	2305

Table 2. Intensity of 5 layers for variation of solution making process

Solution Treatment	Intensity (Count)
Heated and Stirred [$BaTiO_3$ (0 0 1)]	3547
Heating [$BaTiO_3$ (1 0 1)]	2305

The crystallinity level of a material could be determined by following equation (1). The $BaTiO_3$ sample with the number of layers was 5 layers with the preparation of the heated solution has a crystallinity rate of 55%. Samples 3 layers made with heated have a crystallinity level of 45%. Sample 5 layers made with heated and stirred solution has a crystallization rate of 65%. From the calculation of the level of crystallinity, sample 5 layers, preparation with heated and stirred solution have the highest level of crystallinity, the second largest sequence in the 5 layer sample with heated. The lowest crystallinity value in the sample was 3 layers made by the heated solution. This proved to be stirring and heating during manufacture. Heating and stirring cause the atom to diffuse more easily with other atoms so that the bonds between atoms are more orderly and stronger.

The crystallite size of the $BaTiO_3$ thin layer can be calculated using the Scherrer equation in the equation (2) showing the connection between FWHM and crystallite size (D) of each sample with the variation number of different layers and the variation made of the solution. The FWHM value of the $BaTiO_3$ sample is calculated by the origin software used to calculate the crystallite size. The result of $BaTiO_3$ crystallite size calculation can be seen in table 3.

Based on the table. 3 it can be seen that the size of the crystal with the variation number of layers of 3 layers is greater than the 5 layers. The value of crystallite size with the variation of preparation the solution 5 layers with stirred and heated is bigger than the sample with heated so that the making of the solution with heated is better than the preparation of the solution by stirred and heated.

Table 3. Crystallinity, FWHM and Crystallite size of $BaTiO_3$ thin films

Sample	Crystallinity (%)	FWHM	Crystallite Size (nm)
5 Layer Stirred and Heated	65	0,356449	3,960
5 Layer heated	55	0,366766	3,927
3 layer heated	45	0,331099	4,347

3.2 Optical properties

Figure 2 shows the absorption spectra of the $BaTiO_3$ films, (a) variation number of layers (b) variation of solution making process. The results of the absorbance characterization in Fig. 2 (a) show that the sample with maximum layers of 5 layers absorbs light in the range 300-345 nm has a peak of 0.891 and minimum the absorbs light in the range 770-800. Samples with a maximum number of layers of 3 layers absorbing light in the range 300-335 nm have a minimum peak of 0.72999 absorbing light in the range 770-800.

Figure 2 (b) shows that the sample with the number of layers 5 layers made by heated solution, maximum absorbing the light in the range 300-345 nm has a peak of 0.8691 and the minimum absorbance the light in the range 770-800. Samples with 5 layers made by stirred and heated solution have maximum absorbance light in the 360-380 nm range and have peak 0.14873, minimum absorbing light in the range 200-230. The sample 5 layers preparation with stirred and heated solution shows different absorbance values. The sample 5 layers preparation with the stirred and heated solution have a small absorbance value. This indicates that the stirred at which the solution was made has an effect on the absorbance value. The heating during the preparation of the solution allowed the solution to be homogeneous so that in the sample with the stirred solution it was suspected to be homogeneous.

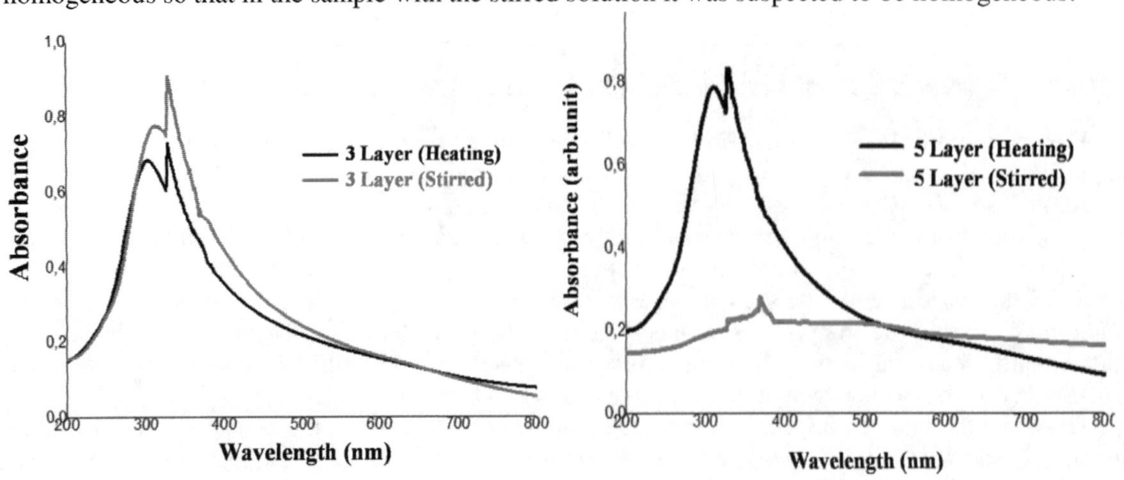

(a) (b)

Figure 2. Graph of absorbance $BaTiO_3$ (a) variation of layers number (b) variation of solution making process

At the point of maximum absorbance shows that the electron cannot absorb energy at that wavelength so that the given energy was simply passed only [13]. As the number of layers increases, the absorbance of the $BaTiO_3$ layer increases, so the absorbance increases with layer thickness. The gap energy is analyzed to determine whether the material are a conductor, a semiconductor or an insulator [13].

4. Conclusion

Based on the result of the penetration done, it can be concluded that the variation of layers number and variation of solution making process has an effect on $BaTiO_3$ microstructure and absorbance. Variation of layers number affects the intensity of the diffraction peaks. The more layers of the intensity are also greater. The variation of solution making the process affects the intensity of diffraction peak. The process of making the solution with stirred and heated has greater intensity than the process of the solution by simply heating it. When stirred at the same time heated to produce atoms diffuses more easily with other atoms so the bonds between atoms are more orderly and strong. The process of making the solution in the heated is larger in the crystallite size of than preparation of solution by stirred and heated. The stirred which the solution is produced influences the appearance of the size of the crystal. Variation number of layers influences the absorbance value of layer. Absorbance increases with increasing number of layers. The absorbance of a sample was made with heated the higher than with stirred and heated.

References

[1] Irzaman, Putra I R, Aminullah, Syafutra H, Alatas H. 2016. *Science Direct*. **33** 607 – 614.
[2] Vijatovic M M, Bobić J D, Stojanović B D.2008. *History and Challenges of Barium Titanate Science of Sintering*. **40** 155-165.
[3] Iriani Y, Jamaludin A, Nurhadi A. 2016. *J. Physics*. **776**
[4] Iriani Y, Saputri D F, Hadiati S, Ramelan A R, Variani V I. 2014. *Adv Mat. Res*. **911** 205-209
[5] Jiang W, Cai W, Lin Z, Fu C. 2013. *Mat. Res. Bull.*. **48** 3092–3097.
[6] Hutauruk F J, Kusumandari, Iriani Y. 2016. *Jurnal Fisika dan Aplikasinya*. **12** 120-123.
[7] Safitri R D, Subarwanti Y, Supriyanto A, Jamaluddin A, Iriani Y. 2016. *Jurnal Fisika dan Aplikasinya*. **12** 112-115
[8] Gao H, Cao J, Liu L, Yang Y. 2011. *J. Mol. Struc*. **1003** 75–81.
[9] Ramakanth S, Raju K C J. 2014. *Solid State Comm*. **187** 59-63.
[10] Hadiati S, Ramelan A H, Variani V I, Hikam M, Soegijono B, Saputri D F, Iriani Y. 2014. *Jurnal Fisika dan Aplikasinya*. **10** 37-43
[11] Zaidah A, Suwarni, Supriyanto A, Jamaluddin A, Iriani Y. 2015. *Jurnal Fisika Dan Aplikasinya.***11** 82-55.
[12] Kwei, G.H., Lawson, A.C., & Billinge, J.L. 1993. *J. Physics Chem*. **97** 2368-2377.
[13] Liman J, Harsono B , Rohman T T, Trimukti U , Khalid M , Roharti E , Irzaman. 2015. *Jurnal Fisika Indonesia*. **111** 1410-2994.

IOP Conf. Series: Materials Science and Engineering **333** (2018) 012036 doi:10.1088/1757-899X/333/1/012036

Acquisition of Co metal from spent lithium-ion battery using emulsion liquid membrane technology and emulsion stability test

Yuliusman[1,*], P T Wulandari[1], R A Amiliana[1], M Huda[1] and F A Kusumadewi[1]
[1] Chemical Engineering Department, University of Indonesia, Depok 16424, Indonesia

*E-mail: usman@che.ui.ac.id

Abstract. Lithium-ion batteries are the most common type to be used as energy source in mobile phone. The amount of lithium-ion battery wastes is approximated by 200 – 500 ton/year. In one lithium-ion battery, there are 5 – 20% of cobalt metal, depend on the manufacturer. One of the way to recover a valuable metal from waste is leaching process then continued with extraction, which is the aim of this study. Spent lithium-ion batteries will be characterized with EDX and AAS, the result will show the amount of cobalt metal with form of $LiCoO_2$ in the cathode. Hydrochloric acid concentration used is 4 M, temperature 80°C, and reaction time 1 hour. This study will discuss the emulsion stability test on emulsion liquid membrane. The purpose of emulsion stability test in this study was to determine optimum concentration of surfactant and extractant to produce a stable emulsion. Surfactant and extractant used were SPAN 80 and Cyanex 272 respectively with both concentrations varied. Membrane and feed phase ratios used in this experiment was 1 : 2. The optimum results of this study were SPAN 80 concentrations of 10% w/v and Cyanex 272 0.7 M.

1. Introduction

Lithium ion batteries consist of heavy metals, organic chemicals and plastics in the proportion of 5-20% cobalt, 5-10% nickel, 5-7% lithium, 15% organic chemicals and 7% plastics, the composition varying slightly with different manufacturers [1]. Some methods to recover cobalt are liquid-liquid extraction, electrowinning, emulsion liquid membrane, vaporization, and hydroxide precipitation. In this study, leaching is used to recover cobalt, to dissolve lithium ion battery wastes with leaching agent and continued with emulsion liquid membrane to extract cobalt metal from leachtant.

Leaching is a solid-liquid extraction method where the separation of solid cobalt mixed with leaching agent so the metal inside can be dissolved into the leaching agent. Strong acids like sulfuric acid and hydrochloric acid were often used as leaching agent. There is a previous study about recovering cobalt metal from lithium ion battery, comparing three reagents as leaching agents which are sulfuric acid, hydroxylamine hydrochloride, and hydrochloric acid [2]. Between three reagents stated, hydrochloric acid is the most suitable and effective with more than 99% cobalt metal leached in a hydrochloric acid 4 M in 80°C temperature, and stirring time of 1 hour.

Emulsion liquid membrane have some attractive features, for example, simple operation, high efficiency, extraction and stripping in one stage, larger interfacial area, scope of continuous operation. By extracting and stripping in one stage, it will reduce reaction time and reagents needed unlike liquid-liquid extraction method [3]. Emulsion liquid membrane is the right choice because it has the highest degree of separation and can be applied in a large scale [4].

IOP Conf. Series: Materials Science and Engineering **333** (2018) 012036 doi:10.1088/1757-899X/333/1/012036

Separation of liquid membrane provides an interesting option for separation process. Compared with conventional process, emulsion liquid membrane (ELM) and surfactant liquid membrane (LSM) have several advantages such as easy operation, high efficiency, extraction and stripping process at the same stage, large interface area, and can run continuously. ELM technique has great potential for recovering and removing metal and hydrocarbon ions from wastes where the efficiency of conventional separation processes is lower. Liquid membrane process uses a selective liquid membrane phase during extraction/stripping process. Separation is achieved by solute permeation through liquid phase of membrane from feed phase to receiving phase. Both phases (feed and receiver phase) are usually soluble but membrane phase is not dissolved with feed phase and receiving phase [4].

Liquid membrane is divided into three types: bulk liquid membrane, liquid supported membrane, and liquid membrane with two emulsions. Of the three types, emulsion liquid membrane has the largest mass transfer area compared to other two types. Emulsion forming system can be divided into two: water-in-oil emulsions dispersed in an external aqueous phase and oil-in-water dispersed in an external organic phase. Membrane phase with type water-in-oil-in-water (W / O / W), undissolved oil phase will separate aqueous phase, whereas oil-in-water-oil (O / W / O) type, undissolved aqueous phase will separate the two organic phases that play role as a membrane. Liquid membrane has two functions which are to move one or more components selectively from external phase to internal and vice versa, and to prevent mixing of external and internal phases. Figure 1 shows configuration in emulsion liquid membrane system with the letter F showing feed phase, E is liquid membrane, and R is receiving phase.

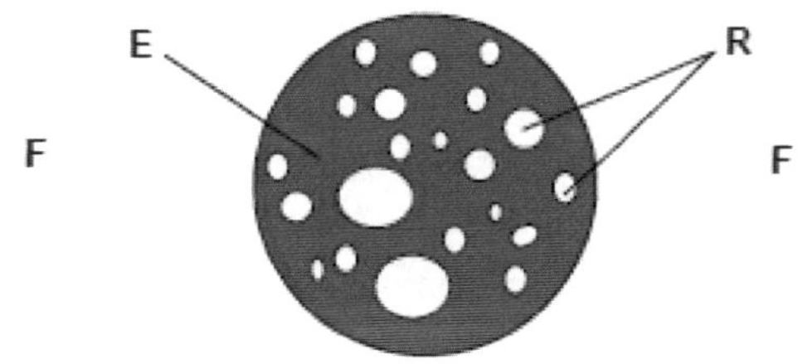

Figure 1. Emulsion liquid membrane system configuration
(Source: Kislik, 2010 [4])

2. Experimental

2.1 Materials

Lithium ion batteries that are used are from cell phone batteries various brand sold at the market (Samsung, iPhone, Xiaomi, and LG). Hydrochloric acid used as leaching agent, Cyanex 272 as extractant. Commercial kerosene, with density of 830 kg/m^3 used as diluent. SPAN 80 used as surfactant, and H_2SO_4 as stripping agent. All the materials were immediately used without reprocessing to purify the materials.

2.2 Experimental procedure

Initial step is preparation of materials. One of the preparation that needed to be done is soaking lithium ion batteries into sodium chloride solution to emptied the remaining electrical charge inside the batteries which may still be inside. Next step is leaching process with varying leaching agent which has been specified before, and will not be discussed further in this study. Solution after leaching (called leachant) then filtered to get rid of solid components which can't dissolve. After filtered, leachant then analysed of cobalt metal content with AAS (Atomic Absorption Spectroscopy). Next step is emulsion stability test using kerosene, Cyanex 272, SPAN 80, and H_2SO_4 with concentration of Cyanex 272 as extractant and SPAN 80 as surfactant being varied.

3. Results and discussion

3.1 Characterization lithium-ion battery waste

To characterize lithium ion battery waste with EDX and AAS analysis, cathode sheet scraped and crushed, after that filtered with 100 mesh filter so the result is cathode powder with size as small as possible so leaching process will be more effective, to make leaching efficiency high, it needs a particle with the size 100 mesh [5]. Figure 2 shows EDX analysis result of battery cathode powder. There is cobalt metal in the powder. EDX analysis is quantitative so AAS analysis is needed to determine how much cobalt metal is in the battery precisely. AAS analysis result with the amount of cobalt metal in cathode battery powder is 11.09%w/w. This goes with previous study by [5], in general range of cobalt metal content in lithium ion battery is 5 – 20% weight, depends on the battery's manufacturer.

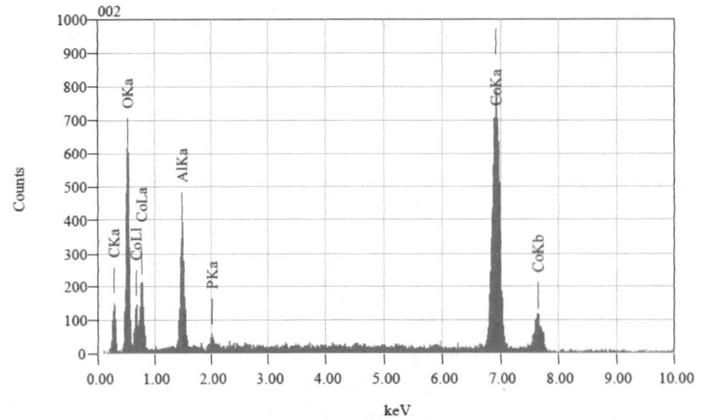

Figure 2. EDX analysis result graph

3.2 Emulsion stability test

Emulsion stability test is done based on eyes visual and percentage of stability based on how the emulsion volume decreases.

3.2.1 Effect of surfactant concentration. Surfactants are usually organic compounds that are amphipathic, meaning they contain both hydrophobic groups (their "tails") and hydrophilic groups (their "heads"). Therefore, they are soluble in both organic solvents and water [6]. Surfactants plays a very important role in the emulsion formation and in the extraction process. It reduces the interfacial tension between oil and water by adsorbing at the liquid-liquid interface [6], maintains the emulsion stability and influences the transport rate of metal ions [7,8]. Wan and Zhang [9] revealed that the selection of a surfactant is the key measure to reduce emulsion swelling and membrane breakage. Therefore, the choice and development of a suitable surfactant [10-12] and establishment of surfactant concentration correctly will determine the success of ELM process.

There are some considerations of choosing surfactant for ELM separations. The hydrophile-lipophile balance (HLB) of surfactant determines the type of emulsion produced. Surfactants with a low HLB, typically ranging from 1 to 10, are more soluble in oil than water and tend to make a water-in-oil emulsion while those with a high HLB, typically ranging from 10 to 20, are more soluble in water than oil and tend to make an oil-in-water emulsion [6]. While to get stable emulsion, surfactant with least hydration capacity and low diffusivity due to its high molecular weight is preferable [9]. However, for overall ELM process, too stable emulsions are unfavorable for the difficulty in demulsification process.

According Figure 3, it is seen that the extractant concentration of 0.7 M and SPAN 80 10% w/v surfactant mixture makes the most stable emulsion and SPAN 80 4% w/v surfactant mixture makes the most unstable emulsion among all variations of surfactant concentration. This is because the emulsion stability depends on the concentration of the surfactant. An excessively high concentration of surfactant will increase resistance to interphase surfaces and this may contribute to the increase in viscosity of emulsion liquid membrane. Using too high surfactant concentrations but not importantly needed will decrease the efficiency of metal extraction due to the formation of high interphase resistances. Conversely, when the surfactant concentration is too low, emulsion leakage may result in a decrease in the volume of the emulsion in this experiment [13]. Beside of surfactant concentration, the emulsion stability is also significantly affected by the types of surfactant. Hasan et al. [14] reported that Span 80 with 12% concentration gave the least breakage compared to Span 85 and Arlacel A.

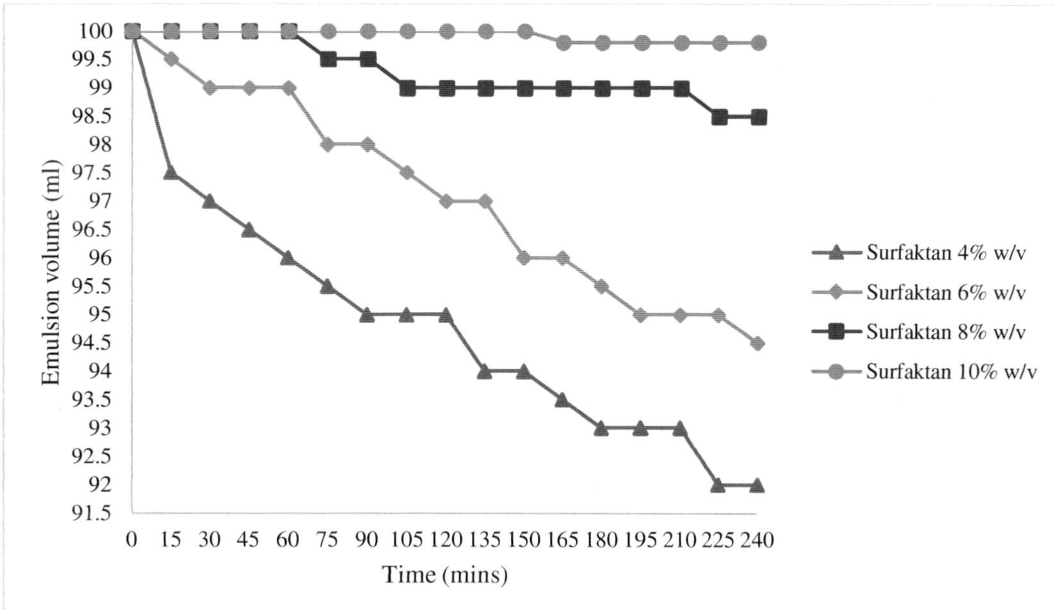

Figure 3. Emulsion stability, surfactant concentration being varied (extractant concentration 0.7 M, stirring speed 1500 rpm, stirring time 60 mins)

3.2.2 Effect of extractant concentration. Extractant acts as a 'shuttle' to carry the metal element through the liquid membrane [15]. However, its presence in membrane phase can decrease the emulsion stability. This is caused by the competitive adsorption with the surfactant as they have opposite behaviour. Interfacial tensions increased with an increase in carrier concentration in the membrane phase leading to formation of larger sized emulsion globules in the dispersed emulsion [16], conversely, interfacial tensions decreased by increasing the surfactant concentration up to a specific value. Gu et al. [17] revealed that the key criterion in selecting a carrier is that it and the complex formed must be soluble in the membrane phase, but not soluble in both the internal and feed phase. Further precipitation within the membrane or at the interfaces must also be prevented [18].

A great number of studies gave evidence that not only the type but also the concentration of extractant affects the emulsion stability. Emulsion breakdown can occur at high concentration of carrier, since the carrier and its complex with the metal present interfacial characteristics that provoke loss of internal aqueous solution [19]. Besides, a very high content of carrier in the membrane does not result in a benefit due to the increase in viscosity, which leads to larger globules.

In Figure 4, it is seen that extractant concentration of 0.7 M is the most stable among the other variations. This shows that the addition of extractant concentration affects emulsion stability.

Interphase tension decreases as the concentration of extractant in organic phase increases, but too high concentration can cause saturation in the phase and will form molecular aggregates called micelles. In non-polar fluids such as kerosene, the solvent used in this study, SPAN 80 alters the micelles, in which the polar head is directed to the center and the nonpolar portion is directed to the dispersed fluid. The extractant used has internal activity in emulsion liquid membrane system so that no surfactant concentration is too high in the organic phase to maintain the emulsion stability. Kerosene solvents also help reduce interphase voltages with feed solutions containing metal ions. This is the most important property of organic solvents for use in emulsion liquid membranes, in addition to having low volatility and insoluble in aqueous solutions [20].

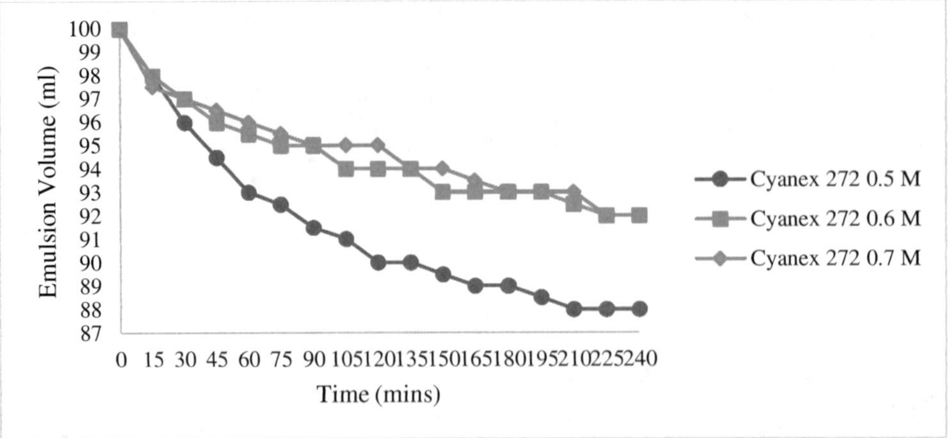

Figure 4. Emulsion stability, extractant concentration being varied (surfactant concentration 4%w/v, stirring speed 1500 rpm, stirring time 1 hour)

Membrane breakage in ELM system includes the rupture of the emulsion and the leakage of the internal reagent and extracted solute through the membrane phase to the external phase, both resulting in the decrease in the volume of the internal phase. As a result, the membrane breakage causes a decrease of driving force for mass transfer and an increase of the raffinate concentration, thereby lowering the extraction efficiency.

3.3 Extraction

Extraction process begins by making an optimum emulsion phase (10% w/v SPAN 80 and 0.7 M Cyanex 272). This extraction process goes in one stage with the stripping process because the extracting technology used is the emulsion liquid membrane. Then the emulsion phases that have been formed are mixed with the optimum leachant that is the result of leaching using 4 M hydrochloric acid which varied the pH of the solution, ie 5 and 6. The pH variation is done because Cyanex 272 is a slightly acidic extractant so the pH of the stripping solution should be less than the pH of leachtant metal during the extraction process with the emulsion liquid membrane. NaOH solids added gradually to change feed phase pH until the pH reaches 5 and 6. The ratio of the mixing volume between the emulsion phase and the feed phase is 1 : 2. After that, the solution is stirred at 250 rpm for 15 minutes. After stirring, the mixture is both transferred to the separatory funnel for extraction and held for 15 - 30 minutes and there is a separation between the feed phase and the emulsion phase. The calculation of the efficiency of the cobalt extraction process is shown in Figure 5.

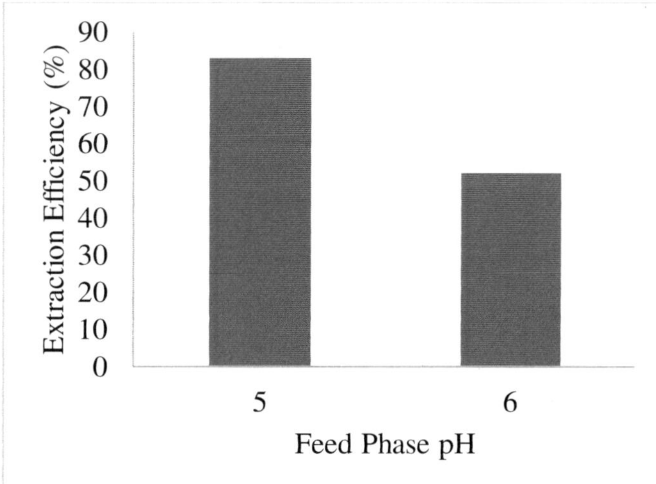

Figure 5. Extraction efficiency with pH 5 and 6 (stripping speed 250 rpm, stirring time 15 mins, room temperature)

Based on Figure 5 shows that extraction efficiency at pH 5 was 83.03% which is higher than pH 6 of 52.1%. This may occur due to emulsion instability. Because of when the pH is high, emulsion stability changes with the emulsion, whereas at too low pH can be explained by competing H^+ ions with the solute due to the release of H^+ ions from extractant to leachant [21]. Extraction rate will decrease when the pH is greater than 5 due to emulsion instability because H^+ ions transfer phenomenon from acidic feed solution to membrane phase so that the membrane can swell and decompose. pH value between stripping solution and feed solution should be different to make a difference of chemical potential for H^+ ions. Physical properties of the extractant will cause a reduction of the ion activity difference on both phases. Cobalt extraction process increases with increasing amount of proton concentration up to a certain concentration [13]. According Figure 5, at pH 6, the emulsion has swelled so the extraction efficiency is lower than that of pH 5.

Type and concentration selection of stripping solution appropriately is one of the determinants of extraction process efficiency with emulsion liquid membrane technology. Reaction by the stripping solution plays an important role in the extraction of solute from feed solution to stripping solution in emulsion liquid membrane process. Based on prior study by [21], performance of Na_2CO_3, H_2SO_4, and HCl as stripping solution was compared. In the first ten minutes, the best stripping solution sequence is Na_2CO_3 > H_2SO_4 > HCl. However, after the first ten minutes, emulsion stability while using Na_2CO_3 is changed so that its extraction efficiency is reduced so that the sequence of extraction efficiency at the end of process from highest to lowest is H_2SO_4 > Na_2CO_3 > HCl.

4. Conclusion

The amount of cobalt metal in cathode Li-ion battery powder is 11.09%w/w. The most stable emulsion is reached when extractant concentration is 0.7 M and surfactant concentration is 10%w/v. The highest extraction efficiency is reached when feed phase's pH is 5.

References

[1] Shin S M, Kim N H, Sohn J S, Yang D H and Kim Y H 2005 *Hydrom.* **79** 172-81
[2] Zhang P, Yokoyama T, Itabashi O, Suzuki T and Inoue K 1998 *Hydrom.* **47** 259-71
[3] Harimu L, Matsjeh S, Siswanta D and Santosa SJ 2010 *Indo. J. Chem.* **10**(1) 69-74
[4] Kislik V S 2010 *Liquid Membranes: Principles & Applications in Chemical Separations & Wastewater Treatment* (Amsterdam: Elsevier B.V.)
[5] Shin S M, Kim N H, Sohn J S, Yang D H and Kim Y H 2005 *Hydrom.* **79** 172-81

[6] Chakraborty M, Bhattacharya C and Datta S 2010 in *V.S. Kislik (Ed.),Liquid Membranes: Principles and Applications in Chemical Separations and Wastewater Treatment* (Amsterdam: Elsevier B.V.) 141-99
[7] Basualto C, Poblete M, Marchese A, Ochoa A, Acosta A, Sapag J and Valenzuela F 2006 *J. Braz. Chem. Soc.* **17** 1374-54
[8] Kakoi T, Goto M and Nakashio F 2000 *Separ. Purif. Tech.* **19** 3-9
[9] Wan Y and Zhang X 2002 *J. Memb. Sci.* **196** 185-201
[10] Nakashio F, Goto M, Matsumoto M, Irie J and Kondo K 1988 *J. Memb. Sci.* **38** 249-60
[11] Goto M, Yamamoto H, Kondo K and Nakashio F 1991 *J. Memb. Sci.* **57** 161-74
[12] Goto M, Kakoi T, Yoshii N, Kondo K and Nakashio F 1993 *Ind. Eng. Chem. Rsch.* **32** 1681-85
[13] Volkan E and Recep K A 2015 *J. Ind. Eng. Chem.* **21** 303-10
[14] Hasan M A, Selima Y T and Mohamed K M 2009 *J. Hzrd. Mat.* **168** 1537-41
[15] Kankekar P S, Wagh S J and Mahajani V V 2010 *Chem. Eng. Proc.* **49** 441-8
[16] Sengupta B, Bhakhar M S and Sengupta R 2007 *Hydrom* **89** 311-8
[17] Gu Z, Ho W S and Li N N in *Ho W S, Sirkar K K (eds) Membrane Handbook, Chapman & Hall* (New York: Elsevier Science BV) 283-352
[18] Perera J M and Stevens G W in *Pabby A K, Rizvi S S H, Sastre A M (Eds.) Handbook of Membrane Separations: Chemical, Pharmaceutical, Food, and Biotechnological Applications* (CRC Press) 709-39
[19] Valenzuela F, Fonseca C, Basulato C, Correa O, Tapia C and Sapag J 2005 *Min. Eng.* **18** 33-40
[20] Valenzuela F, Auspont J, Basualto C, Tapia C and Sapag J 2005 *Chem. Eng. Rsch. Des.* **83**(A3) 247-55
[21] Kumbasar R A 2009 *Separ. Purif. Tech.* **64** 273-79

IOP Conf. Series: Materials Science and Engineering **333** (2018) 012037 doi:10.1088/1757-899X/333/1/012037

Process optimization and leaching kinetics of zinc and manganese metals from zinc-carbon and alkaline spent batteries using citric acid reagent

Yuliusman[1], R A Amiliana[1], P T Wulandari[1], M Huda, and F A Kusumadewi[1]

[1]Chemical Engineering Department, University of Indonesia, Depok 16424, Indonesia

E-mail: usman@che.ui.ac.id

Abstract. Zn-Carbon and Alkaline spent batteries contains heavy metals, such as zinc and manganese, which can causes environmental problem if not handled properly. Usually the recovery of these metals were done by leaching method using strong acid, but the use of strong acids as leaching reagents can be harmful to the environment. This paper concerns the recovery of Zn and Mn metals from Zn-C and alkaline spent batteries with leaching method using citric acid as the environmental friendly leaching reagent. The leaching conditions using citric acid were optimized and the leaching kinetics of Zn and Mn in citric acid solution was investigated. The leaching of 89.62% Zn and 63.26% Mn was achieved with 1.5 M citric acid, 90°C temperature, and 90 minutes stirring time. Kinetics data for the dissolution of Zn showed the best fit to chemical control shrinking core model, while the diffusion controlled model was suitable for the dissolution of Mn kinetics data. The activation energy of 6.12 and 1.73 kcal/mol was acquired for the leaching of Zn and Mn in the temperature range 60°C-90°C.

1. Introduction

Zinc-carbon and alkaline batteries are used in many electronic objects that small quantities of power are required, such as remote control, radios, and watches. Usually, they run out rapidly, producing huge quantity of spent batteries and increasing the environmental problem due to the metallic content, being considered as hazardous waste [1]. Zn-C and Alkaline batteries have the same electrode metal elements, Zn as anodes and Mn as cathodes, so often called as Zn-Mn batteries. Zn content is about 12-28% and Mn content is 26-45% of the total batteries powder mass [2]. It is necessary to recover Zn and Mn metals from these spent batteries to maintain the raw materials and to protect the environment.

Leaching is one of method to extract the metal content from the spent batteries with a liquid solvent or reagent to dissolve the metal in it. Strong acids, such as sulfuric acid, nitric acid, and hydrochloric acid, were often used as leaching reagents, but these acids can release toxic gases like Cl_2, SO_3, and NO_x during the leach process and the waste acid solution after leaching is a threat for environment [3]. Replacing the leaching reagent with more environmentally friendly reagent, like organic acid, can reduce the problem for environment because organic acid more soluble in water and the waste from the leaching process with organic acid is considered more easily to processed.

Citric acid is an organic acid that classified as weak acid, but was often used as a leaching reagent among other organic acid. Citric acid was once used for leaching from low grade oxide ores and able to extract 82% of Zn metal [4]. Citric acid had been also used to recover Li and Co metal from Li-Ion batteries waste [5] and recover Pb, Cu, Zn metals on contaminated soil of plant waste [6]. In this study, the citric acid is used as leaching reagent for recovery the Zn and Mn metal from Zn-Mn spent batteries. The leaching condition is optimized to get the maximum percentage of leaching. The optimum leaching conditions in this study were determined based on three leaching parameters: leaching reagent concentration, temperature, and stirring time.

The leaching kinetics is also investigated in this paper using the variations of temperature and stirring time data. The leaching kinetics are performed to know the rate model of leaching process and to determine the mechanism of leaching. The leaching kinetics of Zn metal in ammonium chloride solution has been studied and showed that the diffusion model was the best kinetic model for this process [7], while chemical reaction model was the suitable leaching kinetics model for Mn metal in sulfur dioxide solution [8]. As the leaching kinetics of Zn and Mn has not been well studied, the objective of this present work is to assess the suitable leaching kinetics models of Zn and Mn metals in citric acid solution.

2. Experimental

2.1. Materials and reagents

The spent Zn-C and alkaline batteries were collected from the local sources. The leaching reagents used in this work were citric acid ($C_6H_8O_7$) and hydrogen peroxide (H_2O_2) used as reducing agent. All solutions were prepared in distilled water and all reagents were analytical grade from Merck.

2.2. Battery dismantling and metal characterization

The spent batteries from several manufactures were manually dismantled. The black powder was a mixture of the cathodic (manganese oxides and graphite) and the anodic (zinc oxides and electrolyte solution). Dismantling products such as plastic film, ferrous scraps and paper pieces were separated. The battery powder then crushed and mashed to expand the surface area and make to increase leaching efficiency. After that, the battery powder was sieved to uniform the size of sample particles. The initial metals content of the powder were characterized using Atomic Absorption Spectroscopy (AAS, spectrometer SpectrAA 200-Varian) in order to evaluate the percentages of Zn and Mn metals.

2.3. Leaching of metals

All leaching experiments were carried out in a 250 mL Pyrex glass reactor with a thermometer to control the temperature and placed at hot plate stirrer to heat up the solution. The reactor was closed to reduce the loss of water by evaporation. A measured amount of battery powder and reagent leaching solution were added to reactor and allowed to reach thermal equilibrium, with agitation provide by the magnetic stirrer at 500 rpm for entire duration of the leaching and 1/20 solid-to-liquid ratio (S/L). All the leach liquor after leaching process were analyzed by AAS (spectrometer SpectrAA 200-Varian) to determine the total amount of zinc and manganese metals after the process.

3. Result and Discussion

3.1. Effect of acid concentration on leaching

The concentration of leaching reagent were varied 0.5 M, 1 M, 1.5 M and 2 M with leaching temperature 80°C and 120 minutes stirring time. The leaching efficiency results from the calculation of AAS analysis in leaching with various concentration of citric acid can be seen in Figure 1.

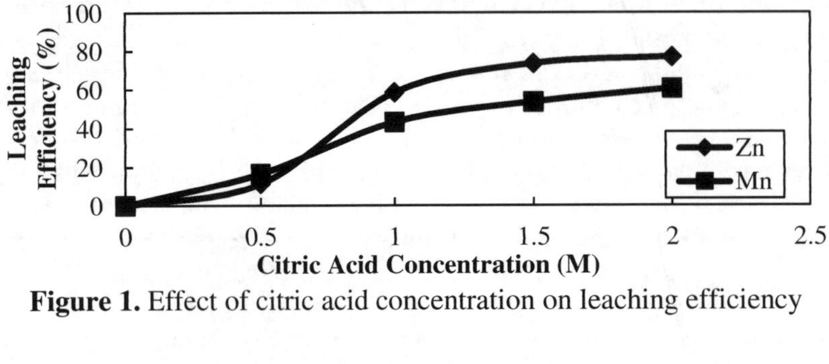

Figure 1. Effect of citric acid concentration on leaching efficiency

From Figure 1, it can be concluded that the greater concentration of citric acid lead to the greater leaching efficiency of Zn or Mn metals. The concentration that produces the maximum leaching percentage from all variation is 2M. The previous research mentioned that the leaching efficiency of metals using citric acid would have a certain optimum point at a certain concentration. As the citric acid concentration was higher and already surpassed the optimum point, there would be a decrease in leaching efficiency [9]. Contrast to the use of strong acids (such as sulfuric acid), which in tandem with increasing concentrations, the leaching efficiency will continue to increase to the constant limit [10]. Zinc and manganese metals originally in the form of powdered solids and tires bound to oxidants might be leached or dissolved due to the binding event with anions of citric acid. When the citric acid concentration increased, the ability and amount of citric acid anion formation also increases, thus it increasing the leaching efficiency of Zn and Mn metals.

Citric acid has 3 carboxyl groups in every one molecule and when dissociation occur one mol of citric acid in water, theoretically 3 mol H^+ will be formed as shown in these following equations [9].

$$C_6H_8O_7 \rightleftharpoons C_6H_7O_7{}^- + H^+, \quad K_{a1} = 7{,}4 \times 10^{-4} \tag{1}$$
$$C_6H_7O_7{}^- \rightleftharpoons C_6H_6O_7{}^{2-} + H^+, \quad K_{a2} = 1{,}7 \times 10^{-5} \tag{2}$$
$$C_6H_6O_7{}^{2-} \rightleftharpoons C_6H_5O_7{}^{3-} + H^+, \quad K_{a3} = 4 \times 10^{-7} \tag{3}$$

The equations above show that the dissociation of citric acid yields 3 Ka values. This proved that citric acid is classified as a good organic acid in the leaching process because it could maintain the content of hydrogen ions (H^+) remain high. The dissociation reaction of citric acid is a reversible reaction, so it must have an equilibrium state. The formation reaction of zinc citrate and manganese citrate can be accelerated when the concentration of hydrogen ions increases, so zinc and manganese ions could bind to the anions of citric acid more.

Based on Figure 1, the increase of leaching efficiency from concentration 1.5 M to 2 M is not significant, it is because the concentration of formed H^+ ions has started to saturated. The dissociation form of citric acid which is a reversible reaction (equation 1 to 3) causes when the citric acid concentration is increased further, the equilibrium will shift to the left. When the equilibrium shifts to the left, it will lead to reduced H^+ ions formed that will reduce the leaching efficiency. Based on that, the optimum concentration for leaching using citric acid in this study is 1.5 M, because the increase of leaching percentage above 1.5 M isn't effective anymore.

3.2. Effect of temperature on leaching

Temperature of leaching is the important parameters to get the high efficiency of recovery. In general, organic acids require higher temperature than strong acids to more ionized, because the high temperature can help to break the H^+ chains that were bound to weak acids. In this study with citric acid, the temperature varied was 60°C, 70°C, 80°C, and 90°C with 1.5 M citric acid and stirring time 120 minutes. The results of the leaching efficiency with varying temperature can be seen in Figure 2.

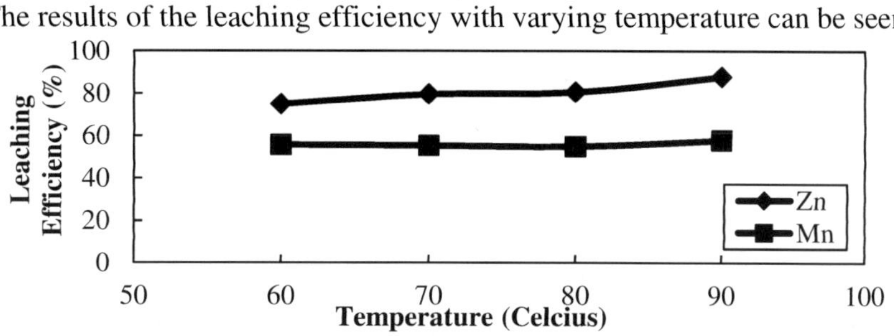

Figure 2. Effect of temperature on leaching efficiency

In the graph on Figure 2 shows that the higher temperature of leaching causes the greater percentage of leaching Zn and Mn metals. The highest leaching percentage was at 90°C with leaching efficiency of Zn reach 87.87% and Mn 57.87%. The high temperature allows a reaction to occur faster because there are collisions between molecules that move more quickly in the compound. High collision rate will make the higher kinetic energy. The relationship between the reaction rate and temperature was determined by the Arrhenius equation with activation energy factor. Activation energy is the energy that must be absorbed in the system from the environment to react and its value is influenced by temperature. From this study with temperature variation, it can be concluded that the optimum temperature for leaching of Zn and Mn metals from Zn-Mn spent batteries is 90°C.

3.3. Effect of time reaction on leaching

In the leaching process, it takes a certain time for citric acid to completely dissolve the zinc and manganese metal that present in the Zn-Mn batteries powder. The time that varied in this study was 30, 60, 90, and 120 minutes with 1.5 M citric acid and in operating temperature 90°C. The results of leaching efficiency with varying stirring time can be seen in Figure 3.

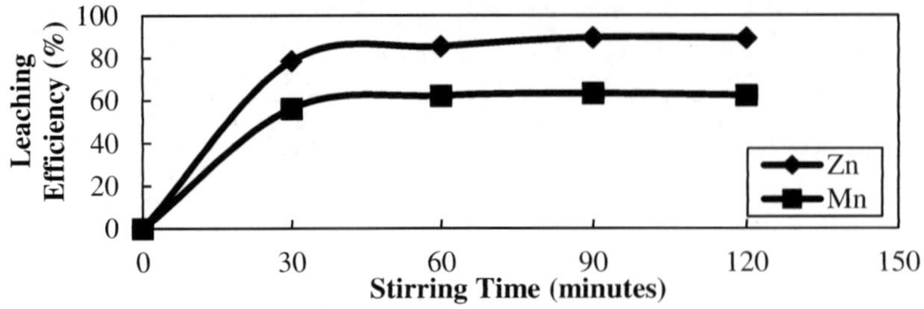

Figure 3. Effect of stirring time on leaching efficiency

Based on the graph in Figure 3, it can be seen that there is an increase of leaching efficiency up to 90 minutes, but the increase of leaching percentage is not significant when heading to 120 minutes. The increase of contact time between citric acid and Zn-Mn batteries powder causes more H+ ions of citric acid that are dissociate and form zinc citrate and manganese citrate compound, so the percentage of leaching Zn and Mn metals were getting bigger. However, the amount of dissociated H+ ions from citric acid will be difficult to increase when it has reached its saturation point, so that longer stirring time will not have a significant effect on the leaching efficiency and tend to be constant when it reaches its optimum time. At this variation of stirring time, it is concluded that the maximum stirring time for leaching of Zn and Mn metals from Zn-Mn batteries is 90 minutes with leaching efficiency of Zn 89.62% and Mn 63.26%. This result is the optimum leaching result with optimized some leaching parameters: citrate acid concentration 1.5 M , temperature 90°C, and stirring time 90 minutes.

3.4. Kinetics of leaching

More data on temperature and stirring time variations are needed to evaluate the kinetics of leaching. Citric acid was used as leaching reagent with concentration 1.5 M at fixed stirring speed and solid-liquid ration. On the previous experiment, we found that the optimum time for leaching of Zn and Mn metals with citric acid from Zn-Mn spent batteries was 90 minutes, then the time variation for kinetic study was only done until the optimum time. In other to determine the kinetic parameters and rate controlling step of leaching process, the popular shrinking core model was utilized [11]. In the solid-liquid reaction system, the reaction may be usually controlled by chemical reaction at the surface of the material or diffusion through the product layer [12]. If the reaction is controlled by diffusion through the product layer, it will have an integrated rate equation as follows [13]

$$1-\frac{2}{3}X-(1-X)^{\frac{2}{3}}=k_d t \qquad (4)$$

If the reaction is controlled by chemical reaction on surface, integrated rate equation as follows [13].

$$1-(1-X)^{\frac{1}{3}}=k_c t \qquad (5)$$

where X is the fraction reacted, k_d is the kinetic parameter for diffusion control, k_c is the kinetic parameter for chemical reaction control, and t is the reaction time (min). To analyze the kinetics of the leaching, data were fitted against both models and reaction time. The higher graphic linearity that formed by the data with the shrinking model showing the more suitable shrinking model for the leaching process. The high linearity can be seen from R^2 value calculated based on the existing graph and showed the smaller data deviation that occurs. The plot of the data of leaching Zn with chemical reaction model can be seen in Figure 4(a), while the diffusion model can be seen in Figure 4(b)

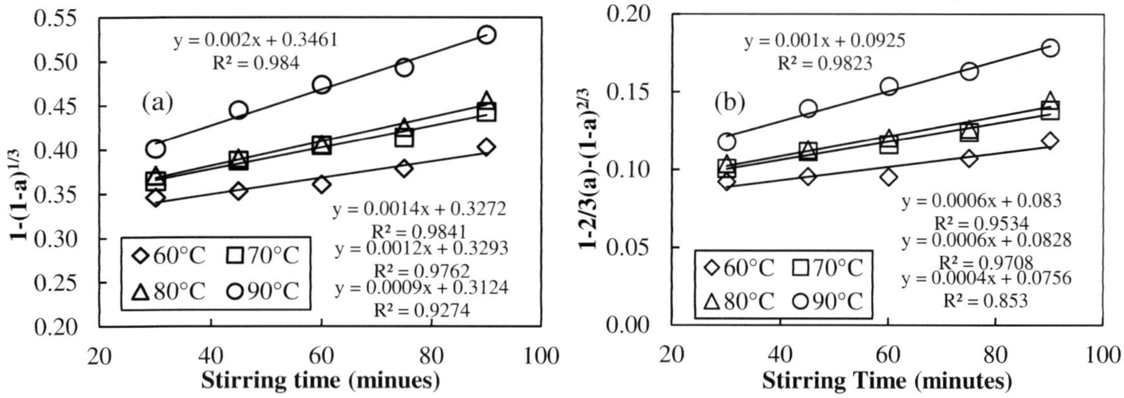

Figure 4. (a) Chemical and (b) diffusion controlled kinetic model for leaching of Zn metal

If the graphs in Figure 4(a) and Figure 4(b) are compared, then the leaching of Zn profile which has higher linearity is chemical reaction on the surface model. It can be seen from the average of R^2 value of chemical reaction model that is bigger than the R^2 value of diffusion model. So, the leaching of Zn with citric acid from Zn-Mn spent batteries is controlled by chemical reaction on the surface. This model is also suitable for leaching of Ni metal from NiCd spent battery using sulfuric acid [14] and also suitable for leaching some metals that contained in Ni-MH spent battery [13]. The plot of the data of leaching Mn with chemical reaction model and diffusion model can be seen in Figure 5.

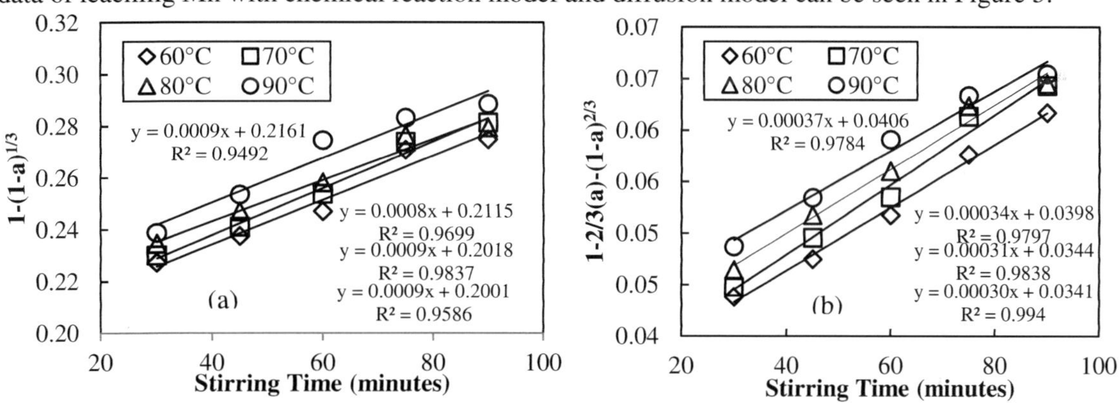

Figure 5. (a)Chemical and (b) diffusion controlled kinetic model for Leaching of Mn metal

Based on Figure 5, the more suitable kinetic model for the leaching of Mn with citric acid from Zn-Mn spent batteries is the diffusion through the product layer, seen from the higher average R^2 values. For the activation energy calculation, leaching kinetic of Zn metal using the chemical reaction model, while Mn metal using the diffusion model.

The calculation to evaluate activation energy was using Arrhenius equation which has been adapted to linear line equation as follows.

$$Ln(k) = Ln\,(A) - \frac{E_a}{R} \cdot \frac{1}{T} \quad (6)$$

where A is the frequency factor, E_a is the activation energy of the reaction, R is the universal gas constant, T is the absolute temperature, and k is the reaction constant for the suitable kinetic model. The activation energy was calculated using the gradient of the linear regression of the equation of the line between *ln (k)* and *1/T*. The graph for leaching of Zn metal using chemical reaction model and diffusion model can be seen in Figure 6.

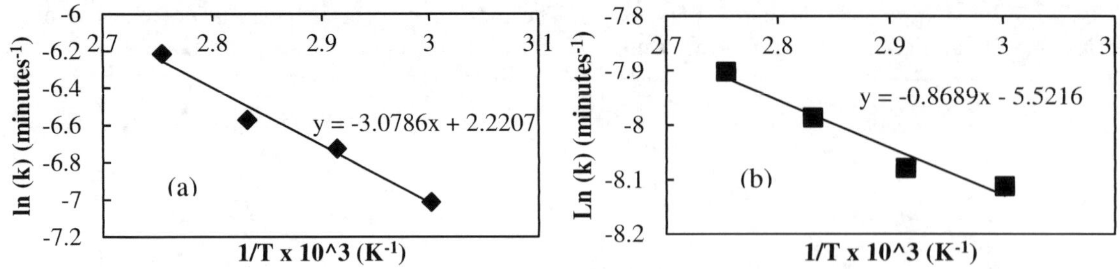

Figure 6. Arrhenius plot for leaching of (a) Zn and (b)Mn metal

Based on Figure 6(a), the calculated activation energy for Zn metal is 25.59 kJ/mol (6.12 kcal/mol), while the calculated activation energy for Mn metal based on Figure 6(b) is 7.22 kJ/mol (1.73 kcal/mol). The control process in the leaching kinetics have been characterized by the magnitude of activation energy, typically, E_a = 1 - 3 kcal/mol denoted diffusion controlled and higher for chemical control kinetics [14]. The activation energy of Zn in this study already in the activation energy range for chemical reaction on the surface, so as the activation energy of Mn in this study that also already in the range of activation energy for diffusion through the product layer, so it can be concluded that the choice of kinetic model for leaching of Zn and Mn metals is correct.

4. Conclusion

The optimum condition for leaching of Zn and Mn metals from zinc-carbon and alkaline spent batteries using citric acid as leaching regent are 1.5 M for citric acid concentration, at temperature 90°C, with stirring time 90 minutes. It can recover until 89.62% of Zn and 63.26% of Mn. The kinetic study show that the leaching kinetic of Zn metal from citric acid solution controlled by chemical reaction on the surface with the activation energy 6.12 kcal/mol and the leaching kinetic of Mn metal controlled by diffusion through the product layer with the activation energy 1.73 kcal/mol.

References

[1] I De Michelis, F Ferella, E Karakaya, F Beolchini, and F Veglio 2007 *J. Power Sources* **172** 975-983
[2] E Sayilgan, T Kukrer, F Ferella, A Akcil, F Veglio, and M Kitis 2009 *Hydrometallurgy* **97** 158-166
[3] Li L, Jing G, Renjie C, Feng W, Shi C, and Xiaoxiao Z 2010 *Waste Management* **30** 2615-2621
[4] Mehdi I, Mohammad M, and Amir R A 2012 Physicochem. Probl. MI **49** 547-555
[5] Li L, Jennifer B D, Xiao X Z, Linda G, Ren J C, Feng W, and Khalil Amine 2013 *J. Power Sources* **233** 180-189
[6] Hongki P, Kyungbae J, Richard D A, and Kyoungkeun Y 2013 *Materials Transactions* **54**(7) 1220-1223
[7] Sheng-hai Y, Hao L, Yan-wei S, Yong-min C, Chao-bo T, ansd Jing H 2016 *Transactions of Nonferrous Met. Soc. China* **26** 1688-1695
[8] Gamini S 2004 *Hydrometallurgy* **73** 215-224
[9] Li L, Jing G, Feng W, Renjie C, Shi C, and Borong W 2009 *J. Hazard. Mater.* **176** 288-293
[10] Manis K J, Anjan K, Amrita K J, Vinay K, Jhumki H, and Banshi D P 2013 *Waste Management* **33** 1890-1897
[11] Shanming H, Jikun W, and Jiangfeng Y 2011 *Hydrometallurgy* **108** 171-176
[12] Xin W, C Srinivasakannan, Xin-hui D, Jin-hui P, Da-jin Y, and Shao-hua J *2013 Sep. Purif. Technol.* **115** 66-72
[13] Pratima M, B D Pandey, and T R Mankhand 2016 *Waste Management* **51** 196-203
[14] Navneet S R, Kalpataru G, and Manoj K 2016 *Hydrometallurgy* **165** 191-198

Recovery of Ni Metal from Spent Catalyst with Emulsion Liquid Membrane Using Cyanex 272 as Extractant

Yuliusman[1], M Huda[1], I T Ramadhan[1], A R Farry[1], P T Wulandari[1], R Alfia[1]

[1]Chemical Engineering Department Universitas Indonesia. Universitas Indonesia, Depok 16425. Indonesia

E-mail: usman@che.ui.ac.id

Abstract. In this study was conducted to recover nickel metal from spent nickel catalyst resulting from hydrotreating process in petroleum industry. The nickel extraction study with the emulsion liquid membrane using Cyanex 272 as an extractant to extract and separate nickel from the feed phase solution. Feed phase solution was preapred from spent catalyst using sulphuric acid. Liquid membrane consists of a kerosene as diluent, a Span 80 as surfactant, a Cyanex 272 as carrier and sulphuric acid solutions have been used as the stripping solution. The important parameters governing the permeation of nickel and their effect on the separation process have been studied. These parameters are surfactant concentration, extractant concentration feed phase pH. The optimum conditions of the emulsion membrane making process is using 0.06 M Cyanex 272, 8% w/v SPAN 80, 0.05 M H2SO4, internal phase extractant / phase volume ratio: 1/1, and stirring speed 1150 rpm for 60 Minute that can produce emulsion membrane with stability level above 90% after 4 hours. In the extraction process with optimum condition pH 6 for feed phase, ratio of phase emulsion/phase of feed: 1/2, and stirring speed 175 rpm for 15 minutes with result 81.51% nickel was extracted.

1. Introduction

Selective separation of metal ions from industrial waste is frequently required in order to prevent environmental pollution. Currently, the application of solvent extraction process with organic extractant for the recovery of nickel metal from catalyst waste has been described. Most of these studies concerned solvent extraction using organophosphorus acid extractants such as D2EHPA (at (2-ethylhexyl) phosphoric), PC-88A (2-ethylhexylphosphonic acid mono-2-ethylhexyl esters acid) and Cyanex 272 (phosphonic acid) [1].

The emulsion liquid membrane (ELM) technique has been regarded as an advance in relation to solvent extraction for separating and concentrating metal ions from aqueous dilute solutions. Liquid membranes have shown potential results in the extraction process, especially in cases where solute concentrations are relatively low and other techniques cannot be applied efficiently since they combine the process of extraction and stripping in a single unit operation. This membrane technology is also highly selective, efficient, and simple because of the extraction and stripping steps that are carried out safely in one continuous process, resulting in less time consuming and less extractant use as in liquid-liquid extraction [2]. This type of emulsion liquid membrane is the best because it produces the highest degree of separation of metals and can be applied on a large scale [3].

In emulsion liquid membranes the extractant is an organic compound have been applied to separate Ni ions in acidic solution. Several studies related to the use of some extractants such as Cyanex 272, D2EHPA, and PC-88A for extracting nickel in liquid-liquid extraction systems have been reviewed. The efficiency of nickel extraction produced when using 0.1 M Cyanex 272 with kerosene solvent at pH of the feed phase 5,8 was 9%, and when using 0.05 M D2EHPA as solvent and the pH feed phase was 6% [4]. Results indicated by 0.15 kmol/m^3 PC-88A with kerosene solvent at pH phase feed at 7.5 also did not show satisfactory results because they were only able to extract 20% of nickel metal [5]. In this study used Cyanex 272 with concentrations varied on the emulsion liquid membrane system because one of the factors that play an important role to produce a good extraction efficiency is extractant concentration.

Surfactants in the emulsion liquid membrane extraction process are used as bridges where the hydrophilic side binds to the water phase and the lipophilic side binds to the oil resulting in a mixture of water and oil [6]. The surfactant also acts as an emulsion stabilizer by forming a protective layer and oriented to the interface and lowering the tension between the dispersed droplets. Emulsion stability is needed to make the extraction process more optimal. Types of surfactants commonly used are derived from the sorbitol ester (Span) or polyoxyethylenated sorbitol ester (Tween) group. Sorbitan mono-oleate (Span-80) is a non-ionic surfactant widely used for water-oil-water emulsion stabilizers [6]. Structure of Span 80 is shown in Figure 1

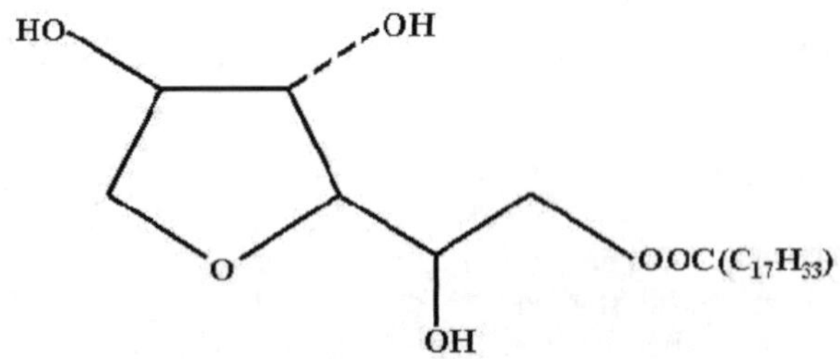

Figure 1. Structure of Span 80

2. Experimental

2.1 Reagents and Apparatus

The liquid membrane phase is composed of surfactant, carrier, and solvent. The non-ionic surfactant used is sorbitane monooleate commercially known as Span-80 which is supplied by Merck Co., Germany. The used carrier is bis (2,4,4-trimethylpentyl) phosphinic acid or commercially known as Cyanex 272 purchased from Cytec Inc. Solvent or organic solvent used is kerosene commercial purchased from Pertamina, Indonesia. Sulfuric acid is used as a stripping phase or internal phase purchased from Merck Co., Germany. Emulsions are prepared using 250 mL Erlenmeyer (Pyrex Inc.) and in homogenization using high-speed homogenizer (IKA, Germany). The pH value was measured using CyberScan 510 pH meters.

2.2.1 Procedure

2.2.2 *Emulsion Preparation and Stability Analysis.* The emulsion is prepared in a 250 mL Erlenmeyer containing a mixture of non-ionic surfactant Span-80, a carrier that has been dissolved with kerosene and sulfuric acid as an internal phase. The mixture stirred for 60 minutes at 1000 rpm. The internal phase and organic phase (I/O) ratios were kept at 1:1 for all experimental variations. The surfactant concentration was varied from 4% -10% w / v and the carrier concentration (extractant) was varied from 0.06-0.2 M to find the optimum concentration in the emulsion stability. The finished primary emulsion was transferred to a 100 mL measuring glass to observe the emulsion volume change for 240 minutes.

2.2.3 *Nickel Extraction.* Prepared emulsions with optimum conditions are poured into another 250 mL Erlenmeyer containing Ni 800 mg / L solution as the feed phase. The emulsion and feed phase ratios are maintained at 1: 1. The emulsion and feed phase mixture is stirred with a stirrer at 250 rpm for 15 minutes. The pH of the feed phase was varied from 5 to 8 to determine the effect of pH of the feed phase on the extraction efficiency. Then the emulsion phase mixture and the feed phase (external phase) are separated by a separatory funnel for 30-60 minutes. Phase feed that has separated the nickel metal content then measured using Atomic Absorption Spectrometry (AAS). The schematic of emulsion liquid membrane extraction is shown in Figure 2.

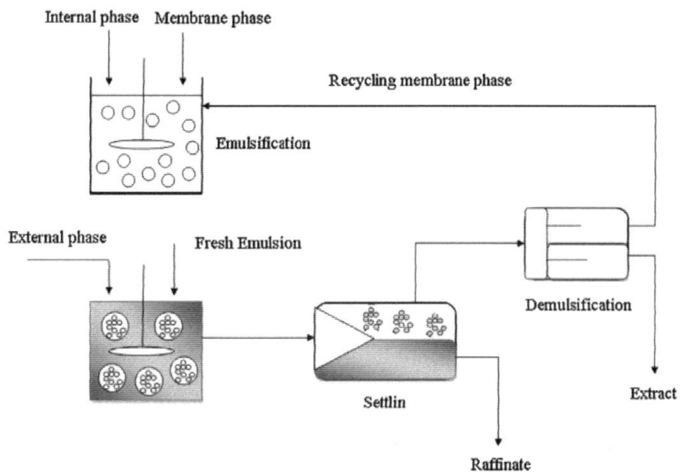

Figure 2. The schematic of emulsion liquid membrane extraction.

3. Result and Discussion

3.1 Effect of Surfactant Concentration

The emulsion was prepared by mixing kerosene as diluent, Cyanex 272 as carrier, H2SO4 0.2 M as stripping agent and Span 80 as surfactant. The internal phase/organic phase ratio was kept 1:1 for all experiment and the concentration of Cyanex 272 was kept 0.06 M for this experiment. After the emulsion was formed, it was removed from erlenmeyer to 100 ml measuring cylinder for immediate stability for 4 hours. The 4 hour time is ideal time to see the stability of the emulsion [7]. Emulsion stability study was observed by visual with the eye of how the emulsion volume decreased. The decrease in the volume of the emulsion may be characterized by a separate organic phase or an internal phase separated from the emulsion. To facilitate readability, the emulsion stability can also be represented by the volume of the emulsion. The Equation 1 shows the percentage formula of emulsion stability dependent on the emulsion volume. The comparison of surfactant variation to emulsion stability is shown in Figure 3.

$$Emulsion\ Stability\ (\%) = \frac{Final\ Emulsion\ Volume\ (ml)}{First\ Emulsion\ Volume\ (ml)} \times 100\ \% \quad (1)$$

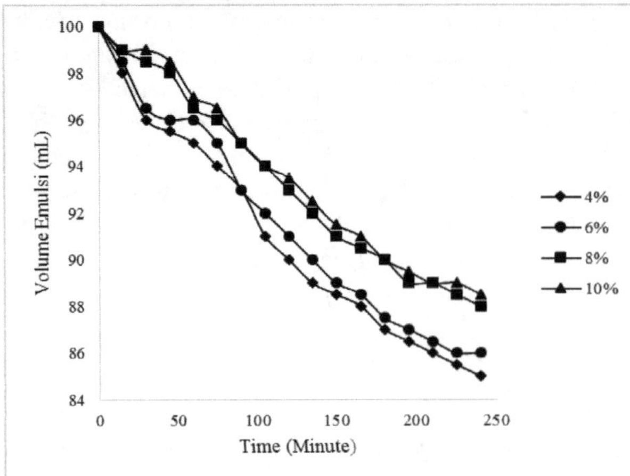

Figure 3. The effect of surfactant Span 80 concentration on the stability of emulsion liquid membrane. Cyanex 272 0.06 M, H_2SO_4 2 M, I/O ratio 1/1

The results of the emulsion stability observation were shown in Figure 3 shows that the Span-80 surfactant concentration increases from 4% w/v to 10% w/v the longer separation between the organic and aqueous phase of the emulsion. This is characterized by a decrease in the emulsion volume that indicates that the effect of surfactant concentration on the emulsion liquid membrane stability is significant.

In Figure 3 was known that with increasing surfactant concentration will increase the emulsion stability. However, in this experiment the use of Span-80 surfactant has not obtained a perfect emulsion stability or in other words there is no emulsion phase which is completely constant or stable in maintaining its volume after 4 hours of observation. This is characterized by the separation of the organic phase and the internal aqueous phase from the emulsion continuously. Although the organic phase separation and the internal aqueous phase of the emulsion take place more slowly at Span-80 concentrations above 8% w/v which is characterized by increasingly sloping curves formed. However, there has not been a constant or stable volume of emulsions for a long period of time.

If surfactant concentration was increased the more surfactants occupy the position in the emulsion interface layer. So the emulsion stability is increased because the surfactant can reduce the surface tension between oil (nonpolar phase) with water (polar phase) by means of the adsorption event in the interface layer of liquids. This is supported by research that has been done by [8], that too low a surfactant concentration can not be a strong enough protector to prevent emulsion droplets from breaking out. While the addition of surfactant levels of more than 8% produce better emulsion stability but not significant. This indicates that the placement of the surfactant in the interface layer through the adsorption event is close to saturation. The event is also due to the higher level surfactant having a limit for the formation of good droplets to produce a good emulsion [9]. Therefore, a further increase in surfactant concentration will only lead to surfactant residues that can not be adsorbed again and accumulate in the membrane/extractant phase. The accumulation of this surfactant residue will increase the viscosity of the whole emulsion phase, giving rise to excess resistance when the extractant wants to transport metal ions from the feed phase to the internal phase.

3.2 *Effect of Cyanex 272 Concentration*

The emulsion is prepared by mixing Span 80 as a surfactant, H2SO4 0.2 M as the internal phase, kerosene as solvent and Cyanex 272 whose concentration is varied as a carrier. The internal phase/organic phase ratio is 1:1 and the concentration of Span 80 is kept constant at 8% w/v. Figure 4 shows the effect of Cyanex 272 concentration on emulsion stability. Effect of Cyanex 272 concentration on emulsion stability The higher concentration of cyanex 272 used stability is decreasing, this is because cyanex 272 also has hydrophilic group and hydrophobic group. Therefore, if extractant quantity is too much it will replace the position of the surfactant at the interface position. Since the concentration and its complex ion are present at the interface it will provoke the loss of the extracted membrane phase and provoke the loss of the internal phase due to swelling [10].

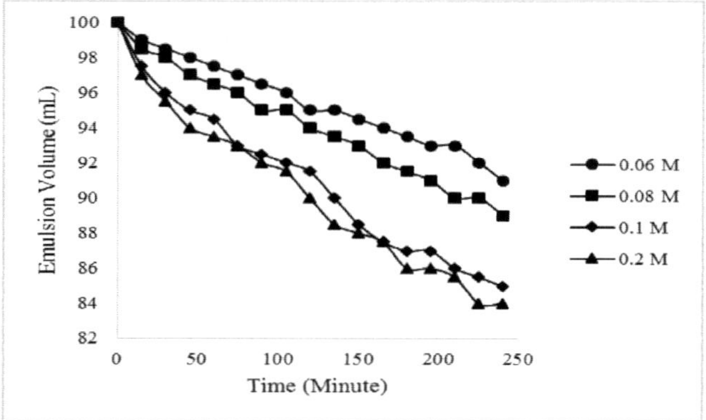

Figure 4. The effect of surfactant Cyanex 272 concentration on the stability of emulsion liquid membrane. Span 80 8% w/v, H_2SO_4 2 M, I/O ratio 1/1.

At low concentrations, the effects of Cyanex 272 concentrations are less visible because at concentrations of 0.06 M this extractant has not been able to compete with surfactants. When the concentration of the extractant was increased to 0.08 M to 0.2 M the effect on the emulsion stability becomes more apparent, as in the graph above, this occurs because the extractant already has a strong enough influence to replace the surfactant position in the interface layer so that it may interfere The stability of the emulsion. When the surfactant position in the interface layer is replaced by the extractant, the surface tension between the polar and polar phases will increase so that the emulsion becomes more susceptible to breakage, since the extractant replacing the surfactant cannot reduce the surface tension as a surfactant function.

The use of high concentration of extractant will also negatively affect the extraction process, because high concentration of extractant will disrupt the emulsion droplets dispersion process, causing the contact surface area between the feed phase (external phase), the membrane phase and the internal phase (Stripping agent) to be reduced which can slow the rate of metal extraction. Meanwhile, from the economic side, the use of extractants at high concentrations is also avoided, because in the manufacture of liquid membrane emulsion extractant price is the most expensive component.

3.3 Effect of Feed Phase pH on Extraction Efficiency

The pH of the feed phase is varied from 5 to 8 as shown in Figure 5. From the figure is known that as the pH of the feed phase increases from 5 to 6, the amount of nickel extraction efficiency produced also increases with the highest value of 81.51% at pH 6. However, when the pH of the feed phase was further increased to 7, the nickel extraction efficiency decreased significantly to only 77.61%. Based on the conditions in the previous study, the pH of feed phase 6 was the optimum condition in the nickel extraction process using Cyanex 272 extractant with an emulsion liquid membrane system [11].

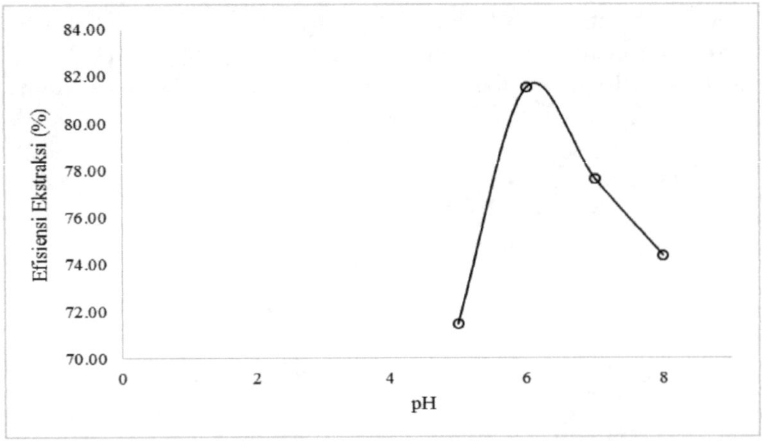

Figure 5. The effect of feed phase pH on nickel extraction

The feed phase pH is too high not can be used in the emulsion liquid membrane system due to the use of SPAN 80 surfactant. Surfactant SPAN 80 is a type of surfactant having a low molecular mass compared to other types of surfactants and has a large hydrophilic group dominantly composed of oxygen atoms with high electronegativity. Therefore, in the alkaline condition SPAN 80 is easily hydrolyzed and causes the emulsion leakege

At pH above 6 (high pH) it is estimated that metal hydroxide precipitation may occur. Metal hydroxide precipitation may occur as the increase in pH means the number of H+ decreases and the amount of OH^- increases to allow the formation of insoluble metal hydroxide species in the distribution process [12].

The effect of pH on the efficiency of extraction is caused by the competition between H3O + and Ni^{2+} in fighting anions. It is therefore desirable to increase the pH of the extraction capacity by increasing the amount of H^+ so that the competition of H^+ and Ni^{2+} for the extractant anion is reduced and Ni2 + can more easily react with the anion than the extractant.

4. Conclusion

As conclusion 81.51% of nickel were extracted under favorable conditions which are feed phase pH 6, Cyanex 272 concentration 0.06 M, Span 80 concentration 8% w/v, 2 M sulphuric acid as stripping agent and 15 min of extraction time.

Acknowledgement

The authors would like to acknoledge the laboratory head and laboratory assistant of Intensification Process Tecnology Laboratory at Chemical Engineering departement, Universitas Indonesia and Universitas Indonesia for making this research possible

References

[1] Pospiech, B. 2015 *Physicochem. Probl. Miner. Process.* **52** 353-364
[2] Harimu L, Sabirin M., Dwi S and Sri J 2010 *Indonesian journal of chemistry* **10**
[3] Kislik V S 2010 *Liquid membranes principles & applications in chemical separation & wastewater treatment*. Amsterdam: Elsevier.
[4] Pospiech B 2015 *Physicochem. Probl. Miner. Process.* **52** 353-364.
[5] Nguyen V T, Lee J C, Jeong J, Kim B S and Pandey B D 2014. *Korean Chemical. Eng. Res* **53** 137-144.

[6] Abou-Nemeh I and Van Peteghem A P 1992 *Journal of membrane science*, **70**.

[7] Zainuddin S 2003 Hg (II) *Ions Extraction using Emulsion Liquid Membrane with Oleat Acid as Carrier, Span 80 and Tween 20 as Mixed Surfactants, and Kerosene as Organik Diluent*. Depok: Universitas Indonesia.

[8] Elsayed F and Farooq A 2014 *International Journal of Chemical and Environmental Engineering* **05** 281-289.

[9] Ahmad A L 2011 *Emulsion Liquid Membrane for Heavy Metal Removal: An Overview on Emulsion Stabilization and Destabilization*. Universitas Negeri Semarang: Semarang.

[10] Valenzuela F, Salinas C, Basualto C, Hagar J S and Tapia C 2003 *Journal of the Chilean Chemical Society*, **48**

[11] Chauduri J B and Pyle D L 1992 *Chemical engineering science*, **47**

[12] Sulaiman R N, Norasikin R, Amin O and Saidina N 2013 *Emulsion Liquid Membrane Stability in The Extraction of Ionized Nanosilver from Wash Water*. Malaysia: Universiti Teknologi Malaysia

IOP Conf. Series: Materials Science and Engineering **333** (2018) 012039 doi:10.1088/1757-899X/333/1/012039

Effect of Post-Braze Heat Treatment on the Microstructure and Shear Strength of Cemented Carbide and Steel Using Ag-Based Alloy

Y Winardi[1], Triyono[1], and N Muhayat[1]
[1]Mechanical Engineering Departement, Engineering Faculty, Universitas Sebelas Maret, Surakarta 57126, Central Java, Indonesia

E-mail : putraalfa86@gmail.com

Abstract. The aim of the present study was to investigate the effect temperature of heat treatment process on the interfacial microstructure and mechanical properties of cemented carbide/ carbon steel single lap joint brazed using Ag based alloy filler metal. The brazing process was carried out using torch brazing. Heat treatment process was carried out in induction furnace on the temperature of 700, 725, and 750°C, for 30 minutes. Microstructural examinations and phase analysis were performed using scanning electron microscopy (SEM) equipped with energy dispersion spectrometry (EDS). Shear strength of the joints was measured by the universal testing machine. The results of the microstructural analyses of the brazed area indicate that the increase temperature of treatment lead to the increase of solid solution phase of enrichted Cu. Based on EDS test, the carbon elements spread to all brazed area, which is disseminated by base metals. Shear strength joint is increased with temperature treatment. The highest shear strength of the brazed joint was 214,14 MPa when the heated up at 725°C.

1. Introduction

In the manufacturing process, the function of cutting tools is very important. However, it needs high cost to make a high quality-cutting tool. It makes the process focuses on selecting and manufacturing cutting tools, so efficiency process is reduced. WC-Co is one of cemented carbide, which is widely used as a cutting tool. Its main compositions are tungsten and cobalt. Tungsten has good strength and resistance, while cobalt serves as a binder and toughness [1, 2]. In order to manufacture a high quality-cutting tool and low-cost production, WC-Co is generally joined with steel. However, the joining of those two materials is difficult. It is because of the differences in properties of both materials. A study shows that joining by welding will produce residual stress. It is because of high difference in coefficient thermal expansion (CTE) [3,4,5].

The joining cemented carbide and steel requires a particular process, such as mechanical assembly, brazing, and diffusion bonding [6]. Brazing is a very potential process and able to joint dissimilar metal. The brazing process consists of several kinds, including vacuum brazing, induction brazing, furnace brazing, and torch brazing. Despite this, brazing is influenced by several elements, such as filler metal selection, characteristics of base metal, heat input, and post brazed treatment. The selection of filler metal is the first step in the brazing process. The filler metal must be compatible with base metals. The good filler metal has a low melting point, high wetness properties, and is able to form bonds with base metals [7].

In recent years, the joining cemented carbide has received real attention. Kaiwa et al [8] examined the effect of adding Ni and Co in the filler metal of induction brazing between the cemented carbides and the martensitic stainless steels. During the brazing process, the specimens were heated to 750ºC with a heating rate of 5ºC / s, and held for 30 seconds. The study results showed that the strength of the joint increased by adding 2.0% mass of Ni and 0.5% mass of Co. In brazing microstructure, Co zone decreases due to dissolution of Co in cemented carbide. Chiu et al [9] studied the effect of temperature and holding time of vacuum brazing between SAE1045 carbon steel with hard alloy WC-Co using copper and brass alloys. Based on the results of the study, the rises in temperature and time of brazing influence on microstructure and mechanical strength. The result of microstructure observation, Fe-Co-Cu alloy layer is formed on the surface of base metal. The higher temperature causes the thickness of Fe-Co-Cu layer to increase. The maximum shear strength is 320 ± 10 MPa was brazed at 1140ºC for 15 minutes.

Besides, Temperature also plays an important role because the filler metal melts at a certain temperature. Based on literature review, temperature rise can improve the properties of the joints. Mousavi et al [10] examine the CK35 steel joint brazed with the cemented carbide using copper filler metal at different temperatures. In this study, it is focused on the properties of wetting, metallographic, and mechanical properties. Wet tests show that brazing time increases, the contact angle decreases. The microstructure in the brazing region indicates a copper-enriched primary phase. The eutectic structure of the silver matrix consists of copper-enriched particles. The shear test results show that the maximum shear strength is 108 MPa at 800°C.

Criss and Meyers [11] investigated the joining of cobalt using silver-copper alloy. The observation results indicate that the weld crack can be minimized, while the bending test indicates the delamination of the filler base does not cause a failure. Effect of holding time on the microstructure and mechanical properties of ZrO2 / TiAl joints brazed with Ag-Cu filler metals using vacuum brazing was reported by Sun et al [12]. The investigation results show that the Cu3Ti3O + TiO layer is formed adjacent to the ZrO2 ceramic while the AlCu2Ti layer is formed on the TiAl substrate. The highest shear strength of 48.4 Mpa was achieved at 880 ° C for 10 minutes. In this study, cemented carbides and carbon steel were brazed using silver filler metal. Then post brazed heat treatment is used to improve the joint properties. The effect of treatment temperature is discussed and studied.

2. Experimental

Silver-based metal alloys are used for filler metals. The chemical composition is shown in Table 1. Carbide tips (YG6) from Zigong Carbide Corp. Ltd. China with size 25x15x8.5 mm and low carbon steel (SS400) with size 40x15x10 mm serve as test specimen. The chemical composition of the two base metals is shown in Table 2. SiC abrasive papers are used to rub the surface of test specimens. Dirt or contaminants on both surfaces of test specimens are cleaned using acetone. Before it is joined, the test specimen is assembled as shown in Figure 1. Thickness gauge 0.1 mm is used to control the spacing between specimen surfaces. To facilitate during the joining process, a jig tool is used, as shown in Figure 2(a).

Table 1. Chemical composition of filler metal

Metal	Ag	Cu	Zn	Ti	P	Other
(wt%)	Main	30	30	0.3	-	-

Table 2. The chemical composition of specimen

Metal	C	Si	Mn	P	S	Ni	Cr	Fe	WC	Co
SS400	0.22	0.59	0.59	0.017	0.02	-	-	Main	-	-
YG6	-	-	-	-	-	-	-	-	94	6

IOP Conf. Series: Materials Science and Engineering **333** (2018) 012039 doi:10.1088/1757-899X/333/1/012039

The brazing process is carried out using torch brazing. After brazing, all specimens are removed from the remaining flux attached to the specimen's surface. Then the test specimen is prepared for the heat treatment process. The induction furnace is used as a heater. The treatment temperature was set at 700, 725, and 750°C respectively and held for 30 minutes. After that, all specimens were cooled to room temperature. All test specimen surfaces are grinded using diamond grinding and pureed using a SiC abrasive paper (grit 320-5000). Then it is polished using a metal polish paste. Microstructures In brazing regions characterized by an electron-scanning microscope (SEM) equipped with X-ray dispersion energy spectrometer (EDS). The mechanical strength of the joint is tested using a universal test machine (UTM-SANS SHT 4106) with a constant velocity of 0.5 mm/min at room temperature. The shear test scheme is illustrated in Figure 2(b).

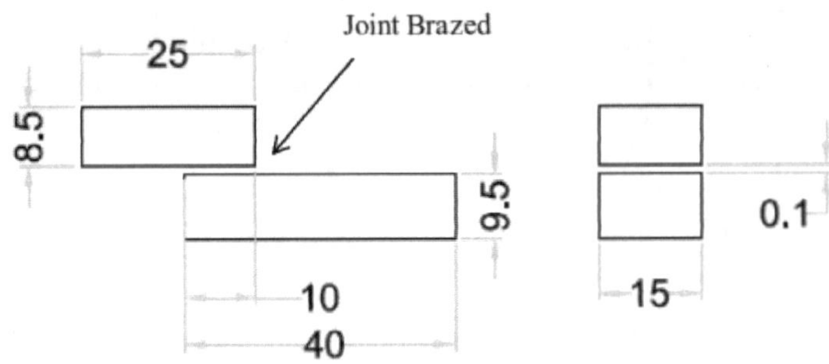

Figure 1. Schematic and types of joints

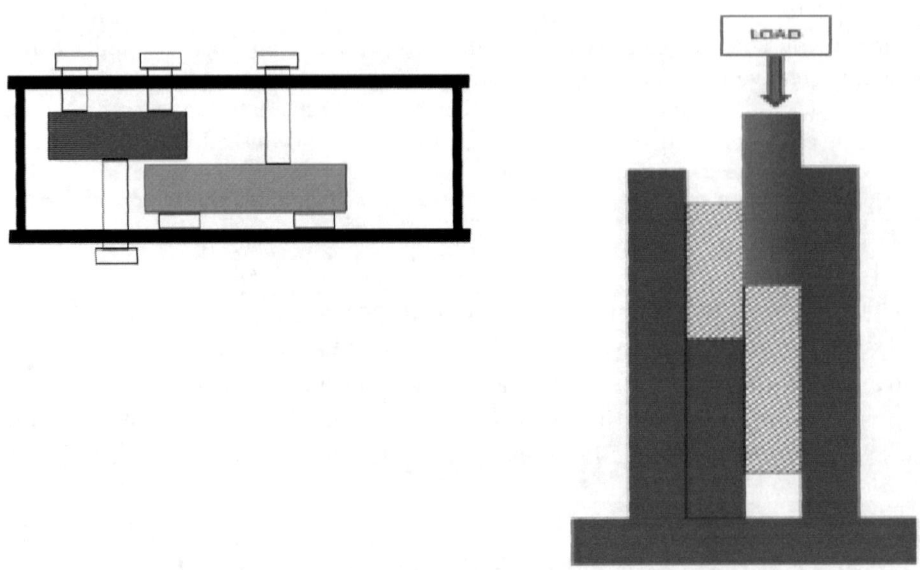

(a) (b)

Figure 2. (a). Jig, (b). Schematic of shear testing

IOP Conf. Series: Materials Science and Engineering **333** (2018) 012039 doi:10.1088/1757-899X/333/1/012039

3. Results and Discussion

The microstructure of brazed joints between carbide tips and carbon steel was observed with SEM, as shown in Figure 3. The results showed that the brazing area consisted of 3 parts: I) Dark colored area with a large texture, II) Light colored areas, and III) Darker areas with smaller textures. Region I is a solid solution phase of filler metal enriched with Cu, Zn, C and Ag. Region II is an Ag matrix enriched with Zn, C, and Cu. Region III is the eutectic phase of Ag-Cu-Zn and C with lower copper elements. The element is widespread in the brazing area. The results of the EDS test on the different region show that the three regions contain C elements released by the parent metal into the area of brazing. This indicates that the C element is capable of diffusing into the filler metal solid solution. The results of the EDS tes are shown in Table 3. Based on the Ag-Cu-Zn phase diagram [13], the eutectic phase occurs at 780 ° C, but with the addition of Zn 10-30 wt% into the alloy may decrease the fill metal liquids to 500°C [14]. Then this can be found at brazed joints by treatment at 700°C for 30 min, as shown in Figure 3(a) but this number of phases gradually decreases with increasing treatment temperature. Figure 3(b) shows the microstructure of the joint brazed at 725°C.

Table 3. Chemical composition on the different region

Region	Chemical Compositions (wt%)				
	Ag	Cu	Zn	Fe	C
I	5.1	55	34.3	-	5.7
II	86.5	3.4	5.6	-	4.5
III	5.6	48.8	39.6	1.4	1.4

It is seen that the eutectic phase spread is less and vice versa the solid solution phase of Cu is greater. Moreover, the enlarged solid phase of Cu-enriched solution is increasing at 750 ° C as shown in Figure 3(c). A previous study [10] states that the cooling rate during solidification has a very important role in the formation of microstructure. It has a slower cooling rate compared to lower treatment temperatures at higher treatment temperatures. Therefore, Cu grainds have more time to grow. So the homogeneity of microstructure can be achieved.

Figure 4. shows the average shear strength of the joints, which are treated at different temperatures for a constant holding time of 30 minutes. It can be seen that the average shear strength increases up to 725°C, and then decrease when the temperature of treatment increases. The maximum shear strength of the joints was 214,14 MPa at 725°C. Increasing of the treatment temperature can improve the fluidity of filler metals and promote the diffusion of elements to achieve a homogeneous microstructure, as reported by the previous literature [9,15,16,17]. Nevertheless, the increasing temperature of treatment can cause the grain growth, and reduce the shear strength.

IOP Conf. Series: Materials Science and Engineering **333** (2018) 012039 doi:10.1088/1757-899X/333/1/012039

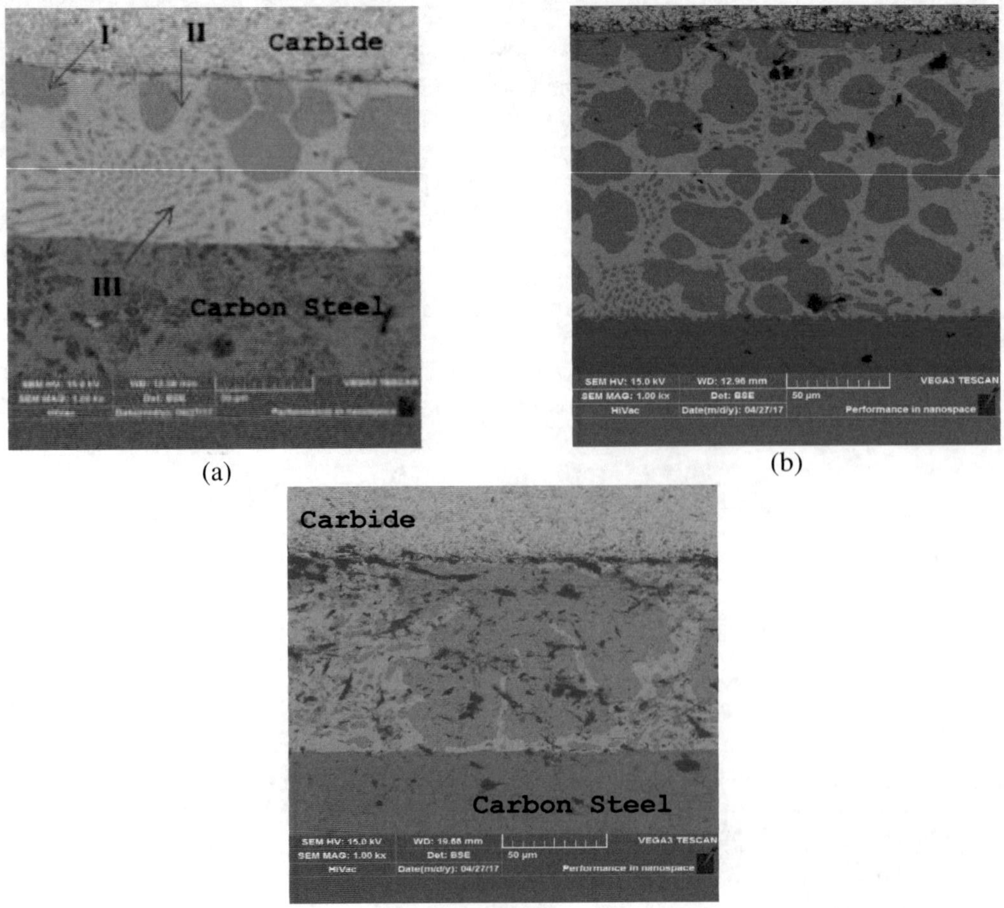

Figure 3. SEM image of brazed joint using silver filler metal (a). 700°C, (b). 725°C, (c). 750°C

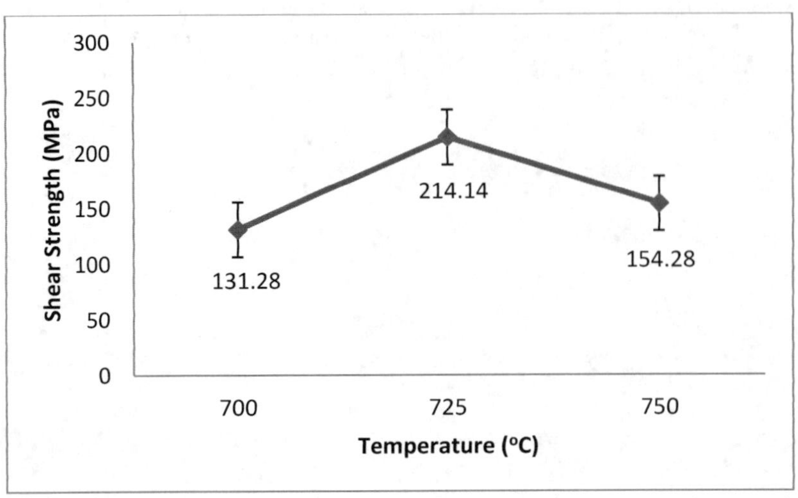

Figure 4. Shear strength of the joints brazed at different temperatures for 30 minute.

4. Conclusions

This study aims to know the effect of post brazed heat treatment on the microstructure and mechanical properties of cemented carbide and steel using silver filler metal. The results obtained are summarized as follows: Microscopic observation results show that the dark area is the phase of Cu-enriched solid solution. Rising treatment temperatures increase the area of Cu solid phase. This is because, during the solidification process, Cu solid phase has more time to grow and develop. The brightest area of the brazing area is the Zn-enriched Ag matrix. EDS test results show that C element is spread to all brazed area. The shear strength of brazing joints increases with the treatment temperature, and then decrease when the temperature up to 750°C. The maximum shear strength reaches 214,14 MPa at 725°C.

References

[1] Chen H, Feng K, Wei S, Xiong J, Guo Z, and Wang H 2012 *Int. J. Refract. Met. Hard Mater.* **33** 70–74
[2] Chen H, Feng K, Xiong J, and Guo Z 2013 *J. Alloys Compd.* **557** 18–22.
[3] Al-Samhan A M 2012 *J. King Saud Univ. - Eng. Sci.* **24** 85–94
[4] Krawitz A and Drake E 2015 *Int. J. Refract. Met. Hard Mater.* **49** 27–35
[5] Kayser W, Bezold A, and Broeckmann C 2015 *Int. J. Refract. Met. Hard Mater*
[6] Miranda R M 2014 *Joining Cemented Carbides.* Elsevier Ltd **1** 527-538.
[7] American Welding Society (AWS) Committee on Brazing and Soldering 1991 *Brazing Handbook 5th edition* (Miami:American Welding Society)
[8] Kaiwa K, Yaoita S, Sasaki T, and Watanabe T 2014 *Adv. Mater. Res.* **922** 322–327
[9] Chiu L, Wang H, Huang C, Hsu C, and Chen T 2008 *Adv. Mater. Res.* **47** 682–685
[10] Akbari Mousavi S A A, Sherafati P, and Hoseinion M M 2012 *Adv. Mater. Res.* **445** 759–764
[11] Criss E M and Meyers M A 2015 *J. Mater. Res. Technol.* **4** 44–59
[12] Dai X, Cao J, Liu J, Su S, and Feng J 2015 *Mater. Des.* **87** 53–59
[13] A. S. M. International 1992 *METALS HANDBOOK VOL 3 Alloy Phase Diagrams* (Materials Park, Ohio:ASM International)
[14] Tsao L C, Chiang M J, Lin W H, Cheng M D, and Chuang T H 2002 *Mat Char.* **48** 341–346
[15] Li L, Li X, Hu K, Qu S, Yang C, and Li Z 2015 *Mater. Sci. Eng.* **634** 91–98
[16] Cao Y, Yan J, Li N, Zheng Y, and Xin C 2015 *J. Alloys Compd.*, **650** 30–36
[17] Jiang C, Chen H, Wang Q, and Li Y 2016 *J. Mater. Process. Technol.* **229** 562–569

IOP Conf. Series: Materials Science and Engineering **333** (2018) 012040 doi:10.1088/1757-899X/333/1/012040

Preliminary Study on the Dissolutions of Ce, Nd, Y and La from Mineral Cassiterite by Acid and Alkaline Leaching

F Firdiyono[1], L Andriyah[1], F N Aini[2], T Arini[1] and L H Lalasari[1]
[1]Research Center of Metallurgy and Materials, LIPI, Puspiptek 470, Indonesia
[2]Chemistry Department, FMIPA of Sebelas Maret University, Surakarta, Indonesia

E-mail: andriyah_yong@yahoo.com, liaa001@lipi.go.id

Abstract. Rare Earth Metal is a rare element that its availability in nature is very small. In Indonesia, the potential of rare earth metals is generally found as the associated mineral in major commodities, especially gold and alluvial tin. These associated minerals can be processed using a particular technology so that the result is a by-product that can increase the added value of the mineral. This purpose of this research was to investigate the dissolution of Cerium (Ce), Neodymium (Nd), Yttrium (Y) and Lanthanum (La) from mineral cassiterite by leaching process using dilute hydrochloric acid (HCl), sulfate acid (H_2SO_4) and sodium hydroxide (NaOH). Firstly, cassiterite was grinded to -100 mesh of particle size and characterized by X-Ray Diffraction (XRD) and X-Ray Fluorescence (XRF) techniques. Secondly, 10 gram of cassiterite was leached in 100 ml solution of 3.26 N HCl, H_2SO_4 and NaOH at variation leaching time of 2, 4, 6, 24 and 48 hours in atmospheric conditions. The products were then filtered to separate filtrate and residue of cassiterite. Finally, to investigate the dissolution of Ce, La, Nd and Y, filtrate from dissolved cassiterite was analyzed by Induced Coupled Plasma-Optical Emission Spectrometry (ICP-OES), while to know the chemical composition of cassiterite leached by dilute HCl, H_2SO_4 and NaOH, residue products of cassiterite was characterized by XRF analysis. The result of ICP-OES analysis showed the dissolution of Ce element higher than Nd, Y and La elements for leaching cassiterite using HCl, H_2SO_4 and NaOH. The increase of leaching time was accompanied by the rise in the amount of dissolved elements from cassiterite. The result of XRF analysis showed the chemistry composition of Ce, Nd, Y and La elements on residue decreased insignificantly from chemistry composition of cassiterite (raw mineral) in all conditions. However, the dissolution of Ce, La, Nd and Y was insignificant in all conditions.

1. Introduction

Indonesia is one of the countries rich in mineral resources. Several studies have shown that some minerals contain undeveloped rare earth metals. Rare Earth Metal is a rare element that its availability in nature is very small. The use of rare earth metals is very broad and closely related to high-tech products such as computer, telecommunication, nuclear and aerospace industries. In the future it is estimated that the use of rare earth metals is increasingly proportional to the development of such high technology. Rare earth metal is the trigger for the birth of new technologies that will continue to grow such as LCDs, magnets and hybrid batteries. This resulted in demand for rare earth metals will

[1] To whom any correspondence should be addressed.

continue to increase. Its incapacity in nature is rare and the level of use for high technology is increasingly widespread, resulting in the price tends to increase.

In Indonesia, the potential of rare earth metals is generally found as the associated mineral in major commodities, especially gold and alluvial tin. These associated minerals can be processed using a particular technology so that the result is a by-product that can increase the added value of the mineral. The potential of gold and tin precipitation as a carrier of rare earth metals is rarely spread across several islands in Indonesia. It is found as the associated mineral of main mineral. The tin ore contains the main minerals of cassiterite (SnO_2) and associated minerals such as columbit-tantalite [(Fe, Mn) $(Ta,Nb)_2O_6$], zircon ($ZrSiO_4$), ilmenite ($FeTiO_3$), rutile (TiO_2), quartz (SiO_2), pyrite (FeS_2), xenotime (YPO_4) and monazite (Ce, La, Y, Th) PO_4. The results of Lalasari et al's study [1] explain that the cassiterite minerals from Bangka, Sumatera, Indonesia contain 3.43% Cerium (Ce), 3.47% Lanthanum (La), 2.18% Neodymium (Nd) and 0.36% yttrium (Y).

Many researchers had reported the result of their investigation, such as : F.Habashi succeeded in extracting of rare earths elements from monazite and xenotime by leaching process with sodium hydroxide (NaOH) and sulfuric acid (H_2SO_4)[2]. Kim et. al used the sulfuric acid leaching after mechanochemical treatment of beneficiated monazite (212 μm size) obtained from the beach sand deposit of Malaysia with sodium hydroxide at room temperature[3]. Kuzmin et. al has reported the nitric acid to leach out the metals from the ground RE ore of Chuktukon deposits, Russia containing 3–7% REOs in the phosphate form (monazite, turnerite, florensite), Fe–Mn oxides and 0.5–1% Nb oxide (chalcolamprite) in a rotating autoclaves. REEs recovery of 87–90% is obtained during leaching in autoclave in the temperature range of 180–220 °C of the ground ore [4]. Kumari and co-worker studied two step leaching of Korean monazite using diluted HCl at moderate temperature for 2 h and pulp density < 100 g/L resulting in only 25% REMs dissolution [5]. Then the concentrate left after phosphate removal was leached in 6 N HCl at 90 °C for 2 h maintaining 60 g/L pulp density resulting in 90% REMs recovery [6].

Various methods have been undertaken to increase the levels of rare earth metals in the main minerals. These include, for example, multilevel precipitation, multilevel crystallisation, ion exchange columns and liquid extraction by attempting various solvents which are selective to one of the desired elements. Separation by way of extraction is mostly done because it has many advantages when compared with other ways that is in terms of process, time and more simple equipment [7]. The purpose of this experiment is to study the solubility of rare earth (Ce, Nd, Y and La) elements in the cassiterite ore in acidic and basic solutions. The acid and base solutions used in this experiment are sulfuric acid (H_2SO_4), hydrochloric acid (HCl) and sodium hydroxide solution (NaOH).

2. Experimental

The raw material used in the experiment was cassiterite ore obtained from Bangka Island, Indonesia. The chemical composition of cassiterite minerals used can be seen in Table 1.

Table 1. The chemical composition of cassiterite from Bangka Island, Indonesia

Element	Sn	La	Ce	Nd	Fe	P	Ti	Ca	W	Al	Y
%Wt	57.82	3.43	3.37	2.18	1.79	1.68	1.45	0.63	0.62	0.52	0.36

This experiment has four stages of activities that are sample preparation, leaching process, filtration and characterization. In the sample preparation, the cassiterite minerals were grinded to a size of -100 mesh using a disc mill, then characterized by XRF to know the chemical composition as shown in Table 1.

In the leaching process, 10 grams of cassiterite were weighed, then dissolved in the three neck flask using variation solvents: 100 ml of HCl 3.26 N, 100 ml of H_2SO_4 3.26 N and NaOH 3.26 N obtained from dilution. The dissolution process was done with variation times of 2, 4, 6, 24, and 48

IOP Conf. Series: Materials Science and Engineering **333** (2018) 012040 doi:10.1088/1757-899X/333/1/012040

hours at temperature 100 ° C-110 °C. For NaOH solvent, leaching process was done without using condenser, with variation times of 2, 4, and 6 hours.

The result of the leaching process is filtered by whatman paper No. 42, and obtained filtrate and residue. The residue is then washed with distilled water until the pH approaches 7 and continued the drying process in the oven at 110 °C for one hour. After one hour, the filter paper containing the residue was weighed in cold conditions and calculated for weight reduction of each experimental variable. The last step was characterization that filtrate product was analyzed using ICP-OES and residue product characterized by XRF.

The reaction of Rare earths elements (Ce, La, Y, Th) PO_4 with H_2SO_4 and NaOH could be described as follows [8]:

$$2\,REPO_4 + 3H_2SO_4 \rightarrow RE_2(SO_4)_3 + 6H^+ + 2\,PO_4^{3-} \quad (1)$$

$$REPO_4 + 3\,NaOH \rightarrow RE(OH)_3 + Na_3PO_4 \quad (2)$$

while the reaction of Rare earths elements (RE)OH with HCl could be described as follows :

$$RE(OH)_3 + 3\,HCl \rightarrow RECl_3 + 3H_2O \quad (3)$$

3. Results and discussion

To investigate the solubility of Ce, La, Nd and Y from cassiterite in variation solvents, it is necessary to observe visually and characterize the product from the leaching process using ICP-OES analysis to identify the solubility of Ce, La, Nd, Y from the filtrate and XRF analysis for know the chemical composition of the residue. The experimental results and characterization are discussed below.

3.1. The Filtrate Color of Leaching Process

Figure 1. The filtrate of leaching process in variation solvent:
(a) H_2SO_4; (b) HCl; (c) NaOH

Figure 1 represented the color of filtrate in the various solvent of leaching process. Fig. 1a, the color of the filtrate of leaching process with H_2SO_4 solution was not changed significantly while Fig.1b, the filtrate changed to be the dark green color with increasing the time of leaching process (2-48 hours) of HCl solution. The change of filtrate color indicated that cassiterite dissolved a lot shown in Fig. 2.

Fig. 1c represented the color of filtrate to be the dark orange color with increasing leaching time (2-6 hours) of NaOH solution. The change of filtrate color indicated that cassiterite dissolved significantly in NaOH solution. The same is done on previous research that F.Habashi [2] where succeeded in extracting of rare earths elements by leaching process with sodium hydroxide (NaOH).

3.2. The Dissolution of Cassiterite

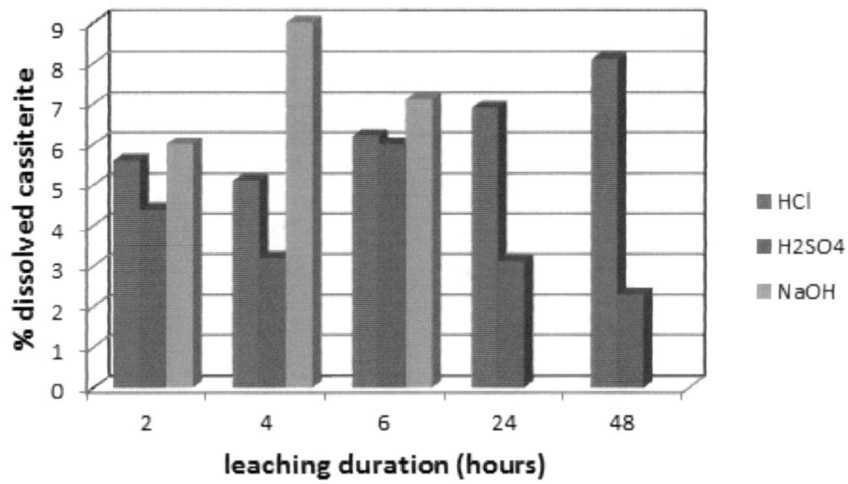

Figure 2. The percentage of dissolved cassiterite in variation solvent (HCl, H_2SO_4, NaOH)

Fig. 2. showed the percentage of dissolved cassiterite in various concentrations. The percentage of dissolved cassiterite rise in proportion to the increase of leaching time. The percentage of dissolved cassiterite was maximum at leaching time 4 hours with NaOH solvent that was about 9 % and the dissolution of cassiterite was not rising significant with hydrochloric and sulfate acid. The same is done on previous research that Lalasari[1] where the dissolution of cassiterite was not rising significant with the increase of hydrochloric acid concentrations.

From the result of a investigation in Fig. 2 could be concluded that the dissolution of cassiterite was not rising significantly with the increase of leaching time in all condition of solvents.

3.3. The Dissolution of Ce, Nd, Y and La from Cassiterite

To study effects of leaching time on the dissolution of rare earth (Ce,Nd,Y, La) elements from cassiterite-Indonesia, 10 gr of cassiterite was dissolved in various solvents (HCl, H_2SO_4,NaOH) with variation of leaching time (2,4,6,24,48 hours) at temperature of 100-110°C. Mixture then was filtrated and filtrate was characterized by ICP-OES analysis. The result of filtrate and residue product were shown in Figure 3 and Table 2.

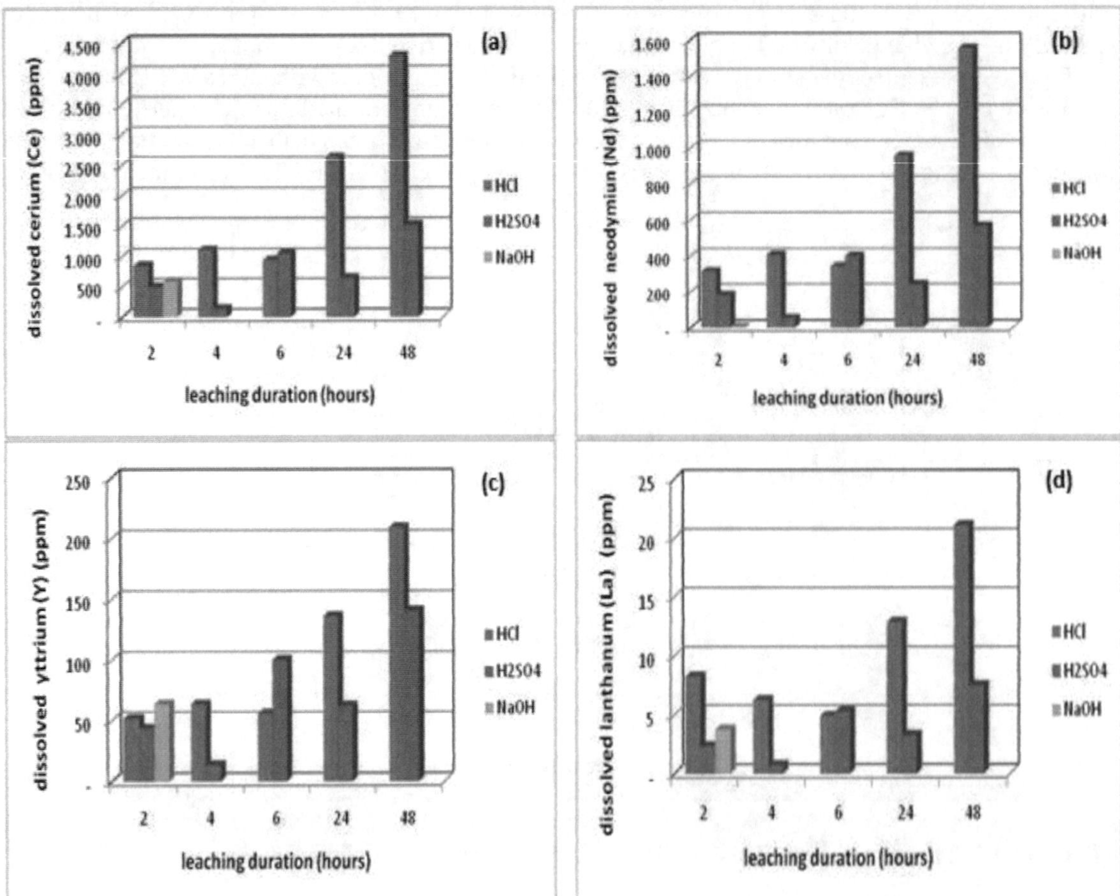

Figure 3. The concentration of dissolved rare earths elements in variation solvent HCl, H_2SO_4 and NaOH (ppm) : (a) Ce ; (b) Nd; (c) Y and (d) La

Figure 3 represented the result of ICP-OES analysis for concentration of dissolved rare earth (Ce,Nd,Y,La) elements. Fig 3 showed the dissolution of Ce element higher than Nd, Y and La elements for leaching cassiterite using HCl and H_2SO_4 solvents. The elements of cerium extracted higher than elements of Nd, Y and La. This condition described that Nd, Y and La was more stable than cerium in acid solution. The increase of leaching time was accompanied by the rise in the amount of dissolved rare earth elements from cassiterite. However, the dissolution of Ce, Nd, Y and La elements was insignificantly in all conditions.

Based on Fig 3, leaching cassiterite using H_2SO_4 solvent more difficult to dissolved than using HCl and NaOH solvent, because H_2SO_4 solvent more unstable than other processes. The same is done on previous research that Kumari[5] and Panda[6] where studied of rare earths elements from Korean monazite using diluted HCl at moderate temperature for 2 h and then the concentrate left after phosphate removal was leached in 6 N HCl at 90 °C for 2 h maintaining 60 g/L pulp density resulting in 90% REMs recovery.

3.4. The Chemical Composition of Residue Product from Leaching Process

Table 2. The chemical composition of residue product in variation solvents

	Percentage (%wt)			
Element	**HCl. 2 hours**	**H_2SO_4. 2 hours**	**NaOH. 2 hours**	**HCl. 48 hours**
Sn	64.85 %	64.70%	64.61%	68.32%
Ce	2.19%	2.29%	2.46%	1.42%
La	2.33%	2.49%	2.67%	1.28%
Si	0.24%	0.22%	0.11%	1.09%
Ti	1.43%	1.36%	1.33%	0.88%
Fe	1.29%	1.27%	1.46%	0.76%
P	1.00%	1.01%	1.08%	0.70%
Nd	1.12%	1.38%	1.33%	0.68%
Cl	0.38%	0.39%	0.41%	0.36%
W	0.375	0.48%	0.19%	0.30%
Zr	0.30%	0.30%	0.29%	0.27%
Y	0.26%	0.29%	0.28%	0.26%
Pr	0.37%	0.44%	0.45%	0.21%
Cd	0.20%	0.16%	0.08%	0.18%
Sb	0.12%	0.08%	0.11%	0.11%
Cr	0.13%	0.13%	0.12%	0.09%
Ta	0.07%	0.10%	0.09%	0.08%
Nb	0.07%	0.07%	0.07%	0.07%
V	0.07%	0.07%	0.06%	0.07%
Te	0.09%	0.04%	0.11%	0.06%
S	0.08%	0.09%	0.09%	0.04%
Pb	0.02%	0.03%	0.03%	0.02%

Table 2 describes the result of XRF analysis from residue products that showed the chemistry composition of rare earth (Ce, Nd, Y and La) elements on residue product decreased insignificant than the chemistry composition of cassiterite (raw mineral) in all conditions. The small change of the chemistry composition indicated that rare earth (Ce, Nd, Y and La) elements dissolved insignificantly in HCl, H_2SO_4 and NaOH solvents.

4. Conclusion

The result of ICP-OES analysis showed the dissolution of Ce element higher than Nd, Y and La elements for leaching cassiterite using HCl, H_2SO_4 and NaOH. The increase of leaching time was accompanied by the rise in the amount of dissolved rare earth elements from cassiterite. Leaching cassiterite using H_2SO_4 more difficult to dissolve than using NaOH and HCl. The result of XRF analysis showed the chemistry composition of Ce, L, Nd and Y elements on residue products decreased insignificantly from chemistry composition of cassiterite (raw mineral) in all conditions.

Acknowledgment

The authors are grateful for the financial support of the Competence Grant from the Research Center for Metallurgy and Materials LIPI.

References

[1] L H Lalasari, A Suharyanto, and F Firdiyono 2017 *IOP Conf. Series: Journal of Physics: Conf.* **817** 1–6
[2] F Habashi 2013 *Canadian Metallurgical Quarterly*.**52** 3
[3] Kim W, Bae I, Chae S, Shin H, 2009 *J Alloys Comp*. **486** 610–614
[4] Kuzmin, V I, Pashkov G L, Lomaev V G, Voskresenskaya E N, Kuzmina V N, 2012 *Hydrometallurgy* **129–130** 1–6
[5] Kumari A., Jha M K, Hait J, Sahu S K, Kumar V, 2013 *J. Ind.Chem. Soc*. **90** 2105–2110.
[6] Panda R., Kumari A., Jha M K, Hait J, Kumar V, Lee J Y, Kumar J R., 2014 *J. Ind. Engg. Chem.* **20** 2035–2042
[7] A Amin 2009 *J. Rekayasa Kim. dan Lingkungan*. **7**
[8] Archana Kumari., Rekha Panda., Manis Kumar Jha., J Rajesh Kumar, Jin Young Lee , 2015 *Minerals Engineering*. **79** 102–115

IOP Conf. Series: Materials Science and Engineering **333** (2018) 012041 doi:10.1088/1757-899X/333/1/012041

Selection of organic acid leaching reagent for recovery of zinc and manganese from zinc-carbon and alkaline spent batteries

Yuliusman[1], R A Amiliana[1], P T Wulandari[1], I T Ramadhan[1], and F A Kusumadewi[1]

[1]Chemical Engineering Department, University of Indonesia, Depok 16424, Indonesia

E-mail: usman@che.ui.ac.id

Abstract. Zinc-carbon and alkaline batteries are often used in electronic equipment that requires small quantities of power. The waste from these batteries contains valuable metals, such as zinc and manganese, that are needed in many industries and can pollute the environment if not treated properly. This paper concerns the recovery of zinc and manganese metals from zinc-carbon and alkaline spent batteries with leaching method and using organic acid as the environmental friendly leaching reagent. Three different organic acids, namely citric acid, malic acid and aspartic acid, were used as leaching reagents and compared with sulfuric acid as non-organic acid reagents that often used for leaching. The presence of hydrogen peroxide as manganese reducers was investigated for both organic and non-organic leaching reagents. The result showed that citric acid can recover 64.37% Zinc and 51.32% Manganese, while malic acid and aspartic acid could recover less than these. Hydrogen peroxide gave the significant effect for leaching manganese with non-organic acid, but not with organic acid.

1. Introduction

Zn-C and Alkaline batteries are primary batteries that often found in the market because of its wide use in various electronic devices and the price is quite affordable. Zn-C and Alkaline batteries have the same electrode metal elements, Zn as anodes and Mn as cathodes, so often called Zn-Mn batteries, differ only in the type of electrolyte used. It is estimated that about 40 billion of Zn-Mn batteries are run out rapidly and thrown away, producing huge quantity of spent batteries which represent major environmental and health threat due to toxic metals such as Zn and Mn [1]. The Zn content are about12-28% and the Mn content reaches 26-45% of the total powder mass in Zn-Mn batteries [2]. Zn and Mn metals are also a valuable metal in industrial world and have been listed as strategic metals by many countries due to the increasing use of these metals with increasingly limited natural resources [3]. Because of the dangers for the environment and the importance of Zn and Mn metals for life, so it is necessary to recover the Zn and Mn metals, especially from the considerable amount of waste batteries in the world.

Leaching is a solid-liquid extraction method in which the process of separating solid metal waste was mixed with leaching reagent, a liquid solvent to dissolve the metal in it. The commonly used leaching reagents are strong acids such as sulfuric acid, nitric acid and hydrochloric acid. However, the use of strong acids as leaching reagents can release toxic gases like Cl_2, SO_3, and NO_x during the leach and the waste acid solution after leaching is a threat to the environment [4]. Reduced adverse effects of the leaching process can be done by replacing the leaching reagent with a more environmentally friendly reagent, which is organic acid because it is more soluble in water and the waste from the leaching process with organic acid is considered more easily processed.

This study was conducted to determine the effectiveness of organic acids as environmentally friendly leaching reagent in the process of leaching zinc and manganese from Zn-C and alkaline spent batteries. The organic acids that used were citric acid, malic acid, and aspartic acid. The three organic acids were selected because of their characteristics including easy natural degradation and the absence of toxic gases in the reaction process, and because of previous reports of other metal leaching processes. In the previous study, leaching on zinc from low grade oxide ores using citric acid was able to extract 82% of Zn metal [5]. Whereas malic acid and aspartic acid had been used as leaching reagents of Li-On batteries and could recover until 90% Co, 100% Li for malic acid and 60% Co, 60% Li for aspartic acid [6].

This study also examined the use of hydrogen peroxide as a manganese reducing agent in the leaching process with organic acids and non-organic acids. The low dissolved Mn element in sulfuric acid caused by the solubility of manganese oxide such as Mn_2O_3 and Mn_3O_4 in sulfuric acid is partial because the produced of MnO_2 is difficult to dissolve on strong acids [7]. The MnO_2 compound is most stable manganese oxide with oxidation number 4+, so an additional reducing agent is required to reduce manganese oxide number to be more easily dissolved in the leaching process. Based on previous research, hydrogen peroxide gave a quite effective result and cheaper than other reducing agents, so it was often used as manganese reducing agent [8].

2. Experimental

2.1. Materials and reagents

The spent Zn-C and alkaline batteries were collected from the local sources. The leaching reagents used in this work consisted of citric acid, malic acid, and aspartic acid as organic acid, then sulfuric acid as non-organic acid, and hydrogen peroxide as reducing agent. All solutions were prepared in distilled water and all reagents were analytical grade (Merck)

2.2. Battery dismantling and metal characterization

The spent batteries from several manufactures were manually dismantled. The black powder was a mixture of the cathodic (manganese oxides and graphite) and the anodic (zinc oxides and electrolyte solution). Dismantling products such as plastic film, ferrous scraps and paper pieces were separated. The battery powder then crushed and mashed to expand the surface area and make to increase leaching efficiency. After that, the battery powder was sieved to uniform the size of sample particles. The initial metals content of the powder were characterized using Scanning Electron Microscopy-Energy Dispersive X-ray Spectroscopy (SEM-EDS, Jeol JSM-6510LA) and Atomic Absorption Spectroscopy (AAS, spectrometer SpectrAA 200-Varian) in order see the characteristic of the powder and to evaluate the percentages of zinc and manganese metals.

2.3. Leaching of metals

All leaching experiments were carried out in a 250 mL Pyrex glass reactor with a thermometer to control the temperature and placed at hot plate stirrer to heat up the solution. The reactor was closed to reduce the loss of water by evaporation. A measured amount of battery powder and 2 M reagent leaching solution were added to reactor and allowed to reach thermal equilibrium, with agitation provide by the magnetic stirrer at 500 rpm for entire duration of the leaching. The leaching operation conditions were made equally to facilitate comparison of the leaching efficiency of each different leaching reagent and the same amount of hydrogen peroxide were added as the reducing agent. The solid-to-liquid ratio (S/L) that used was 1/20 with operating temperature 80°C (353 K) and 3 hours reaction time. All the leach liquor after leaching process were analyzed by AAS (Atomic Absorption Spectroscopy) to determine the total amount of zinc and manganese metals after the process.

3. Result and Discussion

3.1. Characterization of spent battery powder

The powder of Zn-C and Alkaline spent batteries mixture was tested with SEM-EDX and AAS. The SEM-EDX test is performed to determine the morphology and the elements that contained in the initial sample. While the AAS test is performed to determine the amount of concentration zinc and manganese quantitatively as the basis for calculating the percentage of leaching efficiency. The SEM and EDS test results can be seen in Figure 1 and Figure 2.

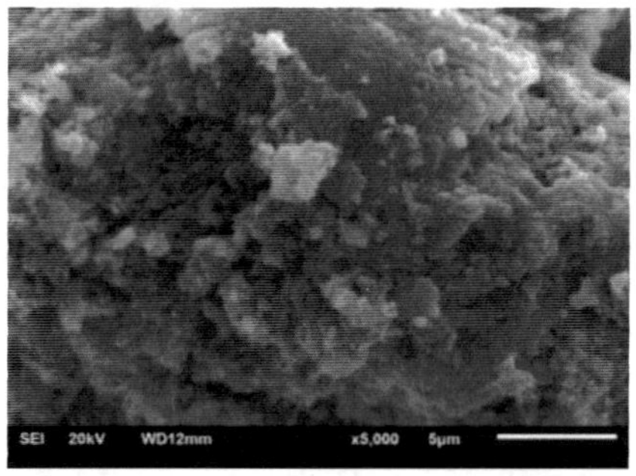

Figure 1. SEM result of Zn-C and alkaline batteries powder

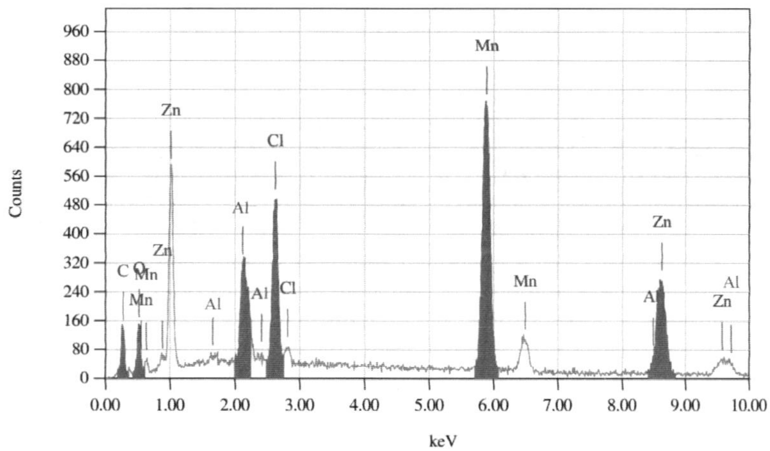

Figure 2. EDS result of Zn-C and alkaline batteries powder

Figure 1 show the SEM image, which reveals the fine nature of powder was identified as irregular and agglomerated. The EDS spectrum in Figure 2 shows the metals contained in Zn-Mn batteries powder are zinc, manganese, and aluminum. In addition, there are also elements of carbon and oxygen detected in the EDS test. This result is in accordance with the research conducted by I De Michelis et al. in 2007 [9] for the metal content of a mixture of Zn-C and Alkaline batteries. The AAS test results can be seen in Table 1.

Table 1. Zn and Mn content of Zn-C and alkaline batteries powder

Element	Content (% w/w)	Unit conversion (mg/g sample)
Zinc	14.52	145.20
Manganese	22.66	226.60

According to Table 1, the concentrations of zinc and manganese metal elements are 14.52% and 22.66% respectively. This result is close to the results of research conducted by De Michelis et al.[7] for the zinc content of the Zn-C and Alkaline battery mixture, which is about 15.46% by weight and the results of research by Peng et al. [9] for the manganese content of Zn-C batteries, which is about 26.3% by weight. The previous esearch had shown that there were four main compounds contained in Zn-Mn battery waste, namely ZnO, MnO, Mn_2O_3, and Mn_3O_4 [2].

3.2. Leaching efficiency with organic acid reagent

There were three organic acid leaching reagents varied in this study, namely aspartic acid, malic acid, and citric acid, and also conducted leaching with non-organic acid, sulfuric acid. These four acids were compared their ability to dissolve the Zn and Mn metals from Zn-Mn batteries powders. The leaching process was carried out using these reagents at temperature 80°C, leaching time 180 minutes, ratio solid/liquid 1/20, and obtained percentage leaching as seen in Table 2.

Table 2. Zn and Mn content in leach liquor and the leaching efficiency

Leaching Reagent	AAS Results (ppm)		Leaching Efficiency (%)	
	Zn	Mn	Zn	Mn
Aspartic acid	3081.46	2941.97	42.44	25.97
Malic acid	4511.69	5332.98	62.14	47.07
Citric acid	4673.43	5814.34	64.37	51.32
Sulfuric acid	6601.10	4850.92	90.92	42.81

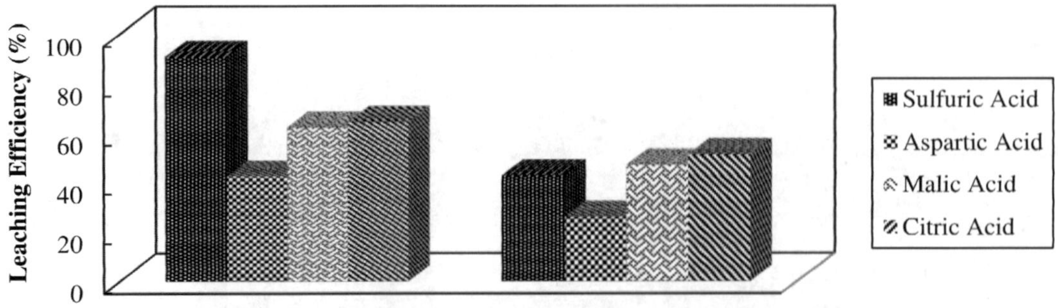

Figure 3. Comparison of leaching efficiency for each leaching reagent

Based on Table 2 and bar graph in Figure 3, the largest percentage of leaching for zinc and manganese metal among organic acid reagent is citric acid which is 64.37% for Zn metal, and 51.32% for Mn metal. The results of leaching by using malic acid for Zn and Mn metals showed almost the same value, although the value is still below citric acid. The lowest leaching result was shown by the use of aspartic acid as reagent, only 42.44% for Zn metal and 25.97% for Mn metal.

The factor affecting the percentage of leaching is the level of acidity (pH) of each organic acid used. Each of the organic acid was prepared into the solution form with a concentration of 2 Molar then measured the acidity level using pH-meter with several repetitions. The result is citric acid having the lowest pH, ranging from 1.1 -1.3. While malic acid has a pH ranging from 1.2-1.5 and aspartic acid pH ranges from 1.5-1.7. So, the highest acidity levels is citric acid, followed by malic acid, and the lowest acidity is aspartic acid. Based on the result of leaching percentage and acidity level, it can be concluded that the higher acidity used, the higher leaching percentage can be reached for Zn and Mn metals. This is consistent with the research of Li Li et al. in 2013 [6], that the reagent acidity level can affect the solubility of metals in the leaching process.

Sulfuric acid as a non-organic acid comparative variable yields leaching percentage of Zn 90.92% and Mn 42.81%. Compared to organic acids, non-organic acids give better results in dissolving Zn metal but show poor results for dissolving Mn metal. High percentage of leaching Zn because the sulfuric acid is a strong acid that permits the perfect ionization, so that the Zn elements that contained in a batteries

IOP Conf. Series: Materials Science and Engineering **333** (2018) 012041 doi:10.1088/1757-899X/333/1/012041

powder will be more easily to dissolved in sulfuric acid than in organic acids. Leaching reaction of Zn and Mn metals in sulfuric acid solution can be seen in the following equation [7].

$$ZnO + H_2SO_4 \longrightarrow ZnSO_4 + H_2O \quad (1)$$
$$MnO + H_2SO_4 \longrightarrow MnSO_4 + H_2O \quad (2)$$
$$Mn_2O_3 + H_2SO_4 \longrightarrow MnSO_4 + MnO_2 + H_2O \quad (3)$$
$$Mn_3O_4 + 2H_2SO_4 \longrightarrow 2MnSO_4 + MnO_2 + 2H_2O \quad (4)$$

The low dissolved of Mn element in sulfuric acid caused by the solubility of manganese oxide such as Mn_2O_3 and Mn_3O_4 in sulfuric acid is partial because it produces an insoluble MnO_2 compound in sulfuric acid [7]. The MnO_2 compound is the most stable manganese oxide with oxidation number 4+, so an additional reducing agent is required to reduce the manganese oxide number to be more easily dissolved in the leaching process.

3.3. Effect of H_2O_2 on leaching

The addition of H_2O_2 solution aims to increase the solubility of manganese in the reagents used. To increase the solubility of the manganese elements, it is necessary to add reducing agent, one of which is H_2O_2 solution. The graphs in Figure 4 and Figure 5 illustrate the effect of adding H_2O_2 solutions to the leaching process by using sulfuric acid reagents as non-organic acids and citric acid reagents as organic acids.

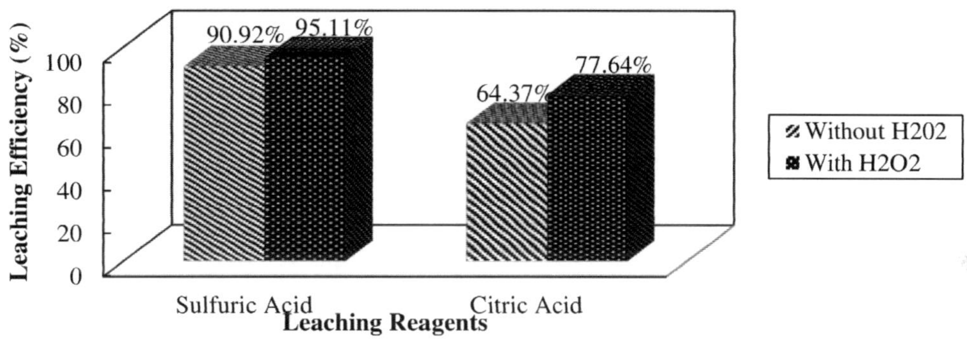

Figure 4. Effect of H_2O_2 on leaching Zn metals

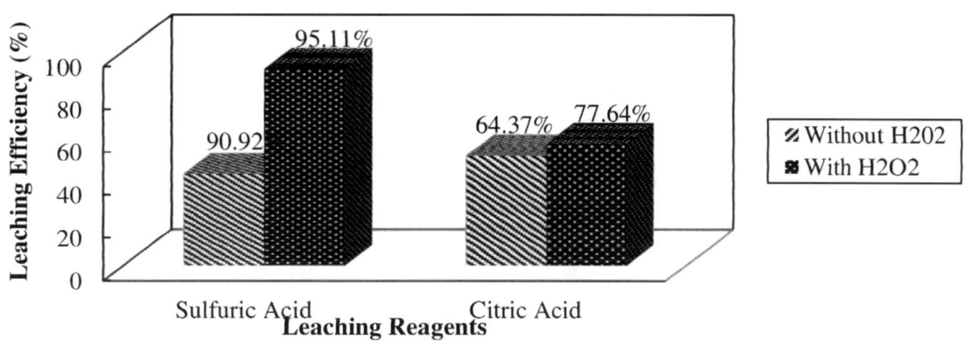

Figure 5. Effect of H_2O_2 on leaching Zn metals

In the Figure 4 and Figure 5, it is shown that the addition of H_2O_2 solution increases the percentage leaching of Mn and Zn metals on the sulfuric acid and citric acid reagents, but the most significant effect is seen in the percentage leaching of Mn metal in sulfuric acid, from 42.81% to 91.75%. This is because hydrogen peroxide can help reduce the Mn metal oxide number in the MnO_2 compound to dissolve it in sulfuric acid. The reduction reaction occurring in the MnO_2 compound can be seen in equation 5.

$$MnO_2 + 4H^+ + 2e^- \longrightarrow Mn^{2+} + 2H_2O \quad (5)$$

The MnO_2 compound is the most stable manganese oxide, so to lower the oxidation number need reducing agents that having a lower redox energy potential than the decomposition basic reaction of the compound, which is 16.82 at 85°C. Hydrogen peroxide has 8.93 redox energy potential at 85°C, so it is possible to become the reducing agent of MnO_2 compound. The effect of adding H_2O_2 solution is not very significant on the leaching of Mn metal by using citric acid, from 51.32% to 56.90%. The reason is the citric acid is an organic weak acid that having a lower energy potential than the decomposition reaction of the MnO_2 compound, so that it has been able to reduce the compound even without the aid of hydrogen peroxide, consequently the addition of H_2O_2 has no significant effect in reducing the MnO_2 compound. However, the percentage leaching of Mn metal using citric acid is still very small compared to sulfuric acid when H_2O_2 is added because citric acid is a weak acid that could not be fully ionized, so that the amount of H^+ ions that can bind to the metal oxide is less than that of sulfuric acid as the strong acids.

The addition of hydrogen peroxide can also increase the percentage leaching of Zn metal although not significantly. The addition of H_2O_2 to sulfuric acid has increase the percentage leaching of Zn metal from 90.92% to 95.11%, whereas in citric acid there was an increase of leaching percentage of Zn metal from 64.37% to 77.64%. This is because hydrogen peroxide is a weak acid that releases H^+ ions that help bind the metal oxides, so that the percentage leaching of Zn metal can increase.

4. Conclusion

Initial metal content of the mixture Zn-C and Alkaline spent batteries powder is 14.52% wt for Zn and Mn metal is 22.56% wt. Leaching reagent of organic acid showing best leaching result for Zn and Mn metal in Zn-Mn spent batteries is citric acid with 64.37% leaching percentage for Zn metal, and 51.32% for Mn metal. Addition of H_2O_2 solution give the significant effect for leaching of Mn metal using sulfuric acid, with 48.94% leaching percentage increase, but didn't give significant effect on leaching of Mn using citric acid.

Acknowledgments

The experimental work of this study was supported by Directorate Of Research and Community Services, University of Indonesia (PITTA Project 2017). The authors would like to thank eh head of laboratory and the laboratory assistant of Intensification Process Laboratory in Chemical Engineering Department, University of Indonesia where this research has been carried out.

References

[1] Yinan S, Qifei H, Zhirui N, Jie M, Baoping X. Shi C, Jiulan D, and Renqing W 2015 *Hydrometallurgy* **153** 66-73
[2] E Sayilgan, T Kukrer, F Ferella, A Akcil, F Veglio, and M Kitis 2009 *Hydrometallurgy* **97** 158-166
[3] Yunqing L and Guoxi X 2005 *J. Hazard Mater.* **127** 244-248
[4] Li L, Jing G, Renjie C, Feng W, Shi C, and Xiaoxiao Z 2010 *Waste Management* **30** 2615-2621
[5] Mehdi I, Mohammad M, and Amir R A 2012 *Physicochem. Probl. MI* **49** 547-555
[6] Li L, Jennifer B D, Xiao X Z, Linda G, Ren J C, Feng W, and Khalil Amine 2013 *J. Power Sources* **233** 180-189
[7] I De Michelis, F Ferella, E Karakaya, F Beolchini, and F Veglio 2007 *J. Power Sources* **172** 975-983
[8] Agnieszka S-T, Wtodzimierz S, and Monika Z-M 2014 *J. Power Sources* **270** 668-674
[9] Peng C H, Bai B S, and Chen Y 2008 *Waste Management* **28** 326–332

Precipitation Rate Investigation on synthesis of precipitated calcium carbonate

E Sulistiyono[1], M Handayani[1], F Firdiyono[1] and E N Fajariani[2]
[1]Research Center of Metallurgy and Materials, Indonesian Institute of Sciences (LIPI), Tangerang Selatan, Banten, 15314, Indonesia
[2]Department of Chemistry, Faculty of Mathematics and Natural Sciences, Sebelas Maret University, Surakarta, 57126, Indonesia

E-mail: eko221068@gmail.com

Abstract. Study on the formation of precipitated calcium carbonate from natural limestone Sukabumi with the influenced of various parameters such as precipitation rate, concentration of $CaCl_2$ and amplitudes were investigated. We also investigated the result with the precipitated calcium carbonate from Merck (p.a) for comparison. The higher concentration of $CaCl_2$ would give effect to the lower of the precipitation rate. It was observed that precipitation rate of calcium carbonate from limestone Sukabumi at concentration of 0.08 molar was 3.66 cm/minutes and showing the optimum condition, while the precipitation rate of calcium carbonate Merck at the concentration 0.08 molar was 3.53 cm/minutes. The characterization of precipitated calcium carbonate was done using X-ray fluorescence (XRF) and scanning electron microscope (SEM). The characterization using XRF showed that CaO content of precipitated calcium carbonate from natural limestone Sukabumi had high purity of 99.16%. The particle distribution using scanning electron microscope (SEM) showed that precipitated calcium carbonate from natural limestone Sukabumi revealed 1.79 μm – 11.46 μm, meanwhile the particle distribution of precipitated calcium carbonate Merck showed larger particles with the size of 3.22 μm – 10.68 μm.

1. Introduction

Sukabumi district has abundant mineral resources, one of the mineral resources is limestone. Potency of limestone in Sukabumi is in Cibadak area, Gunung Guruh Jampang Tengah, Cisolok, Cikidang and Warung Kiara [1]. Until now, the utilization of limestone in Sukabumi regency is used as building materials such as bricks and cement materials [1].

Calcium carbonate is a chemical compound with chemical formula of $CaCO_3$, which is a major component of the shell of marine organisms, snails, pearls and eggshells [2]. The primary forms of limestone is from sedimentation process of snails shell of small fossils, shells, and coral for millions of years [3]. The physical properties of calcium carbonate such as morphology, phase of size and size distributions must be modified according to the applications. Morphological form and phase of calcium carbonate associated with synthesis conditions such as concentration of reactant, temperature, time and natural addictive substances [3].

1 E Sulistiyono, Indonesian Institute of Sciences
2 E N Fajariani, Sebelas Maret University

IOP Conf. Series: Materials Science and Engineering **333** (2018) 012042 doi:10.1088/1757-899X/333/1/012042

Besides as building materials, calcium carbonate is widely used as main raw materials in various industries of paints, inks, paper, plastic, cosmetics, nutritional supplements, polymer pipes and fillers for composit materials [3]. In technological developments, nano-size calcium carbonate materials can be utilized as filler of natural rubber which can provide superior properties of rubber materials[4].

Precipitated calcium carbonate is made through two paths. First path is the process path of dissolution using chloride acid and sodium carbonate precipitation [5]. Second path is calcination process of limestone and then dissolved the calcination result in water and added by carbon dioxide gas [6]. Preparation of calcium carbonate by dissolution and precipitation can be modified by the addition of dodecyl dimethyl betaine material which can produce nano-size calcium carbonate [7]. This article presents experiment result of making precipitated calcium carbonate of limestone from Cibadak, Sukabumi. The pocess used was dissolution process of limestone in solution of hydrochlorida to form calcium chloride and a little impurities. After that, solution of calcium chloride was purified from impurities and then re-precipitation was done using solution of sodium carbonate to form precipitated calcium carbonate.

2. Experimental

2.1. Raw materials

The raw materials for this experiment were limestone from Cibadak, Sukabumi. Samples were taken from hill locations of traditional mining area of limestone which adjacent to limestone combustion units. The limestone that has been selected was subsequently crushed into a fractional shape with a size of approximately 1 cm so that the dissolution process using hydrocloride acid does not take place so quickly. The limestone subsequently partially crushed until smooth for doing analysis of raw materials. In this experiment, the limestone used was raw limestone without heating or calcination. The raw materials of limestone which has the best quality based on X-Ray Fluoresence (XRF) analysis with CaO content around 93% and has white colour was used for further experiments.

2.2. Experimental procedure

The natural limestone for experiment was obtained from Cibadak, Sukabumi district. 200 g of limestone was added in dilute HCl solution. Dilute HCl solution was prepared by mixing 350 ml of concentrated HCl 37% (Merck) with aquadest to a volume of 2 liters. Reaction process will form turbid solution with brown color. Turbid solution was filtered forming a clear yellow solution and then the solution was neutralized with ammonium hydroxide to get solution with pH of 7. Solution then was filtered again using Whatman filter paper No. 42 obtaining clear yellow solution. The solution was subsequently diluted to obtain 1 molar of calcium carbonate solution.
Calcium carbonate solution was then diluted into 0.04 mol, 0.08 mol, 0.12 mol, 0.16 mol and 0.2 mol. After that, weight of Na_2CO_3 needed for precipitation was determined and obtained 4.70624 g; 9.4116 g; 14.1185 g; 18.82496 g dan 23.5312 g. After that, the precipitation process was conducted so that white solid calcium carbonate with very fine grain size was resulted. Precipitation process was done by mixing and stirring 100 ml of calcium carbonate solution with 10 ml of sodium carbonate solution at room temperature. The precipitate obtained was then filtered and washed with aquadest, it did not need to be dried. Precipitation rate investigation was performed by taking precipitate of calcium carbonate added in 200 ml of aquadest. After that it was shaken until evenly and then poured in a large-sized reaction flask. Investigation for decreasing of height level of sedimentation was done by palcing the flask in the upright stand. The proccess was repeated with different variable of concentration of calcium carbonate. Besides that, we measured precipitation rate of calcium carbonate from commercially available calcium carbonate (Merck) for comparison. After optimum point of measurement was obtained, the sample was then analyzed using scanning electron microscopy (SEM) to know grain size of precipitated calcium carbonate.

3. Result and Discussion

3.1. Analysis of dissolution process with HCl and precipitation process of calcium carbonate

A series of experiments of dissolution process of calcium carbonate from Cibadak, Sukabumi with HCl was performed and compared with calcium carbonate from commercially available (Merck). Experiment result showed that calcium carbonate from natural materials of limestone contained a lot of impurities. From the observation, it can be seen that the limestone produced the turbid solution due to the presence of impurities that did not dissolve in dilute HCl solution. After filtering and cleaning process of the impurities, it was seen that the solution was yellow colour which indicated the impurity contained metal ion. This result was different with calcium carbonate from merck which gave a brighter colour. The colour difference of calcium carbonate from natural limestone and commercially available (Merck) could be seen in Figure 1.

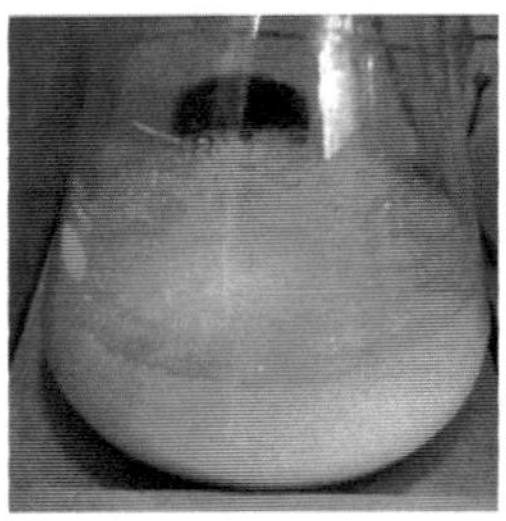

Figure 1. Dissolution process of calcium carbonate with HCl. Dissolution process of natural limestone (left), Colour solution of sodium chloride which showed yellow colour for natural limestone and clear for Merck (right)

The result of analysis using XRF for natural limestone was shown in Table 1. It was depicted that dissolution process of calcium carbonate could increase the content of CaO from 93.44 % (before process) into high purity of CaO content of 99.16 % (after process). This result indicated that impurities were considerably reduced impurities, even several impurities were undetected such as magnesium, silica, alumina and iron. For the comparison, we measured calcium carbonate from Merck using XRF and the analysis result showed that CaO content was 99.40 %. This observation revealed that the difference of CaO content of natural limestone was not large compared with CaO from Merck so that the process was efficiently reduced the impurities and obtained high purity of precipitated calcium carbonate.

Table 1. The analysis result of natural limestone using XRF (before and after process) and compared with calcium carbonate from Merck

Component	Before Process (% wt)	After Process (% wt)	Merck (% wt)
CaO	93.44	99.16	99.40
MgO	2.06	Nd	Nd
SiO_2	0.85	Nd	Nd
Al_2O_3	0.62	Nd	Nd
P_2O_5	0.61	0.14	Nd
SO_3	0.54	0.14	0.16
K_2O	0.44	Nd	Nd
Cl	0.42	0.30	0.34
Fe_2O_3	0.19	Nd	Nd
SrO	0.13	0.14	0.03
BaO	0.08	Nd	Nd

[a] laboratorium : Faculty of MIPA-UNS . Nd = Not Detected

IOP Conf. Series: Materials Science and Engineering **333** (2018) 012042 doi:10.1088/1757-899X/333/1/012042

3.2. Analysis of precipitation rate

Precipitated calcium carbonates were obtained from precipitation process of calcium chloride solution with sodium carbonate. In this research, variations of calcium carbonate concentration were performed. The result of precipitation process was used for measuring precipitation rate on the particle density of calcium carbonate. The result of the experiment showed that the more concentrated calcium carbonate would result precipitation rate slower. Decreasing of precipitation rate was seen in increasing the concentration from 0.04 molar to 0.12 molar. However, precipitation rate would diminish after 0.12 molar. It is possible that decreasing of grain size of calcium carbonate was not linear with the increase of concentration, and only gave effect for the arrangement of precipitate in the solution.

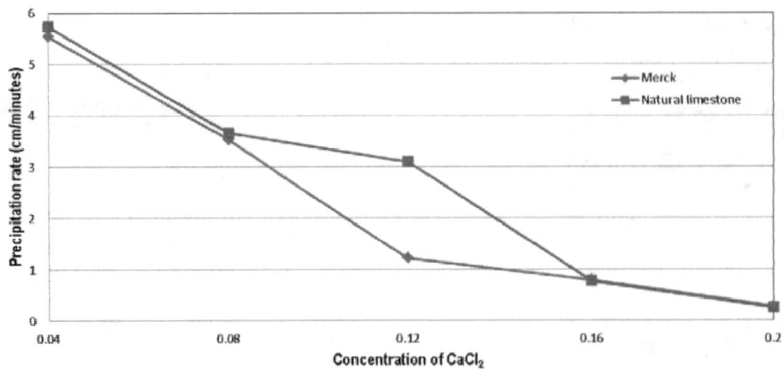

Figure 2. Correlation between precipitation rate of $CaCO_3$ with concentration of $CaCl_2$ solution.

Figure 2 showed the similarity of tendency of decreasing the precipitation rate to the increase of concentration on natural limestone from Cibadak, Sukabumi with calcium carbonate from Merck. However, decreasing of precipitation rate started to slow down at 0.12 for calcium carbonate from Merck, whereas decreasing of precipitation rate of precipitated calcium carbonate from natural limestone Cibadak at concentration of 0.16 molar.

3.3. Analysis of distribution size using scanning electron microscopy

Size distribution of perticipated calcium carbonate from natural limestone was done using scanning electron microscopy (SEM) showed that grain size of calcium carbonate was still quite large when referring to the standard of nanoscale material, where the grain size was still above 100 nm. The particle distribution using scanning electron microscope (SEM) showed that precipitated calcium carbonate from natural limestone Sukabumi revealed 1.79 μm – 11.46 μm, meanwhile the particle distribution of precipitated calcium carbonate Merck showed larger particles with the size of 3.22 μm – 10.68 μm. The result of SEM measurement was depicted in Figure 3

Figure 3. SEM measurement result of precipitated calcium carbonate for Merck (left) and for natural limestone from Cibadak, Sukabumi (right)

From the result of SEM analysis with various enlargement (see figure 4), it can be seen that precipitated calcium carbonates have homogenous crystalline form. From the observations at magnification 500 times and 2.000 times, it indicates that the grain arrangement is in the form of cubic shape which shows that the compound is calsium carbonate[2]. By having a homogeneous arrangement of particle shape, the material is already qualified as a good filler material because the structure is uniform.

Figure 4. The form of calcium carbonate crystals from the dissolution process of HCl and the precipitation process of natural limestone using Na_2CO_3 at magnification 500 times (left) and magnification 2,000 times (right).

4. Conclusion

From the results of the above experiments, it can be concluded that limestone from the Cibadak, Sukabumi Regency has the potential to be developed into advanced materials based on calcium carbonate. Through the process of dissolving with chloride acid and precipitating with sodium carbonate solution into product of precipitated calcium carbonate having a very high purity and uniform grain shape and the same molecular structure. From the result of the analysis of the precipitation rate, it can be seen that the point of optimization of the precipitation rate for natural limestone is 0.16 M and the optimum point of calcium carbonate from Merck is 0.12 M. By comparing natural limestone with calcium carbonate from commercially available, the same process produced precipitated calcium carbonates which have almost the same properties. By having a homogeneous arrangement of particle shape, the precipitated calcium carbonate materials are already qualified as a good filler material.

References:

[1] Anonymous 2015 *Year-End Report of Mining and Environment Office of Sukabumi Regency*
[2] Necmettin E, Haci A E 2016 *Physicochem. Probl. Miner. Process*. **53 (1)** 57
[3] Oates J A H 1998 *Lime And Limestone Chem. Tech.* 130-153
[4] Adi C and Norma A K 2012 *Warta Perkaretan* **33 (2)** 113
[5] Kirboga S, Oner M. 2013 *Chem. Eng. Trans.* **32** 1974
[6] Lailiyah Q, Baqiya M, Darminto 2012 *Journal of Science and Arts ITS* **1** 69
[7] Wang C, Liu Y, Bala H, Pan Y, Zhao J, Zhao X, Wang Z 2007 *Colloids and Surfaces A: Physicochem. Eng. Aspects* **297** 179

IOP Conf. Series: Materials Science and Engineering **333** (2018) 012043 doi:10.1088/1757-899X/333/1/012043

Synthesis of Calcite Nano Particles from Natural Limestone assisted with Ultrasonic Technique

M Handayani[1], E Sulistiyono[1], F Firdiyono[1] and E N Fajariani[2]
[1]Research Center of Metallurgy and Materials, Indonesian Institute of Sciences (LIPI), Tangerang Selatan, Banten, 15314, Indonesia
[2]Department of Chemistry, Faculty of Mathematics and Natural Sciences, Sebelas Maret University, Surakarta, 57126, Indonesia

E-mail: murni.handayani@lipi.go.id

Abstract. This article represents a precipitation method assisted with ultrasonic process to synthesize precipitated calcium carbonate nano particles from natural limestone. The synthesis of nanoparticles material of precipitated calcium carbonate from commercial calcium carbonate was done for comparison. The process was performed using ultrasonic waves at optimum condition, that is, at temperature of 80oC for 10 minutes with various amplitudes. Synthesized precipitated calcium carbonate nanoparticles were characterized using X-Ray Diffraction (XRD), Scanning Electron Microscope (SEM) and Particle Size Analyzer (PSA). The result of PSA measurements showed that precipitated calcium carbonate nano particles was obtained with the average size of 109 nm.

1. Introduction

Calcium carbonate ($CaCO_3$) is an inorganic compound which has been widely studied owing to its abundance in nature as a mineral and biomineral. Calcium carbonate in the form of flour with particle size in nano-scale and a very high degree of purity is an indispensable material in the industrial field. In the industrial field, calcium carbonate is utilized as raw materials for filler on paper, rubber, paint and plastic [1]. Along with technological developments, the use of calcium carbonate is not only for fillers but also for biocompatible materials. The use of calcium carbonate as a biocompatible material is as a coating material because it has the advantage that is easily adaptable in the human body [2]. Calcium carbonate particles have three molecular crystal morphologies, which are Calcite, Aragonite and Vaterite, where calcium carbonate in aragonite crystal is the best form for the coating material [3].

Nano particle material of calcium carbonate for biocompatible coating materials can be made from shellfish material using chitosan as a precursor [3]. The experiments on synthesis of calcium carbonate from the shellfish material are capable to produce calcium carbonate with aragonite crystals, but the grain size is quite large around 5.000 nm [3]. Nanoparticle material of calcium carbonate can be prepared by precipitation process which capable to make calcite nanoparticles with a particle size of 140 nm [1]. The results of another study by precipitation process using sodium silicate with sol-gel system produced calcium carbonate in the form of calcite crystals with particle size of 40 nm [4].

1 M Handayani, Indonesian Institute of Sciences
2 E N Fajariani, Sebelas Maret University

In addition to precipitation methods, synthesis of calcium carbonate could be performed with carbonatation system that is by blowing carbon dioxide gas in a solution of 5% $Ca(OH)_2$. This process used ultrasonic method to control the particle size. The result obtained was nanometer-sized calcium carbonate materials with a range of 45 nm to 72 nm [5]. The process of formation of calcium carbonate nanoparticles by ultrasonic process was also conducted by using surface particle modification with the used of Trioxy Vinyl Silane (TEVS). Through this surface modification process, precipitated calcium carbonate product with nano-size of 50-90 nm was obtained [6]. Ultrasonic process is a process of destruction of particles in a medium with ultrasonic waves that produce cavitation effects [7]. The cavitation effect produces energy propagation in the form of high and low pressure effects simultaneously and repeatedly resulting in a microbubble explosion emanated in a solution with very large quantities [7]. The microbubble, if collided with particles floating in liquid media, causes the particles to break into smaller particle sizes.

In this study, ultrasonic technique was employed to synthesis calcium carbonate nano particles from natural limestone. The synthesis of calcium carbonate nano particles from natural limestone is greatly promising process to be performed because limestone has extensive industrial application and the most common natural form of calcium carbonate. In addition, generally calcium carbonate nano particles were synthesized from commercial calcium carbonate materials, synthesis of calcium carbonate nano particles from natural resources was still very few. By mastering the synthesis technology of calcium carbonate nanoparticles, this will increase the added value of natural resources significantly.

2. Experimental

2.1. Ultrasonic technique

In this research, particle destruction process was performed using ultrasonic process that is sonication equipment Q-1357. The equipment has a maximum power of 1.375 watts by working at a frequency of 20.000 Hz and amplitude with a variation of 10% to 100%. The capacity of the equipment is capable of performing an ultrasonic process of up to 20 litres of fluid in every single process. The ultrasonic equipment is equipped with a temperature gauge to control temperature process. In this research, the variable used is the amplitude associated with ultrasonic power, while the frequency and temperature are fixed. The probe is suitable for the ultrasonic process capacity of 400 ml of fluid.

2.2. Experimental procedure

The natural limestone for experiment was obtained from Cibadak, Sukabumi. Calcium carbonate solution was prepared from dissolution process of limestone with chloride acid. In this research, ultrasonic process was performed on the optimum point of measurement result of precipitation rate in previous experiment. It was known that the optimum point of measurement result for precipitation rate on the concentration of calcium chloride at 0.16 molar for natural limestone and 0.12 molar for calcium carbonate commercially available (Merck). The sonication process and the addition of sodium carbonate solution were carried out for 10 minutes. At the sonication process, the temperature was maintained with maximum $80^{O}C$. The experimental variables used in the ultrasonic process were amplitude 20%, 30%, 40%, 50% and 60%. Characterization was performed using particle size analyser (PSA) equipment to observe the particle size distribution, scanning electron microscopy (SEM) to know grain size and particle shape and also X-ray Diffraction (XRD) to know the molecular structure of calcium carbonate produced.

3. Result and Discussion

3.1. Precipitation result of calcium carbonate of natural limestone using ultrasonic process

Precipitation result using ultrasonic process was conducted with experiment variables of amplitude of 20%, 30%, 40%, 50% and 60%, and obtained the result of a fine solid of calcium carbonate. A fine solid of calcium carbonate was then washed and dissolved in aquadest to measure precipitation rate. From the experimental results, it can be seen that with the ultrasonic process, the particle precipitation rate slows down and particle size is getting smoother and smaller. The result of precipitation rate associated with the amplitude is shown in Figure 1.

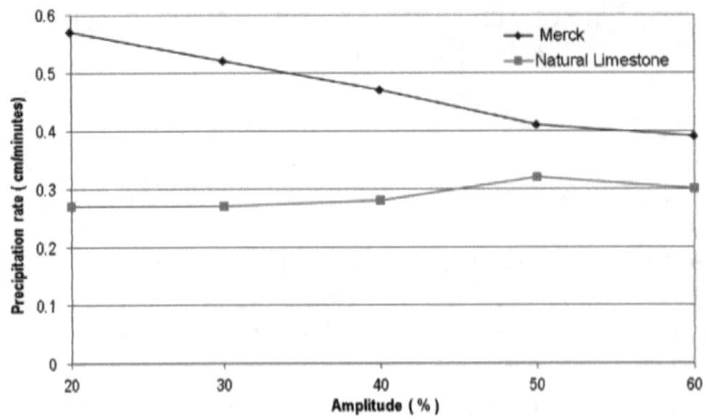

Figure 1. Graphic of correlation between amplitude of ultrasonic with the precipitation process for calcium carbonate.

It can be seen that the optimization point occurs in the ultrasonic process with 20% of amplitude for natural limestone. At amplitude above 20%, the rate of precipitation process decreases slightly, whereas the higher amplitude causes the power required to perform the ultrasonic process becomes greater. In the ultrasonic process of calcium carbonate from Merck, it is seen that the optimum point occurs at 50% of amplitude which is higher compared to the optimum point of natural limestone.

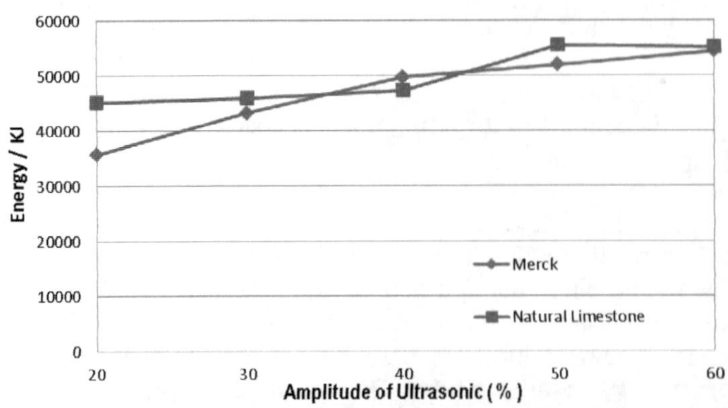

Figure 2. Graphic of correlation between amplitude with energy in ultrasonic process for calcium carbonate

Based on the calculation of energy requirements, it is seen that natural limestone need lower amplitude, this means that the energy for crushing calcium carbonate to make smaller particles is lower (Figure 2). At higher amplitude above 50%, the required energy requirement is greater for calcium carbonate from commercially available (Merck). Overall, the amount of energy required for the ultrasonic process is lower for natural limestone compared with calcium carbonate from Merck. From the calculation of precipitation rate, it is observed that optimum point of calcium carbonate is 0.16 molar with 20% of amplitude and 0.12 molar with 50% of amplitude for natural limestone and Merck, respectively.

3.2. Analysis of surface morphology of calcium carbonate using SEM

From the optimization point, the calcium carbonate solids are dried to get $CaCO_3$ powder and then the powder are filtered to separate the coagulated granules. Analysis of surface morphology of particles using SEM was conducted for calcium carbonate resulted from ultrasonic process compared with calcium carbonate without ultrasonic process. The result of SEM measurements is depicted in Figure 3 which shows that ultrasonic process produces smaller grain size.

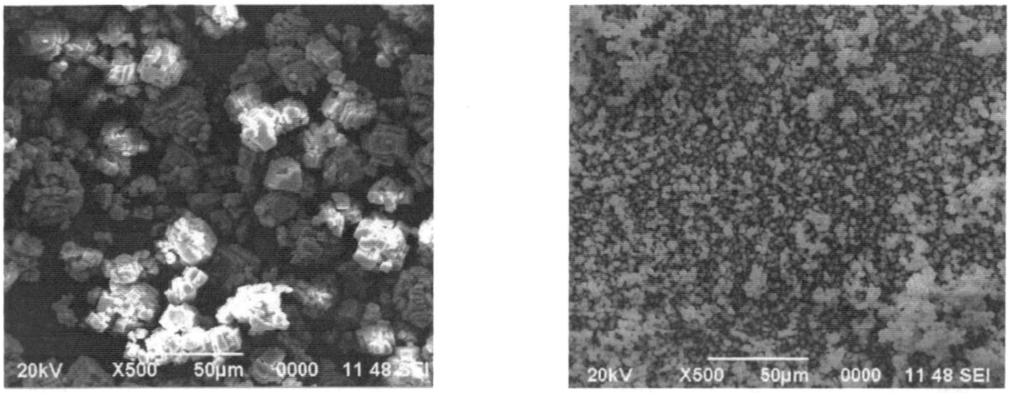

Naturural Limestone Non Ultrasonic **Natural Limestone With Ultrasonic**

Figure 3. The result of SEM analysis comparing the grain size and surface morphology at the same magnification of 500 x .

3.3. Analysis of particle size using particle size analyzer(PSA)

Particle size distribution was measured using particle size analyser. Using this method, particle size could be measured quantitatively. We measured 4 samples for PSA analysis and then the averages were calculated. The result of the PSA measurement is depicted in Table 1

Table 1. Analysis of PSA (units are in nanomater)

Samples	1st measurement	2nd measurement	3th measurement	Averages
Natural Limestone Non Ultrasonic	989	26.049	811	9.283
Natural Limestone Ultrasonic	81	132	115	109

The result shows that $CaCO_3$ sample of natural limestone from Cibadak, Sukabumi is easier crushed into smaller particles with assisted by ultrasonic technique resulting significant decrease in particle size. It is observed that particle size $CaCO_3$ sample of natural limestone without ultrasonic showing particle size of 811 nm to 26,049 nm decreases to grain size of 81 nm to 132 nm when we employed ultrasonic process. The result of the analysis of particle distribution using PSA by taking the smallest measurement results on limestone from Cibadak, Sukabumi is shown as Figure 4.

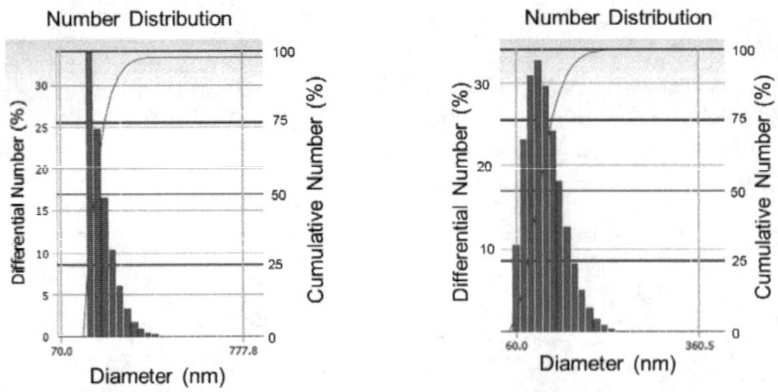

Figure 4. Analysis result of particle distribution by PSA Delsa Nano on $CaCO_3$ from Cibadak – Sukabumi without ultrasonic process (left) and with ultrasonic technique (right).

The results of the PSA analysis show that the particle distribution of precipitation process without ultrasonic method, the smallest particle size starts at 102 nm with a fraction of 33.9%, then continues with the number of fractions down to 270 nm. In the particle size with range 102 nm - 270 nm, there is a fraction count of 97.9%. There is a large grain of 2.2% with a size range between 24,933 nm to 48,933 nm. Iit can be concluded that without the ultrasonic process the resulting calcium carbonate has not qualified as a nano particle materials of $CaCO_3$.

In the precipitation process with ultrasonic technique, it is seen that the smallest fraction occurs at 60 nm with a fraction of 5.3%. Most fractions occur at 75 nm. The number of particles has reached 90.7% in a particle size range of 60 to 100 nm, and particles of a size above 100 nm are about 9.3%. Based on this PSA analysis, it is seen that the calcium carbonate particles after the ultrasonic process have particle size below 100 nm that reaches 90.7%, thus fulfilling the requirements as calcium carbonate nano particles.

3.4. Analysis of crystal structure of calcium carbonate using X-Ray diffraction (XRD)

Analysis of crystal structure using XRD of calcium carbonate nano particles obtained from HCl dissolution and precipitation process with sodium carbonate assisted by ultrasonic technique of natural limestone was conducted. Besides that, calcium carbonate from commercial was also analysed for comparison. The result of XRD analysis shows that the products of calcium carbonate gives the same result for the crystal structure as shown in Figure 5, where two compounds of $CaCO_3$ have identical peaks. This means that even though the raw materials are different but after HCl dissolution and precipitation process with sodium carbonate assisted by ultrasonic technique give the same crystal structure of $CaCO_3$.

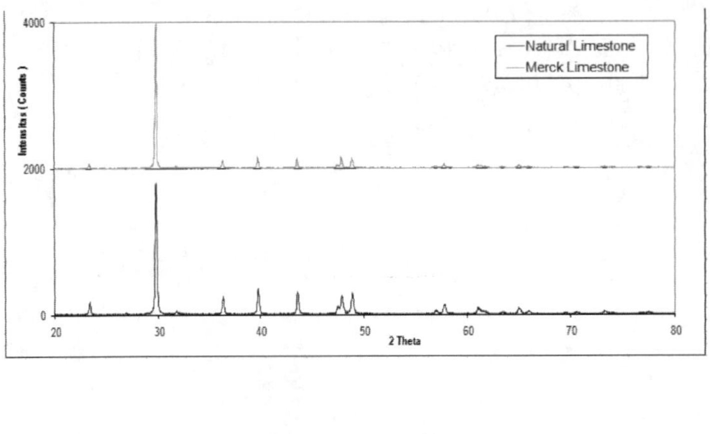

Figure 5. Comparison of XRD analysis of precipitated calcium carbonate resulted from ultrasonic process. Peak with pink color is crystal structure for $CaCO_3$ commercial (top) and peak with blue color is crystal structure for $CaCO_3$ nano particles obtained from natural limestone (down)

It is known that there are three types of crystal structure of calcium carbonate, those are calcite, aragonite and veterite. In order to observe the type of crystal structure of $CaCO_3$ nano particles obtained from natural limestone, the result in Figure 5 is compared with the standard of $CaCO_3$ using MATCH program as shown in Figure 6.

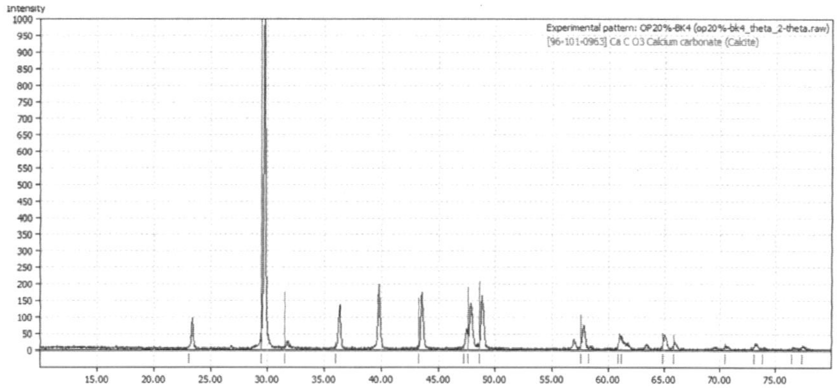

Figure 6. Analysis result of MATCH program on crystal structure of $CaCO_3$ nano particles obtained from natural limestone after ultrasonic process.

In this study, the standard of $CaCO_3$ which match with $CaCO_3$ nano particles obtained from natural limestone is Calcite. We can see that almost all of the peaks obtained in the XRD pattern of calcium carbonate nano particles from natural limestone match very well with the standard of calcite pattern. There are several unidentified peaks at 2Θ of 39.76°, 57.72° , 63.40° and 70.08° which is also observed in commercial calcium carbonate (see Figure 5). This illustrates that the result still contains impurities, even though the intensities of the impurities are small. The possibility of the impurities is because the compound is still wet or carbonization occurs.

4. Conclusion

Synthesis of calcium carbonate nano particles of natural limestone from Cibadak, Sukabumi was performed using ultrasonic process. The result shows that calcium carbonate nano particles fulfill the requirements of calcium carbonate nano material with averages of the particle size obtained from PSA analysis is 109 nm, with the number of particles has reached 90.7% in a particle size range of 60 to 100 nm. From XRD analysis, the crystal structure of calcium carbonate nano particles suits very well with calcite pattern.

References:

[1] Takuya T, Kellie P and Paul G M C 2000 *J. Nanopart. Res.* **2** 375
[2] Dinamani M, Kamath PV and seshadri R 2002 *Mater. Res. Bull.* **37** 661
[3] Majusha H, Neethumol V, A. Benny C, Sreenivasan PV, Jenish P, Asmy A KA 2014 *Int. J. Sci. Res. Pub.* **4** 2250
[4] Shirish H S, Sarang P G, Kunal H K, Satish P M, Kshitij J K, Laxminarayan R, Candrashekhar M M , Madan G P and Muthupandian A 2010 *Intl. J. Chem. Eng* **2010** 1
[5] Sonawane S H, Kadapnis C V, Sathis M, Gumvekar S P and Khanna P K, 2010 *Intl. J. Green. Nanotechnol: Physic and Chemistry* **32** 69
[6] Navichandra S, Ananda M, Hansora D P, Satyendra M, 2015 *Nanoscience and Nanoengineering* **3** 8
[7] Aharon G, 2004 *J. Ultrason. Sonochem* **11** 47

IOP Conf. Series: Materials Science and Engineering **333** (2018) 012044 doi:10.1088/1757-899X/333/1/012044

Effect of Al-TiB Addition on the Mechanical Properties and Microstructure of Al-ADC12/NanoSiC Produced by Stir Casting Methods

D Dhaneswara[1], A Zulfia[1], T P Pramudita[1], D Ferdian[1], and B W Utomo[2]
[1]Department of Metallurgy and Materials Engineering, Universitas Indonesia, Depok 16422, Indonesia
[2]Astra Honda Motor, Cikarang Barat, 17530, Indonesia

Email: donanta.dhaneswara@ui.ac.id

Abstract Addition of Ti-B grain refiner in Al-ADC12/nanoSiC composite results in improvement of tensile strength, hardness, and wear resistance through grain refinement. In this research, composite of Al-ADC12/nano SiC (0.15 %vf) with variations of TiB respectively (0.0), (0.02), (0.04), (0.06), dan (0.08) wt% were produced by stir casting. 10% of Mg were added to promote wettability between reinforce and matrix. It was found the best addition of TiB is 0.04 wt% Ti-B which results 135,9 MPa in tensile strength, 46 HRB in hardness, and 1.47×10^{-5} mm^3/s as wear rate. The increase in mechanical properties of composites mainly because of Al_3Ti acts as nucleants which initiates the grain refinement and the existence of $MgAl_2O_4$ phase indicates an interphase between nano SiC and ADC12 matrix. However, the increase of Ti-B addition after optimum number gives no significant results. High composition of iron and magnesium addition will form intermetallic phase β-Fe, π-Fe, and Mg_2Si.

1. Introduction

Aluminum which covers about 8% of the earth crust is known as one of the three most abundant metals in the earth after iron and copper[1]. It has a good mechanical properties and corrosion resistance both produced by casting or forging, therefore it has a wide applications in transportation. One being studied recently for the purpose of mass transportation is changing cast iron into composites as brake shoe[2]. ADC12 is aluminum with a main alloying element of silicon, copper, and magnesium[3]. These elements improved its casting properties: silicon will increase its flowability, copper reduce shrinkage and combination between silicon and magnesium will form precipitates that increase wear resistance[4,5].

Adding reinforce could enhance yield, tensile strength, and hardness without reducing ductility due to grain refinement mechanism called Orowan Looping[6]. Nano sized particle results in strong multidirectional interphase, good distribution of particles, and low porosity which will increase the ability to transfer applied load compared to micro sized particles[7]. Eventhough nano particles gives refinement effect, the addition of grain refiner gives more significant refinement effect due to innoculation mechanism[8–12]. The refinement takes place during the solidification stage and two phases which act as innoculant are TiB_2 and Al_3Ti which settle at α-Aluminum grain boundaries[13–14]. To achieve the full potential of composite, the SiC/Al interface must possess a strong adhesion through interdifussion by adding magnesium as wetting agent[15,16]. The good interphase of matrix and reinforce indicated by the existence of phase $MgAl_2O_4$ in the composite system[17]. In this work the effects of Ti – B addition were studied to find the ideal compositions provide the best mechanical properties.

2. Experimental

2.1. Synthesis of composite

The alloy was produced through stir casting method. Al-5Ti-1B master alloy were added as grain refiner. ADC12 with nominal composition (mass fraction): Si 10.5%; Fe 0.8%; Cu 2.33%; Mn 0.22%; Mg 0.22%; Zn 0.64% and Cr 0.04% was selected as the matrix, and combined with 0.15% volume fraction of nano silica as reinforce. In the making, aluminum was heated to its molten state in 800°C, then magnesium was added to improve wettability between matrix and reinforce. Titanium boron was added in the form of rod, and nano silica was added the last after degassing process using Argon. Before poured in the molten aluminium nano silica was preheated in muffle furnace for an hour in 500°C to remove moist and increase surface tension. The molten aluminum then stirred 600 rpm for 2 minutes. Before pouring, the mold had been preheated in 300°C for 2 minutes and coated by zircon to reduce thermal shock.

2.2. Microstructural characterization

All cast were microstructurally characterized in the terms of grain size of matrix alloy using optical microscope. They were cut from the cast cylinder, polished, and etched using Keller reagent with standard metallographic technique. The specimen showing highest tensile strength then re-polished, and tested using SEM and XRD to check the phases and knowing the interphase between matrix and reinforce.

2.3. Mechanical properties characterization

All cast were prepared and tested to measure mechanical properties (tensile strength, hardness, wear rate). As another properties, density were tested using Archimedes' Theory. Tensile test was performed using Ogoshi universal testing machine and tested at room temperature. Some parts of sprue were cut into small pieces and grinded to get a fine surface and tested using Brinell. Wear test of the composite were conducted using pin-on-disk machine by sliding a specimen under dry condition.

3. Results and Discussion

3.1. Materials and microstructures

The microstructure of the composite with variation of Ti-B are shown in fig. 1 respectively. Seen in ADC 12 plain (without reinforce and grain refiner) all the structure are needle like, and with the addition of reinforce and grain refiner it all changed into spherical. Significant changes before and after addition of reinforce nanoSiC due Thermal Lag refinement. The nano particles of reinforce forced the α-Al to split and as the result, their grain size become smaller. As the proportion of grain refiner increase, the grains becoming finer and more spherical as seen in figure 1. (b-f).

The decreasing size of grain also proved by the calculation of *Single Dendrites Arm Spacing*, the number of SDAS decreased when the size of grain also decreased (figure 2). As seen in the microstructure there are few morphologies which indicates certain intermetallic phases, and it is predicted using phase diagram which is plotted from its chemical composition test result (OES) in table 1.The excess magnesium in the system derivates the ternary phase diagram Al-Mg-Si. The composition of each composite takes place around *quasi-binary* line reaction which results in formation of intermetallic phase primary Mg_2Si and eutectic Mg_2Si (binary and ternary).

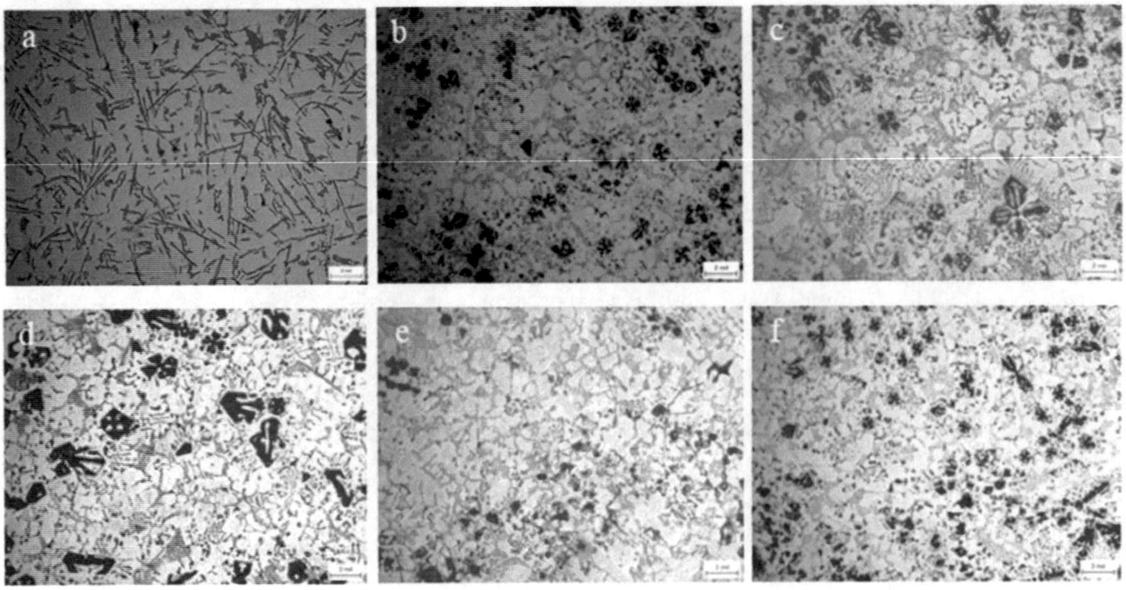

Figure 1.(a) ADC12, (b) 0.0 wt% Ti, (c) 0.02 wt% Ti, (d) 0.04 wt% Ti, (e) 0.06 wt% Ti, (f) 0.08 wt% Ti

Table 1. Chemical composition of composites

Composition	Silicon (Si)	Magnesium (Mg)	Iron(Fe)
Comp 1 (0.0 wt%)	7.21	>12	0,821
Comp 2 (0.02 wt%)	7.23	>12	0.913
Comp 3 (0.04 wt%)	7.20	>12	0.958
Comp 4 (0.05 wt%)	8.17	>12	0.950
Comp 5 (0.06 wt%)	7.20	>12	0.861

As the temperature decrease there will be a change in the morphology of the intermetallic phase. Primary Mg_2Si formed in the early reaction and mostly blocky and compact or snow-flakes shaped. Eutectic Mg_2Si will form after hand and the binary will form a chinese script and the ternary will form white dots around chinese script (figure 2). As the grain get finer, the SDAS reduced (figure 3). The formation follows reaction 1 and 2:

$$L \rightarrow L + Mg_2Si_p \rightarrow Mg_2Si_p + (\alpha\text{-}Al + Mg_2Si)_E \quad (1)$$

$$L \rightarrow L + Al + \beta\text{-}Al_5FeSi \rightarrow \beta\text{-}Al_5FeSi + Al + Si \quad (2)$$

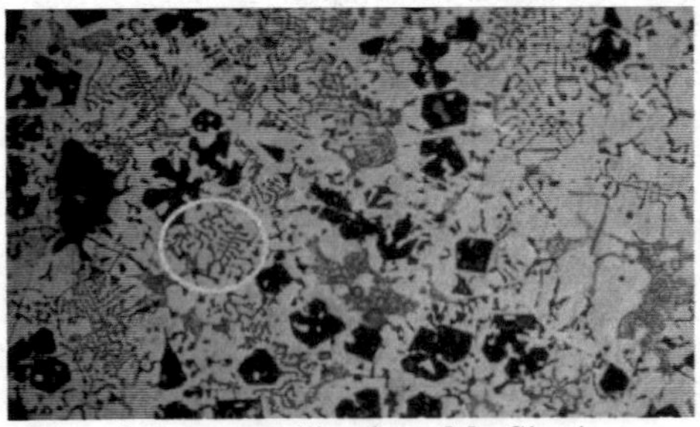

Figure 2. Intermetallic phase Mg_2Si primary (arrow) and eutectic (circle)

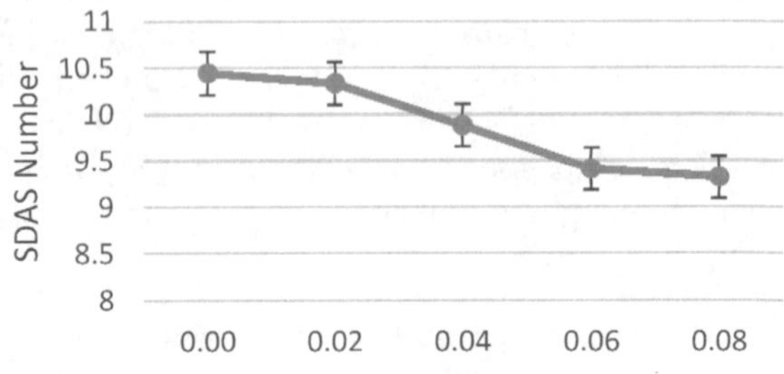

Figure 3. Effect of TiB addition to SDAS composite

The formation of intermetallic phases Al_5FeSi takes form needle like and sharp. This morphology reduces its toughness, and affects tensile strength. The other form of Fe intermetallic is π-Fe which looked like chinese script (figure 4).

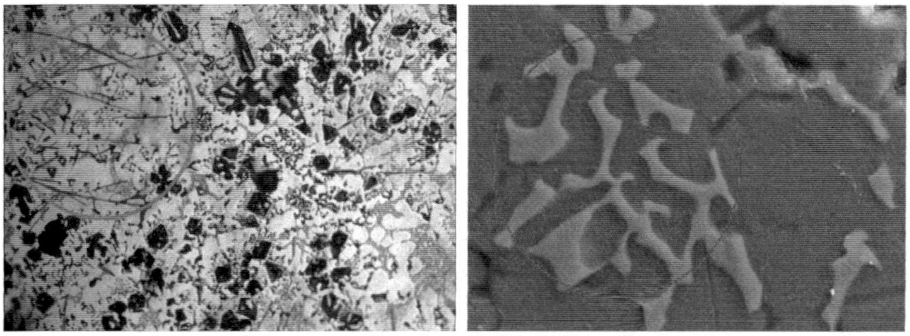

Figure 4. Intermetallic phase β-Fe (left) and π-Fe (right)

SEM and XRD results (figure 5) shows there are intermetallic phases and interphase between matrix and reinforce. Spinel ($MgAl_2O_4$) and titanium in the form TiB_2 and Al_3Ti confirmed by XRD test. The high percentage of oxygen compared to other elements in EDS result (table 2) indicated there is indeed a bonding between matrix and reinforced. There is no evidence on the existence of titanium or boron in the SEM, this is due to very small amount of them added to the system, but all of them showing high peak in the various location in XRD test. It indicates all of them fine dispersed in the matrix.

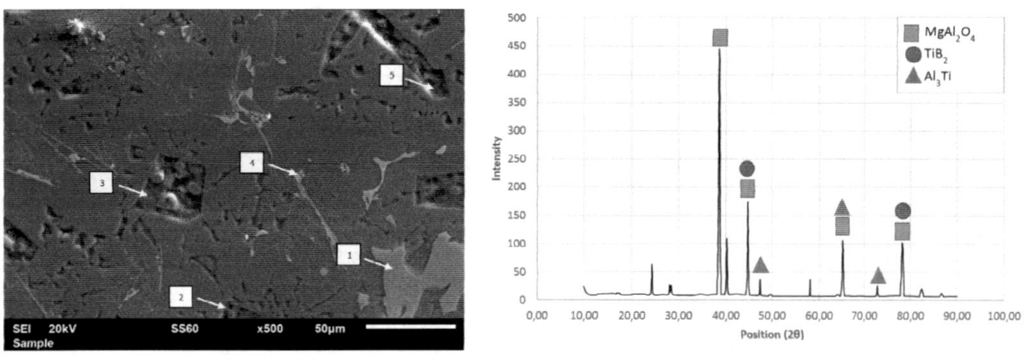

Figure 5. SEM image cast composite (left) and XRD result (right)

Table 2. EDS result cast composite

Region	Element (wt%)							Predicted Phase
	Al	Mg	Si	Fe	Cu	O	C	
1	95.75					4.25		α-Al
2	41.84	16.05	18.57			16.73		Al + Mg_2Si
3	1.41	24.72	26.73			30.82	12.18	Mg_2Si
4	68.29		8.71	9.15		3.69		AlFeSi (β-Fe)
5	54.88	0.93			16.64	3.57	17.28	Al_2Cu

3.2. Density and Porosity

All casts composites density are close to the pure Al (2,59 gr/cm^3). This is because the proportion of Ti-B added was too small, so there will be no significant differences in the term of density. The factor affecting differences between each composition is porosity. An increase of Ti-B addition results in reduction of porosity in cast composite as seen in figure 6. The addition of 0.06 wt% Ti-B gives the highest porosity among due to excess hydrogen which enter the molten aluminum during stirring. The existence of porosity in the system reduce tensile strength significantly[18].

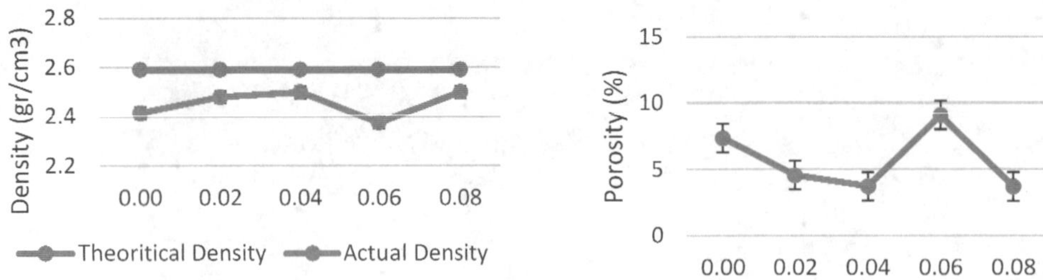

Figure 6. The effect of Ti-B to the density (left) and porosity (right) of composite

3.3. Tensile and Elongation

Porosity in the cast relates to tensile strength, higher percentage of porosity results in low number of tensile strength. Therefore the addition of 0.06 wt% Ti-B gives the lowest UTS (figure 7). Porosity does affect the ability of material to transfer applied stress.

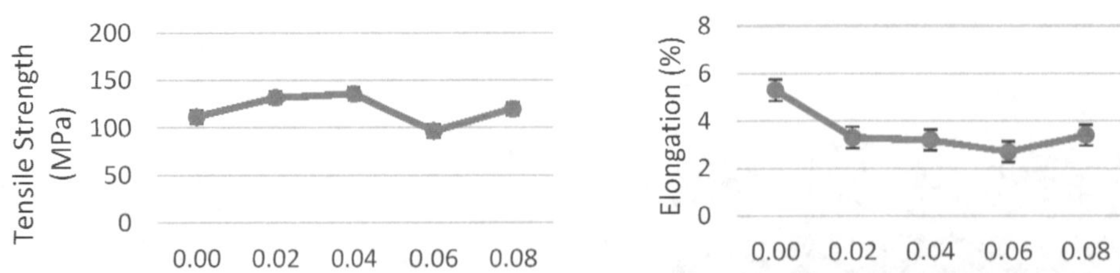

Figure 7. Effect of Ti-B to tensile strength (left) and elongation (right)

3.4. Hardness and wear rate

As the Ti-B added increase, the hardness of cast increase as well (figure 8), but after certain addition, it will give no significant result. The ideal addition is 0.04 wt% therefore addition 0.06 and so on, they give no increase in hardness. Hardness of material relate to the wear rate of the material, as seen variation 0.04 gives highest hardness thus gives the lowest wear rate.

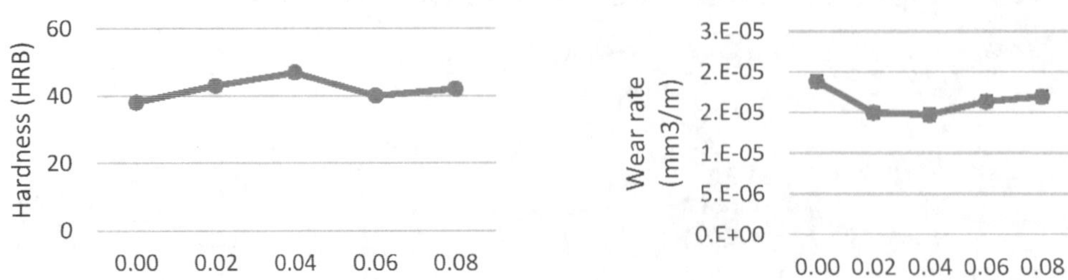

Figure 8. Hardness (left) and wear rate all cast (right)

4. Conclusion

Metal matrix composite can be produced using stir casting method, and the optimal addition of Ti-B which results in highest mechanical properties is 0.04 wt%. It reach highest tensile 135.97 MPa, 46 HRB in hardness, and 3.72% porosity. The addition of grain refiner changes the morphology of primary phase α-Aluminum from coarse dendritic to fine spherical grain which is attributed to grain refinement. Segregation of TiB_2 and Al_3Ti in the system act as inoculant near primary phase α-Aluminum.Addition of magnesium may create intermetallic phases Mg_2Si, and the high percentage of iron creates intermetallic phases β-Fe and π-Fe. The existence of spinel ($MgAl_2O_4$) indicates an interphase between matrix and reinforce which optimized the tensile stress distribution.

Acknowledgement
This research is supported by University of Indonesia through its Research Grant Program (HIBAH PITTA tahun 2017, Nomor: 767 UN2.R3.1/HKP.05.00/2017).

References

[1] Totten G E and MacKenzie D S 2000 *Handbook of Aluminum: Alloy Production and Materials Manufacturing 2nd ed* (New York: Marcel Dekker, Inc)
[2] Kurniawan I 2014 *J. Info. Tek. Mesin* **7(1)** 78-87
[3] Fuxing Y E, Takuya T, Toru K and Kazuhiro N 2006 *JWRI* **35(1)** 53-56
[4] Rana R S, Purohit R and Das S 2012 *Int. J. Sci. Res. Publ.* **2(6)** 1-7
[5] A Zulfia, T Zhakiah, D Dhaneswara and Sutopo, **2017** *IOP Conf. Ser.: Mater. Sci. Eng.* **202** 012089
[6] Mazahery A and Shabani M O 2012 *J. Trans. of Non ferrous Met. Soc. China* Elsevier **22(2)** 275-280
[7] Mondal D P and Jha N 2014 *J. Mat. Eng. App.* Elsevier. **22(5)** 1001-1011
[8] Wannasin J, Canyook R, Wisutmethangoon S and Flemings M C 2013 *J. Acta Metallurgica et Materialia* Elsevier **61(10)** pp 3897-3903
[9] Zhong-wei C, Zhi H E and Wan-qi JIE 2008 *J. Trans. of Nonferrous Met. Soc. China Elsevier* **6326** 410-413
[10] ESamuel, B Golbahar, A M Samuel,H W Doty, S Valtierra and F H Samuel2014 *J. Mat. Mat. Eng. App. Elsevier* **56** 468-479
[11] Bihari A and Das S2014 *J. of Materials Research and Technology* science direct **4(2)** 171-179
[12] Amerioon A, Emamy M and Ashuri G 2015*Procedia Mater. Sci.*Elsevier **11** 32-37
[13] Zhang J, Ke W, Ji W, Fan Z, Wang W and Fu Z 2015 *Mater. Sci. Eng. A* Elsevier ***648*** 158-163
[14] Pai B C, Ramani G, Pillai R M and Satyanarayana K G 1995 *J. Mater. Sci.***30(8)** 1903-1911
[15] Sangghaleh A and Halali M 2009 *J.Appl. Surf. Sci.***255(19)** 8202-8206
[16] Iizuka T and Ouyang Q 2014 *J. Trans. of Nonferrous Met. Soc. China vol. 24* Elsevier **24** 2337-2345
[17] Liu L, Samuel A M, Samuel F H, Valtierra S, Nemak C and Bosques P O B 2003 *J. Mater. Sci.***8** 1255-1267

IOP Conf. Series: Materials Science and Engineering **333** (2018) 012045 doi:10.1088/1757-899X/333/1/012045

Microstructures and Hardness of the High Chromium Oxide Dispersion Strengthened Alloy Fe-25Cr-Y_2O_3 Sintered by the Arc Plasma Sintering (APS)

Bandriyana[1], Arbi Dimyati[1], Agus Sujatno[1], Rohmad Salam[1], Sumaryo[1], Pudji Untoro[2], Bambang Suharno[3]

[1]Center for Science and Technology for Advanced Materials, BATAN-Indonesia
[2]Department of Physics, Surya University
[3]Department of Metallurgy and Materials, Indonesian University

E-mail: bandri@batan.go.id

Abstract. High chromium ODS alloy has been developed for application as structural material in high temperature nuclear reactor. In the present study, Fe-25Cr-Y_2O_3 with dispersed 0.5 wt.% Ytria (Y_2O_3) were synthesized and characterized by means of various techniques as a function of milling time 1, 2 and 3 hours. The alloy synthesis was carried out by the Mechanical Alloying (MA) process and subsequent sintering by means the new plasma technique using the APS apparatus. Scaning Electron Microscopy (SEM) and X-ray diffraction (XRD) were conducted for morphology and phase analysis. Evaluation of the mechanical properties was studied based on the Vickers hardness measurement. SEM examination revealed that the sample after sintering by APS method at different milling duration exhibited some particle aglomeration and homogenized oxide dispersion that obviously strengthened the alloy. The XRD test, however, proved the formation of the main phase Fe-Cr. The alloy showed exceptionally high hardness of 193 VHR which is mainly due to the grain refining that increase by the increasing of the milling time.

1. Introduction

Oxide dispersion strengthened (ODS) ferritic alloy exhibits excellent strength, corrosion, hardness and radiation resistance at elevated temperatures. It is considered to be fitted for structural materials on the advanced and high temperature nuclear systems [1]. The excellent characteristics of this alloy has the origin in the special microstructures mainly consisting of fine grains and tiny oxide particles dispersed homogenously with high number density [2]. The nano oxide particle dispersoids that stable in high temperature can act as pinning point to inhibit dislocation movement that provide excellent creep strength and irradiation resistance [3-5]. It was found that the alloy of Fe-25 Cr steels will have excelent in high temperature oxidation by the formation a protective Cr_2O_3 layer on the surface that exhibited very satisfactory adhesion to the metallic core for all samples. It is clear that Chromium plays an important role in improving the high temperature oxidation resistance of the ODS alloy [6]. In our previous study reported, for achieving sufficient corrosion resistance in severe environment, a new series of high-Cr ODS ferritic steels have been developed to meet the requirements for advanced nuclear fuel claddings. ODS alloys are produced commonly by powder metallurgy and subsequent densification via sintering at high temperature for several hours. However, mechanical alloying (MA) is a useful powder metallurgy processing technique involving mixing, coalescing, fracturing and remixing of powder particles in high-energy ball [3,7]. Densification process by sintering using conventional electric furnace is well-known to be very high energy and time consuming. At least since mid 80's Spark Plasma Sintering (SPS) becomes popular for substituting the use of conventional furnace for the densification process. SPS uses high-pulsed DC current (~5000 Amperes) by 10 Volts to generate heat inside the sample body during relatively moderate time of period which is in the range of 30 minutes to one hour. However, beside of high energy demand SPS does not allows the application on samples with irregular shapes. In this place, APS is recognized as economically good alternative to SPS for solidification of powder green materials such as ODS alloy especially due to its simplicity, low power consumption and flexibility in using defined shape of specimens. APS uses plasma plume produced by dissociation of Argon gas which is exothermic process generating temperature higher than 2000°C. The plasma can be easily controlled in term of its diameter that allow process with high speed. Consolidation of ODS sample using the APS is much faster than the conventional way using heating furnace. For example to consolidate a 15mm diameter Zn-Mg alloy it was needed only 1 second at 1 Amperes. Detailed about APS is described elsewhere [8].

In the present study, Fe-25Cr based ODS alloy via powder metallurgy and sintering by APS method is produced with the aim to study the microstructure and mechanical properties of the alloy and their dependency to the preparation specially the milling time. The results were characterized by means of electron microscopy to reveal the characteristic microstructure.

2. Experimental

For this work, the Fe-Cr ODS alloy was produced with the composition Fe, 25 wt.% Cr and 0.5 wt.% of the dispersed particles Y_2O_3. This refers to standard ODS alloy exhibiting high temperature oxidation resistant with sufficient creep properties. The samples were prepared by mixing and milling all of the materials (Fe, Cr and Y_2O_3 powder) for 1, 2 and 3 hours in a high energy milling (HEM). After milling the samples were isostatic pressed with the compression load of 20 tons to get the sample coins as green material. The coins were then consolidated by sintering in the APS at 25 Amperes and 12 Volt for 4 minutes. The APS apparatus is shown schematically in Figure 1.

Figure 1. The Arc Plasma Sintering Device (APS).

The microstructure and composition of the alloys were studied by means of Scanning Electron Microscopy (SEM) coupled with the Energy Dispersive Spectroscopy (EDX). The EDX spectrums illustrated the chemical composition of the samples. The EDX mapping was performed to show and evaluate the distribution of the alloying element of Fe, Cr and the dispersion of Yttria inside the matrix. The XRD test was conducted to study the phases formed during the APS sintering process. The mechanical properties of the consolidated samples were then studied using the Vickers micro hardness (HV) measurement.

3. Results and Discussion

Microstructures of the sintered ODS alloy samples with dispersion of 0.5 wt.% Yttria were presented in Figure 2. The figures were taken from top view as secondary electron images to increase the contrast of the different phases occur. The alloy generally consists of large Fe-Cr alloy matrix phase seen as bright areas and the dark Cr dominated second phases as also confirmed by the EDX measurement. The small bright particles distributed almost homogenously throughout the image area are Yttrium oxides which appear as individual particle but also as agglomerates of some particles. As also observed in our recent investigation, an area of inter-diffusion zone was detected in the matrix-Cr-rich phase interface This elucidates the fact that the alloying process has been begun with the formation of two single primary Fe and Cr phases followed by the process of interdiffusion the same elements in both direction [9]. Obviously, one can see the grain refining in a function of milling time. The mean grain size of the sample milled for 3 hours is in the range between 2 and 15 um, whereas those of the sample milled for 1 hour between 2 and 50μm. The part of fine grains definitely increases with the milling time.

The number of porosity seems to decrease with the grain refining following the rise of the milling time. The porosity falls consequently from about 15 % on the sintered sample milled for 1 hour to circa 10 % on the sintered sample after 3 hours milling. This is by far acceptable because the APS method is press less sintering process which lefts some porosity on its product.

The presence of large single Cr-rich phase can be explained as a result of large different of particle size of the start material. As a Cr-source was used Cr-powder purchased from Aldrich company with initial grain size in average about 60 μm while Fe powder much more smaller (less than 30 μm). Because the alloying process is known to be diffusion driven phenomena, it may happen that during 4 minutes sintering even at high temperature not enough time for Cr atoms to escape their bonding on the Cr particle surface to diffuse toward Fe matrix to at last form Fe-Cr alloy.

Figure 2. SEM images of the sintered ODS alloy milled (a) 1 hour, (b) 2 hour and (c) 3 hours.

SEM-EDX area mappings in Figure 3-5 show the elemental distribution of Fe, Cr, Y and O atoms on all the samples with different milling times. The presence of phases in the alloy can be determined by superimposing the EDX contrast given by each element. Therefore, Yttria are mainly observable in the diffusion zone of the matrix and the Cr-rich phases. The mapping can clearly image the dispersion of the Ytrria particles in the matrix and in the grain boundary of both Fe and Cr phase. Increasing of milling time has caused homogenize of Yttria presumable due to particle size reduction and intensive mixing during milling. The mean grain size and oxide distribution were similar to those found in the samples sintered by the Spark Plasma Sintering method, which is often reported as a crucial requirement for producing a compact materials with strength and good ductility [10].

EDX mapping again reveals the retained Cr-particle, which remains exist after the sintering process as mentioned before. Figure 3 demonstrate this in undisputed manner the existent of retained Cr particle. The Cr-K line intensity image in the Figure 3a which represents the Cr-rich phase meets deflected zone in the Fe-K intensity in the Figure 3b.

The excess of O-K intensity as present in the images on all the samples correlated directly to Yttrium oxides, which forms agglomeration mainly in the grain boundaries. The most part of O-intensity seems to distributed homogenously on the entire observable surface area inside the grain and grain boundaries. This indicates to the homogenous distribution single Yttrium particles.

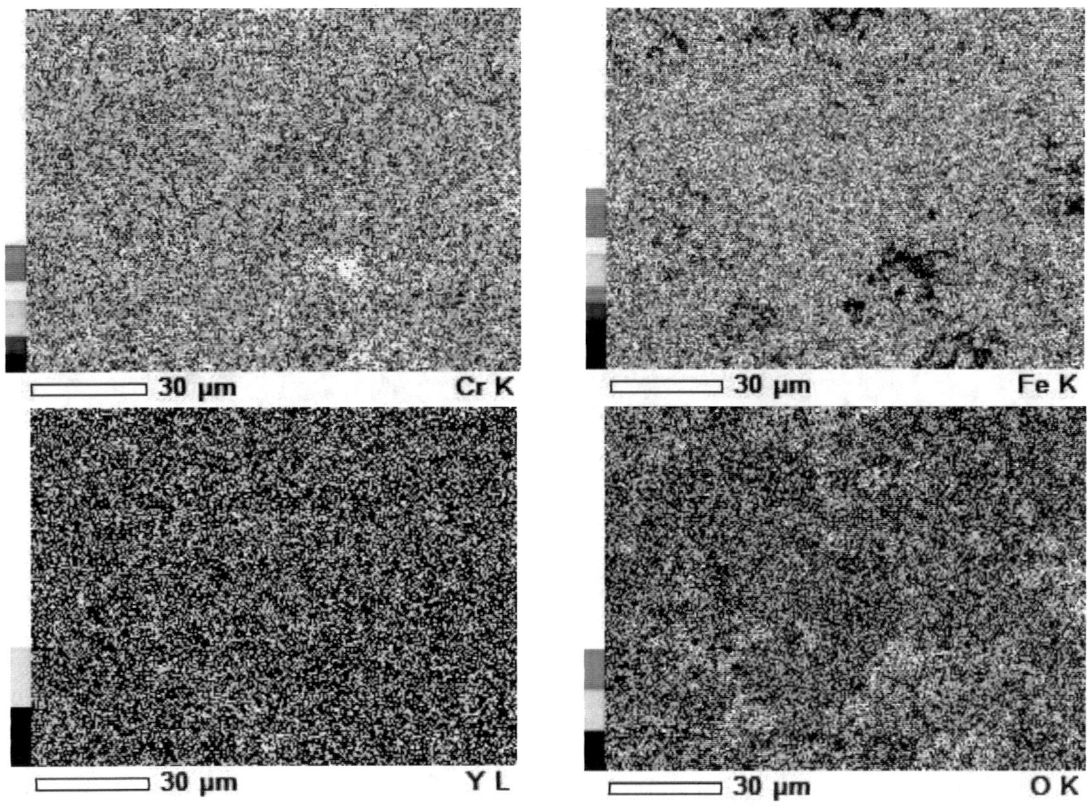

Figure 3. EDX mapping of the ODS alloy prepared by 1 hour milling.

IOP Conf. Series: Materials Science and Engineering **333** (2018) 012045 doi:10.1088/1757-899X/333/1/012045

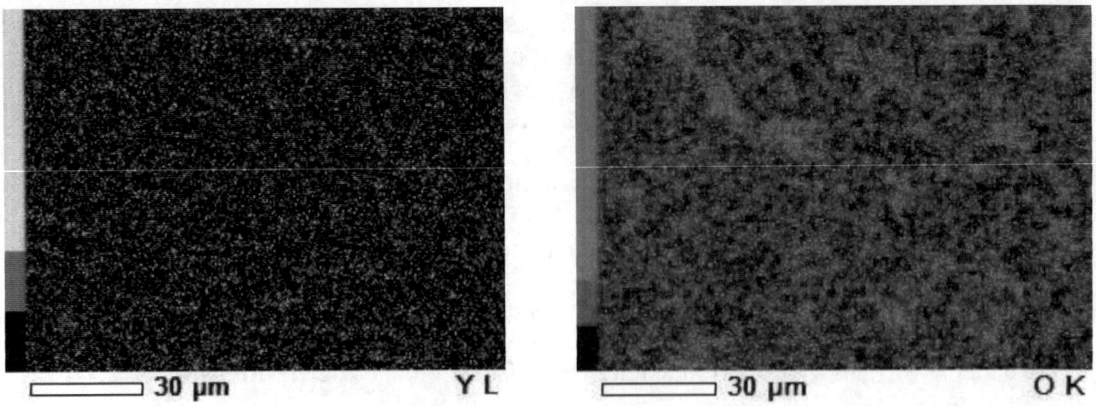

Figure 4. EDX mapping of the ODS alloy prepared by 2 hours milling.

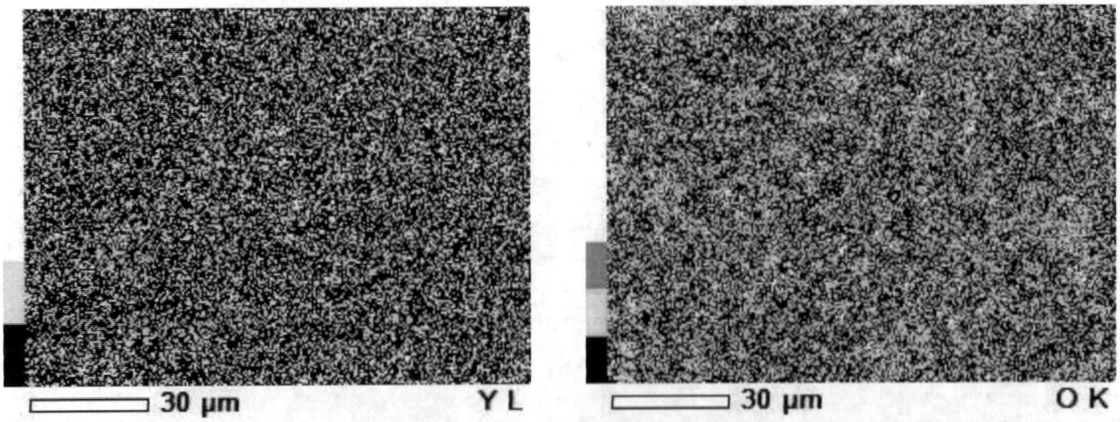

Figure 5. EDX mapping of the ODS alloy prepared by 3 hours milling.

The XRD investigation was performed to study the phase formation during the ODS alloy sintering process. Figure 6 shows the diffraction pattern of sintered ODS alloy samples after 1 hour milling. In general, during time set the alloy was already dominated by the Fe-Cr phase as confirmed by the observation of the microstructure and measurement of elemental composition by the SEM-EDX presented above. Detailed analysis of the occurrence of this Fe-Cr phase can be taken in the easiest way by observation of the first and most intense peak of the XRD pattern which is cropped from the image and presented in Figures 6b. By increasing of milling time the full-width at half-maximum (FWHM) of the peak become wider. This confirmed the grain refining that strengthened the microstructure observation by the SEM described before. More detailed analysis in the first peak of all samples showed that this is actually a superimposing of Fe-Cr and Fe peaks due to the similarity of the crystal parameters. But the peak of the Fe part is quite small so that it can be concluded the most phase dominated by Fe-Cr phase. The alloying process which drive the formation Fe-Cr phase is mostly completed even though the very short sintering time. These development caused the improvement of the mechanical properties that can be observed on the results of the hardness measurement.

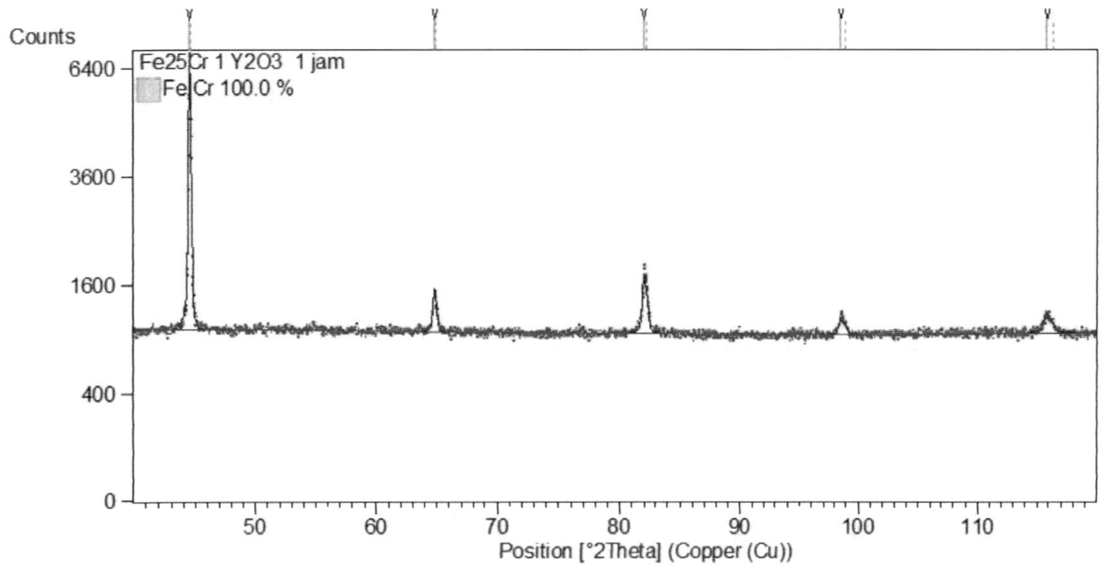

(a)

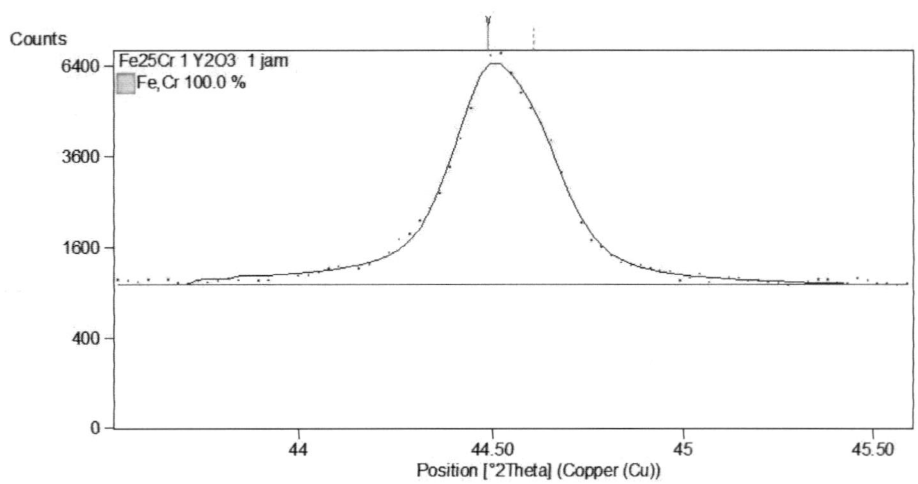

(b)

Figure 6. XRD patterns of the ODS alloy (a) full plot, (b) first peak analysis

The hardness tests on samples with variation of milling time is given in Figure 7. The ODS steel alloy sample showed significant increase of hardness value from around 170 VHN after 1 hour to 176 VHN after 2 hours milling. But the hardness raises to ca. 193 HVN after milling 3 hours. Increasing the milling time will increase the hardness with not linear shape (depletion to linear path as indicated in the image) which is assumed to be caused by the grain refining in one hand and by the more homogenous distribution of oxide particles at longest milling on the other hand.

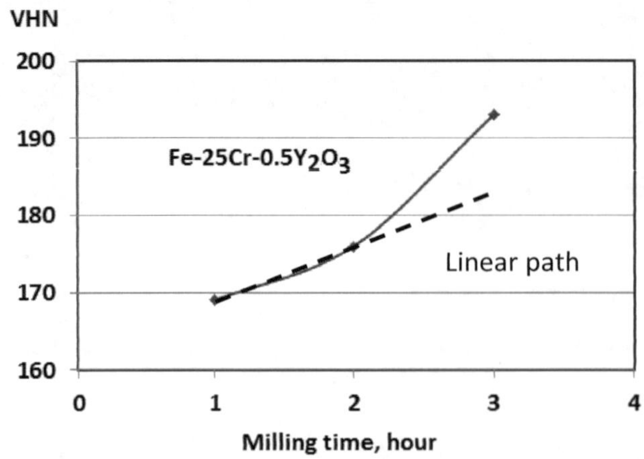

Figure 7. Vickers Hardness of the ODS alloy Fe-25Cr-0.5 Y_2O_3 with milling time variation.

4. Conclusion

From this study, it can be concluded that with APS method high Chromium ODS alloy can be produced with sufficiently quality. Due to the low DC current plasma generation, the heat applied to the sample can be exactly defined and controlled. This significantly simplify the handling of the sintering process compared to the conventional method using furnace even using commercially available SPS method. Electron microscopy investigation found that the sample after sintering with APS at different milling duration exhibited some particle aglomeration and homogenized oxide dispersion that obviously improve the strength of the alloy. The XRD test, confirmed the complete alloy formation of the Fe-Cr matrix phase even after sintering process for only 4 minutes. The alloy showed exceptionally high Vickers Hardness of 193 VHR which is mainly due to the grain refining and homogenous distribution of Yttrium oxides following the milling time.

Acknowledgment

The authors thank to Ministry of Research, Technology and Higher Education (Kemenristekdikti) for financial support under the InSINas 2017 research program, RT2017-0287. Special thanks also given to Mr. Bambang Sugeng for intensive discussion on XRD measurement.

References

[1] Raman L, Gothandapani K, and B.S. Murty B.S 2016 *Defence Sci. J.* **66** 4 316-322.
[2] Noh S, Kasada R, Kimura A, Park S H, Hirano S 2011 *J. Nucl. Mater.* **417** 245-248.
[3] Ukai S, Ohnuki S, Fujisawa T 2013 *Mater. Trans.* **54** 2 215-221.
[4] Iwata R K, Kasada R, Kimura A, Okuda T, Inoue M, Abe F, 2009 *ISIJ International,* **49** 12 1914–1919.
[5] Bischoff J, Motta A T 2012 *J. Nucl. Mater.* **424** 261-276.
[6] Brylewski T, Aleksander Gil, Rakowska A, Chevalier S et.al. 2013 *J. Oxidation of Metals,*
[7] Taguchi M, Sumitomo H, Ishibashi R, Aono Y 2008 *Mater. Trans.* **49** 6 1303-1310.
[8] A. Dimyati 2016 *Proc. Asia- Pacific Microscopy Conference* **11.**
[9] Bandriyana. B, Sujatno A, Salam R, Dimyati A, and P Untoro P 2017, *Int. Conf. ICGRER IOP Conf. Series: Earth and Environmental Science* **75** 1-5.
[10] Hongtao Zhang et al., 2015 *J. Nuclear Materials,* **464** 61-68.

IOP Conf. Series: Materials Science and Engineering **333** (2018) 012046 doi:10.1088/1757-899X/333/1/012046

Characteristics of ADC12/nano Al_2O_3 composites with Addition of Ti Produced By Stir Casting Method

A Zulfia[1], Krisiphala[1], D Ferdian[1] B W Utomo[2] and D Dhaneswara[1]

[1]Departement of Metallurgy and Materials Engineering, Universitas Indonesia, Depok, 16424, Indonesia
[2]Astra Honda Motor, Cikarang Barat, 17530, Indonesia

E-mail: anne@metal.ui.ac.id

Abstract. The mechanical properties and microstructure of ADC12/nano Al_2O_3 matrix composites have been studied in this work. The composites were produced by stir casting method. ADC 12 as matrix composites was combined by Mg and Ti. The addition of Ti was varied from 0.02 to 0.08 wt-% as grain refinement wetting to improve mechanical properties such as tensile strength, hardness and wear resistance, while Mg addition was to promote wetting between ADC 12 and nano Al_2O_3. The optimum tensile strength was found at 0.04 wt-% addition of Ti with value of 132.5 MPa, further adding more Ti cause a poisoning mechanism that will hindered the grain refining process and reduce the tensile strength. The hardness and wear resistance of composites would also increase because of the refinement process. and the added Magnesium in the material that will form Mg_2Si primary phases who have a high hardness value.

1. Introduction

In present day the use of a composites especially with a matrix of metal or called MMC have been increasing in various application. For example in the automotives region the need of a lightweight and high strength material for the brake pads can be realize by the use of a lightweight metals reinforced with a high strength ceramics instead of the conventional cast iron that is heavy and more expensive[1-2]. So this study focused on a lightweight metals that is Aluminium (ADC 12) reinforced with Al_2O_3 nano particles and was fabricated by a convetional stir casting to see the improvement of its mechanical properties[3]

The addition of high strength Al_2O_3 particles can gradually increase the mechanical properties of a low strength material like the ADC12 alloy. In addition, the Al_2O_3 can improve the grain quality as it can also be a nucleating agent for the α-Al and make the grain more finer and uniform [4]. The particle size can also influence the results because the addition of Al_2O_3 will make the material more brittle as the volume fraction of the addition increase. However, with smaller size of the particle it will reduce the volume fraction need to be added which in this study will use a nano size particle for the reinforcement[5].

One of the problems in the fabrication of metal matrix composite is the wetting between the reinforced with their matrix. The addition of Magnesium can increase the wettability of the matrix. The presence of oxide layer in the Al_2O_3 can make the penetration of the matrix difficult. So with Magnesium added it will react and make the spinnel layer ($MgAl_2O_4$) that can make the penetration of matrix

IOP Conf. Series: Materials Science and Engineering **333** (2018) 012046 doi:10.1088/1757-899X/333/1/012046

easier[6]. Also Magnesium can increase the hardness of a material by forming a Mg_2Si phase that have a high mechanical properties[7].

ADC12 which is an Al-Si alloy that have Si flakes microstructural morphology, the adding of Al-5Ti-1B master alloy as a grain refiner can further improve the mechanical properties of the material. The master alloy can produce a more uniform microstructure and a much finer grain than a plain Al-Si alloy[8] by forming an Al_3Ti phases that will be acting as a nucleating sites for the α-Al[9] Nevertheless, high Ti content in a Al-Si system can trigger a poisoning mechanism in which the Ti will form a Ti_5Si_3 layer that can hinder the grain refining process[10]. So by adding variation of the master alloy it can show at what percentage it will have the optimum effect at refining the grain size and improving the mechanical properties.

2. Experimental

2.1 Making of ADC12/Nano Al_2O_3 Composites

Aluminum ADC12 with composition in Table 1 was melted at 800 °C in an electric furnace in the other hand nano Al_2O_3 particles were also preheated at 500 °C for 1 hour in a muffle furnace to remove moisture from the particles. After the aluminum melted, degassing was conducted to remove gas from inside the molten aluminium and followed by removing the slag from the surface of the molten aluminium. Then 10 wt-% of magnesium and variation of Ti from 0, 0.02 to 0.08 wt-% were added into the molten aluminium and stirred for 2 minutes. Meanwhile nano Al_2O_3 with 0.03% Vf was heating for 1 hour then poured into molten aluminium followed by stirring for 2 minutes to maximize the dispersions of the particles in the matrix. The cast composites then poured into tensile metal mold and underwent solidification.

2.2 Characterization of ADC12/Nano Al_2O_3 Composite

The mechanical properties of the composites were characterized by tensile, hardness, wear resistance and impact testing. GOTECH AI-7000 LA 10 machine was used to perform the tensile test with JIS Z2241 standard. Rocky Rockwell B was used to do hardness testing with ASTM E18-16 standard. Ogoshi machine was used to conduct the wear test. Samples for metallography observation were prepared by grinding with sand paper, polishing with TiO_2 and polish cloth, and etching with keller reagent. Keller reagent has composition of 2 ml HF, 5 ml HNO_3, 3 ml HCl and 190 ml aquadest. Samples were observed using OLYMPUS BX41M-LED optical microscope, scanning electron microscope (SEM) link to energy dispersive spectrum (EDS). For further analysis, XRay Diffraction was conducted to identify the phase present in composites.

Table 1 The Chemical Composition of ADC 12

Elements	Al	Si	Fe	Cu	Mn	Mg	Zn	Cr	Ni
%wt	84.8	10.5	0.864	2.33	0.233	0.221	0.637	0.389	0.0759
Elements	Pb	Sn	V	Na	Bi	Zr	B	Ga	Ti
%wt	0.0342	0.0798	0.0102	0.0020	<0.0050	0.0068	<0.0005	0.0118	0.0462

3. Results and Discussion

3.1 Microstructural Observation and Phase Indentification of ADC12/nano Al_2O_3 Composite

The optimum grain refinement process occurs in the addition of 0.04 wt-%, the refining process occur because the Al_3Ti phases which will become a nucleation site for the α-Al and produce a finer and much more uniform grain see Fig. 1. However, with the addition of 0.06 and 0.08 wt-% of Ti the grain become

IOP Conf. Series: Materials Science and Engineering **333** (2018) 012046 doi:10.1088/1757-899X/333/1/012046

larger because of the poisoning effect that is common occured in the Al-Si system with a high Ti content. Ti will form Ti_5Si_3 layer with Si which have a higher affinity than Al.

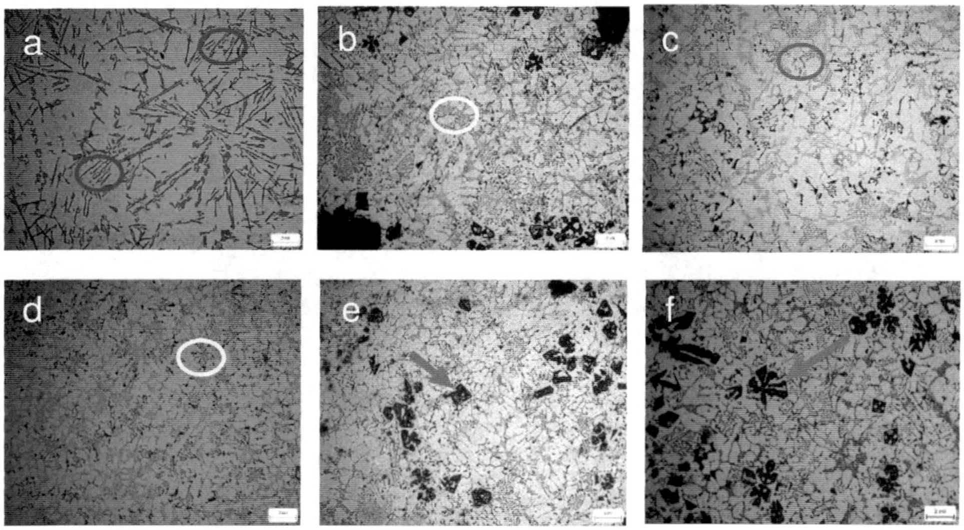

Figure 1 Microstructure of ADC12/nano Al_2O_3 composites (200x): (a) ADC12; (b) ADC12/nano Al_2O_3; (c) 0.02 wt-% Ti ; (d) 0.04 wt-% Ti; (e) 0.06 wt-% Ti ; (f) 0.08 wt-% Ti

As it can be seen by the solidification path of the materials in Figure 2, the microstructure of the composites mainly contains of α-Al, Si eutectic (blue circle), Mg_2Si primary (red arrow) Mg_2Si quasibinary (yellow circle), and Al-Fe-Si intermetallics (green circle). The microstructure in Figure 1 can be seen Mg_2Si phases formed in the system because of 10 wt-% Mg the addition. it was shown in figure 1e and 1f Mg_2Si primary will form in the microstructure and have a blocky like morphology. Mg_2Si primary have a high hardness value that can improve the hardness of the material but can become a site for crack initiation, so more of the Mg_2Si primary phases form the more brittle the material can become[11].

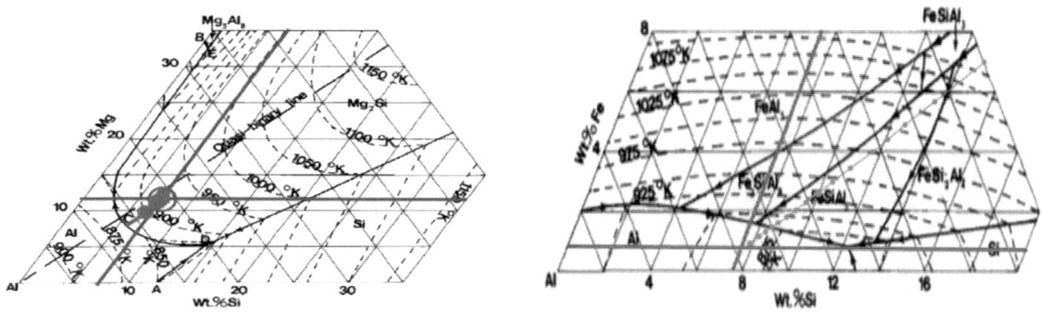

Figure 2 (a) Ternary diagram of Al-Mg-Si alloys and the soldification path (b) Ternary diagram of Al-Fe-Si and the soldification path

ADC12 is a material that commonly have a high Fe content, so in this study it can be seen that the material have the Al-Fe-Si intermetallics. This intermetallics can cause poor mechanical properties because it have low elongation and can reduced the ductility of the material[12]. The Mg_2Si phases of primary and quasibinary were also found in the SEM-EDS examination in Fig.3. Besides the Mg_2Si phase, Al-Fe-Si phases are also found and this phases can lower the output of the mechanical properties.

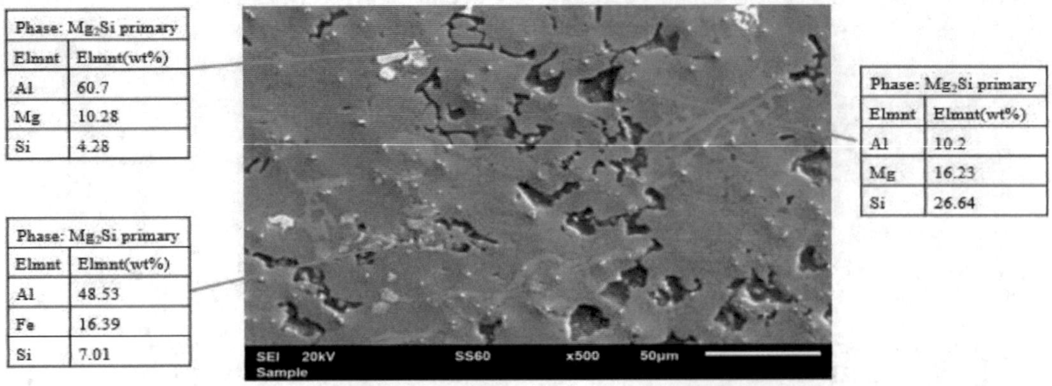

Figure 3 Microstructure of ADC12/nano Al_2O_3 composites with 0.04 wt-% Ti analysed by SEM and EDS Examination

The phases present in ADC12/Nano Al_2O_3 with 0.04 wt-% Ti analysed by XRD is shown in Fig. 4, the phases are $MgAl_2O_4$, Mg_2Si, and Al_3Ti. The spinel layer $MgAl_2O_4$ proven good interface bonding between the matrix and the reinforcement, it is formed by the addition of Mg in this study at 10 wt-%. Al_3Ti is also detected and prove that a refining process occurs as the Al_3Ti became a nucleation sites[13].

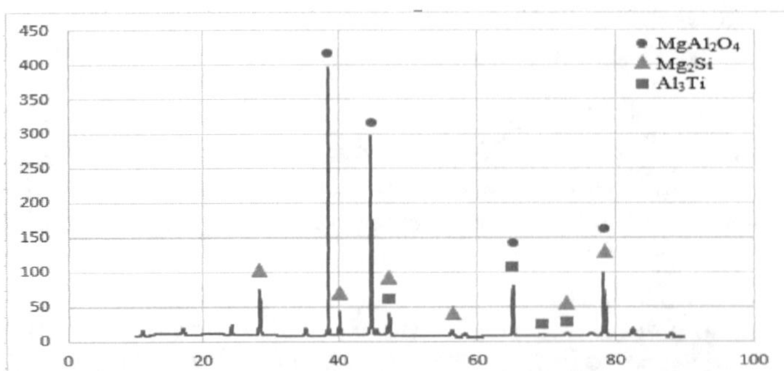

Figure 4 XRD Pattern of ADC12/Nano Al_2O_3 compsoites with 0.04 wt-% Ti addition

3.2 Mechanical Properties of ADC12/Nano Al_2O_3 Composite

The results of the tensile test is shown in Figures 5 and 6. It can be seen that the addition of 0.03 % Vf Al_2O_3 nano particle can improve the tensile strength of the material from 98.36 MPa to 114.87. This is caused by the high strength of Al_2O_3 particles that will increase the mechanical properties of the ADC12 matrix[14]. The addition of Ti content as was discussed before can also improve the tensile properties of a material by making the grain smaller and more uniform. The optimum tensile strength was achieve at 0.04% of Ti addition with a value of 132.48 MPa, it is also have the smallest grain compared to the others. Furthermore addition will decrease the tensile strength because of the poisoning effect occuring in the material, and it was seen in Figure 1e and 1f many Mg_2Si primary were form because of the excess Mg. Mg_2Si primary phases can lowered the elongation of a material by being a site for crack initiation.

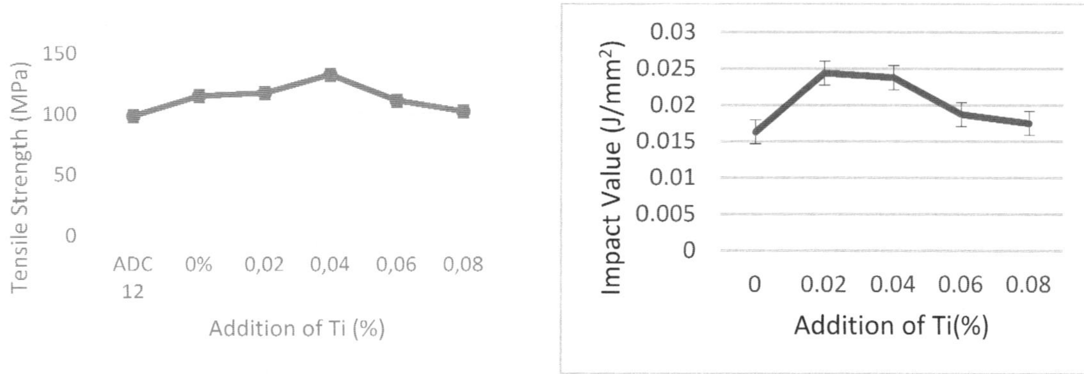

Figure 5 Effect of Ti addition on (a) Ultimate Tensile Strength (UTS) and (b) Impact strength of composites

The addition of nano Al_2O_3 increases the hardness of the material, it is because Al_2O_3 has a high hardness so the hardness of the composites will also improved. Ti addtion can also increase the hardness of a material because of the smaller grain it will produce. In the addition of Ti at 0.06 and 0.08 wt% occur the poisoning mechanism and the grain it produces is larger, but the hardness of the material still increases because of the Mg_2Si primary that forms[15-16]. The highest value of hardness 41.3 HRB. The effect of adding Ti in the composites can be seen in Figure 6(a).

The addition of Al_2O_3 and Ti can also improve the wear resistance of material[17]. It can be seen also in Figure 6(b), that the harder the composite the lower its wear rate becomes. The optimum value of wear rate is at 0.96 x10^{-5}.

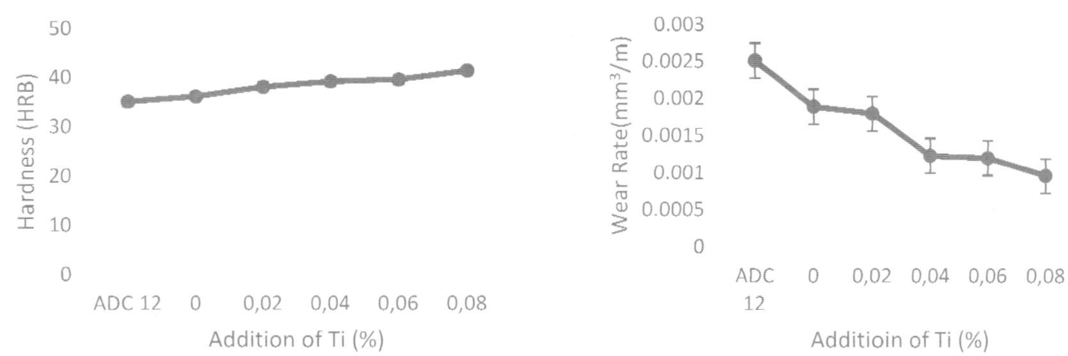

Figure 6 Effect of Ti addition on (a) Hardness (UTS) and (b) Wear rate

The impact value that was obtained are 0,0163; 0,024375; 0,02375; 0,01875; 0,017 corresponding to the increase of Ti addition. The addition of Ti slightly increase the impact value because of the finer grain increasing the energy the material can absorbed. The more brittle the material the impact value also decreases.

4. Conclusion

1. The addition of Al_2O_3 nano particle and Ti improves the mechanical properties of the material such as its tensile strength, hardness, wear resistance, and impact value.
2. The addition of Ti will form Al_3Ti phases in the microstructure that will become a nucleation site for α-Al and reduce the size of the grain so the mechanical properties will also improved.

3. Poisoning effect will occur by adding more than 0.04% wt Ti because of the high Si content in the material.
4. By adding 10 wt-% Mg will create a thin layer of $MgAl_2O_4$ that contributes to the improvement of mechanical properties of the composites but also form the Mg_2Si phases that can make the material brittle.

Acknowledgement

This research is supported by University of Indonesia through its Research Grant Program (HIBAH PITTA tahun 2017, Nomor: 767 UN2.R3.1/HKP.05.00/2017).

References

[1] G. E. Totten and D. S. MacKenzie 2000 *Handbook of Aluminum: Alloy Production and Materials Manufacturing, 2nd ed.* (New York: Marcel Dekker, Inc)
[2] A Zulfia, T Zhakiah, D Dhaneswara, Sutopo 2017 *IOP Conf. Ser.:Mater. Sci. Eng.* **202** 012089
[3] Campbell, F. C. 2010 *Structural Composite Materials*. (Ohio: ASM International)
[4] Shash, Ahmed Y., E. Amer, El-Saeed, Moaztaz 2015 *Advanced Structure Materials*. **70** 13-24
[5] Sajjadi, S. A., Torabi P, Ezatpour, H. R. & Sedghi, A 2012 *J. Alloys Compd.* **511** 226–231
[6] Geng L, Hong-wei, Zhang; Li H., Li-na G, Lu-jun H 2010 *Trans. Nonferrous Met. Soc.* **20** 1851-1855
[7] Feng Y, Shouxun J, Zhongyun F 2013 *Materials Science Forum.* **765** 64-68. 2013.
[8] O. Fakhraei, M. Emamy, and H. Farhangi 2013 *Mater. Sci. Eng. A.* **560** 148–153
[9] S. A. Kori, B. S. Murty and M. Chakraborty 2000 *Science and Engineering.A.* **280** 58-61
[10] S. A. Kori, V. Aruradi, B.S. Murty ,M. Chakaraborty 2005 *M. Materials Forum* **29** 387-393
[11] T. Grostad 2014 *Nucleation of Primary Mg2Si in Al-Mg-Si Alloys* (Trondheim: Norwegian University of Science and Technology)
[12] Zhijun Z, Hiroyasu T, Equo K, Tatsuo S 2013 *Materials Transactions*. **54** 1484-1490
[13] H. Li, Z. Zhang, Y. Yu, dan T. Xie 2012 *Materials under the Magnetic Field.* **1** 1554–1557
[14] Li, Pengting; Liu, Sida; Zhang, Lili; Liu, Xiangfa 2013 *Materals and Designs.* **47**. 522-528
[15] M. Emamy, A. Razaghian, H. R. Lashgari, dan R. Abbasi 2008 *Mater. Sci. Eng. A.* **485** No. 1–2 210–217
[16] K.G. Basavakumar, P.G. Mukunda, M Chakaraborty 2008 *Materials Characterization.* **59** 283-289
[17] P.C. Meena, A.Sharma. S.Singh 2015 *La Metallurgia Italiana-n.* **1** 25-34

IOP Conf. Series: Materials Science and Engineering **333** (2018) 012047 doi:10.1088/1757-899X/333/1/012047

Fabrication of barium titanate doped strontium using co-precipitation method

Y Iriani, M A Yasin and R Suryana*

Department of Physics, Faculty of Mathematics and Natural Sciences, Sebelas Maret University, Jl. Ir. Sutami 36A Kentingan Surakarta 57126 Indonesia

*E-mail: rsuryana@staff.uns.ac.id

Abstract. Fabrication of barium titanate ($BaTiO_3$/BT) doped strontium (Sr) using co-precipitation method has been successfully conducted. The research aim is to get the best of mole variation of Sr doping to ferroelectric material properties. Doping Sr was varied at 1%, 2%, 3%, 4% and 5% in $BaTiO_3$. Each sample was sintered at temperature of 1100°C with holding time for 6 h and temperature rate at 10°C/min. They were then characterized by XRD instrument to investigate the crystal structure, LCR meter to measure the dielectric constant, and Sawyer Tower circuit to reveal the hysteresis curve. The peaks of XRD shift towards larger angle when mole doping Sr increase. The crystallinity of all samples is above 90% and the crystallite size is in the range of 27 nm to 34 nm. Hysteresis curve from Sawyer Tower testing confirms that all samples are ferroelectric material. The RLC measurement results reveal that the less frequency leads to the higher dielectric constant while the highest dielectric constant belongs to the BT doped 3% of Sr. Therefore, it is the best variation obtained in this research.

1. Introduction

Ferroelectric materials can be used as a dielectric material. It is able to change the electric polarization spontaneously when an external electric field is applied [1-2]. That condition is marked by the formation of a hysteresis curve which is the curve between the electric polarization and the electric field. Ferroelectric material has perovskite structure with ABO_3 formula where A is cation with valence of 1 and 2, B is cation with valence of 4 and 5, and O is oxygen [3].

Ferroelectric material has three types of property those are pyroelectric, dielectric and piezoelectric properties. Pyroelectric property is able to be used as a sensor that will produce voltage when it is applied in certain temperature. The dielectric property, by utilizing the ferroelectric properties, is widely used as a memory storage application that is Feroelectric Random Access Memories (FRAM) [4].

One of ferroelectric materials is barium strontium titanate ($Ba_{1-x}Sr_xTiO_3$/ BST) obtained from BT doped by strontium (Sr) atoms. The dielectric constant of BST has been attained between the range of 578 to 2854, and the highest dielectric constant is produced by $Ba_{0.71}Sr_{0.29}TiO_3$ [5]. One of methods to fabricate BST is co-precipitation method [6-9]. The advantage of this method is simply and only required the room temperature in the reaction experiments. In this research, BT doped by Sr with mole variation of Sr concentration has been synthesized by co-precipitation method. This research aims to find the best result of Sr mole variation in BT to ferroelectric properties.

IOP Conf. Series: Materials Science and Engineering **333** (2018) 012047 doi:10.1088/1757-899X/333/1/012047

2. Experimental Method

BT doped Sr was fabricated by co-precipitation method. The mole composition of Sr in BT was varied i.e. 1%, 2%, 3%, 4% and 5%. The raw material involved barium hydroxide (Sigma Aldrich, 95%), strontium nitrate (Sigma Aldrich, ≥ 99%), titanium tetrabutoxide (Sigma Aldrich, 97%), oxalic acid (Sigma Aldrich, ≥ 99%), and isopropanol (IPA). 0.2 M solution of oxalic acid was dissolved into 35 ml of IPA in beaker glass then stirred continuously (named solution 1). The solution of titanium tetrabutoxide (0.1 M) was dissolved in 18 ml IPA (named solution 2). The solution 2 then was poured into the solution 1 then obtained clear solution of oxalotitanic acid (HTO) (named solution 3). Finally, barium hydroxide was added into the solution 3 then dried to get powder sample.

The powder sample was then pressed by a hydraulic pump with the preasure of 5 Pa. The pellets of sample were sintered using furnace at temperature of 1100°C with holding time for 6 h and the temperature rate at 10°C/min. The microstructure of sample then was characterized via XRD with λ Cu of 1.5406Å. The XRD data were then matched with ICDD database to investigate the crystal structure. The crystallinity was calculated by Eq. 1. $I_{peak\,max}$ is the maximum intensity and $I_{peak\,min}$ is the minimum intensity.

$$\text{Crystallinity} = \frac{I_{peak\,max} - I_{peak\,min}}{I_{peak\,max}} \quad (1)$$

The crystallite size of $BaTiO_3$ doped Sr was obtained by using Scherrer equation in Eq. 2 [10]. In which, β is FWHM value, θ is the diffraction angle, k is Scherrer constant that is 0.9; and λ is the wavelength of X-ray.

$$D = \frac{k\,\lambda}{\beta\cos\theta} \quad (2)$$

LCR meter was then performed in the samples to measure the dielectric constant. In addition, to reveal the hysteresis curve, the samples were then test by Sawyer Tower circuit.

3. Results and discussion

Fig. 1 is the diffractogram of BT doped Sr. ICDD database confirms that the peaks belong to BT (#831880) with tetragonal structure. It indicates that the Sr doping process in BT has been successfully conducted. The peaks are dominated by crystal orientation of (101), (111), (002) and (211). An undesirable $BaCO_3$ (*) at diffraction angle about 24° appears in all samples. $BaCO_3$ was usually present at lower temperature of synthesis [11]. The more Sr mole causes the peaks shift the right side, in other word leading to the larger angle diffraction, but it does not change the crystal structure (tetragonal).

Table 1 is the crystallinity and the crystallite size of BT doped Sr with mole variation. All samples have a high crystallinity confirmed by the value that is above 90%. The highest crystallinity is showed by BT doped 2% and 5% of Sr mole. All samples show the similar crystallite size except BT doped 3% of Sr mole and BT doped 4% and 5% of Sr mole show the largest crystallite size.

Fig. 2 is the hysteresis curve of BT doped Sr. This hysteresis curve was applied using Sawyer Tower method. The hysteresis curve formed indicates that all samples of BT doped Sr are ferroelectric material.

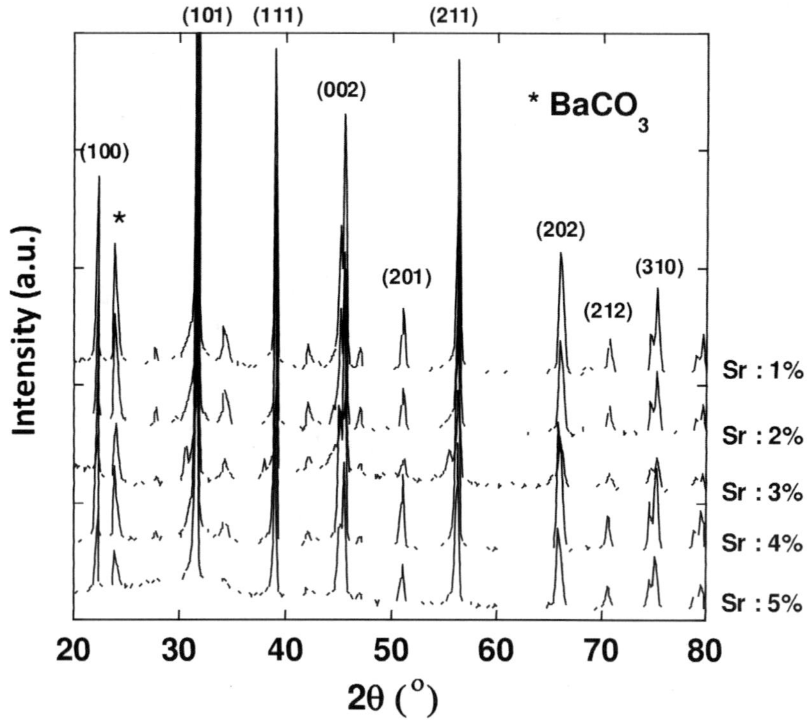

Figure 1. Diffractogram of BT doped Sr at variation of Sr mole

Table 1. Crystallinity and crystallite size of BT doped Sr

Percentage of Sr mole in BT	Crystallinity (%)	Crystallite size (nm)
1%	94	30
2%	96	32
3%	91	27
4%	95	34
5%	96	34

Fig. 3 is the graph of dielectric constant in various frequencies. The greater the frequency decreases the dielectric constant. Each sample has the high value of dielectric constant at the frequency of 0.1 kHz. Fig. 4 is the graph of the dielectric constant of BT doped Sr at frequency of 0.1 kHz. Variations of mole doping Sr will tend to increase the dielectric constant. The highest dielectric constant was obtained in 3% mole doping Sr. This is caused by the radius Sr smaller than Ba, so that will reduce the volume of atoms that affect dielectric constant. However, in the number of moles of doping Sr 4% the dielectric constant decreases. This is possible because Sr doping results in crystal defects.

IOP Conf. Series: Materials Science and Engineering **333** (2018) 012047 doi:10.1088/1757-899X/333/1/012047

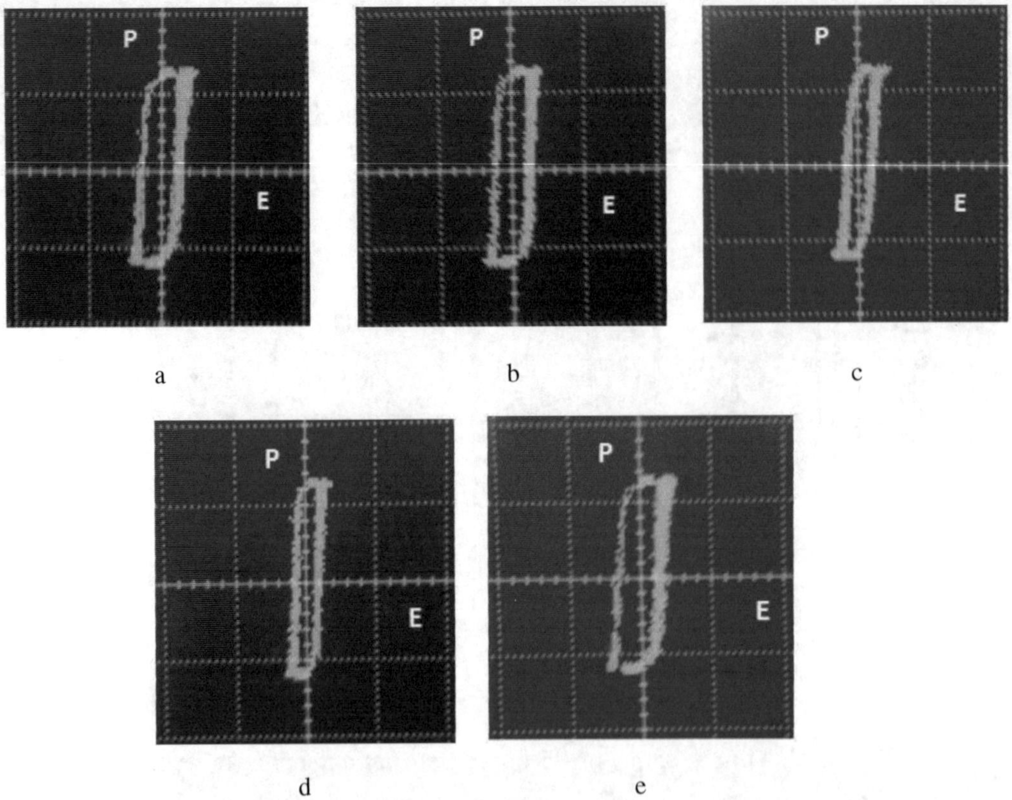

Figure 2. Hysterisis curve of BT doped Sr, a. 1% b. 2% c. 3% d. 4% e. 5%

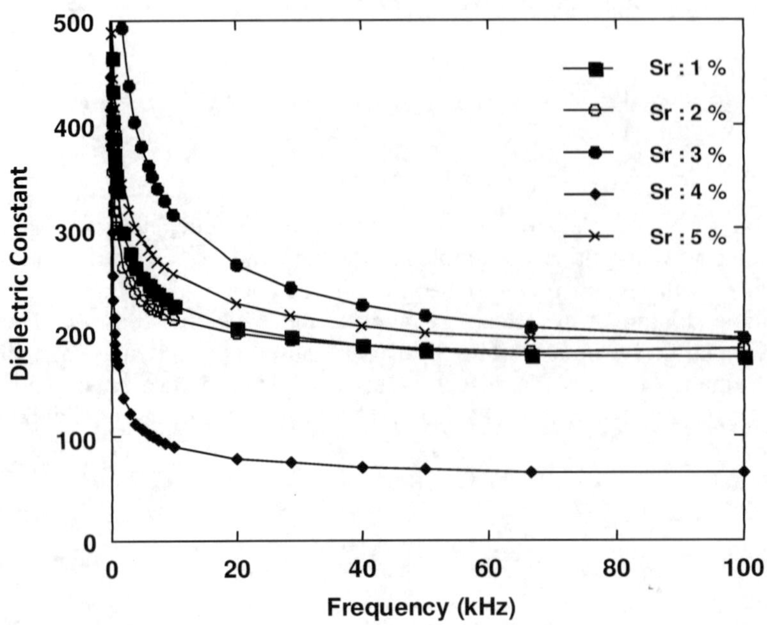

Figure 3. The dielectric constant for each frequency at percent doping variation Sr

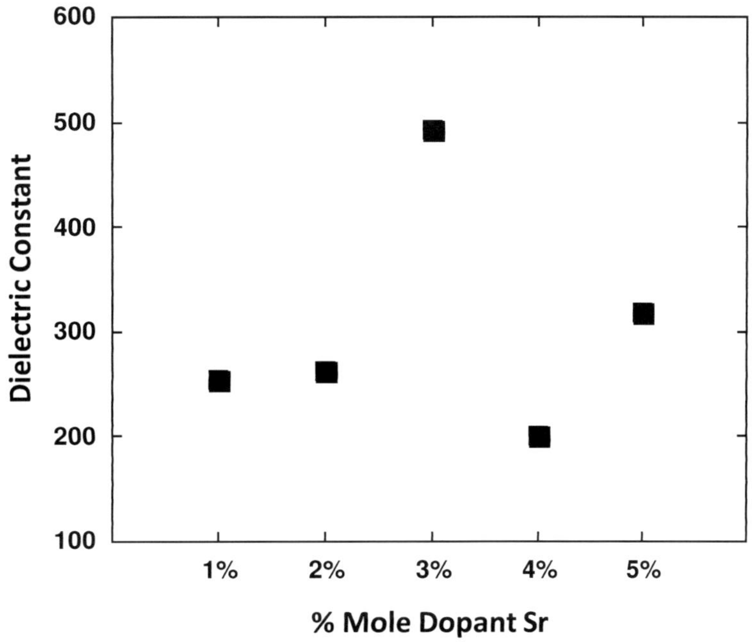

Figure 4. Dielectric constant of BT doped Sr at frequency of 0.1 kHz

4. Conclusion

BT doped Sr ferroelectric material has been successfully fabricated. The more of Sr mole resulted in the larger diffraction angle. The crystallinity of the sample is above 90% and the crystallite size is in the range of 27 nm-34 nm. The hysteresis curve confirms that all samples are ferroelectric material. The greater frequency tends to decrease the dielectric constant. The highest dielectric constant at frequency of 0.1 kHz occurs at doping 3% Sr. Therefore, the best variation obtained in this research is BT with doping of 3% Sr

Acknowledgement
The authors would like to thank PUPT UNS Grant of 2017.

References

[1] Istiqomah M, Jamaludin A, and Iriani Y 2014 *Jurnal Fisika Indonesia* **18(53)** 59-61
[2] Hikam M and Anggraini PWK 2006 *Indonesian Journal of Materials Science (Special Ed.)* 211-216
[3] Jona F and Shirane G 1993 *Ferroelectric Crystal* New York: Dover Publication
[4] Seo J Y and Park S W 2004 *J. Korean Phys. Soc.* **45(3)** 769-772
[5] Sudano H A, Romero J, Zhi Z, and Lin Y 2011 *18th Int. Conf. Composite Mater.*
[6] Khollam Y B, Bhoraskar S V, Deshpande S B, Potdar H S, Pavaskar N R, Sainkar S R, and Date S K 2003 *Mater. Lett.* **57** 1871-1879
[7] Ming-li L and Ming-xia X 2009 *Mater. Res. Bull.* **44** 937-942
[8] Pusit P and Sukon P 2010 *J. Ceramic Process. Res.* **11(3)** 384-387
[9] Y Subarwanti, R D Safitri, A Supriyanto, Y Iriani, A Jamaludin 2017 *IOP Conf. Series: Mater. Sci. Eng.* **176** 012044
[10] Maharsi R, Jamaludin A, Iriani Y 2015 *Adv. Mater. Res.* **1123** 123-126
[11] Simoes A Z, Moura F, Onofre T B, Ramirez M A, Varela J A and Long E 2010 *J. Alloys Compd.* **508** 620-624

IOP Conf. Series: Materials Science and Engineering **333** (2018) 012048 doi:10.1088/1757-899X/333/1/012048

Modification of Montmorillonite with Cetyl Trimethylammonium Bromide and Tetra Ethyl Ortho Silicate

D M Widjonarko[1], O D Mayasari, S Wahyuningsih and K D Nugrahaningtyas
[1]Inorganic Material Research Group, Sebelas Maret University, Ir. Sutami Street 36A Surakarta, Central Java, Indonesia 57126

E-mail: dianmipa@staff.uns.uns.ac.id

Abstract. Modification of montmorillonite (MMt) with cetyltrimethylammonium bromide (CTAB) and tetraethyl ortosilicate (TEOS) has been done. The aim of the research is to study the effect of TEOS and CTAB into MMt. This research is a preliminary step to invent material that can be modified with other functional materials. The study was conducted by recting TEOS with MMt and varying CTAB concentration on MMt previously modified with TEOS. The TEOS concentration was 4.72 M while CTAB concentration was 0.25; 0.5; 2; 3.5; and 5 mmol/g in MMt which has been reacted with TEOS. Material characterization was done by X-Ray Diffraction (XRD), Fourrier Transform Infra-Red Spectrophotometer (FTIR) and Scanning Electron Microscope (SEM). Cation exchange capacity (CEC) of materials was analyzed by titration method. The results show that TEOS and CTAB successfully modified. TEOS adsorbed onto MMt. It was identified from increased basal spacing, specific group and also by its elemental composition, originally having basal spacing 16 Å. After modification with CTAB, basal spacing increased to 28.45 Å or 77.64%. This indicates that CTAB is intercalated within the MMt layers. The CEC of new material is 0.93 meq/g, increase from 0.83 meq/g.

1. Introduction

Inorganic materials are often explored for interact with organic materials. One of the most commonly used inorganic materials is clay. Clay is a natural material easily found in minerals, particle size less than 0.002 mm which are generally dominated by secondary phillosilicate, has large surface area and a negative charge.

The commonly used phillosilicate clay is the smectite which composed of 2: 1 layer structure [1]. One example of smectite is montmorillonite which characterized with its high aspect ratios, large surface area, abundant inorganic compounds, large expansion ability, and easy to modify, abundant inorganic compounds, good expansion capacity, and easy to modify [2,3].

However, montmorillonite has limitation due to the presence of inorganic cations that are electropositive, which tends to attract water molecules to form hydrates, resulting in the surface being hydrophilic and the small cation size causing the intercellular spacing of aluminosilicates to be low. These properties cause montmorillonite rarely used directly to interact with hydrophobic organic molecules, so that surface modification is required [4]. One attempt to increase the value of montmorillonite is to modify montmorillonite with silanes. Therefore, MMt need to be modify to increase the efficiency of montmorilllonite by increasing the surface area and increasing its organophilic properties.

Modification of MMt with Tetraethoxysilane (TEOS) (silanization technique) is not only used to enhance the interaction of the atar surface but also as a link between the chemical reactions between clays and functional groups in macromolecules. The effect of silane on clays is indicated by the reaction between the alkoxy group of the silane and the alkoxy group of the. In other words, silanization potentially modify the surface of the material by increasing the functional group interaction in the material [5].

On the other side, clay modification with surfactants, especially cationic surfactants, will change the nature of clay that was originally hydrophilic to organophilic. Modified Montmorillonite CTAB can be applied in various fields such as adsorbent, heterogeneous catalyst, nanocomposite synthesis, and drug delivery [6]. CTAB can act as a good intercalator agent, since it can insert into montmorillonite molecules that have inter-layered spacing while maintaining its layered structure [7]. However MMt modif with TEOS and Cetyl Trimethyl Ammonium Bromide (CTAB) simultantly is one type of cationic surfactant [8,9]. Research on modification of montmorillonite with CTAB has not been developed.

Therefore, in this study focused on modification of montmorillonite with CTAB to know the chemical characteristics and the material physics of the modified results. The presence of a CTAB at a certain concentration may alter basal spacing (d), surface area, montmorillonite thermal properties, and montmorillonite cation exchange capacity (CEC). The larger the surface area of the modified material, the more efficient the montmorillonite is applied, the greater the cation exchange capacity, the more interchangeable ions it can be applied more widely.

2. Experimental

2.1. Reagent

The MMt nanoclay was K10 grade purchased from Sigma–Aldrich with a surface of 279.28 Æ 0.846 m2/g. Cetyltrimethylammonium Bromide (CTAB) with chemical formula of $C_{19}H_{42}BrN$ and molecular weight of 364.46 g/mol as the surfactant was purchased from Merck, Germany. Its critical micelle concentration (CMC) is 328 mg/L. All others chemicals and reagents were of analytical grade purchased from Merck, Germany.

2.2. MMt Sylanisazed TEOS

An amount of 10 g MMt was first dispersed in 8 mL TEOS and 100 mL ethanol for 12 hours using a magnetic stirrer to swell and to reach homogenity. Then stirring was stopped and the product was filtered and washed with distilled water by several times until became free solvent and dried at 80°C in an oven. The product sylanization was ready to the next step.

2.3. Modification of MMT sylanisazed TEOS with CTAB

The synthesis of CTAB modified MMt was conductedby the following procedure. An amount of 2 g sylanization product was dispersed in 50 mL distilled water and stirrer to swell and to reach homogenity. Then a desired amount of CTAB (0.364 g) which is 1.0 times CEC of MMt was slowly added. The reaction mixture was stirred for 6 hours, then stirring was stopped and the resultant organoclay was filtered and washed by distilled water for several times until it became free from excess salts and dried at 80°C in an oven. The product of CTAB intercalated MMt was ready.

2.4. Characterization Instrument

The *X-Ray Diffraction* (XRD) studies were XRD instrument (Rigaku Miniflex 600) using Cu Ka radiation (λ=1,5406 Å) at wide angle range (2θ = 0-90°), the accelelerating voltage 40 kV and the emission current 15 mA. The functional groups were analyzed by using *Fourrier Transform Infra-Red* (FTIR) spectrophotometer of Shimadzu 8201 PC under a dry air at room temperature by the KBr pellets method. The spectra were collected over the range from 4000 to 400 cm^{-1}. Surface area and pore size were observed by using *Surface Area Analyzer* (SAA) with BET (*Bruneaur-Emmet-Teller*) method of NOVA. Morphology of surface particles and elemental composition were identified with *Scanning Electron Microscope* (SEM) of JEOL JED 2300.

3. Result and Discussions

3.1 Characterization of the MMt and Modified MMt

XRD patterns of MMt shown in Fig. 1. Montmorillonite clay sampel has 2:1 layered structure of smectite class. As seen in Fig. 2, with an addition of surfactant, the basal spacing of the resultant organoclay increasses indicating location of CTA^+ ions between layers of clay and resulting in a decrease of the hydartion water content. So, the surface property changes from hydrophillic to hydrophobic. As known, the amount of added surfactant has a direct effect on the interlayer expansion of smectite. The main peak of MMt in 2θ = 5.512° (d=16.018 Å). In CTAB intercalated MMt the peak of MMt experienced a shift at 2θ = 5.089° (d=17.349 Å), 2θ = 4.791° (d=18.429 Å), 2θ = 4.672° (d=18.897 Å), 2θ = 3.102° (d=28.455 Å), 2θ = 5.223° (d=16.905 Å)

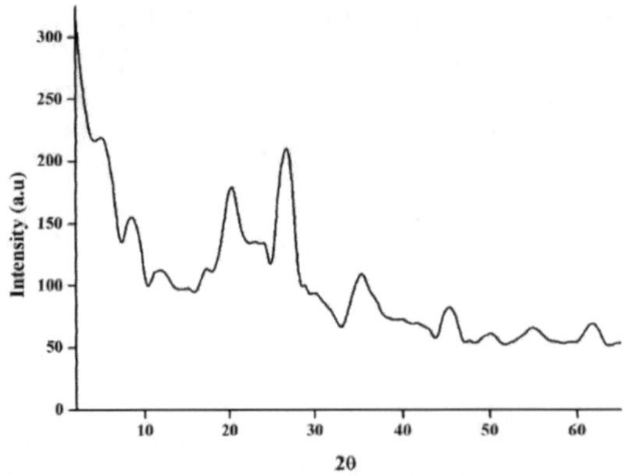

Figure 1. Difractogram of MMt

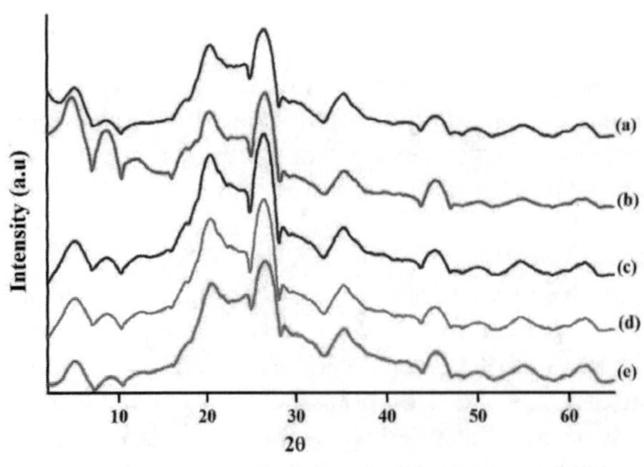

Figure 2. Difractogram (a) Intercalation 0.25 mmol/g. (b) Intercalation 0.5 mmol/g. (c) Intercalation 2.0 mmol/g. (d) Intercalation 3.5 mmol/g. (e) Intercalation 5.0 mmol/g.

The FT-IR spectra of pure MMt and MMt containing CTAB were shown in Fig 3. A relatively small band arround 3626 cm^{-1} was attributed to Al2OH group of the octahedral layer from the IR spectra of MMt. A broad band arround 3643 cm^{-1} was ascribed to the overlapping symmetric and asymmetric hydroxyl group stretching vibration of water molecules on the external layer. The band around 1053 cm^{-1} was attributed to asymmetric stretching vibration of Si–O–Si tetrahedra in the MMT. Comparison of FT–IR spectra of CTAB/MMT with that of pure MMT exhibits significant changes in some of the peaks (Fig. 1). In particular, the shift in the siloxan peak from 1053 cm^{-1} to 1037 cm^-

[1] after loading of CTAB highlights the role of siloxan group in the adsorption of surfactant onto the MMT surface. From the IR spectra of the organocaly (CTAB/MMT), the adsorption band observed around 3631 cm^{-1} corresponding to the stretching vibration of the hydroxyl group and the interlayer water molecules was observed. Bending vibration of water molecules causes peak 1629 cm^{-1}. Also, the additional peaks at 3000 cm^{-1}, appointed to –CH– stretching vibration, could be observed in CTAB/MMT. Additionally, the band at 3460 cm^{-1} was disappeared after the modification of MMT nanomaterial with CTAB, which indicates the removal of water molecules and the change in the hydrophobicity of MMT nanomaterial. It has been reported that the ordered conformation of the alkyl chains in the confined system has a strong dependence on amine concentration and orientation. In relatively high concentration range, the confined amine chains adopt an essentially all-trans conformation and the frequency of the asymmetric CH_2 stretching absorption band locates at a relatively low frequency. However, in the relatively low amine concentration range, the frequency shifts significantly to high frequency and the confined amine chains adopt a large number of gauche conformations. Only when the chains are highly ordered (all-trans conformation), the narrow absorption bands appear around 2918 cm^{-1} asymmetric CH_2 and 2850 cm^{-1} symmetric CH_2 in the infrared spectrum. If conformational disorder is included in the 3460 cm^{-1} was disappeared after the modification of MMt with CTAB, which indicates the removal of water molecules and the change in the hydrophobicity of MMT nanomaterial. It has been reported that the ordered conformation of the alkyl chains in the confined system has a strong dependence on amine concentration and orientation. In relatively high concentration range, the confined amine chains adopt an essentially all-trans conformation and the frequency of the asymmetric CH_2 stretching absorption band locates at a relatively low frequency. Only when the chains are highly ordered (all-trans conformation), the narrow absorption bands appear around 2918 cm^{-1} asymmetric CH_2 and 2850 cm^{-1} symmetric CH_2 in the infrared spectrum.

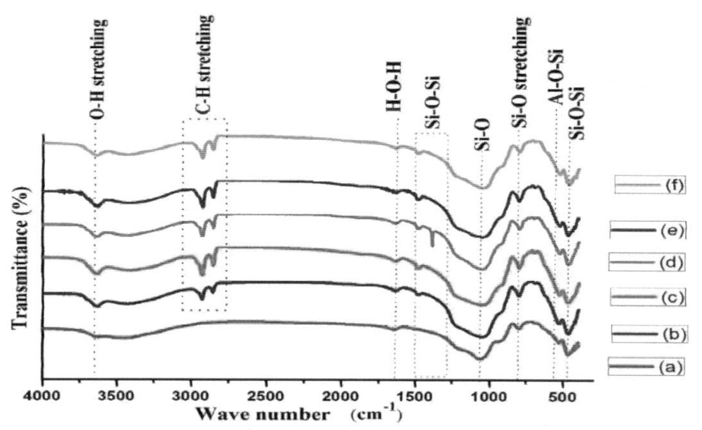

Figure 3. FTIR spectra (a) Modified MMt-TEOS. (b) Intercalation 0.25 mmol/g. (c) Intercalation 0.5 mmol/g. (d) Intercalation 2.0 mmol/g. (e) Intercalation 3.5 mmol/g. (f) Intercalation 5.0 mmol/g.

SEM analysis was carried out to evaluate surface morphology of the unmodified and modified MMt and the results are depicted in Fig. 4. As it is shown, the figure reveals that the surface morphologies of booth MMt samples are different and both unmodified and modified MMt have uneven structure with non-uniform size distribution. The MMt shows massive, aggregated morphology, and some large flakes were observed in some instances. After modification with polymeric species, the clay surface was changed to a non-agregated morphology and there are a large number of small flakes with severely

IOP Conf. Series: Materials Science and Engineering **333** (2018) 012048 doi:10.1088/1757-899X/333/1/012048

crumpled structures. Furthermore, the surface of modified MMt was expanded due to the chemical intercalation.

Figure 4. Morphology (a) Pure MMt. (b) Intercalation 0.25 mmol/g. (c) Intercalation 0.5 mmol/g. (d) Intercalation 2.0 mmol/g. (e) Intercalation 3.5 mmol/g. (f) Intercalation 5.0 mmol/g.

Surface area analysis using BET (*Brunaeur-Emmet-Teller*) method for analysis surface area of MMt and CTAB intercalated MMt. As seen in Table 1, with an addition of surfactant, the surface area of the resultant organoclay decrease indicating that molecule of CTAB has succesfully logged into MMt, so the N_2 adsorption at organoclay decrease. Distribution pore of modified MMt wih CTAB have uniform size distribution. As seen in Fig. 5, distribution pore size of modified MMt with CTAB is at about 1.8-1.9 nm. This shown that the dispersion of modified MMt with CTAB is in teh mesoporous region, which is 2-50 nm.

Table 1. Surface Area of MMt and Modified MMt with CTAB

Sample	Surface Area (m^2/g)
MMt	186.282
MMt-CTAB (0.25 mmol/g)	107.539
MMt-CTAB (0.5 mmol/g)	57.735
MMt-CTAB (2 mmol/gr)	50.629
MMt-CTAB (3.5 mmol/gr)	47.545
MMt-CTAB (5 mmol/gr)	40.589

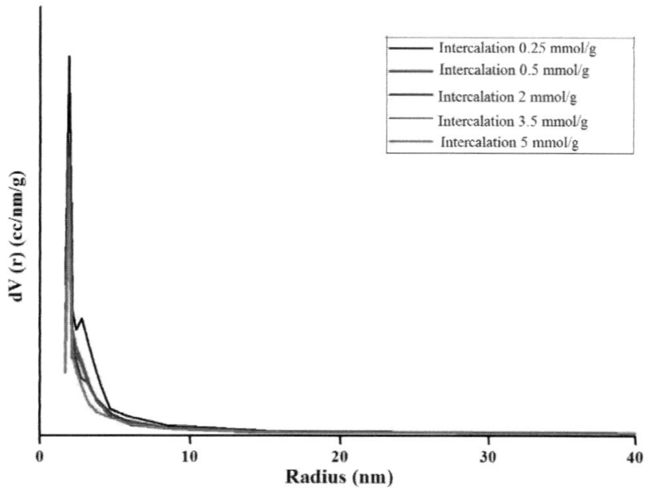

Figure 5. Distribution pore of Modified MMt with CTAB

Cation exchange capacity (CEC) for physico-chemical analysis of MMt and modified MMt with CTAB increases indicating CTA^+ could to replaced cations in interlayer of MMt. MMt had CEC amount 0.83 meq/g, modified MMt with CTAB at concentration (0.25 mmol/gr) had CEC amount 0.83 meq/g, modified MMt with CTAB at concentration (0.5 mmol/g) had CEC amount 0.84 meq/g, modified MMt with CTAB at concentration (2 mmol/g) had CEC amount 0.86 meq/g, modified MMt with CTAB at concentration (3.5 mmol/g) had CEC amount 0.91 meq/g, and modified MMt with CTAB at concentration (5 mmol/g) had CEC amount 0.93 meq/g.

4. Conclusion

Modification of MMt with TEOS and CTAB could to increase basal spacing, influenced of functional group, distribution pore, and cation exchange capacity.

Acknowledgement

This work was supported by the grant of PNBP of UNS and Kemenristek Dikti.

References

[1] Zidelkheir B, Abdelgoad M 2008 *J. Therm. Anal. Calorim.* **94** 181
[2] Bailey L, Lekkerkerker H N, Maitland G C 2015 *Soft. Matter.* **11** 222
[3] Shattar S F, Zakaria N A and Foo K 2016 *Desalin. Water. Treat.* **57** 13673
[4] Xi Y, Frost R L and He H 2007 *J. Colloid. Interf. Sci.* **305** 150
[5] Khajehpour M, Gelves G A and Sundaraj U 2015 *Clay Miner.* **63** 1-4
[6] Varadwaj G B, Parida K, Nyamori V O 2016 *Inorg. Chem. Front* **3** 1100
[7] Moslemizadeh A, Aghdam S K, Shahbazi K, Aghdam H K, Alboghobeish F 2016 *Appl. Clay. Sci.* **127** 111
[8] Zhuang G, Zhang Z, Fu M, Ye X, Liao L 2015 *Appl. Clay. Sci.* **116** 257
[9] Mahesh K V, Murthy H N, Swamy B K, Sharma S C, Sridhar R, Pattar N, Krishna M and Sherigara B S 2011 *Front. Mater. Sci.* **5** 401

IOP Conf. Series: Materials Science and Engineering **333** (2018) 012049 doi:10.1088/1757-899X/333/1/012049

The Effects of Leaching Process to the TiO_2 Synthesis from Bangka Ilmenite

S Wahyuningsih[1*], A H Ramelan[1], E Pramono[1], P Argawan[1], A Djatisulistya[1], F Firdiyono[1], E Sulistiyono[2], P P Sari[1]

[1]Inorganic Material Research Group, Faculty of Mathematics and Natural Sciences, Sebelas Maret University
[2]Metallurgical Research Center, LIPI, Serpong

*E-mail: sayekti@mipa.uns.ac.id

Abstract. Ilmenite mineral is a naturally occurring iron titanate ($FeTiO_3$) and is abundant in nature. The separation of components into TiO_2 and Fe_2O_3 must be expand. The purpose of this research is to synthesis TiO_2 nanoparticles from the filtrate of Bangka ilmenite leaching process. Leaching of ilmenite was done with H_2SO_4 and HCl at various concentrations. The formation of TiO_2 crystal determined by hydrolysis conditions and condensation reaction. TiO_2 synthesized from the filtrate of sulfuric acid leaching that produced from TiO_2 anatase phase when hydrolyzed in an aquaregia solvent and low concentrations of HCl (0.1M). Hydrolysis conditions at higher concentrations of HCl (1M) was produced TiO_2 anatase-rutile phase. The synthesis of TiO_2 from the filtrate of hydrochloric acid leaching was produced anatase phase. While the condition under the alcoholic solvent (2-propanol: H_2O (v/v) = 9: 1) anatase phase crystallites grow in the temperature range up to 550 °C, above this temperature, TiO_2 transform into rutile phase.

1. Introduction

Titanium dioxide (TiO_2) is a semiconductor that has been widely used as raw materials of paint, paper and plastic industry [1]. TiO_2 has also a potential for applications of a gas sensor, environmental cleanup and photovoltaic cells [2]. TiO_2 is widely studied as the base material of DSSC (Dye-sensitized solar cell). Two important phases of TiO_2 are anatase and rutile. TiO_2 P25 commercial from Degussa is produced from these two phases with an exact precise ratio of 70% anatase phase and 30% rutile phase. The optical properties and electrical properties of anatase and rutile are different. Studies have shown that the preparation methods [4], dopant concentrations [5], atmospheres [6] and particle size [7] can affect the anatase-rutile phase transformation, although each influence is unclear explained. Crystal size (grain size) is also very important to notice. Anatase synthesis would obtained less than 15 nm crystal size, whereas 100% rutile is not possible with small crystals size [8].

Anatase-rutile phase content, as well as crystal size are important characters of TiO_2 crystals. The two variables are connected to each other. The transformation of anatase phase into rutile phase is believed to be spontaneous, but not suitable for very low temperatures [9]. Anatase is a tetragonal system that strongly influenced by the size of the crystals. At high temperatures, a tetragonal system is more easily formed, whereas at lower temperatures monoclinic more easily formed. This is influenced by the TiO_2 rutile phase that more thermodynamically stable [9].

TiO_2 anatase is widely produced through sulfate process. Hydrolysis of Titanium (IV) Sulfate, purification of titanium dioxide hydrate and calcination, are factors that affect TiO_2 formation. Polymeric transformation is common in wide area temperatures between 400-1100 °C. Solid phase reactions have been observed to occur at 700-900 °C temperature region [10]. Anatase-rutile transformation influenced by the synthesis procedure of TiO_2, crystal size, temperature and time of calcination [8]. These factors can be used as a control of rutile-anatase transformation and plotting the composition of TiO_2 products and their properties.

2. Experimental

2.1 Pre-oxidation and Characterization of Bangka Ilmenite

Ilmenite was crushed and sieved using 180 mesh sieve. Then ilmenite was dried at 120 °C for 2 hours. Pre-oxidation is done by taking 5 grams of each ilmenite in 20 mL porcelain crust and roasting at 900 °C. The preparation process for more details has been published [10].

2.2 Leaching process

5 grams ilmenite from the roasting result in each experiment then added by iron powder (Fe^0) with the ratio of ilmenite: Fe0 = 5: 1 (w/w). The concentration of HCl used for leaching process is 4M, 6M, and 8M. While for the sulfate process, is using H_2SO_4 in several concentration (2, 3, 4, 5, 6, 7, 8 and 9 M). The acid solution was heated at ± 105 °C until a boiled. 5 gram of ilmenite was inserted and heated for 20 minutes. 1 gram of Fe^0 is added and the leaching process is continued for 2 hours. Procedures without the addition of Fe^0 were also observed.

2.3 Synthesis of TiO_2 from filtrate ilmenite leaching

The synthesis of TiO_2 was carried out by hydrolysis and condensation process using aquades and 2-propanol. The filtrates are hydrolyzed and condensed by:

a) 200 mL aquades was heated at 60 °C, then the filtrate is added into warm aquades to obtain a white precipitate. TiO_2 xerogel washing was done with HCl 1M and aquaregia.

b) The filtrate is added to the 2-propanol and H_2O mixture with the ratio of 2-propanol:H_2O (v/v) 9:1, 8:2, 7:3, 6:4 and 5:5. Therefore, it's allowed to precipitate and decanted. TiO_2 xerogel then washed with HCl 0.1 M.

3. Result and Discussion

3.1 Synthesis of TiO_2 Anatase-Rutile with Sulfate Process

The synthesis of TiO_2 from the filtrate of sulfate leaching at high concentration (H_2SO_4 9M) was showed not rutile phase (Fig. 1). This is due to the using of high concentrations H_2SO_4 that able to dissolve ilmenite and natural rutile. Fig. 1 shows the diffractogram of TiO_2 microcrystalline and $TiOSO_4$ hydrolysis results before calcination. By calculation with Scherrer formulation, relatively small crystallite size (<10 nm) was obtained.

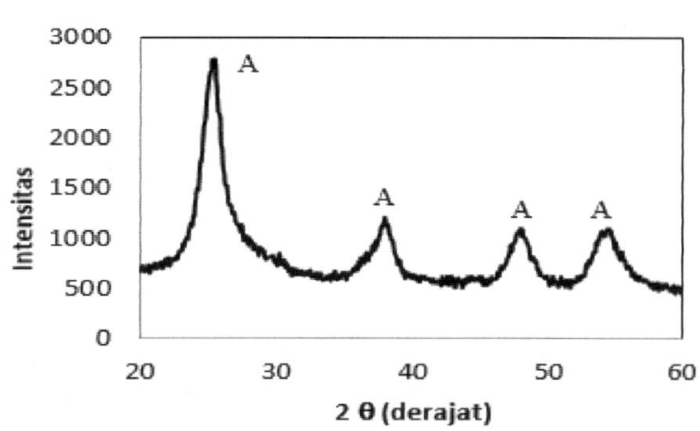

Figure 1. Diffractogram of TiO_2 anatase (H_2SO_4 9M) after washing $HCl:HNO_3$ (v/v) = 3:1

TiO_2 washing is also affect the determination of a crystal phase. TiO_2 xerogel washing with aquaregia shows no formation of rutile phase (Fig. 1). The rutile phase can be produced from the preparation of TiO_2 xerogel by HCl 1M washing (Fig. 2). Peaks at 27,43°(d_{110}); 36,08°(d_{101}) was the characteristic peaks of rutile in accordance with the standard JCPDS No. 89-0920. This is influenced by a faster hydrolysis process at higher concentrations. Although, it appears that the TiO_2 microcrystalline is still amorphous with crystal size 8 nm, TiO_2 xerogel washing with HCl 1M able to form rutile phase. The highly acidic conditions in the TiO_2 show as a rutile phase with crystalline size above 15 nm, as has been done by previous researchers [9].

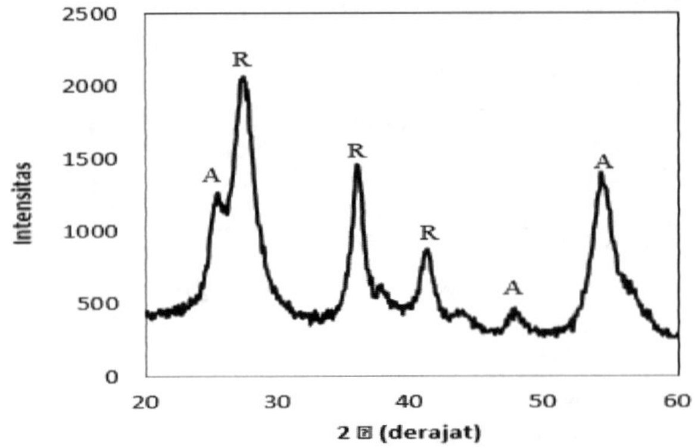

Figure 2. Diffractogram of TiO_2 anatase-rutile after washing HCl 1M

The hydrolysis and condensation reactions of TiO_2 formation shown in the reaction equation (1) and reaction equation (2) :

$$TiOSO_{4(aq)} + H_2O_{(l)} \longrightarrow TiO(OH_2)_{2(aq)} + H_2SO_{4(aq)} + 2\,Cl^- \quad (1)$$

$$TiO(OH_2)_{2(aq)} \xrightarrow{H_2O} TiO_{2(s)} + H_2O_{(l)} \quad (2)$$

The other experimental conditions, washing treatment with HCl solution (HCl 0.1 M) indicates a greater ability to suppress rutile phase Fig. 3. Condensation in this process runs better. HCl 0.1 M solution has also acted as a catalyst for acid hydrolysis process. This annealing process is performed at relatively low temperatures (450°C). Anatase-rutile phase has obtained with anatase phase more dominant. Peaks at 25.31°(d_{101}); 48.04°(d_{200}); and 37.79°(d_{004}) was the characteristic peaks of anatase in accordance with the standard JCPDS No. 78-2486.

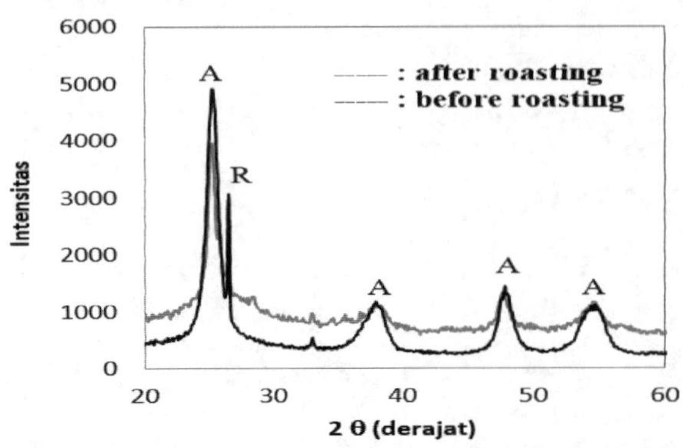

Figure 3. Diffractogram of TiO_2 from leaching residue (H_2SO_4 2 M) after washing HCl 0.1 M. before and after annealing at 450 °C for 2 h.

3.2 Synthesis of TiO_2 Anatase Phase with Chloride Process

The condensation process at very acidic conditions was running slowly. The addition of 2-propanol may substitute the titanyl (TiOH) group to produce titanium isopropoxide. The synthesis of TiO_2 from this process was carried out using filtrate from leaching process with HCl 6M. The condensation process begins by changing the solvent conditions by addition of 2-propanol alcoholic solvents to various comparisons. The ratio of 2-propanol: H_2O(v/v) = 9: 1 showed the condensation of titania sol gel. The hydrolysis and filtrate condensation reaction of the hydride solvent that dissolved in 2-propanol : H_2O (v/v) = 9:1 solvent possibly passes the reaction equation (3) and (4) :

$$TiOCl_2 + H_2O \xrightarrow{HOiPr} TiO(OiPr)_2 + 2HCl \quad (3)$$

$$TiO(OiPr)_2 \xrightarrow{H_2O} TiO_2 + H_2O \quad (4)$$

The titania sol-gel then separated, dried, washed with HCl 0.1 M, and calcinated at 450 °C. Fig. 4 showed TiO_2 that hydrolyzed by filtrate condensation of hydrated solution from the chloride process that using H_2O:2-propanol(v/v) = 9: 1 as solvet. This process followed by HCl 0.1 M washing. The chloride process can not produce rutile until the annealing temperature reach 450 °C. The crystal size then calculated by the Sherrer formula, and obtain 60nm as the result. But that was not indicated a change to the rutile phase.

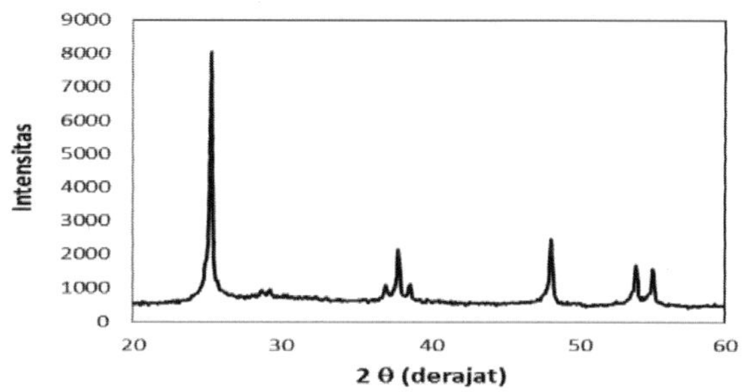

Figure 4. Diffractogram of TiO_2 from *leaching* HCl 6M washing HCl 0,1 M. Annealing at 450 °C 2h.

TEM analysis is used to determine the crystal size (grain size), crystallinity, and sample morphology. TEM results from TiO_2 nanopowder with high anatase abundance is shown in Figure 5 (a), while the lower anatase phases shown in Figure 5 (b). The bright field of TEM result in TiO_2 nanopowder exhibited a dominant anatase phase (Fig. 5 (a)), in line with the XRD results in Fig. 3. The morphological appearance of the anatase nanopowder was spherical (<10nm). Fig. 5b. is a synthesized TiO_2 nanopowder from $TiOSO_4$ ilmenite that dissolved in H_2SO_4 3M after annealing at 600 °C. The rutile phase is more abundant (darker areas) compared to TiO_2 nanopowder synthesized from $TiOSO_4$ using lower H_2SO_4 concentrations (Fig. 5b.), with a spherical morphology of 10-15 nm. The crystal size results from the TEM analysis shows an equivalent value after calculated by Scherrer formula.

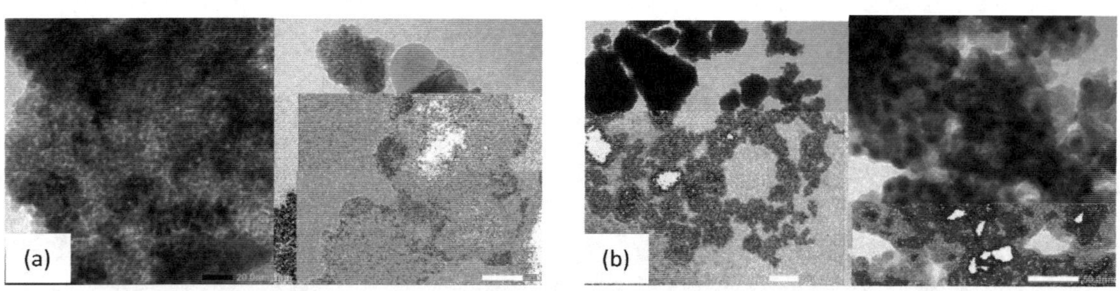

Figure 5. TEM of titanium dioxide from leaching filtrate with H_2SO_4 2M, annealing at (a) 450 °C (b) 600 °C

3.3 Thermal Process of Formation of Rutile Anatase Phase

The crystallization of TiO_2 anatase phase occurs over a relatively wide temperature range up to 550 °C. TGA Thermogram (Fig. 6.) shows a very significant mass decrease that up to 400 °C heating temperature. The mass degradation in the early process is related to the dehydration and the polymerization condensation process. While the crystallization process that occurs in the next stage does not cause a significant mass loss that indicated by the endothermic process on a very clear DTA thermogram (Fig. 6.).

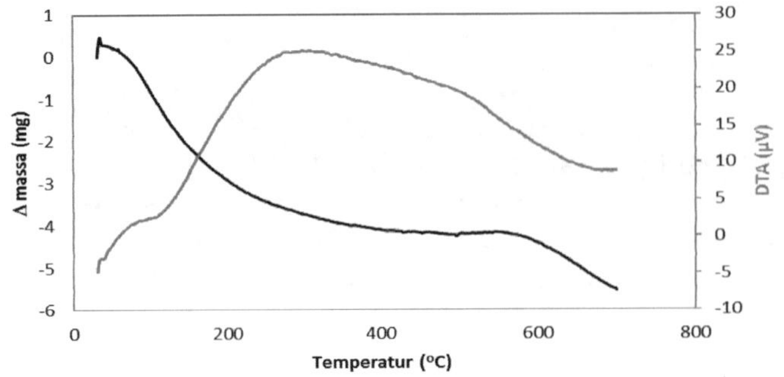

Figure 6. Thermogram of the TGA/DTA sample of the condensed TiO_2 xerogel from leaching filtrates with H_2SO_4 measured up to a temperature of 700 °C. Line - is TGA curve profile and line - is DTA curve profile.

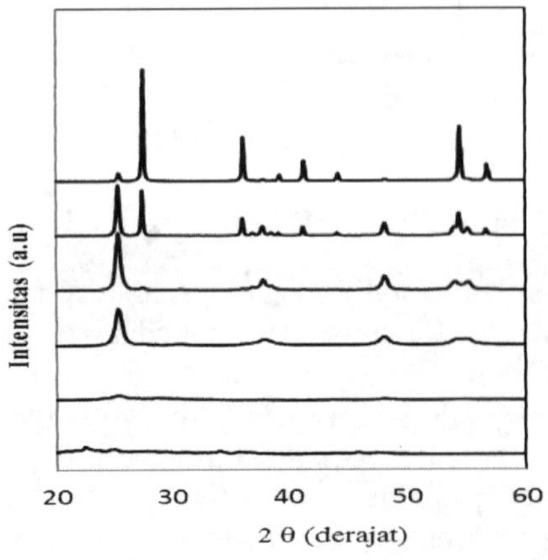

Figure 7. XRD TiO_2 as measured from the variation of annealing TiO_2 at (a) 150 °C, (b) 300 °C, (c) 400 °C, (d) 500 °C, (e) 600 °C, (f) 700 °C. TiO_2 is synthesized from the starting material of titanium tetraisopropoxide [10].

The data of TGA/DTA and XRD to the temperature variation (Fig. 7) of TiO_2 synthesized using the starting material of titanium isopropoxide as pure TiO_2, it appears that anatase phase crystallites exist in the temperature range, up to 500 °C. In XRD TiO_2 annealing temperature of 600 °C and 700 °C, there has been anatase phase change into rutile. In the TGA/DTA thermogram of this temperature range, appears to be a slight decrease in mass. The density of the crystal occurs when anatase phase transform into rutile phase.

The formation of anatase and rutile phase comes through different approaches. This difference in approach and orientation has been shown to be influenced by the environmental state of the solvent and the pH system associated with the energy that accompanies the process. The very high acid concentration is believed to cause the formation of anatase nuclei impaired and leads to the formation of a thermodynamically more easily formed rutile nucleus. Post and Burnham in 1986 [12] have demonstrated thermodynamic calculations that rutile is more stable than anatase with a Gibbs-free energy difference of ~4 kJ/mol and more stable than brookite with a Gibbs-free energy difference of ~20 kJ/mol.

4. Conclusion

Synthesis of TiO_2 can be worked out from filtrate of leaching ilmenite after roasting process. The formation of the TiO_2 crystalline phase is strongly determined by the current conditions of hydrolysis and condensation reactions which include acidity (pH), solvents, and environmental conditions for crystalline growth. The synthesis of TiO_2 from filtrate of diluted H_2SO_4 pathway resulted in anatase phase TiO_2 when hydrolyzed at aqua regia solvent condition and relatively low concentration HCl (0.1M), while hydrolysis at higher HCl concentration conditions (1M) yielded anatase-rutile TiO_2 phase. While the synthesis of TiO_2 from filtrate of diluted HCl solution yields TiO_2 anatase phase when hydrolyzed at 2-propanol:H_2O(v/v) = 9:1. Anatase phase crystallites grow in the temperature range up to 550 °C, at the above temperature undergoing phase transformation into rutile phase.

References

[1] Nayl A A, Awward N S, dan Aly H F 2009 *J. Hazard. Mat.* **168** 793–799
[2] Zhang Y, Tao Q, dan Zhang Y 2009 *Hydrometall.* **96** 52–56
[3] O'Regan B and Grätzel M. 1991 *Nature* **353** 737-739
[4] Wilska, Seppo 1954 *Acta Chemica Scandnavica* **8**
[5] MacKensize K J D 1975 *Trans. J. Brit. Ceram. Soc.* **74** 29-34
[6] Banfield, Jillian F, Bishoff, Brian L, Marc A, Anderson 1993 *Chem. Geol.* 110
[7] Wahyuningsih S, Rahardjo S B, Pramono E, Hidayatullah H, Anatolia F, Sulistiyono E, Firdiyono F 2013 *Asian J. Chem.* **25**(12) 6791-6794
[8] Wahyuningsih S, Kartini I, Narsito 2010 *Eksakta* **11**(1) 1-8
[9] Post J E, Burnham C W 1986 *American Mineralogist* **71** 142-150
[10] Wahyuningsih S, Pramono E, Firdiyono F, Sulistiyono E 2013 *Asian J. Chem.* **25**(12) 6791-6794

IOP Conf. Series: Materials Science and Engineering **333** (2018) 012050 doi:10.1088/1757-899X/333/1/012050

Phosphorus Elimination at Sodium Silicate from Quartz Sand Roasted with Complexation using Chitosan-EDTA

S Wahyuningsih[1*], A H Ramelan[1], N S Suharty[1], M Handayani[2], F Firdiyono[2], E Sulistiyono[2], H Munawaroh[1], P P Sari[1], Y R Kristiawan[1]

[1]Inorganic Material Research Group, Faculty of Mathematics and Natural Sciences, Universitas Sebelas Maret, Surakarta, Indonesia
[2]Metallurgical Research Center, LIPI, Serpong, Tangerang, Indonesia

E-mail: sayekti@mipa.uns.ac.id

Abstract: A phosphorus elimination from sodium silicate solution has been studied. Phosphorus elimination was performed by adding chitosan-EDTA to remove cation phosphorus. Characterization of chitosan-EDTA material was performed using FT-IR, while the decreasing level of phosphorus content was analyzed by quantitative analysis using spectrophotometer UV-Vis refers to SNI 06-6989-2004. The results showed that the content of the sodium silicate can be reduced up to 67.1% through Chitosan-EDTA complexation with phosphorus.

1. Introduction

The development of the photovoltaic industry greatly improved after the 2000s. In 2005 the production of the world's photovoltaic devices has reached 1787 MW. It is planned that by 2020 the production of photovoltaic devices will be increased to 18 GW. In 2005, over 90% of the photovoltaic devices production (solar cells) was made using crystalline silicon. So, now the supply of solar cell grade is an urgent need. Development of low cost silicon production and purification techniques is essential.

One of the source cheap silicon solar cell grade is quartz sand. In Indonesia, high quality quartz sand is available in Belitung (SiO_2 = 98%), Kendawang (SiO_2 = 99%), Samarinda and Kutai Kertanegara (SiO_2 = 98%) and Mandor (SiO_2 = 96%). Lower quality quartz sand is found in Tuban and Banten. Through of the roasted process, the quartz sand is processed into high purity silica to form sodium silicate, but in the sodium silicate there are still minor impurities that must be removed in order to be used as a solar panel material [1].

The common impurities found in natural silicon are Fe, Al, Ca, Ti, P and B. These impurity elements can be reduced by metallurgical methods. Physically, the metal elements with the price of small segregation coefficients are more easily separated from silicon through a sedimentation process, such as Ca and some Fe, Ti and Al. P and B are still in the silicon material will affect the efficiency of solar cell conversion. The minimum requirement of silicon purity of quartz sand silica for solar panels must reach SiO_2 99.999%, with the content of boron and phosphorus elements zero [2].

Phosphorus elements that have speciation from PH_3 to P_2O_5 with oxidation numbers ranging from -3 to +5 make phosphorus having its own technique for the reduction process. In this research, the decrease of phosphorus impurities on sodium silicate solution of roasted sand quartz was done by adding chitosan-EDTA and coagulant. The chitosan-EDTA will bind the cation phosphorus species while the coagulant phosphorus anion species [3,4]. The process of coagulant addition using two coagulants is aluminum sulphate [$Al_2(SO_4)_3.18H_2O$] and calcium hydroxide [$Ca(OH)_2$] [5,6,7,8,9]. Then a variation of coagulant concentration was used to decrease the optimal phosphorus content.

2. Experimental

2.1. Process of sodium silicate solution

100 grams of sodium silicate salt dissolved in 1000 mL of boiling water. The mixture is placed inside the closed system and stirer for 2 hours. The condition maintained at the distilled boiling water. The sodium silicate solution is filtered and diluted in a 1000 mL measuring flask tough filtration. Three sodium silicate solutions (solution A, B, C) were collected at 100 mL for each solution to identified the phosphorus content using UV-Vis spectrophotometer.

2.2. Synthesis and characterization of chitosan-EDTA

2 grams of chitosan were dissolved in 200 mL of 0.1% acetic acid until dissolved, 10 grams of EDTA was added. The chitosan-EDTA solution distilled for 24 hours, and centrifuge to form precipitate. The precipitate is then dried and in a desiccator to remove the water. The dry precipitate then weighed, and characterized by FTIR. The centrifuged solution concentrated and characterized by FTIR and TGA/DTA.

2.3. Reduction of phosphorous using Chitosan-EDTA

The optimal comparation results of chitosan-EDTA were dissolved in 100 mL of 0.1% acetic acid and distilled for 24 hours. Each of 100 ml Sodium Silicate solution (solution A, B, C) were added with 50 mL of chitosan-EDTA solution and stired slowly for 30 minutes to form chitosan gel-EDTA. Separate filtrate with chitosan gel-EDTA formed by filtering process. Then filtrate sodium silicate in the analysis of its phosphorus content with UV-Vis spectrophotometry.

3. Result and Discussion

Sodium silicate from roasted quartz sand dissolved with aquades that have been boiled. Boiled aquades used to soluble the sodium completely. After dissolved and filtered, a clear-colored sodium silicate solution was obtained. After the sodium silicate sample solution was prepared, chitosan-EDTA synthesis was performed. The ratio of chitosan-EDTA in this process is = 1: 5 (chitosan-EDTA) [10]. The reaction that may occur between chitosan and EDTA (Fig. 1).

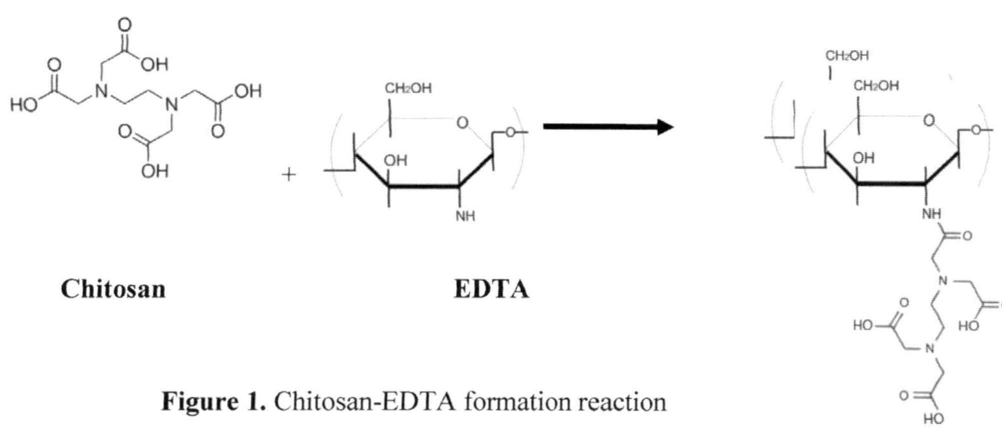

Figure 1. Chitosan-EDTA formation reaction

Characterization of Chitosan-EDTA using FT-IR were showed in Fig. 2.

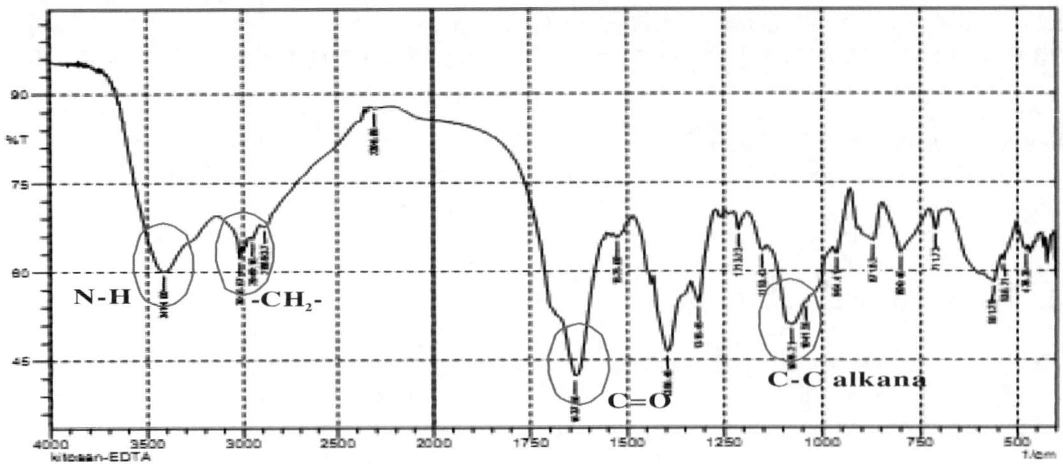

Figure 2. FTIR chitosan-EDTA

FT-IR spectroscopy of chitosan-EDTA in Fig. 2, shows that chitosan has an uptake of N-H group in the 3300-3500 cm^{-1} range, it refers to the chitosan-EDTA formation. In addition, the presence of C=O groups in the range 1600-1700 cm^{-1}, shows the existence of EDTA on chitosan-EDTA. Then in the range 2900-3000 cm^{-1} shows the CH_2 group on chitosan-EDTA that has been successfully formed.

Thermogram data from DTA/TGA characterization (Fig. 3) shows the difference between chitosan and chitosan-EDTA. With the addition of EDTA, the thermal properties of chitosan will increase. Chitosan degradation process starts from 100-200°C, but for chitosan-EDTA starts degraded 200-300 °C with an upward DTA peak (exothermic reaction). These different thermogram characteristics indicate the changing properties of the material that will increase the phosphorous binding of EDTA.

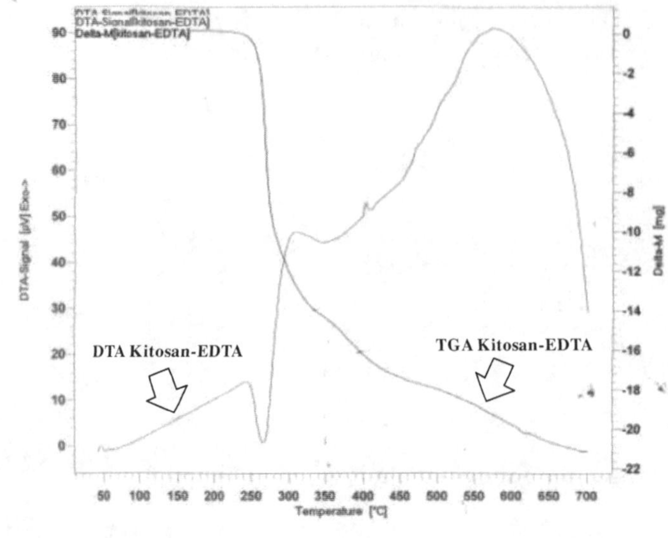

(a)

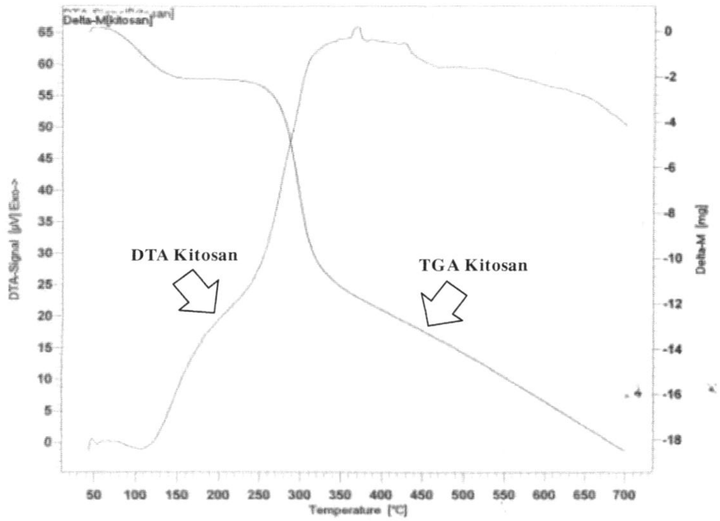

(b)

Figure 3. Thermogram of DTA/TGA (a) Chitosan-EDTA (b) Chitosan

By using the absorbance measurement method with UV-Vis spectrometer that refers to SNI 06-6989-2004, the average concentration of phosphorus is shown in Table 1.

Table 1. The results of measurement of phosphorus content in sodium silicate solution after chitosan-EDTA addition

Sodium Silicate Solutions	Mean of Initial Concentration (ppm)	Mean of Final Concentration (ppm)	Mean of Effectivities (%)
Solution A Solution B Solution C	0.0589 ± 0.00327	0.0194 ± 0.00119	67.063

From the data of initial phosphorus concentration and chitosan-EDTA concentration can be calculated the decreasing of phosphorus level up to 67.063%. This can happen because sodium silicate solution contain several phosphorus that may not converted into phosphorus cation (phosphonium). Actually chitosan-EDTA is able to attract positively charged ions with three active carboxylate groups (COO^-) such as phosphonium (PH_4^+) in sodium silicate solution. However, phosphorus is more easily deposited into phosphate ions (PO_4^{2-}) by forming strong bonds with O atoms on silicate compounds rather than being pentated into phosphonium (PH_4^+).

4. Conclusion

Chitosan-EDTA can be used to decrease the phosphorus content in sodium silicate. The decrease of phosphorus level up to 67,063%. The decrease of phosphorus content showed very low levels due to the chitosan-EDTA ability to bind the dissolved phosphorus in sodium silicate solution.

References

[1]. Zhihao Y, Honglie S, Ye J, Weilong C, Quantao T, Jiale J, Tian P, Jingwang L, Fanjian K, Chunbao R, Jibo C 2014 *Appl. Phys. A* **116** 267-273
[2]. Bart G 2005 *20th European Photovoltaic Solar Energy Conference and Exhibition* Barcelona Spain 2
[3]. Inoue K, Yoshizuka K, Ohto K 1997 *Bull. Chem. Soc. Jpn* **70** 2443-2447
[4]. Bakar M A, Boey H T, Ismail J, Tan W L, and Bakar N H H A 2007 *J. Phys. Sci.* **18** 87–101
[5]. Metcalf and Eddy 1993 *Mc Graw Hill* USA
[6]. Banu R J, Do K U, Yeom I T 2008 *Int. J. Environ. Sci. Tech.* **1** 93-98
[7]. Georgantas D A and Grigoropoulou H P 2006 *Global NEST J.* **8** 121-130
[8]. Han Y U, Park S J, Lee C G 2009 *Environ. Eng. Res* **14** 164-169
[9]. Hirasawa I, and Toya Y 1990 *Am. Chem. Soc.*
[10]. Oktaviana A T W 2011 *Thesis* Faculty of Natural Science Universitas Sebelas Maret Surakarta

Use of *Tectonagrandis* Leaf Extract in Colouring Silk Cloth Material Based on pH and Mordant Variations

A Rosyida[1,2] and Suranto[1]

[1]Department of Environmental Science Graduate Program, Universitas Sebelas Maret, St. Ir. Sutami 36 A Kentingan Jebres Surakarta 57126, INDONESIA,
[2]Textile Chemistry Program. Akademi Teknologi Warga, St. Raya Solo Baki, Km 2, Kwarasan Solo Baru Sukoharjo 57552, INDONESIA.

E-mail: kenur.rosida@gmail.com

Abstract. One of the very potential use of plant natural colour in the industrial activities was Anthocyanin. This substance contains in a number on plant organs such as *Tectona Grandis* (Fabaceae) particularly in their leaf materials. In order to get this secondary metabolism, the aim of this research was to find out the optimum condition of mordant and the pH combination in producing natural colour of *Tectonagrandis* leaf extract. To get this chemical substance, 1 kilogram of young leaves were cut into smaller size and soaked in 7 litre of H_2O for 12 hour[9]. The solution was then boiled for 15 minutes. To get the best colour, variations of pH (5, 7 and 9) respectivelly were applied. The results showed that purple to redness were resulted when pH 5 and no additional mordant was added. Conversly the greyness to black, were produced when ferrosulphate were added. When the addition of alum was made the redies purple will appear. These colours were considered quite good quality under examination of grey and staining scales (4-5) after treated by both wet and dry rubbing methods. The test for wash fastness resulted in sufficiently good value of color change (value of 3 on Grey Scale standard), while test for staining demonstrated good staining (value of 4 on Staining Scale standard). This early finding was quite promosing to be applied in the near future in providing natural colours for industrial purposes.

1. Introduction

As one of countries with abundance of sufficiently potential natural resources, Indonesia presents itself as a powerful country in the provision of nature-derived raw materials. However, the management of the natural resources has not yet been maximal, whereas in fact, they were traditionally managed by our ancestors. Natural dye plants belong to natural resources potential to be used for textile dye in Indonesia, particularly for development of natural, imitive, cultural, and exclusive products, as well as for raw materials with high economic value in textile industries. Now days, sources of natural dyes have been being obtained from such parts of plants as barks, fruit, flowers, timbers, leaves, and roots. In addition, natural dyes can also be derived from fungi, algae, bacteria [1]. However, after synthetic dyes had been invented in 1856 by W.H. Perkins, they started to be widely used [2]. Synthetic dyeing techniques are shifting to the use of natural dyes due to easier process and various yielded colors.

It is estimated that more than 10,000 different synthetic dyes all this time has been being used in industries and more than 7 x 10^5 tons of synthetic dyes is produced in all over the world every year [3]. Based on sequence and production volume, azo dyes are the most widely produced coloring substances, with the percentage of 60-70% of all coloring substances produced all over the world [4], but they have toxic effects, particularly carcinogenicity and mutagenic occurrence [5]. The increase in needs of synthetic dyes will be comparable to the rise in water and soil pollution, as well as in environmental pollution [6] since in textile industries, ± 200,000 tons of dyes, or ranging between 10-50% of the use, are lost to effluents every year during the dyeing and finishing operations due to the inefficiency of the dyeing process [7]. The generated liquid waste contains heavy metal which presents as the primary pollutant [8]. Several dyes are very toxic and mutagenic in nature, and their presence in aquatic environment may resist penetration of sunlight and prevent vegetative photosynthesis, leading to O_2 deficiency and limiting beneficial uses of rivers such as drinking and irrigation [9,10,11]. The use of aromatic and heterocyclic dyes with complex and stable chemical structures causes the dyes to be difficult to degrade when mixed with other pollutants in wastewater [11,12] There has been no method which is able to efficiently remove colors and toxic substances from dye containing waste water [13].

Anthocyanin contents in teak leaves serve as natural dyes which produce more various and attractive colors. Figure 1 presents a chemical structure of anthocyanins, while Table 1 demonstrates anthocyanin components [14].The molecular structures of anthocyanins demonstrate that natural dyes extracted from teak leaves belong to unsaturated organic compounds and flavonoids. Their main structures are characterized by two benzene rings (C_6H_6) connected to three carbon atoms [15]. The three carbon atoms are bonded by an oxygen atom to form a ring between two benzena rings. In a solution, anthocyanins exist in equilibrium forms depending on pH condition. Such equilibrium forms include flavylium cation, carbinol base, chalcone, quinoidal base, and anionic quinoidal base. At the most acidic pH, anthocyanins exist predominantly as flavylium cation contributing to the most stable condition and colorful form. As pH increases above 4, the anthocyanins turn either yellow (chalcone), blue (quinoidal base), or colorless (carbinol base).

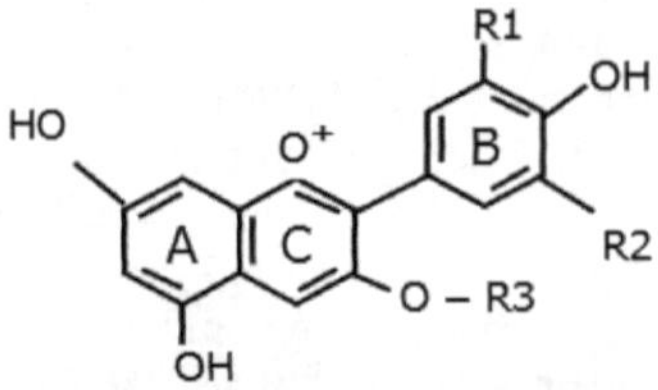

Figure 1. Chemical structure of anthocyanins.

The purpose of this research was to apply young teak leaf extracts as natural dyes on silk, and to find out the colors resulted from the use of various pH and different mordant. The dyeing was done by soaking at room temperature using different methods to obtain more optimal results in coloring.

2. Experimental

2.1 Materials and Methods

Plant materials: Young teak leaves were collected from Baki Sukoharjo.

Research locations: The research was conducted at finishing laboratory and evaluation laboratory of textile chemistry of AkademiTeknologiWarga of Surakarta, as well as finishing laboratory of PT. Danliris, Sukoharjoduring Mei - September in 2014.

2.2 Extraction process of natural dyes

The extraction process was done by preparing 1 kg of young teak leaves collected from the forth or fifth leaf. The leaves were then whased using flowing water and chopping down into small sizes and than 7 L of water was added and boiled for 15 minutes. The solution was the leaved to dry and straining the solution to an extract.

2.3 Preparation of Mordant Solution
Mordant solution was obtained by dissolving alum/ ferrous sulfate in cold water with concentration of 5g/l, and stirring constantly to perfectly dissolve the alum/ ferrous sulfate.

2.4 Dyeing process
The dyeing process was done in glass beaker with vlot of 1 : 30. The extract solution of teak leaves was firstly put into the glass beaker and 1 cc/l of wetting substance (TRO) was added and stirred until homogeneous. After that, the silk was put into the glass beaker and processed at room temperature for 30 minutes. Next, 5 g/l of salt/NaCl was added into the solution. The dyeing was done for 45 minutes at room temperature, and then acid/ alkali was added according to various pH. The dyeing was continued at room temperature for 30 minutes. The colored silk was then squeezed and fixated in the mordant (alum/ ferrous sulfate) solution for 15 minutes. For nonmordant silk, after being dyed, the silk was directly washed, but for mordant silk, after mordanting, the silk was soap, washed and dryed.

2.5 Testing dyeing result
Several tests carried out to samples of dyed fabrics included the test for wash fastness (SNI. 080285-98), the test for rub fastness (SNI.080288-89), as well as the test for darkness of color.

3. Result and Discussion

3.1 The Results of Dyeing
The results of the experiment indicated that teak leaves contribute to purple, reddish-purple, and grayish-black colors of silk fibers due to the from anthocyanin contents. Anthocyanins give permanent colors to silk since the dyed fabrics were not proved to be discolored after frequent washing. Therefore, they could be considered to fulfill requirements to be utilized as textile dyes.

During the coloring, silk fibers are soaked in extract solution of teak leaves to stimulate the swelling of fibers, and therefore pores in the fabrics will open and anthocyanins enter the fibers. The anthocyanins will be then absorbed by and bonded with such reactive groups in fibers as amino (NH_2), hidroxyl (OH) and carboxyl (COOH) groups. Such bonds with silk fibers will get be more abundant and stronger after fixation using alum/ ferrous sulfate than those in dyed fabrics without fixation. It is proved by the fact that fixated fabrics will have darker color and be uneasy to be discolored when being washed. Dyed fabrics without fixation will turn light purple or pink.

3.1.1. Light Fastness. The test for light fastness was performed using spectrophotometer. The results of the test revealed several measurement results, one of which is DL value, a lightness value indicating comparison between darkness/ lightness of color and the used standard values. DL value can be either negative or positive. The positive value denotes that batch has lighter color than standard fabrics, while the negative value signifies that batch has darker color. The standard fabrics refer to white undyed silk.

Table 2 presents DL value of dyed silk. It is clear from the testing results that on dyeing with the addition of fixative agent/ alum, purple to reddish-purple colors were obtained. On fabric samples dyed using acidic pH (5), purple silk with higher DL value than those dyed using neutral pH (7) and basic pH (9) was obtained. The fixation using alum does not result in color changes since alum ($Al_2(SO_3)_4$)belongs to colorless chemical compound. The results of fixation itself will only strengthen previous color resulted from dyeing.

The dyeing with the addition of fixative/mordant agent ferrous sulfate resulted in grayish-black color. Different darkness levels of color of silk dyed in extracts of teak leaves will be obtained if the silk is dyed in solutions with different pH. On silk dyed in solution with acidic pH (5), a fabric with the highest DL value was obtained, while on silk dyed in solution with either neutral (7) or basic (9) pH, similar color with different darkness levels of color was obtained. On silk with no fixation, higher DL value than acidic pH was obtained.

Table 1. Reflectance of silk fabric dyed with *tectonagrandis* leaf extract.

Type of plants	pH	Fixation treatment	Results of the test for darkness of color (DL) Testing					
			1	2	3	4	5	6
Teak leaves	5	without fixation	-36.23	-38.64	-38.54	-29.12	-29.15	-29.38
	7		-34.84	-37.97	-37.84	-29.07	-28.41	-29.02
	9		-34.00	-37.69	-37.82	-28.32	-26.91	-27.56
	5	Fixation using alum	-45.64	-49.86	-48.76	-42.30	-38.86	-35.79
	7		-44.13	-45.13	-44.92	-38.52	-37.37	-34.89
	9		-36.97	-44.64	-44.51	-35.27	-36.91	-34.07
	5	Fixation using ferrous sulfate	-62.82	-63.73	-63.93	-58.22	-59.69	-54.73
	7		-57.58	-62.23	-62.32	-53.73	-58.96	-52.67
	9		-57.48	-62.04	-62.16	-50.11	-58.71	-48.57

Description: L = Lightness
DL+ = Batch has brighter color than fabrics standard
DL- = Batch has darker color than fabrics standard

Higher DL value than acidic pH is obtained since the basic solution may lead to the formation of groups allowing to form cross bonds on protein structures, in which serine residues will form a residues [16]. This may lead to a reduction in reactivity of silk fibers towards anthocyanins during the dyeing. Although the formed residues are small in number, but they exert an influence on DL value of the dyeing results.

On dyeing without fixative mordant agents, light purple or pink was obtained although the dyeing was done in different pH (5,7, and 9). Lighter color is obtained when most of the anthocyanins entering silk fibers after dyeing leave from fibers during washing process. It happens since there is no fixative agent which is able to strongly bind anthocyanins in fibers. As a consequence, the release of pigments of teak leaves can easily occur during washing.

Table 2. Color properties and shades of the silk fabrics dyed with t*ectonagrandis* leaf extract.

pH	Fixation treatment	Results of testing using spectrophotometer Testing						Testing					
		1	2	3	4	5	6	1	2	3	4	5	6
		Da*	Da*	Da*	Da*	Da*	Da*	Db*	Db*	Db*	Db*	Db*	Db*
5	Without fixation	+17.54	+14.75	+14.37	+16.02	+14.25	+12.99	-3.18	-8.34	-8.50	-2.76	-0.11	-0.71
7		+17.00	+14.34	+14.70	+16.67	+14.26	+13.56	-3.83	-9.12	-8.92	-3.30	-0.08	-1.45
9		+18.39	+15.05	+15.17	+16.74	+14.97	+13.65	-5.13	-9.15	-9.06	-3.44	-0.91	-2.22
5	Alum	+18.90	+16.43	+16.30	+16.27	+14.31	+13.05	+4.36	+0.38	+0.50	+5.84	+8.20	+6.57
7		+17.59	+14.94	+14.83	+15.10	+14.06	+12.31	+1.80	+0.31	+0.12	+2.12	+8.14	+4.60
9		+14.58	+14.91	+14.75	+13.92	+13.52	+12.12	+0.94	+0.64	+0.72	+4.39	+6.81	+1.45
5	Ferrous sulfate	+1.38	+1.26	+1.22	+1.50	+0.54	+1.04	-2.96	-2.79	-2.85	-2.46	-1.72	-2.10
7		+2.05	+1.44	+1.35	+1.58	+0.47	+1.33	-3.01	-2.15	-2.60	-2.28	-1.68	-1.55
9		+2.08	+1.53	+1.62	+1.85	+0.31	+1.35	-3.11	-2.06	-1.94	-2.21	-1.47	-1.40

Description : Da+ = red color direction Db+ = yellow color direction
Da- = green color direction Db- = blue color direction

Table 3 indicates color properties and shades of the dyed silk. The results of testing using spectrophotometer revealed that all of color directions indicate red (Da+). Da* value of fabrics treated either without fixation or with alum is greater, this means that the red color direction of dyed silk is stronger than that of dyed silk with smaller Da*, which is dyed silk with fixative agent of ferrous sulfate. Db* values are either negative or positive. All of dyed silk either without fixation or with fixation using ferrous sulfate have negative values, indicating that the color direction of the dyed silk is blue. On dyed silk with fixation using alum, Db+ value shows yellow color direction.

The all experiments of dyeing silk with extract teak leaves that if fixated using different fixative agents but similar pH, silk dyed in extracts of teak leaves will result in different color and color direction. However, if fixated using similar fixative agents but different pH, same colors with different DL values will be obtained.

3.1.2 Wash Fastness. Colors resulted from dyeing are permanent. Table 4 details that the test for washing fastness on silk dyed in extracts of teak leaves with different pH and fixative agents generally resulted in almost similar Grey Scale values. On silk with fixation using alum and ferrous sulfate with different pH (5, 7, 9), value of 3 (sufficiently good) was obtained, while on silk with no fixation, colour changes value of 2-3 (less good). The mean score colour staining on silk dyed either without fixation, with fixation using alum, or with fixation using ferrous sulfate with different pH (5, 7, 9) is 4 (good), meaning that dyed silk has good value of colour staining.

Sufficiently good Grey Scale values on the test washing fastnes appear due to the entrance of pigments of teak leaves (anthocyanins) contained in natural dyes during dyeing to fibers. They are then bonded and together form complex compounds. Although the bond is in the form of hydrogen bond, fixative agents enable to lock and precipitate anthocyanins in silk fibers to avoid them to release and to discolor batch during the test for washing fastness. Therefore, sufficiently good value of washing fastness was obtained. Meanwhile, on silk dyed without fixation, smaller value of color change/ Grey Scale was obtained (2-3/ less good). It is no surprising because there exists no fixative agent which binds and locks anthocyanins in silk fibers. As a consequence, during washing the release of some anthocyanins occurs due to less strong fiber-anthocyanin bonds (only in the forms of hydrogen bonds) and causes discoloration.

Table 3. Washing fastness of dyed silk fabrics with extract teak leaves.

Test Sample	Variance of dyeing		Washing Fastness			
	pH	Fixation treatment	Colour changes		Colour Staining	
			GS	CD	SS	CD
1.	Acidic (pH : 5)	without fixation	2-3	4.2	4	4.0
2.		alum	3	3.0	4	4.0
3.		ferrous sulfate	3	3.0	4-5	2.0
4.	Neutral (pH : 7)	without fixation	2-3	4.2	4	4.0
5.		alum	3	3.0	4	4.0
6.		ferrous sulfate	3	3.0	4	4.0
7.	Basic (pH : 9)	without fixation	2-3	4.2	4	4.0
8.		alum	3	3.0	4	4.0
9.		ferrous sulfate	3	3.0	4-5	2.0

Rubfastness. Table 5 points out results of the test rubbing fastness. It indicates that the test on test rubbing fastness silk dyed either with fixation or without fixation in different pH resulted in Staining Scale value of 4-5 (good).The table shows that the dyed silk had good resistance to rubbing fastness. Similarly, all fabrics treated either with or without fixation in different pH had Staining Scale of 4 (good). Thus, both dry and wet rubbing resulted in good Staining Scale values. The good staining value appears since pigments of teak leaves (anthocyanins) entered to thus fibers and bonded. As a consequence, it is difficult for anthocyanins to escape despite an attempt of rubbing surfaces of fabrics/fibers. This gets rid of discoloration during the test for rubbing fastness.

Table 4. Rubbing fastness of dyed silk fabrics with extract teak leaves.

Test Sample	Variance of dyeing		Rubbing Fastness			
	pH	Fixation treatment	Dry Rubbing		Wet Rubbing	
			SS	CD	SS	CD
1.	acidic (pH : 5)	without fixation	4-5	2.0	4-5	2.0
2.		alum	4-5	2.0	4	4.0
3.		ferrous sulfate	4	4.0	4	4.0
4.	neutral (pH : 7)	without fixation	4-5	2.0	4	4.0
5.		alum	4	4.0	4	4.0
6.		ferrous sulfate	4-5	2.0	4	4.0
7.	basic (pH : 9)	without fixation	4-5	2.0	4-5	2.0
8.		alum	4	4.0	4	4.0
9.		ferrous sulfate	4	4.0	4	4.0

Description: GS = Grey Scale SS = Staining Scale CD = Color Difference

4. Conclusion

One of the very potential use of plant natural colour in the textil industrial activities was anthocyanin. The dyeing silk fabrics with the addition of fixative agent ferrous sulfate resulted in grayish-black color, but the dyeing with the addition of fixative agent alum resulted in purple to reddish-purple colors. On dyeing without fixative mordant agents resulted light purple or pink. Fixation is proved to improve wash fastness and rubfastness on dyed silk fabrics and the use of different pH on dyeing will influence fixation of dyes and light fastness. The use of acidic pH proved to fabric treated with fixation/ mordant (alum and ferrous sulfate) resulted in dyed fabric with the highest light fastness value.

References

[1] Samanta AK and KomarA 2011*In Tech* Croatia
[2] Sanjeeda I and Taiyaba N A 2013 *Environ. Res. Develop.* **8**3
[3] Ogugbue CJ and Sawidis T 2011*Biotech. Res. Int.* DOI 10.4061/967925
[4] Bafana A, Devi S and Chakrabarti T 2011 *Environ. Rev.* **19**350-70
[5] Chung KT and Cerniglia CE 1992 *Mutat. Res.* **3**(277)201-220
[6] Dos, SAB, Cervntes F.J and Van LJB. 2007*Biores. Tech.* **12**(98)2369-85
[7] Ben M, Houas I, Montassar F, Ghedira K, Barillier D, Mosrati R and Chekir G 2012 *Environ. Sci. pollut. Res. Int.* DOI10.1007/s11356-012-0802-7
[8] GyanendraT, MukeshKY, Prabhat U and Shardendu M 2015 *Int. J. Phar. Tech Res.* **8**(1) 96-100
[9] Forgacs E, Cserhati T and Oros G 2004 *Environ. Int.* **30**(7)953-71
[10] PrzystasW, Zablocka GE and Grabinska SE 2012*Water Air Soil Pollut.* **223**(4)1581-92
[11] Hubbe M A, Beck KR, O'Neal WG and Sharma YC 2012*Bio. Res.* **7**(2)2592-687
[12] Ding S, Li Z and Wangrui 2010 *Water Resour. Prot.***26** 73-78
[13] Chequer FMD, Oliveira GAR, Ferraz ERA, Cardoso JC and Zanoni MVB 2013*INTECH*http://dx.doi.org/10.5772/53659
[14] Sea Fast Center 2013 *Merah Ungu Antosianin* (Bandung: Ebook Institut Pertanian Bogor) Ebookdig.biz/ebook/q/pdf/antosianin/html
[15] Koswara S 2009 *Pewarna Alami: Produksi Pangan dan Penggunaannya* ebook pangan.com
[16] Lubis A 1994 *Teknologi Persiapan Penyempurnaan* (Bandung: STTT) p 127

α-keratin/Alginate Biosorbent for Removal of Methylene Blue on Aqueous Solution in a Batch System

G Fadillah[1], E N K Putri[2], S Febrianastuti[2], H Munawaroh[3], C Purnawan[2], S Wahyuningsih[3]

[1]Analytical Chemistry, Faculty of Mathematics and Natural Sciences, Institut Teknologi Bandung, Jl. Ganesha 10 Bandung Indonesia
[2]Analytical Chemistry, Faculty of Mathematics and Natural Sciences, Universitas Sebelas Maret , Ir. Sutami street 36A Kentingan Surakarta Indonesia
[3]Inorganic Material Research Group, Faculty of Mathematics and Natural Sciences, Universitas Sebelas Maret, Ir. Sutami street 36A Kentingan Surakarta Indonesia

E-mail: ganjarfadillah52@gmail.com

Abstract. Methylene Blue (MB) is a cationic dyes which is commonly used in textile industries for coloring agent. The precence of MB in water caused some negative effect on the environment and human health. Many common technologies such as membrane filtration, electrophoresis and adsorption have been widely empolyed for removal of MB in water, but the adsorption technique still has advantages than the others. In this study, removal of MB used a biosorbent α-keratin/alginate (KA). The biosorbent KA was prepared by using the encapsulation technique in $CaCl_2$ 2 % (w/v) solution. The biosorbent was characterized by Fourier Transform Infrared (FTIR) and Scanning Electron Microscope (SEM). The effect of composition of α-keratin and alginate, the pH of solution and contact time on the adsorption were investigated. The optimum adsorption of MB in aqueous solution was found at the composition of α-keratin and alginate of 1:2 (w/w), the pH at 5.0 and contact time at 4 hours. The adsorption of MB on KA biosorbent was comparatively higher than α-keratin and alginate only. Adsorption of MB dyes in aqueous solution followed the Langmuir adsorption isotherm, and the dynamic adsorption model could be described through a pseudo-second order kinetics.

1. Introduction

Methylene Blue (MB) is the most cationic dyes used for industrial applications such as a textile, printing, and plastics industries. This compound has complex aromatic molecular stucture so the compound is stable and difficult to biodegradation in nature and dangerous for the environment [1,2]. Besides, the contamination of MB dyes is dangerous to the environment, the dye contaminants are also dangerous to human life. It can cause irritation of the gastrointestinal tract if ingested, causing cyanosis if inhaled, and irritation of the skin if touched by skin [3]. Therefore, it is necessary to remove the MB dyes from the industrial wastes. Many common methodologies such as membrane seperation [4], flocculation-coagulation [5], aerobic or anaerobic treatment [6], ion exchange and adsorption have been widely used for removal these dyes from the industrial wastes. But, The remove dyes by the adsorption technique appeared to be an efficient, economical, simple and used for numerous years [7].

Various adsorbents such as activated clay and activated carbon [8], carbon nanotubes (CNT) [9], polymer and zeolite [9] have been reported for these dyes adsorption. But these adsorbent materials have deficiencies such as difficult preparation and availability of the raw materials. Therefore, in this study the adsorbent materials are used based on organic materials with the principle of waste for waste such as α-keratin. The α-keratin sources are as numerous as in chicken feathers and human hair, even the main chemical content of the hair is 88-96 % α-keratin protein [10]. This shows that α-keratin has the potential to be further developed. α-keratin is capable of acting as an adsorbent because it has a binding amino and hydroxyl functional group which causes the cation polyelectrolyte properties to act as ion exchangers and potentially developed as adsorbents such as for removal of cationic dyes [11]. Nevertheless, in very recent work, α-keratin is to difficulty of separating from the waste solution may result in the loss of the adsorbent. These advanteges may limit its practical application of α-keratin as a adsorbent in the future.

The material that can be developed as a composite is alginate. Alginate is natural occuring carbohydrate polymer composed of α-L-guluronate and β-D-mannuronate residues and has hydrophilic [12]. Some study has been reported that alginate can encapsulate activated carbon, CNT, maghemite nanoparticles to create adsorbent to remove dyes from aqueous solutions. However, to our knowledge, no study on preparing alginate encapsulated α-keratin (KA) for adsorption of dyes has been reported.

In this study, biosorbent was synthesized and used to develop new series of calcium alginate encapsulated α-keratin to increase the adsorption capacity and physical properties. Alginate with α-keratin will have cross-reactions in the group which results in the formation of amide bonds between carboxylates and amines so that their physical strength becomes better. The developed materials biosorbent were investigated as potential biosorbent for efficient removal of MB dyes from aqueous solutions.

2. Experimental

2.1 Preparation of α-keratin powder, biosorbents KA and characterization

Hair which obtained from some barbershop in Wonogiri and Surakarta, Indonesia, was washed with detergent in several times and dried in the sun to dry. Then, the hair was milled to fine at 180 mesh. The obtained hair powder was washed with acetone for 30 minutes and filtered through whatman filter paper. Then the residue was dried with an oven at 40 °C for 30 minutes. Cleaned hair powder and technical alginate were weighed with the variations of α-keratin:alginate weight composition are 1:1, 1:2, 1:3, 2:1, and 3:1. Then a mixture of hair and alginate with some composition, dissolved with aquadest of 50 mL. The obtained α-keratin-alginate solution, encapsulated with $CaCl_2$ 2 % (w/v) solution immersed in ice cubes. The result of the encapsulation of α-keratin-alginate (KA) composite was filtered and dried at 60 °C for 8 hours. Then, the optimal composition was carried out by adsorption test on MB dyes in aqueous solution at the optimum condition of pH (5.0) and contact time (4 hours). The biosorbent was characterized using FTIR spectra and SEM micrograph. FTIR spectra of biosorbent were recorded in a Perkin-Elmer-238B FTIR spectrophotometer over the wave range 4000-400 cm^{-1}. SEM photographs were taken with JSM 6700F Scanning Microscope to examine the morphology of biosorbents.

2.2 Equilibrium adsorption and kintecis study

Adsorption isotherm, MB dyes solution with various designated concentrations and optimum pH (5.0) was prepared. The biosorbent KA (100 mg) was added into the MB dyes solutions. During the experiments, the mixtures were shaken on a shaker with stirred speed of 200 rpm for 60 minutes. The dye concentrations of final solutions (after 4 hours stirring) were analyzed by Spectrophotometer UV-Vis at λ_{max} 665 nm. All experiment were carried out at 298 K.

Adsorption kinetics, the biosorbent KA (100 mg) was added into the MB dyes solutions while being stirred with a speed of 200 rpm and optimum pH (5.0) at 298 K. The sample solutions were taken at various of time intervals and the concentrations of the dyes were measured by Spectrophotometer

UV-Vis at λ_{max} 665 nm. The common mathematical model used to describe adsorption kinetics absorption kinetics included the Lagergen's pseudo-first-order (**Eq. 1**) and pseudo-second-order (**Eq. 2**).

$$\ln (q_e - q_t) = \log q_e - \frac{k_1 t}{2{,}303} \quad (1)$$

$$\frac{t}{q_t} = \frac{1}{k_2 q_e^2} + \frac{t}{q_e} \quad (2)$$

Where q_e is the equilibrium capacity (mg/g), q_t is the adsorption capacity at (mg/g)t, k_1 is the pseudo-first-order kinetic rate constant (min^{-1}), k_2 is the is the pseudo-second-order kinetic constant (g/(mg.min)).

3. Result and Discussion

3.1 Synthesis and Characterization

The synthesis of KA biosorbent was carried out by encapsulation method. Encapsulation method is carried out in $CaCl_2$ solution which serves as a crosslinking agent to form a α-keratin-alginate composite. The principle of encapsulation is the exchange of Na^+ ions from Na-alginate with Ca^{2+} from $CaCl_2$ when dropped into 2% $CaCl_2$ solution, forming a Ca-alginate gel. The illustration of synthesis biosorbent is shown in **Figure 1**.

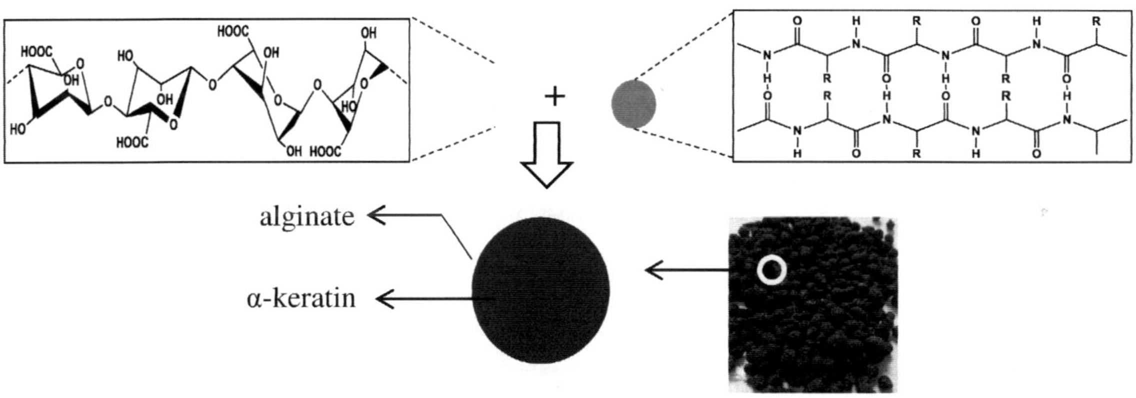

Figure 1. Illustration of synthesis of KA biosorbent

The FTIR spectra of KA, alginate, and α are shown in **Figure 2(a).** The specific spectrum of α-keratin-alginate (KA) biosorbent shows absorption bands at 3423 cm^{-1} (OH stretching), 1639 cm^{-1} (COO^- symmetric stretching). In addition, there is a characteristic absorption of peptide bonds (-CONH-) with three absorption areas which is Amide I, Amide II, and Amide III. The absorption area of Amide I shows the absorption of C=O stretching vibrations at wave numbers 1600 and 1622 cm^{-1}, Amide area II shows the absorption of C-N stretching and N-H bending vibrations at wave numbers 1450 and 1420 cm^{-1}. Meanwhile, the amide III region indicates the absorption of N-H bending, C-N stretching on O=C-N bonds with wave numbers 1367and 1300 cm^{-1}. In keratin and α-keratin-alginate composites, there is also a disulfide group absorption (S-S) in the wave number 1074 and 1038 cm^{-1}. The specific of absorption wavenumber proves that the α-keratin-alginate composite has been formed. The SEM morphology of KA (**Figure 2(b)**) shows that alginate is attached to the α-keratin surface. This proves that α-keratin has been protected by an alginate matrix so that α-keratin has a stronger physical resistance to environmental damage and impact on the adsorption process of MB.

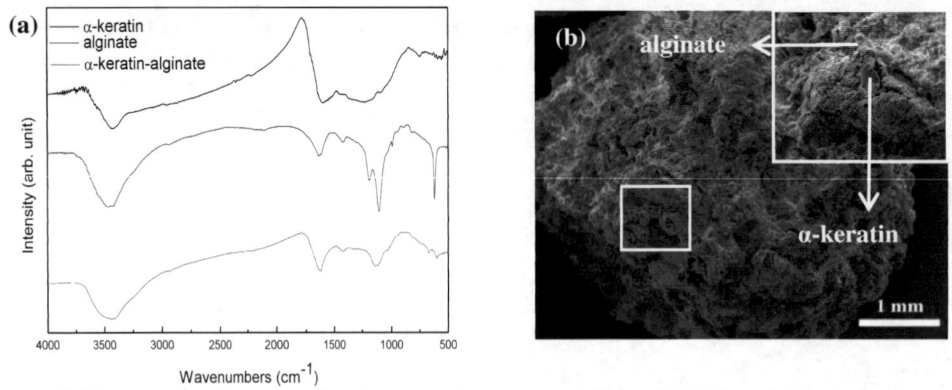

Figure 2. (a) FTIR characterization of α-keratin, alginate and KA ; (b) SEM characterization of KA biosorbent

3.2 Influence of composition biosorbents, pH and contact time

The composition between α-keratin and alginate needs to be studied because it will affect the density of the resulting composite. The results of composition testing are shown in **Table 1**. The composition of α-keratin and alginate (1:2) was found optimum results.

Table 1. Influence of composition KA biosorbents to adsorpion MB dyes

α-keratin : alginat	C_i (mg/L)	C_e (mg/L)	mass of biosorbents (mg)	adsorption capasity (mg/g)
1 : 1		90.52		409.48
1 : 2	500	30.67		469.33
1 : 3		140.89	50	359.11
2 : 1		170.08		329.92
3 : 1		200.28		299.72

In addition, the comparison of composition biosorbent, the other parameters influencing the adsorption capacity are the pH of solution and contact time of adsorption process. The result of the influence of pH and contact time to adsorption MB are shown in **Figure 3**. The optimum pH for KA biosorbent was found to be 6.0 with the maximum contact time at 5 hours for MB dyes adsorption. The effect of pH on adsorption capacity may be discussed on the basis of the nature of the chemical interactions of MB dyes with the biosorbent. The carboxylic (–COOH) from alginate and amino ($-NH_2$) groups present from α-keratin on the biosorbent are responsible for the binding of MB dyes. At lower pH, may be due to the protons competition with the MB dyes molecules for the adsorption sites. While at the optimum pH, it may be due to the more binding sites formed on the surface KA biosorbent to increase its surface complexation capability.

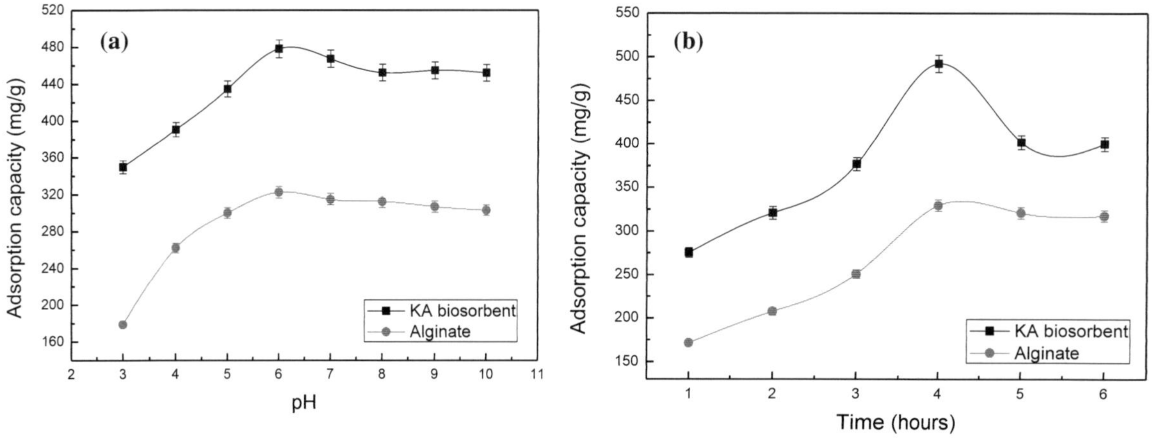

Figure 3. (a) Influence of pH solution and (b) contact time to adsorption of MB dyes

3.3 Adsorption isotherms and Adsorption kinetics

For the adsorption isotherms study, the Freundlich and Langmuir isotherms were used to fit the experimental data. The results from adsorption isotherms study are given in **Table 2**. From the experimental results showed that the adsorption process of MB dyes by KA biosrobent follows Langmuir isotherm. This is because the value of R^2 obtained is greater than Freundlich isotherm and the value of q_e calculated near the q_e experiment. Langmuir isotherm model assumes that adsorption occurs monolayer on a homogeneous surface. If one side of active adsorption has been filled by the adsorbate then there is no further absorption on the active side so that the adsorbent will reach saturation point when maximum adsorption has been reached

Table 2. Langmuir and Freundlich isotherm constants for adsorption of MB dyes

Adsorbent	**Langmuir**			**Freundlich**			**q_e experiment (mg/g)**
	q_m (mg/g)	K_L (L/mg)	R^2	K_F (L/mg)	n	R^2	
KA	426.68	0.0612	0.9991	89.85	3.8712	0.8714	400.58
Alginate	312.83	0.0246	0.9755	35.87	2.6123	0.7143	342.15

The reaction kinetics model which studied in this research is the pseudo-first order and pseudo-second order kinetics model. The model of adsorption kinetics is given in **Table 3**. From the calculation results show that the adsorption of MB dyes follows the pseudo-second order. the second order kinetic model assumes that the rate-limiting step may be chemical adsorption [13].

Table 3. Pseudo-first order and second order rate constants for adsorption of MB dyes

Adsorbent	**Pseudo-first order**			**Pseudo-second order**			**q_e experiment (mg/g)**
	q_e (mg/g)	k_1 (hour^{-1})	R^2	q_e (mg/g)	k_2 (hour^{-1})	R^2	
KA	33754.02	3.7353	0.7930	460.54	0.0182	0.9977	459.75
Alginate	3214.45	2.7634	0.5339	316.23	0.0099	0.9925	306.27

4. Conclusion

In this study, α-keratin-alginate (KA) were developed as biosorbent and characterized using FTIR spectra and SEM micrograph. Adsorption of MB dyes on KA was investigated using batch systems. The data demonstrated that the biomaterials are effective adsorbents for the removal of MB dyes from aqueous solution. The adsorption of MB dyes depended on the composition of α-keratin and alginate, pH of the solution and contact time of adsorption process. The equilibrium adsorption data were correlated well by Langmuir isotherm equations the adsorption kinetics followed pseudo-second order kinetics model. So, the biomaterial has potential to be developed as a biosorbent material of MB dyes adsorption.

References

[1] Kornaros M and Lyberatos G 2006 *J. Hazard. Mater.* **136** 95–102
[2] Chen M Chen Y and Diao G W 2010. *J. Chem. Eng. Data* **55** 5109–5116
[3] Sheng J Xie Y and Zhou Y 2009 *Appl. Clay Sci.* **46** 422–424
[4] Dutta K Mukhopadhyaya S Bhattacharjee S and Chaudhuri B J Hazard 2001 *Mater.* **84**, 57–71
[5] Canizares P Martínez F Jiménez C Lobato J and Rodrigo M A 2006 *Envir Sci. Tech.* **40** 6418–6424
[6] Slokar Y M and Le Marechal A M 1998 *Dyes and Pigments* **37** 335–356
[7] Rafatullah M Sulaiman O Hashim R and Ahmad A 2010 *J. Hazard Mater* **177** 70–80
[8] Weng C H and Pan Y F 2007 *J. Hazard Mater* **144** 355–362
[9] Martyn-Gullon I and Font R 2001 *Water Res.* **35** 516–520
[10] Aluigi A Tonetti C Tonin C and Mazzuchetti G 2011 *Eur. Polym. J.* **47** 1756-1764
[11] Volkov V and Paulo A C 2016 *Int. J. Biol. Macromol.* **85** 476-486
[12] Fuks L Filipiuk D and Majdan M T 2006 *J. Mol. Struct.* **792** 104–109
[13] Dada A O, Olalekan A P Olatunya A M and Dada O 2012 *J. Appl. Chem.* **3** 38-45

IOP Conf. Series: Materials Science and Engineering **333** (2018) 012053 doi:10.1088/1757-899X/333/1/012053

Determination of Pb^{2+} metal ion level in liquid waste from adsorption process by combination adsorbent of rice husk and water hyacinth charcoal using solid-phase spectrophotometry (sps)

S Saputro[1], M. Masykuri, L Mahardiani, and AN Hidayah
Chemistry Education, Faculty of Teacher Training and Education, Sebelas Maret University, Jl. Ir. Sutami 36A Surakarta, Central Java-57126, Indonesia

E-mail: sulistyo_s@staff.uns.ac.id

Abstract. This research are to find out the influence of adsorbent composition between rice husk and water hyacinth in decreasing of Pb^{2+} ion in simulation liquid waste; the optimum composition of combination adsorbent of rice husk and water hyacinth charcoal on Pb^{2+} ion adsorption; and the effectiveness of SPS as a method to determine the decreasing level of Pb^{2+} ion in simulation liquid waste by combination adsorbent of rice husk and water hyacinth charcoal in μg/L level. Rice husk and water hyacinth carbonization using muffle furnace at 350°C for 1 hour. Rice husk charcoal activation in a 2 N NaOH solution and water hyacinth charcoal activated in a 5 M HCl solution. Contacting the combination adsorbent of rice husk and water hyacinth charcoal with a Pb^{2+} solution with variation of mass composition, 1:0 ; 0:1 ; 1:1 ; 1:2 and 2:1. Analysis of the Pb^{2+} ion level using SPS method. Characterization of rice husk and water hyacinth charcoal using the FTIR. The results showed that the combination adsorbent composition of rice husk and water hyacinth charcoal have an impact on decreasing Pb^{2+} ion level. The optimum composition of combination adsorbent of rice husk and water hyacinth charcoal on the adsorption Pb^{2+} ion is 1:2. SPS is an effective method to determine the decreasing Pb^{2+} ion in simulation liquid waste from the adsorption process by combination adsorbent of rice husk and water hyacinth in μg/L, with *Limit of Detection* (LOD) was 0,06 μg/L.

1. Introduction

Indonesia is an agrarian country that has a large agricultural area and as the consequences, it has many agricultural waste like rice husk and water hyacinth. Rice husk is a waste from rice process and water hyacinth is a weed or wild plant that can grow on the water around the agricultural area. Nowadays, management of rice husk and water hyacinth is yet a maximum. Usage of rice husk is just for rub ash and plants media, while the usage of water hyacinth as handicraft material and medicine. In the other hands, rice husk and water hyacinth can be used as a heavy metal ion adsorbent like Pb^{2+} [1-2].

Pb^{2+} is form of the most Pb metal ion in liquid waste. Pb is one of the dangerous environmental contaminant like water that can cause poisoning and even death in living beings. For human, exposure to the Pb impact to reproduction system, nerve system, intelligence system and systemic like decrease the blood pressure. WHO decided Pb limit in the water is 0.1 mg/L [3], and based on Permenkes RI No. 416 Tahun 1990 in Indonesia, Pb limit in the clean water and drink water is 0.05 mg/L [4]. Very small concentration of Pb^{2+} metal ion in the water can give a lot of loss for living beings. Therefore,

IOP Conf. Series: Materials Science and Engineering **333** (2018) 012053 doi:10.1088/1757-899X/333/1/012053

needed an analysis method to determine Pb^{2+} concentration in the water even in very small concentration.

Solid-phase Spectrophotometry (SPS) is an analysis method that can be used to determine the concentration of a solution in very low concentration, until μg/L level [5]. Determining the concentration of a solution based on the light absorption by the target chemical components concentrated in a solid phase, so this method used UV-Visible spectrophotometer [6]. SPS is better than conventional methods because it has high sensitivity, effective to ion speciation in the water, can minimize the use of reagent and applied green analytical chemistry so this method friendly for environment. SPS is an analysis method that very suitable to analysis the effectivity of an adsorbent in the adsorption process.

SPS can be used to determining the concentration of a solution from adsorption process by rice husk and water hyacinth charcoal in μg/L level. Rice husk and water hyacinth charcoal is chosen to combined become adsorbent because they have good ability in adsorping Pb^{2+} metal ion. Rice husk has low nutrient value, abrasive, and contains high carbon [7], while water hyacinth has smooth and diverge roots system and hollows stem. Both of them also contains high cellulose, rice husk have 32.12% of cellulose [8] and water hyacinth have 72.63% of cellulose after milled [9]. This cellulose make rice husk and water hyacinth have potentially become adsorbent of metal ion in liquid system [9]. In this study, rice husk and water hyacinth used as an adsorbent with various ratio for decreasing the concentration of Pb^{2+} metal ion.

2. Experimental

2.1 Tools and Materials

The tools used were a spectrophotometer UV-Vis brand K-MAC, FTIR spectrophotometer Shimadzu, oven, muffle furnace, 100 mesh sieve, mortar and pestle, magnetic stirrer, desicator, analytical balance, filter paper, universal pH indicator, iron clamp, volumetrick flask, beaker glass, measuring glass, flask, drop pipette, volumetric pipette, spatula glass, funnel glass, saucer glass, porcelain saucer and vial. Materials used were rice husk, water hyacinth, standard Pb^{2+} solution, NaOH crystal, 37% of HCl, 0.5 M H_2SO_4, diphenylthiocarbazone, chloroform, Muromac resin AG 50W X2 100-200 mesh and aquadest.

2.2. Research procedure

2.2.1. Preparation of adsorbent. Rice husk and water hyacinth was cleaned and washed by aquadest and then dried on the oven at 105°C for 1 hour. After that, carbonized them in muffle furnace at 350°C for 1 hour. Mashed the results and then sieved with a 100 mesh sieve.

2.2.2. Adsorbent activation. Rice husk charcoal was soaked in a 2 N NaOH solution for 24 hours at room temperature. Filter and rinse the residue with distilled water until the filtrate became neutral, and then dried in an oven at 105°C for 24 hours. Adsorbent was analyzed by FTIR before and after activation. Water hyacinth charcoal is soaked in 5 M HCl solution for 24 hours at room temperature. Filter and rinse the residue with distilled water until the filtrate became neutral, and then dried in an oven at 105°C for 24 hours. Adsorbent was analyzed by FTIR before and after activation.

2.2.3. Resin preparation. Muromac resin AG 50W-X2 100-200 mesh dissolved in distilled water and left it in a few minutes until the resin became expanded.

2.2.4. Determination of Pb^{2+} calibration curve. Standard solution of Pb^{2+} 0 μg/L, 2 μg/L, 4 μg/L, dan 8 μg/L respectively as much as 20 mL added by 1 mL of 0.5 M H_2SO_4, 2 mL of diphenylthiocarbazone and 0.06 mL resin. The mixture was stirred by magnetic strirrer for 20 minutes and then analyzed it using UV-Vis spectrophotometer at a wavelength of 483,411 nm and 558,111 nm. The difference in absorbance of the two wavelengths was determined, $\Delta A = A_{483,411nm} - A_{558,111nm}$, which was ΔA will be made standard curve (ΔA vs concentration).

2.2.5. Determination of Cr (VI) species in simulated liquid waste. Simulated liquid waste solution of 50 µg/L as much as 20 mL added by 1 mL of H_2SO_4 0.5 M, 2 mL of diphenylthiocarbazone and 0.06 mL of resin. The mixture was stirred by magnetic strirrer for 20 minutes and then analyzed it using UV-Vis spectrophotometer at a wavelength of 483,411 nm and 558,111 nm. ΔA obtained will be substituted in equation of Pb^{2+} calibration curve (ΔA vs concentration) so the Pb^{2+} ions in the simulated liquid waste could be known.

2.2.6. Determination of optimum adsorbent composition. Rice husk charcoal and water hyacinth charcoal was mixed with ratio at 1:0, 0:1, 1:1, 1:2 and 2:1. The mixture was added by 20 ml of simulated liquid waste and then stirred by magnetic stirrer for 30 minutes. Filter the mixture and take 5 ml of filtrate to diluted by distilled water until 50 ml. The 20 ml of it was added by 1 mL of H_2SO_4 0.5 M, 2 mL of diphenylthiocarbazone and 0.06 mL of resin. The mixture was stirred by magnetic strirrer for 20 minutes and then analyzed it using UV-Vis spectrophotometer at a wavelength of 483,411 nm and 558,111 nm. ΔA obtained will be substituted in equation of Pb^{2+} calibration curve (ΔA vs concentration) so the Pb^{2+} ions in the simulated liquid waste after adsorption process could be known.

2.2.7. Determination of detection limit. Distilled water as much as 20 ml was added by 1 mL of H_2SO_4 0.5 M, 2 mL of diphenylthiocarbazone and 0.06 mL of resin. The mixture was stirred by magnetic strirrer for 20 minutes and then analyzed it using UV-Vis spectrophotometer at a wavelength of 483,411 nm and 558,111 nm. ΔA obtained will be substituted in equation of Pb^{2+} calibration curve (ΔA vs concentration). This steps was repeated in five times.

3. Result and Discussion

3.1. Preparation of activated carbon from rice husk

Preparation of activated carbon from rice husk consists of three steps, there were dehydration, carbonization and activation. Dehydration is a process to remove water level in rice husk. Carbonization is a process to breakdown organic cellulose in rice husk into carbon and remove non carbon compounds in form of gas. Activation is a process to remove hydrocarbon deposit that closed the pores on the surface of rice husk charcoal. Activation process of rice husk charcoal was used 2 N NaOH solution to soaked it for 24 hours. NaOH was used because it was a strong base that can clean the residues of organic compound and decrease the silica level on rice husk charcoal. NaOH is suitable activator for rice husk charcoal because rice husk charcoal contains more carbon than oxygen.

3.2. Preparation of activated carbon from water hyacinth

Preparation of activated carbon from water hyacinth consists of three steps, there were dehydration, carbonization and activation. Dehydration is a process to remove water level in water hyacinth. Carbonization is a process to breakdown organic cellulose in water hyacinth into carbon and remove non carbon compounds in form of gas. Activation is a process to remove hydrocarbon deposit that closed the pores on the surface of water hyacinth charcoal. Activation process of rice husk charcoal was used 5 M HCl solution to soaked it for 24 hours. HCl was used because it was a strong acid that can clean the residues of organic compound in water hyacinth charcoal and open the pores with dissolved organic substances and certain metal compound. HCl is suitable activator for water hyacinth charcoal because water hyacinth charcoal contains more oxygen than carbon.

3.3. Fourier transform infrared spectra analysis

Figure 1 shows the presence of the -OH group at wavenumber of 3433.44 cm^{-1} on before activation, 3387.15 cm^{-1} on after activation and 3398.72 cm^{-1} on after contact. C-O group at 1096.58 cm^{-1} on before activation, 1105.26 cm^{-1} on after activation and 1120.69 cm^{-1} on after contact. C=O group at 1580.73 cm^{-1} on after activation and 1601.95 cm^{-1} on after contact.

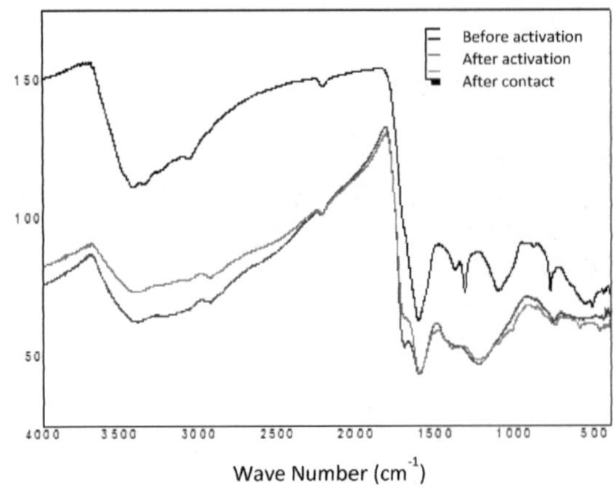

Figure 1. FTIR spectra of rice husk charcoal before activation, after activation and after contact

Figure 2 shows the presence of the -OH group at wavenumber of 3349.53 cm^{-1} on before activation, 3411.26 cm^{-1} on after activation and 3398.72 cm^{-1} on after contact. C-O group at 1109.12 cm^{-1} on before activation, 1225.82 cm^{-1} on after activation and 1227.74 cm^{-1} on after contact. C=O group at 1604.84 cm^{-1} on after activation and 1601.95 cm^{-1} on after contact.

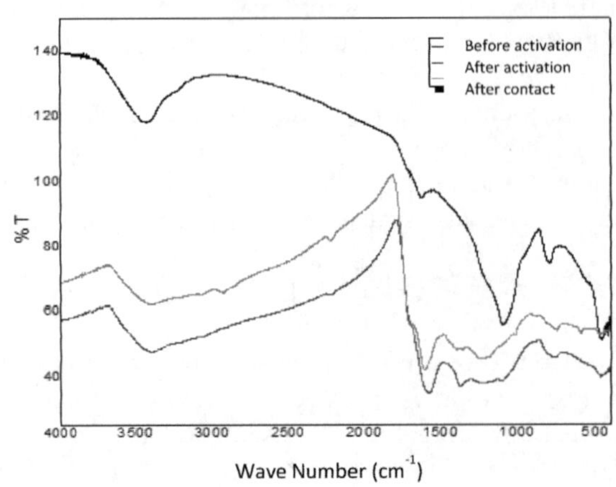

Figure 2. FTIR spectra of water hyacinth charcoal before activation, after activation and after contact

IOP Conf. Series: Materials Science and Engineering **333** (2018) 012053 doi:10.1088/1757-899X/333/1/012053

3.4. Determination of Cr (VI) calibration curve

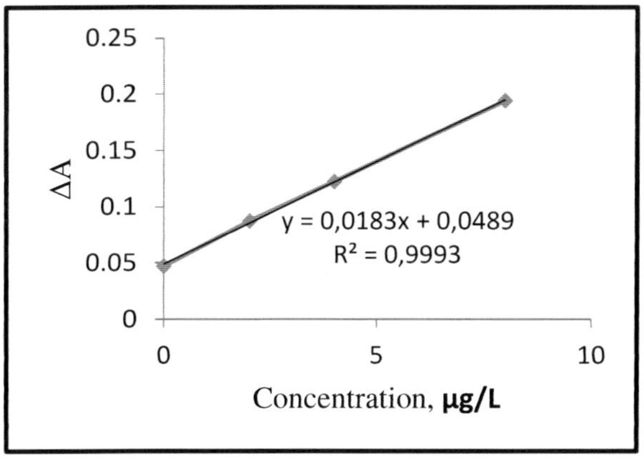

Figure 3. Pb^{2+} calibration curve

Based on the graph in Figure 3, y = 0.0183x + 0.0489 where y is ΔA and x is the concentration. The equation is used to determine the concentration of Pb^{2+} in simulated liquid waste. By using this equation, the Pb^{2+} levels in a simulated waste water was 49,68 µg/L.

3.5. Determination of optimum adsorbent mass composition activated carbon from rice husk and water hyacinth charcoal in a decrease levels of Pb^{2+}

Table 1. Determination of optimum composition

Mass Ratio of Rice Husk and Water Hyacinth Adsorbent	**[Pb^{2+} Initial (µg/L)]**	**[Pb^{2+} Last (µg/L)]**	**[Pb^{2+} Adsorp (µg/L)]**	**Percentage of Pb^{2+} Adsorbed (%)**
1 : 0 (A)	49.68	9.39	40.29	81.10
0 : 1 (B)	49.68	9.26	40.42	81.36
1 : 1 (C)	49.68	9.51	40.17	80.86
1 : 2 (D)	49.68	8.24	41.44	83.42
2 : 1 (E)	49.68	8.59	41.09	82.71

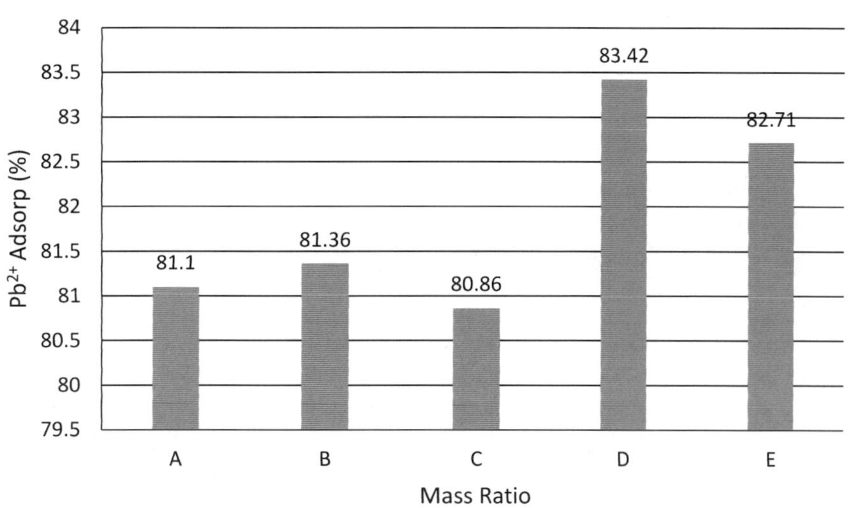

Figure 4. Adsorption percentage chart

Based on the data in Table 3 and Figure 4, the most optimal ratio of combination adsorbent from rice husk and water hyacinth in Pb^{2+} ions adsorption was 1:2. It can decrease Pb^{2+} metal ion concentration until 41,44 μg/L with the adsorption percentage is 83,42%. This ratio showed that the mass of water hyacinth active charcoal was twice from mass of rice husk active charcoal. Its because water hyacinth has a higher cellulose that rice husk. The cellulose showed that the content of carbon in water hyacinth is higher than rice husk.

Cellulose is a polysaccharide that contains D-glucose that connected in uniform by β-glucose bonding. The molecule formula is $(C_6H_{10}O_5)_n$ and it has -OH groups. Based on the result of FTIR spectra, water hyacinth and rice husk charcoal have -OH group. The existence of -OH group allowed the adsorption of Pb^{2+} ions because the ion exchange. H^+ ion in two OH exchange with Pb^{2+} ion, so the metal ion bonded with cellulose. The interaction between cellulose with Pb^{2+} ions happened because the existence of covalent bonding in the interaction [10].

3.6. Determination of detecting limit

Limit of detection (LOD) in this research was 0.06 μg/L, so it can be said that the sample with concentration at 0.06 μg/L can be read. Thus the solid-phase spectrophotometry (SPS) is a sensitive and effective analysis method for use in analysis decreased levels of Pb^{2+} metal ion adsorption by combination adsorbent from rice husk and water hyacinth in μg/L levels.

4. Conclusion

The results showed that rice husk and water hyacinth charcoal can be used as combination adsorbents to adsorb Pb^{2+} metal ions in the simulated liquid waste. Optimum mass ratio of rice husk activated carbon and water hyacinth activated carbon was 1:2 with the percentage of adsorption was 83,42% in the amount of 3 g. Solid-phase spectrophotometry (SPS) is a sensitive method to analyzed the decreased levels of Pb^{2+} metal ions in the level of μg/L with the limit of detection (LOD) was 0.06 μg/L.

Aknowledgement

Authors wishing to acknowledge assistance from Head of Chemistry Laboratory from the Chemistry Study Program, Faculty of Teacher Training and Education, Sebelas Maret University, for using the facilities of the laboratory. This work was partially supported by the Mandatory Grant for S.S (2017), MRG Grant for S.S (2017) and by Competency Grant for S.S (2016-2017) from the Ministry of Research, Technology and Higher Education, Indonesia.

References

[1] Nurhasni, Hendrawati, and Saniyyah N 2014 *Jurnal Valensi* **4** 36
[2] Tosepu R 2012 *Jurnal Manusia dan Lingkungan* **19** 37
[3] Peraturan Menteri Kesehatan No. 416/Men.Kes/Per/IX/1990
[4] WHO Guidelines for Drinking-water Quality WHO/SDE/WSH/03.04/09/Rev/1
[5] Matsuoka S, Nakatsu Y, Takehara K, Saputro S, and Yoshimura K 2006 *Analytical Sciences* **22** 1519
[6] Saputro S, Masykuri M, Mahardiani L, Mulyani B, Qorina I, Yoshimura K, Takehara K, Matsuoka S 2016 *IOP Conf. Ser.: Mater. Sci. Eng.* **176** 012019
[7] S Saputro, M Masykuri, L Mahardiani, S Arini and A H Ramelan 2016 *IOP Conf. Ser.: Mater. Sci. Eng.* **193** 012046
[8] Saputro S, Yoshimura K, Matsuoka S, Takehara K, and Narsito 2009 *Analytical Sciences* **25** 1445
[9] Danarto Y C and Samun T 2008 *Ekuilibrium* **7** 13
[10] Kumar P S, Ramakrishnan K, Kirupha S D, Sivanesan S 2010 *Braz. J. Chem. Eng.* **27** 1
[11] Joejodibroto R 1983 *Berita Selulosa* **29** 3
[12] Abriagni, D 2011 Minithesis of Universitas Negeri Semarang

IOP Conf. Series: Materials Science and Engineering **333** (2018) 012054 doi:10.1088/1757-899X/333/1/012054

The synthesis of corncobs (zea mays) active charcoal and water hyacinth (eichornia crassipes) adsorbent to adsorb Pb(II) with it's analysis using solid-phase spectrophotometry (sps)

S Saputro, M Masykuri, L Mahardiani, D Kurniastuti

Chemistry Education, Faculty of Teacher Training and Education, Sebelas Maret University, Jl. Ir. Sutami 36A Surakarta, Central Java-57126, Indonesia

E-mail: sulistyo_s@staff.uns.ac.id

Abstract. This research aim to examine the effect of the combination between corncobs and water hyacinth to adsorb lead (II), the most effective combination have determined by compared the ratio of corncobs adsorbent and water hyacinth to the increasing adsorption of the Pb(II), prove the effectiveness of the solid-phase spectrophotometry (sps) to determine the levels of Pb(II) as the result of the corncobs active charcoal adsorption and water hyacinth in the level of μg/L.The research method used is experimental method. The data collecting technique is carried out by several stages, which are carbonization using muffle furnace at a temperature of 350°C for 1.5 hours, activation of the corncobs charcoal and water hyacinth using HCl 1M and HCl 5M activator, contacting the adsorbent of corncobs active charcoal and water hyacinth with liquid waste simulation of Pb(II) using variation of corncobns and water hyacinth, 1:0; 0:1; 1:1; 2:1; 1:2, analysis of Pb(II) using an sps, characterization of corncobs active charcoal adsorbent and water hyacinth using FTIR. Research results show that the combined effect of activated charcoal corncobs and water hyacinth can increase the ability of the adsorbent to absorb Pb(II), the optimum adsorbent mass ratio of 1:1 with the absorption level of 90.33%, SPS is an effective method to analyze the decreasing level of Pb(II) as the adsorbtion result of the corncobs active charcoal and water hyacinth in the level of μg/L, with the limit of detection (LOD) of 0.06 μg/L.

1. Introduction

Technological developments have an impact on the development of the industry growing by leaps and bounds. This led to a growing number of industrial waste that pollutes the environment even groundwater that is widely used for the community's daily needs, such as cooking, washing and bathing. One of the heavy metals to become donors in the decline in the quality of human life, namely lead or plumbum [1].

Currently many developed methods to reduce the impact of water pollution by industrial waste containing heavy metals, one is with the method of adsorption. Adsorption is the heavy metal ion absorption method using specific adsorbents. Adsorbents used can be either a corncobs charcoal and water hyacinth.

The corncobs charcoal contain cellulose, hemicellulose, lignin and extractive substances which can be activated carbon [2]. Activated carbon can be used as adsorbent for heavy metals Pb^{2+} with activation first. In addition to the corncob, water hyacinth is also the aquatic plants that grow very rapidly and are considered very detrimental to society because it causes depletion of oxygen in the water and result in more fish that die because of the difficulty gets oxygen. Water hyacinth has good properties, among others, can absorb heavy metals, sulfides, compounds other than protein it contains more than 11.5% and contains cellulose that is bigger than lignin, ash, fat and other substances [3].

To activate charcoal from corn cobs and water hyacinth needed an Activator accordingly. The selection of the appropriate Activator will make the absorption of heavy metals are more optimally.In this research using corncob activator is HCl1 M and the water hyacinth activator is HCl 5 M.

The analysis is used to find out the capabilities of active charcoal adsorbent Hyacinths and corncob absorb heavy metals Pb is solid phase spectrophotometry (SPS).Solid Phase Spectrophotometry is a method of analysis using the spectrophotometer UV rays-looks. This method is based on measuring the sample directly through the absorption phase ion exchanger resin that absorbs the sample in the form of complex [4]. Solid phase spectrophotometry (SPS) has sensitivity better than previous instruments [5]. SPS showed that teak sawdust and zeolite can be used as adsorbents to adsorb Cr(VI) metal ions with the adsorption capacity 1.19 μg/g analyse by solid-phase spectrophotometry (SPS)[6]. The advantages of this method of analysis compared to other methods is the simple operation, high sensitivity and accuracy, which reached the level μg/L, so it can detect the metal ions dissolved in very small concentration [7]. Solid Phase Spectrophotometry are extremely sensitive and highly accurate because it can detect up to the size of the ppb. Higher sensitivity can be obtained by using 20 cm^3 volume of the sample [8]. The advantages of this method of analysis compared to other methods is the simple operation, high sensitivity and accuracy, which reached the level μg/L, so it can detect the metal ions dissolved in very small concentration [8]. The determine the reduction levels of Cd(II) metal ions as an adsorption result of the activated rice husk charcoal in the level of μg/L with the limit of detection (LOD) was0.059 μg/L [9]. Thus, researchers interested in conducting research of the application of activated carbon from corncob and water hyacinth waste as adsorbent of Pb(II) with an analysis using SPS.

2. Experimental

2.1. Tools and materials

The used tools were a UV-visible spectrophotometer from K-MAC, FTIR spectrophotometer from Shimadzu, muffle furnace, oven, pan, analytical balance, volumetric flask, volume pipette, beaker glass, measuring cup, flask, Erlenmeyer, watch glass, stirrer glass, drop pipette, aliquoting devices which are assembled using a syringe, blender, mortar and pestle, a 100 mesh sieve, porcelain bowls, stirrer bar, and a magnetic stirrer. Materials used are rice husk, $Pb(NO_3)_2$, HCl 1 M, HCl 5M, resin AG Muromac 50W-X2, H^+form(100-200 mesh), distilled water, chloroform, Whatman filter paper, blue litmus, and dithizone.

2.2. Research procedure

2.2.1. Production of adsorbent. Corncobs and water hyacinth cleaned and washed with distilled water, then it is dried at 105°C then charred in a muffle furnace at a temperature of 350°C for 1 hours. Results obtained are mashed and then sieved with a 100 mesh size.

2.2.2. Adsorbent activation. Corncobs soaks in a solution of HCl 1 M and water hyacinth soaks in a solution of HCl 5 M for 24 hours at room temperature. Filter and rinse the residue with distilled water until the filtrate become neutral, and then dried in an oven at 110°C for 24 hours. Adsorbent was tested by FTIR before and after activation.

2.2.3. Resin preparation. Muromac resin AG Muromac 50W-X2 H^+form (100-200 mesh) dissolved in distilled water and silence a few moments until the resin becomes more fluffy.

2.2.4. Determination of Ob(II) calibration curve. A 20-mL of standard solution of Pb(II) 0 μg/L, 2 μg/L, 4μg/L, and 8μg/L respectively were added with 1 mL of 0.005% dithizone, and 0.06 mL resin. Stir it for 20 minutes and analyzed using UV-vis spectrophotometer at a wavelength of 483 nm and 558 nm and then absorbance difference of the two wavelengths determined, $\Delta A = A_{483nm}$-A_{558nm}, which ΔA will be made standard curve (ΔA vs concentration).

2.2.5. Determination of Pb(II) species in simulated liquid waste. Simulated liquid waste solution of 50 μg/L was taken 20 mL then added 1 mL of 0.005% dithizone, and 0.06 mL resin. Then, stir it for 20 minutes and analyzed using UV-Vis spectrophotometer at a wavelength of 483 nm and 558 nm. ΔA obtained will be substituted in equation Pb(II) calibration curve (ΔA vs concentration), so that the Pb(II) ions in the effluent can be known.

2.2.6. Determination Of The Most Effective Adsorbents Comparison. Incorporating active charcoal and corn cob charcoal active water hyacinth with a comparison of 1:0; 0:1; 1:1; 2:1; 1:2 into the beaker of solution already contains sewage Pb (II) simulation of as many as 25 mL. Stirer for a solution that contains a mix of active charcoal and waste Pb (II) simulation for 30 minutes. The solution obtained is filtered with a filter paper and diluted from 5 mL filtrate into a 50 mL. Take 20 mL and then adding 0.06 mL resin, 2 mL and 1 mL of H_2SO_4 dithizon. Stirer for 20 minutes. The filtrate obtained tested with solid-phase spectrophotometry with 483 nm wavelength and 558 nm. Test comparison with most effective adsorbents using Fourier Transform Infra-Red (FTIR).Determination of detection limit. Taking five blank solution 20 ml, then each added 1 mL of of 0.005% dithizone, and as much as 0.06 ml resin, then stir it for 20 minutes. Furthermore, it analyzed by UV-vis spectrophotometer at a wavelength of 483 nm and 558 nm. ΔA obtained will be substituted in the equation of the calibration curve.

3. Result and Discussions

3.1 Production of activated charcoal from corncobs and water hyacinth

The process of making corn cob active charcoal and water hyacinth consists of 3 stages, namely the stage of dehydration, and the carbonation stage stage of activation. At the stage of dehydration corn cobs and dried water hyacinth with an oven at a temperature of 105 ° C for 1 hour to eliminate the water levels.The carbonation stage done by inserting the corncob and water hyacinth into the furnace at a temperature of 350 ° C for 1 hour because higher temperatures can increase the rate of reduction of impurity and the volatile compounds that fills pore adsorbents, thus optimizing the establishment of active pore [10].Next up is stage activation charcoal corn cobs and water hyacinth. The main purpose of the activation process is to add or develop pore volume and pore diameter enlarging that had formed on the carbonization process as well as to make some new pore [11].

3.2 Fourier transform infra red spectra analysis

Based on FTIR Spectra analysis against a mix of active charcoal adsorbent corncobs and water hyacinth before activation, after activation and after pengontakkan done, obtained results are not too much difference. It's just that there is no vibration after activation of the C-O-C and C-H amine and vibration CH_3. On the infrared spectra of active charcoal adsorbent blend of corn cobs and water hyacinth after activation or after pengontakkan does not show a difference too much, just that there is no vibration after activation of the C-O-C and C-H amine.While after contacting the absence of C-H vibration amines and CH3. Infrared Spectra analysis results, it can be concluded that the fusion of active charcoal adsorbent and cob of corn contains an aromatic C-H cluster, C-O-C, C = O, C-H,-CH3 amines, C = O, and aromatic C-H.

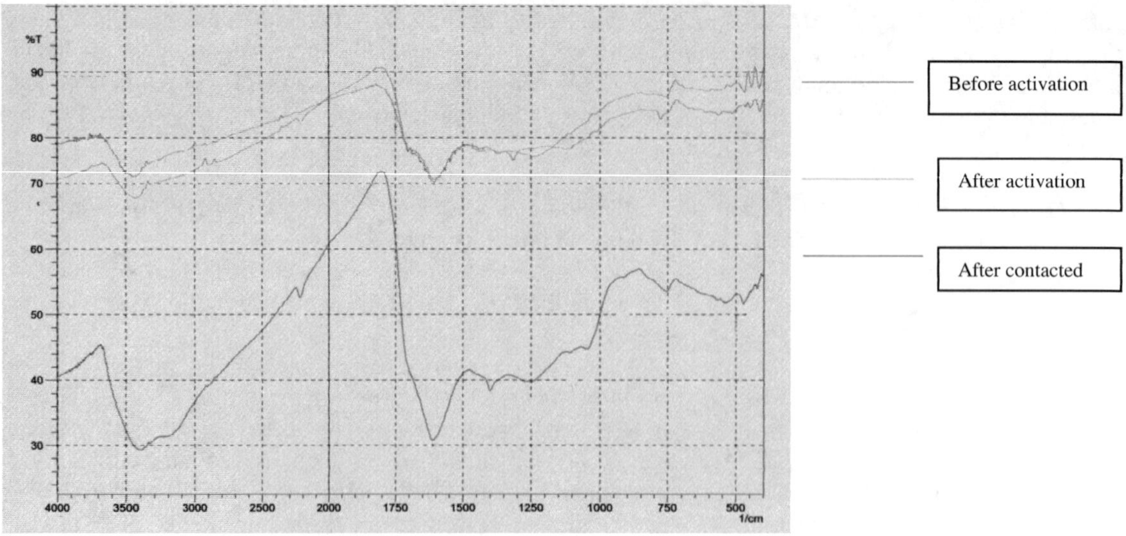

Figure 1. FTIR Spectra

3.3 Determination of Pb(II) calibration curve

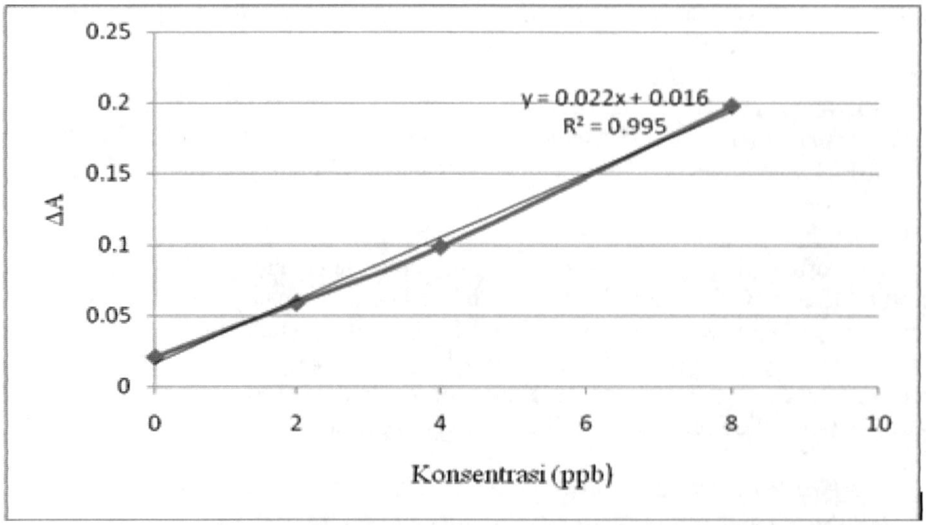

Figure 2. Calibration curve of Pb (II) solution.

From the graph in Figure 2 are obtained one equation that is y = 0.022 x + 0.016, where y is and x is ΔA concentration. The equation which will be used to determine the concentration of Pb (II) contained in the waste liquid simulation. Determination results of Pb(II) levels in simulated liquid waste.

Table 1. Determination results of Pb(II) levels in simulated liquid waste

Sample	**[Pb(II)] (µg/L)**	**ΔA**	**[Pb(II) actually] (µg/L)**
Sample Pb(II)	4.99	0.126	49.99

3.4 Comparison of most effective adsorbents

Table 2. Optimum adsorbent mass comparison

Mass Comparison	Initial [Pb(II)] (μg/L)	ΔA	Final [Pb(II)] (μg/L)	Adsorbed [Pb(II)] (μg/L)	Adsorbed [Pb(II)] (%)
0:1	49.99	0.178	7.39	42.60	85.21
1:0	49.99	0.169	6.96	43.04	86
1:1	49.99	0.122	4.84	45.16	90.33
2:1	49.99	0.139	5.59	44.40	88.81
1:2	49.99	0.139	5.61	44.38	88.77

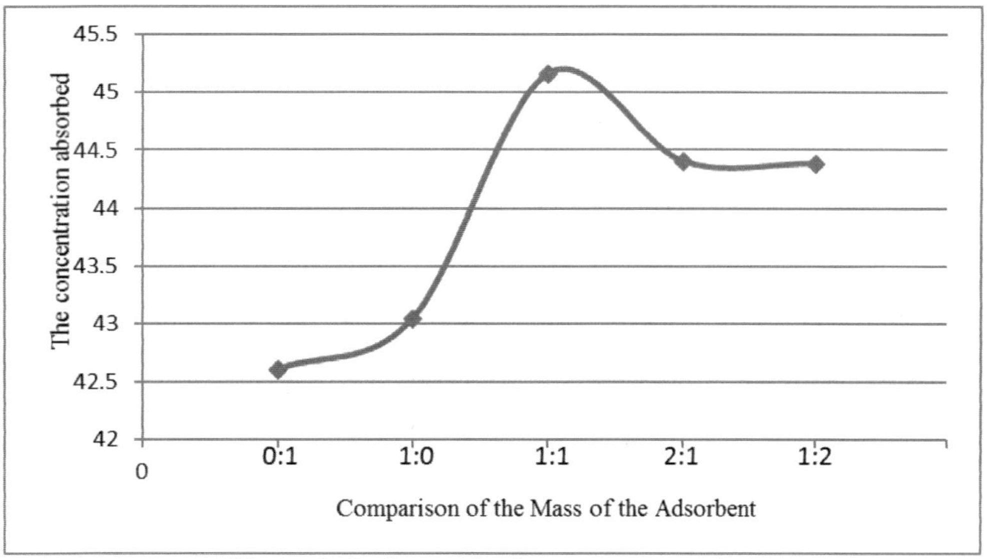

Figure 3. Optimum adsorbent mass comparison

Comparison of the mass of active charcoal adsorbent corncob: copper ion absorption occurs 1:1 of heavy metals Pb (II) highest i.e. 45.16 amounting to μ g/L or 90.33%.Determination of detecting limit.

Table 3. Determination of detecting limit

Blank	ΔA	Concentration	Standard deviation	LOD
A	0.104	4.03		
B	0.104	4.02		
C	0.103	3.98		
D	0.104	4.02		
E	0.104	4		
Average		4.01	0.02	0.06

In the experiments that have been conducted obtained rating LOD 0.06 μ g/L. This shows that at a concentration of 0.06 μ g/L sample is still legible.

4. Conclusion
Influence of combination of active charcoal adsorbent corncob and water hyacinth is able to improve the ability of adsorbents to absorb heavy metals Pb (II), comparison of the mass of active charcoal adsorbent COB corn and water hyacinth are the most effective 1:1 with the absorption of heavy metals Pb (II) of 90.33% and Solid-Phase Spectrophotometry (SPS) is an effective method for determining the levels of heavy metals ions, Pb (II) results of active charcoal adsorbed corncorb and water hyacinth in the level μg/L.

Aknowledgement
Authors wishing to acknowledge assistance from Head of Laboratory Chemistry from the Chemistry Study Program, Faculty of Teacher Training and Education, Sebelas Maret University, for using the facilities of the laboratory. This work was partially supported by the Mandatory Grant for S.S (2017), MRG Grant for S.S (2017) and by Competency Grant for S.S (2016-2017) from the Ministry of Research, Technology and Higher Education, Indonesia.

References
[1] Widowati, dkk 2008 *Efek Toksik Logam Pencegahan dan Penanggulangan Pencemaran.* (Yogyakarta: C.V Andi Offset)
[2] Rahayu A.N, dan Adhitiyawarman 2014 *JKK* **3(3)** 7-13
[3] Tangio, Julhim S 2012 *Skripsi* (Universitas Gorontalo)
[4] Yoshimura K, Waki H, & Ohasi S 1975 *Anal Sci.* **23** 449-454
[5] Matsuoka S, Nakatsu Y, Takehara K, Saputro S, & Yoshimura K 2006 *Anal Sci.* **22** 1519-1524
[6] S Saputro et al 2017 IOP Conf. Ser.: Mater. Sci. Eng. **176** 012019
[7] Sarenqiqige, Saputro S, Kai S, Satoda M, Matsuoka S and Yoshimura K 2013 *Anal Sci.* **29** 677
[8] Saputro S, Yoshimura K, Matsuoka S, Takehara K, & Narsito 2009 *Anal Sci.* **25** 1445-1450
[9] S Saputro 2017 IOP Conf. Ser.: Mater. Sci. Eng. **193** 012046
[10] Siburian A M, Pardede A S D, and Pandia S 2014 *Teknik Kimia USU* **3**(**4**)
[11] Prabarini N & Okayadnya D G 2014 *Jurnal Ilmiah Teknik Lingkungan.* **5**(**2**) 33-41

IOP Conf. Series: Materials Science and Engineering **333** (2018) 012055 doi:10.1088/1757-899X/333/1/012055

Combination of sawdust from teak wood and rice husk activated carbon as adsorbent of Pb(II) ion and its analysis using solid-phase spectrophotometry (sps)

S Saputro, L Mahardiani, D A Wulandari

Chemistry Education, Faculty of Teacher Training and Education, Sebelas Maret University, Jl. Ir. Sutami 36A Surakarta, Central Java-57126, Indonesia

E-mail: sulistyo_s@staff.uns.ac.id

Abstract. This research aimed to know the usage of sawdust of teak wood and rice husk waste as Pb (II) ion adsorbents in simulated liquid waste, the combined optimum mass required adsorbent to adsorb Pb(II) ion, the sensitivity of the solid-phase spectrophotometry (sps) method in determining the decrease of Pb (II) metal ion levels in the μg/L level. This research was conducted by experimental method in laboratory. Adsorbents used in this study were charcoal of sawdust sawdust activated using 15% $ZnCl_2$ solution and activated rice husk using 2 N NaOH solution. The adsorption processes of sawdust and rice husk with Pb(II) solution was done by variation of mass combination with a ratio of 1: 0; 0: 1; 1: 1; 1: 2; and 2: 1. Analysis of Pb(II) ion concentration using SPS and characterization of sawdust and rice husk adsorbent ads using FTIR. The results showed that activated charcoal from sawdust of teak wood and rice husks can be used as Pb (II) metal ion adsorbents with adsorption capacity of 0.86 μg/L, charcoal from sawdust of teak wood and rice husk adsorbent with a combination of optimum mass contact of sawdust and rice husk is 2:1 as much as 3 grams can adsorb 42.80 μg/L.Solid-phase spectophotometry is a sensitive method for analysis of concentration decreasing levels of Pb(II) ion, after it was absorbed by sawdust of teak wood and rice husk with high sensitivity and has the limit of detection (LOD) of 0.06 μg/L.

1. Introduction

Indonesia is a country that is promoting development in all sectors, one of them in the industrial sector. Development in the industrial sector has mixed impacts. These impacts can be generated by the presence of industrial waste and increasing levels of human activity. Industrial waste is generally contained heavy metals therein and the waste is immediately discharged into the environment without any prior processing. The waste will have a negative impact on the environment. The waste comes from a production process using materials or compounds containing such heavy metals. One of the harmful metal ions present in the waste is the Pb metal ion. Pb metal is widely used as packaging material, plumbing such as pipes, household appliances and decoration. In the form of lead oxide is used as pigment/dye in the cosmetic and glace industries and ceramic industries, some of which are used in home appliances [1].

To reduce the heavy metal content of waste is usually used conventional methods such as chemical precipitation, ion exchange (ion exchange), and adsorption. A widely used method is adsorption for many reasons, which are more economical and do not cause harmful effects. Adsorption is defined as the removal of molecules on the outer surface or surface in an adsorbent solid or by the surface of the solution. Adsorption occurs because the molecules on the surface of solids or liquids that have tensile forces in an unbalanced state tend to be attracted inward (the adsorbent cohesion force is greater than the adhesion force). Such tensile gravity imbalances result in solids or liquids used as adsorbents tending to attract other substances that come into contact with their surfaces.

Indonesia's total sawn timber production reaches 2.6 million m^3 per year. Assuming that the amount of waste formed 54.24% of the total production, the resulting sawmill waste is 1.4 million m^3 per year. This figure is quite large because it reaches about half of the production of sawn timber[2]. Sawdust waste generated from the sawmill industry can still be utilized for various purposes, such as planting medium, raw materials of furniture and raw material of charcoal briquettes. One effort to increase the added value of this sawdust is to make activated charcoal as an adsorbent making material.

Indonesia is an agrarian country whose majority population is a farmer. Farmers usually have problems with waste that is a by-product of agricultural products. This is due to the considerable availability of rice husk waste in all places around the rice mill and the limited utilization of the waste. In Indonesia, especially in Klaten district, rice husks are still not getting attention and are limited to a few simple uses such as ash, as a medium for crops and as fuel. In fact, rice husk ash is a very potential material as a heavy metal absorber in water, so it can be an alternative solution to environmental pollution problems. Rice husk has low nutrient value, abrasive, and contains high carbon [3]. Activated rice husk charcoal can be used as adsorbent to adsorb Cd(II) metal ions in the simulated liquid waste[4].

Lately, many adsorbents have been developed to absorb heavy metals. However, many studies are done using a single sorbent as a process. Preparation of adsorbent by combining two materials is still rare, for that in this research will be done merging between sawdust teak and rice husk as absorber of heavy metal ion Pb (II). The combination of coal fly ash, sawdust and coconut shell can increase the adsorption of chromium and cadmium metal ions [5]. Teak sawdust and zeolite can be used as adsorbents to adsorb Cr(VI) metal ions in the simulated liquid waste with the percentage of adsorption was 62.72% in the amount of 0.5 g [6]. Combination of activated charcoal rice husk and zeolite can used as adsorbent in adsorption Cr (VI) metal ion with the adsorption capacity 0,28 μg/g analyse by solid-phase spectrophotometry[7]. This study combining sawdust from teak wood and rice husk is expected to increase the adsorption of Pb (II) metal ions.

Determination of concentration of Pb (II) in water is very important even if the concentration is very small, therefore to know Pb (II) contamination must use instrument that has high sensitivity. Other instruments that have been available can only detect up to ppm, whereas maximum levels of Pb in drinking water according to SNI are 0.005 mg/L. Solid-Phase Spectrophtometry (SPS) is a method of determining the level of a substance in which the absorption of light by the target chemical component is concentrated in the solid phase. This method is more sensitive than conventional methods, and can determine levels up to μg/L level[8].This is evidenced by previous research on the determination of Cr(VI) in natural water by analysis of SPS obtained the limit of detection (LOD) of 0.014 g/dm^3 [9]. Thus, researchers interested in conducting research of the application of activated carbon from sawdust and rice husk waste as adsorbent of Pb(II) with an analysis using SPS.

2. Experimental

2.1. Tools and materials

The used tools were a UV-visible spectrophotometer from K-MAC, FTIR spectrophotometer from Shimadzu, muffle furnace, oven, pan, analytical balance, volumetric flask, volume pipette, beakerglass, measuring cup, flask, erlenmeyer, watch glass, stirrer glass, drop pipette, aliquoting devices which are assembled using a syringe, blender, mortar and pestle, a 100 mesh sieve, porcelain bowls, stirrer bar, and a magnetic stirrer. Materials used are sawdust from teak wood, rice husk, $Pb(NO_3)_2$, $ZnCl_2$ 10%, NaOH 2N, $H_2SO_4$0,5 M, resin AG Muromac 50W-X2H^+ form(100-200 mesh), distilled water, chloroform, Whatman filter paper, blue litmus, and dithizone.

2.2. Research procedure

2.2.1. Production of adsorbent. Sawdust from teak wood and rice husk is cleaned and washed with distilled water, then it is dried at 105°C then charred in a muffle furnace at a temperature of 300°C for 1 hoursfor sawdust and 250°C for 1 hours for rice husk. Results obtained are mashed and then sieved with a 100 mesh size.

2.2.2. Adsorbent activation. Charcoal sawdust soaked in a solution of $ZnCl_2$10% and rice husk charcoal soaked in a solution of NaOH 2 N for 24 hours at room temperature. Filter and rinse the residue with distilled water until the filtrate become neutral, and then dried in an oven at 105°C for 24 hours. Adsorbent was tested by FTIR before and after activation.

2.2.3. Resin preparation. Muromac resin AG Muromac 50W-X2 H^+ form(100-200 mesh) dissolved in distilled water and silence a few moments until the resin becomes more fluffy.

2.2.4. Determination of Pb(II) calibration curve. A 20-mL of standard solution of Pb(II) 0 µg/L, 2 µg/L, 4 µg/L, and 8µg/L respectively were added with1 mL of H_2SO_4 0,5 M, 2 mL of 0.005% dithizone, and 0.06 mL resin. Stir it for 20 minutes and analyzed using UV-vis spectrophotometer at a wavelength of 483 nm and 558 nm and then absorbance difference of the two wavelengths determined, $\Delta A = A_{483nm}-A_{558nm}$, which ΔA will be made standard curve (ΔA vs concentration).

2.2.5. Determination of Pb(II) species in simulated liquid waste. Simulated liquid waste solution of 50 µg/L was taken 20 mL then added 1 mL of H_2SO_4 0,5 M, 2 mL of0.005% dithizone, and 0.06 mL resin. Then, stir it for 20 minutes and analyzed using UV-Vis spectrophotometer at a wavelength of 483 nm and 558 nm. ΔA obtained will be substituted in equation Pb(II) calibration curve (ΔA vs concentration), so that the Pb(II) ions in the effluent can be known.

2.2.6. Determination the optimum combination of adsorbent for sawdust and rice husk. Inserting a 20-mL of simulated liquid waste solution 50µg/L in the 5erlenmeyer. Insert each adsorbent sawdust and rice husk with a ratio of*1:0; 1:1; 1:2; 2:1dan 0:1* and stir it for 30 minutes. Each adsorbent was filtered with Whatman filter paper. The resulted filtrate was taken as much as 5 mL and then diluted with distilled water until 50 mL in volume. Furthermore, the filtrate diluted taken as much as 20 mL added with 1 mL of H_2SO_4 0,5 M, 2 mL of 0.005% dithizone, and 0.06 mL resin then stir it for 20 minutes and analysed using a UV-vis spectrophotometer with a wavelength of 483 nm and 558 nm. ΔA obtained will be substituted in equation of Pb(II) calibration curve (ΔA vs concentration) so species of Pb(II) in a simulated liquid waste after the Pb(II) adsorption is known.

2.2.7. Determination of detection nlimit. Taking five blank solution 20 ml, then each added 1 mL of H_2SO_4 0,5 M, 2 mL of of 0.005% dithizone, and as much as 0.06 ml resin, then stir it for 20 minutes. Furthermore, it analyzed by UV-vis spectrophotometer at a wavelength of 483 nm and 558 nm. ΔA obtained will be substituted in the equation of the calibration curve.

3. Result and Discussion

3.1. Production of activated charcoal from sawdust and rice husk

Production of activated charcoal from sawdust consists of three stages: the stage of preparation, is the removal of impurities in sawdust, stage of carbonization, breakdown of organic matter into carbon by heating it at 300°C for 1 hours for sawdust and 250°C for 1 hours for rice husk, and the activation stage, soaking with an activator solution which is $ZnCl_2$ 15% for sawdust and NaOH 2 N for rice husk for 24 hours. $ZnCl_2$ 15% is used because ZnC12 was able to produce the highest yield with activated

IOP Conf. Series: Materials Science and Engineering **333** (2018) 012055 doi:10.1088/1757-899X/333/1/012055

carbon compared with KOH and H3PO4 [10]. The use of NaOH activator on rice husk can dissolve more and result in a silica surface area became greater than using aquades and HCl Activator.

3.2. Fourier transform infrared spectra analysis

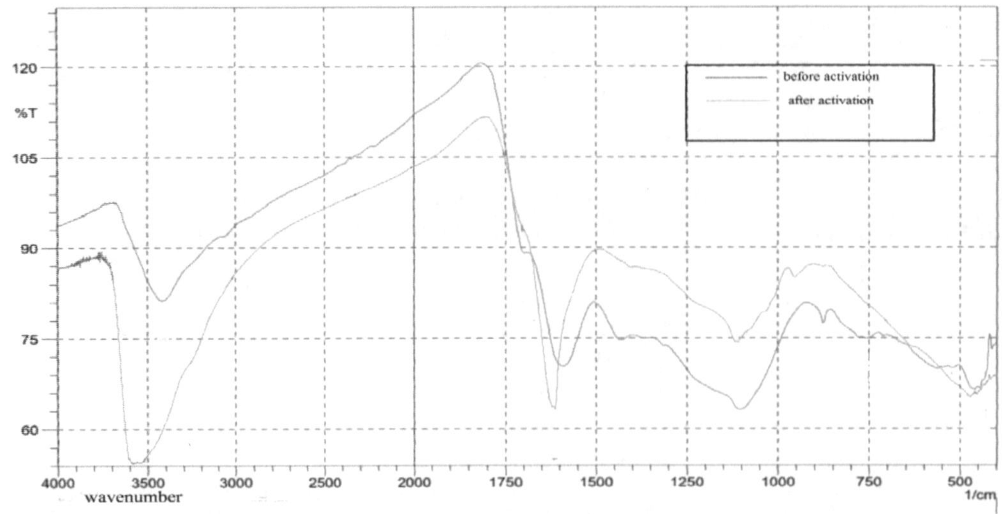

Figure 1. FTIR spectra of sawdust charcoal before and after activation

Based on Figure 1.adsorbent sawdust from teak wood before activation and after activation using the activator $ZnCl_2$ not many changes are happening. From the spectra can be inferred that the sawdust charcoal adsorbent before and after activated contains cluster –CH, C-O,-CH_3,C = C C = O, aromatic, and – OH.

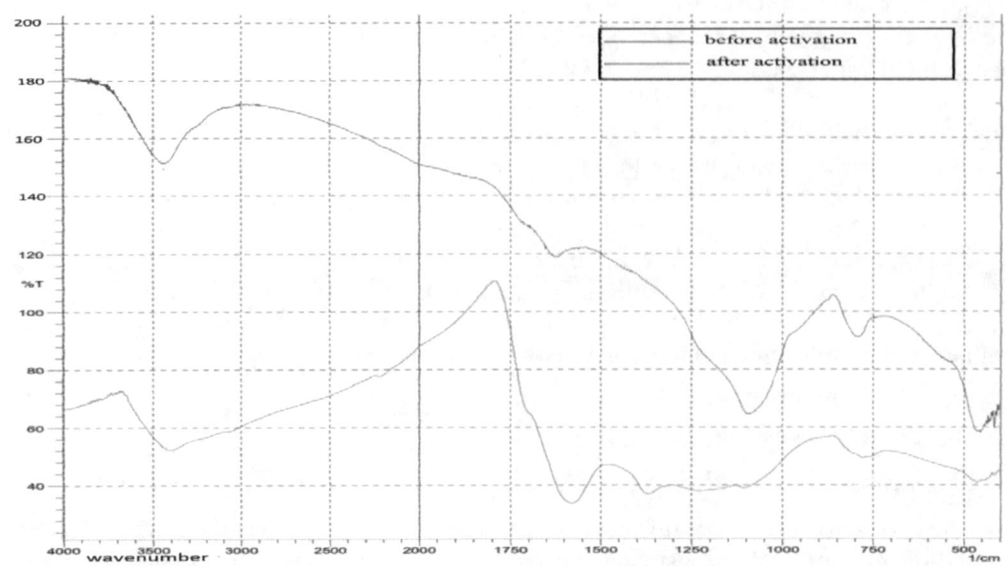

Figure 2.FTIR spectra of rice husk charcoal before and after activation

Based on Figure 2.adsorbent rice husk before and after activation using the activator NaOH not many changes are happening. Based on spectra can be inferred that the rice husk charcoal adsorbent before and after activated cluster containing C=C aromatics, -OH and, Si-O of the Si-O-Si.

3.3. Determination of Pb(II) calibration curve

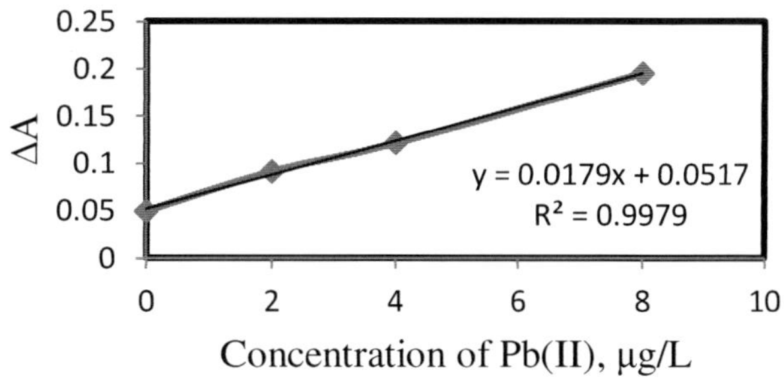

Figure 3. Calibration curve of Pb(II) solution

Based on the graph in Figure 3, y = 0,017x + 0.051 where y is ΔA and x is the concentration. The equation is used to determine the concentration of Pb(II) in simulated liquid waste.

3.4. Determination results of Pb(II) levels in simulated liquid waste

Table 1. Determination results of Pb(II) levels in simulated liquid waste

Sample	[Pb(II)] calculation (ppb)	ΔA	[Pb(II)] actually (ppb)
Sample Pb(II)	50	0.135	49.18

3.5. Determination the optimum combination of adsorbent for sawdust and rice husk in a decrease levels of Pb(II)

Table 2. The influence of the composition of the mass to Ion adsorption of Pb (II)

Composition of the mass	**Final [Pb(II)] (μg/L)**	**Adsorbed [Pb(II)] (μg/L)**	**Adsorbed [Pb(II)] (%)**
1 SG : 0 SP (A)	9.17	40.02	81.36
0 SG : 1 SP (B)	9.63	39.55	80.41
1 SG : 1 SP (C)	8.66	40.52	82.39
1 SG : 2 SP (D)	7.87	41.31	83.99
2 SG : 1 SP (E)	6.39	42.80	87.01

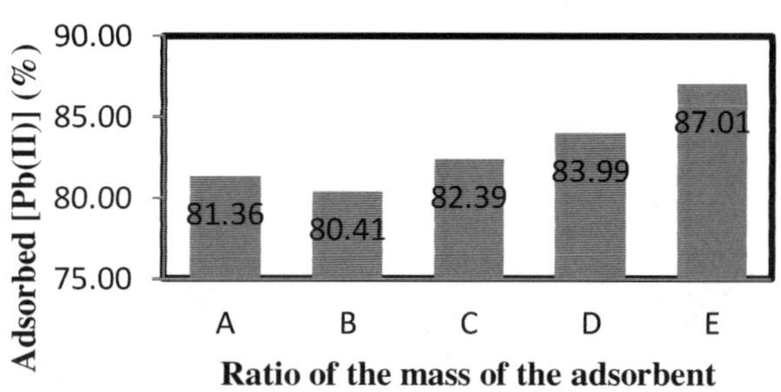

Figure 4. Determination the optimum combination mass of adsorbent

Based on data research in Table 2 and the graph in Figure 4, the optimum combination mass of adsorbent to adsorb Pb(II) metal ions in the simulated liquid waste was 2:1 with adsorbed concentration was 42.80 μ/L, percentage of adsorbed as 87.01% and adsorption capacity was 0.86 μg/g.

3.6. Determination of detecting limit

In this research, limit of detection (LOD) was 0.06 μg/L, so it can be said that the sample at a concentration of 0.06 μg/L can be read, so that the solid-phase spectrophotometry (SPS) is a sensitive and effective method to use in analysis the decreased levels of Pb(II) adsorption results of activated charcoal from rice husk in levels μg/L.

4. Conclusion

Activated charcoal from sawdust of teak wood and rice husks can be used as Pb (II) metal ion adsorbents with adsorption capacity of 0.86 μg/L. Charcoal from sawdust of teak wood and rice husk adsorbent with a combination of optimum mass contact of sawdust and rice husk is 2:1 as much as 3 grams can adsorb 42.80 μg/L. *Solid-Phase Spectophotometry* (SPS) method is a sensitive analysis method for analysis of concentration decreasing levels of Pb(II) ion, after it was absorbed by sawdust of teak wood and rice husk with highly sensitivity and has the *Limit of Detection* (LOD) of 0.06 μg/L.

Aknowledgement

Authors wishing to acknowledge assistance from Head of Chemistry Laboratory, Chemistry Study Program, Faculty of Teacher Training and Education, Sebelas Maret University, for using the facilities of the laboratory. This work was partially supported by the Competency Grant for S.S (2016-2017) from the Ministry of Research, Technology and Higher Education, Indonesia.

References

[1] Gusnita, D 2012 *Berita Dirgantara* ***13(2)*** *95*
[2] Sudarja dan Caroko 2012 *Jurnal Ilmiah Semesta Teknika* **14**(1) 50
[3] Danarto Y C, and Samun T 2008 *Ekuilibrium* **7** 13
[4] S Saputro, M Masykuri, L Mahardiani, S Arini and A H Ramelan 2017 *IOP Conf. Ser.: Mater. Sci. Eng.* **193** 012046
[5] Sahat 2005 *Jurnal Teknik Mesin* ***2(2)*** *75-82*
[6] Saputro S, Masykuri M, Mahardiani L, Mulyani B, Qorina I, Yoshimura K, Takehara K, Matsuoka S 2017 *IOP Conf. Ser.: Mater. Sci. Eng.* **176** 012019
[7] Saputro S, Masykuri M, Mahardiani L, Mulyani B and Wahyuni N.T 2016 *J. Sains Dasar* ***5(2)*** *116*
[8] Sarenqiqige,Saputro S, Kai S, Satoda M, Matsuoka S and Yoshimura K 2013 *Anal Sci.***29** 677
[9] Saputro S, Yoshimura K, Takehara K, Matsuoka S and Narsito 2009 *Anal Sci.* **25** 1445
[10] Esterlita dan Herlina 2015 *Jurnal Teknik Kimia USU* 04(1) 47

IOP Conf. Series: Materials Science and Engineering **333** (2018) 012056 doi:10.1088/1757-899X/333/1/012056

Combination of rice husk and coconut shell activated adsorbent to adsorb Pb(II) ionic metal and it's analysis using solid-phase spectrophotometry (sps)

D N Rohmah, S Saputro, M Masykuri, L Mahardiani

Chemistry Education, Faculty of Teacher Training and Education, Sebelas Maret University, Jl. Ir. Sutami 36A Surakarta, Central Java-57126, Indonesia

E-mail: sulistyo_s@staff.uns.ac.id

Abstract. The purpose of this research was to know the effect and determine the mass comparation which most effective combination between rice husk and coconut shell activated adsorbent to adsorb Pb (II) ion using SPS method. This research used experimental method. Technique to collecting this datas of this research is carried out by several stages, which are: (1) carbonization of rice husk and coconut shell adsorbent using *muffle furnace* at a temperature of 350°C for an hour; (2) activation of the rice husk and coconut shell adsorbent using NaOH 1N and $ZnCl_2$ 15% activator; (3) contacting the adsorbent of rice husk and coconut shell activated adsorbent with liquid waste simulation of Pb(II) using variation comparison of rice husk and coconut shell, 1:0; 0:1; 1:1; 2:1; 1:2; (4) analysis of Pb(II) using *Solid-Phase Spectrophotometry* (SPS); (5) characterization of combination rice husk and coconut shell activated adsorbent using FTIR. The result of this research show that the combined effect of combination rice husk and coconut shell activated adsorbent can increase the ability of the adsorbent to absorb Pb(II) ion then the optimum adsorbent mass ratio required for absorbing 20 mL of Pb(II) ion with a concentration of 49.99 μg/L is a ratio of 2:1 with the absorption level of 97,06%Solid-Phase Spectrophotometry (SPS) is an effective method in the level of μg/L, be marked with the Limit of Detection (LOD) of 0.03 μg/L.

1. Introduction

The disposal of industrial waste effluent into riverine system has given rise to heavily localized pollution and threatens seriously to the environment. The pollution level of the river is increasing sharply and can cause serious problem in near future. For the polluted situation of the river maternal and child health of nearby riverbank slam are in a danger position [1]. The pollution of water is caused of any heavy metals. Most of these metals such as cadmium, lead, arsenic, mercury, copper, chromium, cobalt and nickel are very toxic. They are carcinogenic and at very high concentrations, may lead to brain damage [2]. Industrial potential as a source of pollution of lead (Pb) are all industries that use lead (Pb) as a raw material as well as material support, such as casting industry, the manufacture of batteries, cable, chemical industry and in the manufacture of paint. Whereas the toxicity Pb dangerous for the environment or human being [3]. Environment containing Pb with high concentrations, can affect human health disorders, as well as an increase in the levels of Pb in blood which resulted in disruption of the central nervous system, and can reduce intelligence (IQ) for children.

The impact of the pollution of Pb in the aquatic environment can be dealt with adsorben. Adsorbent of the Pb can be made from agricultural waste in Indonesia, such as rice husk and coconut shell. Rice husk can be used as adsorbent because cellulose and hemicellulose has a considerable potential for absorbing that has tied OH cluster so that it can interact with the component adsorbat. One of the heavy metals that can be absorbed by the rice husk adsorbent is lead (Pb)[5]. On the other hand coconut shell that has been activated shows high ability to adsorb Pb, the active carbon of coconut shell efficient to absorb the metal ion Pb2 + on liquid waste [6]. Adsorbent from coconut shell is cheap and easily adsorbent and obtainable where using them efficiently to absorb metals from water that has been contaminated [7]. The analysis that used to find out the capabilities of active adsorbent from coconut shell is SPS (Solid Phase Spectrophotometry). This method is more sensitive compared to conventional methods that use expensive instrumental, moreover, it can determine the levels up to the level of μ g/L[8]. SPS showed that teak sawdust and zeolite can be used as adsorbents to adsorb Cr(VI) metal ions with the adsorption capacity 1.19 μg/g analyse by solid-phase spectrophotometry (SPS) [9]. And then by using SPS activated rice husk charcoal can adsorbed Cd(II) metal with percentage of adsorbed amount 77.87%[10]. Solid Phase Spectrophotometry are extremely sensitive and highly accurate because it can detect up to the size of the ppb. Higher sensitivity can be obtained by using 20 cm^3 volume of the sample [11]. Solid Phase Spectrophotometry is a method of analysis using the spektofotometer UV rays-looks. This method is based on measuring the sample directly through the absorption phase ion exchanger resin that absorbs the sample in the form of complex [12].

2. Experimental

2.1. Tools and materials

The used tools were a UV-visible spectrophotometer from K-MAC, FTIR spectrophotometer from Shimadzu, muffle furnace, oven, pan, analytical balance, volumetric flask, volume pipette, beaker glass, measuring cup, flask, erlenmeyer, watch glass, stirrer glass, drop pipette, aliquoting devices which are assembled using a syringe, blender, mortar and pestle, a 100 mesh sieve, porcelain bowls, stirrer bar, and a magnetic stirrer. Materials used are rice husk, coconut shell $Pb(NO_3)_2$, HCl 1 M, HCl 5M, resin AG Muromac 50W-X2, H^+form(100-200 mesh), distilled water, chloroform, Whatman filter paper, blue litmus, and dithizone.

2.2. Research procedure

2.2.1. Production of adsorbent. Rice husk and coconut shellcleaned and washed with distilled water, then it is dried at 105°C then charred in a muffle furnace at a temperature of 350°C for 1 hours. Results obtained are mashed and then sieved with a 100 mesh size.

2.2.2. Adsorbent activation. Rice husk in a solution of NaOH 2N and coconut shell in a solution $ZnCl_2$ 15% for 24 hours at room temperature. Filter and rinse the residue with distilled water until the filtrate become neutral, and then dried in an oven at 110°C for 24 hours.

2.2.3. Resin preparation. Muromac resin AG Muromac 50W-X2 H^+form (100-200 mesh) dissolved in distilled water and silence a few moments until the resin becomes more fluffy.

2.2.4. Determination of Pb(II) calibration curve. A 20-mL of standard solution of Pb(II) 0 μg/L, 2 μg/L, 4μg/L, and 8μg/L respectively were added with 1 mL of 0.005% dithizone, and 0.06 mL resin. Stir it for 20 minutes and analyzed using UV-vis spectrophotometer at a wavelength of 483 nm and 558 nm and then absorbance difference of the two wavelengths determined, $\Delta A = A_{483nm}-A_{558nm}$, which ΔA will be made standard curve (ΔA vs concentration).

2.2.5. Determination of Cd(II) species in simulated liquid waste. Simulated liquid waste solution of 50 μg/L was taken 20 mL then added 1 mL of 0.005% dithizone, and 0.06 mL resin. Then, stir it for 20 minutes and analyzed using UV-Vis spectrophotometer at a wavelength of 483 nm and 558 nm. ΔA obtained will be substituted in equation Cd(II) calibration curve (ΔA vs concentration), so that the Pb(II) ions in the effluent can be known.

2.2.6. Determination of the most effective adsorbents comparison. Incorporating active carbon of rice husk and coconut shell with a comparison of 1:0; 0:1; 1:1; 2:1; 1:2 into the beaker of solution already contains sewage Pb (II) simulation of as many as 25 mL. Stir that solution for 30 minutes. The solution obtained then filtered with a filter paper and diluted from 5 mL filtrate into a 50 mL. Take 20 mL and then adding 0.06 mL resin, 2 mL and 1 mL of H_2SO_4 dithizon. Stir it for 20 minutes. The filtrate obtained tested with s with 483 nm wavelength and 558 nm. Test comparison with most effective adsorbents using FTIR. Determination of detectionlimit.taking five blank solution 20 ml, then each added 1 mL of of 0.005% dithizone, and as much as 0.06 ml resin, then stir it for 20 minutes. Furthermore, it analyzed by UV-vis spectrophotometer at a wavelength of 483 nm and 558 nm. ΔA obtained will be substituted in the equation of the calibration curve.

3. Result and Discussions

3.1 Production of activated carbon from rice husk and coconut shell

The process consists of 3 stages, namely the stage of dehydration, and the carbonation stage stage of activation. At the stage of dehydration rice husk and coconut shell with an oven at a temperature of 105 ° C for 1 hour to eliminate the water levels.The carbonation stage done by inserting it into the furnace at a temperature of 350 ° C for 1 hour because higher temperatures can increase the rate of reduction of impurity and the volatile compounds that fills pore adsorbents, thus optimizing the establishment of active pore [13].

3.2 Fourier transform infra red spectra analysis

Based on FTIR Spectra analysis of combination adsorbent between rice husk and coconut shell before activation, after activation and after contacted with Pb ion is just little difference. There is O-H stretching vibrations increasing after activation and after contacting with Pb ion then C = C bonds of aromatics is also getting stronger. In addition the existence of activation and contact with metal Pb lead to the emergence decrease force of C = O and Si. All of that can be represented in that picture below.

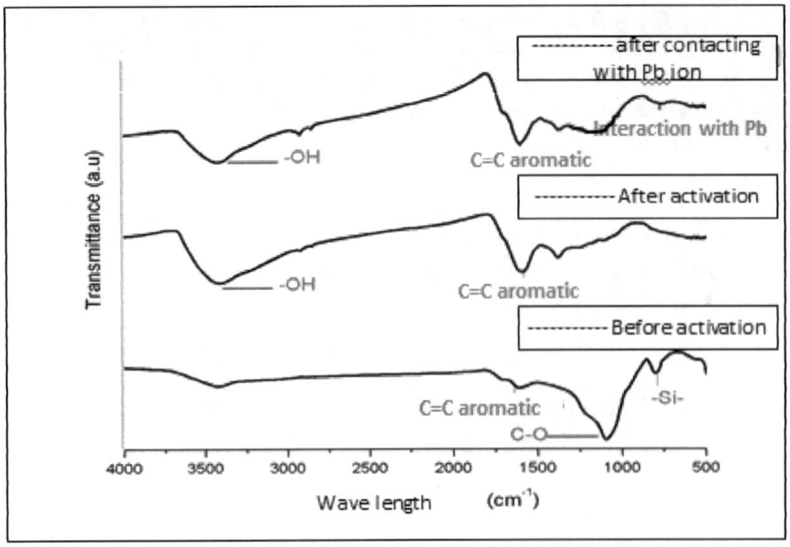

Figure 1. FTIR Spectra

3.3 Determination of Cd(II) calibration curve

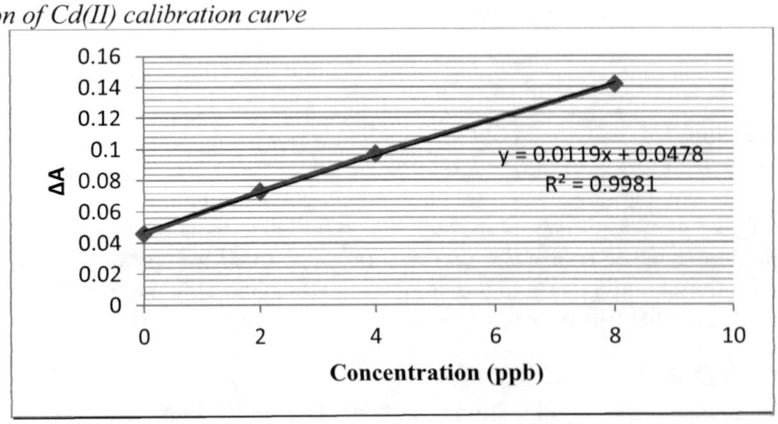

Figure 2. Calibration curve of Pb(II) solution.

From Figure 4. The equation y = 0.011 x + 0.047, where y is and x is ΔA concentration. The equation which will be used to determine the concentration of Pb (II) contained in the waste water simulation.The results of Pb(II) levels in simulated waste water.

Table 1. The results of Pb(II) levels in simulated waste water

Sample	[Pb(II)] (ppb)	ΔA	[Pb(II)] real (ppb)
Sample Pb(II)	4.99	0.101	49.96

IOP Conf. Series: Materials Science and Engineering **333** (2018) 012056 doi:10.1088/1757-899X/333/1/012056

3.4 The comparison of most effective adsorbents

Table 2. Optimum adsorbent mass comparison between rice husk and coconut shell

Mass Comparison	Initial [Pb(II)] (μg/L)	Final [Pb(II)] (μg/L)	Adsorbed [Pb(II)] (μg/L)	Adsorbed [Pb(II)] (%)
0:1	49.96	8.85	41.11	82.27
1:0	49.96	10.08	39.89	79.90
1:1	49.96	3.789	46.17	92.41
1:2	49.96	4.695	45.27	90.60
21	49.96	1.495	48.50	97.06

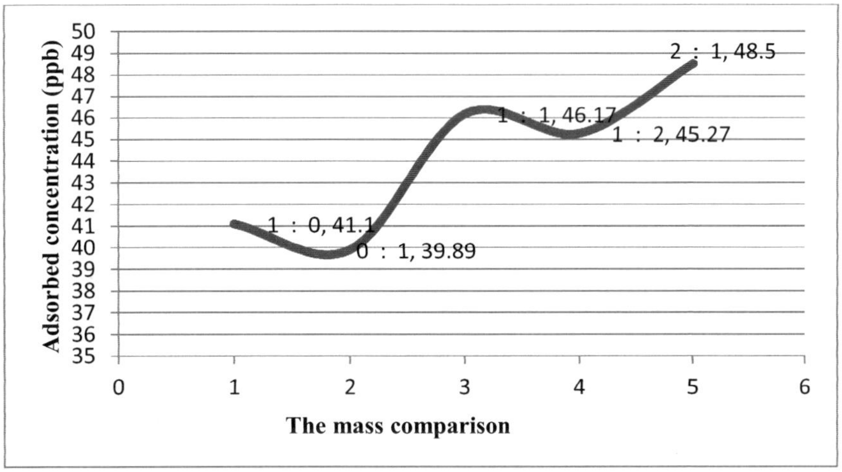

Figure 3. Optimum adsorbent mass comparison

The most effective adsorbent is the combination of active carbon rice husk: coconut shell 2:1 can adsorb the highest Pb ion amount . 48.50 μ g/L or 97.06 % adsorbed.

Table 3. Determination limit of detection

Blank	ΔA	Concentration	Deviation standar	LoD
A	0.1213	6.75		
B	0.1213	6.75		
C	0.1212	6.75		
D	0.1210	6.73		
E	0.1212	6.74		
Average			0.01	0.03

In the experiments that have been conducted obtained rating LOD 0.03 μ g/L.

4. Conclusion

The combination of active adsorbent rice husk and coconut shell can improve the ability of adsorbents to absorb heavy metals Pb (II), the most effective mass comparison of active adsorbent rice husk and coconut shell to adsorb Pb ion is 2:1 with the absorption of heavy metals Pb (II) of 97,06% and Solid-Phase

Spectrophotometry (SPS) is an effective method to determining the levels of heavy metals ions, Pb (II) results of active adsorbent the most effective in the level μ g/L with LoD amount 0,03 μ g/L

Aknowledgement

Authors wishing to acknowledge assistance from Head of Laboratory, Chemistry, Chemistry Study Program, Faculty of Teacher Training and Education, Sebelas Maret University, for using the facilities of the laboratory. This work was partially supported by the Mandatory Grant for S.S (2017), MRG Grant (2017) and Competency Grant for S.S (2016-2017) from the Ministry of Research, Technology and Higher Education, Indonesia.

References

[1] Halder J N, Islam M N 2015 *J. Environ. Human Health* **2**(1)
[2] Yoo H Y, Lee H Y, Jeong W J 2002 *J. Korean Environ. Sci.* **11** 993
[3] Sudarmaji, Mukono J, & Corie I P 2006 *Jurnal Kesehatan Lingkungan* **2**(2) 129
[4] Suksmerri 2008 *Jurnal Kesehatan Masyarakat* **11**(2)
[5] Nurhasni, Hendrawati, & Nubzah S 2014 *Jurnal Valensi* **4** (1) 36
[6] Chengwen S, Shuaiha W, Murong C, Ping T, Mihua S, Guangrul G 2014 *Suistanability* 85
[7] Bernand E, Jimoh A, and Odigure J O 2013 *Res. J. Chem Sci.* **3**(8) 231
[8] Matsuoka S, Nakatsu Y, Takehara K, Saputro S, & Yoshimura K. 2006 *Anal. Sci.* **22**1519
[9] Saputro S, Masykuri M, Mahardiani L, Mulyani B, Qorina I, Yoshimura K, Takehara K, Matsuoka S 2017 *IOP Conf. Ser.: Mater. Sci. Eng* **176** 012019
[10] S Saputro, M Masykuri, L Mahardiani, S Arini and A H Ramelan 2017 *IOP Conf. Ser.: Mater. Sci. Eng* **193** 012046
[11] Saputro S, Yoshimura K, Matsuoka S, Takehara K, & Narsito 2009 *Anal. Sci.* **25** 1445
[12] Yoshimura K, Waki H, & Ohasi S 1975 *Anal. Sci.* **23** 449
[13] Siburian A M, Pardede A S D, & Pandia S 2014 *Jurnal Teknik Kimia USU* **3 (4)**

IOP Conf. Series: Materials Science and Engineering **333** (2018) 012057 doi:10.1088/1757-899X/333/1/012057

Adsorption Effectivity Test of Andisols Clay-Zeolite (ACZ) Composite as Chromium Hexavalent (Cr(VI)) Ion Adsorbent

Pranoto[1], A Masykur, Y A Nugroho
Research Group of Analytical and Environmental Chemistry, Faculty of Mathematics and Natural Sciences
Universitas Sebelas Maret, Ir. Sutami street No. 36 A, Surakarta, Central Java, 57126, Indonesia.

E-mail: pakpranoto@gmail.com

Abstract. Adsorption of chromium hexavalent (Cr(VI)) ion in aqueous solution was investigated. This research was purposed to study the influence of the composition of ACZ, temperature activation, and contact time against adsorption capacity of Cr(VI) ion in aqueous solution. Determination of adsorption effectivity using several parameter such as composition variation of ACZ, contact time, pH, activation temperature, and concentration. In this research, andisol clay and zeolite has been activated with NaOH 3 M and 1 M, respectively. Temperature variation used 100, 200, and 400 °C. While composition variation ACZ used 0:100, 25:75, 50:50, 75:25, 100:0. The pH variation was used 2 – 6 and concentration variation using 2, 4, 6, 8, 10, and 12 ppm. Characterization in this research used such as UV-Vis, Surface Area Analyzer (SAA) and Acidity Analysis. Result of this research is known that optimum composition of ACZ was 50:50 with calcination temperature 100 °C. Optimum adsorption of Cr(VI) at pH 4 with removal percentage 76.10 % with initial concentration 2 ppm and adsorption capacity is 0.16 mg/g. Adsorption isotherm following freundlich isotherm with value K_f = 0.17 mg/g and value n is 0.963. Based on results, ACZ composite can be used as Cr(VI) ion adsorbents in aqueous solutions.

1. Introduction

Chromium hexavalent (Cr(VI)), is toxic and suspected to carcinogenic for all life form [1,2]. Cr(VI) usually found in electroplating, leather tanning, paint, steel fabrication, and textile dyeing [3]. Cr(III) and Cr(VI) state usually exist on environment. Cr(VI) have more high mobility then Cr(III) and is known to capable of permeating cell membrane and is a powerful mutagen for human and animal [4]. According to the Regulation of the Minister of Environment of Indonesia on the Standard of Wastewater Quality that the maximum limit of Cr(VI) ions in waste is 0.1 mg/L.

Conventional methods for the removal of Cr(VI) ion from wastewater include adsorption [5], coagulation-flocculation [6], ion exchange [7], filtration [8], and reverse osmosis [9]. The adsorption process is well known to be one of the most effective techniques and low cost for the removal of Cr(VI) ion in wastewater. The adsorption capacity of the sorbent is related to the radius, the specific surface area and the number of pores [4]. A few materials that can be used as adsorbent for Cr(VI) removal and easy to be found such as zeolite and andisol clay [10,11]. According to Sugiyarto (2013), andisol clay could be good adsorbent for heavy metals removal including Cr(VI) ion and could be

[1] pakpranoto@gmail.com

found in few mountain in java island such as Mt Papandayan, Mt Wilis, and Mt Arjuna. In andisol clay usually found allophane minerals that have high specific surface area, porosity and ion exchange capacity and could be apply to wastewater treatment [10]. Zeolite are the crystalline microporous aluminosilicates that have molecular sieve and adsorbing ability for small molecules such as water. Zeolite are good adsorbent for removal heavy metals ion and contamination/pollution control [12]. Zeolite have SiO_4 and AlO_4 tetrahedral framework that forming stable arrangement through oxygen atom [13].

Andisol clay is still considered to have a lower adsorption capacity against Cr(VI) ions than zeolites. So, the modification of andisol clay by forming andisol clay-zeolit composite is used to improve andisol clay adsorption capacity [14]. This is due to the excess of zeolites that have better adsorption capability than soil andisol may help to increase the adsorption capacity of the andisol soil [15].

In this research, zeolite will be added to the andisol clay to form the ACZ composite as Cr(VI) ion adsorbent and to study the effect of composition, activation, and contact time on Cr(VI) ion adsorption process in aqueous solution . ACZ composite are expected to have a better adsorption capacity that could be applied in industrial wastewater treatment.

2. Experimental

2.1. Adsorbent Preparation

Natural andisol clay from Mt. Lawu and zeolite from Wonosari are dried in the free air. After drying the soil andisol and zeolite mashed and sifted with 150 mesh then washed with aquadest then dried with temperature 105 ^{0}C for 4 hours. Dry andisol clay was activated by soaked in 3 M NaOH solution and stirred with heating 70 ^{0}C for 5 hours then washed with aquades until the pH of andisol clay was neutral then dried at 105 ^{0}C. The dried natural zeolite was activated by soaking in 1 M NaOH solution and stirred with 90 ^{0}C for 1 hour then washed with aquadest until the pH of the zeolite was neutral then dried at 105 ^{0}C.

2.2. ACZ Composite Preparation

Activated andisol soils and zeolite were mixed into 100 ml aquadest with weight variation (g/g) 0: 100 (ZAA), 25:75 (ZAB), 50:50 (ZAC), 75:25 (ZAD), 100: 0 (ZAE). The mixture is then stirred for 1 hour. Each mixture was activated with temperature variations of 100, 200, and 400 ^{0}C for 3 hours. The composite is used as the Cr(VI) ion adsorbent in the aqueous solution.

2.3. Testing Adsorption effectivity of ACZ to Cr(VI) ion

0.1 g of ACZ composite was introduced into 10 ml Cr(VI) solution 2 mg/L with pH 2, 3, 4, 5 and 6 variations. The best pH variations were tested using variation of contact time 20, 40, 60, 80, 100, and 120 minutes. The obtained mixture was filtered and the filtrate was added 1.5-Diphenylcarbazide and was performed by determining Cr(VI) values using UV-VIS Spectroscopy following SNI 6989.71: 2009.

2.4. Isotherm Adsorption analysis of ACZ composite

0.1 g of ACZ composite was introduced into 10 ml Cr(VI) solution with variation of concentration 2, 4, 6, 8, 10, 12 mg/L. The obtained mixture was filtered and the filtrate was added with 1.5-Diphenylcarbazide and determination of Cr(VI) values using UV-VIS spectroscopy.

3. Result and Discussion

3.1. Characterization of ACZ

3.1.1. Surface area, Acidity and Pore Size Characterization. Surface area, acidity and pore size characterization is used to determine effectivity of andisol-clay composite adsorption against Cr(VI) ion. Acidity of Adsorbent is used to determine active site of adsorbent that used to adsorb Cr(VI) ion. Surface area, acidity, pore size characterization of Andisol clay and zeolite will be showed in table 1.

Table 1. Surface area, acidity and pore size characterization of Andisol Clay and Zeolite

Adsorbent	Surface Area (m^2/g)	Acidity (mmol/g)	Pore Size (Å)
Natural Andisol Clay	91.30	0.47	25.06
Activated Andisol Clay	54.36	1.24	51.94
Natural Zeolite	45.95	3.88	76.56
Activated Zeolite	55.68	4.82	86.91
ACZ Composite	48.61	4.94	95.90

Table 1 showed that andisol clay and zeolite activated with NaOH could increase surface area and acidity of andisol clay and zeolite. Increasing surface area and pore size of andisol clay and zeolite showed that impurities on andisol clay and zeolite dissolved during the activation process. Increasing of acidity showed that active site from andisol clay and zeolite increase during activated process and could increase adsorption capacity.

3.2. Adsorption Effectivity of ACZ Composite on Cr(VI). The adsorbent effectivity test on Cr(VI) ion was performed to determine the optimum capability of the ACZ composite adsorbent. This adsorption effectivity test includes several tests such as the effect of contact time, pH, and the activation temperature of the ACZ composite. Adsorption effectivity of ACZ against Cr(VI) ion with variation composition and variation temperature will showed in figure 1 and 2.

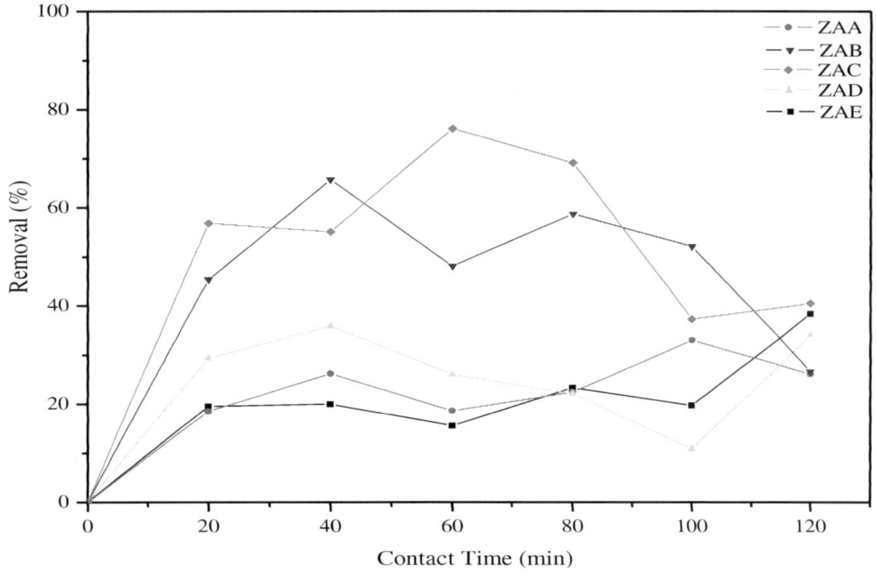

Figure 1. Adsorption effectivity of ACZ Composite against Cr(VI) ion with variation composition

IOP Conf. Series: Materials Science and Engineering **333** (2018) 012057 doi:10.1088/1757-899X/333/1/012057

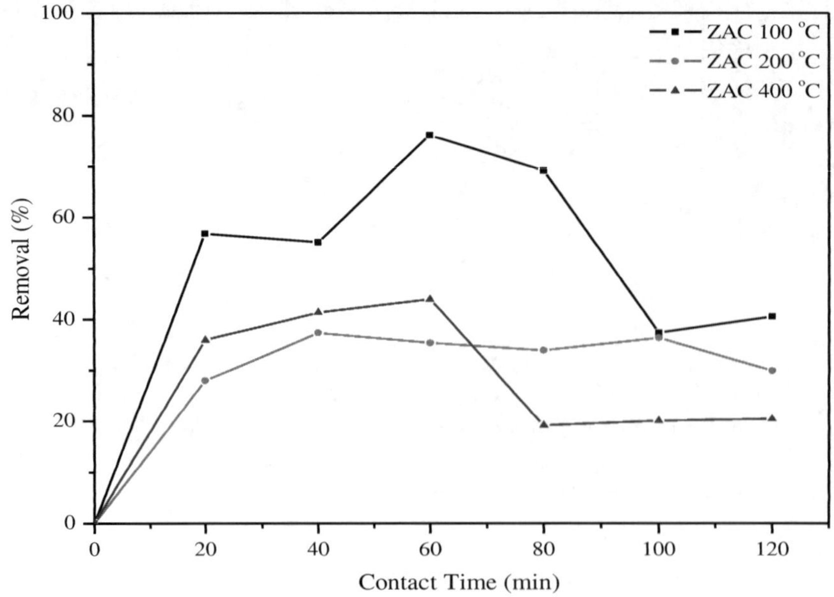

Figure 2. Adsorption effectivity of ACZ Composite against Cr(VI) ion with variation temperature

In figure 1 show that ZAC is optimum composition of ACZ composite with the optimum contact time of adsorption is 60 munites. At this composition, Cr(VI) ion removal is 76.10 % with initial concentration is 2 mg/L. ZAC that activated with different temperature showed in figure 2 that optimum temperature is 100 ^{0}C. It is caused by more higher temperature could broke the structure of ACZ composite. Rouzière *et al.* (2016) state the andisol clay would be broken in temperature above 500 ^{0}C caused –OH group release during calcination process [16,17]. Therefore, the adsorption capacity of ACZ composite would be decrease when temperature increase. The adsorption capacity of ZAC with temperature variation and pH variation will showed in figure 3 and 4.

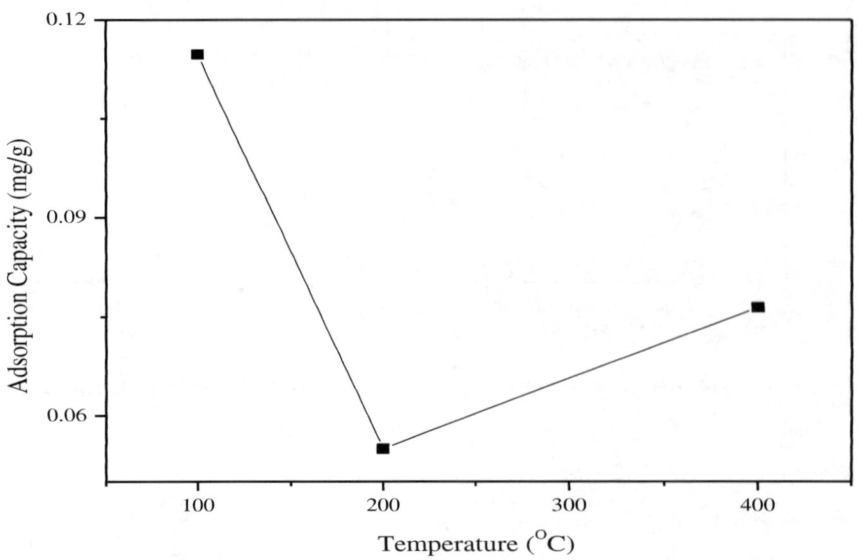

Figure 3. Adsorption capacity of ZAC with temperature variation

IOP Conf. Series: Materials Science and Engineering **333** (2018) 012057 doi:10.1088/1757-899X/333/1/012057

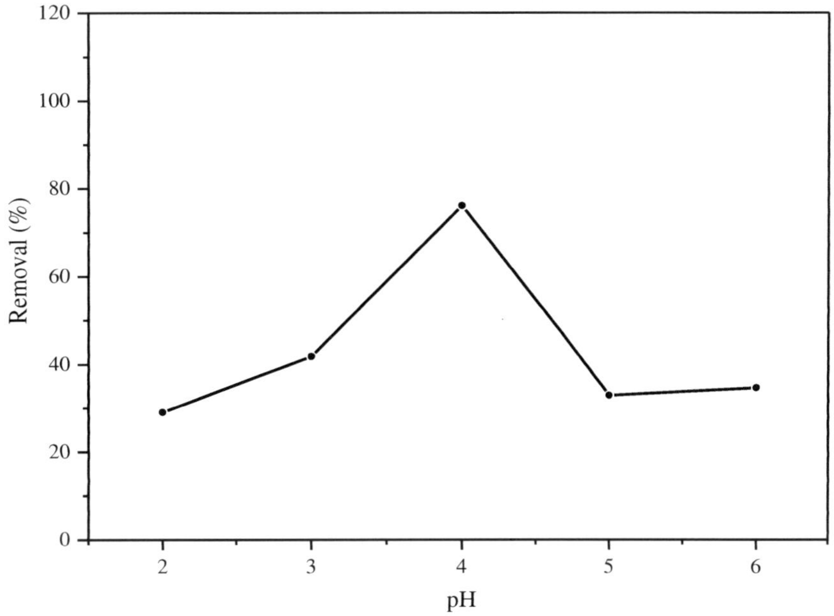

Figure 4. Adsorption of ZAC using pH variation

In figure 3 show that adsorption capacity of ZAC at 100 ^{0}C is 0.16 mg/g. Decrease of adsorption capacity caused by increasing activated temperature of ZAC that could breake the composite structure. In figure 4 show that the optimum condition of Cr(VI) ion adsorption at pH 4. In aqueous solution, Cr(VI) occur in few ion such as $Cr_2O_7^{2-}$, $HCrO_4^-$ and CrO_4^{2-} [18]. In pH 4, the dominant species of Cr(VI) is $HCrO_4^-$ that will interact with active site on ZAC. Reaction mechanism between ZAC with Cr(VI) ion will showed in scheme 1.

$$\equiv Al-OH \xrightarrow{H^+} \equiv Al-OH_2^+ + HCrO_4^-$$

$$\equiv Al-OH_2^+ + HCrO_4^- \longrightarrow \equiv Al-OH_2-HCrO_4$$

Scheme 1. Reaction mechanism between ZAC surface with Cr(VI) ion

In scheme 1 showed that interaction between ZAC surface with Cr(VI) ion that aluminol group will protonized and form $-OH_2^+$ which facilitate to form metal binding with $HCrO_4^-$. In ZAC surface, aluminol group will have different characteristic in different pH. Aluminol group could protonized at pH below 5, while at pH above 5 could relase H^+ ion and make aluminol group have negative charge [19].

3.3. Isotherm Adsorption Analysis of ACZ Composite. Isotherm adsorption analysis is used to determine adsorption perform on ZAC. Type of isotherm adsorption of ZAC against Cr(VI) ion will showed in figure 5 and figure 6.

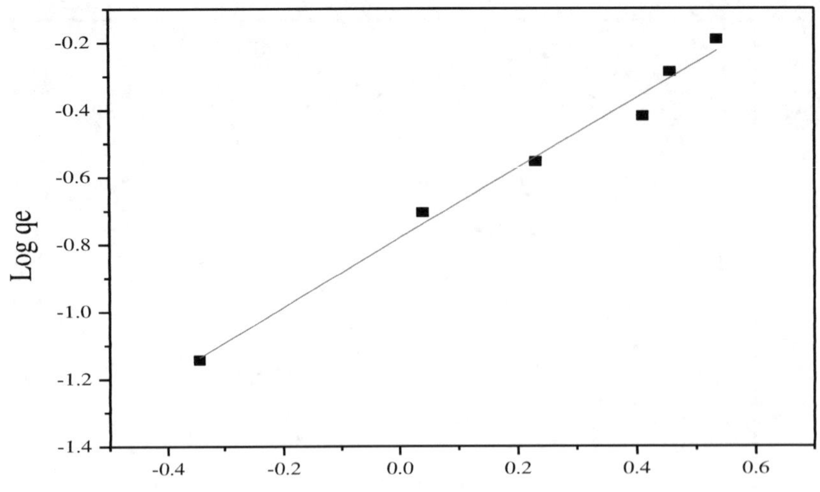

Figure 5. Freundlich isotherm adsorption of ZAC

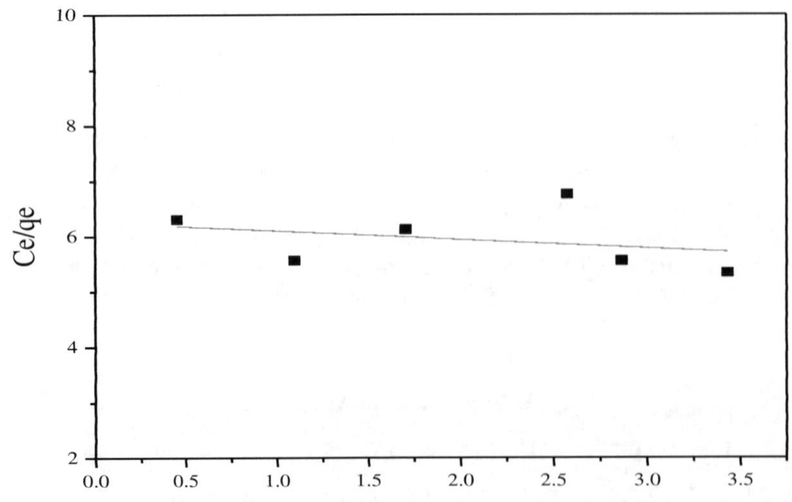

Figure 6. Langmuir isotherm adsorption of ZAC

In figure 5 showed that freundlich isotherm adsorption of ZAC against Cr(VI) ion. Freunclih isotherm analysis based on physically adsorption which assumed formed multilayer of adsorbed molecules on adsorbent surface. Freundlich isotherm of ZAC showed that r value is 0.994 with K*f* value is 0.17 mg/g and n values is 0.963 which mean the adsorption process is favorable [1]. In figure 6 showed that Langmuir isotherm adsorption of ZAC against Cr(VI) ion. Langmuir isotherm analysis based on chemically adsorption which assumed formed monolayer of adsorbed molecules on adsorbent surface. Langmuir isotherm of ZAC showed that r value is 0.327 with Q_{max} 1.25 mg/g and b value is 0.169 L/g. The isotherm adsorption for ZAC will showed in table 2.

Table 2. Isotherm adsorption for ZAC

Isotherm *Langmuir*		**Isotherm *Freundlich***	
Q_{max} **(mg/g)**	1.25	***Kf* (mg/g)**	0.17
***b* (L/g)**	0.169	***n***	0.963
r	0.327	**r**	0.994

In table 2 show that adsorption of ZAC against Cr(VI) ion following freundlich isotherm which means the adsorption process occurs in physically adsorption where the interaction between ZAC surface with Cr(VI) ion is van der waals bonding [20].

4. Conclusion

Result of this research is optimum composition of ACZ composite is 50:50 (ZAC) with activation temperature 100 ^{0}C and optimum contact time is 60 munite. Adsorption capacity of ZAC is 0.16 mg/g with removal of Cr(VI) ion is 76.10 %. Isotherm adsorption result is showed that adsorption process of Cr(VI) ion followed freundlich isotherm or adsorption on physically with freundlich coefficient (Kf) is 0.17 mg/g and r value of freundlich isotherm is 0.994. Physically adsorption based on weak van der waals interaction between ZAC adsorbent with Cr(VI) ion in aqueous solution. From this result we known that ZAC can be used as water treatment in wastewater.

Reference

[1] Bhattacharyya K G and Gupta S Sen 2006 *Ind. Eng. Chem. Res.* **45** 7232–40
[2] Wu X-W, Ma H-W and Zhang Y-R 2010 *Appl. Clay Sci.* **48** 538–41
[3] Ren Z, Xu X, Wang X, Gao B, Yue Q, Song W, Zhang L and Wang H 2016 *J. Colloid Interface Sci.* **468** 313–23
[4] Song W, Shi T, Yang D, Ye J, Zhou Y and Feng Y 2015 *J. Environ. Manage.* **162** 96–101
[5] Mthombeni N H, Onyango M S and Aoyi O 2015 *J. Taiwan Inst. Chem. Eng.* **0** 1–10
[6] Haydar S and Aziz J A 2009 *J. Hazard. Mater.* **168** 1035–40
[7] Sardohan T, Kir E, Gulec A and Cengeloglu Y 2010 *Sep. Purif. Technol.* **74** 14–20
[8] Zhao Y, Kunieda M, Obi N and Watanabe S 2017 *Precis. Eng.* **49** 211–9
[9] George J S, Ramos A and Shipley H J 2015 *J. Environ. Chem. Eng.* **3** 969–76
[10] Sugiyarto K 2013 *J. Environ. Earth Sci.* **3** 48–56
[11] Nguyen T C, Loganathan P, Nguyen T V, Vigneswaran S, Kandasamy J and Naidu R 2015 *Chem. Eng. J.* **270** 393–404
[12] Ates A and Akgül G 2016 *Powder Technol.* **287** 285–91
[13] Erdem E, Karapinar N and Donat R 2004 *J. Colloid Interface Sci.* **280** 309–14
[14] Shukla A and Kumar A 2007 *Sep. Purif. Technol.* **52** 423–9
[15] Churchman G J, Gates W P, Theng B K G and Yuan G 2006 *Developments in Clay Science* vol 1pp 625–75
[16] Rouzière S, Amara M S, Paineau E and Launois P 2016 *Nat. Commun.* **7** 254–78
[17] Wada K 1978 *Developments in Sedimentology* vol 26, ed C Oliver and M C Jamur (Totowa, NJ: Humana Press) pp. 147–87
[18] Tandon R K, Crisp P T, Ellis J and Baker R S 1984 *Talanta* **31** 227–8
[19] Wang G, Hua Y, Su X, Komarneni S, Ma S and Wang Y 2016 *Appl. Clay Sci.* **124–125** 111–8
[20] Atkins P W 1990 *Physical chemistry* (Oxford: Oxford Univ. Press)

IOP Conf. Series: Materials Science and Engineering **333** (2018) 012058 doi:10.1088/1757-899X/333/1/012058

The Variation of Catalyst and Carrier Gas on Anisole Deoxygenation Reaction

D Ariyani, Khoirina Dwi Nugrahaningtyas*, and E Heraldy
Solid State and Catalysis, Department of Chemistry, Faculty of Mathematics and Natural Sciences, Sebelas Maret University, Jl. Ir. Sutami No. 36 A Kentingan Surakarta 57126 Indonesia

E-mail: khoirina@mipa.uns.ac.id

Abstract. This research aims to determine the best catalyst and carrier gas in anisole deoxygenation reaction. The reaction was carried out over a flow system with a variation of catalyst CoMo A (CoMo/USY reduction), CoMo B (CoMo/USY oxidation-reduction), and CoMo C (CoMo/ZAA oxidation-reduction). In addition, variation of carrier gas nitrogen and hydrogen was investigated. The result was analyzed using Gas Chromatography-Mass Spectroscopy (GC-MS). The deoxygenation anisole result showed that CoMo A catalyst with hydrogen as the carrier gas has the highest total product yield (50.72 %), intermediate product yield (38.49 % in phenol and 6.99 % in benzaldehyde), and deoxygenation yield (5.24 %). The CoMo C catalyst exhibited the most selective deoxygenation product. The nitrogen carrier gas with the CoMo C catalyst has the best selectivity of benzene product (93.92 %).

1. Introduction

Bio-oil is an alternative fuel produced from biomass pyrolysis and has the potential to replace conventional fuels. Bio-oil has the advantage of being environmentally friendly when compared to conventional fuels because when combusted, bio-oil produces less pollution than fossil fuels. Bio-oil produces emissions of NO_x and SO_x lower than fossil fuels[1].

Bio-oil contains phenolic compounds (phenol, catechol, guaiacol, syringol, and their derivatives) which are came from the decomposition of lignin fractions of biomass. The major functional groups of lignin derived phenolics are hydroxyl (–OH) and methoxyl (–OCH_3) [2]. Anisole or methoxybenzene has a methoxy group (one of the major functional groups of lignin phenolics) to be used as a bio-oil model compound [3].

Bio oil has many undesirable characteristics such as high viscosity, low stability, high water content, high acidity, and corrosiveness. Those characters caused due to a high of molecular weight and oxygen content. So it cannot be used directly as engine fuel. High oxygen content of bio-oil also causes a low heating value being, which is only half value heating of fossil fuels [4], therefore it is necessary to upgrading bio-oil. Current, bio-oil upgrading techniques include deoxygenation [1].

The process of deoxygenation reaction needs the catalyst to accelerate of the reaction and optimize the product obtained. [5] Reported the product of anisole hydrodeoxigenation reaction with thermal treatment (without catalyst) obtained a lower yield of 0.31% than using a Mo-Co/ USY catalyst of 1.07 %. In the process of deoxygenation reaction, the catalyst serves to absorb the reactant molecule, so reactant reactions occur on the surface of the catalyst and produce the desired product. Different catalyst characters will result in different catalytic activities [2,4,3,6–11]. High temperature (250–400

°C) and high pressure (3–5 MPa) hydrotreating of bio-oil over sulfide CoMo catalyst can effectively reduce the oxygen content of phenolics in the bio-oil [2]. Reference [10] investigates anisole deoxygenation using Pt/Al_2O_3 catalyst. The result showed that the total deoxygenation product is 93.6 %. [12] reported the deoxygenation reaction using $CoMo/Al_2O_3$ catalyst obtained a total product yield of 96.8 %. This indicates that the $CoMo/Al_2O_3$ catalyst has a higher yield than Pt/ Al_2O_3. The effect of CoMo catalysts supported on various materials such as activated carbon, γ-Al_2O_3, HZSM-5, MCM-41, and SBA-15 has been studied by [4]. Among all the supported CoMo catalysts, CoMo/MCM-41 has the best catalytic activity. The size of medium pore diameter (66 Å) leads to a more effective and stable reaction in CoMo catalyst. Based on the studied, in this work used metal Co and Mo with variations catalyst support and methods that have synthesized by some researchers. The catalysts used include CoMo/USY reduction (CoMo A) [13], CoMo/USY oxidation-reduction (CoMo B) [14], and CoMo / ZAA oxidation-reduction (CoMo C) [15]. In the process of testing, the catalyst is formed into pellets to be easily separated from other products.

The catalytic activity may use a flow system [2] and a batch system [9]. The catalytic reaction using the flow system requires a carrier gas to assist the diffusion of the reactant to the catalyst surface. According to [16] the carrier gas also has an effect on the selectivity and stability of the catalyst, as well as being able to as reactants in the formation of the product [1]. [17] found that an increase in H_2 carrier gas pressure could improve the stability of the HZSM-5 catalyst for cracking n-heptane at 270 °C. In addition, [3] reported anisol conversion using H_2 carrier gas is better than using He carrier gas in the same reaction. This is influenced by better catalyst stabilization if the reaction uses H_2 carrier gas compared to He gas. [18] found that higher benzene selectivity was observed when using N_2 compared with H_2 as carrier gas.

2. Experimental

2. 1. Materials

The supported material that used in this research was NH_4-USY which was purchased from Tosoh Inc (Japan), and natural zeolite obtained from Klaten, Indonesia. Other materials were ammonium heptamolybdate and cobalt nitrate hexahydrate purchased from Merck (Germany). Anisole as model compound (methyl phenyl ether, Merck) had a minimum purity of 99%. Nitrogen and hydrogen gas purchase from PT. Samator Surabaya.

2. 2. Methods

Three types of catalysts were prepared. The preparation of the CoMo A catalyst follows [13], the CoMo B catalyst preparation was carried out by [14], and the CoMo C catalyst preparation was carried out by [15].

The deoxygenation catalytic was carried out by drying the catalyst at 110 °C for 3 hours. The catalyst samples are inserted into reactor, then further thermocouple in furnace heated with temperature 350 °C. After a temperature of 100 °C, anisole is included in feed reactor. Then the thermocouple on anisole was turned on at 160 °C and the carrier gas was set at 5 mL/ min. The deoxygenation reaction is carried out for around 1 h. The product is identified using GC-MS to know the type and quantity of the product.

3. Result and Discussions

3.1. Total Product Yield

Catalysts with different components (metal and support) will have different catalyst characteristics and different catalyst activity as well [8]. The catalyst character that influences the catalytic activity can be investigated from the acidity, surface area, the metal contents, and the average pore radius shown in Table 1. Catalytic activity using a flow system requires a carrier gas to react with the surface of the catalyst. The type of carrier gas can affect the deoxygenation process of anisole from its properties.

Table 1. Characteristics of CoMo catalysts [13–15]

Catalyst	Weight (%)		Total acidity (mmol/g)	Specific surface area (m^2/g)	Average pore radius (Å)
	Co	Mo			
CoMo A	0.49	0.69	11.21	497.00	16.23
CoMo B	0.764	0.244	8.61	362.55	19.07
CoMo C	0.229	0.148	0.31	25.48	15.19

Based on figure 1, it can be seen that the highest total product yield is anisole deoxygenation with CoMo A catalyst. The high catalytic performance of CoMo A is estimated because it has the highest acidity compared to other catalysts. The order of total acidity values is among CoMo A> CoMo B> CoMo C (see table 1). This trend is similar to the yield of the total product obtained. The higher acidity value of a catalyst indicates the more acidic sites that contributes to anisol hydrogenolysis [19]. This is according to the research that reported by [8]. [8] says that the higher acidity of the catalyst can increased catalyst activity.

The hydrogen as carrier gas has a higher total product yield than the nitrogen carrier gas in each catalyst, as shown in figure 1. The nitrogen carrier gas has an inert property, so it does not react to the catalytic process and acts only as a carrier gas [18]. The inert property caused by the covalent bonds originating from the three lone pairs electrons in the N_2 molecule. Unlike nitrogen gas, hydrogen gas may react which causes hydrogen not only acts as a carrier gas but also takes a part in hydrogenation and hydrogenolysis reactions to produces deoxygenation products.

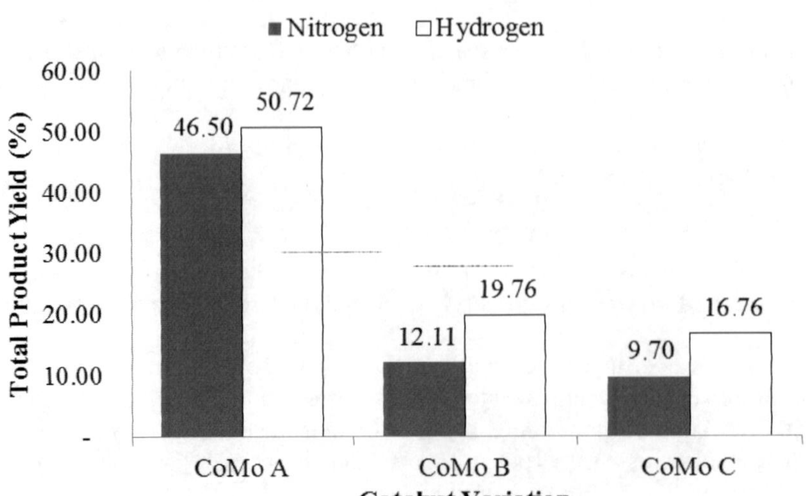

Figure 1. Total product yield on catalyst and carrier gas variations

3.2. Intermediate Product Yield

The yield of intermediate products in this study was represented by the largest intermediate product of phenol and benzaldehyde. [20] also stated in his research that the conversion of anisole over zeolites produces intermediate products such as benzaldehyde and phenol. The CoMo A catalyst has the highest intermediate product yield compared to the other catalysts, as shown in figure 2. This is influenced by Co and Mo metals on the catalyst. The Mo metal is present on the CoMo A catalyst is thought to be a $CoMoO_4$ particle so it is possible to bind the low oxygen, whereas the Co metal that support on the catalyst is not much (see table 1). Consequently, the available hydrogen (both in the reactant and carrier gas) can't be split and reacts for the hydrogenolysis process, so the reaction stops at the formation of intermediate products. Therefore, at the end of the reaction there are more intermediate products obtained. [19].

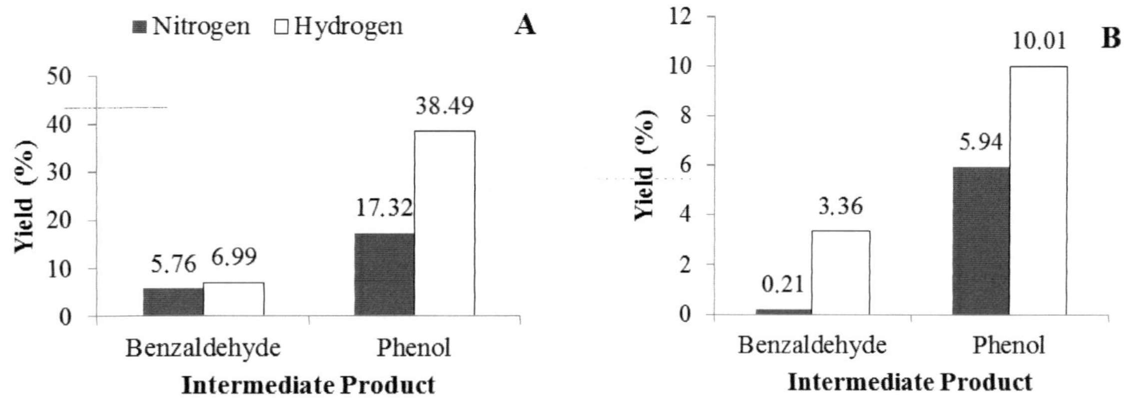

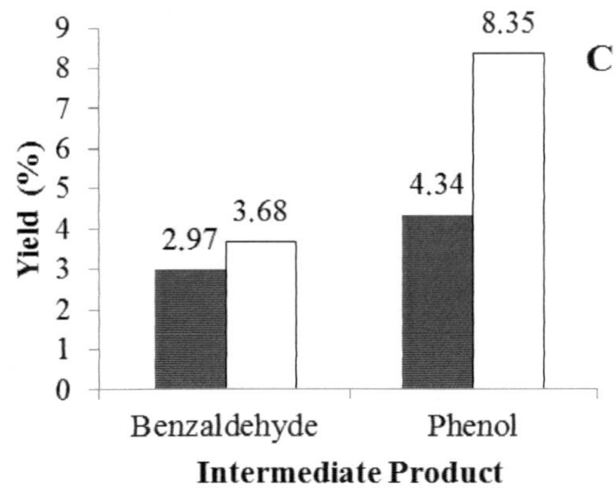

Figure 2. The intermediate product yield with carrier gas variation on the catalyst (A) CoMo A; (B) CoMo B; (C) CoMo C

The catalytic test using three different catalysts produces the highest intermediate product with hydrogen as a carrier gas in both benzaldehyde and phenol products. The use of a hydrogen carrier gas should be capable of providing hydrogen which can be broken down by Co metal and acting as a reactant to the hydrogenolysis reaction. The phenol product should be lower than the benzene product. Excessively low hydrogen flow rates result in high phenol yields.

3.3. Deoxygenation Product Yield

The yield of the anisole deoxygenation product is shown in figure 3. Deoxygenation thermally produces a lower deoxygenation product than catalytic deoxygenation. This case proves that the use of catalysts can improve deoxygenation products. The catalyst can increase the rate of reaction by lowering the activation energy through a different reaction mechanism. So that the reaction can run faster and obtain better results. The CoMo A catalyst has the highest total deoxygenation product yield than any other catalyst. The high value of deoxygenation product yield can be influenced by the large surface area of the catalyst and the amount of Mo metal contain on each catalyst. The order of the surface area and the amount of Mo metal contain are: CoMo A> CoMo B> CoMo C (see table 1). This trend is similar to the yield of anisole deoxygenation product obtained. Large surface area indicates that active site dispersion in the catalyst is increasing [21].

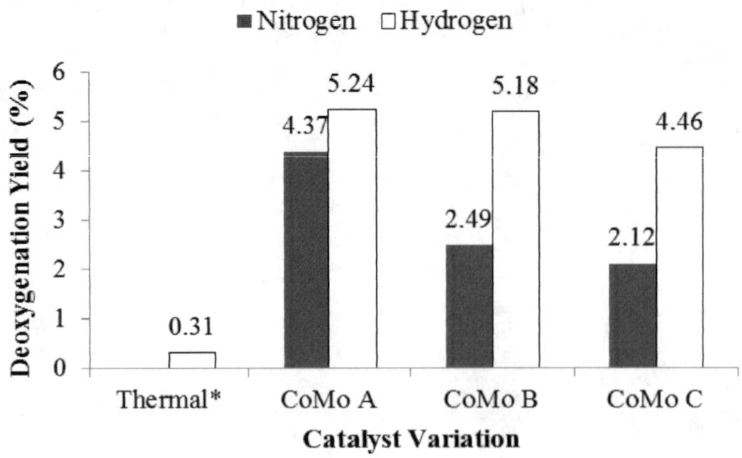

Figure 3. The yield of deoxygenation product by variation of carrier gas on thermal treatment [5] and variation of catalyst

The catalytic test using hydrogen carrier gas results the highest deoxygenation product yield for each catalyst. The hydrogen carrier gas in the catalytic test not only acts as a carrier gas but may also act as a reactant. Metal Co in catalyst can break down hydrogen gas into free hydrogen, then react to form deoxygenation product. In contrast to hydrogen, nitrogen can not participate in the deoxygenation process of anisole and only pure acts as a carrier gas. In the deoxygenation with N2 carrier gas, hydrogen that plays a role in the hydrogenation reaction is only provided by the support. The other hand, hydrogen supplied by the support are limited in number. so, its deoxygenation product yield is low.

3.4. Selectivity Deoxygenation Product

Figure 4 shows the selectivity of the anisole deoxygenation product using CoMo A, CoMo B, and CoMo C catalysts with hydrogen and nitrogen carrier gas. The CoMo C catalyst produces only two oxygen-free products benzene and toluene. This suggests that CoMo C catalyst is the most selective because it is able to inhibit the formation of the other products. It is possible because of the CoMo C catalyst has the smallest catalyst pore (15.19 Å) when compared with catalyst CoMo A and CoMo B . In a small pore radius, the interaction between the reactants and the catalyst is more effective and selective [22].

The selectivity of benzene products was higher when using nitrogen as carrier gas on CoMo A (75.55%) and CoMo C (93.92%) catalyst. However, hydrogen carrier gas has the highest selectivity of benzene product of 76% on CoMo B catalyst. This difference is cannot be explained.

IOP Conf. Series: Materials Science and Engineering **333** (2018) 012058 doi:10.1088/1757-899X/333/1/012058

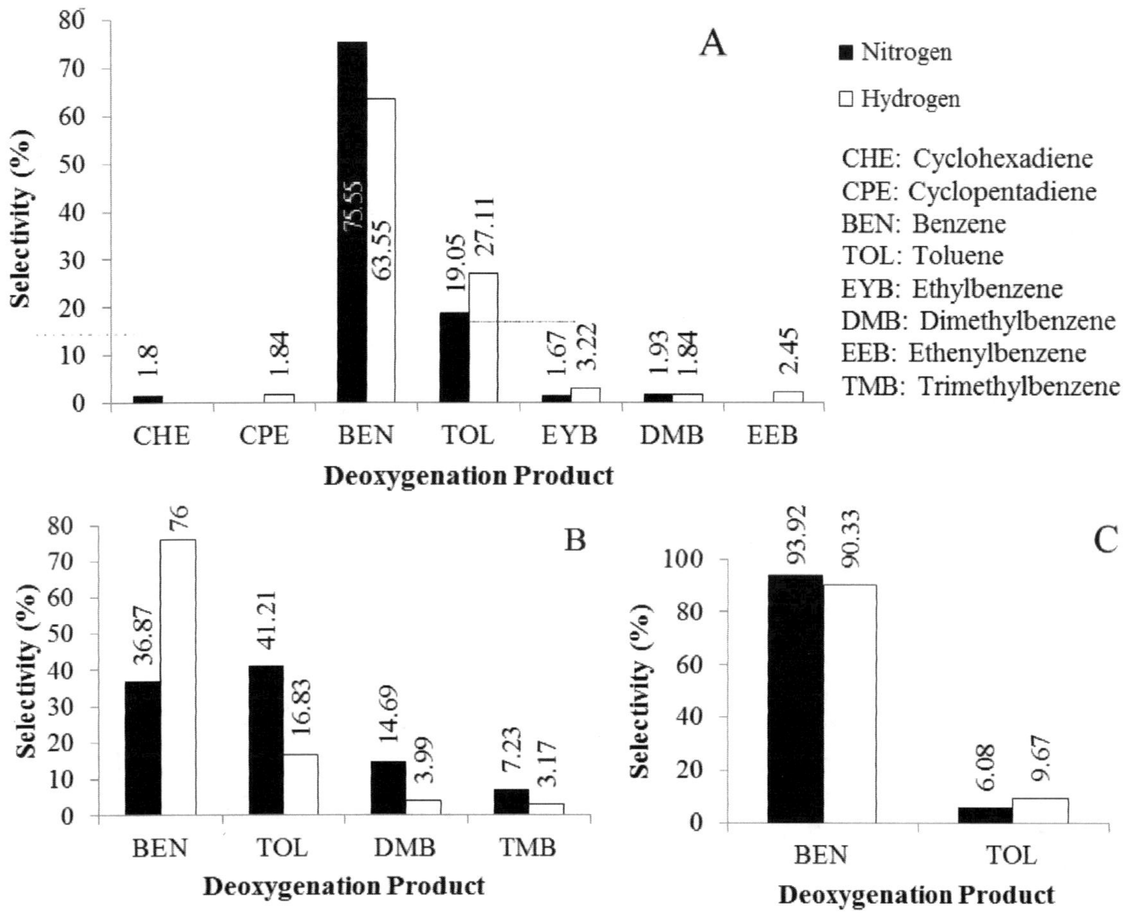

Figure 4. Selectivity of anisole deoxygenation product with CoMo A (A), CoMo B (B), and CoMo C (C) catalysts.

4. Conclusion

The CoMo A catalyst with hydrogen carrier gas has better activity. This is indicated by the result of the highest total product yield and deoxygenation product yield of 50.72 % and 5.24 %, respectively. In addition, anisole hydrodeoxygenation by using CoMo A catalyst and hydrogen carrier gas also has highest intermediate product yield of 38.49 % in phenol and 6.99% in benzaldehyde. The highest selectivity of benzene product is anisole deoxygenation by using a CoMo C catalyst and nitrogen carrier gas, i.e 93.92%.

Acknowledgments

This research is funded by KEMENRISTEKDIKTI under research Competent Grant 2017, number: 873/UN27.21/PN/2017. Therefore, I would like to thank the KEMENRISTEKDIKTI for funding this research.

Reference

[1] Tanneru S K and Steele P H 2015 *Renew Energy.* **80** 251
[2] Zhu X, Mallinson R G and Resasco D E 2011 *Appl. Catal. A Gen.* **379** 172

IOP Conf. Series: Materials Science and Engineering **333** (2018) 012058 doi:10.1088/1757-899X/333/1/012058

[3] Zhu X, Lobban L L, Mallinson R G and Resasco D E 2010 *J. Catal.* **281** 21
[4] Ahmadi S, Yuan Z, Rohani S and Xu C 2016 *Catal. Today* **269** 182
[5] Nugrahaningtyas K D, Hidayat Y and Prayekti P S 2015 *Jurnal Penelitian Saintek* **20** 19
[6] Deliy I V, Vlasova E N, Nuzhdin A L and Bukhtiyarova G A 2011 *Recent Res. Eng. Autom. Control* 24
[7] Boullosa-Eiras S, Lødeng R, Bergem H, Stöcker M, Hannevold L and Blekkan E A 2014 *Catal. Today* **223** 44
[8] Echeandia S, Pawelec B, Barrio V L, Arias P L, Cambra J F, Loricera C V. 2014 *Fuel* **117** 1061
[9] Sankaranarayanan T M, Berenguer A, Ochoa-Hernández C, Moreno I, Jana P, Coronado J M 2015 *Catal. Today* **243** 163
[10] González C, Marín P, Díez F V., and Ordóñez S 2016 *Ind. Eng. Chem. Res.* **55** 2319
[11] Pan Z, Wang R, Nie Z, and Chen J 2016 *J. Energy Chem.* **25** 418
[12] Viljava T, Komulainen R S and Krause A O I 2000 *Catalysis Today* **60** 83
[13] Rachmadhani 2016 *Pengaruh Proses Kalsinasi Dan Reduksi Terhadap Karakter Katalis Sistem Logam Pengemban (Co Dan Mo)/Ultra Stable Y Zeolite* (Indonesia: Universitas Sebelas Maret)
[14] Nugrahaningtyas K D, Hidayat Y, Patiha, Prihastuti N, Yelvi B, and Kalimah R U N 2017 *IOP Sci.*176
[15] Nugrahaningtyas K D, Trisunaryanti W, and Nurcahyo I F 2013 Indonesia ID paten: 2013/01757
[16] George N, Kamath B V, Basrur A G, and Krishnamurthy K R 1997 *React. Kinet. Catal. Lett.* **60** 79
[17] Meusinger J and Corma A 1995 *J. Catal.* **152** 189
[18] Al-Khattaf S, Ali M A and Al-Amer A 2008 *Energy and Fuels* **22** 243
[19] Yang Y, Ochoa-Hernández C, de la Peña O'Shea V A, Pizarro P, Coronado J M and Serrano D P 2014 *Appl. Catal. B Environ.* **145** 91
[20] Thilakaratne R, Jean-Philippe T and Brow R C 2016 *Green Chem* 2231
[21] Wibisono D 2011 *Pengaruh Kandungan Air Dan Laju Umpan Terhadap Perilaku Deaktivasi Katalis Hzsm-5 Dalam Reaksi Aseton Menjadi Hidrokarbon* (Indonesia: Universitas Indonesia)
[21] Rana M S, Ancheyta J, Rayo P and Maity S K 2004 *Catal. Today* **98** 151

IOP Conf. Series: Materials Science and Engineering **333** (2018) 012059 doi:10.1088/1757-899X/333/1/012059

CoMo/USY Catalysts Prepared by Precipitation Methods: Effect of pH Treatment

A I S Gusti, K D Nugrahaningtyas[1], and E Heraldy
Solid State & Catalysis, Chemistry Department, Sebelas Maret University, Faculty of Mathematics and Natural Sciences, Jl. Ir Sutami 36A Kentingan, Surakarta, 57126 Indonesia

Email: khoirina@mipa.uns.ac.id

Abstract. The Effect of pH treatment in the CoMo/USY catalysts prepared by precipitation was investigated. The pH of the precursor solution was controlled at pH 2, 7, and 8. Catalysts were characterized by XRF, XRD, N_2 adsorption, and IR spectroscopy. It was established that pH treatment in the catalysts significantly influences on the metal loading. Catalysts at pH 7 had maximal of amount total metals with particles type of CoO, MoO_3, Co, and Mo. Metal loading affects the decreasing of surface area, pore volume, and average pore radius. In addition, the number of hydroxyl group in the catalyst was less than USY followed by shifting of its wavenumber.

1. Introduction

Hydrotreatment is a hydrocarbon refining process producing clean fuels with better performance by using H_2 over a catalyst. The catalysts has an important role to providing active sites for the hydrotreatment reactions. The conventional catalysts such as $CoMo/Al_2O_3$ was the most commonly used catalysts in hydrotreatment studies. Although hydrotreatment reactions using these catalyst have a disadvantage, since the Al_2O_3 as supports was not contributed to the catalyst active sites. Therefore, zeolites may become a good choice as catalyst support, due to their important role to improve catalytic activity, selectivity, or stability by imparting shape selectivity [1,2]. In contrast to Al_2O_3, zeolite USY (ultra stable Y zeolite) as supports has acid site serving as active site for hydrogenation reactions. Therefore, the combination of CoMo active site with USY as supports will result in a multifunctional catalyst with better activity.

Catalyst properties and activity strongly depend on the preparation method. Catalysts which were obtained via precipitation method exhibited more high activity rather than via impregnation, due to high metal loading, high metal surface area, small metal particle sizes as well as a high number of acidic sites [3]. Variables of catalyst preparation such as types of precipitant, types of precursor, pH of precursor solution, aging time, and temperature have significant effects on the catalyst properties prepared by the precipitation method. According to Koo *et al.* [4], the pH of the precursor solution plays an important role in adjusting the loading amount and particle size of Ni metal and consequently has effects on the activity of catalyst. Salim *et al.* [5] prepared Ni/Celite catalyst by precipitation method for soybean oil hydrogenation and investigated effects of temperature of precipitation, aging temperature, and aging time on the textural properties of the prepared catalyst. The increases of aging time and aging temperature improve the interaction between the support and precursor of active metal to allow the preparation of catalyst with large surface area and high metal dispersion.

2. Experimental

2.1 Materials.

Commercial USY zeolite was obtained from Tosoh Japan Corporation in its NH_4^+ form. H-USY support was prepared by calcined at 550 °C for 1 h. The precursors of ammonium heptamolybdate tetrahydrate [$(NH_4)_6Mo_7O_{24}.4H_2O$] and Cobalt (II) nitrate hexahydrate [$Co(NO_3)_2.6H_2O$] were obtained from Merck in p.a grade. The solution of 25 % ammonia (NH_3) was used as precipitant agent in this investigation and it was obtained from Merck in p.a grade.

2.2 Preparation of catalysts.

CoMo/USY catalysts were prepared by precipitation method with some pH treatments. CoMo-USY-x catalysts where "*x*" is the pH treatment in the preparation process. H-USY (10 g) was added into a two neck flask with several distilled water. The pH was maintained by either (1) without treatment so the pH was at 2; (2) a solution of 25% ammonia was added hence the pH has changed into 7; And (3) pH at 8. Ammonium heptamolybdate tetrahydrate (0.981 g), cobalt (II) nitrate hexahydrate (0.777 g), and distilled water was added simultaneously hence the total volume has reached 50 mL. The slurry was kept stirred slowly and heated at temperature 60 °C for 2 h. The precipitate was filtered and dried at room temperature overnight. The prepared catalysts were dried by rotary evaporator, calcined in N_2 flow at 550 °C for 3 h, and then reduced in H_2 flow at 400 °C for 2 h. The atomic concentration of prepared catalysts were analyzed by energy dispersive X-ray fluorescene (Brüker S2 Ranger). The crystal structure of catalysts were analyzed by X-ray diffraction (XRD, Rigaku miniflex 600 benchtop, CuKα radiation). The particles type was estimated by Le Bail method using Rietica software. The textural characteristics was determined by SAA (quantachrome E 1600) through the nitrogen adsorption at -195.7 °C after the sampels were pretreated at 280 °C for 3 h under vacuum. The specific surface area was calculated using the BET method at a relative partial pressure *P/Po* of 0.05 - 0.3. The total pore volume and average pore radius were determined using the BJH model. The function group was analyzed by infrared spectroscopy (FTIR Shimadzu 8201 PC, KBr powder) with a resolution of 2 cm^{-1} in the range at 400 - 4000 cm^{-1}.

3. Result and Discussion

3.1 Physical properties of CoMo/USY catalysts.

The atomic concentration, specific surface area, total pore volume, and average pore radius of the CoMo/USY catalysts upon pH variation in the precursor solution are summarized in the Table 1. The atomic concentration of cobalt and molybdenum were influenced by the pH of precursor solution, due the pH treatment has changed precursor salts into hydroxides or other salts [4,6,7]. The metal atoms may attach to the pore surface of support, hence the metal concentration may affect the surface area, pore volume, and pore size of the catalysts [8].

Table 1. Physical properties of CoMo/USY catalysts

Sample	Atomic concentration (%)				Specific surface area (m^2/g)	Total Pore Volume (cm^3/g)*	Average pore radius (Å)
	Co	Mo	Si	Al			
CoMo/USY-2	1.96	4.99	22.42	5.88	497.90	0.31	13.36
CoMo/USY-7	3.98	3.39	20.72	5.45	476.15	0.34	15.21
CoMo/USY-8	4.35	0.04	34.64	7.47	500.34	0.38	16.22

*Pores with radius less than 1079.74 Å

The Co concentration was increased, whereas Mo concentration reduced with an increase in pH. The increase in pH has changed Co salts into hydroxides, consequently Co was easier to precipitate on the support surface. Molybdenum salts may change into hydroxide at pH 3 – 6, whereas at pH 7 molybdenum start to be dissolved and form MoO_4^{2-} ions. Therefore, when catalysts were filtered, the soluble Mo containing species was washed out, which would lead to the decrease in Mo concentration in the catalysts under pH 7 and 8. Both Co and Mo were showed high concentration in the CoMo/USY-7 catalysts, due to formation of hydroxides during the preparation process. However, Co and Mo hasn't changed into hydroxides at pH 2, but they were obtained in the catalysts CoMo/USY-2. It was indicated that Co and Mo were trapped in USY pores by other supporting method, such as impregnation under pH 2 [9,10].

Specific surface area, total pore volume, and average pore radius were reduced with an increase in metal concentration (Co and Mo). Metal atoms has entered and attached into USY pores during the preparation process of the catalyst, thus lead to the decrease of pore size and surface area. However, different phenomena occurred at CoMo/USY-2, which has higher specific surface area rather than CoMo/USY-7, whereas total pore volume and average pore radius were smallest rather than others. Typically, physical properties such as metal concentration and pore volume have a significant effect on catalytic performance [11]. Therefore, the CoMo/USY-7 catalysts is expected to have good catalytic activity due to the high metal concentration and pore volume.

3.2 Particles type of CoMo/USY catalysts.

Figure 1 illustrates the XRD pattern of the USY supports and CoMo/USY catalysts prepared by precipitation method. According to Figure 1, the XRD pattern of catalysts CoMo/USY under pH 2, 7, and 8 were similar to USY supports. That was indicated that metal loading was not changed the crystal structure of USY, significantly. The XRD pattern of both Co and Mo were not observed since its crystallite size or intensity were extremely small.

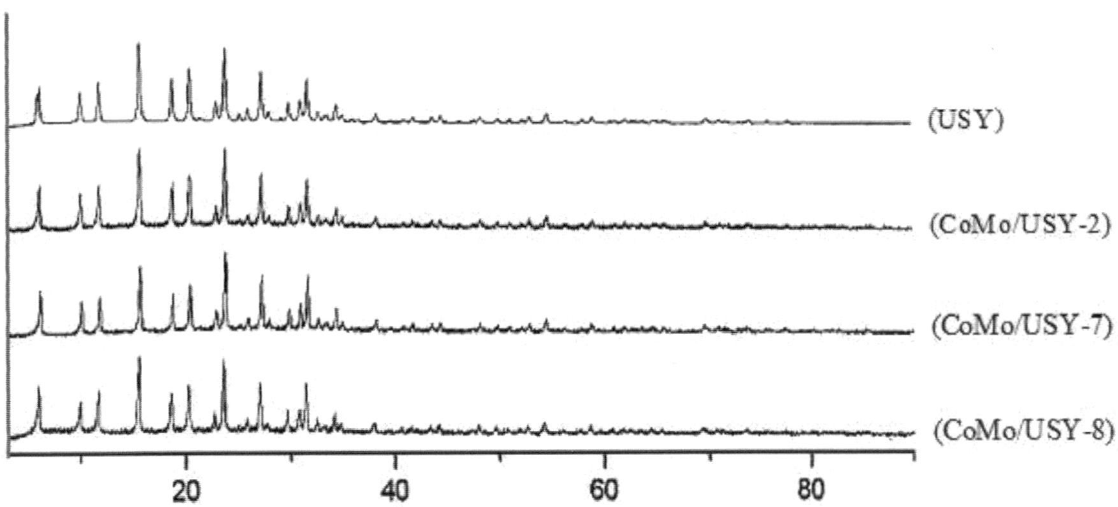

Figure 1. XRD pattern of USY and CoMo/USY catalysts

The particles type of the CoMo/USY catalysts upon pH variation in the precursor solution are summarized in the Table 2. The pH of precursor solution were not significantly influenced the particles type of catalysts, due the precursor salts and hydroxides were relatively easy to decomposed by calcination. According to Table 2, zeolite NaY and faujasite H particles type were dominant (more than 97 %) in the catalysts CoMo/USY, thus explained XRD pattern in Figure 1. In small number MoO_3, Co, and Mo particles type were obtained on the catalysts based on refinement analysis of X-ray diffraction. Molybdenum has interacted with more oxygen atoms, ammonia, and OH^- ions in the form of salts or hydroxides compared to cobalt. Therefore, some molybdenum were not clearly activated and obtained as MoO_3 particles type. CoO particles type was obtained only in the CoMo/USY-7 catalysts. It might be occurred due to the metal agglomeration at CoMo/USY-7 catalysts, which has higher metal concentration [8]. Particles type of metal have a significant effect on catalytic performance [12]. Augustine *et al.* [13] stated that the active metal on the cracking reaction was pure metal particles, therefore Co and Mo particles type estimated to be contribute as active metal which is increase catalytic activity.

Table 2. Particles type of CoMo/USY catalysts

Standard phase	Weight molar particles (%)		
	CoMo/USY-2	CoMo/USY-7	CoMo/USY-8
Zeolit Y ICSD #33599	✓	✓	✓
Zeolit Y ICSD #26920	✓	✓	✓
CoO ICSD #245320	-	✓	-
MoO_3 ICSD #644068	✓	✓	✓
$CoMoO_4$ ICSD #281235	-	-	-
Co ICSD #622439	✓	✓	✓
Mo ICSD #173127	✓	✓	✓
CoMo ICSD #624215	-	-	-

3.3 Function group of CoMo/USY catalysts

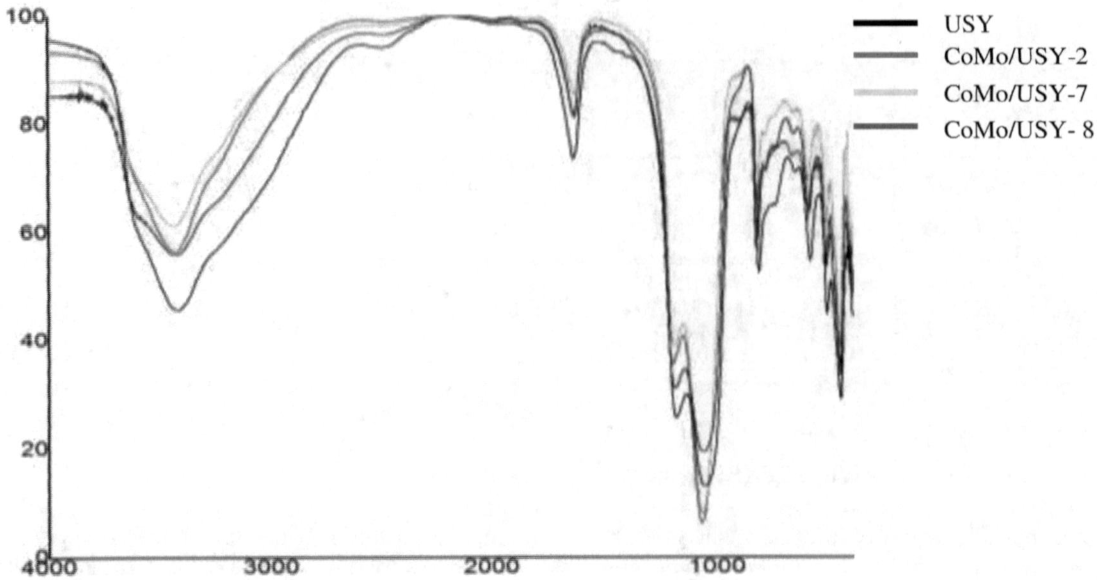

Figure 2. Infrared spectrum of USY and CoMo/USY catalysts

IOP Conf. Series: Materials Science and Engineering **333** (2018) 012059 doi:10.1088/1757-899X/333/1/012059

According to Figure 2, metal loading was not influenced the infrared spectrum of CoMo/USY catalysts, significantly compared to USY supports. It was indicated that metal loading was not changed the function group on the catalysts. The infrared spectrum of catalysts shows several bands attributed to hydroxyl group, internal tetrahedral structure, and external tetrahedral structure of USY, as presented in Table 3.

Cobalt and molybdenum loading was influenced OH stretch band at 3400 cm^{-1}, that attributed to Brönsted acid site on the catalysts. The transmittance of OH stretch band was increased with an increase of metal concentration on the catalysts. That was indicated that metal atoms was replaced H^+ ion, hence the number of Brönsted acid site were reduced on the catalysts. The transmittance of OH stretch band was highest in CoMo/USY-7, due to highest metal concentration. CoMo/USY-8 and CoMo/USY-2 has similar transmittance of OH stretch band, whereas CoMo/USY-8 has lower metal concentration. It was indicated that Brönsted acid sites were reduced significantly in the CoMo/USY-8 catalysts [14,15]. OH stretch band was slightly shifted to high wave number in an increase of metal concentration on the catalysts, which may confirm the proximity between metal atoms and OH groups [15]. That was indicated that metal loading may increase the OH group strength on the catalysts. The wavenumber of OH stretch was highest in CoMo/USY-7 catalysts, due to highest metal concentration. The amount and strength of OH group have a significant effect on catalysts acidity. Therefore, CoMo/USY-7 is expected to have a good catalytic performance in hydrogenation reaction.

Table 3. Infrared spectroscopy absorption of USY and CoMo/USY catalysts

Function group	Wave number (cm^{-1})			
	USY	CoMo/USY-2	CoMo/USY-7	CoMo/USY-8
O-H stretch	3396.79	3421.87	3426.69	3416.08
O-H bend	1634.74	1630.88	1626.06	1628.95
T-O-T asym stretch	1178.56	1195.92	1186.27	1184.34
T-O asym stretch	1051.25	1067.65	1062.82	1058.97
T-O-T sym stretch	816.89	828.46	826.53	822.68
T-O internal sym stretch	657.75	680.90	669.33	663.54
T-O internal bend	454.26	456.18	455.22	455.22

4. Conclusions

CoMo/USY catalysts were prepared by precipitation method with various pH treatment. Metal loading was strongly influenced by the pH of precursor solution. Both concentration of cobalt and molybdenum were highly obtained in the catalysts under pH 7. As the metal concentration increases, the specific surface area, total pore volume, and average pore radius were reduced on the catalysts. The pH of precursor solution were not significantly influenced the particles type of catalysts, since pure metal particles of Co and Mo were obtained on the catalysts for all pH treatments. Metal loading was influenced OH stretch band at 3400 cm^{-1}, that attributed to Brönsted acid site on the catalysts. The highest OH stretch wavenumber was occurred on the CoMo/USY-7 catalysts, indicating the strongest Brönsted acid site. Therefore, CoMo/USY-7 catalyst is considered a promising catalyst with good activity in hydrotreatment reaction.

Acknowledgments
This research is funded by KEMENRISTEKDIKTI under research Competent Grant 2017, number: 873/UN27.21/PN/2017. Therefore, I would like to thank the KEMENRISTEKDIKTI for funding this research.

References

[1] Maesen T L M, Calero S, Schenk M, and Smit B 2004 *J. Catal.* **221** 241
[2] Ding L, Zheng Y, Zhang Z, Ring Z, and Chen Z 2007 *Appl. Catal. A.* **319** 25
[3] Ahmad R, Schrempp D, Behrens S, Sauer J, Döring M, and Arnold U 2014 *Fuel Process. Technol.* **121** 38
[4] Koo K Y, Park M G, Jung U H, Kim S H, and Yoon W L 2014 *Int. J. Hydrogen Energy* **30** 1
[5] Salim V M M, Cesar D V, Schmal M, Duarte M A I, and Frety R 1995 *Stud. Surf. Sci. Catal.* **91** 1017
[6] Song S, Sheng Z, Liu Y, Wang H, and Wu Z 2012 *J. Environ. Sci.* **24** 1519
[7] Van Dillen A J, Terörde R J A M, Lensveld D J, Geus J W, and Jong K P 2003 *J. Catal.* **216** 257
[8] Barton R R, Carrier M, Segura C, Fierro J L G, Escalona N, and Perreti S W 2017 *Appl. Catal., A.* **540** 7
[9] Anderson A and Boundart M 1981 *Catalysis Science and Technology* (Heidelberg: Spinger Verlag)
[10] Alsobaai A M, Hameed B H, and Zakaria R 2006, *Proc. 1st Int. Conf. on Natural Resources Engineering and Technology* (Putrajaya) p 243
[11] Nikulshin P A, Mozhaev A V, Pimerzin A A, Konovalov V V, and Pimerzin A A 2012 *Fuel* **100** 24
[12] Bose S and Das C 2016 *Appl. Catal., A.* **512** 15
[13] Augustine R L and Tanielyan S K 1996 *J. Mol. Catal., A.* **112** 93
[14] Mhamdi M, Zine S K, Ghorbel A, Marceau E, Che M, Taarit Y B, and Villain F 2005 *Stud. Surf. Sci. Catal.* **158** 749
[15] Mendes P S F, Lapisardi G, Bouchy C, Rivallan M, Silva G M, and Ribeiro M F, 2015 *Appl. Catal., A.* **504** 17

IOP Conf. Series: Materials Science and Engineering **333** (2018) 012060 doi:10.1088/1757-899X/333/1/012060

Preparation, Characterization, and Catalytic Activity of MoCo/USY Catalyst on Hydrodeoxygenation Reaction of Anisole

K D Nugrahaningtyas[1], R S R Suharbiansah[2], F Rahmawati[1]
[1]Solid State & Catalysis, Chemistry Department, Faculty of Mathematics and Natural Sciences, Sebelas Maret University, Jl Ir. Sutami No. 36 A, Kentingan, Surakarta, 57126, Indonesia
[2]Chemistry Department, Faculty of Mathematics and Natural Sciences, Sebelas Maret University, Jl Ir. Sutami No. 36 A, Kentingan, Surakarta, 57126, Indonesia

E-mail: khoirina@mipa.uns.ac.id

Abstract. This research aims to prepare, characterize, and study the catalytic activity of Molybdenum (Mo) and Cobalt (Co) metal with supporting material Ultra Stable Y-Zeolite (USY), to produce catalysts with activity in hydrotreatment reaction and in order to eliminate impurities compounds that containing unwanted groups heteroatoms. The bimetallic catalysts MoCo/USY were prepared by wet impregnation method with weight variation of Co metal 0%, 2%, 4%, 6%, 8%, and Mo metal 8% (w/w), respectively. Activation method of the catalyst included calcination, oxidation, reduction and the crystallinity was characterized using X-ray diffraction (XRD), the acidity of the catalyst was analyzed using Fourier Transform Infrared Spectroscopy (FT-IR) and gravimetry method, minerals present in the catalyst was analyzed using X-Ray Fluorescence (XRF), and surface of the catalyst was analyzed using Surface Area Analyzer (SAA). Catalytic activity test (benzene yield product) of MoCo/USY on hydrodeoxigenation reaction of anisole aimed to determine the effect of Mo-Co/USY for catalytic activity in the reaction hydrodeoxigenation (HDO) anisole. Based on characterization and test of catalytic activity, it is known that catalytic of MoCo/USY 2% (catalyst B) shows best activities with acidity of 10.209 mmol/g, specific area of catalyst of 426.295 m^2/g, pore average of 14.135 Å, total pore volume 0.318 cc/g, and total yield of HDO products 6.06%.

1. Introduction

The conventional petroleum component comprises of 50-98% hydrocarbon compounds and non-hydrocarbon compounds with 83 – 87% carbon content, 11 – 14% of hydrogen, 3% of sulfur, 2% oxygen, and 1% of nitrogen [1]. Hydrotreating is a cracking process by the addition of hydrogen gas to saturate the hydrocarbons and eliminate sulfur, oxygen, nitrogen and metal impurities. One of the hydrotreating processes is hydrodeoxigenation (HDO). Hydrodeoxigenation (HDO) aims to remove the oxygen contaminant compounds that attached to hydrocarbon compounds. HDO can be done in two ways, which include the conventional thermal cracking and the catalysis systems. The catalysis systems, will conduct great results due to lower temperature and pressure usages [2]. These properties are formed by a supported material and addition of metal in a solid carrier such as silica-alumina, alumina and zeolite [3].

One type of the supported material that can be used for the process is the Ultra Stable Y-zeolite (USY). There was recent studies about Kazuhisa, et al. studied Si/Al_2O_3 catalysts with several variations of supporting materials, such as Al_2O_3 gamma, mordenite, USY and ZSM-5 [4]. The supporting material USY is a type of zeolite-Y can be synthesized so that it has a higher pore structure and crystallinity. Types of metals commonly used in the petroleum industry include Ni, Co, Mo, Pt and Pd with carrier USY or other synthetic zeolites.

Mo metal has a vital role for deoxygenation reaction [5]. Mo metal will bind the hydroxyl group of phenol to the hydrogenolysis process. The hydrogenolysis process occurs the H_2O released and formed the benzene products. Metal Co plays a role for the hydrogenation reaction, the content of Co metal on Mo-Co/USY catalyst were variated because the higher Co content the greater the ability to break the hydrogen bond [6].

Methods for the incorporation of metals onto the supporting material include ion exchange methods [7], impregnation methods [8] and precipitation methods [9]. Each of these methods will give the results of different catalyst characters. The parameter condition of catalyst preparation is influenced by temperature, time of reflux, pressure as well as ratio of the added metal. Due to these factors will affect the character of the catalyst, such as the crystallinity properties, the amount of metal that was carried [10], specified surface area [11], particle size [12], pore size distribution, volume pore, mechanical strength, and the acidity properties [13].

Based on the results of these studies, the catalytic activity of MoCo/USY was tested on anisole for hydrodeoxygenation reaction. The main reason anisole as a feed compound based on the simplicity of the molecular structure that already represent the compound of the petroleum constituent component, since it bond with the oxygen atom of methoxy representing the main reaction and the benzene ring which may be represented for the hydrogenation reaction. HDO anisole products are cyclohexane, cyclohexene, benzene, toluene, and phenol [6].

2. Experimental

MoCo/USY was prepared by using wet impregnated method. First, 10 g of H-USY were soaked in aquabidest with addition of solution *ammonium heptamolybdate tetrahydrate* $[(NH_4)6Mo_7O_{24}.4H_2O]$ and solution *Cobalt (II) nitrate hexahydrate* $[Co(NO_3)_2.6H_2O]$ with weight variation of Co metal 0%, 2%, 4%, 6%, 8%, and Mo metal 8% (w/w), which signed later as sample of A, B, C, D, and E, respectively. The solution was mixed in total volume of 50 mL aquabidest. The solution was refluxed at 60 °C for 2 hours, the solution is filtered and the filtrate was obtained. Filtrate were dried using rotary evaporator at pressure of 200 mBar, with constant temperature of 48 °C until turned to dried powder. Dried powder samples were thermally calcined, oxidated and reducted. The following procedure for catalytic activity amount of 0.5 g sample was set into the reactor column. The hydrogen gas flow rate was adjusted to 5 mL/min for 1 H and the feed through the reactor containing with the catalyst. Same step procedure was repeated for thermal reaction (without samples), USY, and $PtAl_2O_3$ as commercial catalyst. Several characterization was determined by Fourier Transform-Infra Red (FT-IR) type shimadzu 8201 PC, X-Ray Diffraction (XRD) Phillips type X'Pert, Surface Area Analyzer (SAA) NOVA 1200e.

2.1. Catalytic activity product analysis.

Gas Chromatography – Mass Spectroscopy (GC-MS) type 2010 SE data were analyzed based on the amount of peak band on chromatogram showing the number of compounds contained in the sample. From the GC data, the product distribution was obtained and total product conversion, HDO product conversion, product intermediate conversion and product selectivity were calculated with the following formula percentage of total product shown in equation (1):

$$Percentage\ of\ total\ product = Product\ Area\ (\%)\ X\ \frac{Product\ Volume}{Initial\ Volume} \quad (1)$$

IOP Conf. Series: Materials Science and Engineering **333** (2018) 012060 doi:10.1088/1757-899X/333/1/012060

The activity (yield) of HDO product was the amount of conversion from anisole to benzene and toluene products. Yield of benzene product and total yield of HDO were shown in equation (2) and (3):

$$Yield\ of\ benzene\ product = \%\ Area\ of\ Benzene\ Product\ X\ \frac{Product\ Volume}{Initial\ Volume} \quad (2)$$

$$Total\ yield\ of\ HDO = yield\ of\ benzene\ Product + yield\ of\ toluene\ Product \quad (3)$$

The calculation of the yield of toluene product is the same as the calculation of the yield of the benzene product.

3. Result and Discussion

3.1. Functional Groups

The FT-IR analysis of the MoCo / USY samples was conducted qualitatively to determine the type of functional groups present in the MoCo/USY samples (Figure 1). Based on Figure 1, The FT-IR results show no significant differences in functional groups for each samples. The variation of the addition of Co metal on the catalyst shows no any change in the resulting FT-IR spectra. The absorptions in the spectra was matched with previous results of Bhatia [14], such as the absorption at wavenumber region of the 3450 – 3400 cm^{-1} signified the presence of hydroxyl –OH stretch, adsorption at wavenumber region of 1650 – 1600 cm^{-1} indicated the presence of a bending Si – OH vibration. The presence of functional group adsorption bands of a wavenumber of around 3400 cm^{-1} in H-USY was estimated as a Brønsted acid site of H-USY. That is in accordance with Weitkamp [15]. The FT-IR spectra results analysis shows the O-H stretch appeared in the 3450-3400 cm^{-1} wavenumber. In addition, in the wavenumber area of 1650 – 1600 cm^{-1} indicated the existence of –OH bending accordance with Korkuna [16]. Then the spectra occur in the wavenumber region of the 1100 – 1050 cm^{-1} was detected as a bond of Si – O – Si stretching. Therefore further analysis of other catalytic characters such as acid site values was required.

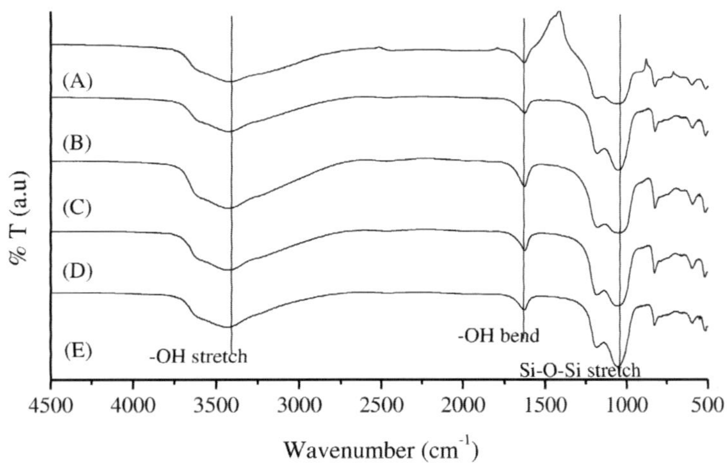

Figure 1. FT-IR spectra of MoCo/USY: sample of A (A), sample of B (B), sample of C (C), sample of D(D), and sample of E (E) after impregnation process.

Based on Figure 2, it can be seen that the FTIR spectras show the relationship between the adsorbed ammonia in the solid surface and the bonding vibrations spectra. The change in vibrational wavenumber can be used to differentiate Brønsted or Lewis acidic sites. The spectra after ammonia adsorption showed the effect of acidity on Brønsted and Lewis acid characters in the samples. The initial absorption of the image spectra remains, such as absorption in the 3450-3400 cm^{-1} wavenumber region indicating the -OH stretch, then the 1650- 1600 cm^{-1} denoted as a bending -OH vibration. After the adsorption of ammonia, the estimated absorption area of Brønsted acid site increased in the wavenumber area of 3593.54-3229.94 cm^{-1} showing the absorption -O-H stretch and attached to the uptake of -N-H amine. This phenomenon occurs in every sample produced. Variations in the addition of Co and Mo metal do not indicate the presence of new functional groups, but produce different absorption rates. The FT-IR spectra analysis in Figure 2 shows that the absorption present in the wavenumber region of 1100-1050 cm^{-1} is the Si – O – Si stretching bond [15]. In the new absorption the amine -N-H stretching function group was found in the wavenumber region of 3229.94 cm^{-1} and in the wavenumber region 1402.31 cm^{-1}.

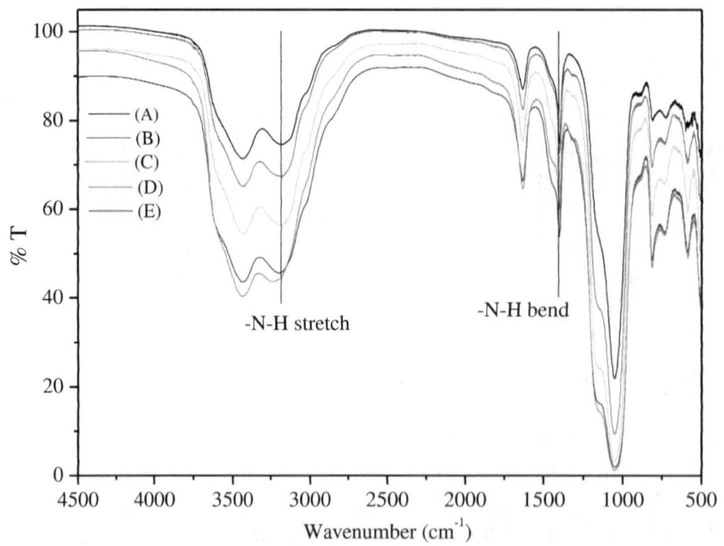

Figure 2. FT-IR spectra of MoCo/USY: sample A (A), sample B (B), sample C (C), sample D(D), and sample E (E) after ammonia adsorption.

3.2. The Powder X-Ray Diffraction and Structural analysis

XRD analysis was performed to determine the type of particle that was carried on the catalysts which can be used to support the previous catalyst characterization. XRD analysis is qualitatively performed on the angle range (2θ) = 5-90°. Based on the characterization, it appears that USY and catalyst of the research results was almost the same diffraction pattern. This almost identical pattern shows that metal impregnation does not alter the cell parameters on each catalyst significantly. When changing the cell parameters, it will affect the peak position in the diffractogram, for example the replacement of the Al-O bond with the shorter bond Si-O causes the cell unit to contract. This would caused the decreased d value and shifts the diffraction peak toward the higher 2θ [7]. Further analysis to find out the compatibility of XRD diffraction data with crystal structure that was generated based on comparison with Inorganic Crystal Structure Database (ICSD) data, and then performed refinement by using Le Bail method with RIETICA software. The refinement of all catalysts carried out by entering the standards were USY according to ICSD #31542 [17] data with cubic crystal structure and space group (F d group -3 m z), CoO (ICSD #9865) [18], MoO_3 (ICSD #36167) [19], $CoMoO_4$ (ICSD #281235) [20], Co (ICSD #41507) [21]. The overall results of refinement with USY standards are presented are presented in Figure 3.

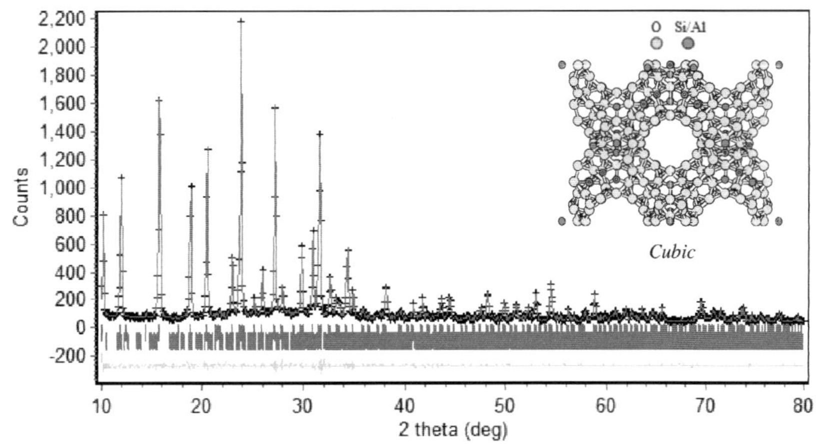

Figure 3. Refinement results of CoMo/USY (sample C) diffractogram with 5 standards phases i.e Y zeolite (ICSD #31542), CoO (ICSD #9865), MoO_3 (ICSD #36167), $CoMoO_4$ (ICSD #281235), and Co (ICSD #41507) + : experimental data ----- : calculation results, ----- : difference between experimental data and the calculation results.

The residual factor is the acceptable limit of the analyzed phase. The more phases that can be analyzed, the residual factor value will be smaller. The unanalyzed peaks are estimated to be the peaks of the metal compounds that emerge on the catalyst. Therefore, further analysis is required by comparing ICSD data from other standards phases are presented in Figure 3. Based on the results of refinement in Figure 3, it is known that the residual factor values (Rp and Rwp) was decreased from the value of one USY standard (the values of Rp and Rwp from 8.86 and 8.61 to 4.48 and 5.77, respectively). Indicates a detected peak of CoO (ICSD # 9865), MoO3 (ICSD # 36167), CoMoO4 (ICSD # 281235), Co (ICSD # 41507) and in accordance with RIETICA calculations.

3.3. Specific surface area, Total pore volume and Average pore size

Further analysis was an analysis with SAA to determine the effect of the ratio of Co and Mo metals on the specific surface area of the catalyst, pore radius average and total pore volume. It shown Catalyst B was relatively highest in specific surface area than the other samples. It is predicted that Co and Mo metals exhibit relatively unequal dispersion on the surface of the catalyst so that it has a specific surface area smaller than the samples of A, C, D, and E. The surface area of A, B, C, D, and E catalysts were 399.745 m^2/g, 426.295 m^2/g, 418.965 m^2/g, 406.028 m^2/g, and 416.852 m^2/g, respectively. The catalyst undergoing a specific surface area increase is assumed that the Co and Mo metals carrying the dispersion are uniformly distributed, so that in the end the catalyst-specific surface area is the total surface area between USY and the Co Mo metal). However, catalytic activity needs to be tested to prove the nature of all the characterization that has been done.

3.4. Catalytic Activity

3.4.1. Product Distribution.

Based on the overall data, an activity test was required on the Catalyst sample to prove its catalytic properties. This test reaction needs to be compared with thermal test treatments, Commercial Catalyst ($PtAl_2O_3$) and USY. The HDO anisole reaction using sample B was carried out using a flow model reactor. The products ware analyzed using GC-MS from the liquid fraction. It was found that from thermal obtained four products, USY five products, whereas with catalyst B made seven products, and $PtAl_2O_3$ made 10 products. In general, it shown that sample B catalyst has better selectivity than $PtAl_2O_3$ catalyst.

3.4.2. Catalytic Activity of HDO

HDO product activity is the amount of anisole conversion into benzene and toluene products. The value of anisole conversion to benzene and toluene is calculated based on GC-MS data. The catalytic activity of HDO refers to the yield of HDO products. The relationship of products yields and different treatment types will be presented in Figure 4.

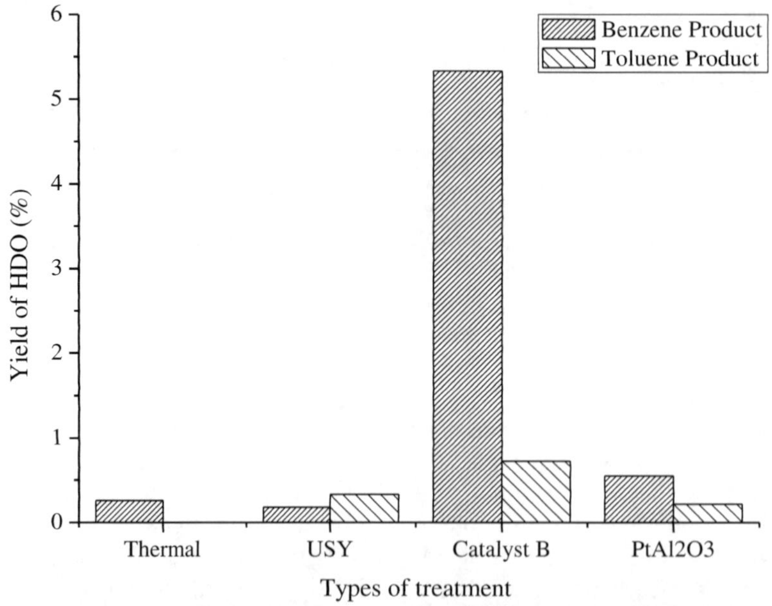

Figure 4. Yields of HDO product of various treatment types.

Based on the Figure 4, the reaction of HDO anisole using catalyst B was obtained benzene of 5.33%, which was higher than by thermal (without catalyst) or USY. In a sequence of benzene yields i.e Catalyst B> $PtAl_2O_3$> thermal> USY. On the other hand, catalyst B obtained toluene products of 0.73%. This phenomenon of hydrodeoxygenation conversion from anisole to benzene and toluene mostly influenced by the presence of Co and Mo metals that incorporated onto supporting material surface (USY). This is due to the Lewis acid sites of Co metal and Brønsted acid sites from USY. In this HDO reaction, the first reaction mechanism is the simultaneous reaction of demethylation and methyl transfer [6]. Demethylation occurs in Co metal of catalyst B. The Co metal is an active site where H2 gas adsorption. Metal Co will break the H2 bond, followed by the formation of bonds between hydrogen atoms with methyl form methane. Meanwhile, the hydrogen atoms from the Brønsted acid site of USY will bind to the oxygen atoms of the anisole that have lost the methyl group. At this stage occurs the formation of phenol compounds. Mo metals will bind the oxygen from the phenol hydroxyl group so that when there is a hydrogen supply, there will be a hydrogenolysis reaction. From the process of hydrogenolysis can occur the release of H_2O and form benzen products.

4. Conclusion

In this study, the acidity of the catalyst will decrease with increasing Co to Mo metal ratio. In the XRF analysis result, Co metal content recorded increased with the addition of metal ratio. The porosity analysis showed that the optimum specific surface area was 426,295 m^2 / g, the optimum pore size was 14,135 Å and the optimum pore volume of 0.318 cc / g was obtained for catalyst B. Furthermore, the MoCo / USY catalyst (Catalyst B) can be used as HDO catalyst anisole with 6.06% HDO product.

Acknowledgment
Authors are grateful to Sebelas Maret University for the financial support through Grant of Maintenance Research Group with contract number 1073/UN27.21/PP/2017.

Reference

[1] Ratna 2010 *Oil Petroleum Fraction* (Depok: UI)
[2] Gates B C 1992 *Catalytic Chemistry* (New York: John Wiley and Son. Inc.)
[3] Sofia A K 2014 *Applied Cat. A: General* **470** 261
[4] Kazuhisa, Inaba M 2007 *J. of the Japan Petroleum Inst.* **51** 234
[5] Kubicka D 2010 *Applied Cat. A: General* **372** 199
[6] Viljava T R 2001 *Cat. Today* **60** 11
[7] Hamdan H 1992 *Introduction to zeolites: synthesis, characterization and modification* (Kuala lumpur: UTM)
[8] Augustine R L 1996 *Heterogenous Catalysis for the synthetic chemist* (New York: Marcel Deklar Inc.)
[9] Hegedus L L 1987 *Catalyst Design progress and perspective* (New York: John Willey and Sons)
[10] Zeghayer Y 2005 *J. of Applied Catalysis* **282** 163
[11] Okamoto Y 2003 *J. of the Japan Petroleum Institute* **46** 343
[12] Rodica Z, Cristina R 2001 *Int. Chem. Eng. Conf.* **1** 399
[13] Maria U 2012 *Reactor* **14** 151
[14] Bhatia V L, Arias J F, Encheandia 2011 *Conf. Series: Material science and eng.* **17** 1
[15] Weitkamp J 1999 *Catalysis fundamentals and applications* (Berlin: Springer)
[16] Korkuna O 2006 *Mic. and Mesoporous Materials* **87** 163
[17] Gallezot, Beaumont P, Barthomeuf R 1974 *J. of Physical Chemistry* **78** 1447
[18] Sasaki S, Fujino, Takeuchi K 1979 *Proceed. of the Japan ac.* **55** 43
[19] Braekken H 2002 *Golden book of phase transitition* (Poland: Wroclaw)
[20] Livage, Hynaux C, Marrot A 2002 *J. of Materials Chemistry* **12** 1370
[21] Haglund, Fernandez J, Grimvall F 1993 *Phys. Review* **48** 11673

IOP Conf. Series: Materials Science and Engineering **333** (2018) 012061 doi:10.1088/1757-899X/333/1/012061

Synthesized of 2,7 dihydroxyxanthone from xanthone and antimalarial activities

Amanatie[1*], Jumina[2], Mustofa[3], Hanafi[4]

[1] Department of chemistry, Faculty of Mathematics and Natural Sciences, Yogyakarta State University, Yogyakarta. 55281 Indonesia.

[2] Department of Chemistry, Faculty of Mathematics and Natural Sciences, Gadjah Mada University, Yogyakarta. 55281 Indonesia.

[3] Faculty of Medicine, Gadjah Mada University, Yogyakarta, 55281 Indonesia.

[4] P3K LIPI Serpong Tangerang, 15314 Indonesia.

E-mail: amanatie@uny.ac.id.

Abstract: The purpose of the research is to synthesize 2,7- di-hydroxyxanthone compounds from xanthone and to evaluate antiplasmodial against activities. The synthesize of 2,7-di-hydroxyxanthone compounds worked with chromatogramphy methods including Thin Layer Chromatography (TLC), Vacuum Liquid Chromatography (VLC). A compound structures were determined based on the spectroscopic evidences including, Infrared (IR), one dimension (1-D) and two dimension (2-D) Nuclear Magnetic Resonance (NMR) spectra and comparison the spectroscopy data with related data from references. The biological properties of compounds are evaluated towards antiplasmodial against activity. The result of the product was obtained as white solid in 63.49% yield. The IR spectrum showed the absorption at 3433 cm^{-1} Which was reinforced with a sharp attack at 1087cm^{-1} indicating the stretching of OH, while the stretching of aromatic C=C appeared at 1620 cm^{-1}. The ^{1}H-NMR (500MHz, and DMSO –d6) spectrum showed that the aryl protons appeared in the region of δ12.98 ppm. In this region, there were 2 singlet at δ_H 12.98 ppm (1H, 2-OH) and (1H,7-OH) and shows the presence of two OH groups. Based on spectroscopy analyses, it could be started that the reaction of 2.7 di-aminoxanthone with $NaNO_2$/HCl and H_3PO_4 produced 2.7- di-hydroxyxanthone. *In vitro* antiplasmodial assay of the product synthesized 2,7 di-hydroxyxanthones against.*Falciparum* strain of 3D7 showed that the IC_{50} values of 2,7-di-hydroxy xanthone, were 0.31 μg/mL, respectively.

1 Introduction

Malaria is global health problem in developing countries. The efforts to eliminate this disease have been doing in many ways. However, the expected results is still not given. Evenmore, malaria becomes one of threatening disease in the world. This is indicated by the increase of malaria incidence especially in the endemic area. There are several problems against malaria. The main problems is the presence of malaria vectors (mosquito) which are resistant to insecticide and the resistant malaria parasites (such as

IOP Conf. Series: Materials Science and Engineering **333** (2018) 012061 doi:10.1088/1757-899X/333/1/012061

plasmodium falciparum) to anti-malaria drugs (such as chloroquin). The parasites are widely spread almost in all endemic area at whole over the world. Therefore, such anti-malaria drugs are not effective and sensitive anymore. These reasons lead the researchers to find the new anti-malaria drugs. One of strategies is based on the development of active compounds obtained from medicinal plants, which are traditionally used by people to cure malaria [1]. From 400 of *Garcinia* plants, it was found that xanthone was the major component, beside terpenoid, benzophenone and biflavonoid. 2-hydroxy xanthone, 2,7 dihydroxyxanthone had the potential biological activies. The 2.7-di-hydroxyxanthone as antimalaria agent has not been reported.

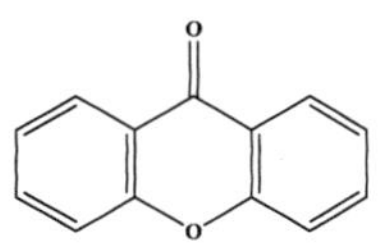

Figure 1. Structure xanthone

The efforts to find the new antimalarial drugs could be done in several ways, such as: a) synthesis of 2-hydroxyxanthone compound which could inhibit the specific metabolism of parasite, b) isolation of active compounds of natural products which are traditionally employed to cure malaria, and c) synthesis of analog compound of anti-malaria drugs, [2]. This research was initialized by isolating and identifying the xanthones from the root of *G. Dulcis* as well as performing the anti-malaria assay [3]. However, the yield obtained was very low. Several xanthone derivatives were reported to display antimalaria activities [4]. Xanthones could selectively inhibit the growth of *P. falciparum* in culture. Study on anti-malaria activity of xanthone derivatives showed that there was correlation between the structure and the anti-malaria activity (IC_{50}). Preliminary study on IC_{50} value (the inhibition value of *P.falciparum* growth) with semi empirical method of PM3 showed that the anti-malaria activity was correlated with the electronic properties of the substituents [5]. Xanthone and its derivatives were commonly obtained from isolation of natural products. Isolation of xanthone has been conducted from the leave [6] and bark [7] of *Garcinia dulcis*. [8] has obtained new xanthone derivatives of 7-O-methyl garcinon-E from *G. cowa* with IC_{50} of 1.50-3.00 μg/mL. Other xanthone derivatives of 1,3,7-trioxygenated and prenylated xanthone have been isolated from Calophylum caledonicum [9]. The originality of this research could be seen as the synthesized for 2.7-dihydroxy xanthones have been conducted. The *in vitro* antiplasmodium assay of the product synthesized 2,7-dihydroxyxanthones to *P. falciparum* strain 3D7 has not been reported. This research was synthesized of 2,7 di-hydroxyxanthone and analyzed of the product synthesized 2.7dihydroxyxanthones using UV-Vis, IR, ^{1}H-NMR, ^{13}C-NMR spectrometers and Test *in vitro* anti-malaria assay of the synthesized 2,7-dihydroxy xanthones against *P.falciparum.* This research was conducted with the main aims of synthesizing the 2,7-dihydroxyxanthones and performing the *in vitro* antimalaria assay of the 2.7 di-hydroxyxanthone. The specific aims were 1) To synthesized the 2,7dihydroxy xanthone from 2,7-diaminoxanthone; 2) To analyzed the product synthesized 2,7di-hydroxyxanthones using spectroscopy method (FTIR, ^{1}H-NMR, ^{13}C-NMR, spectrometer); and 3). To conduct new antimalarial activity of the 2,7-di-hydroxyxanthone.

IOP Conf. Series: Materials Science and Engineering **333** (2018) 012061 doi:10.1088/1757-899X/333/1/012061

2. Experimentals

2.1. Material chemical compounds:

2-Hydroxy xanthone, Hydrogen Chloride (HCl), Natrium Nitride ($NaNO_2$), phosphoric acid, ethanol-aquades. HEPES buffer, P. *falciparum*, Gentamicin sulfate, $NaHC0_3$, serum and red blood cells, Giemsa dyes.

2.2. General procedure:

Spektra of Infra Red (IR): Perkin-Elmer Spectrum One FT-IR spectrophotometers. Spektra of ^{1}H and ^{13}C NMR: spektrofotometer JEOL LTD ECP400, operated at 400 MHz (^{1}H) and 100.53 MHz (^{13}C), use aceton-d_6 as solvent and TMS as internal standard. Separation and purification used Thin Layer Chromatography (TLC), Vacuum Liquid Chromatography (VLC).

2.3. Prosedure of the synthesized of 2,7di-hydroxyxanthone

The 2,7diaminoxantone compound of 0.01 gram (0.00005 mol) was introduced into the 3-neck flask, suspended in Hydrogen Chloride (HCl) and added 10 mL of Natrium Nitride ($NaNO_2$) 2.8 M at 5 ° C until the solution changed entirely to yellow. The mixture was stirred at 5° C. for 30 minutes and then added 15 ml of 1 M Hydrogen Chloride (HCl) solution and be cooled. The resulting mixture was stirred at 5°C for 5 hours, acidified with phosphoric acid. The product is recrystallized with ethanol-aquades. The obtained product was analyzed by Infrared (IR) spectrometer, one dimension (1-D) and two dimension (2-D) Nuclear Magnetic Resonance (NMR) spectra.

2.4. Prosedure Testing the effect of antiplasmodium in vitro

Testing the effect of antiplasmodium in vitro is required P. *falciparum* culture. The cultures used were 3D7, bred by the Trager and Jensen method modified by Waruyanti, **[10], [11]** by candle jar with RPMI 1640, HEPES buffer, Gentamicin sulfate, NaHC03, serum and red blood cells. The breeding is carried out in a sticked glass candle and incubated glass exchanger in the incubator at 37°C. The medium is replaced periodically every 24 hours. The *P. falciparum* stage required for this test was a ring shaped young *trophozoite* obtained by synchronization in a 5% w/v sorbitol solution. The anti-plasmodium effect test of the 2,7-di-hydroxyxanton compound is performed in a microbial well. Into the micro well plate which has been given the test compound with various concentrations, given 50 μl of P. falciparum suspension. Incubate in incubator at 37°C for 24 hours. The results were evaluated by making the dosage form with Giemsa dyes. The number of living schizons accounted for 200 asexual parasites, used as a criterion for the effects of antiplasmodium. From the observation results obtained IC_{50} value indicates that the test compound has activity inhibiting the growth of P. falciparum in vitro. Then continued data analysis with probit analysis.

3 Result and Discussion

3.1 The strategy is via Fries rearrangement of diaryl ester derivatives (Figure 2) [12].

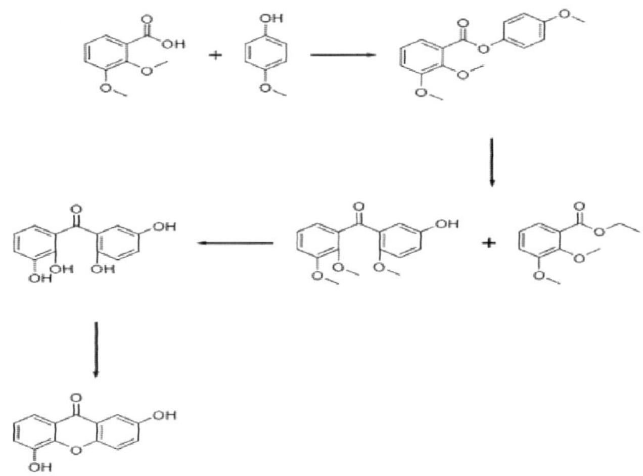

Figure 2. Synthesized of xanthone via benzophenone route

3.2 Synthesized of 2-hydroxyxanthone

The synthesized 2-hydroxyxanthone from monoaminoxanthone was the reacted with $NaNO_2$, HCl, and H_3PO_4 (Figure 3).

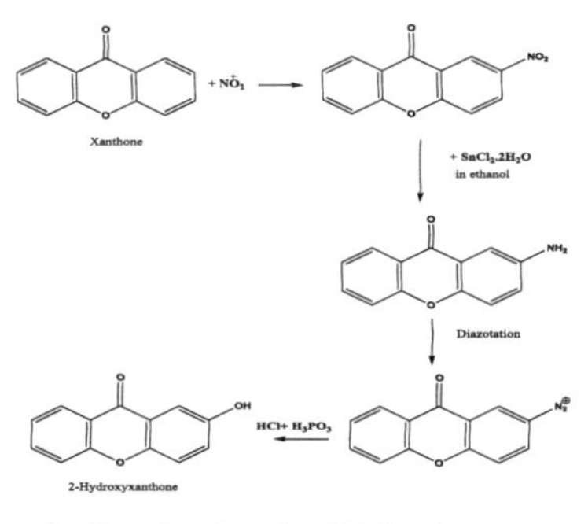

Figure 3. Synthesized of 2-hydroxyxanthone

3.3.Analyzed of 2-hydroxyxanthone

2-Hydroxy xanthone was obtained by reacting 2-aminoxanthone with $NaNO_2$ to produce diazonium salt. Then, hydrolyzed of the salt produce Produced 2-hydroxyxanthone. The product was obtained as white solid in 69.80% yield. The IR spectrum showed the absorption at 3433 cm^{-1} indicating the stretching of OH, while the stretching of aromatic C=C appeared at 1620 cm^{-1}. The ^{1}H-NMR spectrum showed that the aryl protons appeared in the region of δ 6,97 -7,78 ppm. In this region, there were 4 doublet at δ_H 6.97 (H, J = 8.3 Hz), 7.26 (H, J=8.3 Hz),7.37 (H, J=8.3 Hz) and 7.78(2HJ=23Hz) ppm as well as one doublet of doublet peak at δ 7.50 ppm (2H,*dd*,J and 8.3 Hz). One singlet peak from hydroxyl proton appeared at δ_H 12.53 ppm. Identification of the product using ^{13}C-NMR showed aryl carbons at δ106, 110, 117, 137 and

155ppm. The peak at δ 113, 116, 118, 135, 136, 161 dan δ 179,2 from the carbonyl group while the peak at δ 175 ppm was the peak for the carbon (C_2) next to hydroxyl group. Based on spectroscopy analyses, it could be stated that the reaction of 2-aminoxanthone with $NaNO_2$/HCl and H_3PO_4 produced 2-hydroxyxanthone. The reaction mechanism was presented on Figure 4.

3.4. Synthesized 2.7- di-hydroxyxanthone

The product was obtained as white solid in 63.49% yield. The IR spectrum showed the absorption at 3433 cm^{-1} Which was reinforced with a sharp attack at 1087cm^{-1} indicating the stretching of OH, while the stretching of aromatic C=C appeared at 1620 cm^{-1}. The ^{1}H-NMR (500MHz, and DMSO –d6) spectrum showed that the aryl protons appeared in the region of δ12.98 ppm. In this region, there were 2singlet at δ_H 12.98 ppm (1H, 2-OH) and (1H, 7-OH), shows the presence of two OH groups. An aromatic proton appears in the intermediate region of δ 7.30-7.69ppm Ie 7.69 for H_1 and H_8 (2H.d.*J* =2.4Hz), 7.43 for H_3 and H_6 (2H, *dd, J* = 2.4 and 8.2Hz) and 7.30 ppm for H_4 and H_5 (2H.d.j= 8.2Hz).

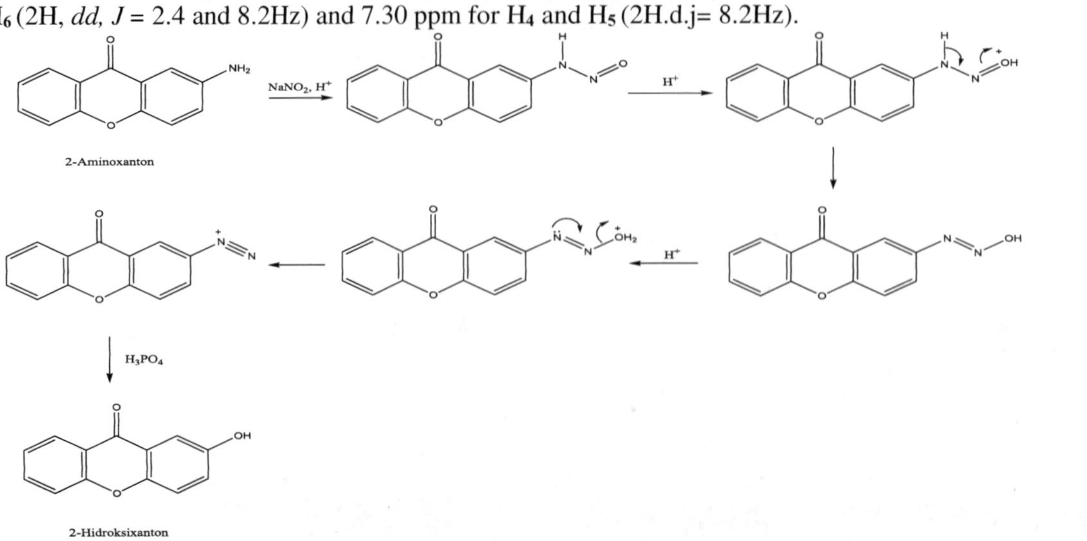

Figure 4. Mechanism of synthesized of 2-hydroxyxanthone

Identification of the product using ^{13}C-NMR(125MHz, $CDCL_3$) Shows the presence of three peaks of the CH group aryl carbons at 119.56 (2C, C_4 and C_5, 11.8.74 (2C, C_3 and C_6), and 117.41ppm(2C, C_1 and C_8). 3 peaks of quaternary carbon groups appear on the region 151.44 (2C. C_9 and C_{12}), 148.58 (2C, C_2 and C_7) and 122.45 ppm (2C, C_{10} and C_{11}), 1 peak of carbonyl group C=O at 179.22 ppm at C_{13}

3.5. Anti-malarial Activity Test

Malaria is a disease caused by the infection of protozoa from the genus of *Plasmodium*. There are 4 species of *Plasmodium* which could cause malaria. They are *P. falciparum, P. vivax*), *P. malariae, P. ovale* ([13], [14], [15], [16], [17]). Among them, *P. Falciparum* is the most responsible to the malaria caused death. Moreover, patients could be infected by more than one species of *Plasmodium*, for example *P. falciparum* and *P.vivax* in subtropical area, *P. falciparum* and *P. malariae* in tropical country in Africa. Results of survey showed that there are 15 million case of malaria, where 70 million citizens live in the malaria endemic area. In Jawa and Bali, the case of malaria increased from 0.12 per 1,000 citizens to 0.81 per 1,000 citizens [18].90% in Africa, 4% in South East Asia, 4% in Mediterania and 2% in other

area)[19]. Resistance of malaria parasite to standard anti-malaria drugs of chloroquin was found in South America (Columbia and Venezuela) in 1960 [20], and followed in Thailand in 1961. the resistance of *P. falciparum* to chloroquin was reported in Kalimantan in 1974 and spread in all provinces in 1996 [21].. The situation is getting worse due to the case of resistance of *P. falciparum* to drugs of sulfadoxin-pirimetamine in 10 provinces and kina in 5 provinces in Indonesia [22]. Another resistance to meflokuin has also been reported. In fact the drug has not been utilized in Indonesia [23].Chloroquin as the first line anti-malaria drug in the world has important role in controlling and medication of malaria. The other important factors are pharmacological and transmission factors[21].In addition, the genetic recombination on sexual stadium between gametocyte and different strain in the mosquitoes could lead the genetic mutation [24].Chloroquin has been considered to have activity in inhibiting the production of hemozoin on the vacuole of malaria parasites. However, the resistant mechanism of *P. falciparum* to chloroquine has not been known yet [25]. There were several hypotheses regarding the resistant mechanism, such as the change in metabolism pathway, thus the anti-malaria drug which enter the *Plasmodium* could not be metabolized properly. Characteristic of resistant parasites to chloroquin is the fast efflux, while that of sensitive parasites is it could survive against radioisotope-label-drug for longer period [26]. The resistance has forced the researchers to find new anti-malarial drugs to substitute the non sensitive anti-malaria drugs against *P. falciparum*.
The study give to synthesized 2.7-dihydroxyxanthone and biological activity as well as new antimalarial activity with IC_{50}=0.31µg/mL

4. Conclutions

According to results and discussion, it could be concluded that: 1) 2,7-di-Hydroxy xanthone which theoretically displayed antimalarial activity. 2) Reaction of 2,7-di-aminoxanthone with $NaNO_2$, HCl and H_3PO_4 produced 2,7- di-hydroxyxanthone in 63.49 % yield. 3) *In vitro* antiplasmodial assay of the product synthesized 2.7-dihydroxyxanthone against P.falciparum strain of 3D7 showed that the IC_{50} values were 0 31µg/mL, respectively.

REFERENCES

[1] Mustofa 2009 *Professor Speech Inaguration* (Yogyakarta: Gadjah Mada University)
[2] Sholikah EN 2010 *Dissertation Doctoral of Program, Faculty of Medicine* (Yogyakarta: Gadjah Mada University)
[3] Amanatie, Jumina M, Hanafi M 2009 *Development of new compound xanton derivatives from Garcinia dulcis root as antiplasmodium* (Yogyakarta: Research Institute of Yogyakarta State University)
[4] Likhitwitayawuid K, Chanmahasathien W, Ruangrungsi N, Krungkrai J 2000 *Plana Med.* **64** 70-77
[5] Likhitwitayawuid K, Ruangrungsi N, Krungkrai J 2001 *Planta Med.* **64** 70-72
[6] Kosela S, Hu IH, Rahmatia T, Hanafi M, Sim K.Y 2000 *J. Nat. Prod.* **63** 406-407
[7] Hanafi M, Soemiati A, Kosela S, Leslie J. H 2004 *Proc. Int. Conf. UGM*
[8] Likhitwitayawuid K, Chanmahasathien W, Ruangrungsi N, Krungkrai J 2000 *Plana Med.* **64** 70-72
[9] Hay AE, Helesbeux J J, Duval O, Labaied, M, Grellier P, and Richomme P 2004 *Life Sci.* **75** 3077-85
[10] Wijayanti MA, Supargiyono, Mustofa, Sholikhah EN, Jumina, Tahir I., Hadanu R 2007 *Proc. Int. Conf. Chem Sci. (ICCS-2007)* 237-242

[11] Neetu T, Shiv V S, Chottes L J, Gaurav V, Anirban P, Vineeta S 2015 *Asian J. Phar. Clin. Res.* **8(2)** 47-50
[12] Naidoo J M 2009 *Novel Methodology for The Synthesis of Xanthones Thesis (*Johannesburg: University Withwatersrand)
[13] Siswandono, Sukardjo B 2000 *Principles of Drug Design* (Surabaya: Erlangga University Press
[14] Ramya TNC, Suroliya N, Surolia A 2002 *Curr. Sci.* **83(7)** 818-25
[15] Zein U 2005 *Recent Handling of Malaria Falciparum Repsository* (Medan: University of Sumatra Utara)
[16] Ashley E, McGready R, Proux S, Nosten F 2006 *Malaria*, *Travel Med. Infect. Dis.* 159-173
[17] Daily JP 2006 *J. Clin. Pharmacol.* **46** 1487-97
[18] Syamsudin, Dewi R M, Hernita 2004 *Airlangga Phar. Mag.* **4(**3) 101-104
[19] WHO 2013 *Malaria* http://www.who.int/mediacentre/factsheets/fs094/en/ accessed on March 23, 2013
[20] Krogstad DJ 1996 *Epidemiol. Rev.* **18(1)** 77-89
[21] Zein U 2005 *Recent Handling of Malaria Falciparum Respository* (Medan: University of Sumatra Utara)
[22] Achmad M.F, Sutanto I 2003 *Indonesian Med. Mag.* **53(2)** 72
[23] Emiliana T, Tuti S, Renny M, Arbani P R, Harijani AM 1993 Cermin *Dunia Kedokteran* 82
[24] Black RH, Canfield CJ, Clide DF, Peters W, Wernsdorfer WH 1968 *Chemotherapy of Malaria* 2nd *Ed* (Geneva: WHO) **136** 36-37
[25] Zaranz B, Jaso, A, Lima LM, Aldana I, Monge A, Maurel S, Sauvain M 2006 *J. Pharm. Sci.* **42(3)** 357-361
[26] Ramya TNC, Suroliya N, Surolia A 2002 *Curr. Sci*, **83(7)** 818-825

IOP Conf. Series: Materials Science and Engineering **333** (2018) 012062 doi:10.1088/1757-899X/333/1/012062

Synthesis and Characterization of Allophane-Like as Chromium (Cr) Ion Adsorbent

Pranoto[1], C Purnawan, A N Husnina
Research Group of Analytical and Environmental Chemistry, Faculty of Mathematics and Natural Sciences
Universitas Sebelas Maret, Ir. Sutami street No. 36 A, Surakarta, Central Java, 57125, Indonesia.

E-mail: pakpranoto@gmail.com

Abstract. The synthesis and characterization of allophane-like as chrom (Cr) ion adsorbent has been studied. The objectives of this study is to determine the characteristics of allophane-like and determine ratio of Al/Si, chromium solution pH, and contact time to get the best decreasing metal ion chrom (Cr) adsorption condition. The study was conducted with the ratio of Al/Si ratios 0.5; 0.75; 1.0; 1.25 and 1.5 from Tetraetyl Orthosilicate (TEOS) solution and Aluminium Nitrate Nonahydrate [$Al(NO_3)_3.9H_2O$] in pH 3-4. The result of synthetic was characterized on functional groups and cristallinity. Experiment of adsorption ability using variation of Cr solution pH 3-7, contact time 30, 60, 90 and 120 minutes with batch method. The results by FTIR shows that functional groups -OH, the asymmetry groups O-Si-O or O-Al-O, relatively weak absorption which stronger then the presence of OH and *bending* vibration Si-O or Al-O on allophane-like. The best conditions of chromium metal adsorption with adsorbent allophane-like was obtained at pH 5, contact time 90 minutes, and the ratio Al/Si 1.5. Types of adsorption in this study follows Freundlich and Langmuir isotherm.

1. Introduction

Allophane is one of the reactive soil minerals because it has a very large surface area and many active functional groups [2]. The allophane surface has properties such as cations and anions exchange, the uptake of organic and inorganic compounds, and the acidity source from silanol (Si-OH) and aluminol (Al-OH) functional groups [11,15], so it can be utilized as an adsorbent. However, the presence of natural allophane as an adsorbent is considered less effective and optimum in adsorbing compounds such as heavy metals. The formation of natural allophane derived from the volcanoes also takes a long time. In addition, natural allophane will change with geographic location differences [9,14]. Uncontrolled and continuous use of natural allophane can damage the environment. Therefore it is necessary to use a new alternative by making allophane synthetic or allophane-like materials.

Allophane-like synthesis can be prepared in a Tetraethyl Orthosilicate (TeOS) and aluminium nitrate nonahydrate solution with an Al / Si ratio of 0.5; 1; 1.5 and 2 [10]. According to Peelletier et al. (2005), the surface area of allophane-like is greater than natural allophane so it can also be utilized as heavy metal adsorbent [12]. One of the heavy metals that often pollute the environment is chromium

[1] pakpranoto@gmail.com

(Cr). Chromium (Cr) is a heavy metal that is harmful to health and the environment. If chrome is used excessively, chrome will result in acute poisoning. Other impacts such as lung cancer [8], kidney failure, anemia, skin allergies, asthma and stomach cancer may also be mutagenic and carcinogenic [5].

There have been many developed methods to reduce Cr heavy metals ion in the environment, one of them using adsorption. Adsorption becomes the preferred method because the process is relatively simple [6]. Factors affecting the adsorption process of heavy metals are the pH of the solution, contact time, initial concentration of solution, temperature, and adsorption method [7]. Based on the description above, then conducted research on the characterization of Allophane-like as adsorbent of chromium metal ion (Cr).

2. Experimental

2.1. Synthesis Allophane-like

100, 66.67, 50, 40 and 30 mL of Tetraethyl Orthosilicate (TeOS) solution were mixed into 50 mL Aluminum Nitric Nonahydrate ($Al(NO_3)_3.9H_2O$] solution at room temperature to obtain an Al/Si ratio of 0.5 ; 0.75; 1; 1.25 and 1.5. Furthermore, 150 ml of 0.1 N NaOH added to the mixture for 3 hours. NaOH was added to solution using ratio of OH/Al by 3. The obtained mixture measured the pH of the solution and made the pH of the solution to acid (pH 3-4), then precipitated for a night until the mixture became gel. Then, the gel dried with a temperature 95 °C for 7 days. The solid is then heated in a furnace at 400 °C for 3 hours. The allophane-like obtained were characterized by FTIR, XRD, SEM, Specific Surface Area and Acidity.

2.2. Determination of Cr ion in Aqueous Solution

0.01 g of allophane-like added to flacon. Then, 10 mL Cr artificial solution was added using pH variation 3, 4, 5, 6 and 7. The mixture was shake with contact time variation 30, 60, 90 and 120 minute. The obtained mixture was filtered and ion Cr was determinated using AAS for determination of Cr ion in Aqueous Solution.

2.3. Determination of Isotherm Adsorption

0.01 g of allophane-like added to flacon and 10 mL Cr artificial solution was added with concentration 2, 4, 6, and 10 ppm. Then mixture was shake with constant temperature. The obtained mixture was filtered and analysed using AAS for determining Cr ion in solution. Then AAS analysis result was used to determine isotherm adsorption.

3. Result and Discussion

3.1. Synthesis and Characterization of Allophane-Like

Allophane-like was made using Tetraetil Ortosilikat (TeOS) solution as Si group source and Aluminium Nitrat Nonahidrat [$Al(NO_3)_3.9H_2O$] solution as alumina group source. Allophane-like using volume ratio Al/Si 0,5; 0,75; 1; 1,25 and 1,5. Obtained allophane-like was characterized using Sufare Area Analysis

3.2. Surface Area and Acidity Analysis

Surface area is important factor in adsorption process because the greater surface area then the greater adsorption capacity. Acidity analysis was used ammonia adsorption using gravimetry methods. Specific surface area and acidity characterization was showed in table 1.

Table 1. Surface Area and Acidity Characterization

Adsorbent	Surface Area (m^2/g)	Acidity (mmol/gram)
Natural Allophane [4]	85.529	1.725
Allophane-Like	201.558	2.728

Table 1. was showed that allophane-like surface area was greater than natural allophane surface area. According SAA analysis, obtained allophane-like pore volume 0.2395 cc/g with pore diameter 4.6 nm. Acidity value of allophane-like is greater then natural allophane. It show that active site of allophane-like is much more than natural allophane so cation exchange in allophane-like more occur.

3.3. Adsorption effectivity of Allophane-like against Cr ion

The ratio of Al / Si ratio can influence the adsorption capacity, whereas the higher the Al/Si ratio of the adsorption capacity will rise to the maximum. The diagram of the adsorption capacity at each ratio can be seen in Figure 6. The higher Al/Si ratio, the adsorption capacity will increase, but at the Al/Si ratio of 1.25 adsorption capacity was decreases, it happen because of the amount of Si contained. From synthesis result, the higher ratio Al/Si content of Al content in Allophane-like is increasing, so that more active sites could increase the surface area. The largest ratio of Al/Si is 1.5 with adsorption capacity of 0.61 mg/g.

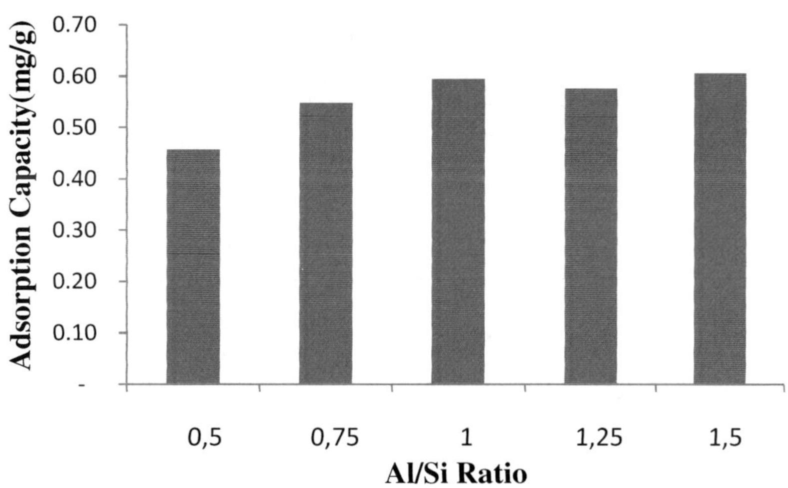

Figure 1. Adsorption capacity diagram of Allophane-Like based on Al/Si Ratio

The pH value and contact time greatly affect the adsorption process. The increase pH can cause the adsorption process to increase but will decrease again if it passes through the optimum point reached. At low pH (<5) protonation will occur resulting in the formation of H_3O^+. This will cause competition between Cr_3^+ and H_3O^+ to bind to the adsorbent negative sites so that Cr ion absorption will be impaired. At an alkaline pH or pH above 5 the number of protons H^+ is small resulting in the possibility of metal binding by a relatively large adsorbent, so that metal ions can form hydroxide precipitates and result in decreased adsorption ability [1]. The adsorption capacity diagram based on the effect of pH can be seen in Figure 2 and the largest adsorption capacity is at pH 5 of 0.87 mg/g.

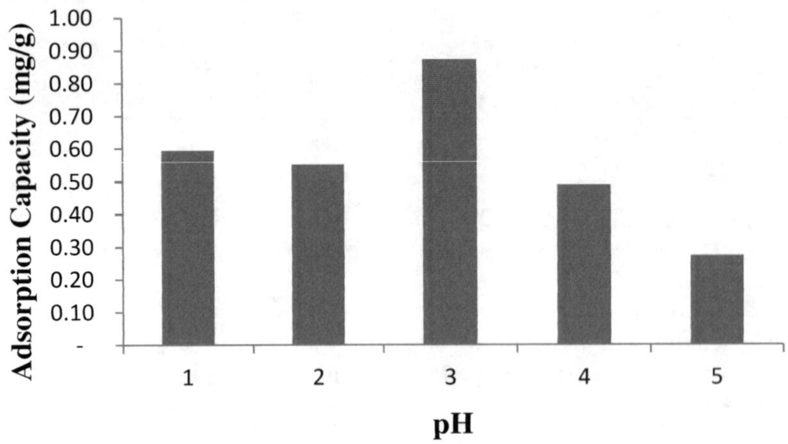

Figure 2. Adsorption Capacity of Allophane-like based on pH variation

The contact time is the time it takes for the adsorbent to adsorb metal ions. The longer contact time between the adsorbent and Cr solution, the adsorption capacity will increase until the optimum condition. However, prolonged contact time causes the adsorption condition to become saturated and the bond between the adsorbent and the adsorbate will be released. From the data obtained has happened adsorption equilibrium between the concentration of Cr ion adsorbed by the Cr ion concentration remaining in solution at the contact time of 90 minutes, while the contact time of 120 minutes in general adsorption ability Allophane-like decline. It can be caused by the surface of the adsorbent has reached a saturation point so that Cr ion could desorbed or escape back into the solution, whereby the release of metal ions of Cr that has been tied to the adsorbent active group. The largest adsorption capacity of Cr metal occurs at a contact time of 90 minutes with an adsorption capacity of 0.71 mg/g. The diagram of adsorption capacity based on pH variation can be seen in Figure 3.

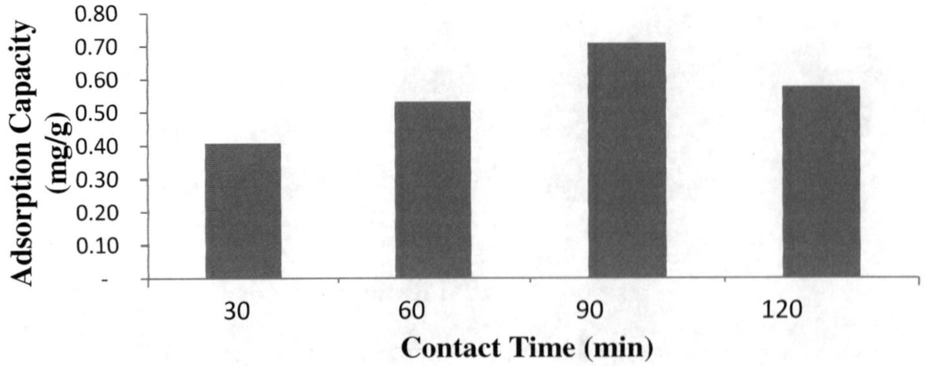

Figure 3. Adsorption Capacity Diagram of Allophane-like based on contact time

The optimum composition of Al/Si ratio is 1.5, then analysed using FTIR for determining possibility of Cr metal ion adsorbed to functional group of allophane-like. FTIR result was showed in Figure 4.

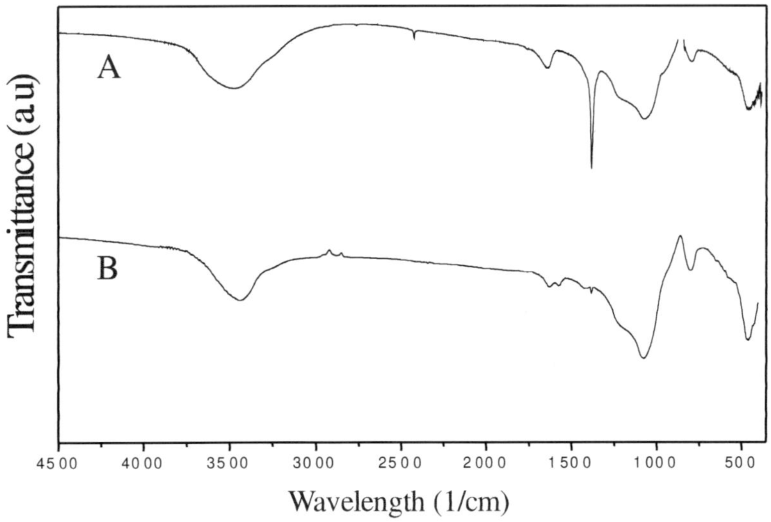

Figure 4. FTIR Spectra of Allophane-like with Al/Si ratio 1,5 before (a) and after (b) adsorption

According in figure 4 showed that there is a change of spectra where in the absorption -OH stretching occur the shift of the wavelength from 3463 cm^{-1} to 3437 cm^{-1}. This is thought to be due to the formation of a hydrogen bond between the adsorbent and Cr metal ion resulting in a decrease in the binding energy of the -OH group on Si-OH or Al-OH [12,13]. The decrease of the wavelength in the hydroxyl group is also due to the bonding that occurs between the Cr metal with oxygen in the hydroxyl group thereby causing the bond length to increase and the -OH binding energy becomes weak.

3.4. Isotherm Adsorption

Isotherm adsorption is used to determine adsorption process happen on chemically or physically process. Langmuir isotherm curve obtained showed in figure 5.

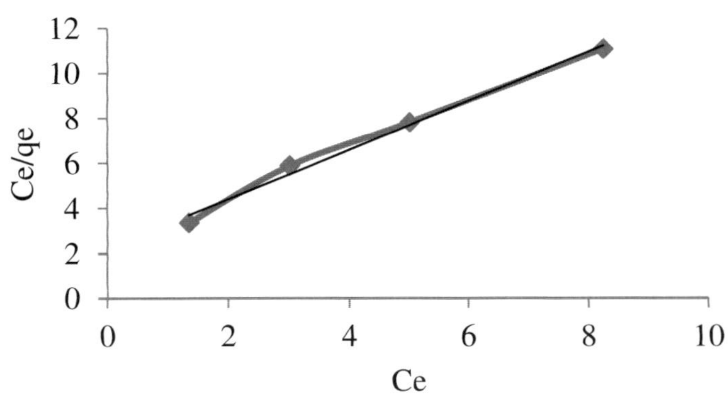

Figure 5. Isotherm Langmuir Curve of Allophane-like

In figure 5 obtained value R^2 = 0.9905. Langmuir isotherms indicate that the adsorption process occurs chemically that the active site of the adsorbent will interact with Cr metal ions in the solution by forming a coordinate covalent bond with the heavy metal ion group in the solution by forming a compound. The adsorption capacity of allophane-like according to Langmuir isotherm is 0.914 mg/g. The determination of the Freundlich isoterm shown in Figure 6.

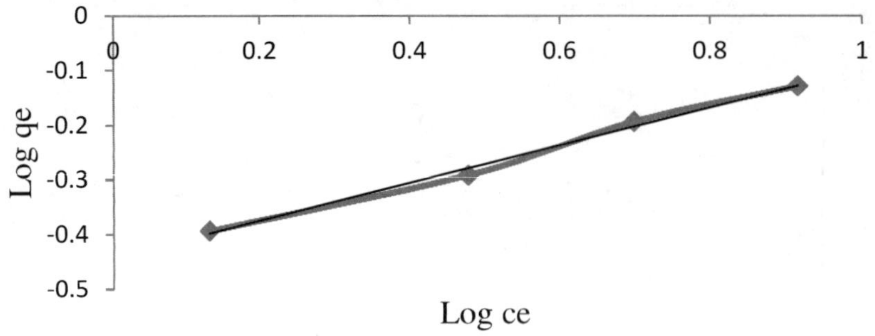

Figure 6. Isotherm Freundlich Curve of Allophane-like

In figure 6 obtained value R^2 = 0.9932. Isotherm Freundlich is an isotherm that describes the process of adsorption physically. Physical adsorption occurs when the intermolecular force is greater than the attraction between molecules. Interaction occurs when heavy metals ion entering the pores of the allophane-like without forming bonds or interacting with weak binding energies through van der waals bonds. This weak pull force causes the adsorbate to move from one part of the surface to the other surface of the adsorbent. The allophane-like adsorption capacity according to freundlich isotherm is 0.353 mg/g. Allophane-like isotherm adsorption data will be shown in Table 2.

Table 2. Isotherm adsorption for Allophane-like ratio Al/Si 1,5

Isotherm Langmuir		**Isotherm Freundlich**	
Q_{max} **(mg/g)**	0.914	***Kf*** **(mg/g)**	0.353
b **(L/g)**	0.494	***n***	2.897
r	0.991	**r**	0.993

Based on table 2, R^2 value Freundlich isotherms equation higher than the R^2 value of the Langmuir equation even though they are close to 1 with a difference that is not significantly different. It can be concluded that this study follows Freundlich isotherms and Langmuir isotherms or adsorption processes occur physically and chemically.

4. Conclusion

Based on the result, Surface area of Allophane-like 201.558 m^2/g and more large then natural allophane. The acidity of Allophane-like is 2.728 mmol/g. The best absorption condition of Allophane-like to chromium (Cr) metal ion occurs in variation of Al/Si ratio 1.5 buffer solution pH 5, contact time 90 minutes. The type of adsorption isotherms follow the isotherms Langmuir with Q_{max} 0.914 mg/g and Freundlich with *Kf* 0.353 mg/g.

Reference

[1] Cordero B, Loidero P, Herrero R, and Vicente 2004 *J. Environ. Chem. I* **1** 180-187
[2] Devnita R, Yuniarti H 2005 *Fakultas Pertanian UNPAD*
[3] Eliopoulus M E, Antivachi D, Vasilatos C, dan Megremi I 2012 *Geoscience Frontiers* **3** 523-539
[4] Husna M N 2012 *Skripsi* Kimia FMIPA UNS Surakarta.
[5] Kaszycki P, Gabrys H, Appenroth K J, Jaglarz A, Sedziwy S, Walczak T, and Koloczek H 2005 *Plant Cell. Environ.* **28** 260-268
[6] Khasanah E N 2009 *Oseana* **34** 1-7
[7] Mc Cabe 1999 *Operasi Teknik Kimia Jilid 2* (Jakarta : Erlangga)

[8] Palar, H. (2008). *Pencemaran dan Toksikologi Logam Berat.* PT Rineka Cipta. 133-144, Jakarta.
[9] Parfitt, R.L., 1990, *Proceedings of the Six International Soil Classification Workshop. Chile and Equador 9 – 20 January 1984.*
[10] Pelletier, E.M., Bogenez, S., Pelletier, M., Razafitianamaharavo, A., Ghanbaja, J., Lartiges, B., and Michot,L., 2005, *Colloids and Surfaces,* **255**, pp **1-10**.
[11] Sukmawati., 2011, *Media Litbang Sulteng IV,* **2**, pp **118-124.**
[12] Iyoda, F., Hayashi, S., Arakawa, S., John, B., Okamoto, M., Hayashi, H., Yuan, G. 2012 *Applied Clay Science* **56** pp **77-83**
[13] Opiso, E., Sato, T., Yoneda, T., 2009 *Journal of Hazardous Materials* **170(1)** pp **79-86**
[14] Pranoto, Suranto, Sugiyarto K. H., Ashadi 2013 Journal of Environmental and Earth Science **3(5)** pp **48-56**
[15] Reinert L., Ohashi F., Kehal M., Bantignies J-L., Goze-Bac C., Duclaux L. 2011 Applied Clay Science **54(3-4)** pp **274-280**

IOP Conf. Series: Materials Science and Engineering **333** (2018) 012063 doi:10.1088/1757-899X/333/1/012063

The Application of Active Paper Incorporated with Oleoresin of Cinnamon Leaf (*Cinnamomum burmanii*) Distillation Residues on Maintaining Dragon Fruits (*Hylocereus costaricensis*) Quality during Storage

M S H Aziz, G J Manuhara, R Utami, L U Khasanah

Department of Food Science and Technology Agriculture Faculty Universitas Sebelas Maret, Jl. Ir. Sutami 36 A, Kentingan, Surakarta 57126, Indonesia

E-mail: mshidiqhaziz@gmail.com

Abstract. The purpose of this study was to determine the effect of active paper placement methods on super red dragon fruits quality during storage at ambient temperature. The active papers were incorporated with oleoresin of cinnamon leaf distillation residues. Various active paper placement methods were applied such as wrapping, placed on the cardboard wall, placed cardboard pad, and scrap of paper on the sidelines. Weight loss, peel color, surface and flesh hardness, total titratable acid, soluble solid total, pH flesh fruit, and total plate count (TPC) of super red dragon fruits samples were investigated during 9 days storage. The result shows that active paper placement methods significantly affected the weight loss, surface firmness and color peel change of super red dragon fruits samples. However, active paper placement methods insignificantly affected the titrable acid total, soluble solid total, pH, flesh firmness and microbial spoilage of super red dragon fruits samples. The best method to maintain the super red dragon fruits quality was wrapping method.

1. Introduction

Dragon fruit is a kind of cactus plant originating from Mexico, Central America, and South America. Dragon fruit contains bioactive substances that are beneficial to the health of the body such as antioxidants in the form of ascorbic acid, beta-carotene, anthocyanin and contain fiber in the form of pectin. Dragon fruit flesh also contains vitamin B1, B2, B3, vitamin C and minerals such as calcium, phosphorus, iron, and others [1]. Super red dragon fruit *(Hylocereus costaricensis)* is a type that is often cultivated because it has a sweet taste [2], and has a high economic value. The red color on the dragon fruit shows the presence of beta-carotene compounds, the redder the color of the flesh indicates higher beta-carotene content [3].

Due to the high water content of dragon fruit, the shelf life is short. At a temperature of 20°C dragon fruit can only storage for 1 week [4]. Harvested dragon fruit will undergo taste changes caused by the changes of titrated acids and soluble solid, physical changes of fruit such as color and texture, weight loss, the changes of pH value, or the changes caused by microbial spoilage during storage [5]. Handling is necessary to minimize dragon fruit damage during storage and distribution.

Active packaging is an alternative to maintain the quality of horticultural products. One form of active packaging is the addition of the active compounds to a paper packaging material which is then known as active paper [6]. In the previous research, the active paper could maintain the quality of plum fruits, peaches and mushrooms *(Agaricus bisporus)* [7] [8].

Oleoresin cinnamon leaves contain linalool as active compounds, which showed antimicrobial activity to inhibit microbial growth [9]. Active paper with the addition of a distillation residues cinnamon leaves oleoresin can provide an inhibitory effect on *Aspergillus niger* and *Pseudomonas fluorescens* [10]. However, the study related to the application of active paper incorporated with oleoresin cinnamon *(Cinnamomum burmanii)* distillation residues on dragon fruit has never been reported. Furthermore, it has the potential to maintain the quality of super red dragon fruit *(Hylocereus costaricensis)* during storage. The purpose of this study was to determine the effect of active paper placement methods on super red dragon fruits quality during storage at ambient temperature.

2. Experimental

2.1. Materials
Cinnamon leaves *(Cinnamomum burmanii)* obtained from Bubakan village, Wonogiri (Indonesia). The active paper materials were Whatman filter paper No.41, chitosan, and tween 80, acetic acid "MERCK", tapioca starch (Rose Brand), and distilled water. The secondary packing materials were cardboard-shaped boxes with length 64 cm, width 18 cm, and height 11 cm which obtained from CV. GIAT, Karanganyar (Indonesia). Cardboard also used as inside partition (8 part for each box) of the box. The super red dragon fruit *(Hylocereus costaricensis)*, harvested after 50 days the flowers bloom, obtained from CV. Purnama Jaya Farm, Sumedang (Indonesia). Testing materials used include distilled water, NaOH "MERCK" 0.1N, phenolphthalein indicator, Plate Count Agar (PCA) "MERCK", sterile normal saline, and alcohol 70%.

2.2. Oleoresin extraction
The extraction of cinnamon leaf distillation residue was conducted by using ethanol 70% (1:6) for 5 hours at 780C. The extract was filtered to separate the pulp and the filtrate. The separation between the solvent and oleoresin by using a vacuum rotary evaporator (IKA RV10) with a pressure of 500-700 mmHg (64 0C/120 rpm) until the solvent evaporated entirely [10].

2.3. Active paper manufacturing
The filter paper was cut to 2x2 mm. 15 g of filter paper immersed in 250 ml of distilled water for 24 h. Immersion filter paper and 250 ml of distilled water were blended for 5 minutes. The 4.5 g of tapioca in 50 ml of distilled water was mixed into the paper pulp and blended for 5 minutes. Furthermore, 0.45 g chitosan in 100 ml of 1% acetic acid was mixed in a blender with pulp for 5 min. 4% (w/w) of oleoresin was put into 50 ml of distilled water and stirred with a magnetic stirrer. During stirring, Tween 80 was added until an emulsion was formed. The emulsion was slowly added to the pulp and blended for 5 minutes. Paper moulding was done by using a 30 x 20 cm mould. Pulp poured evenly into the mould, and on the top covered with a fabric for pressing until flat. Wet paper sheets were dried for 48 hours at room temperature and reversing every 24 hours [10] [11] [12].

2.4. Fruits packaging
The fruits were packed using cardboard-shaped boxes. A total of 7 dragon fruits were placed into cardboard and given a bulk head for each fruit. The applied active paper placement methods were wrapping (Wrap), the placement of paper on the cardboard wall (Wall), the paper on the cardboard pad (Pad) and scrap of paper on the sidelines of the cardboard (Sidelines). The active paper was used for each treatment weighing at 16 g. Dragon fruits that have been packed subsequently analyzed on days 0, 3, 6, and 9 after storage at room temperature (28°C, RH 83% during the day and 31°C, RH 68% in the evening). The data were collected such as weight loss [13], peel color [14] using RGB Color Analyzer -1002 Lutron , surface and flesh hardness [13] using Fruit Hardness Tester FR-5105 Lutron ,

titrable acid total [15], soluble solid total [15] using Hand refractometer ATAGO, pH flesh fruit using pH meter ,and total plate count (TPC) [16].

2.5. Data analysis
The data were statistically analyzed by using *One Way Analyzes of Variance* (ANOVA) at a significance level α = 0.05. If the result shows the difference between the variable then followed by *Duncan's Multiple Range Test* (DMRT) at a significance level α = 0.05.

3. Result and Discussion

3.1. Weight loss
The test result showed that the use of active paper can reduce the weight loss rate. Weight loss caused by reduced water content in the fruits. Postharvest fruit no longer receives water intake from plants. Otherwise, damage to epidermal and periderm tissue causes acceleration of water out of the tissue. In previous research using peaches, plums [7] and mushroom [8] also showed the ability of active paper to inhibiting weight loss. The active paper incorporated with oleoresin of cinnamon distillation leaf residues is known to have antimicrobial compounds that can inhibit the growth of destructive microbes in the fruit that caused the demage to cell tissue [10].

3.2. fruits hardness
The data showed different results between the effect of active paper on the hardness level of fruit surface (fruit peel) (table 2) and fruit flesh (table 3). The level of hardness decline on the fruit surface was higher than the flesh fruit. Although it seems that the active paper was able to slow the hardness decline of the fruit surface. There is no significant decrease in the level of fresh fruits hardness and application of active paper does not show an effect on the hardness level. One of the factors that accelerate the decline in the hardness level is the extracellular enzyme-producing microbes that degrade polysaccharide, hemicellulose, and pectin [17] layers in the skin layer of the fruit. Active paper with the addition of essential oils is known to have the ability to inhibit microbial growth [8]. The active paper will release the compound to headspace. Inhibition effect possible because of the compound released into the headspace has antimicrobial effect.

3.3. Color (red) intensity on fruit surface
Table 4 showed that the use of active paper inhibits the color change that occurs on the surface of the fruit surface. Discoloration in non-climatic fruits indicates the presence of fruit shaking and the process leading to decay [18]. The color change is probably caused by the destruction of the cell and causes the cytosol content out to spur enzyme activation that causes browning [5]. In previous research using active paper with the addition of cinnamon oleoresin was able to inhibit the enzymatic oxidation effect in mushroom. It is possible that the active paper incorporated with oleoresin of distilled cinnamon leaves residue also gives a similar effect.

3.4. pH, total titrable acid, and total soluble solid
Data showed that the use of active paper has no significant effect on pH, total titrable acids, and total soluble solid of fruits flesh compared to controls (table 5, table 6, and table 7). This is thought to be due to the properties of the active paper used. active paper incorporated with essence oil known has antimicrobial activity [8] so that the effect of active paper was only shown in the damage caused by microbes, such as cell damage that causes weight loss, decreased hardness and color. Another possibility is related to the ability of active paper to transmit its compound on the headspace [8] so that the effect does not get into the flesh. in the another study showed that applies edible coating chitosan on cantaloupe did not show any effect on the chemical properties of the fruits during storage [19].

Table 1 Weight Loss during Storage

Treatment	Weight Loss (%)			
	Day 0	Day 3	Day 6	Day 9
Control	0.00^{Aa}±0,00	1.56^{Ab}±0,07	11.78^{Ec}±0,21	24.80^{Ed}±0,00
Wrap	0.00^{Aa}±0,00	1.60^{Bb}±0,00	3.66^{Ac}±0,01	6.08^{Ad}±0,07
Wall	0.00^{Aa}±0,00	6.66^{Eb}±0,07	9.07^{Dc}±0,07	13.62^{Cd}±0,07
Pad	0.00^{Aa}±0,00	1.69^{Cb}±0,00	6.39^{Cc}±0,00	16.16^{Dd}±0,07
Sidelines	0.00^{Aa}±0,00	1.92^{Db}±0,07	4.66^{Bc}±0,00	13.02^{Bd}±0,00

Table 2 Surface Fruits Hardness

Treatment	Hardness Level (Pa)			
	Day 0	Day 3	Day 6	Day 9
Control	14668Ab±465	14408Ab±923	1113Aa±60	1044Aa±17
Wrap	15924Bd±48	14475Ac±352	9259Cb±124	1386Ba±30
Wall	14371Ac±235	13841Ac±125	10107Db±329	1812Ca±52
Pad	14652Ac±490	13922Ac±380	7420Bb±480	1288Ba±116
Sidelines	14954Ac±98	14626Ac±415	7265Bb±195	1394Ba±22

Table 3 Flesh Hardness

Treatment	Hardness Level (Pa)			
	Day 0	Day 3	Day 6	Day 9
Control	947.4ABa±60	926.3Aa±30	841.4Aa±50	898.0Aa±30
Wrap	926.2Aa±30	1060.6Bb±60	947.5Bab±00	961.6ABab±60
Wall	1039.4ABa±50	997.0ABa±70	1110.1CDa±10	1046.5Ba±60
Pad	1050.0Ba±55	1028.8ABa±25	1046.5Ca±40	1000.5ABa±5
Sidelines	1032.3ABa±20	1025.2ABa±10	1166.7Db±10	1025.2Ba±29

Table 4 Color Intensity

Treatment	Intensity Red Color Fruit			
	Day 0	Day 3	Day 6	Day 9
Control	340.88ABd±19.62	278.88Bc±7.60	148.50Ab±4.60	79.25Aa±5.30
Wrap	395.50Bc±8.48	345.75Ec±6.01	296.00Eb±1.77	174.38Da±7.60
Wall	334.09ABd±3.42	319.62Dc±3.71	266.38Db±7.60	112.50Ba±2.12
Pad	322.66Ad±3.77	277.38Bc±4.77	195.92Cb±7.90	136.67Ca±4.72
Sidelines	322.50Ad±2.12	261.00Ac±8.02	165.83Bb±3.06	109.62Ba±6.89

Table 5 pH Value

Treatment	pH Value			
	Day 0	Day 3	Day 6	Day 9
Control	5.29Aa±0.14	5.10Ca±0.03	5.54Aa±0.36	5.34Ba±0.06
Wrap	5.30Aab±0.03	5.07Ca±0.06	5.89Ad±0.01	5.39Bc±0.16
Wall	5.31Ab±0.04	5.02CBa±0.01	5.72Ac±0.04	4.98Aa±0.11
Pad	5.24Aab±0.20	4.94BAa±0.01	5.99Ac±0.19	5.34Bb±0.01
Sidelines	5.16Ab±0.10	4.86Aa±0.02	567Ad±0.22	5.24Bc±0.04

Table 6 Total Titratable Acid

Treatment	Total Titratable Acid (%)			
	Day 0	Day 3	Day 6	Day 9
Control	10.20ABc±0.28	8.45Ab±0.21	9.30Abc±0.71	6.15ABa±0.63
Wrap	10.07BCb±0.42	10.20Cb±0.56	10.80ABb±0.28	4.70Aa±1.27
Wall	9.15Ab±0.07	8.90ABb±0.34	10.00ABb±0.85	5.32Aa±0.17
Pad	11.42Cb±0.60	9.74BCb±0.08	10.95ABb±1.20	5.60Aa±0.00
Sidelines	10.50BCb±0.56	8.65Ab±0.49	11.35Bb±0.21	7.75Ba±0.07

Table 7 Total Soluble Solid

Treatment	Total Soluble Solid (oBrix)			
	Day 0	Day 3	Day 6	Day 9
Control	10.07Aa±0.39	11.16Ba±0.62	10.38Aa±0.25	11.05Aa±0.35
Wrap	9.83Aa±0.24	11.38Bb±0.53	10.12Aab±0.53	11.20Aab±0.56
Wall	9.25Aa±0.35	9.95ABa±0.28	9.50Aa±0.71	11.60Ab±0.28
Pad	9.10Aa±0.85	10.20ABab±0.71	9.50Aa±0.71	11.70Ab±0.14
Sidelines	9.17Aa±0.04	9.40Aa±0.71	9.22Aa±0.32	11.80Ab±0.00

Table 8 Total Plate Count

Treatment	Total Plate Count (CFU/ml)			
	Day 0	Day 3	Day 6	Day 9
Control	3.12Aa±0.39	5.02BCb±0.52	5.54Ab±0.74	7.31ABc±0.37
Wrap	2.97Aa±0.16	3.64Aa±0.24	4.60Ab±0.51	7.08ABc±0.04
Wall	3.16Aa±0.04	4.06Ab±0.18	5.27Ac±0.16	7.36Bd±0.41
Pad	3.01Aa±0.01	4.32ABb±0.24	5.44Ac±0.06	6.34Ad±0.05
Sidelines	3.30Aa±0.02	5.20Cb±0.34	5.51Ab±0.12	7.05ABc±0.60

The number followed by superscript (A, B, C, D, E) in the same column show no significant difference at 5% significant level.
The number followed by superscript (a, b, c, d, e) in the same line show no significant difference at 5% significant level.

3.5. Total plate count

Most of the damage caused to fruit products is caused by decaying microorganism. Test results showed that the use of active paper by wrapping methods and placement on the wall able to withstand the growth rate of microorganisms although other treatments did not show a significant effect. microorganisms found in the fruit of the post-harvest dragon include Xanthomonas campestri, Dithiorella spp, Fusarium lateritium, Aspergillus niger and Aspergillus flavus [20]. Active paper with the addition of tested oleoresin is able to inhibit the growth of niger aspergillus and pseudomonas fluorescence and is known to have linalool compounds capable of inhibiting the growth of microbes [10]. The active compounds contained in the paper will be transferred to the headspace [8]. Treatment with the wrapping method gives a better effect than other treatments is possible because it has a larger surface of the fruit. A larger contact area will provide better protection on packed product

4. Conclusion

The active paper placement methods significantly affected the weight loss, surface firmness and color peel change of super red dragon fruits samples. However, active paper placement methods insignificantly affected the titrable acid total, soluble solid total, pH, flesh firmness and microbial spoilage of super red dragon fruits samples. The best method to maintain the super red dragon fruits quality was wrapping method.

Acknowledgments
This publication supported by research project PNBP Universitas Sebelas Maret

References

[1] Farikha, I N, C Anam, and E Widowati *J. Teknologi Pangan* **2** 30-38
[2] Wahyuni, Rekna 2011 *Jurnal Teknologi Pangan* **2** 69
[3] Markakis, P 1982 *Antocyanin as Food Colors* (New York : Academic Press)
[4] Nerd, Avinoam, F Gutman, and Y Mizrahi 1999 *Postharvest Biol. Technol.* **17** 39-45
[5] Valero, Daniel and Maria Serrano 2010 *Postharvest Biology and Technology for Preserving Fruit Quality* (London : CRC Press)
[6] Akrami F, A Rodriguez-Lafaunte, K Bentayeb, D Pezo, S R Ghalebi, and C Nerin 2014 *LWT-Food Sci. and Tech.* **XXX** 1-5
[7] Rudra, S G, V Singh, S D Jyoti, and U S Shivhare 2013 *Food Biosci.* **3** 49-58
[8] Echegeyen, Y and C Nerin 2015 *Food Chem.* **170** 30-36
[9] Khasanah, L U, Kawiji, R Utami, and Y M Aji 2015 *Jurnal Aplikasi Teknologi Pangan* **4** 48-54
[10] Khasanah, L U, G J Manuhara, D C Pravitama, W Atmaka, R Utami, and Kawiji 2015 Karakterisasi Pengemas Kertas Aktif dengan Penambahan Oleoresin Ampas Destilasi Daun Kayu Manis *Prosiding Seminar Nasional dan Pameran Produk Pangan* (Semarang: Perhimpunan Ahli Teknologi Pangan Indonesia)
[11] Hadi, M 2012 *BIOMA* **6** 12-18
[12] Manuhara, G J, L U Khasanah, and R Utami 2016 *IOP Conf. Series: Materials and Engginering* **107**012031
[13] De Freitas, ST, and E J Mitchan 2013 *Scientia Agricola* **70** 257-262
[14] DeMan, J M 1997 *Teknologi Makanan* (Bandung : Penerbit ITB)
[15] El-Ramady, H R, E Domokos-Szabolcsy, N A Abdalla, H S Taha, dan M Farl 2015 *Postharvest Management of Fruits and Vegetables Storage* (Switzerland : Springer International Publishing)
[16] Fardiaz, S 1993 *Analisis Mikrobiologi Pangan* (Jakarta : Raja Grafindo)
[17] Barth, M, T R Hankinson, H Zhuang, and F Breldt 2009 *Compedium of the Microbial Spoilage on Food and Baverages* Springer Science+Busines Media 135-183
[18] Istianingsih, T and D Efendi 2013 *J. Hort Indonesia* **4** 54-61
[19] Martinon, M E, R G Moreira, M E Castell-Perez, and C Gomes 2014 *J. Food Sci and Tech* **56** 341-450
[20] Paul, R E 2014 *Fruit, Nut and Baverage* Cropa Honolulu

Analytical method development of nifedipine and its degradants binary mixture using high performance liquid chromatography through a quality by design approach

S Choiri[1,2], A Ainurofiq[2], R Ratri[3], M U Zulmi[3]
[1] Department of Pharmacy, Sebelas Maret University, Ir. Sutami 36 A, Surakarta, Indonesia 57126
[2] Faculty of Pharmacy, Gadjah Mada University, Sekip Utara, Yogyakarta, Indonesia, 55281
[3] Department of Pharmaceutical Science and Technology, Setiabudi University, Surakarta, Indonesia, 57127

Email: syaiful.apt@student.uns.ac.id

Abstract. Nifedipin (NIF) is a photo-labile drug that easily degrades when it exposures a sunlight. This research aimed to develop of an analytical method using a high-performance liquid chromatography and implemented a quality by design approach to obtain effective, efficient, and validated analytical methods of NIF and its degradants. A 2^2 full factorial design approach with a curvature as a center point was applied to optimize of the analytical condition of NIF and its degradants. Mobile phase composition (MPC) and flow rate (FR) as factors determined on the system suitability parameters. The selected condition was validated by cross-validation using a leave one out technique. Alteration of MPC affected on time retention significantly. Furthermore, an increase of FR reduced the tailing factor. In addition, the interaction of both factors affected on an increase of the theoretical plates and resolution of NIF and its degradants. The selected analytical condition of NIF and its degradants has been validated at range 1 – 16 µg/mL that had good linearity, precision, accuration and efficient due to an analysis time within 10 min.

1. Introduction

Nifedipine (NIF) as a dihydropyridine derivate is a potent and clinically active as calcium channel blockers which has a low photostability under UV light or sunlight [1]. The degradant products of NIF depend on the light exposure which become dehydronifedipine or nitroso analogs and these are clinically meaningless [2–4]. Several methods were carried out to improve the NIF stability e.g. solid dispersion [5], encapsulation using cyclodextrin [6], polymeric nano encapsulation or formulated in dosage form [7]. However, the degradation process could not be avoided in these systems. In addition, because of a high similarity of structure between NIF and its degradants, a selective analytical method has been purposed and recomended to analyze the NIF and its degradants [8,9].

High performance liquid chromatography (HPLC) is a first choice and selected method to analyze degradant products or related substances in the bulk pharmaceutical or dosage form due to cheap, selectivity, and sensitivity [9,10]. Several works have been reported in the use of liquid chromatography to analyze NIF degradant products e.g. ultra-performance liquid chromatography [8] or HPLC

[2,7,11,12]. However, in these methods did not implement the design experiment (DoE) to determine the most effective and efficient conditions of HPLC analysis.

In order to improve the efficiency and cost e.g. time to analysis in the HPLC method, the DoE was implemented in this study using a full factorial design [13]. Curvature in the design was added to improve the predictive power of model [14]. The DoE was used to determine the optimized region at a range of factor depending on the quality target product profiles [15,16]. The use of DoE was not only to determine the optimized condition but also to elucidate the effect and interaction of each factor [9,17].

The aim of this research was to develop and optimize the HPLC condition using 2^2 full factorial design with a curvature to improve the predictive power of model. Therefore, an effective, efficient, and validated analytical method was achieved to analyze the NIF and its degradant in the force degradation study.

2. Experimentals

2.1. Materials

Nifedipine (NIF) (Lot No. 400196067) was obtained from Dexa Medica (Palembang, Indonesia) Acetonitrile (ACN), methanol (MeOH), chloroform and water was obtained from Merck (Darmstadt, Germany) as a chromatography grade.

2.2. Preparation and characterization of degradation of nifedipine

Pure NIF was exposed to the sunlight during 2 min and 2 days. The degradation was characterized by a thin layer chromatography (TLC). The TLC condition was performed by a silica plate 60 $GF_{254\ nm}$ and chloroform:MeOH (9:1). An identification was performed visually or assisted by a UV lamp at a wavelength of 254 nm. A Shimadzu IR-21 Prestige Fourier transform infrared (FTIR) spectrophotometer (Kyoto, Japan) was used to confirm the degradation process of NIF. A potassium bromide method using a 1% of sample was used to characterize the NIF degradation. The sample was weighed and pressed using a hydraulic presser at 6 kN for 5 min. Samples were scanned using a 32 times iteration and resolution of 2 cm^{-1} from a range of wavenumber of 400-4000 cm^{-1}.

2.3. Design of Experimental of HPLC analysis

A 2^2 full factorial design using an addition of center point as a curvature was used to obtain the design space of HPLC analysis. Flow rate (FR) and mobile phase composition (MPC) i.e. ACN to MeOH ratio were used as factors using different levels of factors which are presented in Table 1. A 4 runs was constructed based on a two factors and levels, respectively and a one run was added in the center point as the curvature. The design space was determined by several parameters i.e. time retention (tR), tailing factor (tf), theoretical plates (N), and resolution (Rs).

Table 1. Factors and levels of HPLC analysis using 2^2 full factorial design with a curvature

Factor	**Level**		
	-1 (low level)	**0 (curvature)**	**+1 (high level)**
Flow rate (mL/min)	0.8	1.05	1.3
ACN to MeOH ratio	0.25	2.125	4

2.4. HPLC condition

A Shimadzu HPLC (Kyoto, Japan) using a LiChrosphere 100 RP-18 (250x4.6; 5μm) column (Merck; Darmstadt, Germany) and ACN to MeOH under several ratios at determined FR depending on DoE (Table 1) with isocratic system was used to assay of NIF from a mixture with degraded NIF. NIF and degraded NIF were mixed and dilute with a mobile phase. The sample was filtered using a 0.22 μm membrane filter and a volume of 20 μL of sample was injected into the HPLC instrument at a column temperature of 40°C.

2.5. Parameter analysis and determination of design space

Each parameter was analyzed by a multiple linear regression analysis (MLRA) model. Several model criteria were used to determine the best fitting model including significant model, not significant curvature, coefficient of determination (R^2) more than 0.7, and difference between adjusted R^2 and predicted R^2 (Adj. R^2-Pred R^2) of 0.2. The best models were used to construct the contour plot. The design space was determined by a desirability function of the superimposed contour plot at several requirements i.e. tR less than 10 min, tf less than 2.5, and Rs higher than 2.

2.6. Validation of optimized condition

Several validation parameters were performed in the optimized condition i.e. linearity, limit of detection (LoD), limit of quantification (LoQ). Cross validation using a leave one out technique was carried out to validate especially the accuration and precision.

Linearity was prepared by several point calibrations. The range concentration from 1 to 16 µg/mL. Each concentration was filtered using a 0.22 µm membrane filter and 20 µL was injected into the HPLC instrument. A calibration model was constructed based on calculated and observed data.

LoD and LoQ were calculated based on a noise to signal ratio and slope of regression coefficient (b). The noise was determined by standard deviation of noise (s) in the baseline of HPLC chromatogram. LoD and LoQ were calculated based on following equations.

$$LoD = 3.3 \times s/b \quad (1)$$

$$LoQ = 10 \times s/b \quad (2)$$

3. Results and Discussion

A good performance of analytical method i.e. selective, effective and efficient was needed to a routine analysis especially for an easily degradable drug [8]. NIF is a photodegradable drug that produce more than two degradation products depending on its mechanism and light exposure [1,2,4]. The NIF degradant product was confirmed by TLC and FITR spectra (data not shown). The TLC results showed that a similar pattern in chromatograms during 2 min and 2 days exposures was observed. It proved that a short time exposure of sunlight promoted the degradation of NIF. Meanwhile, the amount of degradation product increased during exposure. The FTIR spectra confirmed that a new peak was observed due to a C=N bonding vibration at a wavenumber of 1726.29 cm^{-1}. It showed that NIF degradants had a similar structure and distinct in a one functional group. Therefore, a good performance of analytical method to analyze the NIF and degradants must be developed [8]. FR and MPC were the most factors affecting the suitability test system on the HPLC analysis [10,13,18].

In order to develop validated, cheap, effective and efficient analytical methods, a 2^2 full factorial design with a curvature was applied. This system enhanced the validity and predictive power of the model due to a presence of a curvature. The curvature implied the value of intercept of MRLA depending on the observed data [14]. A not significant term of curvature was required to select the suitable model.

All models had significant models ($p<0.05$) and not significant of curvatures ($p>0.05$). The significant model indicated that the model could be used to describe the effect of independent variables on responses. Therefore, it was a first requirement in the model selection [16]. The curvature was used to verify the intercept of each model, a not significant term of curvature showed a not significant difference between the observed value in the middle level of each factor and intercept value. tR was selected as a response due to an efficient reason of analytical process in the HPLC. The shorter analysis time, the more efficient of analytical process using HPLC [19,20]. Hence, the effect and interaction of ACN to MeOH ratio and FR on tR can be shown in Table 2 and Fig. 1a. An increase the ACN to MeOH ratio and FR had significant effect ($p<0.05$) on reducing the tR. Although, an interaction of both factors increase the tR. An increase the ACN fraction comparted to the MeOH fraction reduced the tR due to ability of ACN to increase the elution power [21]. As same as the FR, increasing the FR reduced the time analysis due to reducing the interaction between compound and stationary phase [10,19]. Depending on contour plot of tR (Fig. 1a) showed that the highest tR was obtained at a low level of

ACN to MeOH ratio and FR. On the other hand, the lowest tR was obtained at a high level of both factors.

In addition, the effect and interaction of ACN to MeOH ratio and FR on tf are presented in Table 2 and Fig 1b. Only the ACN to MeOH ratio had a significant effect on an increase the tf ($p<0.05$). Increasing the tf was caused by an interaction between NIF and stationary phase. The higher interaction, the higher tf value. Fig. 1b showed that only the change of MPC affected the tf value, meanwhile the change of FR had no significant effect ($p>0.05$). There was no significant interaction between two factors ($p>0.05$). On the other hand, only FR significant affected on N ($p<0.05$) and had no interaction between both factors. The effect and interaction of ACN to MeOH ratio and FR on the N parameter can be deeply elucidated using N's contour plot (Fig. 1c). A significant color graduation could be observed at a change of FR. Although, an alteration of MPC had no significant effect ($p>0.05$). There was no interaction of two factors, an interaction was observed by a non-parallel or intersect patterns. Rs was selected as a factor due to important parameter to distinguish between main and degradant peaks of NIF. A high resolution was required to determine the design space of degradant analysis [10]. The effect and interaction of both factors are presented in Table 2 and Fig. 1d. MPC and FR had a significant effect on reducing the Rs ($p<0.05$), although there was no significant interaction of both factors ($p>0.05$). An increase the ACN fraction in mobile phase increased the elution power but the resolution was decreased. As similar as the flow rate, an increase the flow rate reduced the tR and reduced the resolution. Depending on the contour plot of resolution (Fig. 1d), it proofed that the MPC had a higher effect than flow rate on Rs and there was no specific interaction between each factor.

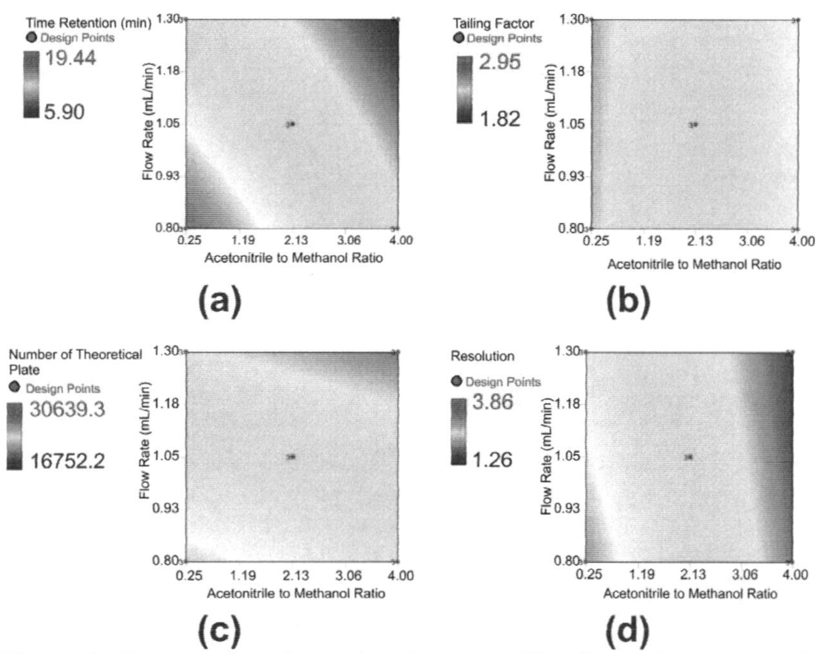

Figure 1. Contour plot of retention time (a), tailing factor (b), number of theoretical plates (c), and resolution (d)

Selection adequate model depended on how the model fit well. Goodness of fit parameters was selected as criteria to determine an appropriate model which could be used to design the optimum region [15]. R^2 measured the effect of independent variables on responses, R^2 more than 0.7 was a criterion in model selection to minimize the error and lack of effect. All models had R^2 more than 0.7. The Adj. R^2-Pred R^2 measured how adequate of model to predict the response, the model can be used to predict the response without a misleading if the Adj. R^2-Pred R^2 was less than 0.2 [15]. The Adj. R^2-Pred R^2 of tf and N responses had higher than 0.2, thus it indicated there was a potential miss-leading in the model to predict the response. Therefore, a verification step was required to obtain an adequate model. PRESS measured the residual between predicted and observed data, the PRESS could be compared between each factor using a relative value with a mean. Adequate precision measured the signal to noise ratio. The adequate precision more than 4 indicated that the model could be used to estimate or predict the response adequately [14,15]. Depending on the all fitting parameters, all models had a potential ability to predict and estimate the response, thus the optimum region could be determined by these models.

Table 2. Coefficient regression, statistical and fitting parameters on responses

	Time retention	**Tailing Factor**	**Number of Theoretical Plates**	**Resolution**
Coefficient Regression				
Intercept	11.96	2.35	23053	2.38
A (ACN to MeOH ratio)	-3.90	0.29	-1124.21*	-0.87
B (Flow rate)	-2.75	-0.006*	-3523.84	-0.18
AB	0.72	0.012*	181.67*	0.083*
Fitting parameters				
R^2	0.9937	0.6858	0.7803	0.9491
Adjusted R^2	0.9918	0.5915	0.7144	0.9339
Predicted R^2	0.9858	0.2930**	0.5057**	0.8856
PRESS	3.99	1.02	1.04×10^8	1.16
Adequate precision	54.72	4.85	7.48	16.108
p-value (CI=95%)				
Model	<0.0001	0.0071	0.0013	<0.0001
Curvature	0.1702*	0.6285*	0.6493*	0.1977*

* = no significant different (p>0.05)
** = difference between adjusted and predicted R^2 more than 0.2

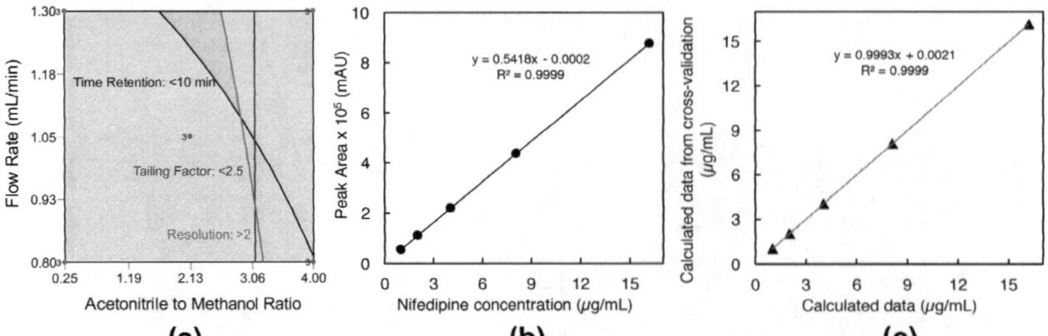

Figure 2. Superimposed of contour plot (a), calibration model (b), and cross-validation calibration model (c)

The optimum region as a controlled space was determined by several criteria. Furthermore, it is presented in the superimposed of each contour plot (Fig. 2a). In order to determine the controlled space, each parameter had a priority to determine the design space. Priority of tR and Rs responses was higher twice than that of the tf and N responses. ACN to MeOH ratio of 2.06 and flow rate of 1.30 mL/min was the optimized condition. This condition was used to construct the calibration model (Fig. 2b). The calibration equation had high R^2 0.9999 and adjusted R^2 of 0.9999. The linearity of concentration was obtained at a range from 1 to 16 μg/mL. Furthermore, the calibration model of cross validation (Fig. 2c) had a high predicted R^2 of 0.9999. There was an error of cross validation of 0.07% and had precision of 100.01% (RSD=0.27%). Based on the noise to signal ratio of baseline, LoD and LoQ were 8.9 and 27.0 ng/mL, respectively. The optimized condition conformed the requirement of validation parameter.

4. Conclusion
A 2^2 full factorial design with a curvature was successfully implemented to optimize the HPLC condition of NIF and its degradant analysis. FR and ACN fraction in MPC had a positive effect on an increase the efficiency of analytical process. However, an increase of both factors reduced the separation ability and an appropriate selection of region in the design space as the controlled strategy had the better results. The optimized condition was validated using cross-validation with a leave one out technique and had good linearity, sensitivity, precision and accuration.

Acknowledgement
This research was supported by "Lembaga Pengelola Dana Pendidikan" (Indonesian Endowment Fund for Education). Syaiful Choiri would like to thank Dexa medica (Palembang, Indonesia) for providing Nifedipine.

References
[1] Grundy J S, Kherani R and Foster R T 1994 *J. Pharm. Biomed. Anal.* **12** 1529–35
[2] Handa T, Singh S and Singh I P 2014 *J. Pharm. Biomed. Anal.* **89** 6–17
[3] Maafi W and Maafi M 2013 *Int. J. Pharm.* **456** 153–64
[4] Shamsipur M, Hemmateenejad B, Akhond M, Javidnia K and Miri R 2003 *J. Pharm. Biomed. Anal.* **31** 1013–9
[5] Keratichewanun S, Yoshihashi Y, Sutanthavibul N, Terada K and Chatchawalsaisin J 2015 *Pharm. Res.* **32** 2458–73
[6] Bayomi M A, Abanumay K A and Al-Angary A A 2002 *Int. J. Pharm.* **243** 107–17
[7] Tagliari M P, Granada A, Segatto S, Stulzer H K, Zanetti-Ramos B G, Fernandes D, Silva I T, Simões C M O, Sordi R, Assreuy J and Soldi V 2015 *Quimica Nova* **38** 781–6
[8] Galan-Rodriguez C, González-Álvarez J and Valls-Remolí M 2015 *Biomed. Chromatogr.* **29** 233–9
[9] Krishna M V, Dash R N, Jalachandra Reddy B, Venugopal P, Sandeep P and Madhavi G 2016 *J. Saudi Chem. Soc.* **20** S313–22
[10] Khan A, Imam S S, Aqil M, Sultana Y, Ali A and Khan K 2016 *Beni-Suef Univ. J. Basic Appl. Sci.* **5** 402–8
[11] Gil-Agustí M T, Carda-Broch S, Ll. Monferrer-Pons and Esteve-Romero J S 2006 *Biomed. Chromatogr.* **20** 154–60
[12] Ohkubo T, Noro H and Sugawara K 1992 *J. Pharm. Biomed. Anal.* **10** 67–70
[13] Sahu P K, Ramisetti N R, Cecchi T, Swain S, Patro C S and Panda J 2017 *J. Pharm. Biomed. Anal.*
[14] Ainurofiq A, Choiri S, Azhari M A, Siagian C R, Suryadi B B, Prihapsara F and Rohmani S 2016 *Adv. Pharm. Bull.* **6** 399–406
[15] Ainurofiq A and Choiri S 2016 *Pharm. Dev. Technol.* Article in Press 1–12
[16] Choiri S, Sulaiman T N S and Kuncahyo I 2014 *Indonesian J. Pharm.* **25** 255–64
[17] Yu L X, Amidon G, Khan M A, Hoag S W, Polli J, Raju G K and Woodcock J 2014 *AAPS J.* **16** 771–83
[18] Ganorkar S B, Dhumal D M and Shirkhedkar A A 2017 *Arab. J. Chem.* **10** 273–82
[19] Kalariya P D, Namdev D, Srinivas R and Gananadhamu S 2017 *J. Saudi Chem. Soc.* **21** S373–82
[20] Kawabe T, Tomitsuka T, Kajiro T, Kishi N and Toyo'oka T 2013 *J. Chromatogr. A* **1273** 95–104
[21] Zendelovska D, Simeska S, Sibinovska O, Kostova E, Miloševska K, Jakovski K, Jovanovska E, Kikerkov I, Trojačanec J and Zafirov D 2006 *J. Chromatogr. B* **839** 85–8

IOP Conf. Series: Materials Science and Engineering **333** (2018) 012065 doi:10.1088/1757-899X/333/1/012065

Understanding the drug release mechanism from a montmorillonite matrix and its binary mixture with a hydrophilic polymer using a compartmental modelling approach

S Choiri[1,2], A Ainurofiq[1]
[1] Department of Pharmacy, Sebelas Maret University, Ir. Sutami 36A Surakarta, Indonesia, 57126
[2] Faculty of Pharmacy, Gadjah Mada University, Sekip Utara, Yogjakarta, Indonesia, 55281

Email: syaiful.apt@student.uns.ac.id

Abstract. Drug release from a montmorillonite (MMT) matrix is a complex mechanism controlled by swelling mechanism of MMT and an interaction of drug and MMT. The aim of this research was to explain a suitable model of the drug release mechanism from MMT and its binary mixture with a hydrophilic polymer in the controlled release formulation based on a compartmental modelling approach. Theophylline was used as a drug model and incorporated into MMT and a binary mixture with hydroxyl propyl methyl cellulose (HPMC) as a hydrophilic polymer, by a kneading method. The dissolution test was performed and the modelling of drug release was assisted by a WinSAAM software. A 2 model was purposed based on the swelling capability and basal spacing of MMT compartments. The model evaluation was carried out to goodness of fit and statistical parameters and models were validated by a cross-validation technique. The drug release from MMT matrix regulated by a burst release mechanism of unloaded drug, swelling ability, basal spacing of MMT compartment, and equilibrium between basal spacing and swelling compartments. Furthermore, the addition of HPMC in MMT system altered the presence of swelling compartment and equilibrium between swelling and basal spacing compartment systems. In addition, a hydrophilic polymer reduced the burst release mechanism of unloaded drug.

1. Introduction

Drug release from a controlled release formulation was a complex mechanism and several mechanisms involved e.g. diffusion, erosion, swelling, and interaction between drug and matrix [1,2]. Drug release mechanism and kinetics could be recognized by a model of drug release [3–5]. A simplification of drug release kinetics and mechanisms had been reported by several models i.e. zero-order kinetics, first-order kinetics, Higuchi, Korsmeyer Peppas, Peppas Sahlin, Weibull, and Hixcon Crowel. However, this model explained based on a simplification of drug release mechanism [4,6,7].

Clay was widely used to modify the drug release from sustained or controlled release formulations [8,9]. Montmorillonite (MMT) was a major component in a bentonite clay which had been approved as an excipient [10,11]. MMT that consist of octahedral silica and alumina in a specific ratio had a basal spacing which can be used to incorporate the drug in the MMT-drug nanocomposite system. Meanwhile,

MMT had a swelling ability which can be applied to control the drug release [9]. Application of MMT in the drug delivery had been reported by several researchers [8,10–12]. In addition, the use of polymers in a controlled release formulation was a fundamental aspect. A hydrophilic polymer was conventionally used to retard the drug release [13,14]. The mechanism to control the drug release in the hydrophilic polymer was distinct from MMT. Although, in the MMT system, there were no reported data i.e. the mechanism and kinetic of drug release depending on a compartmental analysis.

Compartmental analysis was used to analyze the mass transport based on a compartmental system [15,16], this compartment can be designed hypothetically. The mechanism of mass transport from each compartment was distinct in kinetics and mechanism e.g. drug transport in the swelling or leached matrix region [7]. The drug transport kinetics in each compartment can be elucidated by the compartmental analysis [17]. Furthermore, the purpose of this study was to explain the suitable model of drug release mechanism from the MMT matrix and binary mixture between MMT and hydrophilic matrix depending on the compartment modelling approach.

2. Experimentals

2.1. Material
Montmorillonite (MMT) was prepared from purification of bentonite clay which obtained from Wonosegoro (Boyolali, Indonesia), Teofilin (THP) (Shandong Pharm, China) was purchased from Brataco (Surakarta, Indonesia), Methocel K4M (MK4M) was obtained from Colorcon (Herlyesville, PA).

2.2. Purification of montmorillonite
Bentonite clay was washed using demineralized water and colloidal phase was collected. Furthermore, it was centrifuged and washed two times followed by centrifugation. The sediment was dried in oven for 24 hours and passed through a 200 mesh sieve.

2.3. Incorporation of theophylline in montmorillonite and binary matrices
An equal weight ratio of THP and MMT was kneading using ethanol and water (7:3) and dried in oven for 24 hours. A different matrix using HPMC was used at level of MMT:MK4M (1:1). In addition, an equal weight ration of THP and a binary mixture MMT:MK4M (1:1) was kneaded using ethanol and water (7:3), then followed by drying in oven for 24 hours. All sample was passed through a 18 mesh sieve.

2.4. Preparation of tablet
An equivalent to 200 mg of THP in THP/MMT or THP/MMT-MK4M was weighed. Furthermore, THP/MMT or THP/MMT-MK4M, 1% of magnesium stearate as lubricant, and lactose to obtain 500 mg of tablet weight were mixed and compressed into tablet at hardness of 80-90 N.

2.5. Dissolution of tablet
Tablet contained THP/MMT or THP/MMT-MK4M was tested the drug release using an Electrolab DT-08 (Mumbay, India) dissolution tester. A 900 mL of phosphate buffer pH 7.2 was used as a medium at a temperature of 37±0.5°C. Aliquots at 5, 15, 30, 45, 60, 90, 120, 180, 240, 300, and 360 min were withdrawn and filtered using a 0.45 μm membrane filter. Samples were analyzed by a Shimadzu UV2100 (Kyoto, Japan) spectrophotometer at a wavelength 272 nm.

2.6. Analysis of drug release modelling
The drug release was modelled using a WinSAAM (University of Pennsylvania; Kenneth Square, PA). Cumulative of drug release data were plotted using two models depending on distinct of the compartmental approach. Model 1 consisted of 3 compartments i.e. tablet core, swelling front, and dissolution medium. Model 2 consisted of 4 compartment i.e. tablet core, swelling front, equilibrium compartment between MMT basal spacing and swelling front, and dissolution medium. The model was

defined using a WinSAAM notation: IC(1) i.e. initial dose at core compartment (100%); L(A,B) i.e. constant of rate at first order transport from compartment a to the compartment b (e.g. L(2,1), L(3,2), L(3,1), L(4,1), and L(4,2)); DT and DN(C) i.e. delay time and delay number of C compartment, respectively. Specific models are depicted in Fig. 1.

These parameters were calculated using a WinSAAM and the model was evaluated using goodness of fit (GoF), Akaike's information criterion (AIC) and root mean square error (RMSE). The model was validated using a leave one out cross-validation technique using fitting parameters i.e. coefficient of determination (R^2), adjusted coefficient of determination (Adj. R^2), predicted coefficient of correlation (Pred R^2), and RMSE cross-validation (RMSECV).

3. Results and Discussion

The designed model is presented in Fig 1, it principally obtained from the theoretical drug release mechanism from a hydrophilic matrix system [7]. Transport mechanism of drug naturally followed the diffusional release system thus it depends on the driving force of this system i.e. a concentration gradient [2]. In addition, MMT has ability to swelling, this mechanism can be used to control the drug release as a physical barrier [9]. Furthermore, the model 1 was designed based on these mechanisms. Meanwhile, the model 2 was designed depending on the presence of basal spacing of MMT which can be used as a nano-porous system and drug loaded into nano-porous then had interaction via a hydrogen bonding or other interactions with a Si-O-Si group function. This interaction caused releasing of the drug as a controlled manner. Each model had a compartmental system e.g. 1: tablet's core, 2: swelling front, 3: dissolution medium, and 4 as a basal spacing compartment. The mass transport from each compartment followed the direction which was depicted in each model scheme.

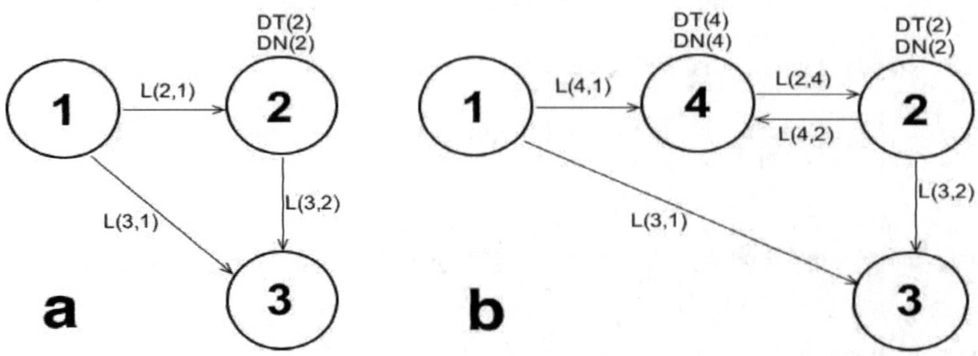

Figure 1. Schematic compartment model of drug transport from montmorillonite and binary mixture with a hydrophilic polymer matrix. The compartment model without an equilibrium system (a) and compartment model with an equilibrium system between swelling and montmorillonite basal spacing (b)

The drug release from MMT matrix was modelled using model 1 and model 2 that is presented in Fig. 2. The 50% of drug released during 6 hours (Fig. 2i). The drug release from MMT matrix hypothetically controlled by swelling and interaction between drug and Si-O-Si functional group in the basal spacing [9]. These mechanisms governed and modulated the drug release. In order to understand the swelling and basal spacing interaction, the drug release was modelled by a system. The model 1 implied that the drug release controlled based on a swelling mechanism, although the model 2 controlled by swelling and interaction between drug and basal spacing of MMT. The result showed that, a L(3,1) was about three times than L(2,1) and L(3,2). Depending on the rate constant of each compartment, the drug release form MMT depending on model 1 is dominantly controlled by a burst release mechanism. The burst release mechanism was proved with no delay time in the swelling compartment due to unloaded drug. Theoretically, the swelling of material needed more time to produce a good enough viscosity to control the drug release [18]. The burst release mechanism can be seen in the initial time in the Fig. 2i. Meanwhile, the drug release based on the model 2 not only was controlled by the burst release mechanism but also the interaction in the basal spacing. There was an equilibrium mass transport between swelling and basal spacing compartments and the mass transport delayed in these compartments. It proved by enabling the rate constant transport between basal spacing (compartment 2) and swelling compartment (compartment 4).

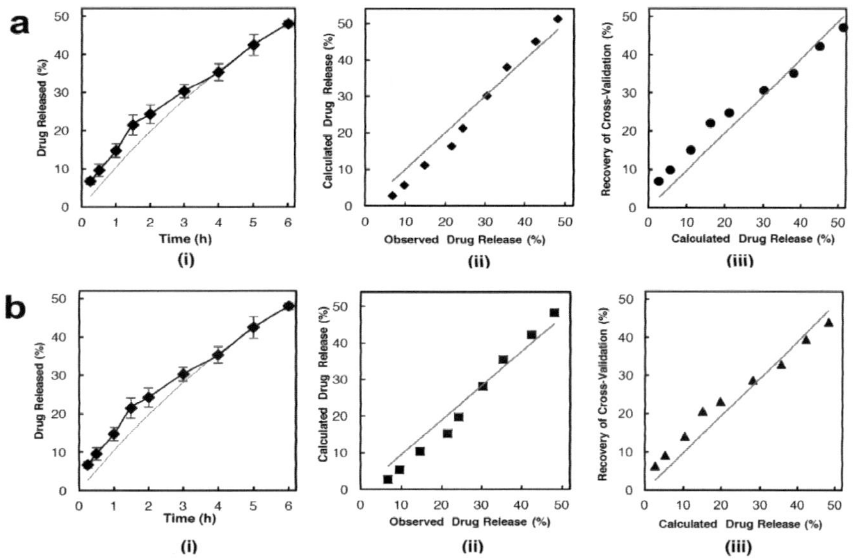

Figure 2. Drug release profile (i), calibration model (ii), and cross-validation model (iii) of montmorillonite matrix system. Model 1(a) and Model 2 (b)

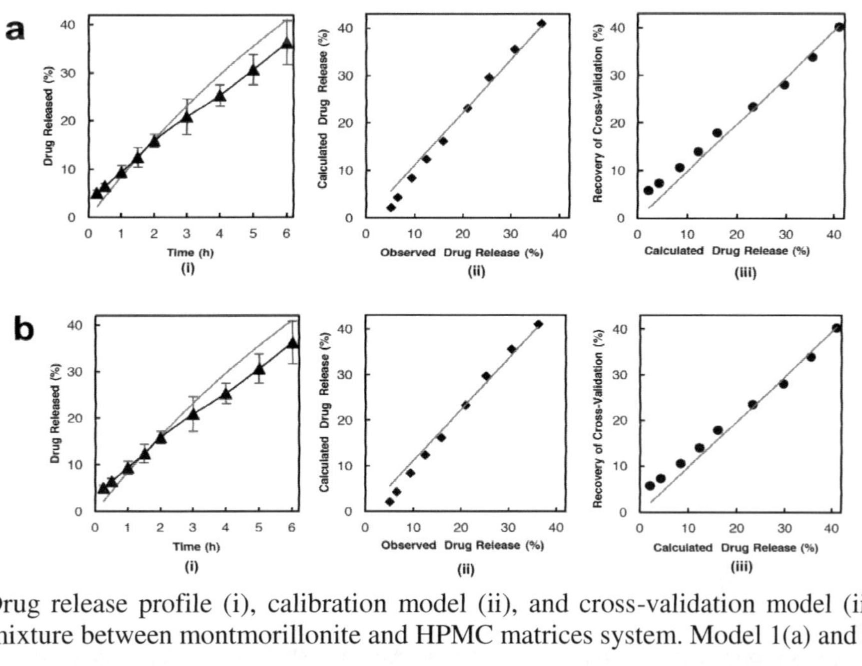

Figure 3. Drug release profile (i), calibration model (ii), and cross-validation model (iii) of binary mixture between montmorillonite and HPMC matrices system. Model 1(a) and Model 2 (b)

In order to reduce the burst release effect, a hydrophilic polymer, MK4M was added in this formulation as a binary mixture of MMT-MK4M. The observed and modelled drug release is presented in Fig. 3(i). The drug release showed the reduction of initial drug release compared to the single MMT system, although the amount of drug release reduced after 3 hours. This system altered the kinetics and mechanism transport from each compartment. Based on two purposed models, it proved that addition of hydrophilic polymers had an absence of the burst release effect and the equilibrium between basal spacing and swelling compartments was disappear. These phenomena showed by a zero constant rate of L(2,1) and L(4,2). Therefore, this system had no mass transport from compartment tablet's core to swelling or no equilibrium phase from basal spacing to swelling compartment. Hence, it can be concluded that addition of hydrophilic polymer reduced the number of the compartment system i.e. three or four compartments to two compartments.

The best fitting model could be used to describe the kinetics and mechanism of drug release [3]. Therefore, the selection of an appropriate model was preformed statistically. Furthermore, fitting parameters were used to estimate how adequate the model to predict the drug release which are presented in Table 1. Compared to the both models in MMT matrix, model 2 was selected to describe the drug release from MMT matrix because AIC and RMSE was lower than that of model 1. These parameters depict the distinct between predicted and observed data [6,17]. Although, GoF parameters, the R^2 and Adj R^2 was lower than that of model 1 but these parameters were not significant difference and the residual was mainly considered compared to the coefficient of determination. Residual of each model can be observed in calibration of model which is presented in Fig 2 (ii) or 3 (ii). Cross-validation technique using an one leave out technique was successfully used to validate the model based on a calibration model [19,20]. The calibration model of cross-validation of each model is presented in Fig 2(iii) or 3(iii). In addition, the selection of model 2 to describe the drug release considered the cross-validation results, the RMSECV was lower than that of model 1. Hence, this model was adequate to describe the model due to high coefficient of determination (more than 0.9) and low residual error (RMSE and Adj R^2-Pred. R^2 less than 0.2)[21].

Table 1. Goodness of fit parameters of model and cross-validation model

Parameter	Montmorillonite System		Montmorillonite and HPMC binary system	
	Model 1	**Model 2**	**Model 1**	**Model 2**
R^2	0.9577	0.9560	0.9784	0.9785
Adjusted R^2	0.9517	0.9497	0.9753	0.9754
RMSE	3.58	3.44	2.05	2.05
AIC	43.69	42.96	33.61	33.62
Cross-validation				
Pred. R^2	0.9245	0.9199	0.9688	0.9689
(Adj. R^2-Pred. R^2)	0.0332	0.0361	0.0065	0.0065
RMSECV	3.87	3.72	2.13	2.13

4. Conclusion

The drug release kinetics and mechanism from MMT and binary mixture MMT-hydrophilic polymer has been studied. Model 2 was the suitable model to describe the drug release form the montmorillonite and its binary mixture with the hydrophilic polymer. The modeling depicted the MMT system had ability to controlled the drug release by swelling ability and interaction with basal spacing i.e. drug intercalated. In addition, a hydrophilic polymer in this system altered the kinetic and mechanism of drug release, although it has been successfully proved to reduce the initial drug release.

Acknowledgement

This research was supported by "Lembaga Pengelola Dana Pendidikan" (Indonesian Endowment Fund for Education). Syaiful Choiri would like to thank Colorcon (West Point, PA) for providing the Methocel K4M.

References

[1] Siepmann J and Peppas N A 2012 *Adv. Drug Deliv. Rev.* **64**, Supplement 163–74
[2] Siepmann J and Siepmann F 2012 *J. Controlled Release* **161** 351–62
[3] Ainurofiq A and Choiri S 2015 *Lat. Am. J. Pharm.* **34** 1328–37
[4] Costa P and Sousa Lobo J M 2001 *Eur. J. Pharm. Sci.* **13** 123–33
[5] Peppas N A and Narasimhan B 2014 *J. Controlled Release* **190** 75–81
[6] Ainurofiq A and Choiri S 2015 *Trop. J. Pharm. Res.* **14** 1129
[7] Siepmann J, Kranz H, Bodmeier R and Peppas N A 1999 *Pharm. Res.* **16** 1748–56
[8] Iannuccelli V, Maretti E, Montorsi M, Rustichelli C, Sacchetti F and Leo E 2015 *Int. J. Pharm.* **493** 295–304
[9] Jayrajsinh S, Shankar G, Agrawal Y K and Bakre L 2017 *J. Drug Deliv. Sci. Technol.* **39** 200–9
[10] Oliveira A S, Alcântara A C S and Pergher S B C 2017 *Mater. Sci. Eng. C* **75** 1250–8
[11] Thakur G, Singh A and Singh I 2015 *Sci. Pharm.* **84** 603–17
[12] Bera H, Ippagunta S R, Kumar S and Vangala P 2017 *Mater. Sci. Eng. C* **76** 715–26
[13] Tiwari S B and Rajabi-Siahboomi A R 2009 *Drug Deliv Tech* **9** 20–7
[14] Zhou D, Law D, Reynolds J, Davis L, Smith C, Torres J L, Dave V, Gopinathan N, Hernandez D T, Springman M K and Zhou C C 2014 *J. Pharm. Sci.* **103** 1664–72
[15] Nugroho A K, Della-Pasqua O, Danhof M and Bouwstra J A 2005 *Pharm. Res.* **22** 335–46
[16] Nugroho A K, Pasqua O D, Danhof M and Bouwstra J A 2004 *Pharm. Res.* **21** 1974–84
[17] Nugroho A K 2014 *Indonesian J. Pharm.* **25** 31
[18] Borgquist P, Körner A, Piculell L, Larsson A and Axelsson A 2006 *J. Controlled Release* **113** 216–25
[19] Kurniawati E, Rohman A and Triyana K 2014 *Meat Sci.* **96** 94–8
[20] Rohman A, Sismindari, Erwanto Y and Che Man Y B 2011 *Meat Sci.* **88** 91–5
[21] Ainurofiq A, Choiri S, Azhari M A, Siagian C R, Suryadi B B, Prihapsara F and Rohmani S 2016 *Adv. Pharm. Bull.* **6** 399–406

Acute and Subchronic Toxicity of Self Nanoemulsifying Drug Delivery Systems (SNEDDS) from Chloroform Bay Leaf Extract (*Eugenia Polyantha* W.) with Palm Kernel Oil as A Carrier

F Prihapsara[1], Mufidah[1], A N Artanti[1], M Harini[2]
[1]Department of Pharmacy, Faculty of Mathematics and Natural Science, SebelasMaret University
[2]Department of Biology, Faculty of Mathematics and Natural Science, SebelasMaret University

E-mail: feapri87@gmail.com

Abstract. The present study was aimed to study the acute and subchronic toxicity of Self Nanoemulsifying Drug Delivery Systems (SNEDDS) from chloroform bay leaf extract with Palm Kernel Oil as carrier. In acute toxicity test, five groups of rat (n=5/groups) were orally treated with Self Nanoemulsifying Drug Delivery Systems (SNEDDS) from chloroform bay leaf extract with doses at 48, 240, 1200 and 6000 mg/kg/day respectively, then the median lethal dose LD_{50}, advers effect and mortality were recorded up to 14 days. Meanwhile, in subchronic toxicity study, 4 groups of rats (n=6/group) received by orally treatment of SNEDDS from chloroform bay leaf extract with doses at 91.75; 183.5; 367 mg/kg/day respectively for 28 days, and biochemical, hematological and histopatological change in tissue such as liver, kidney, and pancreatic were determined. The result show that LD_{50} is 1045.44 mg/kg. Although histopathological examination of most of the organs exhibited no structural changes, some moderate damage was observed in high- dose group animals (367 mg/kg/day). The high dose of SNEDDS extract has shown mild signs of toxicity on organ function test.

1. Introduction

Bay leaf *(Eugenia polyantha)* is one of the plants used to treat diabetes mellitus, beside that it is also known to have potency to cure high blood pressure, high cholesterol, hypertension, gastritis and diarrhea [1]. Ethanolic extract of bay leaf have some active compound such as saponins, triterpenoids, flavonoids, polyphenols, alkaloids, tannins and essential oils consisting of sesquiterpenes, lactones and phenols [2]. Flavonoids lead to decreasing the blood glucose levels [3].

Non polar extract of bay leaf has low solubility and low bioavailability profile. It will influence the therapeutic effect, so that its require a large dose to achieve the optimized theurapeutic effect [4]. SNEDDS *(Self-Nanoemulsifying Drug Delivery System)* is one of dosage form to improve bioavailability and solubility profile in the body.

According to previous research reported that SNEDDS of bay leaf chloroform extract using *Palm Kernell Oil* as carrier with compound of surfactant-cosurfactant using Tween 80-PEG (polyethylene glycol) 400 showed characterization of nanoparticle with transmittance values of 81.81% and the average size droplet 218.90 nm [5]. Antidiabetic activity showed that SNEDDS of bay leaf chloroform extract with Palm Kernell Oil as a carrier to rats at a dose of 91.75 mg/kg BW can lower blood glucose levels equivalent to metformin [6].

The toxicity test consists of 2 types: general toxicity tests include acute toxicity, subchronic and chronic toxicity, and special toxicity tests such as teratogenic test. Acute toxicity test was determined value of LD_{50} [7]. Herbal medicine requires time to achieve therapeutic effects due to slow drug reactions. In addition, its use repeatedly and in the long term, can allow the emergence of unexpected side effects too, it is necessary to test the subcronic toxicity to know harmful effects that may occur in the long-term use of the drug. Research on the toxicity of bay leaf extract and its dosage form has not been widely studied in Indonesia. One study suggested that the LD_{50} extract of bay leaf *(Eugenia polyantha W.)* to zebrafish embryos at 0.060 mg/ml [8].

Based on the above description, an acute and subchronic toxicity test of SNEDDS extract of chloroform bay leaf extract is required to determine the level of the dosage in a single dose and repeated dose. The aim of this research was to find out LD_{50} of SNEDDS bay leaf extract, its toxic symptoms, and to know the effect of SNEDDS preparation of bay leaf extract with repeatable dose for 28 days, profile of weight, SGPT (Serum Glutamic-Pyruvic Transaminase), billirubin and histological profile in liver, pancreas, and kidney of rats.

2. Experimental

2.1 Materials

Materials which were used in this research are bay leaf, male wistar rat, distilled water, chloroform, Palm Kernel Oil (PKO), PEG 400, tween 80, formalin 80%, ethanol 70 % ethanol 95%, ethanol 99.5%, xylol solution, paraffin liquid, Mayer's egg albumin, hematocyclines-xylene, Harri'shematoxylene, and entellan®.

2.2. Procedure

2.2.1. Preparation SNEDDS Bay Leaf Extract with Palm Kernell Oil as a Carrier. Bay leaf simplicia powder was extracted by using maceration method with chloroform solvent for 5 days. Macerate was filtered through glass funnel to separate macerate from simplicia powder. Macerate have been evaporated with rotary evaporator in 55^0C temperature and 6 rotary speeds until it became concentrated. Each material is weighed and prepared. Chloroform extract of bay leaf (0.15 grams) was added to 5 ml of carrier component, then homogenized with magnetic stirrer for 30 minutes. It was sonicated for 15 minutes, then incubated in water bath at a temperature 45°C for 10 minutes. Finally, preparations settling at a temperature of 27°C for 24 hours.

Table 1. Formula of SNEDDS

Material	Amount
Bay leaf chloroform extract	0.15 g
Tween 80	1.6 ml
PEG 400	2.4 ml
Palm Kernell Oil	1.0 ml

2.2.2 *Animal Preparation.* Before testing begin, animals are adapted in the experimental room for seven days. Animals are observed health and behavior. Animals used in the experiment were healthy animals, no weight loss exceeded 10% and showed no behavioral abnormalities and deviations from normal circumstances [9]. Principles of laboratory animal care guidelines were followed and prior permission was sought from the Health Research Ethics Committee for conducting the study (Ref. No. 432/V/HREC/2017).

2.2.3 *Acute Toxicity Test.* Test animals were used male white rats Wistar strain, age 2-3 months, initial weight 113-171 grams. There were 5 groups (n=5) consisting of 4 test groups with 4 dose variations and 1 control group. Test groups consist of SNEDDS from chloroform bay leaf extract. Four test group were administered 4 doses (48, 240, 1200, and 6000 mg/kg BW) while control group was only given vehicle. Then observed the behavior of animal test after 3-4 hours of treatment, the observation was done again after 24 hours of treatment, observed the presence or absence of animal death test. Observations were made 14 days later to see the reversibility of toxic effects. Noted all behavioral changes that occur in animal testing and determination of LD_{50} [10].

2.2.4 *Subchronic Toxicity Test.* Test animals were used male white rats Wistar strain, age 2-3 months, initial weight 90-170 grams. There are 4 groups (n=6) consisting of 3 treatment groups and 1 control group. Determination of doses of subchronic toxicity was based on the lowest dose administered to decrease blood glucose levels in the SNEDDS dosage activity test of 91.75 mg/kg BW [6]. The control group is solution of Na CMC 0.5% comparable with a dose of 91.75 mg/kg BW (KP 91.75); 183.5 mg/kg (KP 183.5); and 367 mg/kg (KP 367). The weight of each animal was measured on days 0, 7, 14, 21 and 28 on subchronic toxicity test, then the percentage of increase weight of each group was calculated. On days 0, 14th, and 28th of animal blood serum test is taken via retro-orbital plexuc for SGPT and billirubin assay using a capillary tube. At the end of the test (day 29), liver, kidney and pancreas of animal has been taken for histological observation [11].

2.2.5 Histopatological Profile. The organ liver, kidney, and pancreas were observed in this study. The organ damage parameter is characterized by permanent cell damage (irreversible). Histologic coloring used HEE type (Hematoxylin eosin). The cell nucleus and cytoplasm so that it can observed the damage of cell components on a specific part. The presence or absence of cell damage depends on the duration and type of necrosis. The stages of necrosis include pycnosis, caryoecsis, and caryolysis.

3. Result and Discussions

3.1 Determination LD_{50} on Acute Toxicity Test

Test animals that have been previously adapted for one week, were given 5 types of treatment each group with a single dose. After a 14-day observation, 100% dead test animals were only shown at the highest dose of dose of 6000 mg/kg BW. Then 60% of dead animals were shown at a dose level of 1200 mg/kg BW. While in the other group there was no death of the test animal. The number of animal mortality tests was used to calculate LD_{50} obtained results of 1045.44 mg/kgBW were included in the slightly toxic category based on the range of the toxicity compound. Slightly toxic means that temporary minor injury will result from contact with, or absorption of a small to moderate amount by a healthy adult. Larger quantities of exposure could, however, cause greater damage. The results of this acute toxicity test show the overall toxicity of the SNEDDS preparation, both the active substance and the carrier as a whole. The high level of toxicity in this study can not be separated from the role of the concentration of each component of the preparation in harmony with the higher dose given. In addition, the volume of dosage also plays a role in causing death at high doses. If there are overvolume, it will cause overhydration of the test animal cells consequently the cell will expand and over time will rupture causing cell death.

Table 2. Percentage of Mortality in Acute Toxicity Test

Groups	Number of rats	Number of Mortality	Percentage of Mortality
Control	5	0	0
48 mg/kg	5	0	0
240 mg/kg	5	0	0
1200 mg/kg	5	3	60%
6000 mg/kg	5	5	100 %

3.2 Subchronic Toxicity Test

3.2.1 Weight profile

Table 3. Precentage Increase of Weight

Day	Control (%)	KP 91.75 (%)	KP 183.5 (%)	KP 367 (%)
0	-	-	-	-
7	2.09	10.24	7.79	3.78
14	2.84	12.77	12.70	6.06
21	17.37	6.75	13.70	21.20
28	8.01	4.39	5.73	6.82

There are many factors can cause weight gain such as limited activity and food which contain carbohydrates and fats. Intake of feed is not converted into energy but will be converted into body fat [12].

3.2.2 Profile of SGPT (Serum Glutamic-Pyruvic Transaminase)

Table 4. Value of SGPT Each Groups

Groups	Day		
	0	14	28
Control (Unit/L)	58.917±6.503	67.950±16.216	58.517±15.726
KP 91.75 (Unit/L)	55.600±9.967	66.483±9.408	59.483±13.703
KP 183.5 (Unit/L)	53.050±5.731	60.417±12.647	72.033±8.536
KP 367 (Unit/L)	51.750±13.779	80.720±12.920	68.920±18.830

Normally, value of SGPT for rate male between 42.9-67.4 U/l. The result showed that the value of SGPT has increased in all groups. Measurements of SGPT were conducted in order to support the observation histopatologic of liver. SGPT is a specific enzyme in the liver. Its present only in the liver so it can be a specific parameter to determine the condition of liver function. Generally, increasing value of SGPT until exceed upper limit, showed likelihood of liver damage.

3.2.3 Profile of Billirubin

Table 5. Value of Billirubin Each Groups

Groups	Day		
	0	14	28
Control (mg/dl)	0.540±0.127	0.492±0.331	0.345±0.055
KP 91.75 (mg/dl)	0.522±0.094	0.590±0.199	0.455±0.109
KP 183.5 (mg/dl)	0.538±0.175	0.527±0.169	0.368±0.077
KP 367 (mg/dl)	0.458±0.179	0.454±0.059	0.512±0.115

Normally, value of billirubin between 0.2-0.55mg/dl. The value of bilirubin in each group can be affected by the magnitude of red blood cell reshuffle. Differences in red blood cells are reorganized in each individual to produce different values of bilirubin, this is not independent of the adaptation of animals tested against foreign substances. Overall, the presence of foreign substances that were pawned on test animals did not affect bilirubin levels.

3.2.4 Histopatological Profile

Table 6. Percentage of Total Damage Cells

Groups	Total Damage in 100 cells			Percentage of Damage (%)
	Hepar	Pancreatic	Kidney	
Control	20	19	7	15.33 %
KP 91.75	12	12	11	11.67 %
KP 183.5	23	27	15	21.67 %
KP 367	45	39	21	35 %

The toxicity of nano-sized material came from the size, surface wide, composition, and nanomaterial size [13] . The most important mechanism from nanomaterial toxicity in vitro and in vivo is related to the oxidative stress induction by free-radical formulation [14] . Increasing of Reactive Oxygen Species (ROS) can cause oxidative stress condition that causes the occurrences of tissue damage in the whole biological membrane by protein attack, lipid or fat, nucleate acid, and *glico-conjugat* [15]. Non-ionic surfactants are often used as carriers of nanomedicine preparations because it is less toxic, not cause hemolytic, not irritation and can maintain the pH of a liquid [16]. The study mentions surfactant surface tension above 35 mN/m has a low toxicity, whereas the surface tension of tween 80 at the study 43.2 mN/m [17]. Tween 80 has value of LD_{50} 25 g/kg orally and 4.5 g/kg intravenous in rat [18]. Previous study has been conducted that LD_{50} of PEG 400 orally >10g/kg and did not cause significant side effects with repeated administration for two years in mice at a dose of 20 mg/kg/day, but with higher doses showing the low effect on growth and swelling of the liver [19]. Basically, each compound is non-toxic but in use in the long term will still cause side effects. The observation of histopathology blood smear on pancreas, hepar, and kidney after giving for 28 days shows the minor damage in form of cell degeneration. Based on various sources mentioned each organ largely suffered only minor damage (control group, KP 91.75 and KP 183.5) and moderate damage (KP 367).
Cell degeneration can be cellular swelling, picnosis, caryorecsis, and caryolisis. The occurrence of cell degeneration can cause the changes of structure and cell shape that cause inability of cell to function normally. Damage of cells in each organ is still at cellular swelling and picnosis phase. Swelling of the endoplasmic reticulum from increased cell water, one of the earliest ultrastructural changes observed in injured cells, is reversible. Picnosis or core corrugating is a cytoplasm homogenization and eosin cordage increase by the damage intercytoplasmic protein. In caryoecsis, the core cells that experience picnosis is fragmentized. In caryolisis, chromatine become pale. This change is the reflexion of DNA activity because of the cell acid degree reduction [20]. In contrast, picnosis, caryorecsis, and caryolisis are all nuclear signs of cell death and represent irreversible changes.

4. Conclusions

Overall the study of acute and subchronic toxicity test preparation SNEDDS bay leaf chloroform extract showed LD_{50} have amounted in 1045.44 mg/kg BW. Subchronic toxicity test results showed a minor damage to the liver, kidneys and pancreas in 183.5 mg/kg rat weight and moderate damage in a dose of 367 mg/kg as indicated by the increased value of SGPT and billirubin. In conclusion, SNEDDS bay leaf chloroform extract used in this study is possible developed as a future drug, because the damage to the body caused by SNEDDS extract is minor and not permanent, considering the dose.

Acknowledgments
This work was supported by the grant of Penelitian Unggulan Perguruan Tinggi (PUPT) 2017 from Indonesian Ministry of Research and Technology and High Education.

References

[1] Dalimartha S 2003 *Encyclopedia of Indonesia Plants* (Jakarta: Agriwidya)
[2] Sudarsono 2002 *Plant Drugs* (Yogyakarta: Center for Traditional Medicine)
[3] Nublah 2011 Identification Group Compound Blood Glucose Lowering Blood Rats (Rattus norvegicus Berkenhout) Hyperglycemia in Leaves Bread Fruit (*Artocarpusaltilis* (park.) Fosberg) *Thesis* Gadjah Mada University
[4] Bansal A B Munjal and Patel S 2010 Self-Nano-emulsifying Curcuminoids Composition with Enhanced Bioavailability *WO 2010/010 431*
[5] Prihapsara F Harini M Nurona A Zilla S N and Intan M W 2015 *Optimization Self Nanoemulsifying Drug Delivery System (SNEDDS) Ethyl Acetate Fraction of Bay Leaf (Eugenia polyanta Wight)* (Surakarta : International Conference on Biodiversity)
[6] Harini M, Widiyani T and Prihapsara F 2016 Development Nanoherbal of Bay Leaf as an Antidiabetic Agent *Research Report* Sebelas Maret University
[7] Lu C F 1995 *Toxicology Basic Principles, Target Organ and Risk Assessment* (Jakarta: Indonesia University Press)
[8] Ismail H F, Zanariah H, Wong T N, Nur S R, Zainudin A N and Majid F A A 2017 *J.Tradit. Complement. Med.***1** 14
[9] Sulistiyorini D 2016 Subchronic Toxicity Test Herbal Product "X" By In Vivo *Thesis* University of Muhammadiyah Purwokerto
[10] Sari W P 2010 Acute Toxicity Test Mixture Ethanol Extract Betel leaf (*Piper betle* L.) and Dry Extract Gambir (*Uncaria Gambir* R.) Mice against White Males *Thesis* University of Islamic State Jakarta
[11] Krysanti A and Widjanarko S B 2014 *J. Food and AgroIndustry* **2** 1
[12] Yanti R K, Yoyon D A, Martino and Dini S D 2015 *J. Commun. Med.* **3** 1
[13] Rejman J V, Oberle I S, Zuhorn and Hoekstra D 2004 *Biochem. J* **37** 7
[14] Chang Y N, Zhang L, Xia M, Zhang J and Xing G 2012 *Materials* **5** 1
[15] Sharma A, Bansal S and Nagpal R K 2003 *Indian Journal of Pediatrics* **70** 9
[16] Kuma G P and Pogaku R 2011 *Acta Pharmaceutica Sinica B* **1** *14*
[17] Imai T, Satoshi K M and Takane F 1994 *Myzus Persicae. Appl. Entomol. Zool.* **29** 3
[18] Rowe R C, Sheskey P J and Quinn M E 2009 *Handbook of Pharmaceutical Excipients 6th Edition* (London :Pharmaceutical Press)
[19] Webster R, Eric D, Harris P, Siegel N, Stadler J, Tilbury L and Smith D 2007 *ASPET* **35** 1
[20] Sudiono J 2003 *Patologi* (Jakarta : Buku Kedokteran EGC)

IOP Conf. Series: Materials Science and Engineering **333** (2018) 012067 doi:10.1088/1757-899X/333/1/012067

Synergistic Cytotoxicity Effect by Combination Treatment of Polyketide Derivatives from *Annona muricata* Linn Leaves and Doxorubicin as Potential Anticancer Material on Raji Cell Line

A N Artanti[1*], O P Astirin[2], A Prayito[3], R Fisma[2], F Prihapsara[1]
[1]Department of Pharmacy, Faculty of Mathematic and Natural Sciences, Sebelas Maret University, Surakarta, Indonesia
[2]Department of Biology, Faculty of Mathematic and Natural Sciences, Sebelas Maret University, Surakarta, Indonesia
[3]Department of Pathobiology, Faculty of Medical, Sebelas Maret University, Surakarta, Indonesia

Email: antantinuranif@gmail.com

Abstract. Nasopharynx cancer is one of the most deadly cancer. The main priority of nasopharynx cancer treatment is the use of chemotherapeutic agents, especially doxorubicin. However, doxorubicin might also lead to diverse side effect. An approach recently develop to overcome side effect of doxorubicin is to used of combined chemotherapeutic agent. One of the compounds found effication as an anticancer agent on nasopharynx cancer is acetogenin, a polyketide compound that is abundant in *Annona muricata* L. leaves. This study has been done to examine polyketide derivatives was isolated from *Annona muricata* L. which has potency to induce apoptosis by p53 expression on raji cell line. The determination of cytotoxic combination activity from polyketide derivative and doxorubicin was evaluated using MTT assay to obtain the value of CI (*combination index*). Data analysis showed that combination of polyketide derivative from *Annona muricata* L. (14,4 µg/ml) and doxorubicin with all of concentration performed synergistic effect on raji cell line with CI value from 0.13 – 0.65.

1. Introduction

Monotetrahydrofuran annonaceous acetogenins is one of the active compound were isolated from *Annona muricata* L leaves. It was one of Indonesia's native plants widely used in traditional medicine, especially cancer therapy and parasitic infections [1]. Chloroform extract and fraction of *Annona muricata* L leaves acetogenins resulted in higher apoptosis rate compared to ethyl acetate solvent which was 91.86% at 2000 ng/mL concentration, while in ethyl acetate extract 23.79% [2]. Annonaceous acetogenin often referred to inhibit the growth of cancer cell. Acetogenin is a polyketide compound with structure C-35 or C-38 compounds, possessing one or two tetrahydrofuran rings and γ-lactone (either saturated or unsaturated) ususally involving three carbon chain attached to long aliphatic chain and having unbranched aliphatic region [3]. Incidence of cancer caused by virus is increasing every year, with estimation of new cases of cancer caused by Epstein Barr Virus ranges from 3.1 cases per 100,000. Most (83%) nasopharyngeal cancer are diagnosed at regional or distant stages. The overall 5 year relative survival rate is 60% [4]. There are several ways to treat cancer caused by virus such as surgery, radiation therapy, chemotherapy, and combination of these. To date, the use of chemotherapeutic agent is the mostly used treatment beside neoadjuvant chemotherapy and concurrent chemoradiotherapy [5] but it cause many side effects, such as severe nausea and vomiting, acute and chronic nephrotoxicity, anemia acute cochlear toxicity (ototoxicity) [6] and also developed resistance. Many studies was conducted on natural compound as an anticancer agent that has shown potentially of anti-cancer that have toxicity selectively without damaging in normal cells. One of the natural compound which has potential effect as anticancer agent is *Annona muricata* L. Liaw et. al., 2002 reported that four compound of *Annona muricata* L. have a potential anticancer namely are monotetrahydofuran acetogenins, muricin H, muricin I and cis-annomontacin [7]. The main focus of this study was to observe the synergistic cytotoxic activity of polyketide compound from *Annona muricata* L after treatment with doxorubicin on raji cell line. Chemotherapeutic combination could reduce doxorubicin side effects and cellular resistance by decreasing the concentration of doxorubicin for treatment [8]. Herbal medicines have been used as anti-cancer agents to overcome side effects and drug resistance [9].

2. Experimental

2.1 Sampel preparation

Annona muricata L. leaves isolate was obtained from previous research belongs to Professor. Okid Parama Astirin, Sebelas Maret University. It was obtained from Karanganyar, Central Java, Indonesia. Dried sample of *Annona muricata* L leaves was extracted with methanol and evaporated at 80°C. Thus methanol extract was partiotioned with chloroform, n-hexane and ethyl acetate using vaccum liquid chromatography (VLC). Polyketide derivatives from *Annona muricata* L.leaves was isolated by chloroform and each sample from TLC process which have one spot is joined and evaporated again to FTIR and GC-MS analysis.

2.2 Cell culture

Raji cells were obtained from Faculty of Medicine, Gadjah Mada University and were grown in RPMI (Gibco), supplemented with 10% Fetal Bovine Serum (FBS Qualified, Gibco, Invitrogen USA), 1.5% (w/w) penisilin-streptomisin (Gibco, Invitrogen USA), and 0.5% fungizone (Gibco, Invitrogen USA). Cells were incubated at 37°C and 5% CO_2.

2.3 MTT assay

Cytotoxic activity of polyketide isolation from *Annona muricata* L. leaves was evaluated using MTT assay to obtain the value of IC_{50}. $12x10^3$ Raji cells/well were grown in 96-well plate before being exposured to sample treatment. For cell viability assay, cells were treated for 24 hours with increasing concentration of isolate alone and in combination with doxorubicin. As negative control, only growth medium was added. At 100 μg/ml of MTT solution (0.5 mg/mlin PBS) was added to each well continued with incubation for 3 hours at 37°C. There action was stopped by dilution with 10% (w/v) Sodium Dodecyl Sulphate in 0.01 N HCl, and cells were incubated overnight. The absorbance was determined by using ELISA reader at λ 595 nm.

2.4 Statistical analysis

Single citotoxic assay was determined by linier regression between concentration and % cell viability giving the equation $y = Bx+A$ were used to calculate IC_{50} value. It was analyzed statistically by using Microsoft Excell and statistical significance was estimated by using ANOVA test. Statistical significance was placed at $p<0.05$. Synergistic effect was evaluated based on the Combination Index (CI).The interpretation of CI value was done based on classification listed in Table 1.

3. Result and Discussions

3.1 Effect of polyketide derivative from Annona muricata L. leaves on Raji Cells growth

Polyketide isolation from *Annona muricata* L. leaves shows potent cytotoxicity on Raji cells. Furthermore, IC_{50} value acquired as the parameter of concentration that needed to yield 50% cells' growth inhibition of isolate on Raji cells. The effect of Polyketide isolation from *Annona muricata* L. leaves to cell viability of 24-hours treatment. Treated cells showed cytotoxic effect with change of cell morphology and decreased cell viability. Viable cells had epithelial shape, but after being treated with certain concentration of samples they gave spherical-shape and shrunk cell wall. Polyketide isolation from *Annona muricata* L. caused inhibition of cell growth in a concentration-dependent manner. It has performed cytotoxic effect on Raji cells with IC_{50} of 71.96 μg/ml. Cell viability after treatment with polyketide isolation from *Annona muricata L.* was significantly decreased by IC_{50} value is 71.96 μg/ml on Raji cells (Fig.1). According to Prayong et al. (2008) this active compound exhibited potent cytotoxicity ($IC_{50} \leq$ 100μg/ml) [10]. These results above suggested that polyketide isolation from *Annona muricata* L. leaves indicate as the most potential concentration which has potential cytotoxic effect.

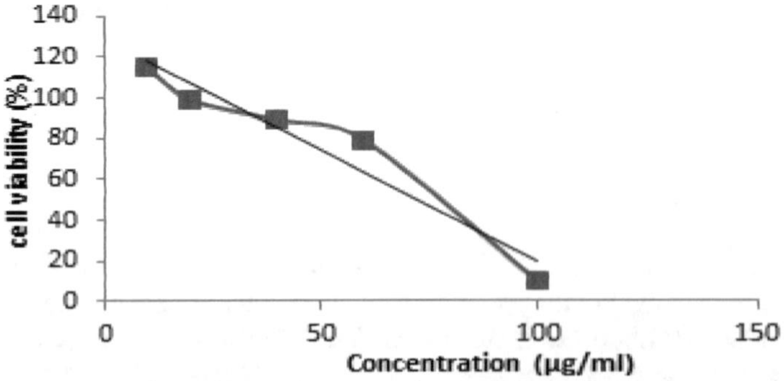

Figure 1. The effect of polyketide derivative from *Annona muricata* L. on the growth Raji cells. The cells were incubated in the 96 well plates and given the treatment the sample in the range 10 μg/ml - 100 μg/ml.

3.2 Synergistic effect of polyketide derivatives from Annona muricata L. combined with doxorubicin on Raji cell growth

Synergistic effect was done by combining ½ IC_{50}, ¼ IC_{50}, $^{1}/_{5}$ IC_{50} of polyketide derivative from *Annona muricata* L. with ½ IC_{50}, ¼ IC_{50}, $^{1}/_{5}$ IC_{50} of doxorubicin. The strongly synergistic effect obtained were combaining $^{1}/_{5}$ IC_{50} and ½ IC_{50} of polyketide derivative from *Annona muricata* L. with ½ IC_{50} and $^{1}/_{5}$ IC_{50} of doxorubicin with CI value in range 0.1-0.30 (Fig. 2). The synergistic effect obtained were combaining ½ IC_{50} of polyketide derivative from *Annona muricata* L. with ½ IC_{50} and $^{1}/_{4}$ IC_{50} of doxorubicin repectively with CI value in range 0.3-0.7. The synergistic effect of doxorubicin combine polyketide derivative from *Annona muricata* L. probably through the apoptotic pathway. This result indicates that polyketide derivative from *Annona muricata* L. can be used in combination with doxorubicin on Raji cells. Ability of it to increase apoptotic and necrosis cell death induced by doxorubicin.

Table 1. Combination Index (CI) value of polyketide derivative from *Annona muricata* L. combined with doxorubicin on Raji cells

Polyketide derivative from *Annona muricata* L (μg/mL)	**Doxorubicin (μM)**		
	1.4	**1.75**	**3.5**
14.4	0.21	1.08	0.17
18	1.37	0.22	0.03
36	0.28	0.35	0.40

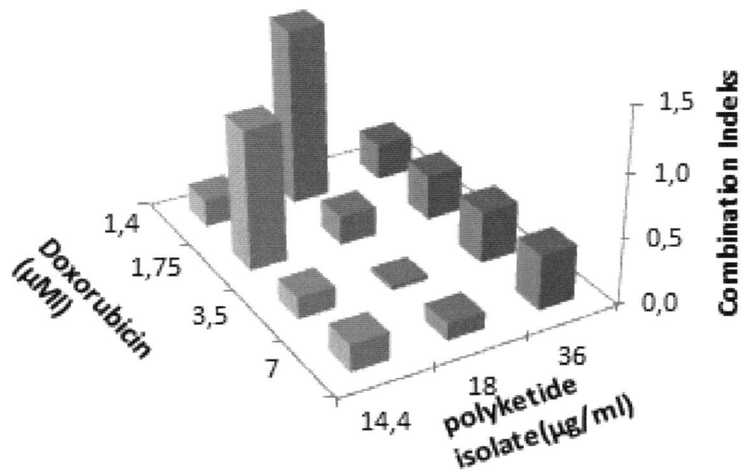

Figure 2. The synergistic effect of polyketide derivative from *Annona muricata* L. after treatment with doxorubicin on Raji cells. The cells were incubated overnight in the 96 well plates and given the treatment of these two compound by combining ½ IC_{50}, ¼ IC_{50} and 1/5 IC_{50}. The graph shows CI value on Raji cell after treatment ($p<0.05$)

Annona muricata L. reported to be utilises as remedies againts cancer, however is not especially for cancer caused virus. Our result demonstrated that polyketide isolation from *Annona muricata* L. have growth inhibitory and cytotoxic effect on nasopharing cancer cell line. Decreasing cell viability may be because of either cell death or cell cycle arrest, so that the observation of these should be done. Therefore, polyketide derivative may be potential to develop as viral inhibitor agent and competitor of vaccine to prevent nasopharinx cancer. However, this speculation still needs further investigation by in vitro and in vivo study.

4. Conclusions
Polyketide derivative from *Annona muricata* L. has synergistic effect with doxorubicin. It means that polyketide derivative from *Annona muricata* L. potential to be developed as a co-chemotherapeutic agent on Raji cell lines. Further molecular target detection to investigate its cellular pathway needs to be conducted.

Acknowledgment
We thank to acknowledge Universitas Sebelas Maret for Hibah Peneliti Utama 2017 with contract number No.623/UN27.21/PP/2017 which has funded this research.

References
[1] Moghadamtaosi S Z, Rouhollahi E, Karimian H, Fadaeinasab M, Firoozinia M, Abdulla M A, Kadir H A 2015 *PLoS ONE* **10**(4) e0122288
[2] Astirin O P, Anif N A, Meutia S F, Eva A P, dan Adi P 2013 *J. Cancer Therapy* **4** 1244-1250
[3] Rupprecht J K, Hui Y H, McLaughlin J 1990 *J. Nat. Prod.* **53**(2) 237-278
[4] American Cancer Society 2017 Williams Street Atlanta
[5] Pan X B and Zhu X D 2012 *Chin. J. Cancer* **31**(12) 573-578
[6] Mukherjea D and Rybak L P 2011 *Pharmacogenomics* **12**(7) 1039–1050
[7] Liaw C C, Chang F R, Lin C Y, Chou C J, Chiu H F, Wu M J, Wu Y C 2002 *J. Nat. Prod.* **65**(4) 470–475
[8] Zhao L, Wientjes M G, and Au J L-S 2004 *Clin. Cancer Res.* **10** 7994-8004

[9] Zhang Y, Wang C, Wang H, Wang K, Du Y, and Zhang J 2011 *Cancer Lett.* **304**(1) 21-32
[10] Prayong P, Barusrux S, and Weerapreeyakul N 2008 *Fitoterapia* **79**(7-8) 598-601

IOP Conf. Series: Materials Science and Engineering **333** (2018) 012068 doi:10.1088/1757-899X/333/1/012068

Acute And Subchronic Toxicity Studies Of SNEDDS (Self Nanoemulsifying Drug Delivery Systems) From Ethyl Acetate Extract Of Bay Leaf (*Eugenia polyantha W.*) with Virgin Coconut Oil As Oil Phase

F Prihapsara[1], R I Alamsyah[1], T Widiyani[2], A N Artanti[1]
[1]Department of Pharmacy, Faculty of Mathematics and Natural Science, Sebelas Maret University
[2]Department of Biology, Faculty of Mathematics and Natural Science, Sebelas Maret University

E-mail: feapri87@gmail.com

Abstract. Bay leaf (Eugenia polyantha) is widely used as an alternative therapy for diabetic and hypercholesterol. However, the administration of the extract has a low oral bioavailability, therefore it is prepared by Self Nanoemulsifying Drug Delivery Systems (SNEDDS) ethyl acetate extract of bay leaf. Therefore, acute and subchronic toxicity test is required. The toxicity test performed was an experimental study, including acute and subchronic toxicity tests. Animal experiments were used using Wistar strain rats. Acute toxicity test using 5 groups (n=5) consisted of 1 control group and 4 groups of SNEDDS dose with 48 mg/kgBW 240 mg/kg, 1200 mg/kg, and 6000 mg/kg, while for subchronic toxicity test with 1 group control and 3 groups of doses of SNEDDS with dose group variation 91.75 mg/kgBW, 183.5 mg/kg, and 367 mg/kg. Duration of observation at acute toxicity test for 14 days while for subcronic toxicity test for 28 days with continuous SNEDDS dosage. The results of the acute toxicity test showed toxic symptoms and obtained median lethal dose (LD_{50}) values from SNEDDS from ethyl acetate extract of bay leaf 1409.30 mg/kgBW belonging to slightly toxic category. Subchronic toxicity studies show that the test drug has minor damage in liver and kidneys and moderate damage in pancreas.

1. Introduction

Bay leaf has been used for the treatment of high cholesterol, diabetes (diabetes mellitus), high blood pressure (hypertension), gastritis, diarrhea and gout [1]. Bay leaf extract has a large molecular size and low stability in the body. Tween 80 has been used because it has high value of Hydrophilic Lipophilic Balance (HLB), a good wetting agent and a good emulsifying agent [2]. PEG 400 was selected as cosurfactant which is help solubilize hydrophilic surfactants as well as drugs in an oil base [3]. Cosurfactant determines the emulsification time in the medium as well as the size of the nanoemulsion due to cosurfactant molecule to place its position between surfactants. Cosurfactant is an amphiphilic compound such as propylene glycol, polyethylene glycol, and glycol esters having affinity to the water and oil phases [4]. The carrier which often used in the manufacture of SNEDDS is vegetable oil. Vegetable oil was chosen because it was more easily degraded by microorganisms so it is more

[1]Coressponding autor : feapri87@gmail.com

IOP Conf. Series: Materials Science and Engineering **333** (2018) 012068 doi:10.1088/1757-899X/333/1/012068

environmentally friendly [5]. Virgin Coconut Oil (VCO) has long-chain fatty acids that contain a lot of linoleic acid. Long-chain triglycerides have the advantage of being able to increase the transport of drugs through the lymphatics thereby reducing the first pass metabolism. SNEDDS ethyl acetate extract of bay leaf with VCO carrier has a transmittance value of 90.95% and these results indicated the high clarity of nanoemulsion. That has a droplet size range from 50-100 nm. SNEDDS ethyl acetate extract of bay leaf with VCO carrier has activity for antidiabetic type II insulin resistant. It can decrease blood glucose level equivalent to metformin [6] [7].

Toxicity studies are required to assess potential hazards to humans through the acute, subchronic, and chronic exposure of laboratory animals to pesticides. Acute toxicity studies provide information on the potential for health hazards that may arise as result of short-term exposure. Determination of acute oral, dermal, and inhalation toxicity is usually the initial step in evaluating the toxic characteristics of a drug. In each of test, the animal is exposed to the test material once on one day. One study of acute toxicity test showed that the LD_{50} bay leaf extract to zebrafish embryos at 0.060 mg/ml [8].

Subchronic exposures do not elicit effects that have a long latency period. However, they do provide information on health hazards that may result from repeated exposures to a drug. Subchronic tests also provide information necessary to select proper dose levels for chronic studies, especially for toxicity studies for which an maximum tolerated dose (MTD) must be selected.

Based on the description, an acute and subchronic toxicity test of SNEDDS ethyl acetate extract of bay leaf is required to determine the level of the dosage in a single dose and repeated dose which caused toxicity. The aim of this research was to determined LD_{50}, toxic symptoms, the effect of repeated dose for 28 days, profile of weight animal test, SGPT (Serum Glutamic-Pyruvic Transaminase) value, billirubin value, and histological profile of liver, pancreas, and kidney.

2. Experimental

2.1 Materials

Materials which were used in this research are are bay leaf obtained from Sukoharjo, male wistar rat (125-170 grams), distilled water (Brataco), chloroform (Brataco), ethyl acetate (Brataco), Virgin Coconut Oil (Agung Jaya), PEG 400 (Agung Jaya), tween 80 (Agung Jaya), formalin 80% (Brataco), ethanol 70 %(Brataco), ethanol 95% (Brataco), ethanol 99,5% (E Merck), xylol solution (E Merck), paraffin liquid (Brataco), Mayer's egg albumin (E merck), hematocyclines-xylene (E Merck), Harri's hematoxylene (E Merck), and entellan® (E Merck).

2.2. Procedure

2.2.1. Preparation of SNEDDS. Bay leaf simplicia powder was extracted by using maceration method with chloroform solvent for 5 days. Macerate was filtered through glass funnel to separate macerate from simplicia powder. The powder then remaceration with ethyl acetate. It was evaporated in 55^0C temperature and rotary speed at 60 rpm until it became concentrated. Each material is weighed and prepared. Ethyl acetate extract of bay leaf (0.15 grams) was added into 5 mL of carrier component, then it was homogenized with magnetic stirrer for 30 minutes. It was sonicated for 15 minutes, incubated in waterbath 45 °C for 10 minutes. Finally, it was kept at 27 °C for 24 hours.

Table 1. Formula of SNEDDS [6]

Material	Amount
Bay leaf ethyl acetate extract	0.15 g
Tween 80	1.33 mL
PEG 400	2.67 mL
Virgin Coconut Oil	1.0 mL

2.2.2 *Animal Preparation.* Before proceeding for toxicity studies animals, Wistar strain rats (125-170g) reared at animal house of Sub Biology Laboratory Sebelas Maret University, were selected and housed separately and kept under strict observation for a period of one week with free access to food and water. Any animal showing sluggish movement or any sign of illness was rejected. Principles of laboratory animal care guidelines were followed and prior permission was sought from the Health Research Ethics Committee for conducting the study (Ref. No. 433/V/HREC/2017).

2.2.3. *Acute Toxicity Test.* Test animals were used male white rats Wistar strain, age 2-3 months, initial weight 125-170 grams. There were 5 groups (n=5) consisting of 4 treatment groups (SNEDDS ethyl acetate extract of bay leaf) with 4 dose variations and 1 control group. For treatment groups, group I with dose of 48 mg/kg BW; Group II 240 mg/kg; Group III 1200 mg/kg and group IV 6000 mg/kg while the control group is SNEDDS without extract. The reversibility of toxic effects was observed for 14 days. All behavioral changes of animal was noted [9].

2.2.4. *Subchronic Toxicity Test.* Test animals were used Wistar male white rats, age 2-3 months, initial weight 125-170 grams. There are 4 groups consisting of 3 treatment groups and 1 control group. Determination of doses of subchronic toxicity was based on the lowest dose administered to decrease blood glucose levels in the SNEDDS dosage activity test of 91.75 mg / kgBW [7]. Three test group were administered 3 dose (91.75, 183.5, and 367 mg/kg BW) while control group was only given vehicle. The weight of each animal was measured every week on subchronic toxicity test, then the percentage of increase weight of each group was calculated. Animal blood serum test was taken via retro-orbital plexuc for SGPT and billirubin assay using a capillary tube on days 0, 14th, and 28th. Histological observation has been done at day 29th [10].

3. Result and Discussions

3.1 Determination LD_{50} on Acute Toxicity Test

Test animals that have been previously adapted for one week, were given 5 types of treatment each group with a single dose. After a 14-day observation, 100% dead test animals were only shown at the highest dose of dose of 6000 mg / kgBW. The median lethal dose (LD_{50}) was 1409.30 mg/kg BW for male rats. The result showed that *SNEDDS* is minor toxic for oral administration. Minor toxic means that temporary minor injury will result from contact with, or absorption of a small to moderate amount by a healthy adult. Rats and humans have similarity, which is both of them are mammals. If SNEDDS extract is minor toxic in rats, hopefully so is in human. The others plant which have antidiabetic effect are minor toxic such as crude extract of *Vernonia amygdalina* (LD_{50} 1265 mg/kg in rat, oral) [11], and *Physalis peruviana L.* (LD_{50} 1280 mg/kg in guinea-pig, oral) [12].

The maximum tolerated dose is higher than the dose required having antidiabetic effects. However, like crude extract which has been recognized toxic, the use of SNEDDS should also be carefully adjusted or discouraged.

Table 2. Percentage of Mortality in Acute Toxicity Test

Groups	**Percentage of Mortality**
Control	0
I	0
II	0
III	40%
IV	100 %

IOP Conf. Series: Materials Science and Engineering **333** (2018) 012068 doi:10.1088/1757-899X/333/1/012068

3.2 Subchronic Toxicity Test

Table 3. Body Weight Profile

Day	**Body weight of Rats (gram)**			
	Control	**91.75 mg/kg**	**183.5 mg/kg**	**367 mg/kg**
0	140.83±12.70	120.33±21.45	131.17±13.99	123.67±10.31
4	143.83±13.17	128.33±21.62	135.50±13.65	126.50±8.50
7	146.50±13.31	135.33±23.02	139.33±14.39	128.33±7.66
11	150.00±14.30	141.83±24.31	145.00±13.70	128.60±7.44
14	152.23±15.56	148.33±26.68	149.50±13.81	132.80±6.02
18	162.83±16.24	158.17±33.14	154.33±13.00	140.60±10.97
21	170.50±17.03	161.17±34.31	155.83±13.73	143.20±14.57
28	179.83±19.72	168.33±37.25	159.17±14.82	155.25±15.41
Increase of weight (%)	27.69	39.89	21.34	25.53

Based on the table 3, dose 91.75 mg/kg give highest of increase weight than other groups. The alteration of body weight gain of the treatment groups compared to the control group would reflect the toxicity of the substance [13]. Significant difference in body weights between the treatment and control group may occur in the absence of any morphological changes [14].

Table 4. Value of SGPT each groups

Groups	**Day**			**Criteria**
	0 (Unit/L)	**14 (Unit/L)**	**28 (Unit/L)**	
Control	57.25±11.00	63.63±4.60	75.88±8.90	**42.9-67.4 U/L**
91.75 mg/kg	53.32±3.07	51.33±7.14	69.08±16.45	
183.5 mg/kg	58.57±15.60	70.00±16.01	71.60±8.98	
367 mg/kg	51.23±4.76	72.55±14.53	79.10±10.20	

Table 5. Value of Billirubin each groups

Groups	**Day**			**Criteria**
	0 (mg/dl)	**14 (mg/dl)**	**28 (mg/dl)**	
Control	0.45±0.06	0.47±0.05	0.44±0.15	**0.2-0.55 mg/dl**
91.75 mg/kg	0.49±0.12	0.57±0.14	0.32±0.06	
183.5 mg/kg	0.50±0.11	0.55±0.05	0.38±0.06	
367 mg/kg	0.49±0.05	0.58±0.09	0.36±0.09	

Elevation in SGPT and billirubin is an indicator of liver and heart damage[15]. Based on the table 4 and 5, after 28 days of extract administration there were no significantly different in the values of SGPT and billirubin in male rats test groups compared to control groups. It indicates the extract was not toxic for liver and did not affected lipid metabolism.

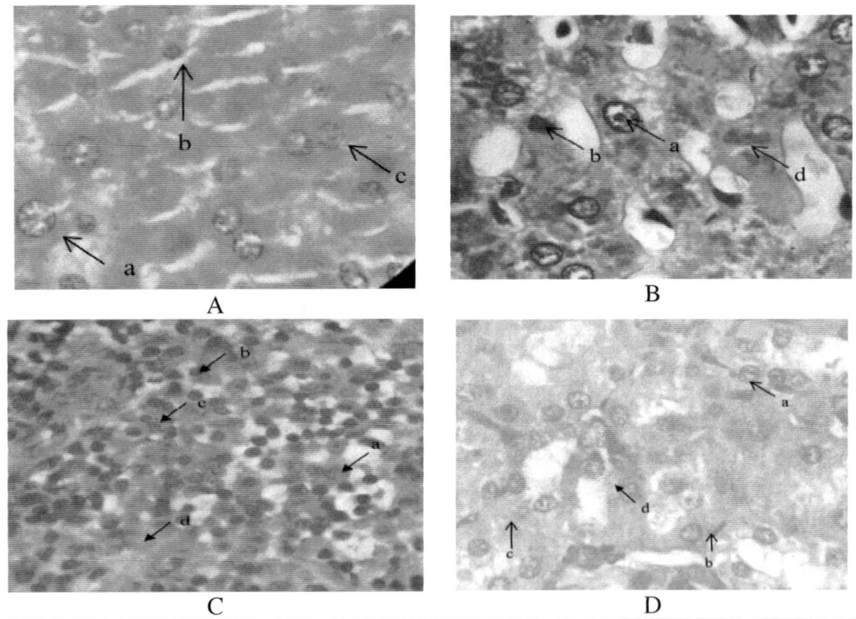

Figure 1. Photomicrograph of tissues in 28 days of SNEDDS ethyl acetate extract of bay leaf treatment. (A) control groups, (B) SNEDDS 367 mg/kg BW in hepar tissues, (C) SNEDDS 367 mg/kg in pancreatic tissues, (D) SNEDDS 367 mg/kg in kidney tissues
400x

Table 6. Percentage of Total Damage Cells

Groups	Total Damage in 100 cells			Percentage of Damage
	Kidney	Pancreas	Liver	
Control	13	25	18	18.67 %
91.75 mg/kg	15	33	9	19.00 %
183.5 mg/kg	22	46	21	29.67 %
367 mg/kg	30	44	28	34.00 %

The four weeks daily treatment with the SNEDDS at three doses did not show any toxicity related mortalities and changes in general health, behaviour, motor activities, and growth. Hence, it can be concluded that SNEDDS has no effect on growth and normal functions of the rat at low doses [16].

The present histopathological examination indicated that some of the pancreas sections obtained from rat treated SNEDDS 367 mg/kg body weight test dose showed picnosis and distorted general architecture. Picnosis, which appeared chiefly in pancreas sections of rat treated at SNEDDS 367 mg/kg dose, occurs when a cell receives a signal to initiate apoptosis [17-20]. Damage from toxic or immunologic insult may cause hydropic degeneration of hepatocytes in which cells take on a swollen, edematous appearance with irregularly clumped cytoplasm and vacuolations [20]. The presence of picnosis and protoplasmic changes like vacuolation in swollen hepatocytes possibly points potential morphologic evidence of degeneration [17-20].

In subchronic toxicity study, the general histological architecture of kidney and liver sections of rat in all experiment groups was not affected. However, some sections of kidney obtained from rat treated with the SNEDDS 367 mg/kg body weight test dose were showed cellular swelling and picnosis. Being focal and minor, the observed changes are not necessary indicators of renal disease or injury [21]. However, further toxicological investigation is recommended on other vital internal organs as well as on non-rodent species to confirm this finding.

SNEDDS makes macrophages entering it in cells and degradation processes in the cell may have a cytotoxic effect [22]. Small-sized nanomaterials will accumulate in certain tissues (the spleen, liver, lungs, and kidneys) which will cause impaired biological function of each of these tissues [23]. The mechanisms that occur in nanomaterial toxicity are related to oxidative stress induction by free radical formation [24]. Free radicals are released by phagocyte cells in response to foreign bodies, lack of antioxidants, and environmental factors. The organ that is most frequently affected by the nanomarterial is the liver and spleen due to phagocyte cell prevalency in the endothelic reticulum. In addition, the organs with high blood flow such as kidneys and lungs [25]. Intracellular nanoemulsions may interact with different components, especially with mitochondria and nucleus, which are the main sources of cell toxicity, induced apoptosis, and ROS formation [26]. SNEDDS at relatively lower doses does not produce obvious toxic effects after acute and prolonged oral administration in rat. However, further investigation is needed to confirm this.

4. Conclusions

The high dose of SNEDDS ethyl acetate extract of bay leaf may cause intoxication. The use of the SNEDDS should also be carefully adjusted or discouraged.

Acknowledgments
This work was supported by the grant of PNBP PU UNS 2017 No. 623/UN27.21/PP/2017.

References

[1] Utami P and Tyas D E 2013 *The Miracle of Herbs* (Jakarta: Agro Media)
[2] Salager J L 2002 *Surfactants Types and Uses Laboratory Of Formulation, Interfaces, Rheology and Process* (Colombia:Universidad De Los Andes)
[3] Amrutkar C, Salunkhe K and Chaudhari S 2014 *World Journal of Pharmaceutical Research* **3 (4)**: 2137-2151
[4] Makadia H A, Bhatt A Y, Parmar R B, Paun J S and Tank H M 2013 *Asian J. Pharm. Res.* **3** 21–27
[5] Patel J Kevin G Patel A Raval M and Sheth N 2011 *Int. J. Pharm. Investigation* **1** 112–118
[6] Prihapsara F, Harini M, Widiyani T, Artanti A N and Ani I L 2017 *IOP Conference Series: Materials Science and Engineering* **176 (1)**
[7] Harini M, Widiyani T and Prihapsara F 2016 Development Nanoherbal of Bay Leaf as an Antidiabetic Agent *Research Report* Sebelas Maret University
[8] Ismail H F, Zanariah H, Wong T N, Nur S R, Zainudin A N and Majid F A A 2017 *Journal of Traditional and Complementary Medicine* **1** 14
[9] Sari W P 2010 Acute Toxicity Test Mixture Ethanol Extract Betel leaf (*Piper betle* L.) And Dry Extract Gambir (*Uncaria Gambir* R.) Mice Against White Males *Thesis* University of Islamic State Jakarta
[10] Krysanti A and Widjanarko S B 2014 *Journal of Food and Agro Industry* **2** 1
[11] Michael U A, David B U, Theophine C O, Philip F U, Ogochukwu A M and Benson V A 2010 Journal of Basic and Clinical Pharmacy **1 (3)**
[12] Kasali F M, Kadima J N, Mpiana P T, Ngbolua K and Tshibangu D T 2013 *Asian. Pac. J. Trop. Biomed.* **3 (11)**
[13] Michael J D and Manfred A H 2014 *Handbook of Toxicology 3rd Edition* (United States : CRC Press Inc.)
[14] Bailey S A, Zidell R H and Perry R W 2004 *Toxicol. Pathol.* **32(4)** : 448-466
[15] Crook M A 2006 *Clinical Chemistry and Metabolic Medicine* (London : Hodder Arnold) p 426
[16] OECD 2001 *Guidelines for Testing of Chemicals, Acute Oral Toxicities up and down Procedure* 425: 1-26
[17] Vanden B T, Vanlangenakker N, Parthoens E, Deckers W and Devos M 2010 *Cell Death and Differentiation* **17**: 922-930.

[18] Trump B F, Berezesky I K, Chang S H and Phelps P C 1997 *Toxicol. Pathol.* **1 (8)**
[19] Kerr J F, Wyllie A H and Currie AR 1972 *Br. J. Cancer* **4 (57)**
[20] Kumar V, Cotran R and Robbins S 2002 *Robbins Basic Pathology 7th Edition* (Philadelphia: Elsevier Saunders)
[21] Rose B D 1987 *Pathophysiology of Renal Disease* 2nd *Edition* (New York : McGraw-Hill)
[22] Rawat, M, Singh D and Saraf S 2006 *Biological and Pharmaceutical Bulletin* ***29*(9) :** 1790-1798.
[23] Nel A, Xia T, Madler L and Li N 2006 *Science* 311: 622-627
[24] Chang, H W, Hsu P C and Tsai Y C 2012 Ag/Carbon Nanotubes for Surface-Enhanced Raman Scattering (Berlin: Springer) p119-135
[25] Clichici S and Filip A 2015 *License In Tech.* 93-121
[26] Aillon K L, Xie Y, El G N, Berkland C J and Forrest M L 2009 *Adv. Drug Deliv. Rev.* **61(6)** : 457-466

IOP Conf. Series: Materials Science and Engineering **333** (2018) 012069 doi:10.1088/1757-899X/333/1/012069

Aqueous Extract Composition of Spent Ginger (*Zingiber officinale var. Amarum*) from Essential Oil Distillation

G J Manuhara[1]*, G P Mentari[1], L U Khasanah[1], R Utami[1]
[1] Department of Food Science and Technology, Faculty of Agriculture, Sebelas Maret University, Jl. Ir. Sutami no.36A, Surakarta, Central Java, Indonesia

*E-mail: godrasjati@yahoo.com

Abstract. Ginger (*Zingiber officinale var Amarum*) is widely used as raw material for essential oil production in Indonesia and contain high functional compounds. After producing essential oil, distillation leave less valuable spent ginger. This research was conducted to determine the bioactive compounds remained in aqueous extract of the spent ginger. The extracts were produced at various combination of temperature (55, 75, 95°C) and duration (15, 30, 45 minutes). The extract composition was observed using Gas Chromatography - Mass Spectrometry analysis. The temperature and time of maceration extraction affected the content of compounds in spent ginger aqueous extracts. The extracts contained four largest components of α-curcumene, α-zingiberene, β-sesquiphellandrene and β-bisabolene. The aqueous extracts from spent ginger contained the compounds which may contribute to distinctive flavor of ginger and also bioactive function.

1. Introduction

Ginger rhizome is widely used as a kitchen spice or medicine around the world. It has a distinctive spicy flavor and a pleasant aroma so that since many centuries, it has been extracted by boiling in water for beverage. The Koran, as moslem scripture, mentions (chapter 76/ Al-Insaan, verse 17) that the drink of "zanjabil" (which is an Arabic word for ginger) is served to the inhabitants of heaven. Among other varieties, in Indonesia, *Zingiber officionale var Amarum* is the most popular spice used by people to cook and produce essential oils, because although it has small rhizome size, but has soft fiber, and strong aroma. Its aroma is not too spicy like red ginger. Ginger contain zingerone, shogaol, gingerol, paradol, β-phellandrene, curcumene, cineol, geranyl acetate, terpineol, terpene, borneol, geranyl, limonene, zingiberol, linalool, α- zingiberene, β-sesquiphellandrene, β-bisabolene, zingiberenol and α-farnesene [1].

Ginger compounds act as antioxidants which are phenolic compounds (such as 6-gingerol and 6-shogaol), alanine, and vitamin C. Antioxidant compounds have an important role in the human health and are also widely used as food additives to prevent food damage [2, 3].

High value commodity from ginger processing is essential oil produced by distillation. The ginger oil industries also produce spent ginger as by product. Generally, these by products are dried and directly used as boiler fuel of distillation, whereas the bioactive compounds allegedly remain in the spent ginger.

The extraction of the bioactive compounds from spent ginger using organic solvent is the most popular method recently. However, concern about food safety related with solvent residue remaining in the extract probably limit consumer acceptance when the extract applied into food processing. The use of ethanol as a solvent in the extraction for food purposes also rises doubtness of halal state among Moslem consumer. Water is the cheapest solvent available, neutral, and harmless so it is safe when used in food manufacturing without any negative concern about remaining solvent, but its higher boiling point than the other solvent result in longer evaporation to concentrate the extract.

Time and temperature affect extraction rate. The longer time of extraction, the bigger chance of contact between solute and solvent, thus the more solute were extracted. Maceration provide a simple method of extraction, but it takes a longer time when the processing was held at room temperature. In order to accelerate the extraction, modifications in the method by using heating and stirring are effective. Increasing extraction temperature is very effective in accelerating the extraction process because the temperature exhibit mobility of solvent molecules and also cause pores of the solid raw material getting bigger. However some functional compounds of ginger are sensitive to the temperature fluctuation. Gingerol is converted into shogaol when ginger or its derivatives is heated with very high temperature. Therefore, the extraction temperature should be employed carefully to keep the important functional components in the extract.

Previous research concluded that highest total phenol and antioxidant activity in the extract was produced by brewing ginger powder (from fresh ginger/ not spent ginger) in water at 56.12°C [4]. Research on extraction of spent ginger from essential oil distillation using water has never been done. Furthermore, variations in temperature and extraction time probably affect the composition of spent ginger extract obtained. It is therefore necessary to study the effect of temperature and time of the spent ginger on the composition of the extract by the mean of Gas Chromatography - Mass Spectrometry (GC-MS) analysis. The GC-MS analysis had been used to determine components in air dried ginger root extract by other researchers [5]. The benefit of this study is to find a method for producing aqueous flavor extract from the spent ginger based on evaluation of components observed in the extracts.

2. Experimental

2.1. Materials.

Fresh ginger (*Zingiber officinale var. Amarum*) rhizome harvested after two years cultivation in December from Ngunut Village, Jumantono, Karanganyar, Central Java, Indonesia.

2.2. Methods

2.2.1. Ginger oil distillation. The ginger rhizome were sorted then washed and brushed to release soil and other impurities attached to its surface. Clean fresh ginger rhizomes were chopped using a knife by cutting cross with a thickness of ± 0.6 cm. Furthermore, the rhizomes were air dried until its visual appearance is wilted and not easily broken (moisture content 23.08 ± 0.21%). Essential oil distillation of ginger rhizomes was conducted in CV. Orizho, Bantul, Yogyakarta, Indonesia using steam distillation method. The distillation lasted for 9 ± 0.25 hours until produced ginger essential oil and spent ginger as the by product.

2.2.2. Spent ginger extraction. The spent ginger rhizomes were then dried using a cabinet dryer at a temperature of 50-60°C for 27 hours until the water content is 5-7%. Dried spent ginger were grinded using a disk mill, then sifted (40 mesh). Furthermore, the spent ginger powder was extracted by maceration using distilled water as a solvent (the powder to the water ratio was 1: 5). The extraction the was performed at various temperature (55°C, 75°C and 95°C) which were combined with various extraction time (15 minutes, 30 minutes and 45 minutes). Furthermore, filtration using filter paper was conducted to separate filtrate / extract from the solid waste. Before tested, the spent ginger extracts were stored in dark amber bottles at 10°C.

2.2.3. GC MS analysis. The component identification was carried out using GC - MS equipped with mass Restek RxiTM-1ms column (30 m length). Helium was used as carrier gas at a constant flow of o.44 ml/min in and an injection volume of 1 μl was employed, injector temperature 280 °C; ion - source temperature 250 °C. The oven temperature was programmed from 100°C for 4 min, with an increase of 5°C/min, to 250°C. Total GC running time was 34 min.

3. Result and Discussion

Table 1. Composition of spent ginger aqueous extract (in relative percentage for each compund)

Components	Extraction condition								
	55°C			75°C			95°C		
	15'	30'	45'	15'	30'	45	15'	30'	45'
Decanol	0.93	-	-	-	-	-	-	-	-
α-zingiberene	0.99	9.92	-	-	5.65	10.80	10.07	9.19	-
β-sesquiphellandrene	0.73	12.44	-	-	4.43	11.32	10.81	10.98	4.95
Nerolidol	10.07	-	-	-	-	-	-	-	-
1-buthyl 2-octyl phthalate	0.89	-	-	-	-	-	-	-	-
Geranyl acetate	0.74	-	-	-	-	-	-	-	-
Phenol,4-ethyl-2-methoxy	0.73	-	-	-	-	-	-	-	-
Oxiranemethanol	-	1.77	-	-	7.95	-	-	2.42	-
Borneol	-	3.18	-	-	-	-	-	-	-
α-curcumene	-	12.16	-	-	5.92	13.28	12.80	9.43	7.61
Germacrene D	-	4.05	5.89	-	-	-	-	-	-
β-bisabolene	-	8.12	-	-	5.23	7.35	7.02	7.03	-
Furazan	-	-	6.44	-	-	-	-	-	-
2-camphonone	-	-	-	-	5.54	-	-	-	-
γ-muurolene	-	-	-	-	-	2.89	-	-	-

The α-zingiberene, β-sesquiphellandrene, α-curcumene, and β-bisabolene performed similarity index ≥ 90 in comparison with the library of masss spectra, while the others ranged 80-89.

Based on the result, the extraction should not be carried out at 55°C – 45 minutes, 75°C – 15 minutes, or 95°C – 45 minutes due to lack of functional or flavoring compounds (see table 1). At 55°C – 45 minutes or 75°C – 15 minutes, ginger starches might started to swell before they were gelatinized finally. The swelling starches probably absorbed most of the compounds, while the earlier gelatinized form of starches obstructed the constituents to be extracted by the water. Previous research observed the gelatinization temperature of ginger starch at 78 ± 0.1°C [6]. At 95°C – 45 minutes, the aqueous spent ginger extract demonstrated only two detected compounds (α-curcumene and β-sesquiphellandrene) but in low percentage. The longest extraction at high temperature (95°C) seemed to degrade most of the functional compounds.

The α-zingiberene, β-sesquiphellandrene, α-curcumene, and β-bisabolene were most frequently found in most of the extracts (see table 1). Those four compounds are sesquiterpenes commonly found in ginger essential oil, methanolic extract of ground ginger and also aqueous extract of roasted/boiled ginger [7, 8, 9, 10]. At longer extraction, the compounds performed a tendency of percentage increase, but α-zingiberene, β-sesquiphellandrene and α-curcumene decreased as the longer extractions were carried out at 95°C. Each compounds perform unique odor as follows: α-zingiberene (warm-spicy, weak woody), β-sesquiphellandrene (spicy, mild), α-curcumene (spicy, herbal), and β-bisabolene (warm, spicy, sweet-balsamic) [5].

Nerolidol and borneol (see table 1) were also reported in research on ginger extracted by boiling ginger rhizome in water at 100°C [10]. It should be noted that nerolidol, which perform many bioactive function, were present in high percentage as the result of 55°C - 15 minutes extraction, but not found in other extracts. Borneol is monoterpene contributed to camphoraceous, peppery, and earthy odor [5]. Borneol compound was present in the extract from 55°C - 30 minutes treatment, while geranyl acetate was only present in the extracts of 55°C - 15 minutes treatment. The germacrene D, which contributed to dry-spicy and weak woody odor, was also found in the study of ginger powder extracted using a high pressure CO_2 at 12-14°C and 4.6-5.0 bar for 120 minutes [5]. The germacrene is sesquiterpene compound [11] but not all extracts contained this compound. The germacrene D was present in the result of 55°C - 30 minutes extraction and also 55°C - 30 minutes extraction.

Other compound found in the extract was phenol, 4-ethyl-2-methoxy (see table 1). This was also detected on the maceration extract of ginger powder at 50°C for 6 hours with ethanol solvent [12]. Camphonone was known as ginger extract component giving fresh and camphoraceous odor [5].

The key nonvolatile constituents in ginger are gingerol and zingerone [13]. However, the compounds were not found from the GC-MS analysis result (see table 1). This was probably due to degradation of the compounds as the result of high temperature employed during distillation. Phenolic compounds are sensitive, unstable and highly susceptible to degradation. The most important cause for phenolic degradation is the temperature [14]. In addition, drying of wet spent ginger probably affect gingerol compounds. The drying during material preparation also lower gingerol and increase terpene hydrocarbons in ginger rhizome [15].

Ginger rhizome contains compounds that act as antioxidants, including terpenoids and polyphenols [16]. Terpenoid is a type of compound derived from a combination of two or more units of isoprene. The terpenoids include monoterpenes, sesquiterpenes, terpenes, triterpenes, tetraterpenes and polymeric terpenoids. Sesquiterpenes are terpenoids that have 15 carbon atoms and composed of 3 isopropen units. Some of the aqueous spent ginger compounds, i.e. nerolidol (see table 1) belonging to sesquiterpenes which are medically important as bioactive compounds and perform antioxidant activity [17].

Based on each composition, the extract obtained at 75°C – 45 minutes, 95°C – 15 minutes or 95°C – 30 minutes indicated strong ginger flavor, while at 75°C – 30 minutes indicated mild ginger flavour with nuances in fresh and camphoraceous odor. The extract obtained at 55°C – 30 minutes indicated not only strong ginger flavour but also nuances in camphoraceous, peppery and earthy odor. All extracts mentioned above denoted same characteristic: warm, spicy, woody, herbal, sweet balsamic.

4. Conclusion

The temperature and time of maceration extraction affected the content of compounds in spent ginger aqueous extract. The spent ginger aqueous extracts contained four largest components of α-curcumene, α-zingiberene, β-sesquiphellandrene and β-bisabolene. The aqueous extract from spent ginger contained compounds which contribute to distinctive flavor of ginger. Although GC-MS analysis revealed many key compounds, the using of more sensitive equipment such as High Performance Liquid Chromatography (HPLC) was sugested to cope with the absence of gingerol and shogaol as important flavoring agents and antioxidant compounds in the extract.

Acknowledgements

This work was the part of research project financially supported by research grant of PNBP UNS 2017 from Universitas Sebelas Maret, Indonesia.

References

[1] Obeten K E, Patrick O E, Victoria N I and Patricia P O 2015 *International Journal of Medical and Health Sciences Research* **2** 25-35
[2] Tsai T H, Tsai P J and Ho S C 2005 *J. Food Sci.* **70** C93-7
[3] Suhaj M 2006 *Journal of Food Composition and Analysis* **19** 531-37
[4] Kishk Y F M and Hemat E E S 2010 *World Journal of Diary and Food Sciences* **5** 188-96
[5] Stoyanova A, Denkova Z, Nenov N, Slavchev A, Jirovetz L, Buchbauer G, Lien H N, Schmidt E and Geissler M 2006 *Electronic Journal of Environmental, Agricultural and Food Chemistry* **5** 1615-23
[6] Kolawole A A, Igwemmar N C and Bello H A 2013 *International Journal of Science and Research* **2** 71-75
[7] Kamaliroosta Z, Kamaliroosta L and Elhamiradc A H 2013 *Journal of Food Biosciences and Technology* **3** 73-80
[8] Buang F Jantan I Zawani Amran A Z and Arbain D 2014 *Research Journal of Applied Sciences, Engineering and Technology* **7** 5098-105
[9] Hasan HA, Rasheed Raauf AM, Abd Razik BM and Rasool Hassan BA 2012 *Pharmaceut Anal. Acta* **3** 1-5
[10] Purnomo H Jaya F and Widjanarko S B 2010 *International Food Research Journal* **17** 335-47
[11] Suresh V N, Venugopalan V V, Joy B, Sreekumar M and Menon A N 2012 *Asian Pacific Journal of Tropical Biomedicine* **2** S1347-50
[12] Jayanudin, Rochmadi, Wiratni, Yulvianti M, Barleany D R, and Ernayati W 2013 *Proceeding of International Conference on Chemical Engineering* Parahyangan University pp: 72-7
[13] Zick S M, Ruffin M T, Djuric Z, Normolle D and Brenner D E 2010 *International Journal of Biomedical Science* **6** 233-40
[14] Vatai T Skerget M and Knez Z 2009 *Journal of Food Engineering* **90** 246–54
[15] Bartley J P and Jacobs A L 2000 *Journal of the Science of Food and Agriculture* **80** 209-15
[16] Ghasemzadeh A Jaafar H Z E and Rahmat A 2011 *Journal of Medicinal Plants Research* **5** 1147-54
[17] Chan W K, Tan L T H, Chan K G, Lee L H and Goh B H 2016 *Molecules* **21** 529

The Effect of Edible Coating Enriched With Kaffir Lime Leaf Essential Oil (*Citrus hystrix* DC) on Beef Sausage Quality During Frozen Storage (-18°±2^{0}C)

R Utami[1], Kawiji[1], L U Khasanah[1], and R Solikhah[1]
[1]Department of Food Science and Technology, Agriculture Faculty, Universitas Sebelas Maret

Email: rohulautami@staff.uns.ac.id

Abstract. The aim of this study was to determine the effect of edible coating enriched with kaffir lime (*Citrus hystrix* DC) leaves essential oil at various concentration on beef sausage quality during frozen storage (-18°±2^{0}C). The concentration of kaffir lime leaves essential oil enriched in edible coating were varied at 0%; 0.2%; 1.4%. Microbiological, physical and chemical characteristics (TPC, color, TBA, TVB, and pH) were investigated at 0, 1, 2, 3, and 4 months of storage. The result showed that edible coating with the addition of kaffir lime leaves essential oils decreased the microbial growth, TVB value, and TBA value of beef sausage. The color and pH of samples can be stabilized during storage. The selected kaffir lime leaves essential oil concentrations based on microbial, physical, and chemical characteristics of beef sausages during frozen storage at -18°C was 0.2%.

1. Introduction

Meat is one of livestock foodstuffs production which has good nutritional content for human consumption. Nutritional content of meat are water (66%), protein (18.8%), and fat (14%) [1]. In food industries, meat also processed into food product such as sausage. According to SNI 01-3020-1995, sausage is a food produced from the mixture of ground meat (not less than 75% of meat) with flour or starch, and with the addition or no addition of spices and other food ingredients that permitted and put into a sausage casing.

Beef sausage has high enough of nutrient content and food processed practically so many people like to consume it. Nutrient content of 100 gram of beef sausages ingredients is 27.98 g of fat, 0.35 gram of carbohydrate, and 18.21 g of protein [2]. Fat and protein content of beef sausages were high enough easier to spoil during storage. The sausages damaged due to microbial growth and lipid oxidation. Beef sausage was suitable media for pathogenic microbial growth and decaying that caused the decreasing of nutritional content. Lipid oxidation will cause the beef sausage flavor damaged as rancidity. Therefore, preservation is needed to inhibit the spoilage.

Beef sausages shelf life was extended by low temperature or frozen storage. According to *United States Department of Agriculture* [3], the shelf life of the uncooked and cooked beef sausage was one up to two months at frozen storage (-18°C). However, chemical changes will occur during frozen storage such as protein denaturation, lipid oxidation, flavor decreased and degradation of pigments and vitamins [4]. *Pseudomonas* often resulted in mucus and pigment formation on frozen meat surface [5]. Therefore, alternative preservation is needed to inhibit other frozen resistant microorganisms growth especially for *Pseudomonas* genus and chemical changes especially lipid oxidation during frozen storage, by using an antimicrobial edible coating.

The edible coating is thin layer coating foodstuffs and safe to eat. Edible film or edible coating show ability to inhibit the moisture, oxygen, aroma and solute transportation [6]. To improve edible coating function, the addition of antimicrobial compound in the packaging materials is needed. The combination of edible coating and antimicrobial could inhibit the growth of pathogenic and spoilage microbial that could extend the shelf life of food.

The spices contain some substances that could inhibit the microbial growth. One of those spices is kaffir lime leaf essential oil. The main compounds of kaffir lime leaf essential oil (*Citrus hystrix* DC) were *Citronellal* (64.15%), *Beta-citronellol* (10.71%), *Trans-caryophyllene* (5.54%), and *Linalool* (5.31%) [7]. *Citronellal* is monoterpene compounds that showed high antifungal properties [8]. Those compounds as an active compound that could inhibit and destruct the microbial cell.

Previous research has been utilized extract or essential oil of plant as a beef sausage natural preservative agent. Garlic essential oil [9], galangal essential oil, and lemongrass essential oil [10] were the essential oils have been used as a sausage preservative agent. However, no research has been studied about edible coating enriched with kaffir lime leaf essential oil as frozen sausage preservative agent. Furthermore to investigate the effect of edible coating enriched with kaffir lime leaf essential oil on the quality of beef sausage, the microbiological, physical and chemical characteristics of beef sausage during frozen storage were determined.

2. Experimental

2.1. Preparation of kaffir lime leaf essential oil

Kaffir lime leaf (obtained from a local market in Surakarta, Indonesia) was stored for one day [7]. Before storage stage, the leaf was kneaded and tored to break the leaf cells, sprinkled by water (200ml of water for 3 kg of material), and covered with a wet cloth. Essential oil of kaffir lime leaf was produced by water vapor distillation. Distillation conducted for 3h after the first oil drop.

2.2. Preparation of edible coating solution enriched with kaffir lime leaf essential oil

The edible coating was made by dissolved 5 g tapioca flour in 100 ml distilled water. Therefore flour solution was heated at 60°C until gelatinized. The gelatinized solution was mixed with a plasticizer (2 ml glycerol (Merck)) and heated at 60°C for 30 minutes. The essential oil was added and mixed in edible coating solution when the solution cooled [11].

2.3. Application of edible coating enriched with kaffir lime leaf essential oil on beef sausage

Edible coating solution poured into beef sausages "Besto". Beef sausages with coating were hanged and dried by a dryer. Beef sausages were packaged in polypropylene plastic 0.88 mm of thickness. The coated Beef sausages were stored in the freezer (-18 ± 2^{0}C) and analyzed for chemical (TVB, TBA, pH, and color) and microbiology properties (TPC). The value of TVB, TBA, and TPC were determined by followed the methods of Kasmadiharja *et al* [12], Apriyantono *et al* [13] and Fardiaz [14] respectively. pH analysis used pH meter (Eutech Instrument Handheld Series). The color analysis used Chromameter Konica Minolta CR-400/410. Each analysis was done at 0, 1st, 2nd, 3rd, and 4th month. The analysis of the 0-month sample was done after the sample was stored for 2 days at -18 ± 2^{0}C and the next month analysis were done after 30 days of storage.

2.4. Statistical analysis

The statistical analysis which used this research was randomized design (RAL) with one factor (concentration of kaffir lime leaf essential oil variation (0%, 0.2%, 1.4%)). Each sample prepared in duplicate. Obtained data were analyzed by ANOVA (Analysis Of Variance). If there were the difference, it will continue by *Duncan Multiple Range Test* (DMRT) analysis to know whether there was a difference of each sample at 0.05 of significant.

3. Result and Discussion

3.1. Total Plate Count (TPC)

Table 1. Total Plate Count (TPC) of beef sausages coated with edible coating enriched with kaffir lime leaf essential oil during frozen storage -18±2^{0}C (log CFU/g)

Oleoresin concentration	Storage time (month)				
	0	1	2	3	4
0%	$4.21\pm0.38^{A}_{a}$	$4.08\pm0.31^{A}_{a}$	$4.48\pm0.57^{AB}_{a}$	$4.76\pm0.19^{AB}_{a}$	$5.16\pm0.17^{B}_{a}$
0.2%	$3.72\pm0.07^{A}_{a}$	$3.88\pm0.41^{AB}_{a}$	$4.45\pm0.58^{AB}_{a}$	$4.63\pm0.46^{AB}_{a}$	$4.91\pm0.23^{B}_{a}$
1.4%	$3.15\pm0.73^{A}_{a}$	$3.88\pm0.49^{AB}_{a}$	$4.19\pm0.16^{AB}_{a}$	$4.31\pm1.08^{AB}_{a}$	$4.93\pm0.38^{B}_{a}$

Note: Notation with same letters in the same column (lowercase) and same row (upper case) showed no significant difference at a significance level of 5%

During storage, the TPC value of all samples increased (Table 1). The increasing of TPC value related to the previous study that reported that beef sausage coated with edible coating enriched with kaffir lime leaf oleoresin also increased during frozen storage [15]. This indicated that at frozen storage still allowed bacterial growth. However, kaffir lime leaf essential oil more able to inhibit the microbial growth on beef sausage during storage. At 4th month storage, the TPC value of beef sausage coated with edible coating enriched with 0.2% and 1.4% of kaffir lime leaf essential oil were 4.91 CFU/g and 4.93 CFU/g, respectively. These values were lower than that of beef sausage coated with edible coating enriched with kaffir lime leaf oleoresin at 4th-month storage (6.52 CFU/g; 6.60 log CFU/g). This due to citronellal content of kaffir lime leaf essential oil was higher than that of kaffir lime leaf oleoresin. The citronellal content of kaffir lime leaf essential oil was 64.15% [7] while citronellal content of kaffir lime leaf oleoresin was 25.66% [16]. The kaffir lime leaf essential content consists of citronellal, beta-citronellol, trans-caryophyllene, and linalool. Citronellal showed an antimicrobial activity [17]. Citronellal was terpenoid compound that could inhibit the microbial growth by disrupting the cell membrane and cell wall formation [18]. The strong antifungal activity of *C. nardus* essential oil associated with synergy effect both of citronellal and linalool [19].

According to previous report [20], the requirement level of microbial content of ready to eat food product for cooked meat product was 10^4 colony/g for satisfied level, 10^4-10^6 colony/g for rather satisfied level, 10^6-10^8 colony/g for unsatisfied (danger potential) level, and more than 10^8 colony/g for not accepted product. While the maximum level of TPC by Indonesian National Standard SNI 01-3820-1995 was 1 x 10^5 colony/g or 5 log CFU/g. The study result showed that the TPC value of 4th-month storage control sample has been passed the SNI standard (5.16 log CFU/g), while 0.2% and 1.4% of kaffir lime leaf essential oil treatment samples still on SNI standard until four months storage. This indicated that kaffir lime leaf essential oil inhibited the microbial growth and maintain the beef sausage quality during frozen storage.

3.2. Total Volatile Base value (TVB)

Table 2. Total Volatile Base Value (TVB) of beef sausages coated with edible coating enriched with kaffir lime leaf essential oil during frozen storage -18±2^{0}C (mg N/100g)

Essential oil concentration	Storage time (month)				
	0	1	2	3	4
0%	$2.51\pm0.01^{A}_{a}$	$10.94\pm1.25^{B}_{a}$	$23.89\pm1.74^{C}_{b}$	$45.18\pm1.47^{D}_{c}$	$52.12\pm0.64^{E}_{c}$
0.2%	$2.48\pm0.06^{A}_{a}$	$7.56\pm0.01^{B}_{a}$	$17.63\pm1.80^{C}_{a}$	$38.32\pm0.92^{D}_{b}$	$42.51\pm3.33^{D}_{b}$
1.4%	$2.50\pm0.03^{A}_{a}$	$8.78\pm1.80^{B}_{a}$	$15.10\pm0.04^{C}_{a}$	$31.50\pm0.08^{D}_{a}$	$35.89\pm0.98^{E}_{a}$

Note: Notation with same letters in the same column (lowercase) and same row (upper case) showed no significant difference at a significance level of 5%

Table 2 showed that TVB value of all beef sausage samples significantly increased during 4 months frozen storage. TVB value of control samples increased from 2.52 mg N/100g to 52.12 mg N/100g. TVB value of beef sausage coated with edible coating enriched by 0.2% of kaffir lime leaf essential oil increased from 2.48 mg N/100g to 42.51 mg N/100g. TVB value of beef sausage coated with edible coating enriched by 1.4% of kaffir lime leaf essential oil increased from 2.50 mg N/100g to.89 mg N/100g.

TVB value related to the bacterial growth on samples. TVB is resulted from protein decomposition by bacterial activities and autolytic enzyme. Resulted protein breakdown was volatile and caused off-flavor compounds such as ammonia, H_2S, mercaptans, phenol, cresol, indole, and skatole [21]. Dimethyl amine and trimethylamine also included as TVB compounds [22]. Trimethylamine was produced by decomposition bacteria, dimethyl amine was produced by autolytic enzyme during frozen storage, while ammonia was produced by amino acid deamination and catabolite of nucleotides [23]. Proteolytic enzymes produced by microorganisms will degrade protein of beef sausage to be amino acids and further will degrade into alkali compound (indole or trimethylamine) [24].

Limit TVB value of fresh meat and processed products were 10 mg N/ 100 g for fresh meats, 20-30 mg N/ 100 g for the early stage of spoilage meat, and 30 mg N/ 100 g for spoilage meat [12]. After three month storage, TVB value of all samples exceeded the limit value. However, TVB value of control sample significantly higher than that of treatment samples. The higher of kaffir lime leaf essential oil enriched on edible coating solution the lower TVB value of beef sausage.

The result was similar to the previous study that incorporated the kaffir lime leaf oleoresin on edible coating solution as the coating material of beef sausage. TVB value of control samples also higher than that of treatment samples [15]. Inhibitory effect of kaffir lime leaf essential oil was caused by disturbing of microbial cell wall formation [25]. Essential oil could inhibit the microbial growth by disturbing the membrane or cell wall formation [18].

3.3. Acidity (pH)

Table 3. Acidity (pH) of beef sausages coated with edible coating enriched with kaffir lime leaf essential oil during frozen storage -18±2^{0}C

Essential oil	Storage time (month)				
concentration	0	1	2	3	4
0%	$6.58 \pm 0.00^{B}_{a}$	$6.24 \pm 0.15^{AB}_{a}$	$5.77 \pm 0.21^{A}_{a}$	$6.22 \pm 0.35^{AB}_{a}$	$6.36 \pm 0.36^{AB}_{a}$
0.2%	$6.21 \pm 0.33^{A}_{a}$	$5.92 \pm 0.23^{A}_{a}$	$6.05 \pm 0.35^{A}_{a}$	$6.28 \pm 0.38^{A}_{a}$	$6.43 \pm 0.05^{A}_{a}$
1.4%	$6.37 \pm 0.01^{B}_{a}$	$6.31 \pm 0.19^{B}_{a}$	$5.96 \pm 0.07^{A}_{a}$	$6.36 \pm 0.14^{B}_{a}$	$6.38 \pm 0.07^{B}_{a}$

Note: Notation with same letters in the same column (lowercase) and same row (upper case) showed no significant difference at a significance level of 5%

The concentration of kaffir lime leaf essential oil enriched on edible solution had no significant effect on pH of samples. The initial pH of beef sausage ranged from 6.21-6.58. During storage, there were slightly increasing and decreasing of pH value. However, pH value of beef sausage coated with edible coating enriched by 0.2% of the kaffir lime leaf essential oil was stable during 4 months of storage. At the end of storage, the pH values of samples ranged from 6.36-6.43. This pH value similar to a previous study [26] that reported the pH value of cooked beef sausage between 6 and 6.5. However, if the pH value under 4.6 to 5.5, indicating the sausage was spoilage.

Increasing and decreasing of pH value of beef sausage were related to the bacterial growth. Decreasing of pH value related to lactic acid bacteria contamination. Decreasing of pH value during storage due to *psychrophilic* bacteria activity which degraded the carbohydrate and spices which produced organic acid especially lactic acid during carbohydrate fermentation [9]. Bacterial which usually contaminated foodstuffs and produced decreasing of acidity related to various bacteria such as *Lactobacillus, Acinetobacter, Bacillus, Proteus, Micrococci, Clostridium*, and *Enterococci* [22].

Increasing of pH of beef sausage during storage was affected by the activity of decomposition bacteria which produced more alkali metabolite.The pH value of anchovy also increased during vacuum frozen storage [27]. Increasing of pH value was affected by the producing of alkali compound such as ammonia, dimethyl amine, trimethyl amine by decomposition bacterial activity.

3.4. Color

Table 4. Color of beef sausages coated with edible coating enriched with kaffir lime leaf essential oil during frozen storage -18±2^{0}C

	Essential oil concentration	Storage time (month)				
		0	1	2	3	4
L*	0%	$55.72\pm2.86^{A}_{a}$	$54.11\pm2.02^{A}_{a}$	$52.79\pm0.52^{A}_{a}$	$54.75\pm1.24^{A}_{a}$	$54.66\pm0.04^{A}_{a}$
	0.2%	$53.08\pm1.30^{A}_{a}$	$54.62\pm0.87^{A}_{a}$	$52.67\pm0.91^{A}_{a}$	$54.59\pm2.14^{A}_{a}$	$56.15\pm2.21^{A}_{a}$
	1.4%	$53.50\pm1.03^{A}_{a}$	$54.68\pm1.50^{A}_{a}$	$53.67\pm0.38^{A}_{a}$	$55.33\pm0.77^{A}_{a}$	$55.28\pm1.19^{A}_{a}$
a*	0%	$22.88\pm0.40^{A}_{a}$	$22.35\pm0.27^{A}_{a}$	$23.26\pm1.15^{A}_{a}$	$23.58\pm0.59^{A}_{a}$	$22.95\pm0.55^{A}_{a}$
	0.2%	$24.38\pm1.16^{A}_{a}$	$23.90\pm0.40^{A}_{b}$	$22.14\pm0.88^{A}_{a}$	$23.61\pm1.16^{A}_{a}$	$22.47\pm0.97^{A}_{a}$
	1.4%	$23.61\pm0.75^{A}_{a}$	$22.78\pm0.42^{A}_{ab}$	$22.06\pm1.41^{A}_{a}$	$22.56\pm0.88^{A}_{a}$	$21.59\pm0.85^{A}_{a}$
b*	0%	$17.99\pm0.41^{A}_{a}$	$18.03\pm0.18^{A}_{a}$	$17.96\pm0.73^{A}_{a}$	$19.98\pm0.33^{B}_{a}$	$19.33\pm0.15^{B}_{a}$
	0.2%	$19.56\pm0.78^{A}_{a}$	$18.82\pm0.81^{A}_{a}$	$17.76\pm0.74^{A}_{a}$	$19.40\pm1.27^{A}_{a}$	$18.72\pm0.42^{A}_{a}$
	1.4%	$17.96\pm0.79^{A}_{a}$	$17.57\pm0.33^{A}_{a}$	$17.53\pm0.91^{A}_{a}$	$18.14\pm0.52^{A}_{a}$	$17.56\pm0.01^{A}_{b}$

Note: Notation with same letters in the same column (lowercase) and same row (upper case) showed no significant difference at a significance level of 5%

During frozen storage for 4 months, the L* (Lightness) value of beef sausage was not affected by the concentration of kaffir lime leaf essential oil enriched on an edible solution (Table 4). The L* (Lightness) value of beef sausage ranged from 52.67 to 56.15. The lightness of beef related to the oxidation reaction of sausage which produced metmyoglobin compound. *Pseudomonas* growth on meat will reduce the partial pressure of oxygen and support *metmyoglobin* (metMb) formation which resulted in gradual discoloration of meat product [28]. The a* (*redness*) color intensity of beef sausage samples were stable ranged from 21.59 to 24.38. The edible coating application could maintain the *redness* of beef sausage during frozen storage. The b* (yellowness) value of control sample increased during frozen storage for four months. Yellowness degree on meat increased with increasing of storage time [29]. This was due to the lower content of metmyoglobin pigment on sausage. Decreasing of myoglobin content into metmyoglobin can cause the pale brown color of meat.

Kaffir lime leaf essential oil contains *citronellal* and *beta-citronellol*. Both of components were can inhibit the oxidative damaged on beef sausage during storage. The main components of essential oil of *Artemisia scoparia* waste were *Citronellal* and *Beta-citronellol* [30]. The essential oil can be a powerful antioxidant and radical activity against hydroxyl (OH) ions and hydrogen peroxide (H_2O_2).

3.5. Thiobarbituric Acid Value (TBA)

Table 5. Thiobarbituric Acid value (TBA) of beef sausages coated with edible coating enriched with kaffir lime leaf essential oil during frozen storage -18±2^{0}C (mg malonaldehyde/kg)

Essential oil concentration	Storage time (month)				
	0	1	2	*3*	4
0%	$0.25\pm0.05^{A}_{a}$	$0.11\pm0.02^{A}_{a}$	$0.13\pm0.07^{A}_{a}$	$0.13\pm0.04^{A}_{a}$	$0.18\pm0.08^{A}_{a}$
0.2%	$0.32\pm0.01^{B}_{a}$	$0.09\pm0.01^{A}_{a}$	$0.13\pm0.04^{A}_{a}$	$0.10\pm0.04^{A}_{a}$	$0.16\pm0.04^{A}_{a}$
1.4%	$0.25\pm0.04^{B}_{a}$	$0.08\pm0.01^{A}_{a}$	$0.12\pm0.04^{A}_{a}$	$0.11\pm0.01^{A}_{a}$	$0.11\pm0.01^{A}_{a}$

Note: Notation with same letters in the same column (lowercase) and same row (upper case) showed no significant difference at a significance level of 5%

The TBA value of all samples ranged from 0.25 to 0.32 at initial storage and 0.11 to 0.18 mg malonaldehyde/kg at the end of storage. The concentration of kaffir lime leaf essential oil enriched on an edible solution has no significant effect on TBA value of the sample. However, the kaffir lime leaf essential oil has high antioxidant activity with Inhibitory Concentration value were IC_{50} 59.98 mg/mL [31]. IC_{50} is number of extract concentration (microgram/milliliter) capable inhibiting 50% oxidation. This is due to kaffir lime leaf essential oil contained citronellal and beta-citronellol [7]. The compound can show the antioxidant and radical activity of hydroxyl ions (OH) and hydrogen peroxide (H_2O_2) [30]. Kaffir lime leaf essential oil also contained Linalool compound [7]. Linalool compound was estimated can trigger to increasing of singlet oxygen. Linalool compound was a compound which caused singlet oxygen increased [32]. When singlet oxygen developed and reacted with unsaturated lipid so the hydroperoxide formed (ROOH) [33].

Based on the TBA value of samples, all samples still acceptable to consume because of the TBA value less than 2 mg malonaldehyde/kg of the sample [34] and have not reached the rancidity level (0.5 mg malonaldehyde/kg of the sample).

4. Conclusion

The result showed that edible coating with the addition of kaffir lime leaves essential oils decreased the microbial growth, TVB value, and TBA value of beef sausage. The color and pH of samples can be stabilized during storage. The selected kaffir lime leaves essential oil concentrations based on microbial, physical, and chemical characteristics of beef sausages during frozen storage at -18°C was 0.2%.

Acknowledgements

This publication supported by research project PNBP Universitas Sebelas Maret.

References

[1] Muchtadi T R, Sugiyono, and Ayustaningwarno F 2011 *Ilmu Pengetahuan Bahan Pangan* (Bandung: Alfabeta)

[2] Fatscret 2013 *Beef Sausage (Cooked)* http://www.fatsecret.com.au/calories-nutrition/generic/beef-sausage-%28cooked%29 Accessed at Thursday, 26 December 2013

[3] United States Department of Agriculture 2011 *Sausages and Food Safety* http://www.fsis.usda.gov/wps/wcm/connect/3e6befc0-b62e-4168-bf4b-b3a 386e3fe19/Sausage_and_Food_Safety.pdf?MOD=AJPERES Accessed at Thursday, 26 December 2013

[4] Sun D W 2006 *Handbook of Frozen Food Processing and Packaging* (Florida: CRC Press)

[5] Buckle K A, Edwards R A, Fleet G H, and Wootton M 2010 *Ilmu Pangan* (Jakarta: Universitas Indonesia Press)

[6] Pranoto Y, Salokhe V, and Rakshit K S 2005 *Food Res. Int.* **38** 267-272

[7] Khasanah L U, Kawiji, Utami R, and Aji Y M 2015 *Jurnal Aplikasi Teknologi Pangan* **4(2)** 49-54

[8] Nakahara K, Alzoreky N S, Yoshihashi T, Nguyen H T T, and Trakoontivakorn G 2003 *JIRCAS* **37(4)** 249-252

[9] Mudawaroch R E and Zulfanita 2012 *Surya Agritama* **1(1)** 71-84

[10] Jirukkakul N 2013 *Int. Food Res. J.* **20(3)** 1199-1204

[11] Utami R, Kawiji, Nurhartadi E, Putra A Y T, and Setiawan A 2014 *IFRJ* **21(1)** 413-419

[12] Kasmadiharja H 2008 *Kajian Penyimpanan Sosis, Naget Ayam, dan Daging Ayam Berbumbu dalam Kemasan Polipropilen Rigid* Skripsi (Bogor: Fakultas Teknologi Pertanian Institut Pertanian Bogor)

[13] Apriyantono A, Fardiaz D, Puspitasari N L, Sedarnawati, and Budiyanto S 1989 *Petunjuk Laboratorium Analisis Pangan* (Bogor: Pusat Antar Universitas Pangan dan Gizi IPB)

[14] Fardiaz S 1993 *Analisis Mikrobiologi Pangan* (Jakarta: Raja Grafindo Persada)

[15] Utami R, Kawiji, Khasanah L U, and Narinda A H 2015 *Jurnal Teknologi Industri Pertanian* **25(2)** 118-119
[16] Kawiji, Khasanah L U, Utami R, and Aryani N T 2015 *AGRITECH* **35(2)** 182-183
[17] Chaudhari L K D, Jawale B A, Sharma S, Kumar H S C D M, and Kulkarni P A 2012 *JCDP* **13(1)** 71-74
[18] Ajizah A 2004 *BIOSCIENTIAE* **1(1)** 31-38
[19] Aguiar R W S, Ootani M A, Ascencio S D, Ferreira T P S, Dos Santos M M, and Dos Santos G R 2014 *The Scientific World Journal* **2014** 1-8
[20] Herawati H 2008 *Jurnal Litbang Pertanian* **27(4)** 1-12
[21] Pandit I G S, Suryadhi N T, Arka I B, and Adiputra N 2007 *Indonesian Journal of Biomedical Sciences* **1(3)** 1-12
[22] Siagian A 2002 *Mikroba Patogen pada Makanan dan Sumber Pencemarannya* Fakultas Kesehatan Masyarakat (Sumatera Utara: Universitas Sumatera Utara Digital Library)
[23] Riquixo C 1998 *Evaluation of Suitable Chemical Methods for Seafood Products in Mozambique* Fisheries Training Programme (Mozambique: The United Nations University)
[24] Wardana A A and Widyaningsih T D 2006 *Edible Film Packaging of Red Cabbage Extract (Brassica Oleracea Var. Capitata L F Rubra) as Bioindicator for Detect Shelf Life of Sausage* Skripsi (Malang: Jurusan Teknologi Hasil Pertanian Universitas Brawijaya)
[25] Utami R, Khasanah L U, Kawiji, Utami A W 2014 *Jurnal Teknologi Hasil Pertanian* **7(2)** 122-123
[26] Korkeala H J and Bjorkroth J 1997 *Journal of Food Protection* **60(6)** 724-731
[27] Gunsen U, Ozcan A, and Aydin A 2011 *Turkish Journal of Fisheries and Aquatic Sciences* **11** 233-242
[28] Zhang Y, Mao Y, Ke Li, Dong P, Liang R and Luo X 2011 *Asian-Aust. J. Anim. Sci.* **24(5)** 713 – 722
[29] Nurdjannah R and Sumarlin R 2010 *Seminar Nasional Teknologi Peternakan dan Veteriner* 645-653
[30] Singh H P, Mittal S, Kaur S, Batish D R, and Kohli R K 2008 *Food Chemistry* **114 (2009)** 642-645
[31] Nurtjahya K, Lenny S, and Zuhra C F 2011 *Komposisi Kimia, Aktivitas Antimukroba Dan Antioksidan Minyak Atsiri Daun Dan Kulit Buah Jeruk Purut (citrus Hystrix) Serta Penggunaannya Sebagai Zat Aditif Pada Adible Film Pati Sukun* Laporan Penelitian Universitas Sumatera Utara
[32] Trabelsi D, Ammar A H, Bouabdallah F, and Zagrouba F 2014 *Journal of Environmental Science, Toxicology and Food Technology* **8(5)** 18-27
[33] Latupeirissa G 2014 *85478781-dp* http://ml.scribd.com/doc/90929791/8547 8781-dp Accessed at Monday, 4 Agustus 2014
[34] Rahayu S, V P Bintoro, dan Kusrahayu 2012 *Jurnal Aplikasi Teknologi Pangan* **1(4)** 108-114

IOP Conf. Series: Materials Science and Engineering **333** (2018) 012071 doi:10.1088/1757-899X/333/1/012071

The Effect of Kaffir Lime Leaves Distillation Residue Oleoresin Concentration on Active Paper Packaging Characteristics

Kawiji[1], R Utami[1], S Ulum[1], L U Khasanah[1], G J Manuhara[1], and W Atmaka[1]
[1]Food Science and Technology Department, Faculty of Agriculture, Universitas Sebelas Maret, Jl. Ir. Sutami 36A, Surakarta, Indonesia.

Email : kawiji_kawiji@yahoo.co.id

Abstract. Oleoresin of kaffir lime leaves distillation residue still contains some active compounds such as *Citronellal*, *β-Citronellol,* and *Linalool* which potential to incorporated on the active paper packaging. The purposes of this study were to determine the effect of kaffir lime leaves distillation residue oleoresin concentration on the physical characteristics, sensory characteristics, and antimicrobial activity of the active paper packaging incorporated with kaffir lime leaves distillation residue oleoresin and to determine the functional groups of active paper packaging. The concentration of kaffir lime leaves distillation residue oleoresin were varied at 0%, 2%, 4% and 6%. The result showed that the addition of kaffir lime leaves distillation residue oleoresin increased the thickness and moisture content of the paper and decreased the tensile strengths and folding endurances of active paper packaging. The microbial inhibition tends to increase along with the higher oleoresin concentration addition. Aromatic CH group were found at a wavelength of 897.90 cm-1 of on paper packaging with 2% oleoresin indicated as functional aromatic functional group allegedly obtained from the kaffir lime leaves oleoresin.

1. Introduction

Packaging has an important role in the food industry. The main role is its ability to extend the shelf life of the packaged product. The basic functions of packaging are to contain and protect the product from damage and thus more easily stored, transported and marketed [1]. In addition, there are some additional functions of the use of packaging for food products, namely as the identity of the product, protect the product from bad influences from outside, increase the attractiveness of potential buyers and increase efficiency. To improve the function of the packaging in extending the shelf life, active packaging that can prevent food damage due to the proliferation of microbes, as well as the gas exchange occurs, could be developed.

Active packaging is a system that can actively change the condition of packaged food to extend shelf life, to improve food safety or to maintain sense quality of foods [2]. Packaging will be classified as active packaging if the packaging that have either one of the following function: absorbent materials O_2 (oxygen scavenger), an absorbent material or CO_2 addition, ethanol emitters, ethylene absorber, water absorbent, antimicrobial materials, heating / cooling, absorber which can secret aroma / flavor, or a protective light. Additionally, active packaging can also equip with an O_2 indicator, CO_2 indicator, or indicators of quality defects which reacts with volatile materials produced from chemical reactions, enzymatic and/or damage from the microbial activity in food. Some examples of active packaging type: MAP (Modified Atmosphere

Packaging), plastic packaging with the anti-fog addition [2], and paper packaging with the addition of an antioxidant and antimicrobial compound (active packaging paper) [3].

Active paper packaging is packaging with the addition of an antioxidant or antimicrobial agent that can extend the shelf life of the packaged product [1]. Some substances were have been used as active compound in the packaging paper including the uses of essential oils of cinnamon, essential oil of cloves and essential oil of oregano [4]; the uses of potassium sorbate and potassium metabisulphite in coating paper [5]; chitosan and DPCT (o, o'-dipalmitoyl chitosan) as an antimicrobial [6]; protein isolate and carvacrol [7][8], cinnamaldehyde [9] and ginger pulp oleoresin [10]. The concentration of the active substance in the paper packaging can also affect the physical properties of the paper produced. Potassium sorbate and sodium permanganate addition can decrease tensile strength and folding endurance of the paper packaging [5].

Spices and herbs in the form of leaves (fresh or dried), stem, bark or root (rhizome) used as a flavoring of foods and drinks. Spices are also known to have nutritional value, antioxidants, antimicrobial and as medicine. Essential oils extracted from herbs have been known have antimicrobial properties that can be used to control microbial growth in foods. Such as a mixture compounds of carvacrol and thymol in different proportions can reduce the growth of *Pseudomonas aeruginosa* and *Staphylococcus aureus* [9]. Kaffir lime has been used by Indonesian, especially in the form of the fruit and leaves as food seasonings. Kaffir lime leaves are also used as ingredients in traditional medicines because contain alkaloids, polyphenols, essential oils, tannins, and flavonoids [11]. Active compounds produced in the essential oil of kaffir lime leaves such as *Citronellal*, *Isopulegol*, *β-Citronellol*, *Geranyl acetate*, *Neryl acetate*, *Acetic citronellil*, *Charyofilen*, *Nerolidol*, *Diethyl carbitol* and *Fitol* [12]. Whereas other study reported that essential oils lime leaves contain the active compounds in the form of *Citronellal* 64.15%, 10.71% *β-Citronellol*, *Linalool* 5.54%, and *Trans-caryophyllene* 5.31% [13].

This study needs to be conducted to determine the addition of a distillation residue kaffir lime leaves oleoresin (Citrus hystrix DC.) at some concentration (0%; 2%; 4%; and 6%) on the characteristics of the active paper packaging include water content, tensile strength, folding endurance, thickness, antimicrobial activity and sensory test and to determine the content of the paper active packaging functional group by the addition of kaffir lime distillation residue oleoresin (*Citrus hystrix* DC.) at selected concentration.

2. Experimental

2.1 Preparation of Kaffir Lime Distillation Residue Oleoresin

The residue of kaffir lime distillation was air-dried using a blower for 8 hours. The prepared materials were extracted by maceration method using 96% ethanol as a solvent in a ratio of materials: solvent = 1: 5 for 5 hours and 14 minutes at 74°C. Furthermore, the extract was evaporated using a rotary evaporator 100 rpm temperature of 60°C [14].

2.2 Preparation of Active Paper Packaging

Filter papers were cut into 2 mm x 2 mm. Then, 5 g of cut paper soaked in 250 mL of distilled water for 24 hours. Soaking pieces of paper added with 250 mL of distilled water and crushed with a blender for 15 minutes to become pulp. 0.45 grams of chitosan powder added beaker glass that containing 100 mL of 1% acetic acid and stirred into a solution of chitosan in acetic acid [9]. In another beaker glass, 50 mL of distilled water added to the kaffir lime leaves distillation residue oleoresin that varied at 0 g; 0.3g; 0.6g; and 0.9g and added with tween 80 at 0, 10 drops, 12 drops and 13 drops (respectively) to form an emulsion of oleoresin and Tween-80. Paper pulp, a solution of chitosan in 1% acetic acid, and emulsion oleoresin with Tween-80 mixed using a blender for 5 minutes. In that mixture, distilled water was added until the volume of 1,000 mL and blend for 5 minutes [9]. Solution of 4.5 g of tapioca in 50 mL of distilled water was added to the mixture and blended again. The last mixture was poured into molding paper with screener (20 cm x 30 cm), trimmed rolled and pressed until wet paper sheet formed. The wet paper sheet was dried at room temperature (30°C) for 48 hours with reversal of paper after 24 hours of drying.

2.3 Analysis of the Active Paper Packaging

Each active packaging paper was analyzed the physical characteristics such as water content with thermogravimetric method [15], thickness (ISO 536: 2010), tensile strength (ISO 1306: 2009), folding

IOP Conf. Series: Materials Science and Engineering **333** (2018) 012071 doi:10.1088/1757-899X/333/1/012071

endurance (SNI 0491: 2009). Antimicrobial activity [16], sensory testing [17] and the functional groups on the packaging paper with the addition of selected oleoresin concentration [18].

2.4 Data Analysis

This research design used to a completely randomized design (CRD), which consists of a single factor, which is a variation of concentration kaffir lime leaves distillation residue oleoresin. Data were analyzed using Oneway ANOVA to determine whether there is a difference of treatment at the 5% significance level was followed by Duncan test DMRT Multiple Range Test at a significance level of 5%.

3. Result and Discussion

3.1 Physical Characterization of Paper Packaging

3.1.1 The water content. The addition of oleoresin not significantly affected the water content of the paper packaging (Table 1). The addition of ginger pulp oleoresin up to 6% also showed insignificant effect on the water content of paper [10]. However, the increasing trend of the water content observed by the addition of oleoresin. Water content increased along addition of oleoresin in the paper. The higher concentration of oleoresin, the higher proportion of the tween-80 as emulsifier were needed at the paper packaging production. Tween-80 can role function as an emulsifier or surfactant that can stabilize the emulsion [19]. Tween-80 is hygroscopic and provide a porous structure [20]. In the packaging paper with the addition of oleoresin have a higher water content than the control paper packaging due to the porous structure was formed allowing the packaging paper re-absorb water prior water content testing process after drying. This is not consistent with the previous study [21], which states that in the manufacture aloe vera powder, the addition of Tween-80 reduced the water content of the flour because tween-80 can expand the surface so that it will speed up the drying process because the transportation system is accelerated in removing water contained in the material on the evaporation process. The results obtained on testing the water content of paper packaging of this study fulfilled the requirement by ISO 0123: 2008 on cardboard duplex that maximum 10% of water content.

Table 1. Physical characterization test active paper packaging with an addition of kaffir lime leaves distillation residue

Oleoresin concentration	Water content (wb)%	Thickness (cm)	Tensile strength (N/mm)	Folding endurance
0%	8.893^{a} ±0.901	0.791^{a} ± 0.005	1.007^{b} ±0.152	0.430^{a} ±0.220
2%	9.921^{a} ±0.245	0.879^{ab} ±0.061	0.492^{a} ±0.155	0.240^{a} ±0.020
4%	9.981^{a} ±0.334	1.014^{c} ±0.023	0.446^{a} ±0.249	0.095^{a} ±0.095
6%	10.142^{a} ±0.433	0.935^{bc} ±0.040	0.298^{a} ±0.040	0.070^{a} ±0.000

Note: Notation with same letters in the same column showed no significant difference at a significance level of 5%

3.1.2 Thickness. There was a significant difference in the thickness parameter (Table 1) of all samples. The thickness of paper packaging ranged between 0.7-1 mm. Paper packaging with the addition of oleoresin tends to have a higher thickness than the control paper packaging (without oleoresin addition). Several studies reported that the higher total solid of the material the higher thickness of the film. The thickness of the edible film increased with the addition of corn starch and white saffron filtrates [22] and several essential oils [23].

3.1.3 Tensile Strength. From Table 1 can be seen that the addition of oleoresin can affect tensile strength value. This is because higher water content value and a porous structure resulting from the addition of tween 80 as an oleoresin emulsifier that can affect the physical properties of the packaging paper becomes more brittle, so that impact to the decreasing of tensile strength. This is consistent with the other study [24] which stated that the addition of tween 80 as a plasticizer can improve the flexibility of the mixture but may decrease the value of Young's modulus in the film. The tensile strength of the pulp sheet derived from several types of wood in Indonesia ranged between 42-61 Nm / g [25]. Lowering of tensile strength value also performed by the addition of ginger pulp oleoresin on paper materials [10].

3.1.4 Folding Endurance. Folding endurance value of paper packaging decrease with the increasing of oleoresin addition (Table 1). The highest folding endurance value was performed by control paper packaging. However, the statistical analysis at a significance level of 5% of the folding endurance analysis results showed no significant difference. Folding endurance value is influenced by the structure of the packaging paper sheet that rigid and porous. Folding endurance decrease because it's affected by moisture content [26]. Active paper with lower water levels tends to show much lower folding endurance values. Water molecules affected stretching on the cellulose fibers, due to the reduction in hydrogen bonding intramolecular while flexibility cellulose fibers affect the durability of paper folding. However, in this study the results obtained in contrast to the results of the research. Higher water content samples showed much lower folding endurance value.

3.2 Sensory Test of Active Paper Packaging

The color of packaging paper with the addition of 2%, 4% and 6% significantly different from the control paper packaging (Table 2). It can be seen that the color of 2% addition of oleoresin concentration showed significantly different with the that of 4% and 6% addition. Paper packaging control showed the highest assessment score of 4.31 which can be interpreted that paper packaging was preferred by the panelists.

Table 2. Sensory test of active paper packaging

Oleoresin concentration	Colour	Texture	Odor	Overall
0%	4.31^{c} ±0.604	3.34^{a} ±0.614	3.66^{b} ±0.769	3.93^{b} ±0.593
2%	3.63^{b} ±0.615	3.33^{a} ±0.547	2.90^{a} ±1.094	3.13^{a} ±0.730
4%	2.93^{a} ±0.828	3.20^{a} ±0.805	2.83^{a} ±1.053	3.00^{a} ±0.695
6%	2.83^{a} ±0.952	3.20^{a} ±0.761	3.13^{ab} ±1.137	3.07^{a} ±0.785

Note: Notation with same letters in the same column showed no significant difference at a significance level of 5%

The texture values of papers were 3.34; 3.33; 3.20; 3.20 for 0%, 2%, 4% and 6% samples, respectively (Table 2). This value can be interpreted that neutral assessment given by the panelists. This was apparently due to the method of making the manual paper, packaging paper, causing the similar texture.

The odor of control active paper packaging showed significantly different values with packing paper with the addition of 2%, 4%, and 6% oleoresin. Control paper packaging has the highest value compared to other papers packaging. The higher oleoresin concentration addition generated stronger odor.

Control paper packaging showed the highest overall value as shown in Table 2. The value of control paper packaging significantly different from the packing paper with the addition of 2%, 4%, and 6% oleoresin. Paper packaging control was preferred by the panelists (3.93). The packaging paper with the addition of 2%, 4%, and 6% oleoresin showed the neutral values those were 3.13; 3.00; and 3.07, respectively. The additional of ginger pulp oleoresin also reduced sensory characteristic (color, texture and overall) of paper packaging [10].

3.3 Antimicrobial Activity of Active Paper Packaging

The result of antimicrobial activity testing against *Pseudomonas fluorescens* and *Aspergillus niger* showed that the addition of 2%, 4%, and 6% oleoresin concentration were not significantly different from the control paper packaging (without the addition of oleoresin) (Table 3). However, it can be seen that the inhibition tends to increase along with the higher concentration addition.

Table 3. Antimicrobial activity of active paper packaging

Oleoresins concentration	*Pseudomonas fluorescens* (mm)	*Aspergillus niger* (mm)
0%	5.72[a] ±0.035	9.47[a] ±0.274
2%	6.09[a] ±0.579	12.65[a] ±2.227
4%	6.46[a] ±0.071	13.58[a] ±0.389
6%	7.07[a] ±1.034	14.27[a] ±3.662

Notes: Notation with same letters in the same column showed no significant difference at a significance level of 5%
Calculation includes paper packaging diameter

3.4. Analysis of Active Paper Packaging Functional Groups

In this study, the cellulose can be demonstrated at a wavelength of 1366.62 cm^{-1} (CH) and 1029.07 cm^{-1} (CO) on the control packaging paper and 1372.41 cm^{-1} (CH), 1058.97 cm^{-1} (CO) in the 2% oleoresin packaging paper (Table 4). Based on FTIR analysis, at 2% oleoresin packaging paper can be found OH groups (3224.76 cm^{-1}), CH (2899.76 cm^{-1}, 1429.31 cm^{-1} and 1372 cm^{-1}). CN group can be known from the uptake 1258.55 cm^{-1} and 1317.44 cm^{-1}, while the CO group showed at 1058.97 cm^{-1} and group C = C absorption showed at 1,639 cm^{-1}. In addition, at the selected paper packaging was performed any aromatic CH group at a wavelength of 897.90 cm^{-1} (Figure 1). This functional aromatic functional group allegedly obtained from the kaffir lime leaves oleoresin. Tween 80 indicated by absorption bands at 2938 and 2869 cm^{-1} to CH, 1730 cm^{-1} to C = O, and 3462 cm^{-1} to OH [27]. In this study, the CH contained in uptake 2899.76 cm^{-1}, and 3224.76 for the OH group. The spectrum of chitosan can be identified at a wavelength of 895 cm (bond NH) and also at a wavelength of 1,654 cm (stretching of C = O group of the amide of N-acetyl) [6]. While on paper packaging with a concentration of 2% chitosan oleoresin can be identified from the wavelengths of 897.9 cm.

Table 4. Functional Groups Analysis

Functional Groups	Wavelength (cm^{-1})	
	Control	2% Oleoresin
O-H	3451.76	3424.76
C-H	1431.24;1366.62; 2887.56	2899.13;1429.31; 1372.41
C-N	1315.51;1168.91; 1058.97	1258.55;1317.44; 1238.35
C-O	1029.07	1058.97
C=C	1636.67;1511.29	1639
C-H (aromatic)	-	897.90

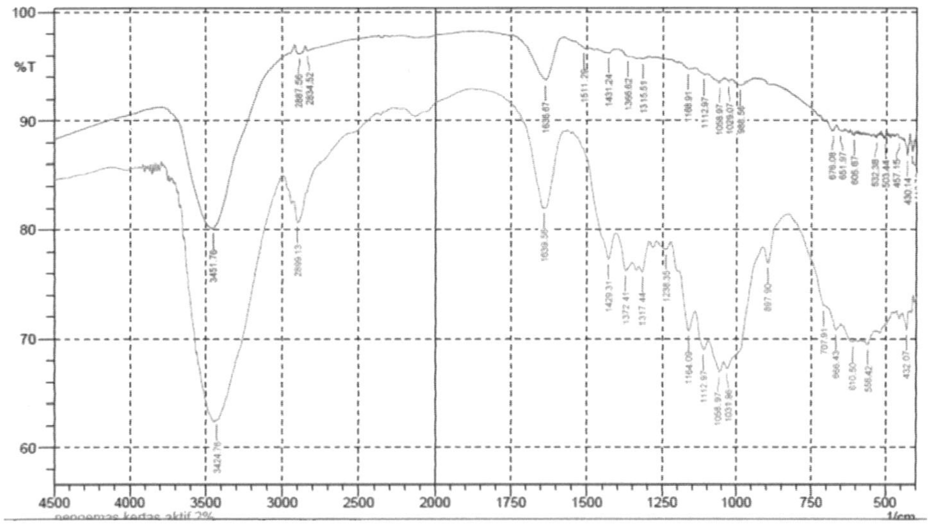

Figure 1 Functional groups of active paper packaging
(Note : Control: top spectrum line; 2% Oleoresin:bottom spectrum line)

4. Conclusion

The result showed that the addition of kaffir lime leaves distillation residue oleoresin increased the thickness and moisture content of the paper and decreased the tensile strengths and folding endurances of active paper packaging. The microbial inhibition tends to increase along with the higher oleoresin concentration addition. Aromatic CH group were found at a wavelength of 897.90 cm^{-1} of on paper packaging with 2% oleoresin indicated as functional aromatic functional group allegedly obtained from the kaffir lime leaves oleoresin. Generally, the paper packaging with 2% oleoresin showed similar physical characteristics with the control samples except for the tensile strengths properties. However, the antimicrobial activity of the paper packaging with 2% oleoresin was slightly higher than that of control sample indicating the improvement of the paper packaging properties.

Acknowledgment
This work financially supported by research project PNBP Universitas Sebelas Maret.

References

[1] Julianti E, Nurminah M 2006 *Buku Ajar Teknologi Pengemasan* (Medan : Universitas Sumatera Utara)
[2] Ismariny 2010 *Deputi Bidang Teknologi Informasi Energi dan Material (TIEM)* 22-117
[3] Rodriguez A, Nerin C, and Battle R 2008 *J. Agric. Food Chem.* **56** 6364-6369
[4] Rodriguez A, R Battle and C Nerin 2007 *Prog. Org. Coat.* **60** 33-38
[5] Rudra S G, V Singh, S D Jyoti U and S Shivare 2013 *Food Biosci.* **3** 49-58
[6] Bordenave N, Greller S, and Coma V 2010 *Biomacromolecules* **11** 88-96
[7] Arfa, AB, Preziosi-Belloy L, Chalier P, and Gontard 2007 *J. Agric. Food Chem.* **55** 2155-2162
[8] Arfa, AB, Chrakabandhu Y, Preziosi-Belloy L, Chalier P, and Gontard N 2007 *J. Food Res. Int.* **40** 22-32.
[9] Manuhara G J, Khasanah L U, and Utami R 2016 *10th Joint Conference on Chemistry: Materials Science and Engineering* **107** 2
[10] Wiastuti T, Khasanah L U, Kawiji, Atmaka W, Manuhara G J, and Utami R 2016 *10th Joint Conference on Chemistry: Materials Science and Engineering* **107** 3-4
[11] Ravindran PN, Pillai GS 2004 *Under-utilized herbs, and spices* (London: Woodhead Publishing)
[12] Mayasari D, Jayuska A, and Wibowo MA 2013 *JKK* **2(2)** 74-77
[13] Khasanah L U, Kawiji, R Utami, and Yoga M A 2015 *Jurnal Aplikasi Teknologi Pangan* **4(2)** 48-55

[14] Khasanah LU, Kawiji, Utami R, Anindito BK, and Achmad F 2013 *Prosiding Seminar Nasional Agribisnis* **2013** 404-408
[15] Sudarmadji S, Haryono B, and Suhardi 2010 *Prosedur analisa untuk bahan makanan dan pertanian* (Yogyakarta: Liberty)
[16] Utami R, Nurhartadi E, and Trisna P A Y 2013 *Jurnal Teknosains Pangan* **2(2)** 51-56
[17] Setyaningsih D, Apriyantono A, and Sari MP 2010 *Analisis sensori untuk industri pangan dan agro* (Bogor: IPB Press)
[18] Anam C, Sirojudin, and Firdausi KS 2007 *Jurnal Berkala Fisika* **10 (1)** 79-85
[19] Lin TJ 1968 *Journal of The Society of Cosmetic Chemist* **19** 683-697
[20] Ratti C and Kudra T 2006 *Drying Technology* **24(9)** 1101-1108
[21] Ramadhia, Mufihah, Sri K and Imam S 2012 *Jurnal Teknologi Pertanian* **13(2)** 125-137
[22] Amaliya R R and Putri W D R 2014 *Jurnal Pangan dan Agroindustri* **2(3)** 43-53
[23] Tongnuanchan P, Benjakul S and Prodpran T 2014 *Journal International Aquatic Research* **6(62)** 1-12
[24] Girdthep S, Punyodom W, Molloy R and Channuan W 2011 *International Conference on Chemistry and Chemical Process, IPCBEE* **10 (2011)** 95-100
[25] Siagian, Rena M, Setyani B L and Yoswita 2004 *Jurnal Penelitian Hasil Hutan* **22(2)** 75-86
[26] Mozina K, Pracek S, and Bukosek V 2013 *Cellul. Chem. Technol.* **47(5-6)** 461-468
[27] Ren W, Tian G, Jian S, Gu Z, Zhou L, Yan L, Jin S,Yin W and Zhao Y 2012 *Tween coated NaYF4:Yb,Er/NaYF4 core/shell upconversion nanoparticles for bioimaging and drug delivery* (China: Electric Supplementary Material (ESI) for RSC Advances)

IOP Conf. Series: Materials Science and Engineering **333** (2018) 012072 doi:10.1088/1757-899X/333/1/012072

The effect of water volume and mixing time on physical properties of bread made from modified cassava starch-wheat composite flour

S Srirejeki[1*], G J Manuhara[1], B S Amanto[1], W Atmaka[1], P W Laksono[2]
[1] Department of Food Science and Technology, Faculty of Agriculture, Sebelas Maret University, Indonesia
[2] Department of Industrial Engineering, Faculty of Engineering, Sebelas Maret University, Indonesia

*E-mail: sucisrirejeki@yahoo.com

Abstract. Modification of cassava starch with soaking in the whey (by product on cheese production) resulted in changes of the flour characteristics. Adjustments of processing condition are important to be studied in the making of bread from modified cassava starch and wheat composite flour (30:70). This research aims to determine the effect of water volume and mixing time on the physical properties of the bread. The experimental design of this research was Completely Randomized Factorial Design (CRFD) with two factors which were water volume and mixing time. The variation of water volume significantly affected on bread height, dough volume, dough specific volume, and crust thickness. The variation of mixing time had a significant effect on the increase of dough volume and dough specific volume. The combination of water volume and mixing time had a significant effect on dough height, bread volume, bread specific volume, baking expansion, and weight loss.

1. Introduction

Wheat flour is the main ingredients in bread making. With the increasing demand of fresh bread causes the needs of wheat flour increase too. Therefore, efforts should be made to reduce the amount of wheat flour needs in bread making. The superiority of cassava flour compared to other root based flour is cheap, abundant raw materials, neutral color and flavour and has a higher swelling power [1].

However, cassava starch application in bread making produces weak dough strength, so that it produces less expand bread, and hard crumb properties [2]. Therefore, it is necessary to modify cassava starch so it has better characteristics and meet industrial requirement [3]. Based on the preliminary research, white bread produced from composite flour of wheat and cassava flour modified with soaking in whey performed the best properties at substitution level of 30%. However, hard texture, bigger and uneven pore were observed in the bread.

It because the water volume and the mixing time used in all formulations are the same. The addition of sufficient water produce a perfectly hydrated dough and higher expansion after fermentation [4]. Inadequate or excessive water volume affected the interaction between components in the dough [5]. Development of gluten matrix during mixing formed dough gradually. The optimum dough is formed at the maximum point of mixing so that the dough resistance collapsed when the mixing continued [6].

2. Experimental

2.1. Materials

High protein flour, cassava starch, refined sugar, water, salt, shortening, milk powder, yeast used in this study were obtained from local store in Surakarta, Indonesia. The whey used in this study was obtained from local producer in Boyolali, Indonesia.

2.2. Modification of cassava starch

Based on the preliminary research cassava starch was modified by soaking starch in the whey with cassava starch - whey ratio was 1: 3, then heated in waterbath at 50°C for 60 minutes with mixing, and dried in cabinet dryer at 50°C during 10 hours. The dried cassava starch, for further was through the size reduction stage using blender and sieving (80 mesh).

2.3. Bread making

The bread was made by straight dough method which all ingredients were mixed in single mixing step after weighing [7]. Water was added into the 250 g composite flour (wheat flour : modified cassava starch ratio was 70 : 30) bit by bit according to the designed formulas then the ingredients were mixed according to the designed treatments. The formed dough was then rounded and rested for 15 minutes. Then the dough was degassed using rolling pin, rolled, and put in a baking pan. The dough height [8], weight, volume and specific volume [9] was then measured. The next processing was fermentation in proofer for 30 minutes at 45-50°C and relative humidity 75-80%. After that, the dough was baked for 30 minutes at 150°C. The bread was further cooled at room temperature. The bread height [8], bread volume and specific volume [9], baking expansion [10], weight loss [11], and crust thickness [12] was then measured.

2.4. Data analysis

The design of this study used a factorial completely randomized design with two factors: water volume (volume at water absorption capacity/ WAC of the composite flour, WAC+5%, and WAC+10%) and mixing time (time at viscoelastic point/ VEP, VEP+10%, and VEP+20%). The combination of two factors was repetead three times. The data was analyzed using two way ANOVA and continued with Duncan's Multiple Range Test (DMRT) at $\alpha = 0.05$ when significant influence was observed.

3. Results and Discusion

3.1. Dough Height

Statistically, combination of water volume and mixing time time had a significant effect on dough height with p value 0.043 ($p < 0.05$). The dough height rised and then decreased, it was due to the effect of gluten formed. At the highest height, the dough is expected to obtain a strong gluten network. The dough height is influenced by the dough fermentation rate through gas formation and formed gluten network [13]. The stronger gluten network, the dough ability to hold the gas produced during fermentation was optimized so that the higher dough obtained. The highest of dough height was found in the combination of VEP+10% - WAC+5%. It was lower when compared to other studies that used the mixture of wheat flour and tapioca and resulted dough height of 8.35 cm [4]. The difference arised because the difference in composition of flour, water volume and yeast resulted in different gas production.

3.2. Bread Height

Water volume variation had a significant effect on bread height with p value of 0.024 ($p < 0.05$). Water is an important factor in bread making. The addition of water in optimum amounts, form optimum viscoelasticity properties of dough so that gluten-formed network is strong [14]. The stronger the gluten hold the carbon dioxide gas formed, the greater the bread volume, the higher the bread produced [15]. The combination of VEP - WAC+5% resulted in highest bread height as well as bread volume and specific volume.

Table 1. Physical Characteristic of Dough and Bread from Modified Cassava Starch-Wheat Composite Flour

	Parameter								
	DH (cm)	BH (cm)	DV (cm^3)	DSV (cm^3/g)	BV (cm^3)	SVB (cm^3/g)	BE (%)	WL (%)	CT (mm)
A	5.03	6.30	320.11	0.79	835.87	2.23	283.63	6.36	2.75
B	5.23	6.80	344.58	0.86	820.23	2.18	238.94	5.72	3.12
C	4.96	6.76	316.70	0.80	970.59	2.66	306.42	8.15	3.38
D	5.03	7.15	371.37	0.91	1084.3	2.85	274.20	5.73	3.03
E	5.45	6.95	468.81	1.15	836.92	2.19	178.52	5.89	3.14
F	5.16	6.93	391.12	0.96	1003.76	2.66	257.10	6.67	3.04
G	5.10	6.25	360.10	0.80	689.57	1.82	192.78	7.91	2.43
H	5.00	6.86	345.19	0.84	970.36	2.53	282.78	6.21	2.82
I	5.16	6.86	313.32	0.76	928.42	2.41	296.72	5.72	2.08

[A] VEP - WAC
[B] VEP+10% - WAC
[C] VEP+20% - WAC
[D] VEP - WAC+5%
[E] VEP+10% - WAC+5%
[F] VEP+20% - WAC+5%
[G] VEP - WAC+10%
[H] VEP+10% - WAC+10%
[I] VEP+20% - WAC+10%
[DH] Dough Height
[BH] Bread Height
[DV] Dough Volume
[DSV] Dough Spesific Volume
[BSV] Bread Specific Volume
[BE] Baking Expansion
[WL] Weight Loss
[CT] Crust Thickness

3.3. Dough Volume and Specific Volume

Statistically, water volume and mixing time had a significant effect on dough volume (with p value 0.0001 and 0.021) and specific volume (with p value 0.0001 and 0.024), but there was no interaction between two factors. The resulting trend of dough specific volume was the same as the trend of dough volume. It was because the specific volume of dough is obtained from the ratio of the dough volume and the dough weight, while the dough weight did not change significantly [9]. Optimum mixing form the gluten structure and improve the extensibility and elasticity of gluten. The addition of water in the right amount can form a dough with optimum viscoelasticity properties so that the resulting gluten is also optimal [4]. This gluten ultimately hold the fermentation gas so that the dough is developed so that the optimum volume is generated [14]. The combination of VEP+10% - WAC+5% resulted in the highest dough volume and specific volume as well as dough weight. The result of the dough volume was greater when compared with other studies. The research of the effect of addition of various variations of hydrocolloids on non-gluten bread making with cassava flour formulations obtained the dough volume of 35.50 - 40.91 cm^3 with the total weight of the material 220.798 - 222.5 g [16]. While the research of making non-gluten bread based on cassava flour using hydrocoloid and enzyme obtained the dough volume of 52,06-55,28 cm^3 with the total weight of the material 108,375 g [17].

3.4. Bread Volume and Specific Volume

The variation of mixing time significantly affected the volume of the bread with p value of 0.009 ($p < 0.05$). The longer of the mixing time causes the greater volume. Longer mixing time allows the development of sufficiently formed gluten [18]. Water volume variation significantly affected the volume of the bread with p value of 0.012 ($p < 0.05$). The optimum addition of water, form a strong gluten network [4]. Reduced gluten network causes lower dough volume so that the volume of bread obtained decreases [14]. While the combination of water volume and mixing time used significantly affected the bread volume with p value 0.0003 ($p < 0.05$). The stronger the gluten holds the carbon dioxide gas formed, the greater the bread obtained [15]. The resulting trend of bread specific volume was the same as the trend of bread volume. In this research the highest of bread volume and specific volume was found in the combination of VEP - WAC+5%. The range of bread volume and specific volume in this research was comparable with other study (study of breadmaking from wheat flour substituted with cassava flour and soybean as an improver) resulting the bread volume and specific volume 803 - 1400 cm^3 and 2,35 – 3,93 cm^3/g at 0-50% substitution of cassava flour with total weight 406 g [9].

3.5. Baking Expansion

Based on statistical results of two way ANOVA, the mixing time and combination of water volume and mixing time significantly affected baking expansion with p value of 0.007 and 0.003 ($p < 0,05$). The dough resistance to mixing or mixing tolerance is usually 10-20% of the optimum mixing peak [19]. In this research, mixing until VEP+20% increased baking expansion. The maximum bread volume is obtained sometime after the point of maximum resistance [19]. In this research, the highest baking expansion was found in the combination of VEP+20% - WAC. The baking expansion was lower when compared with the research of the production of white bread from wheat flour substituted by 10-30% tapioca and added with glycerol monostearate which performed 290 - 345% baking expansion [20].

3.6. Weight Loss

Based on statistical results of two way ANOVA, combination of water volume and mixing time had a significant effect on weight loss of the bread with p value 0.015 ($p < 0.05$). The weight loss of bread was the same trend as the baking expansion. This is related to the formation of gluten, so the dough is ready to hold CO_2 gas from fermentation activity [14]. In this research the highest of weight loss of bread is produced equal to the highest baking expansion of bread. It was found in the combination of VEP+20% - WAC. The resulting weight loss is considerably lower when compared with other studies of baking with tapioca substitution derived from sweet and sour cassava flour [21].

3.7. Crust Thickness

Based on statistical result of two way ANOVA, water volume variation significantly influence to crust thickness of the bread with p value 0.001 ($p < 0.05$). The results showed that the more water volume, the crust was thinner. The crust thickness occured due to the baking process [22]. During baking, heat causes the water vapor pressure from the inner layer (crumb) while the outer layer of bread get dry. Gelatinization during baking caused the formation of layers on the surface of materials that inhibited the transfer of moisture mass [23]. The water vapor pressure from inside the crumb caused the outer structure of the bread to become deflated and solid structure. This structure is called the crust [22].
The resulting crust thickness was lower when compared with other studies, ie, the production of bread with temperature differences (185, 195 and 205 °C) and baking time (25, 30 and 35 minutes) obtained a much larger crust thickness 6.08 - 9.0 mm [12].

The best method of breadmaking was the combination of VEP - WAC+5%. WAC+5% obtained the largest bread volume so that WAC+5% was selected in terms of water volume variations. Meanwhile, in terms of variation in the amount of mixing time, VEP and VEP+10% was not significantly different, but both were significantly different with VEP+20%. So that VEP is chosen due to its cheaper operational costs than VEP+10%. Furthermore this research still needs to be studied further to know the quality of bread made with best method when compared with bread of wheat flour (commercial bread).

4. Conclusion

The variation of water volume significantly affected on bread height, dough volume, dough specific volume, and crust thickness. The variation of mixing time had a significant effect on the increase of dough volume and dough specific volume. The combination of water volume and mixing time had a significant effect on dough height, bread volume, bread specific volume, baking expansion, and weight loss

Acknowledgments
This work was the part of research project financially supported by research grant of PNBP UNS 2017 from Universitas Sebelas Maret, Indonesia.

References
[1] Balagopalan C 1988 *Casava in Food, Feed, and Industry* (Florida: CRC Press)
[2] Ana C B L, Accacia J G P, and Roberto G J 2004 *Brazilian Archives of Biology and Technol.* **47** 66
[3] G J Manuhara, B S Amanto, and T A Astuti 2016 *IOP Conf. Ser.: Mater. Sci. Eng.* **193** 012024
[4] Rusdin R and Dwi S 2015 *J. Agritech.* **35** 325
[5] Shewry P R, Popineau Y, Lafiandra D, and Belton P 2001 *J. Food Sci and Technol.* **83** 826
[6] Govindan V and Ravi P S 2013 *Phenotyping in Wheat Breeding* (New York: Springer)
[7] Sutrisna K 2009 *Teknologi Modifikasi Pati* Ebook.com
[8] Felicia W, Nurwitri C C, and Elisa F R 2008 *J. Ilmu dan Teknologi pangan* **6** 203
[9] Justina N N, Owuamanam C I, Omeire G C, and Eke C C 2014 *American J. of Research Communication* **2** 110-111
[10] Samuel A M 1992 *3rd Bakery Technology and Engineering* (New York: The AVI Publishing Co, Inc)
[11] Victor D M D M Z, Claudia E M M, Ernesto A C A, and Mayela B J 2005 *Revista Salud Publica y Nutricion* **13** 458-460
[12] Jusoh Y M M, Chin N L, Yusof Y A and Abdul R 2009 *J. of Food Engineering* **94** 367
[13] Georgiana G C, Silvia M, Daniela V V, and Costel M 2013 *Czech J. Food Sci.* **31** 222
[14] Ainy J, Evi L and Herman H 2012 *J. Perikanan dan Kelautan* **3** 21
[15] Potter N N and Hotchkiss J H 1995 *5th Food Science* (New York: Champan and Hall)
[16] Angelina R P 2009 *Efek Hidrokoloid Pada Sifat Rheologi Adonan dan Fisikokimia Roti Pada Roti Non Gluten dengan Formulasi Tepung Singkong (Manihot esculenta crantz)*(Semarang: Katolik Soegijapranata University)
[17] Laksmi and Rhani A H 2010 *Peningkatan Kualitas Roti Non Terigu Berbasis Tepung Ubi Kayu (Manihot utilissima) Menggunakan Hidrokoloid dan Enzim* (Semarang: Katolik Soegijapranata University)
[18] Asih Priyati, Sirajuddin H A and Guyup M D P 2016 *J. Ilmiah Rekayasa Pertanian dan Biosistem* **4** 219
[19] Stanley P C and Linda S Y 2006 *Baked Products: Science, Technology and Practice* (UK: Woodhead Publishing)
[20] Enny K B, Ratna Y, and Doni H 2013 *Rekapangan* **5** 133
[21] Sidsel J, Leif H S, Ula K, and Anette K T 2015 *J. Elsevier LWT - Food Science and Technology* **60** 14
[22] Jusoh Y M M, Chin N L, Yusof Y A and Abdul R 2008 *Pertanika J. Sci. & Technol.* **16** 245
[23] G Lisinska and W Leszczynski 1989 *Potatoes Science and Technology.* (New York: Elsivier Applied Science Publishers Ltd.)

IOP Conf. Series: Materials Science and Engineering **333** (2018) 012073 doi:10.1088/1757-899X/333/1/012073

Proximate Nutritional Evaluation of Gamma Irradiated Black Rice (*Oryza sativa L. cv. Cempo ireng*)

Riyatun[1], Suharyana[1], A H Ramelan[1], Sutarno[2], O A Saputra[3] and V Suryanti[3,*]
[1]Department of Physics, Faculty of Mathematics and Natural Sciences, Sebelas Maret University, Jl. Ir. Sutami 36A Surakarta, 57126 Indonesia
[2]Department of Biology, Faculty of Mathematics and Natural Sciences, Sebelas Maret University, Jl. Ir. Sutami 36A Surakarta, 57126 Indonesia
[3]Department of Chemistry, Faculty of Mathematics and Natural Sciences, Sebelas Maret University, Jl. Ir. Sutami 36A Surakarta, 57126 Indonesia

Email: venty@mipa.uns.ac.id

Abstract. Black rice is a type of pigmented rice with black bran covering the endosperm of the rice kernel. The main objective of the present study was to provide details information on the proximate composition of third generation of gamma irradiated black rice (*Oryza sativa L. cv. Cempo ireng*). In respect to the control, generally speaking, there were no significant changes of moisture, lipids, proteins, carbohydrates and fibers contents have been observed for the both gamma irradiated black rice. However, the 200-BR has slightly better nutritional value than that of 300-BR and the control. The mineral contents of 200-BR increased significantly of about 35% than the non-gamma irradiated black rice.

1. Introduction

Rice (*Oryza sativa L.*) is an important cereal grain which consumes nearly half of the world's population. Pigmented rice has been consumed for a long time in Asia and is considered to be a healthy food [1]. Black rice is a type of pigmented rice with dark purple of outer bran layer which appears to be black. It has high nutritional value due to its high content of fiber, minerals, anthocyanin, vitamins B and E [2-3]. This rice is low in sugar, salt and fat and free of gluten and cholesterol. In spite of its beneficial, black rice paddy has some limitations which can hinder the mass production of black rice. The demand for black rice has been increasing due to its high nutritional value and multiple biological activities. Black rice paddy has a long period of cultivation which is more than 145 days [4]. The habitus is very tall (more than 150 cm), as a result the plant falls easily.

On previous work, our group have applied gamma radiation on the black rice grains in order to overcome the limitations of the black rice [5]. The black rice grains were irradiated with gamma rays yielded by core of Cobalt 60 which gamma rays radiation doses used were 200 and 300 Gy. This irradiation was applied to induce mutation in black rice plant to improve its quality and physical appearance. This genetically engineered has been successfully modified the morphological of leaves, stems and grains of the black rice plant. After 74 days of plantation, the irradiated black rice have been flowered faster than non-irradiated one, which were 9.25 and 11.45% for 200 and 300 Gy irradiated black rice, respectively [5].

The doses of exposed gamma sources often also influence the nutrition content of cereal grains [6]. To meet the criteria on the commercial application, the irradiated products must be assayed carefully in term of safety and acceptability [7]. One of the method widely employed to evaluate the nutrition content on the product is proximate analysis following AOAC standard method [8]. The nutritional value of the irradiated black rice with doses of 200 and 300 Gy are reported in this paper.

2. Experimental

2.1. Materials

Third generation of gamma irradiated black rice (*Oryza sativa L. cv. Cempo ireng*) with doses of 200 and 300 Gy were used in this research. The non-irradiated black rice was applied as a control. All chemicals were used from e-Merck. Solvents of analytical grade were utilized without further purification.

2.2. Proximate Analysis

The black rice powdered was sieved through a 100 mesh sifter and stored in refrigerator for further used. Proximate analysis of black rice samples was conducted according to SNI 01-2891-1992 following AOAC 2000 [9].

2.2.1. Moisture content. The black rice samples was weighed and recorded as initial weight (*Wo*). Afterward, each black rice samples was dried in oven until obtaining constant weight (*Wi*). The moisture content was calculated according to Equation 1.

$$Moisture = \left(1 - \frac{Wi}{Wo}\right) \times 100\% \quad \text{.....} \quad (1)$$

2.2.2. Mineral content. Mineral content of each black rice samples was determined through dry method. Black rice samples (5 g) were placed in a weighed porcelain and burned in furnace at 500-600°C. The mineral content was expressed as the final weigh of sample.

2.2.3. Lipid content. The lipid content was examined *via* soxhlet extraction method. Black rice samples (5 g) were extracted using 150 mL of petroleum ether for 6 h. The solvent was then evaporated and the extract lipids were weighed (W_{lipid}). The lipid content was calculated using Equation 2.

$$Lipid(\%) = \frac{W_{lipid}}{Wo} \times 100\% \quad \text{.....} \quad (2)$$

2.2.4. Protein content. The protein content was determined using Kjehdahl method. Black rice samples (0.1 g) was destructed with 1 g of $CuSO_4$, and 2.5 mL of H_2SO_4 for 2 h at 100 °C. Afterward, to the destructed sample was added 50 mL of distilled water and 15 mL of 50% w/w NaOH, then distilled. The distillate was collected in Erlenmeyer containing 10 mL of 0.02 N HCl. The sample was then titrated using standardized NaOH to determine protein content.

2.2.5. Carbohydrate content. The carbohydrate content was calculated using Equation 3.

$$Carbohydrate = 100\% - \%(Moisture + Mineral + Lipid + protein) \quad \text{......} \quad (3)$$

2.2.6. Fiber content. The fiber content determined using hydrolysis method. Free lipid content of black rice (5 g) was hydrolyzed using 50 mL of 1.25% H_2SO_4 solution and boiled for 30 minutes. Then, 30 mL of 3.25% NaOH was added into the system and boiled for next 30 minutes. The sample was filtered in warm condition and washed with 1.25% H_2SO_4, distilled water and 96% ethanol. The sample was dried at 105 °C and weighed. The final weigh was defined as the fiber weight.

3. Result and discussion

The proximate analysis of gamma irradiated black rice is necessary to conduct regarding to the food safety and security regulation. Proximate nutritional evaluation of gamma irradiated black rice has been conducted for quantifying the moisture, minerals, lipids, proteins, carbohydrates and fibers contents. The results obtained for proximal composition of black rice investigated in this study are shown in Table 1.

Table 1. The proximate nutritional data of black rice samples.

No.	Analysis	Black rice samples		
		Control	**BR-200***	**BR-300****
1	Moisture (%)	14.82	12.89	13.22
2	Mineral (%)	1.23	1.67	1.13
3	Lipid (%)	3.00	3.06	2.93
4	Protein (%)	7.15	7.63	7.47
5	Carbohydrate (%)	73.80	74.75	75.25
6	Fiber (%)	1.13	0.97	0.80

*BR-200 : gamma irradiated black rice with a dose of 200 Gy
**BR-300 : gamma irradiated black rice with a dose of 300 Gy

Generally, there is no significant nutrition content changes of gamma irradiated black rice in respects to the control one. These results were in line with other researchers. Taipina *et al.* (2011) reported that the macronutrient of irradiated sample is unaffected by radiation doses up to 10 kGy [7]. Moreover, Diehl and Josephson (1994) reported that the essential nutrition of irradiated samples undergo no significant losses under condition of actual or potential commercial application [10].

The moisture and fiber contents of both gamma irradiated black rice slightly decreased and moderately no changes have been observed for the lipids contents of both irradiated black rice. The moisture content of BR-200 decreased to 12.89% which is meet the acceptable of moisture content of about 12%. This moisture levels is recommended for long term storage as well as avoid microbial growth [8]. The fiber content of black rice obtained from cv. Cempo Ireng was 1.13%. Fiber hold important role on the decreasing of the blood cholesterol and sugar in diabetics, also reduce the bowel disorder as well as fight against constipation. The high content of fiber will prevent all of the mentioned diseases. However, the fiber content also should meet the required standard. Oko and Onyekwere (2010) reported that the standard fiber content of rice is 0.5% - 1.0%. Both of the irradiated black rice is still comply with the fiber content requirement which can gave a positive impact to fulfill the required fiber content to the body [11].

It is noted that the minerals content of BR-200 increased significantly of about 35% in respect to the control. However, slightly decreased in minerals content was observed for the BR-300 sample. Minerals are known as essential nutrients playing important role in the body activity [12]. Mineral content on the rice is mostly influenced by genetic characteristic and environment factors [13].

In respect to the control, the proteins and carbohydrates contents were slightly increased for both gamma irradiated black rice. Carbohydrate content which is higher than 70% can be considered to be a good source of carbohydrate [14]. Hence, these gamma irradiated black rice varieties are good source of carbohydrate. Carbohydrate in rice is primarily starch composed by amylose and amylopectin. The carbohydrate content is strongly influence the rice physical parameter, where at high content of carbohydrate will make the individual grains is easy to sticking to each other, however at low level, it is prevented [15].

4. Conclusion

Although there were no significant changes of moisture, lipids, proteins, carbohydrates and fibers contents have been observed for the both gamma irradiated black rice in respect to the control, the 200-BR has slightly better nutritional value than that of 300-BR. The minerals, lipids, proteins and fibers contents of the 200-BR are higher than those of the 300-BR. Moreover, the minerals, lipids, proteins and carbohydrates contents of the 200-BR is higher than those of the non-gamma irradiated black rice. It is notably increased of the minerals content of 200-BR of about 35% in respect to the control. Results generated in this study could be able to provide adequate information's on the quality of the black rice new variant for alternative cereal grains.

Acknowledgements
Authors would like to thank to Sebelas Maret University through Hibah Penelitian Unggulan UNS (PU-UNS) with contract number 473/UN27/HK/2017 dated 12 May 2017 for financial support.

References

[1] Simachanu C S, Yenagi N B, Math K K 2016 *J. Farm Sci.* **29(1)** 61
[2] Suzuki M, Kimur T, Yamagishi K, Shinmoto H, Yamak K 2004 *Nippon Shokuhin Kagaku Kogaku Kaishi* **51(58)** 424
[3] Kristamtini, Taryono, Basunanda P, Murti R H, Supriyanta 2012 *J. Agric. Biol. Sci.* **7** 982
[4] Kushwara UKS 2016 Black Rice Switzerland (SW): Springer International Publishing DOI 10.1007/978-3-319-30153-2_2
[5] Hartanti R S, Putri T A N, Zulfa F, Sutarno, Suranto 2017 *IOP Conf. Ser.: Materi. Sci. Eng.* **193** 012038 doi:10.1088/1757-899X/193/1/012038
[6] Marcu D, Damian G, Cosma C, Cristea V 2013 *J. Biol. Phys.* **39** 625
[7] Taipina M S, Garbelotti M L, Lamardo L C A, Santos J S, Rodas M A B 2011 *Proced. Food Sci.* **1** 1992
[8] Verma D K, Srivastav P P 2017 *Rice Sci.* **24(1)** 21
[9] AOAC International 2000 Official Methods of Analysis, 17th Ed., Washington DC
[10] Diehl J F, Josephson E S 1994 *Acta Alimentary* **23(2)** 195
[11] Oko A O, Onyekwere S C 2010 *Int. J. Biotechnol. Biochem.* **6(6)** 949
[12] Wang K M, Wu J G, Li G, Zhang D P, Yang Z W, Shi C H 2011 *J. Cereal Sci.* **54(1)** 116
[13] Zimmermann M B, Hurrell R F 2002 *Curr. Opin. Biotechnol.* **13(2)** 142
[14] Thomas R, Wan-Nadiah W A, Bhat R 2013 *Inter. Food Res.* J **20(3)** 1345-1351
[15] Mbatchou V C, Dawda S 2013 *Int J. Res. Chem. Environ.* **3(1)** 308

IOP Conf. Series: Materials Science and Engineering **333** (2018) 012074 doi:10.1088/1757-899X/333/1/012074

Assay for hypoglycemic functional food of cocoyam (*Xanthosoma sagittifolium* (L.) Schott.) tuber

N S Handajani[1], M Harini[1], R Yuliningsih[1], S Afianatuzzahra[1], U Hasanah[1], and T Widiyani[1]

[1]Study Program of Biology, Faculty of Mathematics and Natural Sciences, Universitas Sebelas Maret, Jl. Ir. Sutami 36 A Surakarta 57126, Indonesia

E-mail: noor_handajani@yahoo.com

Abstract. Diabetes Mellitus (DM) type II is a degenerative disease that is a major killer in many countries. It is characterized by an increase of the blood glucose level above normal. It is important to choose an appropriate food sources using glycemic index (GI) concept in order to prevent blood glucose increase. One of Indonesian traditional carbohydrate source is cocoyam (*Xanthosoma sagittifolium* (L.) Schott.) tuber. The tuber is assumed having a higher carbohydrate content with lower GI. The research aims to measure GI of cocoyam tuber (CT) and determine glucose and glycogen level in animal model after CT fed. Experimental research was carried out by using completely randomized design. We used twenty four male rats as animal models. They were grouped in to 4 different treatments. Group I was treated with standard feed, group II was treated with standard feed and glucose, group III was treated with steamed CT, and group IV was treated hypoglicemic agent standard, glibencamide. The research results that GI of steamed CT was low. It was 54. Blood glucose of diabetic rats after fed by CT decreased significantly ($p<0.05$), similar to diabetic rats after treated by glibencamide. Whereas glycogen level in diabetic rats after fed by CT was higher than in diabetic rats after fed by standard feed. Cocoyam tuber increased glycogen level in diabetic rats significantly ($p<0,05$). Glycogen level in diabetic rats fed by CT was as high as in healthy rats. Therefore CT is potential consumed for DM type II patients.

1. Introduction

Nowaday, philosophy on food needs has much shifted. Food does not only overcome hunger, but also the most important food is good to the physical health. In the past nutrients on food were only expected give benefit for basic physical needs and sensory satisfaction. Otherwise now, food is expected to be functional. Food has advantages improving physical health and preventing any chronic diseases. Several chronic diseases caused death and dissability worldwide. One of them is diabetes mellitus (DM).

Diabetes mellitus is the sixth biggest cause of mortality in all age groups. It is a collective symptoms which characterized by hyperglycemic in the blood and glycogenesis failure due to insulin metabolic disorder. In a normal physical condition, excessive glucose will be stored as glycogen. Otherwise glycogen will be broken down into glucose in the glucose lack condition. Long term DM may cause a series of metabolic disorder which causes macrovascular and microvascular pathological disorders [1]. The number of DM patients increase rapidly worldwide and it is now a global epidemic

[2]. Wild *et al.* [3] estimate that in 2030, the number of DM patients in Indonesia reaches out 21.3 million. It is fourth position below China, India, and United States [4].

An action to prevent this chronic disease is healthy diet. It can be performed by selecting carbohydrate sources based on glycemic index (GI) concept. According to Rimbawan and Siagian [5], GI concept emphasizes to understand the ability of carbohydrate source on the blood glucose level speed increase. Generally, food which increases blood glucose level quickly has high GI, while food which increases blood sugar level slowly has low GI. Low GI food is potentially to develop as a functional food. According to Miller *et al.* [6] and Pruett [7], GI score is categorized into three classes, i.e. low (<55), medium (55 to 70), and high IG (>70). Carbohydrate sources with low GI should be explored to use in DM prevention.

One of Indonesian traditional food material is cocoyam (*Xanthosoma sagittifolium*) tuber. It is also known as kimpul (Javanese) or tannia. Cocoyam tuber is consumed as an alternative staple food because of the high carbohydrate content, around 70-80%, and have a digestible characteristic [8]. However, it is less information about CT potency as functional food source. Therefore, it is necessary to study the potency of CT as functional food source. The study aims to determine the GI score of CT and its hypoglycemic activity.

2. Experimental

This study was performed experimentally using completely randomized design. Twenty four adult male rats (*Rattus norvegicus*) aged 2–3 months was used as animal models. Cocoyam tuber originated from Wonogiri. After peeled and cleaned up, it was steamed at temperature 70° for 20 minutes. The study consists of two stages. Firstly, we measured blood glucose level of normal rats and determined GI value of CT. Secondly, we measured blood glucose level and liver and muscle glycogen levels of DM rats.

In the first stage, twelve rats were grouped into three different treatment groups. In the first group rats were treated pellet diet *ad libitum,* in the second group rats were treated 6.75 g/kg bw pure glucose orally, and in third group rats were treated steamed CT *ad libitum.* The blood glucose level of the rats was determined by using glucose oxidase-phenol + aminophenazone (GOD-PAP) method [9]. It was performed three times at the 0th, 1st, and 2nd hours after treatment. The blood was collected from vena orbitalis.

Glycemic index was calculated based on the blood glucose level in individuals treated cocoyam tuber obtained by comparing the Area under Curve (AUC) of individuals treated CT and the AUC of individuals treated standard glucose which had GI 100. Area under curve was determine by using trapezoid method [11]. According to Venn and Green [12], GI was determined by using formula as follow:

$$GI = \frac{AUCcocoyamtuber}{AUCstandardglucose} x100 \qquad (1)$$

In the second stage, twelve DM rats which induced by 40mg/kg BW streptozotocin (STZ) interaperitoneally [13] were used. In the first group, DM rats were treated pellet diet *ad libitum.* In the second group or positive control group, DM rats were treated 0.9 mg/200g bw of glibenclamide. Whereas in the third group DM rats were treated steamed CT *ad libitum.* Blood was collected 4 times on day 0th, 1st , 3rd, and 6th. The blood glucose level was determine by using GOD-PAP method similar to the previous stage [9]. On the last days, rats in the first and third groups were sacrificed and dissected. Each of one gram liver and gastrocnemius muscle samples were preserved to determine the glycogen levels by using *Anthrone* method [14].

Data analysis were performed statistically by using *Statistical Product and Service Solutions* (SPSS) ver.17. Significant difference among treatment groups was examined by using *Analysis of Variance* (ANOVA) and followed by *Duncan Multiple Range Test* (DMRT).

3. Result and discussion

3.1. Blood glucose level of normal rats and glycemic index of cocoyam tuber

Blood glucose level of rats which treated pure glucose increased rapidly within 1 hour after treatment. It was higher than the blood glucose level of rats treated CT (Table 1). The increase level of blood glucose in rats which treated CT was rather similar to the control group. After 2 hours, the blood glucose level decreased slightly in rats treated pure glucose. Nevertheless it was higher than the blood glucose level before treatment. Meanwhile, the blood glucose level of rats which treated CT decreased close to the normal level. Blood glucose data of control group and CT group were used to determine AUC and calculate GI then. Cocoyam tuber had lower GI. It was 54. According to Rimbawan and Siagian [5], this score is categorized as low GI. Low GI food (<55) is considered as slow release carbohydrate. This carbohydrate can be broken down into monosaccharide slowly so the glucose releases into bloodstream last longer. This leads a gradual release of insulin to corrrespond glucose in the blood [15].

Table 1. Blood Glucose Level and Glycemic Index Score

Treatment	Average Blood Glucose Level (mg/dl) at hour			Blood Glucose Level Increase at hour		AUC (mg.hour/dl)	GI
	0	1	2	1	2		
Control	115.8	148.6	188.5	32.8	72.7	293.4	-
Pure Glucose	103.5	275	259	171.5	155.5	456.7	100
Cocoyam Tuber	104.4	141.9	113.4	37.5	9	250	54

According to Lenner et al [16], DM patients should consume low GI food. It would help patients improving their blood glucose control. Comparing to other tubers, CT has lower GI than potato, but higher GI than yam and taro. Glycemic index of steamed potato, yam and taro are 62, 41, and 47.7 respectively [17]. Most DM patients consume potato for daily diet though it has a medium GI.

3.2. Blood glucose level on DM rats

Rats were treated with STZ intraperitoneal in order to induce DM. The blood glucose rose in highly level over 150 mg/dl. It showed that all animals were in hyperglycemic condition. Barrett-Connor et al [18] stated, hyperglycemic condition is attained when blood glucose level is greater than or equal to 109 mg/dl. Blood glucose level increases because beta cells of pancreatic gland which produces insulin are damaged by STZ [19] .

Table 2. Blood glucose, liver and muscle glycogen level on DM rats

Treatment	Blood glucose level (mg/dl)			Blood glucose level increase/decrease (mg/dl)		Glycogen level (µg/25 mg)	
	1st day	3rd day	6th day	3rd day	6th day	Liver	Muscle
Na CMC (placebo)	158.9	347.6	475	188.7	127.4	266.4	92
Glibenclamide (positive control)	288.7	369.7	434.9	81	65.2		
Cocoyam tuber	249	118.3	83.5	-130.7	-34.8	531.2	169.8

Blood glucose were changed in different level after treatments (Table 2). At the end of day 6, the blood glucose level of DM rats treated placebo and glibenclamide still rose. However, the blood glucose increase in placebo group was unsignificant ($p<0.05$). Glibenclamide is a standard drug for DM treatment. Glibenclamide has hypoglycemic effect which stimulates insulin release in any glucose input [20]. Meanwhile, the blood glucose of DM rats treated CT declined in a significantly level back to normal glucose level (< 109 mg/dl). It was similar to the Folasire et al's [21] study. They concluded that blood glucose level of DM rats induced by alloxan declined as well after treated CT.

There are several possibilities why CT has hypoglycemic activity. Firstly it is because of CT has lower GI as previously described. Secondly, CT has antioxidant activity [22]. Vitamin C content on the CT is considered to this activity. Cocoyam tuber contains 2 mg/100 g vitamin C [23]. This is the most important antioxidant soluble in the plasma. It is clearer free radicals in the plasma. Regarding DM, Azrimaidaliza [24] stated that vitamin C acts as an antioxidant. It reduces oxidative damage which caused by free radicals, thus vitamin C prevents to the DM. In the DM which is induced by STZ, there is very reactive free radicals formations, e.g. NO, O_2 and H_2O_2. They damage several parts of pancreatic beta cells, such as cell membrane, protein and cause DNA fragmentation. Vitamin C from CT prevents beta cell damage, thus beta cell produces insulin as normal function.

Cocoyam tuber also contains flavonoid [25]. Any certain flavonoids have an antioxidant function [26] by acting as atom binder or free radicals scavenger [27]. Furthermore, several flavonoids capable to regenerate pancreatic beta cells. These flavonoids can stimulate progenitor cells in the pancreatic gland to differentiate and form new Langerhans Islet cells. Diosgenin is the most predominant phytochemical of CT [28]. It is known have anti-cancer activity which prevent cell proliferation and have hypoglycemic activity [29]. It reduces blood glucose level by decreasing lactase and maltase activity which play a role in carbohydrate breaking into simple sugar (monosaccharide). Breaking down of disaccharide in the intestines will be lower and glucose release into blood will be slower [30]. Diosgenin is an inhibitor for α-amylase and α-glucosidase enzymes which act in starch breaking down. Therefore this compound is the potential in DM treatment [31].

3.3. Glycogen levels of DM rats liver and muscle

In the DM rats which induced by STZ, the liver and muscle glycogen level increased (Table 2). The increase level was higher than normal individuals. The glycogen levels of the livers and muscle in normal rats are 227.56 μg/25 mg and 79.91 μg/25 mg respectively [30]. In this study, glycogen level of placebo group did not increase as high as CT group. The glycogen level of placebo group was rather similar to normal level. Whereas glycogen level of CT group increased significantly ($p<0.05$), twice of normal level.

Liver and muscle have important roles on the body glucose homeostasis by involve in glycogen synthesis. In DM which induced by STZ, the glycogen synthase phosphatase enzyme becomes defective. Therefore its activity is lower [32]. Suarsana et al [33] stated that less glucose enters to the cell in DM patients. Cell lacks of glucose. It interferes glycogenesis and glycogen storage in liver and muscle as well. Otherwise, glycogen in liver and muscle will be catabolized. Glucose will be released to the bloodstream and it results hyperglycemia condition and low glycogen status in hepatocytes and muscle cells [34]. Table 2 shows that diabetic rats treated Na-CMC had lower glycogen level than DM rats treated CT. It might CT has a high glycogenesis activity, but there is no strong evidence supportting study yet. It should be further study to determine on the biochemical mechanism of CT.

4. Conclusion

The research results that blood glucose level of normal rats which treated CT decreased close to the normal level. Blood glucose of DM rats after fed CT decreased significantly ($p<0.05$), similar to DM rats after treated by glibencamide. Whereas glycogen level in DM rats after treated by CT was higher than in DM rats after fed standard feed. Cocoyam tuber increased glycogen level in DM rats significantly ($p<0.05$) as high as in healthy rats. Glycemic index of CT was low. It was 54. Therefore CT is potential consumed as a functional food source for DM type II patients

References

[1] Whitney E, Rolfes SR, Pinna K 2002 *Nutrition and Diabetes Mellitus Understanding Normal and Clinical Nutrition* 7th ed (Belmont: Wadsworth)
[2] Sinaga E, Wirawanni Y 2012 *J. Nutr. Coll.* **1(1)** 312-321
[3] Wild S, Roglic G, Green A, Sicree R, King H 2004 *Diabetes Care* **27(5)** 1047-1053
[4] IDF (International Diabetes Federation) 2013 *Diabetes Atlas* 6th ed (www.idf.org)
[5] Jenkins DJA, Kendall CWC, Augustin LSA, Franceschi S, Hamidi M, Marchie A, Jenkins AL, Axelsen M 2002 *Am. J. Clin. Nutr.* **76(1)** 266-273
[6] Miller JB, Pang E, Bramall L 1992 *Am. J. Clin. Nutr* **56** 1034-1036
[7] Pruett A 2010 *A Comparison of The Glycemic Index of Sorghum and Other Commonly Consumed Grain* Thesis (Kansas: Food Science Kansas State University)
[8] Kusumo S, Hasanah M, Moeljoprawiro S, Thohari M, Subandrijo, Hardjamulia A, Nurhadi A, Kasim H 2002 *Pedoman Pembentukan Komisi Daerah Plasma Nutfah Badan Penelitian dan Pengembangan Pertanian* (Bogor: Komisi Nasional Plasma Nutfah)
[9] Hinkle JL, Cheever KH 2013 *Brunner & Suddarth's Textbook of Medical-Surgical Nursing* (Wolters Kluwer Health)
[11] Allison DB, Paultre F, Maggio C, Mezzitis N, Pi-Sunyer FX 1995 *Diabetes Care* **18(2)** 245-250
[12] Venn BJ, Green TJ 2007 *Eur. J. Clin. Nutr.* **61(1)** 122–131
[13] Susilawati Y, Muhtadi A, Soetardjo S, Supratman 2014 *Jurnal Ilmu-ilmu Hayati dan Fisik* **16(3)** 127-131
[14] Leyva A, Quintana A, Sanchez M, Rodriguez EN, Cremata J, Sanchez JC 2008 *Biologicals* **36** 134-141
[15] Willet W, Manson J, Liu S 2005 *Am. J. Clin. Nutr.* **76(1)** 274-280
[16] Lenner RA, Nils GA, Mette A, Susanne B, Eliina H, Anette J, Brita K, Anne R, Annica S, Inga T, Bengt V 2004 *Scan. J. Nutr.* **48(2)** 84-94
[17] Yang Y, Wang H, Cui H, Wang Y, Yu L, Xiang S, Zhou S 2006 *World J. Gastroenterol.* **12(21)** 3430-3433
[18] Barrett-Connor E, Wingard DL, Criqui MH, Suarez L 1984 *J. Chronic Dis.* **37(9–10)** 773-779
[19] Elsner M, Guldbakke B, Tiedge M, Munday R, Lenzen S 2001 *Diabetalogia* **43** 1528-1533
[20] Tjay TH, Rahardja K 2002 *Obat-obat Penting: Khasiat, Penggunaan, dan Efek-Efek Sampingnya* Ed. VI (Jakarta: PT. Elex Media Komputindo)
[21] Folasire OF, Oridupa OA, Owolabi AJ, Adepoju OT 2016 *Int. J. Nutr. Metab.* **8(4)** 24-29
[22] Nishanthini A, Mohan VR 2012 *Asian Pac. J. Trop. Biomed.* 701-706
[23] Lingga P 1995 *Bertanam Ubi-ubian* (Jakarta: Penebar Swadaya)
[24] Azrimaidaliza 2011 *Jurnal Kesehatan Masyarakat* **6(1)** 36-41
[25] Rita ES, Sophia D 2010 *PJN* **9(8)** 810-814
[26] Winarsi H 2007 *Antioksidan Alami dan Radikal Bebas* (Yogyakarta: Penerbit Kanisius)
[27] Sulistyorini R, Sarjadi, Andrew J, Kis D 2015 *Majalah Kedokteran Bandung* **47(2)** 69-76
[28] Senanayake SA, Ranaweera KKDS, Bamunuarachchi A, Gunaratne A 2012 *TARE* **15(1)** 32-36
[29] Jatmiko GP, Teti A 2014 *Jurnal Pangan dan Agroindustri* **2(2)** 127-134
[30] Patel K, Manoj G, Vijay T, Dinesh KP 2012 *Nat. Prod. Bioprospect.* **2** 46-52
[31] Ghosh S, Piyush M, Abhishek D, Ajay BP, Pramod M, Adersh A, Navanath K, Mahemud LS, Boppana R, Vaishali SS, Dilip DD, Balu AC 2014 *PLOS ONE* **9(9)** 1-9
[32] Niewoehner CB, Nuttall FQ 1986 *Diabetes* **35** 705-711
[33] Suarsana IN, Priosoeryanto BP, Bintang M, Wresdiyati T 2010 *Jurnal Ilmu Ternak dan Veteriner* **15(2)** 118-123.
[34] Dewi UK, Tyas RS 2009 *Bioma* **11(1)** 1-5.

Application of Biosurfactants Produced by *Pseudomonas putida* using Crude Palm Oil (CPO) as Substrate for Crude Oil Recovery using Batch Method

V Suryanti*, D S Handayani, A Masykur and I Septyaningsih
Department of Chemistry, Faculty of Mathematics and Natural Sciences, Sebelas Maret University, Jl. Ir. Sutami 36A Surakarta 57126 Indonesia

E-mail: venty@mipa.uns.ac.id

Abstract. The application of biosurfactants which have been produced by *Pseudomonas putida* in nutrient broth medium supplemented with NaCl and crude palm oil (CPO) for oil recovery has been evaluated. The crude and purified biosurfactants have been examined for oil recovery from a laboratory oil-contaminated sand in agitated flask (batch method). Two synthetic surfactants and water as control was also performed for oil recovery as comparisons. Using batch method, the results showed that removing ability of crude oil from the oil-contaminated sand by purified and crude biosurfactants were 79.40±3.10 and 46.84±2.23 %, respectively. On other hand, the recoveries obtained with the SDS, Triton X-100 and water were 94.33±0.47, 74.84±7.39 and 34.42±1.21% respectively.

1. Introduction

Oil recovery comprises a primary phase which uses natural stored energy and a secondary phase which includes injection of water to force oil recovery. The crude oils trapped in wells secondary phase makes up to 67% of the total reserves [1-2]. Various chemical and physical processes are applied for recovery of this residual oil which are can caused serious bad effects on the environment for long term uses. Enhanced Oil Recovery (EOR) is a tertiary oil recovery process to regain unrecoverable oil from an oil reservoir by chemicals flooding. The use of EOR methods have become an increasing important strategy. In particular, synthetic surfactants are used in chemical flooding process of EOR due to their effects on wettability alteration [3]. However, some synthetic surfactants have the potential toxicity risk or hazard to environment and human health [4].

Biosurfactants are surface-active compounds produced by microorganisms which are good candidates to replace the synthetic surfactants in EOR technology due to their low toxicity and biodegradability [5]. They are able to reduce surface and interfacial tension thus forming stable water-oil emulsions which is important for maximum oil extraction. The salt concentration and pH effect the biosurfactants solubility and activity. The effective pHs are pH ranging between 4-10 [6-7]. Many biosurfactants are precipitate at pH≤4 because their isoelectric point is near pH 4. The surfactant activities have been reported to increases 0-8% NaCl concentration due to salting-in affects [7-8].

The used of biosurfactants for oil recovery has been reported [9]. *Acinetobacter calcoaceticus* RAG-1 produces emulsan which is an extracellular protein-associated lipopolysaccharide. This biosurfactant has been used in chemical flooding process of EOR. The surfactant present in the slug reduces the interfacial tension between the oil/water and oil/rock to large extent, thereby desorbing the oil and emulsan stabilizes the desorbed oil in water, which is removed with the injection water [10]. The flooding experiment using lipopeptidic biosurfactants produced by *Bacillus mojavensis* (PTCC 1696)

IOP Conf. Series: Materials Science and Engineering **333** (2018) 012075 doi:10.1088/1757-899X/333/1/012075

showed 25% oil recovery. These biosurfactants reduce the interfacial tension from 65 to 26.7 mN/m under highly saline conditions [11]. *Rhodococcus* sp strain TA6 produces a mix of extracellular lipids and glycolipids which reduce surface tension of a hydrocarbon-based growth medium from 68 to <30 mN/m. These biosurfactants enhance the residual oil recovery of oil saturated sand packs by up to 70% [12].

On previous work, we reported the production of biosurfactants by *Pseudomonas putida* using crude palm oil (CPO) as substrate. This biosurfactans was identified as rhamnolipids [13]. As one of the world's biggest CPO producer, Indonesia exports 50% of produced CPO in an unprocessed form. Using CPO as a substrate for the production of biosurfactants is one process for diversification products of CPO in order to increase its economic value. In this work, we are reported the application of crude and purified biosurfactants produced by *P. putida* using CPO as substrate for crude oil recovery from a laboratory oil-contaminated sand in agitated flask (batch method). We compare their capacities for crude oil recovery with two synthetic surfactants (Sodium Dodecyl Sulphate and Triton X-100) and water as control.

2. Experimental

2.1. Materials

All chemical were used are analytical grade from e-Merck. The CPO was obtained from Centre for Chemical and Packaging, Jakarta, Indonesia. The crude oil was obtained from PT Pertamina Blok Cepu, Blora Indonesia. The strain used in this work, *P. putida* FNCC 0071, was purchased from IUC Food and Nutrition, Universitas Gadjah Mada, Indonesia.

2.2. Production and Purification of Biosurfactants

The production and purification of the biosurfactants have been carried following the methods of our previous work [13]. Cultures of bacteria were maintained on nutrient agar. The media were used for biosurfactants production composed of nutrient broth (8 g/L), NaCl (5 g/L) and CPO (10 % v/v). Media was sterilized prior used. The flasks containing 125 mL media were inoculated with 3 ml overnight pre-culture of the strain and incubated at room temperature on a reciprocal rotary shaker at 150 rpm.

Culture liquid of *P. putida* was filtered by Buchner filtrations using Whatman filtration paper grade 42. The supernatant was then acidified to pH 2.0 with HCl 6 N and leaved overnight at 4°C. The supernatant was extracted twice with n-hexane for removing of the remaining CPO. The aqueous solution was then extracted twice using ethyl acetate. Sodium sulphate anhydrous was then added to the organic layer. The organic layer which was free form water was then evaporated to obtain the purified biosurfactants.

2.3. Application of Biosurfactants for Oil Recovery using Batch Method

Application of biosurfactants for crude oil recovery using batch method was carried out pursuing the modified procedure reported by Suthar *et al.* (2008) [2]. Sand samples were cleaned and air dried. Sand samples of 100 Mesh particle size (15 g) were washed with NaCl 5% and contaminated with 2 g of crude oil, and then transferred to 250 mL erlenmeyer flasks. The erlenmeyer flask then was added with 30 mL of the biosurfactants and incubated under agitation of 100 rpm at room temperature for 24 h in a rotary shaker. After this period, the solution was separated with the sand. The hexane (15 mL) was added to the sand as the extracting solvent. The evaporation was performed to obtain the residual oil which was then determined by gravimetric analysis. The percentage of crude oil removal was calculated using the equation 1.

$$\text{Crude oil removed (\%)} = (O_i - O_r) / O_i \times 100\% \qquad (1)$$

Where Oi is the initial oil in the sand (g) before washing and Or is the oil remaining in the sand (g) after washing. The SDS (Sodium Dodecyl Sulphate), Triton X-100 and distilled water as a control were also analyzed at the same conditions for comparing the crude oil removal ability of biosurfactants. All experiments have been carried out in triplicate.

3. Results and Discussion

Although chemically synthesised surfactants have been used for enhanced oil recovery and oil spill clean up for over decades, biosurfactants have been studied for a possible replacement of chemical surfactants due to the toxicity and resistance to degradation of the chemical surfactants [14-15]. The contact time is an important parameter affecting efficiency of oil removal, as a sufficient contact time is required for effective oil removal. According to Lai *et al.* (2009), 24 h was sufficient contact time for solubilisation of the hydrocarbon to mobile phase, therefore, in this study, we investigated the oil removal for 24 h [16].

In order to evaluate the use of the crude biosurfactants, the removal ability of the cell-free broth was tested. Table 1 presents the results of the experiments carried out by batch method in erlenmeyer flasks for the removal of crude oil from the oil-contaminated sand by the crude and purified biosurfactants. The biosurfactants capacities for oil recovery were compared with two synthetic surfactants (SDS and Triton X-100) and water as a control. These experiments were performed at the slightly above CMC values of the biosurfactants/surfactants. Removing ability of crude oil from the oil-contaminated sand by purified and crude biosurfactants were 79.40±3.10 and 46.84±2.23 %, respectively. On other hand the recoveries obtained with the SDS and Triton X-100 as chemical surfactants at the same concentration were 94.33±0.47 and 74.84±7.39%, respectively. Water as a control removed only 34.42±1.21% of crude oil present in sand.

Table 1. Crude oil removal ability by biosurfactants, SDS and Triton X-100 using batch method.

Samples	**Crude Oil Removal (%)**
Biosurfactant	79.40±3.10
Crude Biosurfactants	46.84±2.23
SDS	94.33±0.47
Triton X-100	74.84±7.39
Water	34.42±1.21

The obtained results showed that the crude oil removal ability of water was about 34% which is a great economic interest. Therefore, a water washing process can be suggested applied before any other recovery process to diminish the crude oil content and consequently the consumed any surfactant quantity. As comparisons, Khalladi *et al.* (2009) report that water washing of a diesel-polluted soil remove 24% of *n*-alkanes and Chang *et al.* (2000) find that water washing remove 30-80% of PAHs. Our results are in accordance with the reported researches in the ability of positive controls of synthetic surfactants (SDS and Triton X-100) for crude oil removal [17-18]. Chang *et al.* (2000) also observe that 73.6-100% of PAHs were removed in the presence of SDS and Chaprao *et al.* (2015) report that 55-80% of motor oil pollutant was subtracted by Triton X-100 [4, 18].

From the results gained, it can be established that the purified biosurfactants are more effective in oil recovery when compared with the crude biosurfactants. The purified biosurfactants oil removal ability is slightly higher than that of the SDS, however, it is lower than that of the Triton X-100. Biosurfactants begin to form micelle in solution at the CMC value and solubilize petroleum hydrocarbons in soil-water systems at the above CMC. The biosurfactants have crude oil removing ability because they can stimulate the mobilization of oil toward aqueous phase resulting in stable emulsification and micellization, leading its removal [19]. The biosurfactants possessing a combination of surface and emulsification activity have a great ability in oil recovery.

Present studies showed that the ability of crude biosurfactants in removing crude oil were about 46%. This result indicated that the crude biosurfactants has lower ability than that of reported results. The crude biosurfactants from *Candida tropicalis* grown in waste frying oil removed 78-97% of the petroleum and motor oil from sand [20], while the crude biosurfactant from *C. guilliermondii* cultivated in industrial residues extracted about 90% of motor oil from sand [21]. Abu-Ruwaida *et al.* (1991)

reported that the crude biosurfactant produced by *Rhodococcus* sp. removed 86% of crude oil from sand [22].

Our gained results, it can be established that the crude biosurfactants are less effective in oil recovery the when compared with purified biosurfactants. Several studies report differently with our results. Some studies show that the crude biosurfactants have equal ability in removing oil with the purified biosurfactants. Other results report that the crude biosurfactants have better ability removing oil than that of the purified biosurfactants. Chaprao *et al.* (2015) report the ability of biosurfactants produced by yeast *C. sphaerica* and by bacterium *Bacillus sp.* for removing motor oil from soil [4]. The crude biosurfactants and the purified biosurfactants show almost equally effective in the removal of the motor oil pollutant under kinetic conditions (70-90%). Silva *et al.* (2010) report that the crude biosurfactants from *P. aeruginosa* was as effective as the purified biosurfactants in removing about 85% diesel oil from sand [23]. Moreover, the crude biosurfactant from *C. lipolytica* grown in medium containing animal fat and corn steep liquor was more effective in removing motor oil than that of the isolated biosurfactant [24]. Thus, the possible use of the crude biosurfactants without purification steps, which can further reduce the production cost of biosurfactants.

Other method, such as using column, need to be evaluate to for better ability in removing crude oil. Furthermore, other factor which can be affected the oil removal such as the kind of soil also need to be observe, since Silva *et al.* (2014) reported different ability of biosurfactant *P. cepacia* grown in mineral medium supplemented with corn steep liquor and soybean waste frying oil achieved for the removal of motor oil from sand and clay soil. Overall, biosurfactants produced by *P. putida* grown in media containing CPO have good prospect for crude oil removal from sand.

4. Conclusions

Biosurfactants which have been produced by *P. putida* in the nutrient broth medium containing CPO for crude oil recovery has been examined. The crude and purified biosurfactants have been investigated for crude oil recovery from a laboratory oil-contaminated sand using batch method. The purified biosurfactants were more effective than that of the crude biosurfactants in removing crude oil from sand. Furthermore, the purified biosurfactants were able to remove crude oil which were almost equally effective in the oil removal with the commercial available chemical surfactants (SDS and Triton X-100). Thus, the purified biosurfactants produced by *P. putida* could be applied in crude oil recovery by batch method.

Acknowledgments

The authors acknowledge their thanks to Ministry of Research, Technology and Higher Education Indonesia for the financial support to this research in the form of Insentif Riset Sistem Inovasi Nasional (Insinas) 2017 with contract number: 1047/UN27.21/PP/2017.

References

[1] Bryan R S, Stepp A K, Bertus K M, Burchfield T E and Dennis M 1993 *Dev. Petr. Sci.* **39** 289-306
[2] Suthar H, Hingurao K, Desai A and Nerurkar A 2008 *J. Microbiol. Methods* **75** 225-230
[3] Adelzadeh M R, Roostaazad R, Kamali M R and Lotfabad T B 2010 *Transaction C: Chemistry and Chemical Engineering* **17(1)** 46-54
[4] Chaprao M J, Ferreira I N S, Correa P F, Rufino R D, Luna J M, Silva E J and Sarubbo L A 2015 *Electronic J Biotechnol.* **18** 471-479
[5] Zhou W, Wang X, Chen C and Zhu L 2013 *Colloids Surf A Physicochem Eng. Asp.* **425** 122-128
[6] Amani H, Sarrafzadeh M H, Haghighi M, and Mehrnia M R 2010 *J. Petrol. Sci. Eng.* **75** 209-214
[7] Al-Bahry S N, Al-Wahaibi Y M, Elshafie A E, Al-Bemani A S, Al-Makhmari H S and Al-Sulaimani H S 2013 *Int. Biodeter. Biodegr.* **81** 141-146
[8] Darvishi P, Ayatollahi S, Mowla D and Niazi A 2011 *Colloids. Surf. B, Biointerfaces* **84** 292-300
[9] Bachman R T, Johnson A C and Edyvean R G J 2014 *Int. Biodeter. Biodegr.* **86** 225-237
[10] Kosaric N 1992 *Pure Appl. Chem.* **64** 1731-1737

[11] Ghojavand H, Vahabzadeh F, Shahraki A K 2012 *J. Petrol. Sci. Eng.* **81** 24-30
[12] Shavandi M, Mohebali G, Haddadi A, Shakarami H, Nuhi A 2011 *Colloids. Surf. B, Biointerfaces* **82** 477-482
[13] Suryanti V, Handayani D S, Masykur A and Lindasari 2017 *IOP Conf. Ser.: Earth Environ. Sci.* **175** 012013
[14] Mulligan C N 2005 *Environ. Pollut.* **133** 183-198
[15] Kryachko Y, Nathoo S, Lai P, Voordouw J, Prenner E J and Voordouw G 2013 *Int. Biodeter. Biodegr.* **81** 133-140
[16] Lai C C, Huang Y C, Wei Y H and Chang J S 2009 *J. Hazard Mater.* 167 609-614
.[17] Khalladi R, Benhabiles O and Bentahar F 2009 *J. Hazard Mater.* **164** 1179-1184
[18] Chang M C, Huang C R and Shu H Y 2000 *Chemosphere* **41** 1295-1300
[19] Costa S G, Nitschke M, Lepine F, Deziel E and Contiero J 2010 *Process Biochem.* **45** 1511-1516
[20] Batista R M, Rufino R D, Luna J M, Souza J E D and Sarubbo L A 2010 *Water Environ. Res.* **82** 1-8
[21] Coimbra C D, Rufino R D, Luna J M and Sarubbo L A 2009 *Current Microbiol.* **58** 245-251
[22] Abu-Ruwaida A S, Banat I M, Haditirto S, Salem A and Kadri M 1991 *Acta Biotechnol.* **11** 315-324
[23] Silva S N, Farias C B, Rufino R D, Luna J M, Souza J E D and Sarubbo L A 2010 *Colloids Surf. B Bioterfaces.* **79** 174-183
[24] Santos D K F, Rufino R D, Luna J M, Santos V A, Salgueiro A A and Sarubbo L 2013 *J. Pet. Sci. Eng.* 2013 **12** 1-32
[25] Silva N M, Luna J M, Rugino R D, Santos V A and Sarubbo L A 2014 *Biocatal. Agric. Biotechnol.* **3** 132-139

IOP Conf. Series: Materials Science and Engineering **333** (2018) 012076 doi:10.1088/1757-899X/333/1/012076

The Potential of Lignolytic *Trichoderma* Isolates in LDPE (Low Density Polyethylene) Plastic Biodegradation

M Hikmah, R Setyaningsih, A Pangastuti[1]
Biology Department, Faculty of Matematics and Science,
Sebelas Maret University, Surakarta, Indonesia

E-mail: artinipangastuti@gmail.com

Abstract. Plastic is experiencing buildup in the environment. Biodegradation process can be used as an alternative for LDPE plastic degradation because the process is environmentally friendly. Some fungi of the genus *Trichoderma* are known to have a role in plastic biodegradation. This study aims to find out how the potential of that lignolytic *Trichoderma* spp. isolates in LDPE biodegradation. Five isolates were screened by growing on MSMB (mineral salt medium broth) emulsified LDPE powder, with 35 days incubation at 30°C and shaking at 80 rpm. TL1, TL4, and TL5 are the three most potential isolates, indicated by the growth marked by increasing colony size on screening media. They were then tested for biodegradability by growing the isolates in MSMA (mineral salt medium agar) which then inoculated by 4 sheets of sterile LDPE 1x3 cm^2 above the colony surface, incubated for 5, 15, 25 and 35 days. The degredability assessment is done by measuring the weight loss of LDPE sheets after biodegradation treatment. The obtained degradability percentage of TL1, TL4, and TL5 are 4.87%, 7.12%, and 7,51% respectively. The visual micrograph of LDPE film by SEM showed the appearance of damage and unevenness on the surface of the post-degradation film.

1. Introduction

Plastics are synthetic polymers that are difficult to degrade, because they have a high molecular mass density, resulting in a buildup in the environment [1]. Plastics are material that can potentially threatens the survival of living things on earth, because they are made of chemicals that are difficult to degrade by microbes in the environment. One of the most commonly used types of plastic is low density polyethylene or LDPE.

Biodegradation process can be used as an alternative degradation because the process is environmentally friendly and does not cause natural damage. Biodegradation is the process when microorganisms are able to degrade or break down natural polymers (such as lignin and cellulose) and synthetic polymers (such as polyethylene and polystyrene) [2]. Biodegradation can be done by microorganisms such as bacteria, fungi, yeast and algae [3]. Fungi that have the ability of degradation of plastic in the environment among others derived from the genus *Aspergillus*, *Penicillium*, and *Trichoderma* [4]. In many studies, fungus are considered capable of LDPE biodegradation because of their ability to form hydrophobic proteins that can stick to polymer surfaces [5]; ability to produce suitable degradation enzymes against insoluble LDPE [6]; the growing speed of fungi biomass in soil faster than bacteria [7]; extension of body mass and penetration into other locations through the

[1] Author for correspondence: artinipangastuti@gmail.com

distribution of hyphae; and also the fungus can survive in environments with poor nutrition availability, low pH, and low humidity [8].

Prior research which did isolation and selection of the lignolytic fungi from zalacca leaf litter, produces fungal isolates from the genus *Trichoderma*. The lignolytic activity of fungi is due to the presence of lignolytic enzymes, some of which are most laccase, manganese peroxidase, and lignin peroxidase [9]. The polyethylene degredation may be performed by the *Trichoderma harzianum* fungi and produce a high percentage of degradability [10]. From the results of the research known that the enzymes that play a role in the process are laccase and manganese peroxidase, two enzymes are also encountered in the lignolytic test before. This study aims to test the potential of *Trichoderma* spp isolates. lignolytics from previous studies to determine whether there is a capability in LDPE polyethylene biodegradation.

2. Experimental

Isolates of lignolytic *Trichoderma* spp. obtained from the stock of biological laboratory of FMIPA UNS, they are TL1, TL2, TL3, TL4 and TL5. The culture stock is isolated from leaf litter of zalacca plantation in previous research and has been tested to have lignolytic ability.

2.1. Preparation for Media

Media that is used for rejuvenation of *Trichoderma* spp. is potato dextrose agar medium (PDA). While the media used for screening and biodegradation tests is mineral salt medium (MSM), with compositions in 1000 ml of aquades containing K2HPO4 1 g; KH2PO4 0.2 g; NaCl 1 g; CaCl2.2H2O 0.002 g; (NH4) 2SO 1 g; MgSO4.7H2O 0.5 g; CuSO4.5H2O 0.001 g; ZnSO4.7H2O 0.001 g; MnSO4.H2O, 0.001 g and FeSO4.7H2O 0.01 g. For solid media (MSMA) agar is added for as much as 20 grams.

2.2. LDPE Powder Preparation

LDPE polyethylene crystals in amount 5 grams dissolved into 150 mL xylene and then boiled for 30 minutes. To remove xylene from the solution, the LDPE solution was given ethyl alcohol. The mixture is then evaporated and dried by oven at a temperature of 40-50°C during overnight. The rough powder that is formed is then smoothed with a blender, then filtered to obtain the best powder.

2.3. Screening

The five isolates were inoculated on MSMB medium emulsified LDPE in erelnmeyer by 1 cork borer (39.25 mm^2) of inoculums. After that, it was incubated at 30°C in a shaker incubator at 80 rpm for 35 days. Isolates that exhibited growth characterized by increasing colony size or inoculum diameter in the media are then assessed as potentially biodegradable. This growth was observed visually or qualitatively, then measured the colony area with millimeter block.

2.4. LDPE Sheets Preparation

LDPE sheets with a thickness of 0.03 mm cut into 1x3 cm^2 pieces. The pieces are weighed until a constant dry weight is obtained. The LDPE sheet was then disinfected using 70% ethanol for 30 minutes, then transferred to sterile distilled water for 20 minutes and then dried again.

2.5. Biodegradation Test

Two pieces cork borer of 5-day-old Trichoderma spp. inoculum was inoculated on a petri dish containing MSMA, then incubated for 1 week at room temperature. A total of 4 pieces of sterile LDPE that have been known to have their dry weight inoculated on the surface of the growing isolates on the test medium. The culture was incubated at room temperature with a period of 5, 15, 25, and 35 days. In each incubation period, one sheet of LDPE film is taken. The LDPE film was first rinsed with alcohol and distilled several times, then dried with oven at 50°C for 6 hours. The constant dry weight of each LDPE sheets is then weighed. The value of degradation efficiency is obtained from the weight loss of the incubation start and end time with the equation:

$$DE\% = \frac{W0-W1}{W0} \times 100\,\% \quad (1)$$

W_1: the original weight of the sample
W_2: the weight of sample after biodegradation treatment
DE%: percentage of degredation efficiency

2.6. Observation of LDPE Film Morphology with SEM

SEM analysis was done at the UNDIP Integrated Laboratory of Semarang

3. Result and Discussion

3.1. Growth of Low Density Polyethylene Degrading Fungi (LDPE)

Visually, the isolates appeared to be growing quite well on media that do not support the growth of fungus in general. MSMB is a medium containing only mineral salts without addition of sugar or other carbon sources as triggers and growth supporters. In the medium, the only carbon source was the LDPE powder. At screening, the material LDPE appeared to be on the surface of the media and surrounded the isolates, or partly at the bottom of the media and appeared to be trapped, clinging, and covering the mycelium. LDPE powder was attached to the fungal colony (mass of mycelium). This is a sign that the fungal isolates used LDPE as a carbon source in MSMB. The initial size of the isolates inoculated into the medium was 1 cork borer (39.25 mm^2) of 1 week old isolates. From these early colonies it was possible to assess the growth of isolates after 35 days incubation, from the increasing of colony size.

At this screening, growing colonies showed irregular shapes of different sizes in liquid medium. Isolate TL1 colony looked round shaped with uneven edges, brownish green, mycelium looked dense, like there were fine hairs around the colony, a little bit of LDPE powder attached and entangled on the surface of the mycelium. In the TL4 isolate, spherical colonies appeared clumped unevenly, some of the mycelium that came out elongated like a tail and irregular, bright green colony color, only a small amount of LDPE powder attached to the surface. Whereas in the TL5 isolate, visible colonies with long and large, branched and irregular clumps, colony morphology and coloration could not be clearly seen because the colony is totally covered by LDPE powder at the base of erlenmeyer. Growth of these isolates in screening can be seen as in Figure 1.

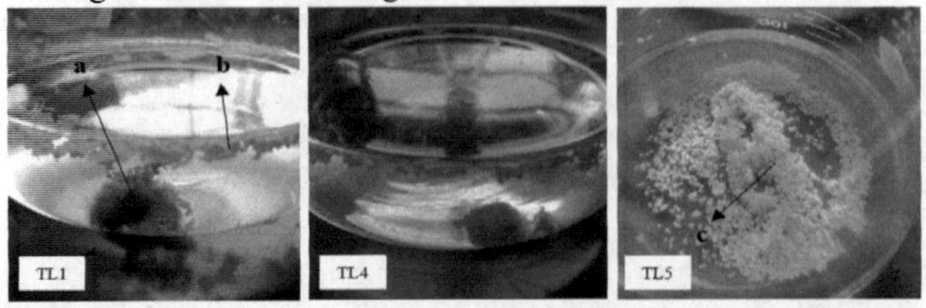

Figure 1. Growth of LDPE degrading potential isolates in screening media. (a) colony. (b) LDPE powdew partially floated on the surface of the media. (c) the colony was shrouded by powder at the base of the media.

After visual observation, the measurement of colony wide area was done by using millimeter block paper placed on the base of erlenmeyer. From the five isolates, three isolates (TL1, TL4, and TL5) showed good growth as seen in Table 1 which shows colony area data, so the three isolates were taken for further biodegradation test.

Table 1. Wide Range Size of Colony by Screening

Numb.	Isolate code	Early broad (cork borer)	Wide range area by screening
1	TL1	39,25 mm^2	3.1cm
2	TL2	39,25 mm^2	1.8 cm
3	TL3	39,25 mm^2	2.8 cm
4	TL4	39,25 mm^2	2.8 cm
5	TL5	39,25 mm^2	0 cm
6	control	-	-

3.2. Biodegradation of LDPE Plastic Sheet by Potential Isolate

Potential isolates from screening were further tested in this stage. Growth of inoculum *Trichoderma* spp. on the MSMA was not as fast as growth on the usual standard media for fungal cultures such as PDA. This MSMA was containing mineral salts only with the addition of plain agar in the absence of glucose and other carbon sources, thus less supporting growth. The colonies that grow on the medium appeared to be small to medium size but spread throughout the petri surface with the size and thickness of colony color varying each isolate. After the inoculums on MSMA are about one week old, the sterile LDPE sheets were prepared to be inoculated on the inoculum surface as seen in Figure 2.

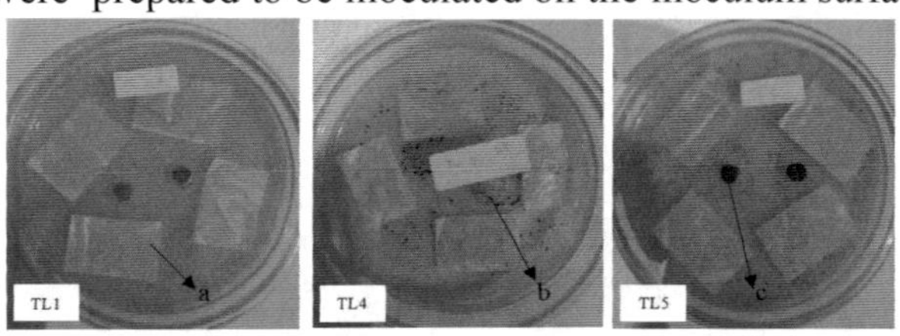

Figure 2. LDPE sheets placed on the surface *Trichoderma* spp. colonies. (a) LDPE sheet. (b) colonies of isolates (c) inoculum.

The growth of fungal isolates TL1, TL4, and TL5 in MSMA with LDPE membrane as a carbon source was relatively slow and there was no significant increase in mass of mycelium during the incubation period. The fungi utilizes complex organic polymers and converts them into simplier molecules by secreting the degrading enzymes to break the polymer chain and producing short chains such as oligomers, dimers, and monomers of a size small enough to pass through a semipermeable membrane [4]. However, the large molecular weight of LDPE, 3-dimensional structure, hydrophobic properties, and lack of functional groups strongly influence microbial attachment efforts in LDPE polymers [11]. This causes the growth of the fungus becomes very slow. During the degradation process, LDPE is first converted into its monomers, then the monomer is mineralized. LDPE is too large to pass through cellular membranes, it must first be depolymerized into smaller monomers before being adsorbed and biodegraded in microbial cells. These low molecular weight compounds will then be utilized by microbes as a source of carbon and energy.

3.3. Decreasing of Dry Weight Film by Presence of Biodegradation

Considering the mass of the polymer before and after the biodegradation process over a period of time is the simplest quantitative method to characterize the biodegradation of a polymer. Fungi that colonize plastic surfaces and utilize their polymers will exhibit visible growth traits with the naked eye, while the integrity of the polymer will decrease and lead to weight loss of LDPE sheets [12]. From the average dry weight gained, we got the graph of dry weight loss from the beginning to the end of the incubation period of each isolate and control as can be seen in Figure 3.

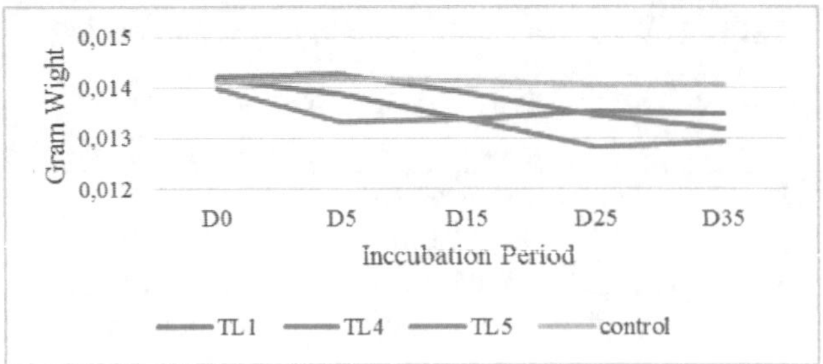

Figure 3. Film Weight Loss by Biodegradation

3.4. Percentage of Isolate Degradability

After 35 days of biodegradation test mass, control LDPE did not show any significant weight change, meanwhile LDPE sheet weight treatment was reduced by TL 1 fungal isolate by 4.87%, TL4 isolate by 7.12%, TL5 isolate of 7.51 %, and the control yielded a 0.28%. The controls showed no significant difference in each period. The control shows that there is a number indicating the percentage value of degradation, so it can be assumed that there is a decrease of plastic biomass due to the physical factor due to shake and rotary shaker during the incubation period. Comparison of percentage value of degradability of each isolate and control can be seen in Figure 4.

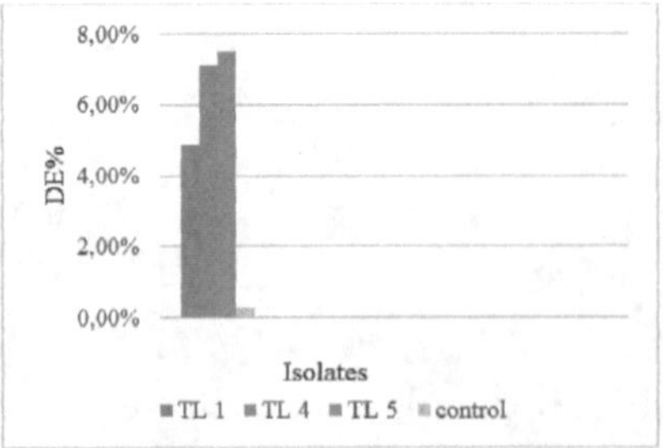

Figure 4. Percentage of Degradation Efficiency (DE%) of Isolates

Das and Kumar cultivated potential fungal isolates in MSM incubated at 33.3°C and shaken on a rotary shaker at 130 rpm for 60 days, with LDPE sheets being the only carbon source. After 60 days incubation it was found that *Aspergillus* sp. capable of reducing about 5% -8%, while *Fusarium* sp. capable of reducing 9% of LDPE sheet weight [13]. Singh and Gupta reported changes in LDPE weight were degraded by *Aspergillus flavus* (30%), *Aspergillus niger* (20%), *Aspergillus japonicus* (36%), *Mucor* sp. (16%), *Penicillium* sp. (24%), and *Fusarium* sp. (32%) after a 30-day inactivity period on synthetic medium (without yeast extract) and LDPE sheet as the only carbon source [14].

Harefa [15] conducted the research by isolating the fungus from TPA Namo Bintang. Two fungal isolates capable of degrading potential LDPE plastics from the study were isolates of the *Aspergillus* genus, with the percentage of degrees are 5.31% and 6.63% respectively after 45 days of biodegradation testing. If the results of this study were compared with those three studies, then the percentage of LDPE weight loss treated by the ligolytic *Trichoderma* spp isolates TL1, TL4, and TL5 were higher compared to the percentage of boot loss obtained by Das and Kumar [13] and Harefa [15], but still much lower with the degradability values obtained by Singh and Gupta [14].

3.5. Observation of Film Morphology After Biodegradation Test

The biodegradation process can cause changes in surface morphology. The result of LDPE sheet micrograph representing the three isolates along with the control at 3000X magnification can be seen in Figure 5. The analysis using SEM (Figure 5) shows that the LDPE surface control is incubated and after 35 days of biodegradation is flat, without cracks and holes, and no particles are attached to the surface. LDPE treatment with isolated TL1 isolates shows that the LDPE surface becomes more rough, forming curves, leading to cracks in LDPE materials. Similarly, in LDPE treated TL4 and TL5 isolates, they also suffered damage, produced pores on the surface, and even visible spore particles left behind during rinsing of LDPE membranes.

The natural properties of LDPE that are resistant to biological attacks are caused by hydrophobic carbon bonds and have high molecular weight polymers [16]. However, once the organism is attached to the polymer surface, it will begin to grow by utilizing the polymer as a carbon source. The filamentary fungi are generally capable of attaching to the hydrophobic surface by forming hydrophobic proteins [17]. The extracellular enzymes are too large for deeper penetration into the polymeric material so that they work only on the surface of the polymer alone. As a result plastic biodegradation is generally only a process of surface erosion.

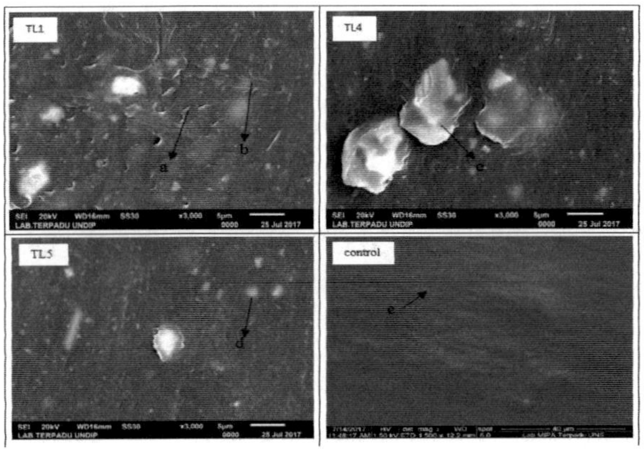

Figure 5. Appearance of LDPE Film Morphology at 3000X magnification. (a) pores or holes. (b) damage and cracking on the surface of the film. (c) SEM electrons reflect the particle on the sample due to uneven surface of the sample. (d) spore particles left on the surface of the film. (e) the control film surface looks flat and clean

4. Conclusion

From this research it can be concluded that 3 of 5 isolates of lignolytic *Trichoderma* spp. isolated from leaf litter of zalacca plantation has potential in LDPE plastic biodegradation, demonstrated by the ability of isolates to grow on the screening media emulsified LDPE powder as carbon source, and from biodegradation test which resulted in shrinkage of dry weight indicated by the degradability value of each isolate, TL 1 of 4,87%, TL4 of 7,12%, and TL5 by 7,51%. Potential degredability is also supported by polymer surface observations with SEM indicating a membrane damage due to biodegradation by fungi.

References

[1] Arutchelvi J *et al* 2008 *Indian J. Biotech* **7** 10
[2] Kaseem M *et al* 2012 *Polymer Science Series A.* **54** 168
[3] Guzman A *et al* 2011 *Chem. Tech.* **5** 119
[4] Bhardwaj H *et al* 2012 *Open Access Scientific Reports* **1** 274
[5] Seneviratne G *et al* 2006 *Current Sci.* **90** 20
[6] Shah A *et al* 2008 *Biotech. Adv.* **26** 247
[7] Kim D Y and Rhee Y H 2003 *Microbiol. and Biotech.* **61** 302
[8] Zahra S *et al* 2010 *Waste Management* **30** 397
[9] Placido J and Capareda S 2015 *Bioresources and Bioprocessing* **2** 25
[10] Sowmya H *et al* 2014 *Environ Monit. Assess.* **186** 6577
[11] Esmaeili A 2013 *PlosOne* **8** 2
[12] Gajendiran A 2016 *Biotech* **6** 52.
[13] Das M P and Kumar S 2014 *Int. J. Chem. Tech. Research* **6** 300
[14] Singh J and Gupta K 2014 *Int.J. Curr.Microbiol. App. Sci.* **3** 443
[15] Harefa R 2016 *Biodegradasi Plastik Low Density Polyethylene Menggunakan Jamur dari Tempat Pembuangan Akhir (TPA)* Skripsi Universitas Sumatera Utara.
[16] Kyaw B 2012 *Indian J. Microbiol.* **52** 412

[17] Kershaw M L and Talbot N J 1998 *Fungal. Genet. Bio.* **23** 20

IOP Conf. Series: Materials Science and Engineering **333** (2018) 012077 doi:10.1088/1757-899X/333/1/012077

Antioxidant Activities of Cinnamaldehyde Derivatives

V Suryanti*, F R Wibowo, S Khotijah and N Andalucki
Department of Chemistry, Faculty of Mathematics and Natural Sciences, Sebelas Maret University, Jl. Ir. Sutami 36A, Surakarta, 57612, Indonesia

E-mail: venty@mipa.uns.ac.id

Abstract. The modification structures of cinnamaldehyde, which was isolated from cinnamon oil, has been carried out. The synthesized compounds were tested their antioxidant activity by using 1,1-diphenyl-2-picryl hydrazyl (DPPH) radical scavenging assay and the IC_{50} was done by spectrophotometric assay method compared with standard vitamin E. The cinnamaldehyde derivatives, e.g. cinnamic acid **2**, methyl cinnamate **3** and cinnamyl alcohol **5** had significantly higher antioxidant activity than that of their starting materials, cinnamaldehyde. However, although cinnamic amine **5** had a hydroxyl group, it gave no antioxidant activity possibly due to its bulky structure.

1. Introduction

Antioxidants are compounds that are able to trap and deactivate damaging oxygen-centred radicals. Antioxidants have been proven for protection against certain diseases and aging since they have capacity for scavenge free radicals, which are responsible for the oxidative damage of lipids, proteins and nucleic acids. Many synthetic antioxidant compounds have shown toxic and/or mutagenic effects, and therefore there has been a constantly rising interest in the antioxidative constituents of various medicinal plants. Natural occurring antioxidants such as flavonoids from fruits, vegetables and leaves have been consumed for protection against diseases, including cancer, cardiovascular and cerebrovascular, due to their potential in controlling the levels of free radicals and the process of lipid peroxidation [1-5].

It has been established that the cinnamon oil and extract possess antioxidant activity, which is ascribed to the presence of phenolic and polyphenolic substances. Cinnamaldehyde (3-phenyl-2-propanal) **1** is the major component of cinnamon oil or cinnamon bark extract. This compound has various substituents on the aromatic ring, which are useful for starting material in synthesizing of its derivatives. Cinnamaldehyde derivatives have been reported as useful compounds for various application [6-11]. Cinnamic acids derivatives have been studied to have antioxidant, anti-inflammatory, anti-tuberculosis and cytotoxic properties. Cinnamate derivatives have been shown to have anti-inflammatory activity. This study is focused on the antioxidant activities of cinnamaldehyde derivatives. The modification structures of cinnamaldehyde, which have antioxidant activity is need to be done to use them in foods and pharmaceutical preparations as safe synthetic antioxidants.

We have synthesised cinnamaldehyde derivatives **2-4**, e.g. cinnamic acid **2**, methyl cinnamate **3** and methyl-3-(2-hydroxy-5-nitrophenyl amino)-3-phenylpropanoate **4** from cinnamaldehyde **1** [12]. The structures of the synthesized compounds have been established on the basis of UV-Vis, IR, ^{1}H-NMR and ^{13}C-NMR spectral. In this paper, we report the antioxidant properties of those synthesized compounds. We have examined for *in vitro* antioxidant activities of the compounds **1-4** using DPPH

(2,2-diphenyl-1-picrylhydrazyl) radical as a contribution to possible future applications. Additionally, we also synthesised and assessed antioxidant activity of cinnamyl alcohol **5** (Figure 1).

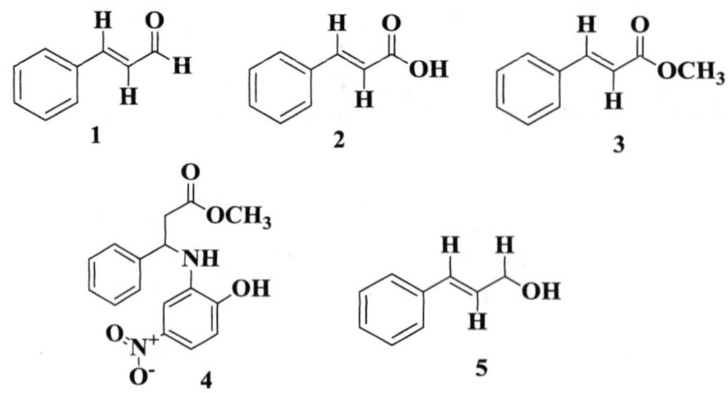

Figure 1. The synthesised compounds.

2. Experimental

2.1 Materials

Cinnamon oil was purchased from CV M & H FARM. Chemicals and solvents were analytical grade, which were obtained from e-Merck or Aldrich.

2.2 Methods

2.2.1 *Cinnamaldehyde **1** Isolation.* Cinnamaldehyde **1** was obtained from cinnamon oil by adding sodium bisulphate through a nucleophilic addition reaction. Cinnamaldehyde was separated by solvent extraction using diethyl ether. After evaporation, the cynnamaldehyde was washed with distilled water until neutral and then dried with anhydrous sodium sulphate.

2.2.2 *Cinnamic Acid **2** Synthesis.* Cinnamaldehyde was converted into cinnamic acid by reflux method for 6 h. Cinnamaldehyde (2.46 g) and CrO_3 (7.9 g) in a mixture of diethyl ether (100 mL) and water (100 mL) in a flask was added four drops of KTF. The mixture was then stirred continuously at a reflux for 6 h. The mixture then was cooled using an ice bath to reach 4-5°C and then allowed to reach room temperature. The mixture was extracted using diethyl ether to separate CrO_3 from the cinnamic acid.

*2.2.3 Methyl Cinnamate **3** Synthesis.* Cinnamic acid was esterified with methyl alcohol to obtain methyl cinnamate. Cinnamic acid (670 mg) in 10 ml of methanol was added 2 drops of sulphuric acid. The mixture was refluxed for 6 h. The mixture was then evaporated and cooled in room temperature to form precipitation.

*2.2.4 Cinnamic Amine (methyl-3-(2-hydroxy-5-nitrophenylamino)-3-phenylpropanoate) **4** Synthesis.* Reaction between methyl cinnamate and 2-amino-4-nitrophenol to give methyl-3-(2-hydroxy-5-nitrophenylamino)-3-phenylpropanoate. Methyl cinnamate was added with 2-amino-4-nitrophenol in a ratio of 1: 1.2 mmol equivalent. The mixture was refluxed in dichloromethane for 72 h and then evaporated the mixture to remove residual solvent.

*2.2.5 Cinnamyl Alcohol Synthesis **5**.* The mixture of 10 mmol cinnamaldehyde and 10 mmol $NaBH_4$ were grinded for 10 mins. A saturated $NaHCO_3$ solution was added. The mixture was then extracted

IOP Conf. Series: Materials Science and Engineering **333** (2018) 012077 doi:10.1088/1757-899X/333/1/012077

with dichloromethane. The organic layer was dried with anhydrous Na_2SO_4 and the solvent was evaporated.

2.2.6 Antioxidant activity test. The 1,1-diphenyl-2-picrylhydrazyl (DPPH) radical was used to evaluate the free radical scavenging capacity of synthesized compounds [13]. Firstly, the antioxidant activity test was performed by determining the maximum wavelength of DPPH in the range of 800-400 nm. The maximum DPPH wavelength was found to be 516 nm.

The aliquot of the different concentrations of the test sample (12.5, 25, 50, 75 and 100 ppm) was added to 3 mL ethanolic solution of DPPH. Test tubes were kept for 30 mins at room temperature and were recorded at 516 nm by UV-Vis Spectrophotometer. DPPH was also applied on the blank test tubes at the same time where only ethanol was taken as blank. The absorbance results were then used to determine the value of IC_{50}. The percentage inhibition was calculated using equation 1, where Ac is the absorbance of the control and As is the absorbance of the sample.

$$\% \text{ inhibition} = (Ac - As/Ac) \times 100 \quad \text{.......} \quad (1)$$

Then percentage inhibitions were plotted against log concentration of each of the test sample. Then IC_{50} value was calculated from the graph. The IC_{50} value denotes the concentration of sample required to scavenge 50% of the DPPH free radicals.

3. Result and Discussion

The starting material used was cinnamaldehyde **1** and the reaction was accomplished in three step reactions to give the final product **4**. Cinnamaldehyde was converted into cinnamic acid **2**, which was then esterified with methyl alcohol to obtained methyl cinnamate **3**. The methyl cinnamate was then reacted with 2-amino-4-nitrophenol to give final product of methyl-3-(2-hydroxy-5-nitrophenylamino)-3-phenylpropanoate **4**. The synthetic compounds have been characterized by ^1H NMR and the spectra showed the right chemical shifts as reported previously [12]. Cinnamaldehyde **1** was obtained as yellowish liquid with the yield of 80%. Cinnamic acid **2** was obtained as white powder with the yield of 64.9%. Methyl cinnamate **3** was obtained as dark brown liquid with the yield of 42%. Cinnamic amine **4** was obtained as brownish-yellow solid with the yield of 68%.

The cinnamyl alcohol showed as yellowish liquid with the yield of 93%. FTIR showed peaks at 3350 cm^{-1} for O-H stretching, 3081 and 3059 cm^{-1} for C-H stretching aromatics, 1449 cm^{-1} for C=C aromatic stretching. The ^1H NMR spectrum of the cinnamyl alcohol compound exhibited 9 protons. A peak at $\delta = 1.56$ ppm indicated for a proton of hydroxy group. Two peaks at $\delta = 4.30$ and 6.57 ppm showed 2 coupled for proton C = C. Peaks at $\delta = 7.23$-7.37 ppm showed the existence of 5 protons for aromatic protons.

The DPPH solution was scanned at the wavelength of 800-400 nm and the maximum wavelength of DPPH was obtained at 516 nm. This maximum wavelength can be applied for determination of the antioxidant activity of the synthetic compounds because the maximum wavelength of DPPH exists within the 515-520 nm. DPPH is a stable free radical containing an odd electron in its structure and commonly employed for evaluation of the radical scavenging activity in chemical analysis. The odd electron in the DPPH free radical gives a strong absorption maximum at 516 nm and is purple in colour. DPPH addition and incubation for 30 mins gave different colour from purple to yellow. The colour turned to yellow as the molar absorptivity (optical density) of the DPPH radical at 516 nm reduces when the odd electron of DPPH radical forms the reduced DPPH-H. The resulting decolourization is stoichiometric with respect to the number of electrons captured [14-16]. The absorbance from the several variations concentration at 516 nm was used to determine the percentage inhibition, and plotted with the concentration to obtain the linear equations that was used to determine the IC_{50} values. Table 1 showed the antioxidant activities of all compounds.

Table 1. Antioxidant potential of the all compounds

Compound	IC_{50} Value	Remarks
Vitamin E	7.48	Highly active
Cinnamaldehyde **1**	95.38	Moderately active
Cinnamic acid **2**	38.52	Highly active
Methyl cinnamate **3**	40.76	Highly active
Cinnamic amine **4**	-37.32	No activity
Cinnamyl alcohol **5**	21.45	Highly active

Cinnamaldehyde as a starting material had IC_{50} value of 95.38, which indicated moderate antioxidant activity. The cinnamaldehyde derivatives **2**, **3** and **5** exhibited IC_{50} values of 38.52, 40.76 and 21.45, respectively. These cinnamaldehyde derivatives showed highly antioxidant activities, which were comparable to the antioxidant activity of vitamin E. Furthermore, these IC_{50} values showed that these compounds had higher antioxidant capacity than that of starting material **1**. Cinnamyl alcohol **5** was found to have the most efficacious antioxidant among all the synthetic compounds. However, compound **4** gave a negative value of IC_{50} which indicated having no antioxidant activity.

As per chemical structural features, there were four different types of compounds synthesized under the study area. It is obvious that structural modification gives about the bioactivity and, of course, structural variations of molecules alters the biological activity in a regular trend [17]. Compound **2** has a hydroxyl group that exerted its antioxidant property. The presence of a hydroxyl group showed upward tendency of antioxidant activity. However, although compound **4** has a hydroxyl group but it does not show antioxidant properties. It might be due to its bulkier structure than that of compound **2**. Although compound **3** does not have hydroxyl group, however, this compound showed highly antioxidant activity. The reason of this finding is still unclear. Compound **5** was found to have the highest IC_{50} value among other compounds. It is clearly understood that the alcohol functional group sharply enhanced the antioxidant potency. This is why compounds **5** was found to have highest activity than the others. This study showed that the cinnamaldehyde derivatives as the potent antioxidants and may be efficient as preventive agents in food and drugs.

4. Conclusion

The present investigation emphasized that the modification structure enhanced the antioxidant activity of the cinnamaldehyde. Cinnamaldehyde derivatives **2**, **3** and **5** had significantly higher activities than their starting materials. However, the modification structure for compound cinnamic amine **4** gave no antioxidant activity possibly due to its bulky structure. These findings marked that the cinnamaldehyde derivatives **2**, **3** and **5** are more desirable antioxidant compound than that of their starting material. The results of DPPH assays also demonstrated that compound **5** has the greatest capacity as free radical scavengers, possibly due to the presence of alcohol functional group, which is essential for its antioxidant capacity. However, the potential application of these cinnamaldehyde derivatives in various foods and regulatory acceptance still needs further investigation.

Acknowledgements

The authors acknowledge their thanks to Sebelas Maret University for the financial support to this research in the form of Penelitian Unggulan Perguruan Tinggi (PUPT) Grant 2017 with contract number: 098/SP2H/LT/DRPM/IV/2017 dated 3 April 2017.

References

[1] Aruoma O I 1998 *J. Am. Oil Chem. Soc.* **75** 199-212.
[2] Torreggiani A and Tamba, M 2005 *Trends Heterocycl. Chem.* **10** 115–137.
[3] Wang H, Cao G and Prior R L 1996 *J. Agric. Food Chem.* **44** 701-705.
[4] Hossain M M, Shaha S H and Aziz F 2009 *Bangladesh Med. Res. Counc. Bull.* **35** 49-52.
[5] Sadhu S K, Okuyama E, Fujimoto H and Ishibashi M 2003 *Chem. Pharm. Bull.* **51** 595-598.
[6] Pontiki E, Hadjipavlou-Litina D, Litinas K and Geromichalos G 2014 *Molecules* **3(19)** 9655-9674.
[7] Godoy M E, Rotelli A, Pelzer L and Tonn C E 2000 *Molecules* **5** 547-548.
[8] Bairwa R, Kakwani M, Tiwari N R, Lalchandani J, Ray M K, Rajan M G R and Degani M S 2010 *Bioorg. Med. Chem. Lett.* **20(5)** 1623-1625.
[9] De P, Baltas M and Bedos-Belval F 2011 *Curr. Med. Chem.* **18(11)** 1672-1703.
[10] De P, Yoya G K, Constant P, Bedos-Belval F, Duran H, Saffon N, Daffé M and Baltas M 2011 *J. Med. Chem.* **54(5)** 1449-1461.
[11] Yoya G K, Bedos-Belval F, Constant P, Duran H, Daffé M and Baltas M 2009 *Bioorg. Med. Chem. Lett.* **19(2)** 341-343.
[12] Suryanti V, Wibowo F R, Pranoto, Isnaeni S R, Sari M R K, Handayani S 2016 *IOP Conf. Ser. Mater. Sci. Eng.* **107** 012062.
[13] Hossain, M M, Shaha S and Aziz F 2009 *Bangladesh Med. Res. Counc. Bull.* **35** 49-52.
[14] Espin J C, Soler-Rivas C and Wichers H J 2000 *J. Agric. Food Chem.* **48** 648-656.
[15] Yu L 2001 *J. Agric. Food Chem.* **49** 3452-56.
[16] Mattha¨us B. 2002 *J. Agric. Food Chem.* **50** 3444-3452.
[17] Jiang H, Sun S-L, Zhang C, Yuan E, Wei Q-Y and Zeng Z 2013 *Trop. J. Pharm. Res.* **12(6)** 1017-1022.

Secondary metabolites of Mirabilis jalapa structurally inhibit Lactate Dehydrogenase A in silico: a potential cancer treatment

R Kusumawati[1], A H Nasrullah[2], R N Pesik[3], Muthmainah[4] and D Indarto[1,5]

[1] Department of Physiology, Faculty of Medicine, Universitas Sebelas Maret, Surakarta, Indonesia

[2] Medical Student, Faculty of Medicine, Universitas Sebelas Maret, Surakarta, Indonesia

[3] Department of Pathological Anatomy, Faculty of Medicine, Universitas Sebelas Maret, Surakarta, Indonesia

[4] Department of Anatomy, Faculty of Medicine, Universitas Sebelas Maret, Surakarta, Indonesia

[5] Biomedical Laboratory, Faculty of Medicine, Universitas Sebelas Maret, Surakarta, Indonesia

E-mail: donoIND323@gmail.com

Abstract: Altered energy metabolism from phosphorylated oxidation to aerobic glycolysis is one of the cancer hallmarks. Lactate dehydrogenase A (LDHA) is a major enzyme that catalyses pyruvate to lactate in such condition. The aim of this study was to explore LDHA inhibitors derived from Indonesian herbal plants.In this study, LDHA and oxamate molecular structures were obtained from protein data bank. As a standard ligand inhibitor, oxamate was molecularly re-validated using Autodock Vina 1.1.2 software and showed binding energy -4.26 ± 0.006 kcal/mol and interacted with LDHA at Gln99, Arg105, Asn137, Arg168, His192, and Thr247 residues. Molecular docking was used to visualize interaction between Indonesian phytochemicals and LDHA. Indonesian phytochemicals with the lowest binding energy and similar residues with standard ligand was Miraxanthin-III (-8.53 ± 0.006 kcal/mol), Vulgaxanthin- I (-8.46 ± 0.006 kcal/mol), Miraxanthin-II (-7.9 ± 0.2 kcal/mol) and Miraxanthin-V (-7.96 ± kcal/mol). Lower energy binding to LDHA and binding site at these residues was predicted to inhibit LDHA activity better than standard ligand. All phytochemicals were found in Mirabilis jalapa plant. Secondary metabolites in Mirabilis jalapa have LDHA inhibitor property in silico. Further in vitro study should be performed to confirm this result.

1. Introduction

Cancer is the first leading cause of death in developed countries while in developing countries, cancer becomes the second cause of mortality after cardiovascular diseases [1]. In 2012, approximately 32.6 million cancer cases were reported globally and 8.2 millions of death was due to cancer [2]. Data from epidemiological studies shows that up to 14.1 millions of new cancer cases are found every year [3]. WHO estimates that in the next two decades, the number of new cancer cases will increase by around 70% worldwide [1].

Cancel cells use glycolysis pathway as the main pathway for energy metabolism even when the oxygen supply is sufficient. Glucose and energy metabolism in cancer cells are mainly through aerobic glycolysis mechanism [4,5]. The shift in energy metabolism from oxidative phosphorylation to glycolysis causes the cancer cells to compensate for increased energy demand by up-regulating glucose transport [6,7]. This glycolysis process creates nucleoside and amino acids which facilitate biosynthesis of macromolecule and cell organelles leading to rapid growth of the cancer cells [8].

Glycolysis process is initiated by glucose phosphorylation resulting in the formation of pyruvate molecule and ATP. The pyruvate is reduced to lactate by an enzyme catalyst called lactate dehydrogenase (LDH) [9]. LDH plays an important role in ATP formation and in maintaining the sustainability of glycolysis process. LDH has two subunits with distinct function. LDHA subunit is responsible to convert pyruvate into lactate while LDHB subunit has a reverse function [10]. It is reported that in invasive cancer cells, the expression of LDHA is up-regulated. Increased expression of LDHA is associated with several factors such as tissue damage, hypoxia and necrosis. These conditions increase glycolysis process giving rise to excessive lactate production and reduced cell dependency toward oxygen [11,12].

Most of the currently available anti-cancer drugs not only affect cancer cells but also normal cells. The most common side effect of anti-cancer drugs is the toxicity to kidney and gastrointestinal organs, bone marrow suppression and immune system suppression [13]. The development of LDHA inhibitor as anti-cancer drug is aimed to get new anti-cancer with less severe side effects. Previous studies showed that LDHA deficiency was not life threatening and selective LDHA inhibitor caused minimal side effects [14]. Reduced LDHA activity will induce mitochondrial respiration and increase intracellular oxidative stress. This condition will affect the ability of cancer cells to proliferate, lower the likelihood of metastasis, and increase the sensitivity to chemotherapy agent [15,16]. Today, bio-computational program has been widely used for drug development. This method requires less time and funding as compared to *in vitro* or *in vivo* research directly. One of the methods is molecular docking used to see interaction between drug and its target receptor [17,18]. This study aimed to explore potential LDHA inhibitor from Indonesian herbal plants.

2. Experimental

Virtual data of all phytochemicals from Indonesian herbal plants were obtained from Herbal Indonesia database (HerbalDB). Approximately 6.776 active compounds from 3.825 plant species in Indonesia can be found in this database [17]. Three dimensional structure of the phytochemicals were downloaded from PubChem NCBI (http://pubchem.ncbi.nlm.nih.gov). The standard ligand used in this study was Oxamate. Oxamate is a potent LDHA inhibitor [19]. The structure of oxamate resembles pyruvate's structure so that it can compete with pyruvate [20]. Three dimensional structure of LDHA which binds with standard ligand's inhibitor (oxamate) was downloaded from Protein Data Bank in http://www.rcsb.org/pdb/ with PDB IDI code : 1I10. Subsequently, structure of oxamate was separated from LDHA structure by using Chimera 1.9. LDHA in *.pdf format file was prepared using AutoDock Tools by removing water molecule and adding hydrogen atom in the polar site. Then they were saved in *.mol2 and *.sdf format. Interaction between oxamate and LDHA was validated five times using AutoDockVina program to measure the binding energy. AutoDock Vina software is the development of AutoDock program which can predict conformation and binding energy better [21].

Molecular docking is a bio-computational program which can predict the structure of a molecule complex. This method comprises of two steps including predicting the binding location of a molecule and predicting the score of binding affinity [22]. The phytochemicals predicted to act as LDHA inhibitor must fulfill the Lipinski Rule of Five criteria, molecular weight less than 500 Dalton, have lower binding energy, and show similar conformation as the standard ligand. Molecular docking between phytochemicals and LDHA enzyme was done at mobile loop of the enzyme's active site (at amino acid residue 163-247 and 267-331) and at the site of LDHA-oxamate binding (at His 192, Asn 137, Arg 105, Arg 168 and thr 247) using software AutoDockVina 1.1.2. This molecular docking result between standard ligand-LDHA was used to analyze the docking between Indonesian

IOP Conf. Series: Materials Science and Engineering **333** (2018) 012078 doi:10.1088/1757-899X/333/1/012078

phytochemicals and LDHA. Interaction between standard ligand inhibitor-LDHA and phytochemicals-LDHA were visualized with Pymol 1.7. Pymol is a software used to visualize the binding site between receptor and its ligand [23].

3. Result and Discussion

3.1. Validation of Docking between Oxamate and LDHA

Before docking the active phytochemical of Indonesian herbal plants with LDHA, we firstly validated the docking between standard inhibitor (Oxamate) and LDHA using Autodock Vina and Pymol 1.7 software. Validation was conducted five times and the final average score for Oxamate-LDHA docking was -4,26 ±0,06 kcal/mol (Table 1). This score becomes the reference to screen for candidate phytochemical of LDHA inhibitor. The phytochemicals must have lower binding energy to LDHA than the standard ligand (Oxamate).

Table 1. Result of docking validation between Oxamate and LDHA

	Docking Score (kcal/mol)	Binding site
Validation 1	-4.3	Arg 105, Asn 137, Gln 99, Arg 168, Thr 247
Validation 2	-4.2	Arg 105, Asn 137, Gln 99, Arg 168, His 192, Thr 247
Validation 3	-4.3	Arg 105, Asn 137, Gln 99, Arg 168, Thr 247
Validation 4	-4.2	Arg 105, Asn 137, Gln 99, Arg 168, His 192, Thr 247
Validation 5	-4.3	Arg 105, Asn 137, Gln 99, Arg 168, Thr 247
Average	-4.26 ±0.06	

Visualisation of Oxamate-LDHA interaction showed that Oxamate binds with LDHA using Hydrogen bond (Figure 1). Validation of standard inhibitor (Oxamate) and LDHA interaction was conducted to get the docking score and binding site. Previous study reported that Oxamate bound with LDHA at 5 residues including Arg 105, Asn 137, Arg 168, His 192, and Th2 247 [24]. Arg 105 and His 192 plays a crucial role in LDHA enzymatic reaction. His 192 has two main functions. In addition to being proton exchanger, His 192 also facilitates pyruvate orientation enabling it to interact with co-factor Nicotinamide Adenine Dinucleotide Hidrogen (NADH). Arg 105 is responsible to stabilize Hydrogen ion transfer reaction [25]. Binding at these two residues is predicted to inhibit LDHA activity.

Oxamate has an identical structure as pyruvate so that it can act as substrate-like inhibitor competitor [20]. During validation process between Oxamate and LDHA, we removed NADH due to the fact that NADH can block Oxamate-LDHA interaction and thus affect the docking process.Validation process revealed that there was one additional binding site at Gln 99 (Figure 1). Gln 99 is part of the mobile loop of the active site that directly influence the catalytic reaction of pyruvate to lactate conversion. Binding of substrate at these residues can inhibit Lactate Dehydrogenase [24,25].

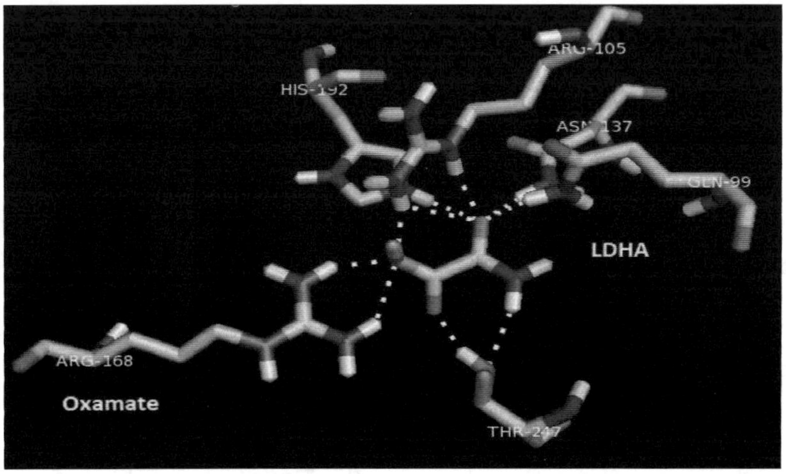

Figure 1. Visualisation of Oxamate-LDHA Binding Site
(white = atom H; blue= atom N; red= atom O; green = atom C)

3.2. Result of Docking between active phytochemicals of Indonesian herbal plant with LDHA

After screening 6.776 active phytochemicals listed in HerbalDB database, we found that 516 of them met the Lipinski Rule of Five criteria. Docking of the 516 phytochemicals with LDHA was done three times resulting in 501 phytochemicals with docking score less than or equal to the standard ligand (Oxamate). Visualisation of the 501 phytochemicals using Pymol software revealed that 23 active phytochemicals bound with at least 5 amino acid residues similar as that of Oxamate. Among these, 4 phytochemicals bound at 5 similar residues as oxamate with one additional binding at Arg 98 which serves as a mobile-loop of the active site (Table 2). Visualisation of these 4 phytochemicals showed that they generally had identical structure and formation (Figure 2). During glycolysis metabolism, NADH serves as a co-factor which induces LDHA enzymatic reaction. Initially, NADH binds with LADH and creates a binding site for the substrate (in this case, pyruvate). Subsequently, pyruvate will bind with LDHA resulting in closing of the mobile-loop of the active site [26,27].

Table 2. Result of docking between phytochemical of Indonesian herbal plant with LDHA

Phytochemical name	Molecular weight (g/mol)	Binding site	Average of Binding Energy (kcal/mol)	Docking Score (kcal/mol)		
				I	II	III
Oxamate (standar ligand)	89.05	Gln 99, Arg 105, Asn 137, Arg 168, His 192, Thr 247	-4.26±0.006	-4.3	-4.2	-4.3
Miraxanthin-III	330.34	Gln 99, Arg 98, Arg 105, Arg 168, His 192, Thr 247	-8.53±0.006	-8.6	-8.5	-8.5
Vulgaxanthin-I	339.304	Gln 99, Arg 98, Arg 105, Asn 137, Arg 168, Thr 247	-8.46±0.006	-8.5	-8.5	-8.4
Miraxanthin-II	326.261	Gln 99, Arg 98, Arg 105, Asn 137, Arg 168, Thr 247	-7.9±0.2	-8.1	-7.7	-7.9
Miraxanthin-V	346.339	Gln 99, Arg 98, Arg 105, Asn 137, His 192, Thr 247	-7.96±0.06	-7.9	-8.0	-8.0

Through molecular docking, we found that 4 phytochemicals potentially act as LDHA inhibitor including Miraxanthin-II, Miraxanthin-III, Miraxanthin-V, and Vulgaxanthin-I because they bound at the mobile loop of the enzyme's active site (Figure 2). Oxamate has been proven to be LDHA inhibitor in cancer cell. Oxamate inhibited tumor growth both *in vivo* and *in vitro*, induced cancel cell

IOP Conf. Series: Materials Science and Engineering **333** (2018) 012078 doi:10.1088/1757-899X/333/1/012078

apoptosis, and reduced tumor size [28,29]. However, cell permeability toward Oxamate is low and non-specific due to its small molecular weight and intracellular penetration power [20]. The molecular weight of Miraxanthin-II, Miraxanthin-III, Miraxanthin-V, and Vulgaxanthin-I are larger than Oxamate (Table 2). It is estimated that with larger molecular weight, these phytochemicals will be more permeable and specific compared to Oxamate.

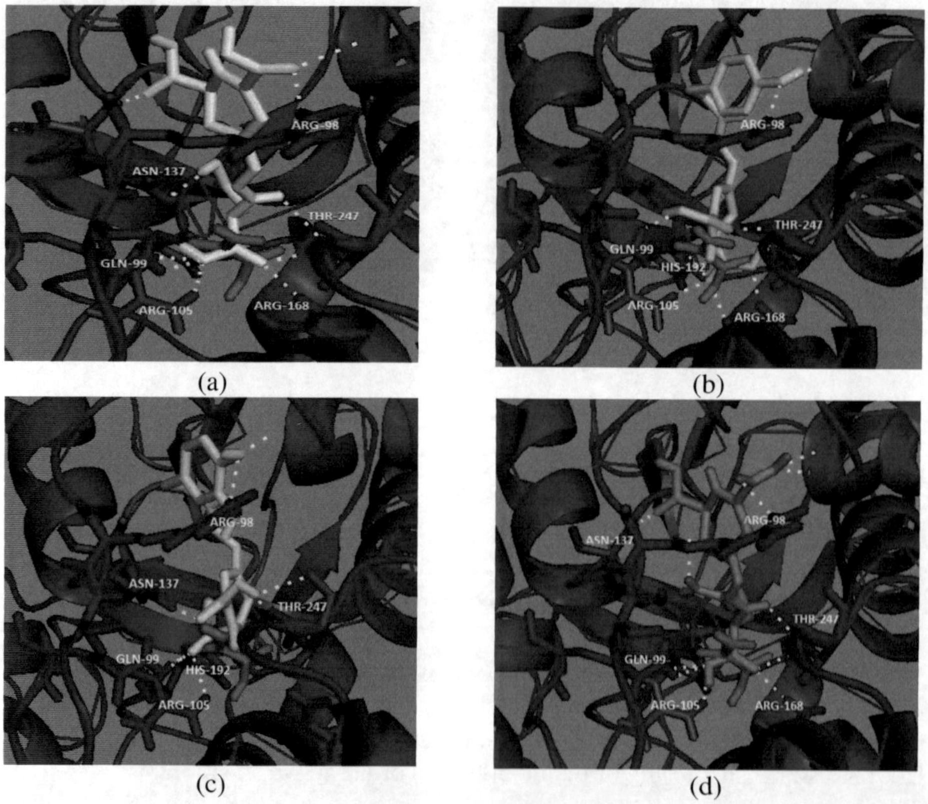

Figure 2. Visualisation of phytochemicals-LDHA binding compare to Oxamate (red). (a) Miraxanthin-II (b) Miraxanthin-III (c) Miraxanthin-V (d) Vulgaxanthin-I.

A phytochemical tends to create stronger hydrogen bond to water than to lypophilic cell membrane as the number of H-bond acceptor increases [30]. Oxamate is a compound that has been proven to be LDHA inhibitor in some study [28, 29]. Of the four phytochemicals we found in this study, Miraxanthin-II has more H-bond acceptor than Oxamate. Thus, we predict that it will be more difficult for Miraxanthin-II to cross cell membrane as compared to Miraxanthin-III, Miraxanthin-V, and Vulgaxanthin-I. The molecular weight of the four phytochemicals we found are three times larger than Oxamate. It is estimated that the phytochemicals will bind NADH co-factor in hydrogen binding. This binding has impact to inhibit the glycolysis metabolism.

Based on HerbalDB database, Indonesian herbal plant which contains Miraxanthin-II, Miraxanthin-III, Miraxnathin-V, and Vulgaxanthin-I is *Mirabilis jalapa* (It is called *Bunga pukul empat* in Indonesia). Previous study reported that *Mirabilis jalapa* leaf extract has cytotoxic effect on cancer cells although the mechanism of its cytotoxicity is not clear yet [31]. It is important to isolate every single compound of *Mirabilis jalapa* to identify the phytochemicals with anti-cancer property.

4. Conclusion

Miraxanthin-II, Miraxanthin-III, Miraxanthin-V, and Vulgaxanthin-I can potentially act as LDHA inhibitor *in silico*. Docking result showed that all these four phytochemicals had lower binding energy than Oxamate and interacted at LDHA inhibitor binding site. Indonesian herbal plant, *Mirabilis jalapa* contains all these four phytochemicals thus raising the likelihood that secondary metabolites of this plant can serve as LDHA inhibitor. Further research is needed to confirm this finding.

Reference

[1] WHO 2011 *Global status report on noncommunicable diseases 2010* (Geneva : WHO press)
[2] Torre L A, Bray F, Siegel R L, Ferlay J, Lortet-Tieulent J and Jemal A 2015 Global cancer statistics 2012 *CA. Cancer. J. Clin.* **65**: 87–108
[3] Jemal A, Bray F, Center M, Ferlay J, Ward E and Forman D 2011 *CA Cancer J. Clin.* **61** 69-90
[4] Warburg O 1956 On respiratory impairment in cancer cells *Science* **124** 269– 70
[5] Hanahan D and Weinberg R A 2011 Hallmarks of cancer: the next generation *Cell* **144** 646–674
[6] Jones R G and Thompson C B 2009 *Genes. Dev.* **23** 537–48
[7] Hsu P P and Sabatini D M 2008 Cancer cell metabolism: Warburg and beyond *Cell* **134** 703–7
[8] Vander Heiden M G, Christofk H R, Schuman E, Subtelny A O, Sharfi H, Harlow E E, Xian J and Cantley L C 2009 *Biochem. Pharmacol.* **79** 1118–24
[9] Yu L, Chen X, Wang L and Chen S 2016 *Oncotarget* **7** 38908-26.
[10] Valvona C J, Fillmore H L, Nunn P B and Pilkington G J 2016 *Brain. Pathol.* **26** 3-17
[11] Miao P, Sheng S, Sun X, Liu J and Huang G 2013 *IUBMB Life.* **65** 904-10
[12] Augoff K, Anita H J and Tabola R 2015 *Cancer. Lett.* **358** 1-7
[13] DeVita V T, Lawrence T S and Rosenberg S A 2008 *DeVita, Hellman & Rosenberg's Cancer: Principles & Practice of Oncology, 8th Edition* (Lipincott Williams & Wilkins)
[14] Gray L R, Tompkins S C and Taylor E B 2013 *Cell. Mol. Life. Sci.* s00018-013-1539-2
[15] Wang Z Y, Loo T Y, Shen J G, Wang N, Wang D M, Yang D P, Mo S L, Guan X Y and Chen J P 2012 *Breast Cancer Res. Treat.* **131** 791-800
[16] Sheng S L, Liu J J, Dai Y H, Sun X G, Xiong X P and Huang G 2012 *FEBS J.* **279** 3898–910
[17] Yanuar A 2012 Penambatan molekuler: praktek dan aplikasi pada virtual screening (Depok: Fakultas Farmasi Universitas Indonesia)
[18] Morris G M and Lim-Wilby M 2008 *Molecular docking. In A. Kukol (Ed)* (Totowa :Human Press) pp 365-82
[19] Thornburg J M, Nelson K K , Clem B F, Lane A N, Arumugam S, Simmons A, Eaton J W, Telang S and Chesney J 2008 *Breast Cancer Res.* **10** R84
[20] Granchi C, Paterni I, Rani R and Minutolo F 2013 *Future Med. Chem.* **5** 1967–91
[21] Trott O and Olson A J 2010 *J. Comput Chem.* **31** 455-61
[22] Meng X Y, Zhang H X, Mezei M and Cui M 2011 *Curr. Comput Aided Drug.* **7** 146-57
[23] Seeliger D and de Groot B L 2010 *J. Comput. Aided Mol. Des.* **24** 417-22
[24] Read J A, Winter V J, Eszes C M, Sessions R B and Brady R L 2001 *Proteins* **43** 175–85
[25] Fiume L, Manerba M, Vettraino M and Di Stefano G 2014 *Future Med. Chem.* **6** 429-445
[26] Shi Y and Pinto B M 2012 *PLoS One* **9** e86365.
[27] Sun R, Lil X, Li Y, Zhang X, Lil X, Lil X, Shi Z and Bao J 2014 *J. Mol. Model* **21** 133
[28] Li X, Lu W, Hu Y, Wen S, Qian C, Wu W and Huang P 2013 *Int. J. Oncol.* **43** 1710-8
[29] Miskimins W K, Ahn H J, Kim J Y, Ryu S, Jung Y S and Choi J Y 2014 *PLoS One* **9** e85576.
[30] Pollastri M P 2010 *Curr. Protoc. Pharmacol.* **49** 9
[31] Wicaksono PA, Name S, Martien R and Ismail H 2016 *Asian Pac. J. Cancer Prev.* **17** 2277-84

IOP Conf. Series: Materials Science and Engineering **333** (2018) 012079 doi:10.1088/1757-899X/333/1/012079

Miraxanthin-V, Liriodenin and Chitranone are Hepcidin Antagonist In silico for Iron Deficiency Anemia

S Yotriana[1], YH Suselo[2], Muthmainah[3], and D Indarto[2,4]
[1]) Medical Student, Faculty of Medicine, Universitas Sebelas Maret
[2]) Department of Physiology, Faculty of Medicine, Universitas Sebelas Maret
[3]) Department of Anatomy, Faculty of Medicine, Universitas Sebelas Maret
[4]) Biomedical Laboratory, Faculty of Medicine, Universitas Sebelas Maret

E-mail: yulianaheri@staff.uns.ac.id

Abstract. Anemia is one of the greatest nutrition problem in the world that is commonly found in children, pregnant women and reproductive women. This disorder is predominantly caused by iron deficiency. Hepcidin, a hepatic hormone, regulates iron metabolism and high serum levels of this hormone are detected in patients with iron deficiency anemia (IDA). Anticalin is a sintetic compound which is able to interacts with hepcidin leading to inhibition of ferroportin-hepcidin binding complexes but its therapeutic effects are still under investigation. Indonesia has various herbal plants which are potentially developed to treat some human diseases. Therefore, the purpose of this study was to identify phytochemicals derived from Indonesian plants that is able to inhibit hepcidin-ferroportin interaction. A bioinformatics study with molecular docking method was used in this study. Three-dimensional structures of human hepcidin and anticalin were obtained from the Protein Data Bank (ID: 1M4F and 4QAE respectively). Because their molecular size was big, each molecule was cut into 2 parts of its binding sites. All phytochemicals structures were obtained from HerbalDB and PubChem NCBI database. Truncated anticalin/phytochemicals were molecularly docked with truncated hepcidin by using AutoDock Vina 1.1.2. and their interactions were visualized using PyMol 1.3. Truncated Anticalin had -4.6 and -4.2 kcal/mol binding affinity to truncated human hepcidin. Truncated anticalin 1 was bound to Cys^{13},Cys^{14}, Arg^{16}, and Ser^{17} residues in truncated hepcidin 1 while truncated anticalin 2 was at Cy^{23}, and Lys^{24} residues in truncated hepcidin 2. Miraxanthine-V, Liriodenin and Chitranone had lower binding affinity (-4.8±0.77, -4.7±0.33 and -5.01±0.30 kcal/mol respectively) than that of anticalin and occupied binding sites as same as anticalin did. There are three phytochemicals that potentially become hepcidin antagonists *in silico*. In vitro assays are required for verification of the antagonist effect of these phytochemicals on iron metabolism.

1. Introduction
Anemia affects a half billion of people in the world and it is generally caused by iron deficiency. In 2011, approximately 29% non-pregnant women who age 15-49 years old and 38% pregnant women suffer anemia [1]. In Indonesia, the prevalence of anemia is 21.7% in baby aged >1 years, 29% in children aged 5-12 years, 22.7% in children aged 13-18 years and 37.1% in pregnant women [2]. The Indonesian Ministry of Health has implemented an iron supplementation program to overcome anemia problem especially in pregnant women since 1975. Based on Indonesian Health Profile (2014), the coverage of this program has reached more than 90% for first visit and more than 80% for fourth visit

to health institutions during 2009 - 2012 [3]. However, this program has not been able to reduce the prevalence of anemia in pregnant women [4].

Many studies have reported that some patients with anemia are unresponsive to oral iron supplementation [5, 6]. High levels of serum hepcidin in these patients increase inhibition of iron absorption, which leads to persistent iron deficiency. In recent years, several approaches which are intended to inhibit hepcidin production has been developed to treat this disorder. Modulation of hepcidin production through iron signaling, inflammation, and erythropoiesis pathways is the most common developed approaches whilst antigen therapy is targeted to inhibit hepcidin mRNA. Another approach is inhibition of ferroportin-hepcidin endocytosis or *de novo* stimulation of ferroportin [7]. Anticalin PRS-080, for instance, is a synthetic molecule which is derived from small extra-cell protein (lipocalin) and able to bind biologically with various endogenous ligands. In addition, anticalin could neutralize human hepcidin hormone in hipoferremia mice [8]. So that anticalin potentially becomes a hepcidin antagonist but efficacy and safety of this drug is further investigated.

Indonesia has approximately 30.000 plant species, which contribute in 80% plant varieties in the world. Around 9.609 plants have therapeutic effects as herbal medicine [9]. Therefore, some phytochemicals from various Indonesian plants may be developed as hepcidin antagonist. Molecular docking is one method of virtual screening which provides valuable information about drug and receptor interaction. So this study aimed to identify Indonesian herbal plants that potentially have hepcidin antagonist by using molecular docking approach.

2. Experimental

All Indonesian herbal plants which were registered in database of HerbalDB were used in this bioinformatics study. Three dimensional structures of hepcidin and anticalin were downloaded from Protein Data Bank (PDB) with accession number 1M4F and 4QAE respectively. While three dimensional structure of phytochemicals were obtained from PubChem NCBI and had to meet Lipinski's rule of five. PyRx version 0.8 and AutoDock Tools version 1.5.6 were used to molecularly dock between anticalin/phytochemicals and hepcidin. Selected hepcidin antagonists were then visualized using PyMol version 1.3 softwares

Before running molecular docking, water content in hepcidin and anticalin molecules were removed by using AutoDock Tools and added with polar hydrogen molecule. However, both molecules were too big for docking process and these molecules were then cut in the binding site position (grid box) using the same software. Truncated hepcidin had 6 chains (P, Q, R, S, T, and U) and R chain directed the binding site of anticalin. Whereas truncated anticalin had 6 chains (A, B, C, D, E, and F). Residues of Tyr78, Gln174, and Cys175 at A chain, Glu40, Glu44, Val126, and Arg130 at C chain and Lys46 at D chain served as binding sites against truncated hepcidin (Table 1). Grid box I which consisted of 174-176 residues of A chain, 39-41, 43-46 and 128-130 residues of C chain, and 43-46 residues of D chain bound to Cys13, Cys14, Arg16, and Ser17 of truncated hepcidin. Secondly, grid box II which contained 78-79 residues of A chain 125-126 residues of C chain interacted with Cys23 and Lys24 residues of truncated hepcidin.

The position of grid box I was at X center = ±122.794 size =16, Y center = ±44.184 size=15, and Z center = ±15.406 size= 12 while grid box II position was at X center= ±132.175 size =11, Y center = ±36.824 size=12, and Z center = ±18.427 size= 9. Molecular docking was done at least three times to get root mean square deviation (RMSD) <2 Å. The results of this docking were used as a standard for screening of the phytochemicals. Phytochemicals were then molecularly docked with truncated hepcidin to get binding energy scores. *In silico* candidates of hepcidin antagonist were selected by comparing their binding energy scores, binding sites and molecular conformation with the standard anticalin.

3. Results and Discussion

Interaction of anticalin and hepcidin was firstly validated to obtain consistent binding energy scores. Truncated anticalin 1 bound to Cys^{13}, Cys^{14}, Arg^{16}, and Ser^{17} residues of truncated hepcidin 1 in grid box I with -4.6 kcal/mol binding energy (RMSD 0.0) and the other bound to Cys23 and Lys24 residues of truncated hepcidin 2 in grid box II with -4.2 kcal/mol (Table 1). Cutting process of anticalin and hepcidin molecules should consider their amino acid charges because it will influence their molecular structure and function. Amino acids can be divided into polar (hydrophilic) and non-polar (hydrophobic). The polar amino acid consists of charged amino acids and uncharged amino acids. Polar amino acid with positive charge includes lysine, arginine, and histidine while the ones with negative charge include aspartate and glutamate. Uncharged polar amino acid consists of methionine, glutamine, asparagines, serine, threonine, and cysteine. The non polar amino acid includes glycine, alanine, valine, leucine, isoleucine, and proline. Amino acids in truncated anticalin should have the same charge as the amino acids at its binding site to prevent alteration of atom interaction.

Table 1. Validation of standard anticalin and hepcidin based on binding energy and binding site

No	Standard	Residues	Chain	Docking Score	*Binding site* on Hepsidin						Molecule Formula	*Lipinski's Rule of Five*			
					Chain A		Chain C		Chain D	Other residues		Molecular weight (<500g/mol)	*H-Bond Donor (<5)*	*H-Bond Acceptor (<10)*	*Compound lipophilicity (Log P<5)*
					Standard 1 (Cys14, Arg16)	Standard 2 (Cys23)	Standard 1 (Cys13, Ser17)	Standard 2 (Lys24)	Standard 1 (Arg16)						
1	Anticalin Standard 1 (Grid Box 1)	Gln174-Xn-Ile176	A	-4.5	Cys14, Arg16	-	-	-	-	-	C14H26N4O4S	346.446			
2	Anticalin Standard 1 (Grid Box 1)	Asn39-Xn-Val41	C	-3.9	-	-	Cys13, Ser17	-	-	-	C14H24N4O6	344.363			
3	Anticalin Standard 1 (Grid Box 1)	Gln128-Xn-Arg130	C	-4.6	-	-	Cys13	-	-	-	C15H28N8O5	400.433			
4	Anticalin Standard	Arg43-Xn-Lys46	C	-3.6	-	-	Ser17	-	-	-	C21H38N8O8	530.575			

IOP Conf. Series: Materials Science and Engineering **333** (2018) 012079 doi:10.1088/1757-899X/333/1/012079

	1 (Grid Box 1)														
5	Anticalin Standard 1 (Grid Box 1)	Arg43-Xn-Lys46	D	-3.7		-	-	-	Arg16	-	C21H38N8O8	530.575			
6	Anticalin Standard 2 (Grid Box 2)	Tyr78-Xn-Leu79	A	-4.2	-	Cys23	-	-	-	-	C15H22N2O3	278.346			
7	Anticalin Standard 2 (Grid Box 2)	Val125-Xn-Val126	C	-3.7	-	-	-	Lys24	-		C10H20N2O2	200.278			

IOP Conf. Series: Materials Science and Engineering **333** (2018) 012079 doi:10.1088/1757-899X/333/1/012079

Table 2. Selected phytochemicals as an *in silico* candidate of hepcidin antagonist

No	Pubchem Code	Compound	Docking score	*Binding site* on Hepsidin						Molecule Formula	*Lipinski's Rule of Five*			
				Chain A		Chain C		Chain D	Other residues		Molecular weight (<500g/mol)	*H-Bond Donor (<5)*	*H-Bond Acceptor (<10)*	*Compound lipophilicity (Log P<5)*
				Standard 1 (Cys14, Arg16)	Standard 2 (Cys23)	Standard 1 (Cys13, Ser17)	Standard 2 (Lys24)	Standard 1 (Arg16)						
1	5281203	*Miraxanthin-V*	-4.8	Arg16	Cys23	Cys13, Ser17	Lys24	Arg16	Cys11	C15H10O6	346.339	5	8	1.1
2		*Liriodenin*	-4.7	Cys14, Arg16	-	-	Lys24	Arg16	Thr25	C17H9N1O3	275.263	0	4	3.4
3		*Chitranone*	-5.1	Cys14, Arg16	-	Cys13, Ser17	-	Arg16	Cys11	C22H14O6	374.348	2	6	3.8

Identification of potential phytochemicals was started from screening Indonesian herbal plants which have complete molecular structure and small size. There were 517 phytochemicals which met the criteria among 6,776 phytochemicals. After running several times, it revealed that 11 phytochemicals had lower binding energy compared to the standard 1 and 39 phytochemicals had lower binding energy than that of the standard 2. All these compounds interacted with the truncated hepcidin at the same location as the standard did. However, there were only three phytochemicals which had the lowest binding energy (Miraxanthin-V with -4.8, Liriodenin with-4.7 and Chitranone with -5.1 kcal/mol), similar binding sites and conformation, compared with standard (Table 2).

Lipinski criteria which consist of molecular weight, H-bond donor, number of H-bond acceptor and octanol-water partition coefficient (LogP) are usually used to identify new drug candidates. The recommended molecular weight is less than 500 Da. Small molecular weight creates more stable bond and requires less energy to bind. H-bond donor and H-bond acceptor are used to evaluate the ability of a drug to cross cell membrane. As the number of H-bond increases, the cell permeability decreases. LogP or lipophilicity is important to measure the degree of a compound to dissolve in fat. A compound becomes more hydrophilic as the lipophilicity decreases resulting in lower toxicity [10].

We have firstly demonstrated that Miraxanthin-V, Liriodenin and Chitranone potentially become a natural hepcidin antagonist in silico. Miraxanthin-V has five H-bond donors, eight H-bond acceptors and LogP of 1.1. It interacts with hepcidin at Cys^{11}, Cys^{13}, Arg^{16}, Ser^{17}, Cys^{23}, and Lys^{24}. This compound can be found in the stem and root of *Mirabilis jalapa* plant. It has been used to treat gastrointestinal diseases such as dysentriae, diarrhea, muscle pain, abdominal colic and peptic ulcer [11, 12].

Liriodenin does not have H-bond donor but has four H-bond acceptor and logP of 3.4. It interacts with hepcidin at Cys^{14}, Arg^{16}, Lys^{24} and Thr^{25}. Liriodenin interacts with hepcidine at Cys^{14}, Arg^{16}, Lys^{24} and Thr^{25}. This compound can be found in several plants including *Annona cherimolia, Annona reticulata, Nelumbo nucifera* and *Polyalthia Longifolia*. Liriodenin was reported to inhibit proliferation of ovarian cancer cell and have various biological activities as antimicrobial, antifungal, anti-tumor, anti-platelet, anti-aritmia, sedative and anti-depressant effect [13, 14, 15].

Chitranone has two H-bond donors, six H-bond acceptors and LogP of 3.8. Chitranone interacts with hepcidine at Cys^{11}, Cys^{13}, Cys^{14}, Arg^{16}, and Ser^{17}. This compound can be found in the leaves and root of *Plumbago Zeylanica* [16]. Chitranone has several therapeutic properties such anti-diabetes, anti-cholesterol, anti-inflammation, anti-cancer, anti-microbial, anti-malaria and anti-fertility property [17].

IOP Conf. Series: Materials Science and Engineering **333** (2018) 012079 doi:10.1088/1757-899X/333/1/012079

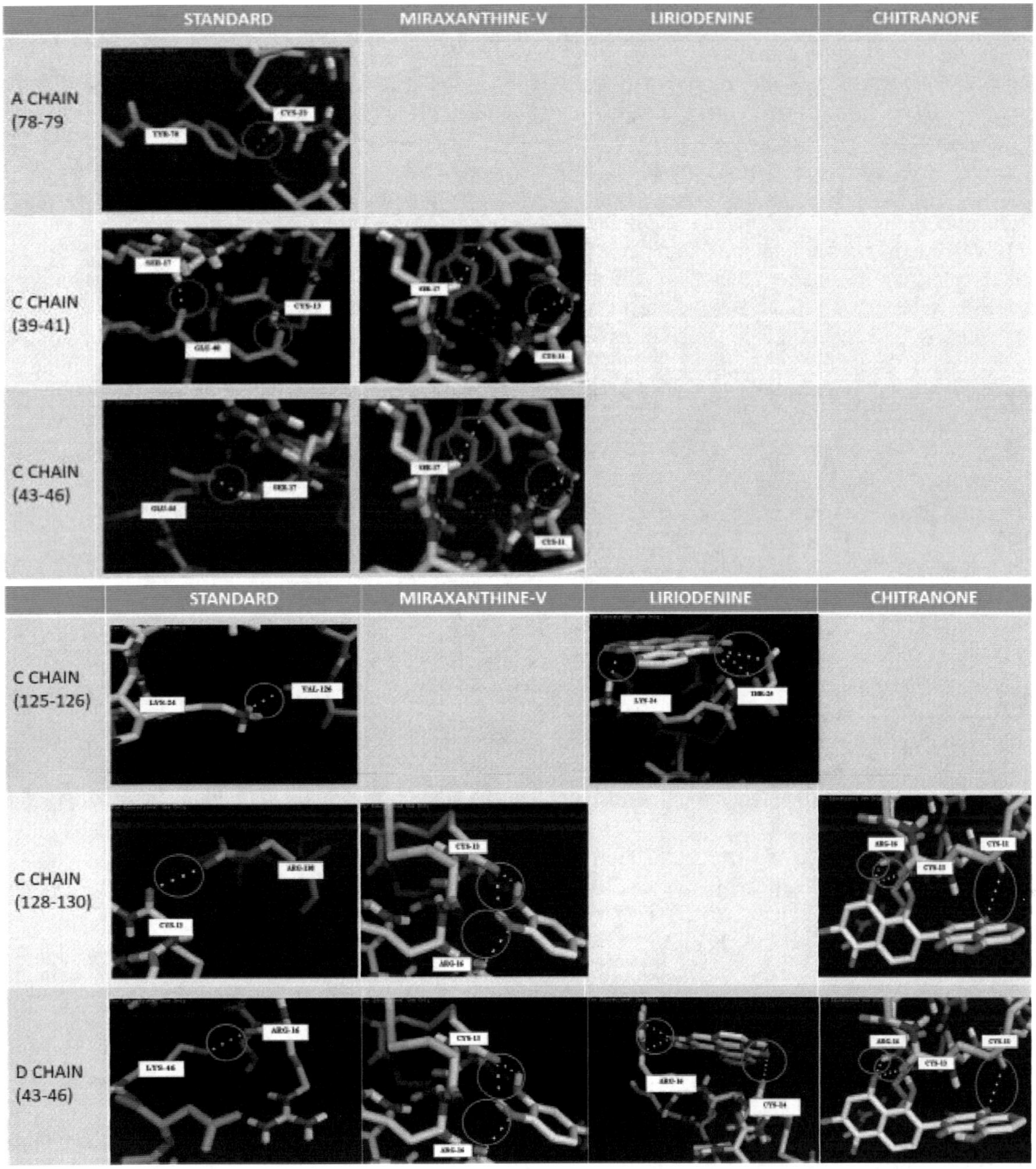

Figure 1. Visualization of anticalin/phytochemicals and hepcidin binding complexes using PyMol. Green: Carbon (C), Red: Oxygen (O), White: Hydrogen (H), Blue: Nitrogen (N), Yellow: Sulfur (S), dashes-line: atomic interactions, white boxes: amino acid residues.

We realized that this study has limitations. One of them is the fact that anticalin molecule that we used was a modified molecule obtained by cutting anticalin-hepcidin molecule complex. In addition, the standard molecule anticalin and target protein, hepcidin, are large protein. AutoDock Vina program can only perform protein-ligand docking. Thus, we modified anticalin and hepcidin molecule by cutting several residues. It is likely that this modification influence the docking result.

IOP Conf. Series: Materials Science and Engineering **333** (2018) 012079 doi:10.1088/1757-899X/333/1/012079

4. Conclusion

Miraxanthin-V, Liriodenin, and Chitranone potentially become hepcidin antagonist *in silico* which have higher binding affinity and molecular similarity, compared with the standard anticalin. In vitro assays should be performed to evaluate the antagonist effects of these phytochemicals on iron regulation.

References

[1] WHO 2014 *Global nutrition target 2025: anaemia policy brief (WHO/NMH/NHD/14.4)* Geneva
[2] Departemen Kesehatan 2013 *Riset Kesehatan Dasar (Riskesdas) 2013* (Jakarta: Badan Litbangkes)
[3] Ministry Health Republic of Indonesia 2014 *Indonesia Health Profile* Jakarta.
[4] Barkley J S, Kendrick K L, Codling K, Muslimatun S, and Pachón H. 2015 *Asia Pac. J. Clin. Nutr.* **24(3)** 452–455
[5] Bregman D B, Morris D, Koch T A, He A, Goodnough L T 2013 *Am J Hematol.* **88**: 97-101 http://dx.doi.org/10.1002/ajh.23354
[6] Suselo Y H, Wulandari S, Ayusari A A, and Indarto D 2017 *Bali Medical Journal* **3(3)**: S61-S64. DOI:10.15562/bmj.v3i3.722
[7] Fung E, and Nemeth E 2013 *Haematol.* **98(11)** 1667–1676
[8] Sun C C, Vaja V, Babitt J L, and Lin H Y 2012 *Am. J. Hematol.* **87(4)** 392–400.
[9] Yanuar A, Mun'im A, Lagho A B A, Syahdi R R, Rahmat M, and Suhartanto H 2011 *Int J Com Sci* **8(5)** 180–183.
[10] de Brito M A 2011 *Br. J. Pharm. Sci.* **47(4)** 788-805
[11] Ganta L, Dogiparthi R K, Pulusu P R, Vallela S R, Andugula M K, and Srinivasan R 2013 *World J. Pharmacy and Pharmaceutical Sci.* **3(5)** 1364-1393
[12] Rozina R 2016 *Int. J. Phar. Res.* **6(5)** 160-168
[13] Nordin N, Majid N A, Hashim N M, Rahman M A, Hassan Z, and Ali H M 2015 *Drug Design, Development and Therapy* **9** 1437-1448
[14] Chen C Y, Wu H M, Chao W Y, and Lee C H 2013 *Af. J. Pharmacy and Pharmacol.* **7(8)**1067-1070
[15] Perviz S, Khan H, and Pervaiz A 2016 *Frontiers of Phamacology* **7(28)**
[16] Jian P, Sharma HP, Basri F, Baraik B, Kumari S, and Pathak C 2014 *Int. J. Phar. Sci. Rev. Res.* **24(1)** 157-163
[17] Ganesan K and Gani S B 2013 *Am. J. Phytomed. Clin. Therapeut.* **3** 313-337

IOP Conf. Series: Materials Science and Engineering **333** (2018) 012080 doi:10.1088/1757-899X/333/1/012080

Antibacterial activity study of *Attacus atlas* cocoon against *Staphylococcus aureus* and *Escherichia coli* with diffusion and dilution method

Aminah[1], E R Nugraheni[1*], A Yugatama[1]
Pharmacy Department, Mathematics and Sciences Faculty, Sebelas Maret University

E-mail: retnaningtyas_n@yahoo.co.id

Abstract. The aim of this study was to evaluate the antibacterial activity from *Attacus atlas* cocoon extract against Gram-negative bacteria (*Escherichia coli*) and Gram-positive bacteria (*Staphylococcus aureus*) with 3 diffferent solvents polar, semi-polar and non polar which was ethanol, ethyl acetate and chloroform, also to determine the Minimum Inhibitory Concentration (MIC) and Minimum Bactericidal *Concentration* (MBC) of the extract. Cocoon was extracted with maceration method using 3 solvents with ratio of sample and solvent 1:10. Antibacterial activity of the Extracts obtained was evaluated with Agar disk diffusion method. The best result was then continued to determine the MIC and MBC of the extract using broth macro-dilution method.The results show that each of the extracts have antibacterial activity with broad spectrum against two different type of bacteria at concentration of 1 g/mL with different clear zone between these extracts. Clear zone from the biggest to the smallest against *Escherichia coli* was ethyl acetate (10.5 mm), chloroform (9 mm) and ethanol (8 mm). While against *Staphylococcus aureus*, was obtained by chloroform (12.5 mm), ethyl acetate (10.5 mm) and ethanol (7 mm). The MIC value of extracts can not be determine. The smallest MBC value against both bacteria was obtained by ethyl acetate with concentration of 3.125% b/v as a bactericidal.

1. Introduction

Infectious diseases are one of the leading causes of death worldwide. During the past few decades, new infectious diseases have appeared and old ones previously thought to be controlled have reemerged [1] and thus, despite of many significant developments in the antibacterial therapy, many problems remain to be solved for most of the antibacterial drugs available [2]. Hence, discovery of novel antibacterial agents with better pharmacological profile is still highly desirable [3]. Antibacterial substance can be from chemical synthesis or substance from plants or animals [4].

Indonesia has a wide range of natural wealth of flora and fauna. Indonesia's natural wealth, especially in fauna, many are not getting the interest from researchers. So far, the use of fauna for traditional medicine is very minimum, compare to flora. One of the natural ingredients that can be used for traditional medicine is *Attacus atlas* cocoon, also known as wild silkworms or *ulat keket* in Indonesia.

Attacus atlas common name is Atlas Moth, which is one of the saturniidae family usually can be found in citrus, cinnamon, guava trees [5]. *Attacus atlas* cocoon is widely used to produce silk, but not every cocoon can be further processed to produce silk, i.e unqualified cocoon. Usually, these cocoons will be removed. Silk worm cocoon also has benefits for health. But, it is still empirical. Previous

IOP Conf. Series: Materials Science and Engineering **333** (2018) 012080 doi:10.1088/1757-899X/333/1/012080

research was conducted against *Antheraeamylitta* cocoon, which is still in one familia with *Attacus atlas*, is known to have antibacterial activity [6].

Staphylococcus aureus is a facultative anaerobic gram positive coccal bacteria. It is found on the skin as a part of normal skin flora and nasal passages. It can cause a range of infections include minor skin infections such as pimples, impetigo, boils, cellulitis, folliculitis, scaled skin syndrome and abscesses and life threatening diseases such as pneumonia, meningitis, osteomyelitis, endocarditis, toxic shock syndrome, bacteremia and sepsis also can cause staphylococcal scaled skin syndrome, a severe disease, in infants [7]. *Escherichia coli*, a gram negative bacterium, can cause food poisoning. The diseases caused by E. *coli* include gastroenteritis, urinary tract infections, neonatal meningitis, septicaemia and pneumonia [7].

Other research [8], concluded that in food which was wrapped with cocoon sheets could inhibit microbes grow. But, the research could not prove the antibacterial activity in *Attacus atlas* cocoon scientifically. The current study investigated the antimicrobial activity of *Attacus atlas* cocoon extracts, MIC and MBC concentration of the extracts.

2. Experimental

2.1 Materials

Attacus atlas cocoons were obtained from PT. Yarsilk, Yogyakarta, Indonesia. Cultures of *Escherichia coli* ATCC 25922and *Staphylococcus aureus* ATCC 25923 were obtained from Microbiology lab, Faculty of Medicine, Universitas Sebelas Maret. Cultures were maintained in nutrient agar in refrigerator at 4^0 C. Ethanol 75%, Chloroform and Ethyl Acetate as solvents. Mueller Hinton Agar, Nutrient Agar and Nutrient Broth as medium for Antibacterial assay.

2.2 Cocoon extraction

Cocoon Attacus atlas included in this study were collected from local company PT. Yarsilk, Yogyakarta, Indonesia. The dried material was grounded into powder, 150 g of the grounded cocoon was macerated 24h in 1500 mL ethanol, ethyl acetate and chloroform to obtain 3 extracts and remacerated twice, then filtered with filter paper. The filtrates were evaporated under rotary vacuum evaporator. The extract yields were weighted, stored in tube in fridge at 4°C and their yield percentages were calculated using the following formula:

$$\text{Extract yield (\%)} = \frac{\text{weight of extracted sample residues (g)}}{\text{weight of raw sample (g)}} \times 100$$

2.3 Bacterial strains

The antibacterial potency of each plant extracts were evaluated using 2 bacterial strains. 1 strain of Gram positive (*Staphylococcus aureus*) and 1 strain of Gram negative (*Escherichia coli*). The bacterial strains were obtained from the culture collection of Sebelas Maret University, Indonesia.

2.4 Inoculums preparation

Each bacterial strain was subcultured overnight at 35°C in Nutrient agar. The bacterial growth was harvested, diluted in sterile saline water and adjusted the suspension turbidity with 0.5 McFarland Standard of 10^8 CFU/mL.

2.5 Antibacterial activity of cocoon extract

The disk diffusion method is used to evaluate antimicrobial activity of each extracts. The extract residues (1 g) were dissolved in 1 mL DMSO. MHA medium was poured into sterile petri dishes to create 4 mm depth, sterilized and swabbed with the bacterial suspension of 10^8 CFU/mL sterile filter paper discs loaded with extracts of 25 µL and placed on top of MHA plates. Filter paper discs loaded with 10 µg Gentamycin was used as positive control. The plates were incubated at 35 C for 18 h. the presence of inhibition zones were measured, recorded and considered as indication for antibacterial activity.

2.6 Determination of the minimum inhibitory concentration (MIC) of the extracts

MIC is defined as the lowest concentration the antimicrobial agent that inhibits the microbial growth after 18 h of incubation. The extract which exhibiting an antibacterial activity was continued to determine their MIC using macro-broth dilution method as recommended by CLSI guidelines. Extract diluted 1/10 in sterile NB and transferred 2 mL in a tube which had already contained 1 mL Nutrient broth and then two-fold dilution by transferring 1 mL from the first tube to the 5th tube which each had already contained 1 mL NB. Inoculum of bacterial suspension of 0.5 McFarland diluted 1/150 in Nutrient broth and transferred 1 mL in each tube, then incubated 35 C for 18 h. 3 additional tubes as Antibiotic control (1 mL medium and 1 mL antibiotic), Growth control (1 mL medium and 1 mL inoculum) and Medium control (2 mL medium).

2.7 Determination of the Minimum Bactericidal Concentration (MBC) of the extracts

Streaks were taken from the two lowest concentrations of the tubes exhibiting invisible growth (clear tube) and subcultured onto sterile MHA plates. The plates were incubated at 35C for 18 h, then eamined for bacterial growth. MBC was taken as the concentration of plant extract that did not exhibit any bacterial growth on the freshly inoculated agar plates.

3. Results and Discussion

3.1 Cocoon extraction yield

The extract of 150 g dried material with ethanol yielded extract residues of 4% w/v, with ethyl acetate of 4.67% w/v and with chloroform of 5.33% w/v. the highest yield of cocoon extract was obtained with chloroform followed by ethyl acetate and ethanol.

3.2 Antibacterial activity of cocoon extract

Attacus atlas cocoon were investigated to evaluate its antibacterial activity against Gram positive bacteria (*S. aureus*) and Gram negative bacteria (*E. coli*) using disc diffusion method. Evaluation of antibacterial activity activity of these cocoon extracts was recorded in Figure 1. and Figure 2. The results revealed that all extracts were potentially effective in suppressing microbial growth of *S. aureus* and *E.coli* with different potency. Ethyl acetate extract was the most effective extract against *E. coli* and chloroform was the most effective extract against *S. aureus*. In positive control the inhibition zone was 28 mm and there was no inhibition zone in negative control.

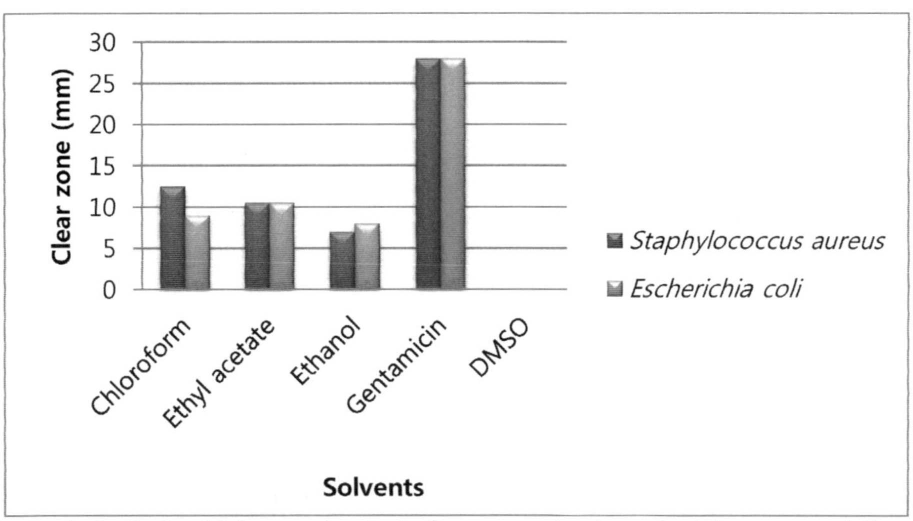

Figure 1. Antimicrobial screening test of cocoon extracts against *S. aureus* and *E. coli*.

a. b.

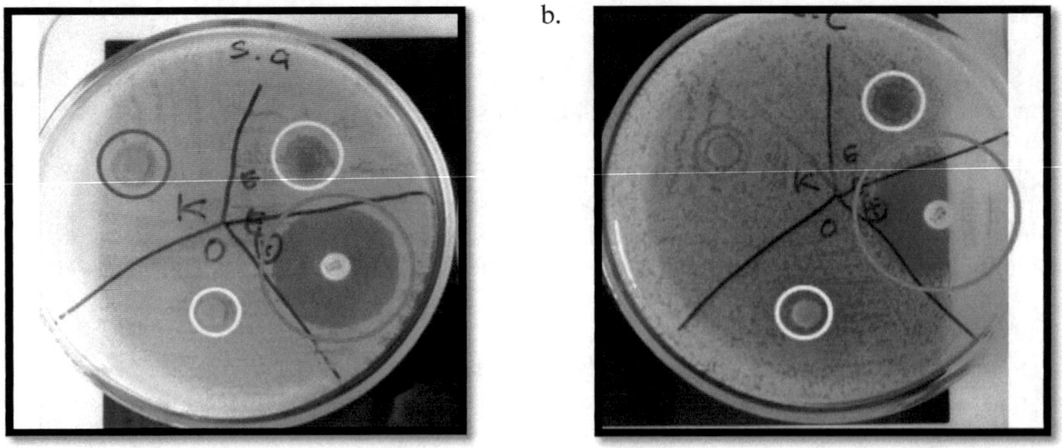

Figure 2. Growth inhibition of *S. aureus* (a) and *E. coli* (b) caused by *Attacus atlas* cocoon

3.3 Minimum inhibitory concentration (MIC's) of the effective cocoon extract

The MIC and MBC of the extracts were employed to evaluate their bacteriostatic and bactericidal properties. The inhibitory effect of extracts can not be evaluated because of biased results. The tubes were cloudy even before incubation because of the extracts color, so we can not determine the MIC and subcultured all of the concentration onto fresh MHA plate to determine the MBC.

3.4 Minimum bactericidal concentration (MBC's) of the effective cocoon extracts

The MBC was confirmed by absence of bacterial growth of the tested strains streaked form inhibition zone corresponding to their lowest MIC's, but we streaked all concentration because the MIC could not be determine. The MBC results were recorded in Table 1. and Figure 3-6. Ethyl acetate and chloroform extracts showed potentially bactericidal activity against the tested pathogenic bacteria (*S. aureus* and *E. coli*). Etyhl acetate had MBC 3.125% w/v for both bacteria while chloroform had MBC 6.25% w/v against *S. aureus* and 12.5% w/v against *E. coli*.

Table 1. Minimum Bactericidal Concentration of each extracts.

Extracts	Concentration (g/100 mL)	Bacterial growth	
		E.coli	*S. aureus*
Chloroform	50	-	-
	25	-	-
	12.5	-	-
	6.25	+	-
	3.125	+	+
Ethyl Acetate	50	-	-
	25	-	-
	12.5	-	-
	6.25	-	-
	3.125	-	-
Ethanol	50	+	+
	25	+	+
	12.5	+	+
	6.25	+	+
	3.125	+	+

a. b.

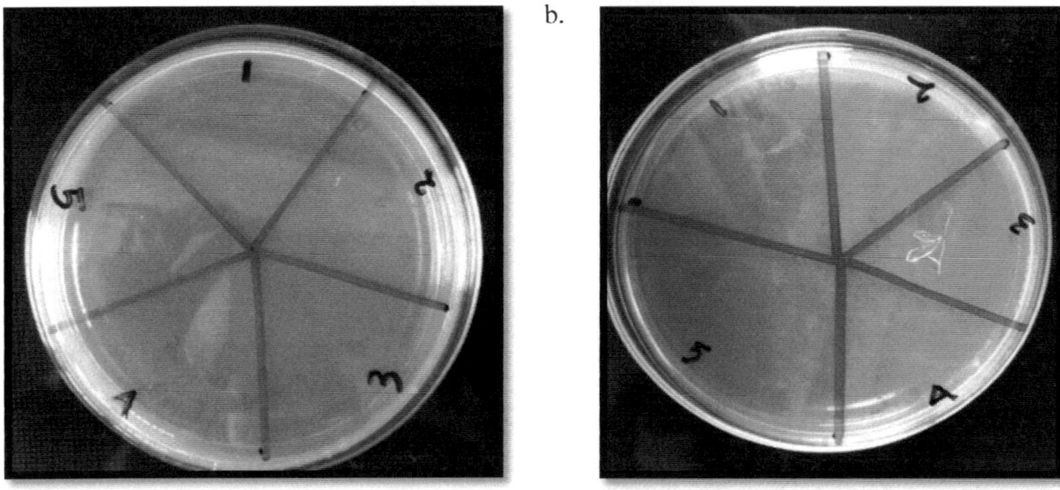

Figure 3. MBC of ethyl acetate extracts against *S. aureus* (a) and *E. coli* (b)

a. b.

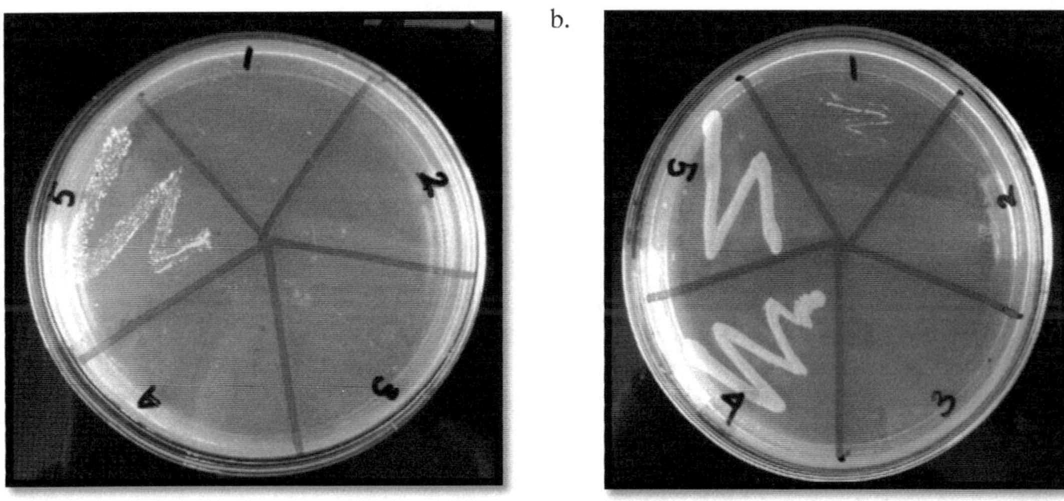

Figure 4. MBC of chloroform extracts against *S. aureus* (a) and *E. coli* (b)

a. b.

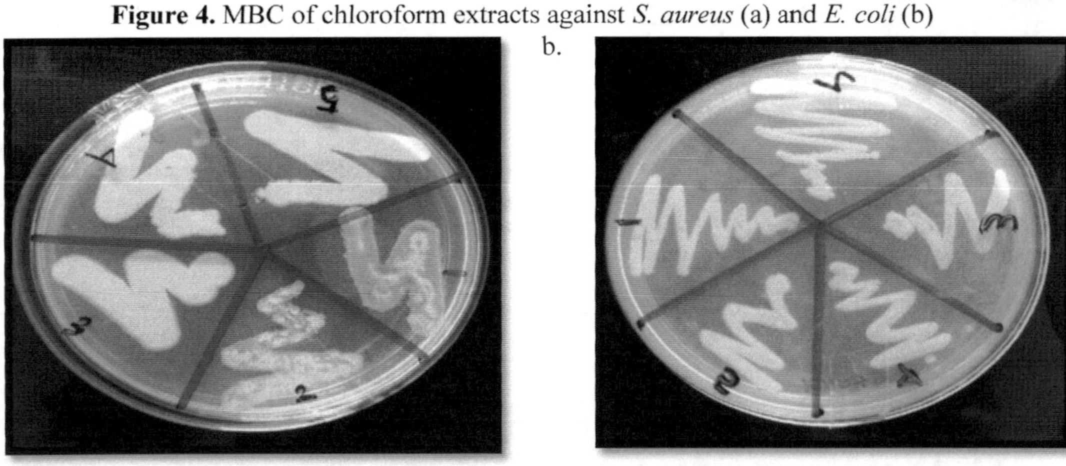

Figure 5. MBC of ethanol extracts against *S. aureus* (a) and *E. coli* (b)

IOP Conf. Series: Materials Science and Engineering **333** (2018) 012080 doi:10.1088/1757-899X/333/1/012080

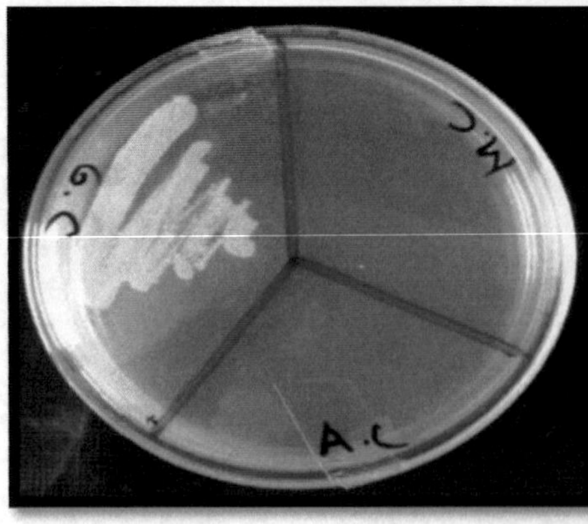

Figure 6. Antibiotic control (A.C); Growth control (G.C) and Medium control (M.C)

4. Conclusion

Attacus atlas cocoon have antibacterial activity against *Staphylococcus aureus* and *Escherichia coli* with agar disk diffusion method. MBC of the extracts against *S.aureus* from lowest to highest was ethyl acetate (3.125%), chloroform (6.25%) and against *E.coli* (3.125%), chloroform (12.5%), respectively. Ethyl acetate and chloroform was bactericidal, while ethanol bacteriostatic.

References

[1] Sharma PC and Jain S. Acta 2008 *Pharm Sci* ***50*** *(*35).
[2] Sharma PC and Jain S. Acta 2008 *Pharm Sci* ***50*** *(*551).
[3] UjjwalSahoo , A.K.Seth , A.Sen , Dhanya B , J.Patel and R.Chawla 2010 *Research Journal of Pharmaceutical, Biological and Chemical Sciences***1** 102-107
[4] Hoeprich PD 1992 *Antifungal Agents in Encyclopedia of Mikrobiology 1st edition* (San Diego: Academic Press Inc)
[5] Dennis S. Hill *Pests of Crops in Warmer Climates and Their Control* (Berlin: Springer)
[6] Akai H Anti-bacteria Function of Natural Silk Materals1997*International Journal Wild Silkmoth & Silk* **3** 79-81
[7] Elsa V, K Leena P, S Sathia N 2012 *Research Journal of Pharmaceutical, Biological and Chemical Sciences* **3** 960
[8] Faatih M 2005 *Jurnal penelitian Sains & Teknologi* **6** 35-48.

IOP Conf. Series: Materials Science and Engineering **333** (2018) 012081 doi:10.1088/1757-899X/333/1/012081

Development and Validation of High Performance Liquid Chromatography Method for Determination Atorvastatin in Tablet

A Yugatama[1], S Rohmani[1], A Dewangga[1]
[1]Department of Pharmacy, Faculty of Mathematics and Natural Sciences, Sebelas Maret University, Indonesia

E-mail: adiyugatama.apt@gmail.com

Abstract. Atorvastatin is the primary choice for dyslipidemia treatment. Due to patent expiration of atorvastatin, the pharmaceutical industry makes copy of the drug. Therefore, the development methods for tablet quality tests involving atorvastatin concentration on tablets needs to be performed. The purpose of this research was to develop and validate the simple atorvastatin tablet analytical method by HPLC. HPLC system used in this experiment consisted of column Cosmosil C18 (150 x 4,6 mm, 5 µm) as the stationary reverse phase chomatography, a mixture of methanol-water at pH 3 (80:20 v/v) as the mobile phase, flow rate of 1 mL/min, and UV detector at wavelength of 245 nm. Validation methods were including: selectivity, linearity, accuracy, precision, limit of detection (LOD), and limit of quantitation (LOQ). The results of this study indicate that the developed method had good validation including selectivity, linearity, accuracy, precision, LOD, and LOQ for analysis of atorvastatin tablet content. LOD and LOQ were 0.2 and 0.7 ng/mL, and the linearity range were 20 - 120 ng/mL.

1. Introduction

Atorvastatin is the statin derivatives that works with inhibiting HMG-CoA reductase in a competitive manner and the most commercially avaliable drugs for the treatment of dyslipidemia [1]. Atorvastatin is the most potent drug to reduce LDL levels because its use with medium dose intensity (10 – 20 mg/day) evident effectively to reduces levels of LDL by 30 – 50% from previous, and with high dose intensity (40 – 80 mg/day) evident capable of reduces LDL levels of > 50% from previous [2].
Due to patent expiration of Lipitor® which is a drug patent atorvastatin in 2011 [3], the pharmaceutical industry makes copy of the drug. Therefore, tablet quality tests involving atorvastatin levels on tablets needs to be performed. To achieve it needs to be developed the method of analysis atorvastatin levels in tablet.

The development method of analysis levels of an active substance in tablet with high performance liquid chromatography (HPLC) has become the the common method because it is easy enough to operated and time of analysis is relatively short. In previous studies a mobile phase used was mixture of a solvent acetonitril-buffer ammonium acetate (60:40 v/v) [4]. But in that method, the use of buffer solution as mobile phase would be easy clogging the columns and longer laundering.
The use of mixture of methanol and water for mobile phase was more efficient than use of buffer because its not clogging the columns, not require mobile phase preparation, and the selectivity of methanol is better than acetonitril [5]. Therefore, it is necessary to develop of atorvastatin analysis method by high performance liquid chromatography (HPLC) with a mixture of methanol and water as

IOP Conf. Series: Materials Science and Engineering **333** (2018) 012081 doi:10.1088/1757-899X/333/1/012081

mobile phase that meets the validation parameters including: selectivity, accuracy, precision, linearity, LOD (Limit of Detection), and LOQ (Limit of Quantitation).
The purpose of this study was to obtain a simple method for analyzing the content of atorvastatin tablets by high performance liquid chromatography (HPLC) that meets the validation parameters including selectivity, linearity, accuracy, precision, LOD (Limit of Detection), and LOQ (Limit of Quantitation).

2. Experimental

2.1. Instrument
The instrument used were High Performance Liquid Chromatography (Lab Alliance) featuring Rheodyne sample injectors (20 μL sample loops), UV SCL-10A VP detector (Lab Alliance) and LC-10AT VP pump (Lab Alliance), Cosmosil C18 Column (150 x 4.6 mm, 5 μm), UV-Vis spectrophotometer (Shimadzu AA 6650), analytical balance scale (Ohaus PA413), glass tools (Pyrex), and other supporting tools.

2.2. Material
The materials used were atorvastatin standard (PT Kalbe Farma, Indonesia), methanol for HPLC (Merck), and aquabidest (PT Ikapharmindo).

2.3. Stock solution Preparation
Atorvastatin stock solution made with a concentration of 1 mg/mL, using 100 mg of standard atorvastatin dissolved in methanol up to 100 mL.

2.4. Determination of the maximum standard atorvastatin wavelength (scanning λ) in the mobile phase
Stock solution of standard atorvastatin was created in concentration of 100 μg / mL. Scanning of maximum wavelength was performed by UV-Vis spectrophotometer at 200-400 nm.

2.5. Optimization of mobile phase composition and flow rate on HPLC systems
A 20 μL standard atorvastatin solution with a concentration of 100 ng/mL was injected into the HPLC system at maximum wavelength and flow rate of 1 mL/min, where variations in mobile phase composition and mobile phase flow rate were performed. Furthermore, the composition and flow rate of the mobile phase was determined by the best separation performance, based on retention time (Rt), resolution (Rs), tailing factor (TF), and theoretical plate (N). Required separation requirements were retention time < 10 min, Rs > 2, TF ≤ 2, and N > 2000 [5].

2.6. System Suitability Test
System suitability test was done by making concentration of standard solution of 120 ng/mL. A 20 μL standard 120 ng / mL atorvastatin solution was injected into the HPLC system. Based on chromatogram obtained, retention time (Rt), peak area, and peak height were set.The system suitability test was meet the qualification if the CV value was 2.0% for the area and peak height, and $\leq$ 1.0% for retention time, TF ≤ 2, Rs> 2 and N > 2000 [5].

2.7. Validation of Analysis Methods
Validation methods including (i) selectivity testing by comparing standard chromatograms of atorvastatin, atorvastatin tablets, and mobile phase solvents, (ii) determination of linearity by making series of concentrations of 20 - 120 ng / mL, (iii) accuracy and precision testing, (iv) determination of LOD and LOQ.

IOP Conf. Series: Materials Science and Engineering **333** (2018) 012081 doi:10.1088/1757-899X/333/1/012081

2.8. Analysis of samples of atorvastatin tablets

Each atorvastatin tablet were weighed and crushed, then dissolved in 10 mL of methanol, homogenized, and was filtered. The solution was diluted in mobile phase by 20,000 times. A 20 µL was injected in the HPLC system.

3. Result and Discussion

3.1. Determination of the maximum standard atorvastatin wavelength (scanning λ)

The results of maximum wavelength scanning of atorvastatin in a methanol-water pH 3 (80:20 v/v) mobile phase with a concentration of 100 µg/mL using a UV-Vis spectrophotometer at a wavelength of 200-400 nm showed that atorvastatin had maximum absorption at 245 nm. Scanning results can be seen in Figure 1.

3.2. Optimization of mobile phase composition and flow rate on HPLC systems

From the optimization result of mobile phase composition in table 1, the 80:20 ratio was chosen because it yielded the best resolution and tailing factor values with a reasonably fast retention time compared to other mobile phase compositions. In addition, the result of N value was also qualified (N > 2000) [5].

Furthermore, by the result of the mobile phase flow rate optimization in table 2, the chosen flow rate was 1 mL/min as it yields optimum resolution, tailing factor, N value, peak area and retention time.

3.3. System Suitability Test

The result of the system suitability test seen in table 3, indicates that the HPLC system used for analysis of atorvastatin levels in tablets meets the requirements of the system suitability test.

3.4. Validation of Analysis Methods

In the selectivity determination, it was resulted a chromatogram with a peak of atorvastatin standard and sample at 3.9 min retention time, and there was no peak seen in mobile phase solvent chromatogram. This showed that the method used was selective for atorvastatin analysis. In the linearity determination, 5 series of standard atorvastatin concentrations of 20 – 120 ng/mL obtained r^2 more than 0.997 so it meets the requirements of linearity [6]. In the determination of accuracy and precision result in table 4, it was indicating that the recovery value and CV has fulfilled the requirements of 80 - 110% recovery value and < 7.3% CV value [7].

In the LOD determination, was used visual instrumental method with the calculation of signal to noise ratio [8]. The results obtained that LOD and LOQ values were 0.2 ng/mL and 0.7 ng/mL.

3.5. Tablet Uniformity Test

The result of uniformity test of tablet sample in Table 5 showed that the atorvastatin concentration in tablet sample was meet the requirements of the content not less than 90.0% and not more than 110.0% than indicated on the label [3].

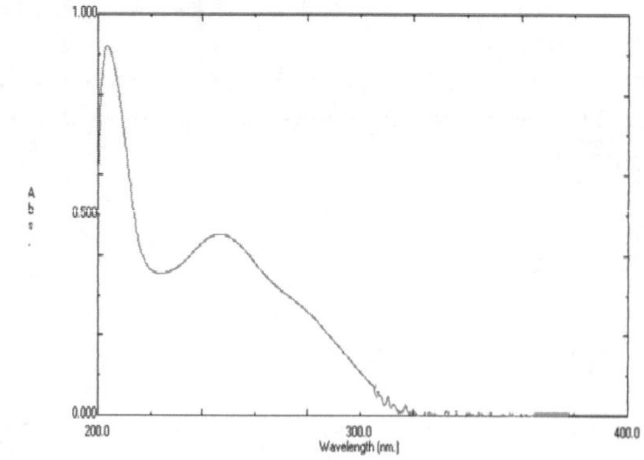

Figure 1. UV Spectrum of Atorvastatin in the Mobile Phase

Table 1. Optimization Result of Mobile Phase Composition

Parameter	Mobile Phase Composition (Methanol-Water pH 3)					Requirement (Snyder et al., 2010)
	70:30	**75:25**	**80:20**	**85:15**	**90:10**	
Resolution (Rs)	1.54	0.83	3.15	1.06	3.00	> 2
Rt (min)	6.24	4.98	3.91	3.25	2.72	≤ 10
Tailing Factor (TF)	1.6	0.89	0.9	0.75	1.2	< 2
Theoretical Plate Number (N)	5784	5377	5184	4210	5980	> 2000

Table 2. Optimization Result of Flow Rate

Parameter	Flow Rate (mL/min)			Requirement (Snyder et al., 2010)
	0,9	**1,0**	**1,1**	
Resolution (Rs)	3.00	3.15	2.63	> 2
Rt (min)	4.49	3.91	3.61	≤ 10
Tailing Factor (TF)	1.12	0.90	0.75	< 2
Theoretical Plate Number (N)	5367	5184	5184	> 2000

Table 3. System Suitability Test Result

Parameter	Average	CV (%)	Requirement (Snyder et al., 2010)
Retention time (Rt)	3.96	0.17	CV ≤ 1.0%
Peak area	2726465	0.77	CV ≤ 2.0%
Peak height	362225	0.51	CV ≤ 2.0%
Resolution (Rs)	2.45	2.02	RS > 2
Tailing Factor (TF)	0.88	1.67	TF ≤ 2
Theoretical Plate Number (N)	6055	0.07	N > 2000

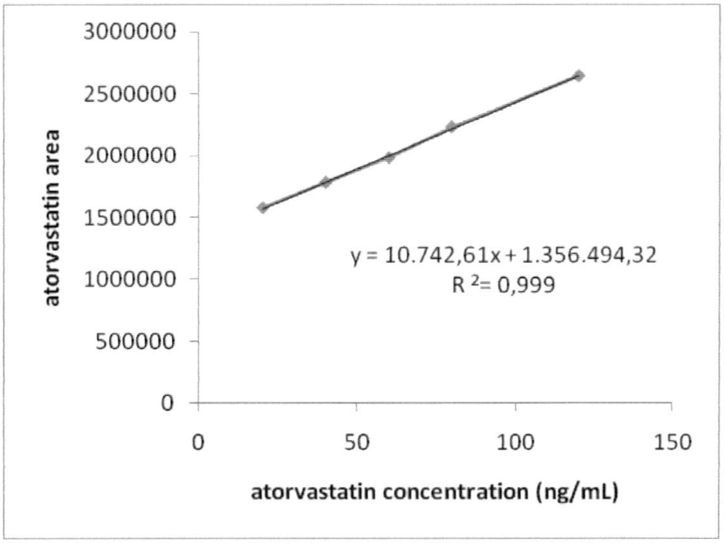

Figure 2. Determination Result of Linearity

Table 4. Accuracy and Precision Result

Atorvastatin Concentration are Known (ng/mL)	Measured Atorvastation Concentration (ng/mL)	Recovery (%)	CV (%)
20	20.22	101.15	0.64
	20.66	103.32	
	19.02	95.09	
	20.96	104.85	
	18.92	94.62	
40	43.59	108.99	1.38
	39.66	99.16	
	38.49	96.22	
	37.53	93.84	
	39.72	99.30	
60	58.28	97.14	1.35
	56.15	93.59	
	56.12	93.53	
	56.73	94.55	
	62.02	103.37	
80	81.58	101.98	1.73
	81.31	101.63	
	73.84	92.31	
	75.04	93.79	
	78.40	98.01	
120	111.69	93.08	2.82
	129.53	107.95	
	119.80	99.86	
	123.18	102.65	
	115.25	96.05	

Table 5. Result of Uniformity Test

Tablet	Concentration (%)	SD	CV
1	105.09		
2	100.52		
3	105.82		
4	102.45		
5	102.55	1.79	1.75
6	101.93		
7	103.60		
8	100.34		
9	102.09		
10	101.51		

4. Conclusion

1. Simple method of analysis by HPLC resulting optimum conditions for determination of atorvastatin tablet content were using a mixture of methanol-water pH 3 (80:20 v/v) as mobile phase; a C18 column as stationary phase, a flow rate of 1 mL / min and detected with a 245 nm UV detector.
2. The simple analytical methods developed have met the validation requirements of the analysis method including selectivity; linearity in the range 20 - 120 ng/mL with $r^2 = 0.999$; accuracy; precision; Limit of Detection (LOD) of 0.2 ng/mL; and Limit of Quantitation (LOQ) of 0.7 ng/mL.

Acknowledgments

Thank you to Sebelas Maret University for the financial support.

References

[1] Novakova L, Vickova H, Statinsky D, Sadilek P, Solichova D, Blaha M, Blana V and Solich P 2009 *J. Chromatogr. B 877 2093*

[2] Stone N J, Robinson J and Lichtenstein AH 2014 *J. Am. Coll. Cardiol.* **63**(25) 889

[3] Aini N S D and Sari I O 2015 *Jurnal Kefarmasian Indonesia* **5**(2) 90

[4] Kumar S G V, Rajendraprasad Y and Chandrashekar S M 2010 *Int. J. Pharm. Tech. Res.* **2**(1) 463

[5] Snyder L R, Kirkland J J and Dolan J W 2010 *Introduction to Modern Liquid Chromatography* (New Jersey: John Wiley and Sons)

[6] Chan C C, Lam H, Lee C Y and Zhang X M 2004 *Analytical Method Validation and Instrument Performance Verification* (New Jersey: John Wiley dan Sons)

[7] Gandjar I G and Rohman A 2012 *Analisis Obat Secara Spektrofotometri dan Kromatografi* (Yogyakarta: Pustaka Pelajar)

[8] Ahuja S and Dong M.W 2005 *Handbook Of Pharmaceutical Analysis by HPLC Volume 6* (Oxford: Elsevier)

IOP Conf. Series: Materials Science and Engineering **333** (2018) 012082 doi:10.1088/1757-899X/333/1/012082

Leaves Waste Composite with Glass Fiber Reinforcement

N Khoiri[1], W N Jannah[2], C. Huda[1], RM Maulana[2], P Marwoto[2], Masturi[2]
[1]Physics Education Department, Universitas PGRI Semarang, Indonesia
[2]Physics Department, Universitas Negeri Semarang, Indonesia

E-mail: nurkhoiri@upgris.ac.id

Abstract. A research has been made to fabricate leave waste composites and Polyvinyl Acetate (PvAc) polymers reinforced with glass fibers. The method used was a simple mixing of leaves powders, PvAc, and glass fibers varied from 0 g to 1 g. Mass of 16 g leaves powder and mass of PvAc 4 g. The mixing result is suppressed by 5 metric-tons for 15 minutes. The composite is dried at room temperature for 1 day then in the oven at 100°C for 1 hour. The compressive strength is measured bu a hydraulic press. The result show that the compressive strength increased to the highest point of 0.8 g and will decrease significantly when the addition of glass fiber mass of 1 g. The highest compressive strength reaches 52.6 MPa when the glass fiber mass is 0.8 g. The result of this research showed that leaves composites with Polyvinyl Acetate polymer reinforced with fiber glass can be used as alternative material of wood substitute.

1. Introduction

The demand for household goods and wood-based buildings continues to increase along with population growth. The massive supply of timber needs has seriously affected the destruction of forests and forests. On the other hand there begins a collective awareness of the need for greening as a guardian of world air quality [1]. This condition causes wood-based industries to experience serious problems of wood raw material scarcity which causes raw material prices to become very expensive and environmental issues that make wood-based industries become the main suspects in the rapid process of environmental damage [2]. At the same time the production of organic and inorganic waste as a consequence of waste of life activity is increasing. One of them is leaves waste that has not been studied and interesting for research. In Indonesia as a tropical country, leaves waste is very abundant. During this leaves waste only in the pile and burned and a small part for compost. The problem in this research is how to use leaves waste as wood substitute [3].

One of the answers to the above problem is by processing leaves waste into wood substitute material, it has a function that is strategic and very promising for our nation that is in tropical area. On the other hand, this idea will be one solution to the abundance of leaves waste and on the one hand will save the use of wood, so hope will be able to maintain environmental sustainability[4].

Since the use of more wood in some sectors triggered the amount of waste sawdust. In addition, the utilization of leave waste still not optimal [5]. It is necessary to support the implementation of appropriate technology that is expected to increase productivity. The use of press machine is helpful for improving product quality and mechanical properties of a product [1]. Based on these problems try to be approached by making composite leaves and Polyvinyl Acetate polymer reinforced with glass fiber.

2. Experimental

The dried leaves waste was used is mahogany leaves. It is destroyed with a blender until it becomes a fine powder. Next PVAc was weighed and dissolved in 8 ml of water and stirred using magnetic stirer for about 15 minutes. The leaves waste powder and PVAc are then added with varying glass fibers, while the mass of leaves powder and PVAc remains 16 grams, the mass of PVAc 4 grams, and glass fiber varied from 0-1 grams. the effect of the addition of glass fiber to improving the strength of the material. Furthermore, from the various variations are sought on how many mixtures will be produced the most powerful materials. After leaves powder, PVAc and glass fibers are evenly mixed, then sample put into the mold and then pressed using a pressure of 5 metric-tons for 15 minutes. The composite is dried at room temperature for 1 day then in the oven at 100°C for 1 hour.

The compressive strength is measured using a hydraulic press tool in the Physics laboratory of Semarang State University.

3. Result and Dsicussion

3.1. The result of compressive strength

The results of the composite sample compression test with fiber glass filler obtained the data such as Table 1.

Table 1. Test results of composite leaves sample with glass fiber filler

Sample	Fiber glass (g)	Compressive strength (MPa)
A	0	17.0
B	0.2	32.3
C	0.4	36.4
D	0.6	48.6
E	0.8	52.6
F	1	37.4

3.2. The Effect of fiber glass

Properties of Powdered Leaves Although it has been done in hot-presing still has a very high porosity, to increase the strength of leaves waste composite materials required polymer as a binder. The addition of PVAc will make the composite material strength stronger and lighter [7].

Furthermore, strength testing was done by adding fiber glass to the powder and PVAc mixture. From the results of sample tests on the sample table 1, where the results of its strength increased after the mixture of leaves powder and PVAc in add 0.2 gram glass fiber. This increase can be explained by the polymer impregnation approach. In the process of infiltration, penetration or polymer insertion into the pivot particles, the interaction between waste and polymer with fibers glass, fibers glass has increased the number of surface interactions that were originally only polymers with leaves particles, now added harder fibers glass so as to further enhance the mechanical strength of the material product. The presence of a shaft in leaves powder and PVAC after hot press allows for increased impregnation after added fiber glass. As a result the pore pores become smaller due to increased interaction between polymer surfaces with leaves particles. This decrease in pores is followed by increased interaction between polymer particles and waste fillers due to the proximity of the distance between them.

The addition of polymers and fibers glass to the leaves powder has a function as a particle binder. Initially the increase in composite compressive strength continues to increase in line with the addition of glass fibers from glassless samples of glass in add 0.2 grams up to 1 gram, while the mass of leaves powders and PVAc always remain in number. The maximum compressive strength is obtained when the fiber glass mixture is 08 grams, it is assumed that the mixed composition is mixing with the most suitable composition, under these optimum conditions, the surface area of the interaction between the particles reaches the maximum point, where all the right particles interact effectively. The addition of either a fraction, either leaves powder, PVAc or fiber glass will have an impact on the addition of an unintended region. This is the cause when the addition of fiber glass with a mass of 1 gram, causing the composite strength decreased [8], [9], [10]. Wherein the increase of the fibers glass tends to increase composite strength of the compound produced through a certain fraction of 0.8 gram. The addition of fiber glass fraction actually decreases composite compressive strength. Fiber Glass fraction 0.8 is called the maximum fraction.

3.3. The influence of hot pressing
The process of impregnating fibers glass into the pores of the powder mixture and PVAc is strongly influenced by temperature and pressure [10]. An increase in pressure and temperature will allow for a more effective contact between the Adhesive surface and the surrounding Adherent surface resulting in an increase in mechanical strength [11]. With a decrease in the Polymer viscosity value, the penetration rate into the pores will become even faster [12]. The introduction of polymers into pores results in increased density and composite density. These conditions have an impact on increasing the mechanical strength of the resulting composite material [13] so that the higher the temperature gives the greater the compressive strength produced.

In addition to temperature, impregnation is also affected by pressure. This pressure affects the process of impregnation of leaves powder and PVAc with fiber glass. Pressure in principle as a speeding penetration glass fibers on leaves powder and PVAc. At the same temperature, increased pressure results in an increase in penetration rate. Pressure also causes the distance between the particles closer, the porosity decreases, the density increases and the particle arrangement becomes more solid so that the overall surface interaction between the particles also increases. This condition results in an increase in mechanical compressive strength of the resulting composite material.

4. Conclusion
Based on the discussion, it can be concluded that the powder of garbage leaves added by PVAc and fiber glass can effectively be processed into a strong and light composite material. The addition of fiber glass to the leaves powder and PVAc has a function of strengthening the resulting composite pressure. The maximum composite pressure is obtained when adding fibers glass is 0.8 grams. Leaves composites with Polyvinyl Acetate polymers reinforced with glass fibers provide an alternative to wood substitutes.

Aknowledgment
Acknowledgments the authors convey to the Ministry of Research, Technology and Higher Education for the support of this research through Research Superior University of 2017.

Reference
[1] Slamet S 2013 *Proc. SNST 4th 2013 Unwahas* 1–9
[2] Masturi M, Aliah H, Aji M P, Sagita A A, Bukit M, Aliah H and Abdullah M 2012 *AIP Conf. Proc.* **90** 89–93
[3] Lyutyy P, Bekhta P, Sedliacik J and Ortynska G 2014 *J. Acta Facultatis Xylologiae Zvolen* **56** 39–50
[4] Intang S H, Maryoto A and Haryanto Y 2008 *J. Litbang Provinsi Jawa Tengah* **6** 23–37
[5] Aigbomian E P and Fan M 2013 *J. TOBCT* **7** 108– 117
[6] Najafi S K, Hamidinia E and Tajvidi M 2006 *J. Appl. Polym. Sci* **100** 3641–45
[7] Masturi, Mikrojuddin, Khairurrijal 2011 *J. of Mater. Cyc. And Waste Management* **13** 225-231
[8] Mikrajuddin 2008 *Pengantar Nanosains* (Bandung: Penerbit ITB)
[9] Starokadomskii D L 2008 *Russ. J. Appl. Chem.* **11** 1987-91
[10] Zhang Y, Zhang S Y and Chui YH 2006 *J. Appl. Polym. Sci.* **102** 2668-69

IOP Conf. Series: Materials Science and Engineering **333** (2018) 012082 doi:10.1088/1757-899X/333/1/012082

[11] Packham D E 2005 *Handbook of Adhesi Second edition* (Chichester: John Wiley & Sons Ltd)
[12] Matyka M, Khalila A and Kozal Z 2008 *Phys. Rev. E* **78** 1026306-08
[13] Sperling L H 2006 *Introduction to Physical Polymer Science Fourth Edition* (New Jersey: John Wiley & Sons)

IOP Conf. Series: Materials Science and Engineering **333** (2018) 012083 doi:10.1088/1757-899X/333/1/012083

EDIBLE FILM FROM POLYBLEND OF GINGER STARCH, CHITOSAN, AND SORBITOL AS PLASTICIZER

N Sariningsih[1], Y P Putra[1], W P Pamungkas[1], T Kusumaningsih[1]
Chemistry Department, Faculty of Mathematic and Natural Science, Sebelas Maret University, Indonesia

E-mail: triana_kusumaningsih@staff.uns.ac.id

Abstract. Polyblend ginger starch /chitosan based edible film has been succesfully prepared and characterized. The purpose of this research was to produce edible film from polyblend of ginger starch, chitosan, and sorbitol as plasticizer. The resulted edible film were characterized by using FTIR, TGA and UTM. Edible film of ginger starch had OH vibration (3430 cm^{-1}). Besides, edible film had elongation up to 15.63%. The thermal degradation of this material reached 208°C indicating high termal stability. The water uptake of the edible film was 42.85%. It concluded that edible film produce in this research has potential as a packaging.

1. Introduction

Packaging is a required material for food quality maintenance in order to keep it in good condition. Plastic is often used as a type of package. The use of plastic as a material of food packaging gave rises variety of problems to environment due to its unability to be recycled and decomposed naturally by microbes. Based on recent data by Jambeck [1] Indonesia is ranked as the world's second producer of plastic garbage to the sea (187.2 million tons) after China (262.9 million tons), in third place was Philippines (83.4 million tons), followed by Viet Nam (55.9 million tons), and Sri Lanka (14.6 million tons) per year. Each year the production of plastic generates about eight percent the production of the oil world about 12 millions barrels of oil or equivalent to 14 million of tree [2]. To resolve this problem, the alternative to reduce packaging material is with changes the using of plastic with the edible film [3]. The ginger production in Indonesia in 2000 amounted to 115.092 tons with an average 0.31% growth per year [4]. Component of ginger is gingerol, shogaol, zingerone and have effect of pharmacology and physiology as antioxidant [5]. The accumulation of ginger production until 2016 has reached approximately 1,401,435 ons and if the starch amount of ginger according to Hermani [6] reaches 82% then the starch of industry waste results more or less amounted to 1,149,178 tons. Given that the magnitude of ginger starch waste industry in Indonesia has not been used optimally and the existing antioxidant components in the ginger starch [7], research need to be done to process the waste ginger starch becoming a beneficial edible film that can be used as an alternative to reduce plastic waste in Indonesia. The main component of edible film is hydrocolloid, fat, and composite [8].

IOP Conf. Series: Materials Science and Engineering **333** (2018) 012083 doi:10.1088/1757-899X/333/1/012083

2. Experimental

2.1 Instruments
Instrument used to characterization physic characteristic of material is TGA- DTA Stanton Redcroft TG-750 and FTIR Merk Shimadzu, Type IRPrestige21. Instrument used to characterization mechanic characteristic is tensile strength test use Universal Testing Machine Merk Time Shijin Type WDW-20-E, morphology test use Scanning Electron Microscopy (SEM) JEOL JSM-6360LA, thickness test used couplers micrometer and water uptake test.

2.2 Materials
The material used to produce edible film is ginger starch, chitosan, sorbitol as a plasticizer, acetic acid 2% (Merck ®, 100%) and aquadest.

2.3 Edible Film Preparation
A number of starch and chitosan was weighed in comparison of mass variation between starch and chitosan (b/b) i.e. 5:5, 6:4, 7:3, 8:2, 9:1 with a total mass of ginger starch and chitosan 1 gram. Ginger starch was dissolved using aquadest, while chitosan is dissolved in acetic acid 2% (Merck ®, DIXI 25%) under stirring for approximately 30 minutes. Both solution had been mixed. The mixture is heated in temperature 80°C [6]. After 25 minute warm-up, the mixture was added 3.7 mL the sorbitol solution 30% and stirred for 5 minutes. The solution is cooled and the air bubbles or the polluter were then ± 150 grams solution was poured into a mold glass and placed in an open room ± 3 days. The molded dry solution was separated as plastic and ready to be characterize.

2.4 Characterization of edible film
2.4.1 The Thickness Measurement. The thickness measurement of edible films was measured using micrometers with accuracy of 0.0001 mm. The measurement was carried out on at least 5 different places to get the average thickness in accordance with the standard thickness [6].

2.4.2 Mechanic Test (Tensile Strength). Tensile strength test was done using Mesdan Lab strength by means the tip of sample was clipped in tensile testing machine. Next, recording the thickness and length of the initial sample. Start button on the computer was pressed would make the tool draw samples with speed of 100 mm/min until the sample is broken. Tensile strength test was conducted 3 times on different samples and the average was measured. Tensile strength edible film is calculated with the following equation

$$\tau = \frac{Fmax}{A} \quad (1)$$

Based on equation above, τ indicated tensile strength (MPa), Fmax showed voltage (N), and A showed broad cross-section transverse (mm^2). Measurement of at the broken elongation was done the same way with powerful pull test. The extension is expressed in percentage with calculation:

$$\text{Elongasi (\%)} = \frac{Lenght\ maximum\ (mm)}{first\ lenght\ (mm)}\ X\ 100\% \quad (2)$$

2.4.3 *FTIR Analysis (Fourier Transform Infrared Spectroscopy).* FTIR analysis of functional groups was aimed to find the process that followed the mixing procedures on samples, whether physically or chemically in each process of making edible film. The sample is placed into a holder set, then the corresponding spectrum was set. The results would be obtained by diffractogram in the relation between the wave numbers and intensity. FTIR spectra was recorded using spectrophotometer at room temperature [9].

2.4.4 Water Uptake Test. Water resistance test procedure was done with water absorption test by weighing the initial weight of sample that would be tested (W_O), sample was put into a container that contains aquadest for 10 seconds. Sample was lifted out from the container and the water in the surface of sample was removed with tissue paper, next the weight was measured again. Sample was put back into the container for 10 seconds. Then the sample was lifted from the container and weighed again. Soaking and weighing procedure carried out back until a constant weight was obtained [10]. The absorbed water by the sample were calculated through equation

$$\text{Water (\%)} = \frac{W-Wo}{Wo} X\ 100\% \qquad (3)$$

Based on the equation above, Wo showed the initial weight of edible films and W is the final weight after water uptake test.

2.4.5 Thermal Analysis (TGA-DTA). TGA-DTA was thermal analysis method in a polymerization. This method was dynamically recorded the weight of sample in heated or cooled conditions in a controlled ratio as a time or temperature function. The sample was placed into a specific place, heated and sought the spectrum that showed a reduced weight of polymer against temperature or time changes due to dehydration or decomposition process. The results would be obtained by diffractogram in relation between the wave number and intensity.

2.4.6 Scanning Electron Microscope Analysis. The morphological analysis of edible film is done by using SEM (Scanning Electron Microscopy). Edible film samples are affixed to the set of holder with double adhesive, then coated with gold medal in a vacuum. After that, the sample is put in place in the SEM, then the topographic image is observed and 5000 times magnified [9].

2.4.7 Antibacterial Test. Antibacterial test on edible film was done by diffusion method in Agar. The diffusion method was often used for testing an antibacterial properties. *E. coli* bacteria was used for gram negative. Antibacterial activity of edible film was showed with the emergence of a clear zone in the vicinity of edible film. The clear zone was calculated every 24 hours during the period of 96 hours.

IOP Conf. Series: Materials Science and Engineering **333** (2018) 012083 doi:10.1088/1757-899X/333/1/012083

3. Results and Discussion

The edible film ginger starch has been successfully prepared and characterized. The data of water uptake test, mechanic test, the thickness test shown in Table 1. The data shown are data from 5 variations with composition ratio of ginger starch and chitosan is 5:5, 6:4, 7:3, 8:2, 9:1.

Table 1. Test result of water uptake, thickness, elongation, and Maximum Drag from edible film

Starch : chitosan	Water Content (%)	Average of thickness (mm)	Elongation (%)	Maximum Drag (N)
0.5 g : 0.5 g	50.00 %	0.01	2.63	13.5
0.6 g : 0.4 g	50.00 %	0.01	1.59	14.6
0.7 g : 0.3 g	50.00 %	0.02	4.09	17.4
0.8 g : 0.2 g	42.85 %	0.02	5.35	18
0.9 g : 0.1 g	52.38 %	0.03	15.63	24.5

From the data, it is known that the best composition of the ginger starch and chitosan in ratio 9:1. In this comparison, the water uptake test shows a good percentage that is equal to 52.38 %. This identifies that in the ratio of ginger starch and chitosan 9:1 will be more easily degraded by water, this is because the starch contains more hydroxyl groups (OH), so it will bind more water because of the hydrogen bond. From the water uptake data then it can be seen that this edible film will be digested well by the body. Water content is smaller than previous researchers is 212.98% [9]. Thickness testing aims to know that edible films can be used as well as food packaging. Data from the average thickness indicates that of the five comparisons have a thickness between 0.01 mm to 0.03 mm. Japanese industrial standard states that edible films used as food packaging should not have a thickness above 0.25 mm. Of the data, edible film ginger starch and chitosan meets the standard for use as food packaging. Edible film has the best elongation on composition ratio 9:1 with elongation percentage of 15.63%. According to Krochta (1997) state that the percentage of edible film elongation is said not good if less than 10% [11]. From the data, it can be seen that the edible film has a good elongation on the composition ratio of ginger starch and chitosan is 9:1. The maximum tensile strength of the edible film at a ratio of 9:1 is 24.5 N, from the data it can be seen that edible film can be used for food packaging.

Functional groups of edible film samples with variations of ginger starch: chitosan (5:5), (6:4), (7:3), (8:2), and (9:1) were successfully performed. The FTIR spectra results are shown in Figure 1. Based on the FTIR spectra there is an uptake in the region of 3500-3300 cm^{-1} indicating the presence of the -OH group. The absorption in the 1700-1600 cm^{-1} region shows the presence of C=O absorption.

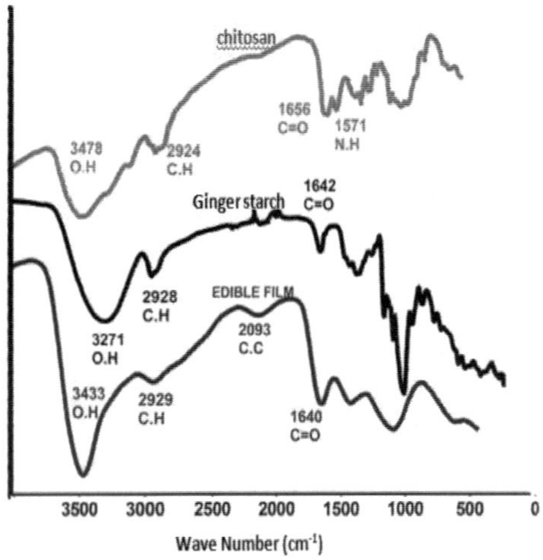

(a)

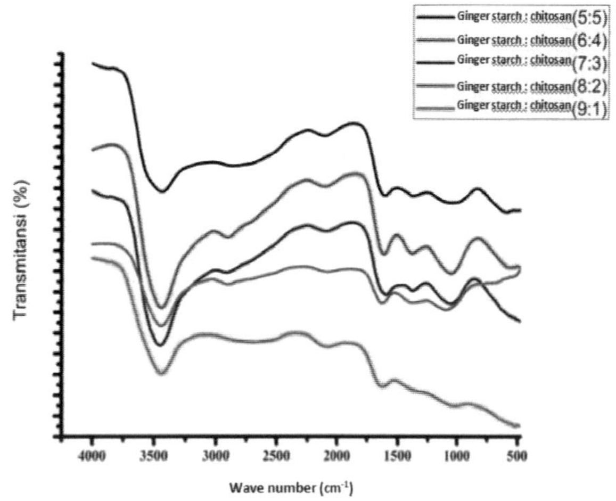

(b)

Figure 1. Spectra FTIR edible film (a) spectra of edible film, ginger starch and chitosan, (b) spectra of edible film on five variations of composition.

The spectra of the edible film indicate the absorption of the wave number 2093 cm^{-1}, the absorption of C-C, this absorption is a new absorption that identifies that new compounds have been formed, and from the five variations that FTIR has tested show the same spectra, it shows that the five variations have formed an edible film. The purpose of FTIR testing is simply to know that the edible film has been successfully synthesized and from the FTIR spectra showing that the edible film was successfully synthesized.

The analysis of thermal properties in the edible film sample was determined by thermogravimetric method in which edible film was heated at room temperature up to 900°C with a heating rate of 10°C/min. The TGA-DTA edible film thermogram is shown in Figure 2. The data show that the edible film experiences three times the thermal degradation that occurs in the heating process, at a temperature of about 100°C, 200 - 360°C, and 360 - 600°C. The degradation of the first stage of edible film at a temperature of 100°C indicated that the process of releasing water molecules in the edible film [12]. This degradation is characterized by a reduction in mass directly as the temperature begins to rise. The second degradation occurs at a temperature of about 200 - 360°C. Degradation at this stage was the breaking of the starch bonds. The third degradation occurs at temperatures around 360 - 600°C. Degradation at this stage was the breaking of the main chitosan chain [12].

IOP Conf. Series: Materials Science and Engineering **333** (2018) 012083 doi:10.1088/1757-899X/333/1/012083

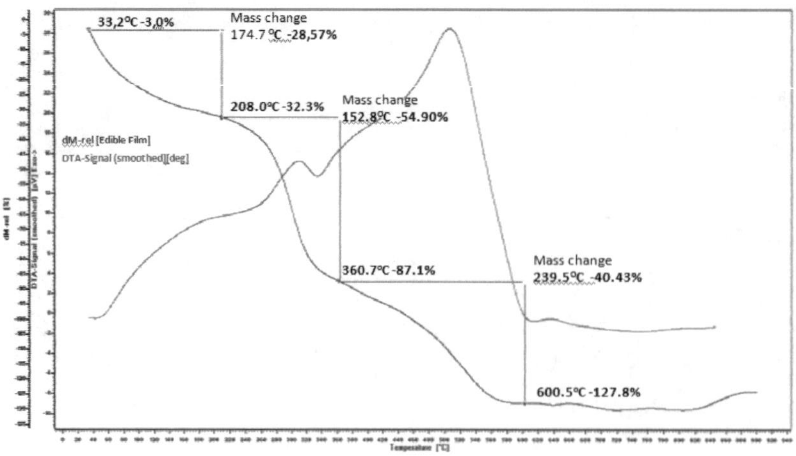

Figure 2. Thermal analysis results (TGA- DTA) of edible film ginger starch and chitosan

SEM surface morphology analysis of edible film has been successfully performed. The SEM test results are shown in Figure 3 which is a SEM morphological image from edible film ginger starch and chitosan compared to edible film sukun starch chitosan [9].

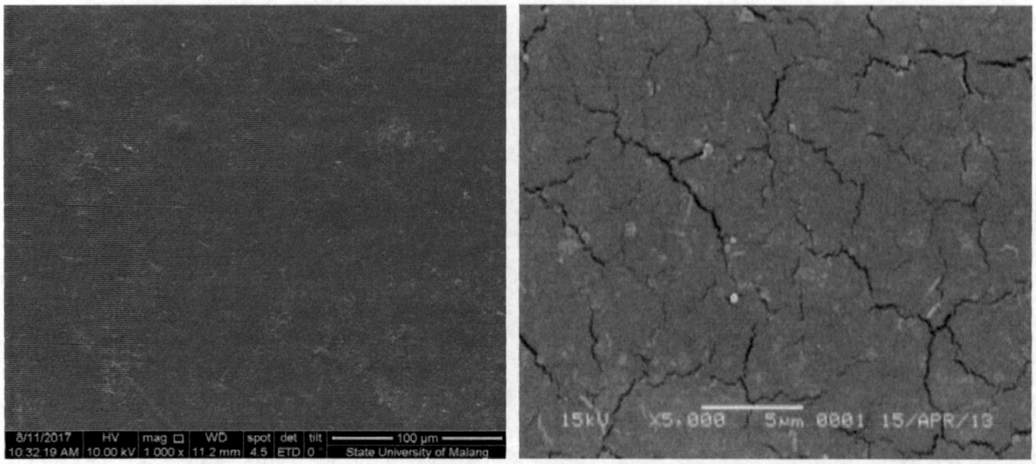

(a) **(b)**

Figure 3. SEM Analysis (a. SEM analysis from edible film ginger starch and chitosan, b. SEM analysis from sukun starch and chitosan)

Based on the SEM test data, edible film ginger starch and chitosan has a finer surface when compared with edible film sukun starch chitosan. This is identifies that this edible film has better physical properties when compared to edible film sukun starch chitosan. Antibacterial testing, it is known that *e.coli* bacteria can not grow in this edible film, so it can be seen that edible film is bacteriostatic.

4. Conclusion

Based on the above data, it can be concluded that:

1. Manufacture of edible films that are strong, elastic, and have antibacterial properties has been successfully synthesized from ginger starch, chitosan, and sorbitol as a plasticizer.
2. Edible film has elongation reach 15.63%. The thermal degradation of this material reaches a temperature of 208°C which indicates good thermal resistance. In addition, the results of water resistance test of edible film was 42.85%.
3. The best comparison between ginger starch and chitosan, which produces edible film with the best mechanical properties, is in the composition ratio of ginger starch and chitosan at 9: 1.

Acknowledgement

Thanks to Ristekdikti who has given us confidence to do this research and to the supervisor, thank you for guiding and giving direction for this research.

References

[1] Jambeck J R, Geyer R, Wilcox C, Siegler T R, Perryman M, Andrady A, Narayan R and Law K L 2015 *Science* **347(6223)** 768-771.
[2] Tuti Wahyuni 2016 *CNN Indonesia* Thursday 23/02/2016.
[3] Henrique C M, R F Teofilo, L Sabino, M M C Ferreira, M P Cereda 2007 *Journal of Food Science* **74** 184-189.
[4] Andriani S, Yunianta 2015 *Food and Agroindustrial Journal* **3(3)** 1128-1135.
[5] Bhattarai S, V H Tran and C C Duke 2001 *Journal of Environmental, Agricultural and Food Chemistry* **90** 1658-1664.
[6] Hermani and Winarti C 2013 *BBPPP* 1-6.
[7] Kikuzaki N, Nakatani 1993 *Journal of food science* Wiley Online Library.
[8] Rodriguez M, Oses J, Ziani K and Mate J I 2006 *Journal of Food Research International* **39** 840-846.
[9] Setiani W, Tety S, Lena R 2013 *Valensi* **3(2)** 100-109.
[10] Ban W, Song J, Argyropoulos D S & Lucia L A 2005 *Journal of Applied Polymer Science* **100** 2542-2548.
[11] Krochta 1997 *Food Technology* **51.**
[12] E. Pramono 2012 *ALCHEMY* **8(1)** 70-78.

IOP Conf. Series: Materials Science and Engineering **333** (2018) 012084 doi:10.1088/1757-899X/333/1/012084

Thermal properties of Bentonite Modified with 3-aminopropyltrimethoxysilane

E Pramono[1,3], W Pratiwi[1], D Wahyuningrum[2], C L Radiman[1*]
[1] Inorganic and Physical Chemistry Division, Faculty of Mathematics and Natural Sciences, Institut Teknologi Bandung, Jalan Ganesha 10, Bandung, Indonesia
[2] Organic Chemistry Division, Faculty of Mathematics and Natural Sciences, Institut Teknologi Bandung, Jalan Ganesha 10, Bandung, Indonesia
[3] Inorganic Materials Research Groups - Faculty of Mathematics and Natural Sciences, Universitas Sebelas Maret, Jalan Ir. Sutami 36 A, Surakarta, Indonesia

E-mail: cynthia@chem.itb.ac.id

Abstract. Chemical modifications of Bentonite (BNT) clay have been carried out by using 3-aminoprophyltrimethoxysilane (APS) in various solvent media. The degradation properties of products (BNTAPS) were characterized by thermogravimetric analysis (TGA). Samples were heated at 30 to 700 °C with a heating rate 10 deg/min, and the total silane-grafted amount was determined by calculating the weight loss at 200 – 600 °C. The thermogram of TGA showed that there were three main decomposition regions which are attributed to the elimination of physically adsorbed water, decomposition of silane and dehydroxylation of Bentonite. High weight loss attributed to the thermal decomposition of silane was observed between 200 to 550 °C. Quantitative analysis of grafted silane results high silane loaded using a solvent with high surface energy, which indicates the type of solvent affected interaction and adsorption of APS in BNT platelets.

1. Introduction

Clay is cheap and an abundant inorganic material resource on Earth. In Indonesia, Ca-Mg bentonite (BNT) is dominant type of clays which is contained in soil and generally used for purification [1]. BNT consists of monmorillonite-MMT as the largest component and other oxides such as Fe_2O_3 in its structure. The BNT structure was covered by two tetrahedral sheets of silica and octahedrons of alumina sheets with exchangeable cations such as Mg^{2+}, Ca^{2+}, Na^+ in the interlayer space. The ions are strongly hydrated in the presence of water, resulting in a hydrophilic environment of BNT [2].

BNT and several clays are widely applied as adsorbent and inorganic components in polymer composites [3-6]. Clay is naturally has polar properties and easily agglomerate due to strong interaction between clay particles compared to clays with polymers [7]. It is necessary for clay mineral to be modified to improve the surface and clay interlayer properties so that the polymer can be intercalated. Clay modification with quaternary ammonium salts compounds have been previously reported to improve interlayer spacing [8]. Another clays modification to improve thermal stability at high temperature, modification with organosilane such as 3-aminopropyltrietoxysilane has been previously reported [9, 10]. The reaction between the clay with modifier can occur on the surface and interlayers of the clay, and it was influenced by the type of solvent. Modification of clays with

IOP Conf. Series: Materials Science and Engineering **333** (2018) 012084 doi:10.1088/1757-899X/333/1/012084

organosilane changes both chemical properties and physical properties of clays such as the thermal stability [11].

In this paper, we studied a chemical modification of BNT with 3-aminopropyltrimethoxyisilane (APS) and thermal properties of modified materials. In order to understand the effect of a dispersion medium, Water, Ethylene glycol and the combination of solvent was used. The modified BNTs (BNTAPS) were characterized with Thermogravimetric analysis (TGA).

2. Experimental

2.1 Materials

Bentonite (BNT) 180 mesh supplied by a local industry in Indonesia. 3-Aminopropyltrimethoxysilane (APS) was obtained from Aldrich Chemical Company, Ethanol (Merck) and Ethylene glycol (Merck) solvents were used without further purification.

2.1.1 Aminosilane functionalization of BNT. Five gram of BNT was dispersed in 250 mL of solvent and stirred at room temperature for an hour. Five gram of APS dissolved in 250 mL of solvent was added to the BNT dispersion and stirring was continued for an hour. The product was then filtered and dried at 40 °C for 24 hours. The products were designated by BNTAPS_x_y, where x and y represent the solvents. Distilled water (W), Ethylene glycol (EG), Ethanol (Et) were used as solvents.

2.1.2 Thermal Analysis. Thermogravimetric analyses were carried out using Simultaneous Thermal Analyzer (STA) LINSEIS PT1600 in the temperature range from 30 to 700 °C at heating rate 20 °Cmin^{-1} with air atmosphere. Using TGA data, silane loaded amount was calculated using equation 1, where $W_{200\text{-}600}$ is percentages of weight between 200 °C and 600 °C [10, 12].

$$\text{Silane load amount } (\%) = \frac{100 \times W_{200\text{-}600}}{100 - W_{200\text{-}600}} \quad (1)$$

3. Results and Discussion

To analyse interaction of BNT with APS and total grafted silane, thermal analysis was conducted. Thermogram TGA and differential thermal gravimetric (DTG) of BNT and BNTAPS were shown in Figure 1 and Figure 2. BNT thermogram shows two major stages of degradation process, that are in the below 100 °C and about 500 °C which is attributed to dehydration and dehydroxylation alumnosilicate [13]. Weight loss at temperatures below 100 C is attributed to evaporation of water molecules that physically interact on the BNT surface and obtain a total degradation about 11%. At a temperature range of 150-450 °C BNT is slowly degraded and the high weight loss observed at 450-550 °C. The degradation peak around 450-550 °C is attributed to the dehydroxylation of alumino-silicate layers.

Figure 1 and Figure 2 show different degradation patterns of BNTAPS especially in areas 150-450 °C and 550-700 °C. Previous studies reported that DTG from aminosilane-modified clays produced four major peaks of water degradation, intercalation, grafted silanes and alumino-silicate layers. In this study is obtained degradation peak around 100-150 °C which is attributed to dehydration of water molecules on the surface of BNTAPS. The peak around 150-450 °C shows some degradation peaks. Previous research has reported that the absorbed and grafted modifier molecule is degraded in 150-450 °C. The degradation around 450-550 °C is observed for BNT and BNTAPS, which is attributed to the degradation of alumino-silicate layers. Meanwhile, BNTAPS is still degraded again at higher temperatures at 550-700 °C.

Table 1 shows mass loss percentages of BNT and BNTAPS. Modified materials showed higher mass loss at 150-450 °C than BNT. BNTAPS modified in water media showed highest mass loss which is about 9.57%, while the lowest mass loss about 6.81% was observed in BNTAPS modified in mixture solution of ethylene glycol and ethanol. At a temperature of 550-700 °C, the degradation is obtained for BNTAPS in various solvents; on the other hand, this is not obtained for BNT. BNTAPS

IOP Conf. Series: Materials Science and Engineering **333** (2018) 012084 doi:10.1088/1757-899X/333/1/012084

modified in water and mixture of water and ethanol showed higher mass loss, 2.83% and 3.56% respectively. The lowest mass loss about 1.34% was observed in BNTAPS_Eg.

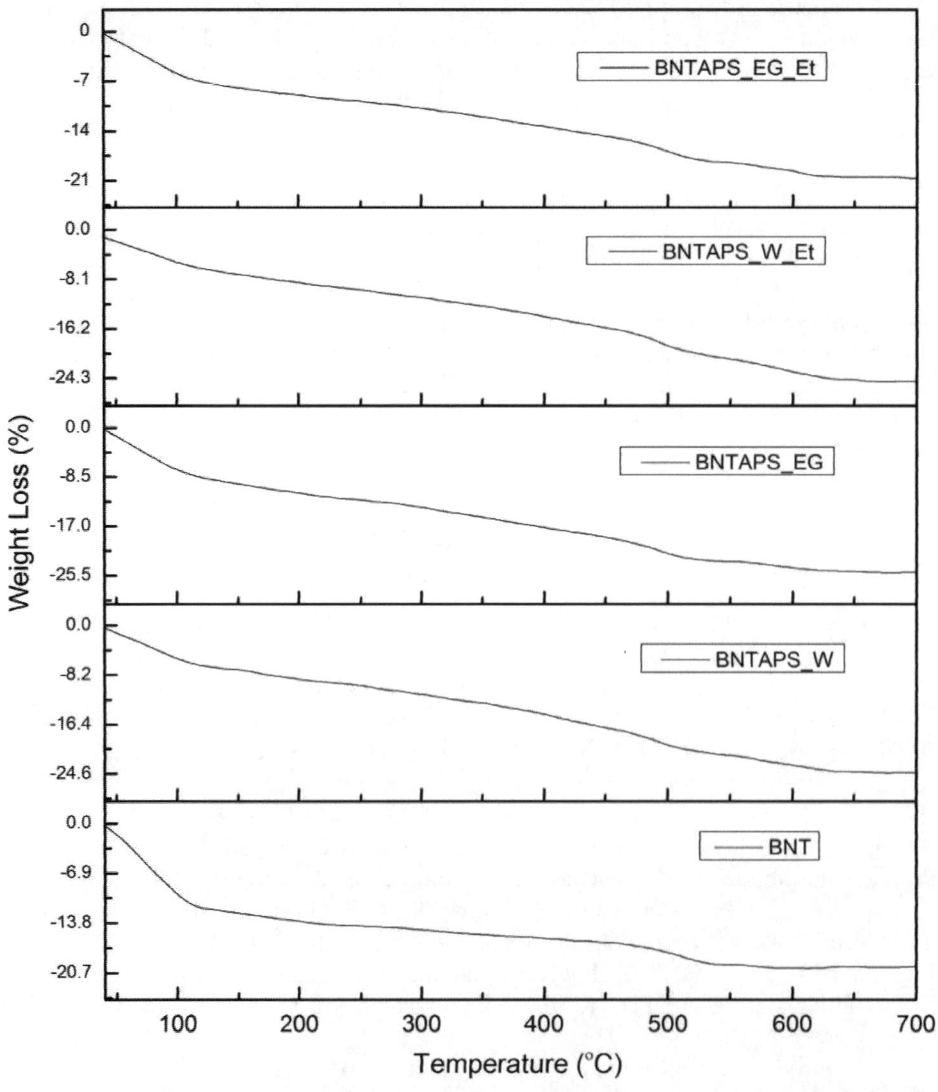

Figure 1. Thermogram TGA of BNT and BNTAPS in any solvents.

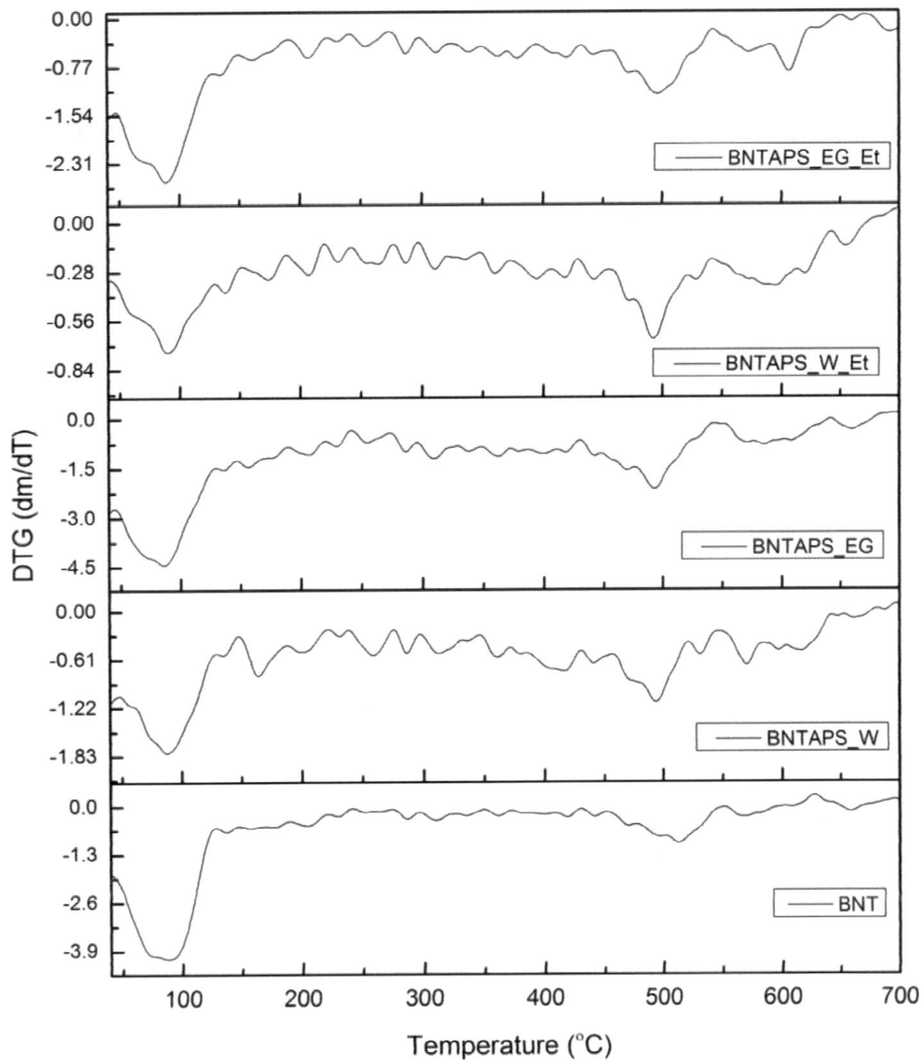

Figure 2. Thermogram DTG BNT and BNTAPS

Table 1. Mass loss characteristics of BNT and BNTAPS obtained from TGA analysis

Sample	Mass loss (%)			
	40-125 °C	**150-450 °C**	**450-550 °C**	**550-700 °C**
BNT	11.8	4.10	3.07	-
BNTAPS_W	6.8	9.57	4.65	2.83
BNTAPS_EG	8.8	9.18	4.17	1.34
BNTAPS_Wet	6.5	8.72	5.15	3.56
BNTAPS_EGEt	7.2	6.81	3.77	2.18

The usage of different types of solvents produces different degradation patterns at 150-450 °C. Water and mixtures of water and ethanol as dispersing medium resulted in degradation at 100-150 °C which is attributed to evaporation of physisorbed water molecules and hydration water of exchangeable cations [9, 11]. The next degradation peaks are obtained at a temperature of about 150-

450 °C and this degradation is also produced on the use of EG and mixtures of EG and Et. The degradation around 150-450 °C is related to the decomposition of absorbed and intercalated APS. In this study, the modification of APS in BNT resulted in degradation peak at 550-700 °C area for all dispersing media, while that degradation did not occur in BNT. Peaks at 550-700 °C are related to the degradation of cross-linked APS on the inter-clay layer BNT and followed by the dehydroxylation of aminosilicate layers.

Previous studies reported that clay modification with amimosilane is influenced by surface energy or surface tension of the solvent [9]. Solvents with lower surface tension or similar to the BNT surface tension (γ-MMT 44 ± 2 mJ / m^2) can wet the BNT and facilitate the APS to react with BNT on the surface. While solvents with higher surface tension can increase basal spacing of BNT so that APS can be intercalated or form a crosslink in the basal spacing of BNT. Figure 3 shows the effect solvent surface tension on the silane load, which calculated using equation 1. The using of solvents with low surface energy produce lower total silane load than higher surface energy solvents, and these result agree with previous study [9]. The use of mixed EG solvent with ethanol produced the lowest silane load which is about 11.5%, meanwhile the usage of EG (γ 52.4 mJ / m^2) and water (γ 72 mJ / m^2) were obtained 14.77% and 16.7%, respectively. The modification using high surface energy such as water not only facilitates the APS reaction on the BNT surface, but also intercalates and interacts with the -OH groups in the basal spacing of BNT and it is produced higher thermal degradation [9, 11].

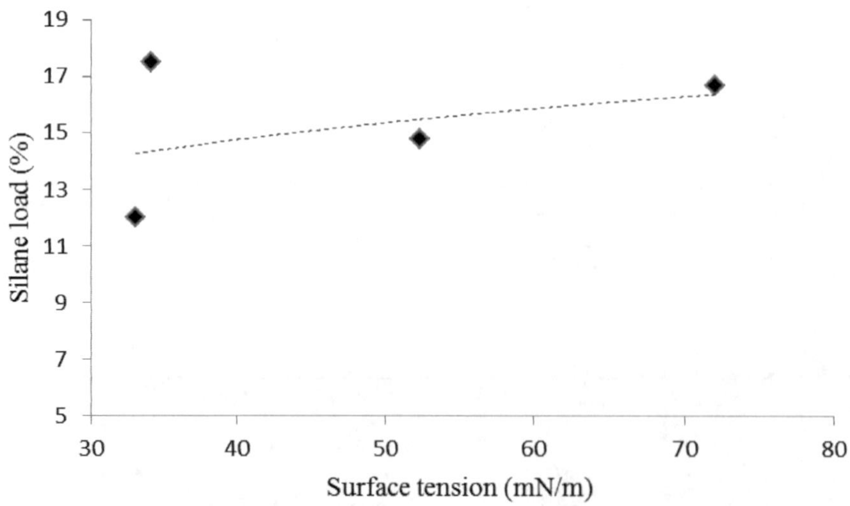

Figure 3. Effect of solvent surface tension on silane load

4. Conclusion

Chemical modifications of bentonite have been prepared with APS under various solvents. Thermal analysis was performed to determine the interaction and total grafted silane on BNT. The TGA and DTG data of BNTAPS shows degradation below 150 °C, 150-450 °C, 450-550 °C, and above 550 °C due to releasing water molecules, degradation of absorbed and intercalated APS, bonded APS in the clay interlayers, and degradation of aluminosilicate layers. The percentage of grafted silane on the BNT, calculated using TGA, showed a higher grafted silane for high surface tension solvent. The usage of water as solvent produces a large grafted silane with a percentage 16.7%.

Acknowledgment
The authors give the gratitude to Lembaga Penelitian dan Pengabdian Masyarakat Institut Teknologi Bandung for providing the financial support for this research through the KK Research Grant, Program A, 2017.

References

[1] Koswojo R, Utomo R P, Ju Y H, Ayucitra A, Soetaredjo F E, Sunarso J and Ismadji S 2010 *Appl. Clay Sci.* **48** 81-86
[2] Özcan A S and Özcan A 2004 *J. Colloid Interf. Sci.* **276** 39–46
[3] Othman N, Ismail H and Mariatti M 2006 *Polym. Degrad. Stabil.* **91** 1761-1774
[4] Tahir S S and Rauf N 2006 *Chemosphere* **63** 1842 - 1848
[5] Wang P, Ma J, Wang Z, Shi F and Liu Q 2012 *Langmuir* **28** 4776 - 4786
[6] Bergaya F and Lagaly G 2001 *Appl. Clay Sci.* **19** 1-3
[7] Zulfiqar S, Kausar A, Rizwan M and Sarwar M I 2008 *Appl. Surf. Sci.* **255** 2080-2086
[8] Paiva L B, Morales A R and Díaz F R V 2008 *Appl. Clay Sci.* **42** 8-24
[9] Shanmugharaj A M, Rhee K Y and Ryu S H 2006 *J. Colloid Interf. Sci.* **298** 854–859
[10] Piscitelli F, Posocco P, Toth R, Fermeglia M, Pricl S, Mensitieri G and Lavorgna M 2010 *J. Colloid Interf. Sci.* **351** 108–115
[11] Bertuoli P T, Piazza D, Scienza L C and Zattera A J 2014 *Appl. Clay Sci.* **87** 46 – 51
[12] Herrera N N, Letoffe J M, Putaux J L, David L and Bourgeat-Lami E 2004 *Langmuir* **20** 1564-1571
[13] Holtzer M, Bobrowski A and Zymankowska-Kumon S 2011 *J. Mol. Struct.* **1004** 102–108

IOP Conf. Series: Materials Science and Engineering **333** (2018) 012085 doi:10.1088/1757-899X/333/1/012085

Influence of Polyvinyl Alcohol (PVA) Addition on Silica Membrane Performance Prepared from Rice Straw

S Wahyuningsih[1*], A H Ramelan[1], D T Wardoyo[2], S Ichsan[1], Y R Kristiawan[1]
[1]Inorganic Material Research Group, Faculty of Mathematic and Natural Sciences, Sebelas Maret University, Surakarta, Indonesia

[*]E-mail: sayekti@mipa.uns.ac.id

Abstract. The utilization and modification of silica from rice straw as the main ingredient of adsorbent has been studied. The aim of this study was to determine the optimum composition of PVA (polyvinyl alcohol): silica to produce adsorbents with excellent pore characteristics, optimum adsorption efficiency and optimum pH for methyl yellow adsorptions. X-Ray Fluorescence (XRF) analysis results showed that straw ash contains 82.12 % of silica (SiO_2). SAA (Surface Area Analyzer) analysis showed optimum composition ratio 5:5 of PVA: silica with surface area of 1.503 m^2/g. Besides, based on the pore size distribution of PVA: silica (5:5) showed the narrow pore size distribution with the largest pore cumulative volume of 2.8 x 10^{-3} cc/g. The optimum pH for Methanyl Yellow adsorption is pH 2 with adsorption capacity = 72.1346%.

1. Introduction

The development of science and technology in Indonesia directly encourages the growth of the industry. The environmental pollution problem such as industrial waste waters into river has been more serious problem of this decade. Liquid waste as a product of industrial activity often creates problems of the environment [1]. Liquid wastes containing hazardous and toxic materials could disrupt the biological processes that occur to municipal environment. Water pollution from several heavy metals and organic wastes originated by industries such as metallurgy, tannery, and textile industries [2]. The textile industry produced large amount of liquid waste such as dyestuffs, suspended solids, oils and fats that need to be treated before the wastes were released into the river [3]. The amount of dye waste in the textile industry would cause environmental pollution, while the environment has a limited ability to degrade the dye, so further research is needed to overcome these problems. Liquid waste treatment was intended to eliminate levels from pollutants that were contained in liquid waste to be disposed directly to the environment.

Several studies of the dye and organic waste removal compounds present in batik industrial have been widely studied, not only by chemical treatment such as, color degradation with oxidation reactions, anaerobic reactions and photocatalysis reactions [4], but also by physical treatment like coagulation, sedimentation, adsorption using activated carbon, silica and biomaterials [5]. The coagulation and sedimentation technique has noted by the good efficiency for waste treatment but generate new waste, such as floc or coagulant. Not only the use of activated carbon still lack of requirement of high enough cost because the price of activated carbon is relatively expensive, but also the use of photocatalyst reactions requires a high cost because the price of photocatalysis reagents such as TiO_2 was quite expensive. Liquid waste waters process by biological method was also widely applied to reduce the organic compound of liquid waste textile industry. However, the efficiency of color removal through this biological process was not really satisfactory, since the dye has the resistance to biological degradation (recalcitrance). To solve that problem, a new alternative is needed to treat the textile industries effective and efficient liquid waste for reducing organic pollutants and dyestuffs [6].

Based on El-Maghraby [7] study about utilized rice hulls for *Methylene blue* absorption, the hidroxyl group in rice hulls increasing with the increases of pH. Increases the number of negatively charge sites could enlarge the attraction between dye and adsorbent surface. The research of Mahatmanti et al in 2016 [8], the utilization of rice hulls as the source of silica in the making of chitosan based membrane with the supporting material Poly Ethylene Glycol (PEG) has been done. Silica/Chitosan/PEG membrane was used to reduce the content of heavy metals such as Ca, Cd, Zn, and Mg in solution. The source of silica on the membrane was obtained from rice straw. The selection of rice straw due to straw has 75% content of silica in one kilogram [9]. The manufacture of silica membranes used a mixture of rice straw silica, Poly Vinyl Alcohol (PVA), and Poly Ethylene Glycol (PEG).

2. Experimental

2.1. Chemicals

Rice straw obtained from local industry (Pacitan, East Java, Indonesia), Polyvinyl alcohol (PVA) (Merck, Indonesia), Polyethylene glycol (PEG) (Merck, Indonesia), Nitric acid (Merck, Indonesia), and Hydrochloric acid (HCl) (Merck, Indonesia) were used in the production of PVA/Silica membrane. Whereas the Methanyl Yellow used in this study is a local material and Fe^{3+} solution is from Merck, Indonesia.

2.2. Process of rice straw ashes and Activation of Rice Straw Ash

The main ingredient of rice straw from Pacitan which is then nested with heating in a cauldron placed on the stove. This process was done until the gray rice straw is gray, and further characterization using STA and chemical group analysis using FT-IR. Furthermore, rice straw charcoal was burned at 800 ° C for 3 hours resulting in rice straw ash. The resulting rice straw ash is reddish-gray, and then analyzed its chemical content using FT-IR.

The rice straw ash was activated by soaking the ash of rice straw on HCl 1 M for 24 hours. Then it was filtered and neutralized with water, after neutral then heated in oven for 5 hours at 110 °C. Then crushed using porcelain cup and sieved with 150 meshes sieve. Then, analysis of functional group content using FTIR and chemical content analysis using XRF.

2.3. Fabrication of Silica membrane

A total of 2, 4, 6, 8, 10 grams of PVA were mixed with 10 mL 1M HNO_3, and 200 mL aquadest, stirred with a magnetic stirrer for 2 hours on 80 ° C heating. Then taken each 1, 2, 3, 4, 5 gram of each PVA solution that has been made, then added ash of activated rice straw as much as 5 grams and stirred until homogeneous. The silica-PVA dough of each PVA / silica composition added PEG 0.3 grams, 5 ml aquades and 4 g of Na_2SiO_4. The mixture was then stirred until homogeneous and poured onto a mold ± 20 mm in diameter and arranged with a membrane thickness of ± 4 mm, then allowed to stand for ± 30 hours.

The next stage after the process of forming the silica membrane includes drying and burning. The drying process was carried out at a temperature of about 60 °C for 1 hour. Then it was burned by raising the temperature to 800 °C. The combustion process was carried out for 5 hours, following the cool down in the furnace to room temperature. Finally we have gotten silica membrane composed by PVA:silica = 1:5; 2:5; 3:5; 4:5; 5:5, respectively.

2.4. Methanyl Yellow Adsorption Process

Silica membrane composed by PVA:silica = 1:5; 2:5; 3:5; 4:5; 5:5, respectively, was placed on a porcelain funnel, and an already prepared Methanyl Yellow solution was placed on the burette of 20 ml. Then the Yellow Methanyl solution was continuously streamed into the silica membrane. Thus, silica membrane with optimal PVA / silica ratio will be obtained to adsorb Methanyl Yellow solution. The Uv-Vis spectrophotometer UV-Vis Shimadzu PC 1601 was used to calculate the maximum wavelength in the 300-600 nm range. Based on maximum wavelength have been obtained we plotted the standard curve of that dye to determined the sample concentration later .

2.5. Determination pH Optimum of Methanyl Yellow and Fe^{3+} Solution

The silica membrane was used to adsorb Methanyl Yellow. However, the Metanyl Yellow solution was prepared at various pH condition from pH 1 to pH 8. The silica membrane were immersed into Fe^{3+} solutions in certain pH conditions in the range 1 – 7. While, the adsorption process of Methanyl Yellow with optimum pH was carried out. In addition, the membrane was placed on a porcelain funnel, and a 20 mL solution of Methanyl Yellow was placed on the burette. Yellow Methanyl solution was continuously streamed into the silica membrane. Based on this procedure we also study the infuence of Fe^{3+} additon in methanyl yellow solutions and also calculate the optimum pH for Fe^{3+} solution will pass to silica membrane.

3. Result and Discussion

3.1. Characteristic of Straw Ash due to Activation and combustion Process

Rice straw is one of the large amount of agricultural waste and has not been fully utilized. Rice straw when burned will produce ash as much as 10-17%, which means that any 1000 kg of rice straw burning will produce 100-170 kg of rice ash, and contain 75% silica in each kilogram [9].

From the Table. 1, it can be seen that the silica content is 82.12%, which is the dominant composition of rice straw ash used in this study. Several studies that have been done previously obtained the silica ash content of rice straw as much as 82.5% [10] and 75% [9].

Table 1. Result of Chemical Composition Content (% w/w) of Rice Straw Ash using XRF

Composition	SiO_2	K_2O	CaO	Al_2O_3	P_2O_5	Cl	MnO	Fe_2O_3	SO_3	TiO_2	Rb_2O	ZnO	SrO
Quantity (%)	82.12	9.72	2.52	1.00	0.99	0.98	0.87	0.82	0.73	0.09	0.06	0.05	0.01

Base on the result (Figure 1), there is no significant difference of wave numbers in each sample, which means that in each treatment does not change the composition of the functional groups in the sample. The sharped absorption number at 1090 cm^{-1}-1095 cm^{-1} was represented Si-O-Si (siloxane) groups. The FTIR spectras (Figure 1) shown that reduction of C content was happened in Activated Rice Straw Ash. HCl solution was able to dissolve impurities, its can be known by the loss of small peaks (very low intensity).

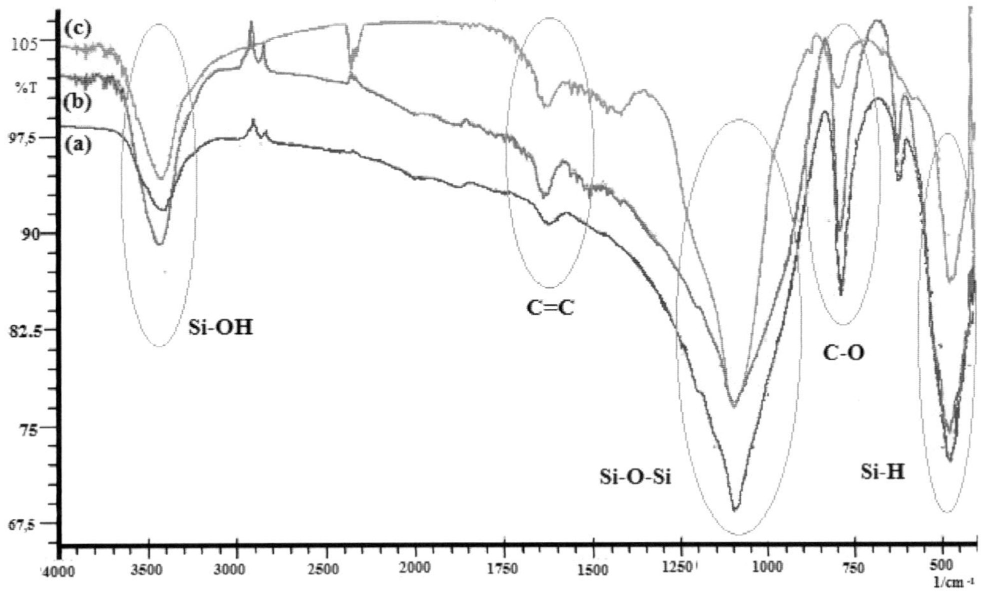

Figure 1. FTIR spectra of (a) Rice straw charcoal (b) Rice Straw Ash without activation (c) Activated Rice Straw Ash

3.2. Characteristic of silica membrane with addition of PVA.

In the manufacture of silica membrane consists of several stages. The initial stage is the manufacturing of PVA (liquid) with the addition of nitric acid as the activating agent of the hydroxyl functional group. PVA will perform a strong secondary bond with siloxane (Si-O-Si) and silanol (Si-OH) groups in rice straw ash. This is because PVA has a hydroxy group in every iteration of the polymer, therefore, there will be a three-dimensional structure between PVA and silica. According to Reed James (1988) [11], PVA will act as a binder between particles that will increase the *green strength* of the silica membrane.

The function of the addition of PEG as a pore-forming agent in the silica membrane. As stated by Kim & Lee (1998) [12], that PEG can be used as a pore-forming and increase membrane permeability. In addition, PEG also has a function as a plasticizer. The only adhesive-containing system is very elastic like rubber, so it takes a plasticizer that will make the dough is viscoelastic. Viscoelastic will make the dough easily formed according to desire [11].

The thermal characteristics of the PVA/silica membrane in the combustion process is performed thermal analysis which can be seen as Figure 2.

IOP Conf. Series: Materials Science and Engineering **333** (2018) 012085 doi:10.1088/1757-899X/333/1/012085

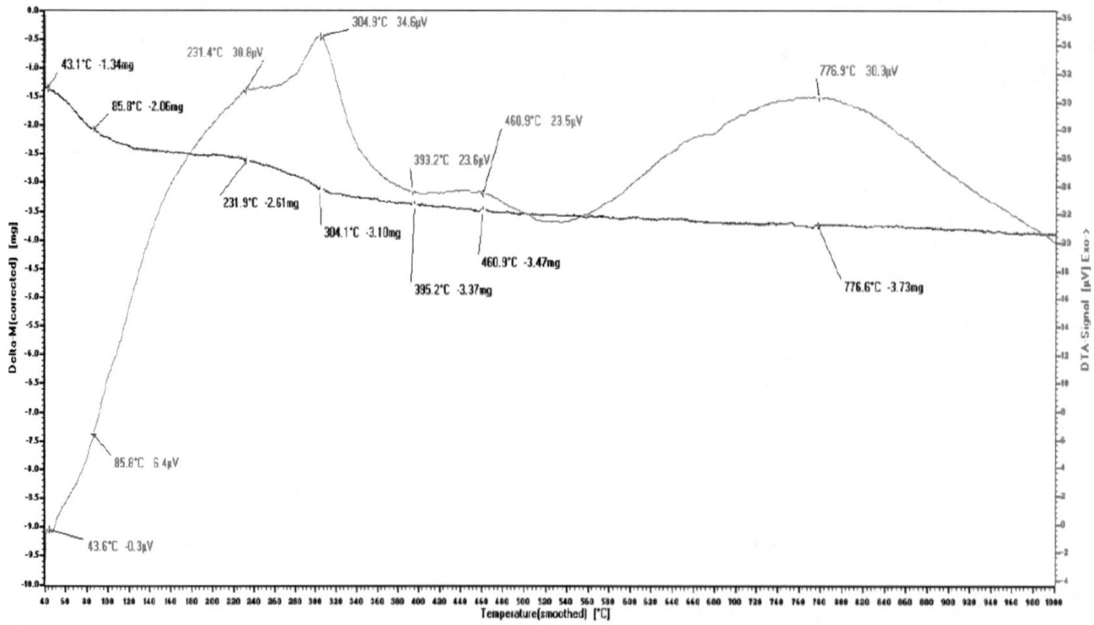

Figure 2. Termogram of silica membrane (PVA/Silica = 5/5)

From the silica membrane thermogram, at 43.6 ° C there is a mass loss of 1.34 mg which means that the water content in the membrane has decreased and up to 85.8 °C has decomposition of 2.06 mg. Transition Glass (Tg) PVA occurs at a temperature of 85.8 °C. While melting temperature (Tm) of PVA occurs at a temperature of 231.9 0C.

At the temperature of 395.2 - 460.9 °C occurs endothermic, where at this temperature PVA and PEG will burn. According to Gilman (1994) [13], at 350 °C, the alcohol fraction, methylene, aliphatic carbon will be lost by ± 70 ppm, ± 45 ppm, ± 15-50 ppm from the original composition up to 776.6 °C most of the PVA and PEG are lost, where up to this temperature the total mass decomposition is 3.73 mg.

At the temperature of 460.9-776.6 °C oxidation reactions, recombination and polymerization occur within the silica membrane. According to Remmey (1994) [14], the temperature of 500 °C-800 °C is the temperature of the filtration, where the glass forming elements will melt and become hard when cooled. In this period the silica changes in volume and crystal structure so that it will expand rapidly. While at temperatures above 776.6 °C there will be recrystallization in the membrane, which takes place exothermically (releasing heat).

With the loss of PVA and PEG will result in empty spaces (pores) in the silica membrane and rearrangements of bonds within the membrane. The pore characteristics of each membrane composition can we seen in SAA result. Comparison of pore size, pore distribution, pore volume, and surface area of each PVA/silica composition can be seen in Table 2.

Table 2. Surface and Pore Analysis Result of Silica Membrane

PVA/Silica	BJH Desorption		BET		
	Pore Diameter (Å)	Cumulative pore volume x10^{-3} (cc/g)	Average Diameter Pore (Å)	Pore Volume x 10^{-3} (cc/g)	Surface Area (m^2/g)
1/5	**18**	**1.5**	34.4747	1.472	0.854
	20	1.3			
	32	1			
	48	0.6			
2/5	**19**	**1.6**	27.7030	1.555	1.122
	39	1			
3/5	**28**	**0.9**	64.2095	1.159	0.361
4/5	**18**	**2.5**	53.7122	2.515	0.937
	25	2.2			
5/5	**18**	**2.8**	37.1623	2.793	1.503
	21	2.3			

The cumulative volume of pore is the volume (cc) of each pore diameter in each 1 gram of sample. From Table 2, it can be seen that in the silica membrane composed from PVA: silica = 1:5; 2:5; 4:5; and 5:5 had the largest pore size distribution on micropore (< 20 Å), whereas in the silica membrane composed from PVA:silica = 3:5, the highest pore size distribution was obtained at mesoporous (20 Å - 50 Å).

From the micropore diameter, it can be seen that the composition of the 5:5 silica membrane has the highest cumulative volume of the micropore. The more micropores in the sample, the greater the total pore volume and total surface area in the sample are known by the BET method. Where BET (Brunau-Emmett-Teller) is a method used to determine the total surface area and pore structure of a solid (Haber et al, 1995) [15].

The 5:5 silica membrane provides the largest total surface area. This proves that, when burned to a temperature of 800 ° C, PVA will be degraded and rearranged to form new pores. Thus, the more PVA the added surface of the silica membrane will be. However, the pore size distribution on the 5:5 silica membrane does not different significantly between micropore and mesoporous size. This gives information that on the 5:5 silica membrane, the micropore is not distributed uniformly (randomly). While the silica membrane with uniform pore size distribution can be seen on the 2:5 silica membrane, which is characterized by a significant difference between micropore and mesoporous size. Thus, the PVA:silica composition which gives a uniform pore size distribution is the PVA:silica = 2:5 composition. However, the optimum surface area was obtained on a silica membrane with a PVA:silica composition = 5:5.

3.3. Performance of Silica membrane on adsorption of Methanyl Yellow

The 10 ppm Yellow Methanyl solution was used as the starting solution which would later be adsorbed with the silica membrane. The variation of PVA/silica composition (b/w) is 1/5; 2/5; 3/5; 4/5; 5/5 adsorbed with 20 mL of Methanyl Yellow solution.

IOP Conf. Series: Materials Science and Engineering **333** (2018) 012085 doi:10.1088/1757-899X/333/1/012085

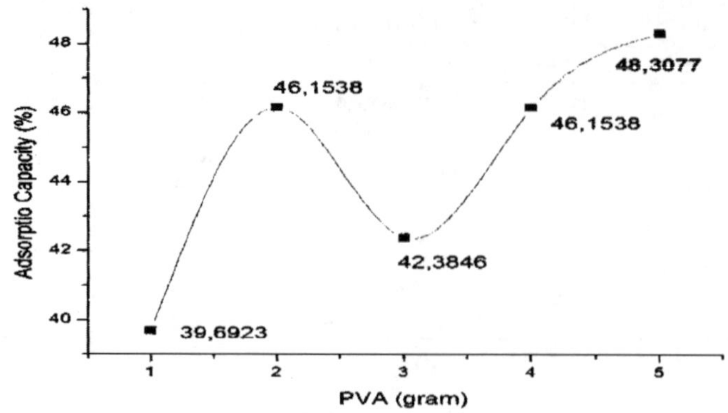

Figure 3. Optimum Composition of Silica membrane

Based on data in Figure 3, it can be seen that the optimum adsorption capacity occurs when use silica membrane composed from PVA:silica = 5:5 (w/w) as 48.3077 %. Methanyl Yellow molecules will be retained more in the silica membrane as the radius (microspores) becomes smaller and the larger the surface area. In addition, Methanyl Yellow is a kind of negatively charged (anionic) textile dyestuff. Thus, the negative group of Methanyl Yellow, the SOOH functional group, will bind to the Si-O-Si group present in the silica membrane.

The determination of pH optimum of Methanyl Yellow was conducted at pH = 1, 2, 3, 4, 5, 6, 7, and 8 the pH solution have effected Methanyl Yellow adsorption process. The optimum Methanyl Yellow can be absorbed maximum in silica membran 72.1346 % at pH 2 condition. At an acidic condition, the NH and -SOOH groups will be formed as $-NH^{3+}$ and $-SOOH^{2+}$. The cluster will bind strongly with the negative group ($-SO^{4-}$, $-SiO^{-}$) on the silica membrane. In addition, the acid conditions will change the nature of Methanyl Yellow to be a hydrophobic. The hydrophobic methane particle of Methanyl Yellow will be readily accumulated on the surface of the solution [16], making it more readily adsorbed by the silica membrane. However, if it is too acidic (pH 1) the adsorption becomes less efficient because at this pH the excess H^+ occurs, so the negative group of silica membranes can no longer bind to the amount of H^+ (saturated), consequently there are still many free H^+ ions in the solution causing competition between dyestuffs with free H^+ ions to be adsorbed by the silica membrane. It is characterized by the color change of Methanyl Yellow into a dark purple.

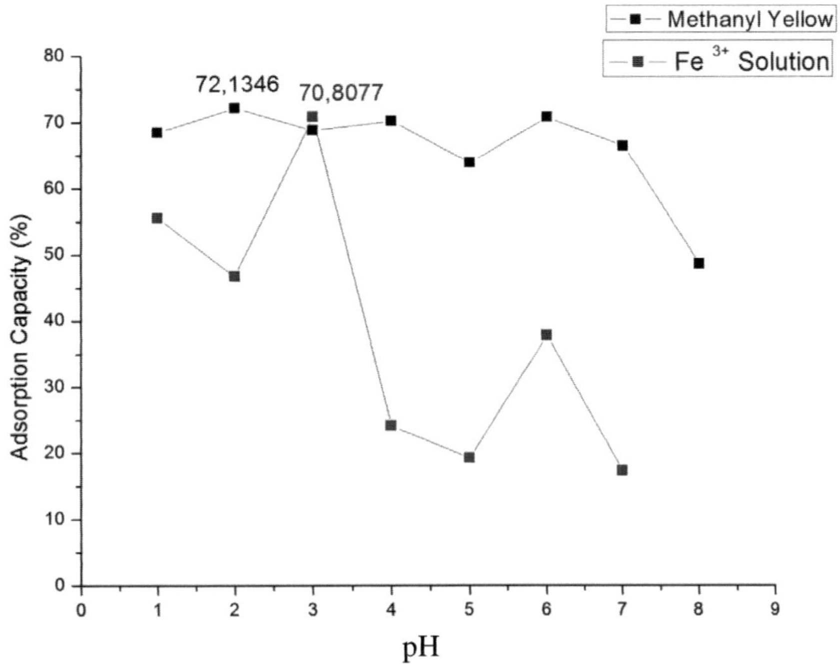

Figure 4. Optimum pH of adsorption on Methanyl Yellow and Fe^{3+}

The determination of the optimum pH of Fe^{3+} solution aims to find out how the influence of Fe^{3+} on Methanyl Yellow optimum pH adsorption. The effect of pH variation of Fe^{3+} solution in Methanyl Yellow adsorption is shown in Figure 4. Based on the Figure 4, the optimum pH of Fe^{3+} solution in Methanyl Yellow adsorption process is pH 3 with adsorption capacity = 70.8077%.. However, more OH^- or more alkaline causes the Fe^{3+} solution over the saturation point (Yellow $FeCl_3$ precipitate). With the formation of this precipitate, $FeCl_3$ covers the pores of the silica membrane. So the adsorption of methanyl Yellow is not optimal. At pH 3, the Fe^{3+} solution will change to $Fe(OH)_3$ which will be bound to the pores of the silica membrane. So that will increase the amount of positive charge on the silica membrane that will bind Methanyl Yellow (anionic compound).

4. Conclusion

Silica membrane have been successfully synthesized. X-Ray Fluorescence (XRF) analysis results showed that straw ash contains 82.12 % of silica (SiO_2). Based on the pore surface area, a silica membrane composed from PVA: silica = 5:5 have highest surface area (1.503 m^2/g) with an adsorption capacity value is 48.3077 %. Besides, based on the pore size distribution of PVA: silica (5:5) showed the narrow pore size distribution with the largest pore cumulative volume of 2.8 x 10^{-3} cc/g. The optimum pH for Methanyl Yellow adsorption is pH 2 with adsorption capacity = 72.1346%.

References

[1] Krim L, Sahmoune N, and Goma B 2006 *Am. J. Environ. Sci.* **2** 31-36
[2] Vijayaraghavan K and Yeoung-Sang Y 2008 *Biotech. Adv.* **26** 266-291
[3] S Dey and A Islam 2015 *Resour. Environ.* **5** 15-44
[4] Rashed, M N dan El-Amin AA 2007 *Int. J. Phys. Sci.* **2** 73-81

[5] Mondal S 2008 *Environ. Eng. Sci.* **25** 383-396
[6] Kuhad R C, Sood N, Tripathi K K, Singh A and Ward O P 2004 *Adv. Appl. Microbiol.* **56** 185-213
[7] El-Maghraby A and El Deeb H A 2011 *Global NEST J.* **13** 90-98
[8] Mahatmanti F W, Nuryono and Narsito 2016 *Indo. J. Chem.* **6** 45 - 52
[9] Kargbo R 2010 *Agric. Bio. J. North Am.* **3** 195-200
[10] Wattanasiriwech S, Wattanasiriwech D and Svasti J 2010 *J. Non-Crystal. Solid.* **356** 1228–1232
[11] Reed, James 1988 *Introduction to the Principles of Ceramic Processing* (Singapore: John Wiley and Sons)
[12] Kim J H, Lee K H 1998 *J. Mebrane Sci.* **138** 153-163
[13] Gilman J W, VanderHart D L and Kashiwagi T *Thermal Decomposition Chemistry of Polyvinyl alcohol)* (Gaithersburg: National Institute of Standards and Technology)
[14] Remmey G B 1994 *Firring Ceramics Advanced Series in Ceramics Vol. 2* (Singapore: World Scientific Publishing)
[15] Haber J, Block J H, and Delmon B, 1995 *Pure and Applied Chemistry* **67** 1257-1306.

IOP Conf. Series: Materials Science and Engineering **333** (2018) 012086 doi:10.1088/1757-899X/333/1/012086

Mechanical and Barrier Properties of Semi Refined Kappa Carrageenan-based Composite Edible Film and Its Application on Minimally Processed Chicken Breast Fillet

D Praseptiangga, B H Maimuni, G J Manuhara, D R A Muhammad

Department of Food Science and Technology, Sebelas Maret University (UNS)
Jl. Ir. Sutami 36A Surakarta 57126, Central Java, Indonesia

E-mail:dpraseptiangga@yahoo.com

Abstract. Kappa-carrageenan (KC) is one of the most interesting biopolymers that is composed of a linear chain of sulfated galactans and extracted from red seaweed, *Kappaphycus alvarezii*. It shows good potential for development as a source of biodegradable or edible films. However, KC films do not have good water vapor barrier properties, as they are intrinsically hydrophilic. Palmitic acid (PA) as hydrophobic material was incorporated into semi-refined kappa-carrageenan (SRKC) edible films in order to improve water vapor barrier properties. In this study, composite films based on SRKC incorporating PA were prepared and their applications on minimally processed chicken breast fillet were evaluated. Composite SRKC-based films with varying concentrations of PA (5%, 10%, and 15% w/w) were obtained by a solvent casting method. Their mechanical and barrier properties were investigated. Results showed that the incorporation of PA in films caused an increase in thickness, but decrease in water vapor transmission rate (WVTR) as the concentration of PA increased (from 5% to 15% w/w). Composite SRKC-based edible film incorporating 15% w/w of PA presented better water vapor barrier properties as compared to other films with 5% and 10% w/w PA incorporation. Thus, formulation containing 15% w/w PA was used as a wrapping material for film application on minimally processed chicken breast fillet. The application results showed that the incorporation of PA in film caused an effect ($p < 0.05$) on preventing of weight loss significantly compare to control (non-wrapping), however it did not significantly ($p > 0.05$) change the color of minimally processed chicken breast fillet.

1. Introduction

Global demand for meat is expected to increase by 44% by more than 400 million tonnes by 2030 with demand for poultry by 39% [1]. While increasing consumer demand for high food quality, fresh, nutritious, and easy to prepare caused minimally processed food is more preferred, including fresh chicken meat. However, the main obstacle for this commercial product is short shelf life due to high nutritional content and moisture content in fresh chicken meat resulting in the rapid growth of decaying microorganisms and pathogens. Chicken breast meat can be stored at a temperature not exceeding 4ºC to preserve its good condition for 4-5 days. While the broiler chicken pieces stored at a temperature of 28°C has a shelf life for 8 hours [2]. Edible films and/or coatings can improve the quality of poultry meat, by reducing the loss of moisture content on the meat surface, inhibiting discoloration due to the combination of lipid oxidation and myoglobin oxidation, and inhibiting the activity of decaying microorganisms and pathogens [3]. Therefore, the application of edible films and/or coatings on chicken breast fillet is expected to be alternative natural ingredients for packaging that can prolong the shelf life of fresh and/or minimally processed chicken breast fillet, so that consumer demand for high quality fresh chicken breast fillets can be fulfilled. Different structural materials have been used for the development of edible films and/or coatings, such as proteins, lipids, and polysaccharides. Film-forming materials can be utilized individually or as mixed composite blends. Proteins and polysaccharides used for their mechanical and structural properties, while hydrophobic substances (lipids (fatty acids), essential oils, and emulsifiers) used for providing good moisture barrier properties [4,5,6,7].

Kappa carrageenan is water-soluble polymer with a linear chain of partially sulphated galactans, which present high potentiality as film-forming material. Shorter and cheaper extraction method of red alga, *Kappaphycus alvarezii*, could be resulted in semi-refined kappa carrageenan (SRKC) of which a renewable resource and commercially available at a reasonable cost. Carrageenan-based edible films are brittle, thus a plasticizer to improve the mechanical properties is needed [6,8,9]. Sorbitol was employed as plasticizer in this work due to its several advantages such as retaining moisture and demonstrating stabile performance at high temperature [10]. Moreover, hydrophobic materials such as fatty acids could be used to enhance the barrier properties of this film. Thus, this present study deals with the formulation and characterization of a semi-refined kappa carrageenan-based composite edible film incorporated with palmitic acid and evaluation its application on minimally processed chicken breast fillet.

2. Experimental

Semi-refined Kappa-carrageenan, as the main component of the film matrix, was purchased from Galic Artabahari, Co., Ltd. (Cikarang Barat, Indonesia). Sorbitol (plasticizer) and palmitic acid (PA) were purchased from Brata Chemical (Surakarta, Indonesia) and Chemix Pratama (Yogyakarta, Indonesia), respectively. Zein was obtained from Laboratory of Food Chemistry and Biochemistry, Gadjah Mada University (Yogyakarta, Indonesia). Minimally processed chicken breast fillet was purchased from local market 'Pasar Demangan' (Yogyakarta, Indonesia). Aquadest was used for all sample preparations and all other chemicals used were in analytical grade.

Composite Semi-refined Kappa Carrageenan (SRKC)-based edible films were produced as described in Manuhara *et al.* with modification [11] and the SRKC concentration was selected based on the best properties of SRKC film as described in our previous study [9]. In this study, two groups of solutions were prepared. The first group, zein solution of 5% (w/w semi-refined kappa carrageenan) was prepared by dissolved in 10 ml of 96% ethanol, and stirred with a magnetic stirrer for 10 minutes at room temperature (28±2°C). Palmitic acid at various concentrations (5, 10, and 15%, w/w dry SRKC) was then added into zein solution, stirred with a magnetic strirrer for 10 minutes until it became completely homogenous (Solution A). The other group was the preparation of the main solution. It was prepared by dispersing 2 g of SRKC in aquadest (90 ml) and stirred under continuous stirring until dissolved and reached the temperature of 60°C. It was maintained for 10 minutes. Sorbitol liquid (1% v/v) was then added as the plasticizer and the Solution A were also poured into a main film solution. The mixture was then heated to 80°C and the temperature of mixture was then maintained at 80°C for 5 minutes on a hot plate with constant stirring to obtain a homogeneous solution. The film-forming emulsions were left for several minutes and stirred manually with stainless steel spatula to naturally remove dissolved air bubbles produced during stirring. The film-forming emulsions were then casted by pouring onto a casting plate (W x L x H: 15 × 23 × 2 cm) evenly and immediately to avoid the solution to turn into a gel. The plates were dried at 60°C in drying oven for 8 hours. The dried film were carefully peeled off from the plates and dried at room temperature (28 ± 2 °C) for 24 hours with 50% RH condition before analysis [6,9,11].

Mechanical (thickness, tensile strength (TS), and elongation at break (EAB)), and barrier properties of composite SRKC-based edible films were analyzed by methods as described in our previous work [6,9]. The water vapor transmission rate (WVTR) of the film specimens was measured according to a modified ASTM E96/E96M-05 method 1997 [12], and its analysis was conducted as mentioned in our previous work [6,9]. Weight loss for film application by wrapping was conducted by a method as described in our previous work and color was analyzed by a method of Andarwulan et al. [13]. A total of 3 samples were examined for each film type, thus, each film sample consisted of two replicate measurements. The experiments of factorial with a completely randomized design were used. The various concentrations of palmitic acid were the independent variable. All data were statistically evaluated by analysis of variance (ANOVA) procedure using SPSS software (version 16.0 for Windows software; SPSS Inc., Chicago, USA). Duncan's multiple range tests were used to compare the difference among the mean values for the film properties. Differences between the mean values were considered significant at $p < 0.05$.

3. Results and Discussion

Table 1. Mechanical and Barrier Properties of Composite Semi-Refined Kappa Carrageenan (SRKC)-based Films

Palmitic Acid (% w/w)	Thickness (mm)	TS (MPa)	EAB (%)	WVTR (g/ h.m^2)
5	$0.063^a \pm 0.00$	$16.82^a \pm 2.15$	$9.89^a \pm 4.54$	$8.93^a \pm 1.89$
10	$0.067^a \pm 0.01$	$13.84^a \pm 2.90$	$9.86^a \pm 3.47$	$8.26^a \pm 2.03$
15	$0.070^a \pm 0.00$	$14.64^a \pm 2.90$	$9.88^a \pm 2.77$	$6.51^a \pm 1.58$

TS: Tensile strength; EAB: Elongation-at-break; WVTR: Water Vapor Transmission Rate

In the present study, the thickness of composite Semi-refined Kappa Carrageenan (SRKC)- based films were ranged from 0.063 to 0.070 mm, and as a comparison (0% PA concentration), the thickness was 0.053 mm [11]. The lowest thickness of composite SRKC-films were obtained at the 5% w/w PA concentration, whereas the highest thickness was obtained at the 15% w/w PA concentration (Table 1.), however they did not show significantly differences ($p > 0.05$) on thickness parameter. Tensile strength (TS) and elongation-at-break (EAB) are included in mechanical properties of the composite SRKC-based films as shown in Table 1. TS indicate the maximum tensile stress that the film can sustain, while EAB is the maximum change in length of a test specimen before it breaks [14]. The low interval concentration (5%) of palmitic acid incorporation to the films did not give any significant effects to TS and EAB of composite SRKC-films. The percentage of TS and EAB were decreased and then increased. The decrease occurred at a 10% concentration of palmitic acid addition, and increased at a concentration of 15%. The TS of composite SRKC-films were ranged from 13.84 to 16.82 MPa, while EAB of composite SRKC-films were ranged from 9.86 to 9.89 %. As comparison (0% PA concentration), TS and EAB of composite SRKC-films were 21.14 MPa and 12.36% [11], respectively. Controlling mass transfer plays important role of food packaging for preservation of food. Hydrophobic compounds with their non-polar molecules can improve moisture migration, thus they could be employed as effective barriers [15]. The WVTR values of the composite SRKC films at 28±2°C and 50% RH are shown in Table 1. The addition of 15% w/w PA was the most effective in improving water barrier properties. However, no significant difference ($p > 0.05$) was observed in comparing to the films produced in this study. Interestingly, the decline in film WVTR is possibly to be associated with the film thickness. For instance, as the palmitic acid concentration increased, the thickness successively increased (Table 1), followed by a decrease in WVTR values (Table 1), probably due to the significant reduction of water mobility through the film. Formulation containing 15% w/w PA was then used as a wrapping material for film application on minimally processed chicken breast fillet.

Table 2. Weight Loss of Minimally Processed Chicken Breast Fillet

Treatment	Weight Loss (g/h)
Control (non-wrapping)	$0.101^{c}\pm0.002$
Wrap plastic	$0.003^{a}\pm0.000$
Composite SRKC-film	$0.067^{b}\pm0.002$

Weight loss in food is generally associated with decreased moisture content. The use of film as a wrapping material of food is determined by the film's ability to reduce moisture loss in food. The weight loss analysis on minimally processed chicken breast fillet was conducted by weighing chicken breast fillets, either wrapped with composite SRKC film, wrapping with cling wrap, or non-wrapping (control) for 8 hours. Chicken breast fillet was placed on acrylic cup and stored in desiccator containing saturated silica gel (RH 0%). Weighing is done from hour 0 to hour 8 every 1 hour. During storage there was weight loss of chicken breast fillet. Table 2. shows that wrapping application by wrap plastic and composite SRKC film inhibited weight loss of minimally processed chicken breast fillet during treatment. It can be seen that control (non-wrapping) demonstrated the highest of weight loss (0.101 g/h), while chicken breast fillet wrapped by composite SRKC film 0.067 g / h and by plastic wrap 0.003 g/h. Thus, packaging of minimally processed chicken breast fillet in wrapping treatment with composite SRKC film can reduce the weight loss of chicken breast fillet by about 2 times and had a significant effect (p <0.05) on reducing weight loss after 8 hours of storage compared to control (non-wrapping). While wrapping with wrap plastic able to minimize weight loss of chicken breast fillet approximately 30 times compared to control. Therefore, it may suggests that the composite SRKC film used in this study provides barrier properties against water vapor, i.e. by inhibiting the loss of water content in chicken breast fillets, but the wrap plastic provides the barrier properties better than composite SRKC film. This is because the wrap plastic is a plastic made of polyvinyl chloride, which is a flexible plastic that is obtained by the addition of plastic materials and has a smaller pore size (density) so as to withstand the rate of water vapor transmission [16].

Table 3. Color Analysis of Minimally Processed Chicken Breast Fillet

Treatment	L*	a*	b*	ΔE
Control (non-wrapping)	$42.06^{a}\pm1.59$	$3.26^{a}\pm1.37$	$8.50^{a}\pm1.82$	-
Wrap plastic	$44.20^{a}\pm1.57$	$1.53^{a}\pm0.87$	$6.10^{a}\pm0.14$	3.66
Composite SRKC-film	$42.95^{a}\pm1.27$	$3.28^{a}\pm0.56$	$7.30^{a}\pm0.98$	1.50

The color of the meat is a visual parameter that is first seen by consumers as an indicator of freshness. It is related to pigment concentrations, especially myoglobin and hemoglobin. The color change of meat during storage is influenced by several factors, including the level of metmyoglobin accumulation. Metmyoglobin produces a brown color in the flesh thus lowering the freshness and the level of consumer acceptance of meat. The rate of metmyoglobin accumulation is related to intrinsic factors, such as pH and fiber types in animal muscle, and extrinsic factors, such as post-slaughter treatment and carcass chilling treatment. Environmental factors also affect meat color changes, such as temperature, oxygen, light, microbial growth, and meat storage packaging. The discoloration of meat during storage occurs as a combination of pigment oxidation and lipid oxidation in meat [17]. Moreover, biopolymer-based edible films used as packaging material of meat products can reduce water loss, minimize lipid oxidation, and prevent discoloration [18]. This color analysis aimed to determine the effect of wrapping treatment by composite SRKC film in maintaining the color of minimally processed chicken breast fillet.

Table 3 shows that the color parameters (L*, a*, and b*) of minimally processed chicken breast fillet by wrapping and non-wrapping treatments were not significantly different ($p > 0.05$). Although the results showed no significantly difference, the lightness (L*) of chicken breast fillet by wrapping with composite SKRC film showed L* value higher than control (non-wrapping), and the highest value of L* was wrap plastic. Color changes of chicken breast fillet during storage may be associated with enzymatic and non-enzymatic reactions that cause degradation of myofibrillar proteins [19]. L* value in meat was associated with protein decomposition in the muscle of the meat [20]. The value a* shows the redness color of minimally processed chicken breast fillet, while the value of b* shows the yellowness color. Moreover, the ΔE value provides quantitative information of the overall color difference in each chicken breast fillet by wrapping treatment using wrap plastic or composite SRKC film, i.e. chicken breast fillet with non-wrapping treatment (control). Thus, the difference between control and wrapping application is symbolized by ΔE. The color difference value (ΔE) of chicken breast fillet wrapped by wrap plastic with control color is 3.66 indicating a moderate color difference (ΔE = 3-6). While, the color difference value (ΔE) of chicken breast fillet wrapped by composite SRKC film with control color is 1.50 indicating a small color difference (ΔE = 1-3). In this present study showed that chicken breast fillet wrapped by wrap plastic and composite SRKC film looked fresher when compared to control (non-wrapping). Non-wrapping chicken breast fillets will more easily change color to brown or dark because of direct contact with air. This is caused by the pigment of myoglobin undergoes oxidation to metmyoglobin. While the wrapped chicken breast fillet, either with wrap plastic or composite SRKC film was able to protect or minimize direct contact from outside. This corresponds to the function of the film as a barrier to the mass (such as moisture, oxygen, light, lipids) [21].

4. Conclusion

SRKC-palmitic acid (PA) composite films were successfully prepared and obtained after solvent casting and drying to the formation of an emulsion film. Composite SRKC-based film incorporating 15% w/w of PA demonstrated better water vapor barrier properties as compared to other films with 5% and 10% w/w PA incorporation. Thus, formulation containing 15% w/w PA was used as a wrapping material for film application on minimally processed chicken breast fillet. The application results showed that the incorporation of PA in film caused an effect ($p < 0.05$) on preventing of weight loss significantly compare to control (non-wrapping), however it did not significantly ($p > 0.05$) change the color of minimally processed chicken breast fillet.

Acknowledgements

This study was financially supported by the research grant of Sebelas Maret University (UNS), Surakarta, Indonesia, and the authors wish to express their gratitude for this research support.

References

[1] Australian Chicken Meat Federation Inc 2011 *The Australian Chicken Meat Industry: An Industry in Profile* 1-48
[2] Fernandez-Pan I, Carrion-Granda X, Juan I M 2014 *Food Control* **36** 69-75
[3] Khan M I, Adrees M N, Tariq M R, Sohaib M 2013 *Pak. J. Food Sci.* **23** (2) 71-79
[4] Espitia P J P, Du W, Avena-Bustillos R J, Soares N F F, McHugh T 2014 *Food Hydrocoll.* **35** 287-296
[5] Tavassoli-Kafrani E, Shekarchizadeh H, Masoudpour-Behabadi M 2016 *Carbohyd. Polym.* **137** 360–374
[6] Praseptiangga D, Fatmala N, Manuhara G J, Utami R, Khasanah L U 2016 *AIP Conf. Proc.* **1746** 020036
[7] Praseptiangga D 2017 *IOP Conf. Ser.: Mater. Sci. Eng.* **193** 012003
[8] Wirawan S, Kompiang A, Prasetyo E 2012 *Reaktor* **14** (1)
[9] Manuhara G J, Praseptiangga D, Muhammad D R A, Maimuni B H 2016 *AIP Conf. Proc.* **1710** 030043
[10] Shoukui He, Wang Y, Sun Y, Chen S, Zhang Y, Ying M 2015 *Adv. J. Food Sci. Technol.* **9** (7) 523-528

[11] Manuhara G J, Marseno D W, Santoso U 2008 *J. Teknol. Hasil Pert.* **1**(2)
[12] Ozdemir M, Floros J D 2008 *J. Food Eng.* **86** 215–224
[13] Andarwulan N, Kusnandar F, Herawati D 2011 *Analisis Pangan* (Jakarta: Penerbit Dian Rakyat)
[14] Bourbon A I, Pinheiro A C, Cerqueira M A, Rocha C M R, Avides M A, Quintas M A C, Vicente A A 2011 *J. Food Eng.* **106** 111-118
[15] Peroval C, Debeaufort F, Despera D, Voilley A 2002 *J. Agric. Food Chem.* **50** 3977-3983
[16] Buckle K A, Edwards R A, Fleet G H, Wootton M 2010 *Ilmu Pangan* (Jakarta: Penerbit UI Press)
[17] Troy D J, Kerry J P 2010 *Meat Sci.* **86** 214-226
[18] Gennadios A, Weller C L L 1990 *Food Technol.* **44** (10) 63-69
[19] Jouki M, Seyed A M, Farideh T Y, Arash K, Naimeh K 2014 *Food Sci. Human Wellness* **3** 65-72
[20] Latou E, Mexis S F, Badeka AV, Kontakos S, Kontominas M G 2014 *LWT-Food Sci.Technol.* **55** 263-268
[21] Krochta J M 1992 *Control of Mass Transfer in Food with Edible Coatings and Films* (Florida: CRC Press)

IOP Conf. Series: Materials Science and Engineering **333** (2018) 012087 doi:10.1088/1757-899X/333/1/012087

Bioplastic from Chitosan and Yellow Pumpkin Starch with Castor Oil as Plasticizer

M Hasan[1], R F I Rahmayani[1], Munandar[1]
Department of Chemistry Education, Faculty of Teacher Training and Education, Syiah Kuala University, Darussalam Banda Aceh 23111

E-mail: muhammadhasan.kimia@unsyiah.ac.id

Abstract. This study has been conducted on bioplastic synthesis of chitosan and yellow pumpkin starch (*Cucurbita moschata*) with castor oil as plasticizer. The purpose of this study is to determine the characteristics of the effect of chitosan and starch composition of pumpkins against solvent absorption, tensile strength and biodegradable. The first stage of the research is the making of bioplastic by blending yellow pumpkin starch, chitosan and castor oil. Further, it tested the absorption capacity of the solvent, tensile strength test, and biodegradable analysis. The optimum absorption capacity of the solvent is obtained on the composition of Pumpkin/Chitosan was 50/50 in H_2O and C_2H_5OH solvent. Meanwhile the optimum absorbency in HCl and NaOH solvents is obtained by 60/40 composition. The characterization of the optimum tensile strength test was obtained on the 40/60 composition of 6.787 ± 0.274 Mpa and the fastest biodegradation test process within 5-10 days occurred in the 50/50 composition. The more chitosan content the higher the value of tensile strength test obtained, while the fastest biodegradation rate occureds in the composition of yellow pumpkin starch and chitosan balanced 50:50.

1. Introduction

The use of plastic from petrochemical ingredients tends to increase. The Plastics have been used for various needs, such as food packaging, bottle drinks, and various household appliances and toys. Plastics from petrochemical raw materials can not be degraded either by the influence of solar radiation or microbial decomposers [1]. As a result, plastic waste from these products is also increasing along with the increasing use of plastic products, especially plastic as a packing material. Possible solutions to this problem have been given, such as recycling scrap plastic and burning plastic waste. Recycle plastic waste is difficult to obtain products with the same strength as plastic of origin because it is difficult to separate the plastic which is from the type of polymer. Burning plastic is also not a wise move, because it can cause air pollution due to the production of toxic gases, such as cyanide gas, chlorine, and carbon monoxide gas [2].

Other efforts continue to be encouraged is the synthesis of biodegradable plastics, namely strong plastic when used and immediately biodegradable or damaged when thrown away as garbage. In addition, the expected biodegradable plastic characteristics are that the resulting biodegradation product is not toxic to the environment. Synthesis of biodegradable plastics can be done through a number of alternatives, including copolymerization [3], biosynthesis [4], and poliblending between natural polymers and synthetic polymers [5-6]. Copolymerization is a chemical microstructure modification. This method has been able to create a variety of polymers as raw material for

IOP Conf. Series: Materials Science and Engineering **333** (2018) 012087 doi:10.1088/1757-899X/333/1/012087

biodegradable plastics, but considered less economical. Biosynthesis is a plastic synthesis by utilizing microbes. The disadvantage is that in addition to the resulting low yield of plastic is also difficult to regulate the resulting polymer chain structure, so the properties of plastic produced is usually difficult to control. The biodegradable plastic synthesis of highly economically viable prospects is through blending of natural polymers, in addition to being biodegradable in nature the polymer feedstock is readily available and renewable. A lot of very up-to-date research has been developed to ecologically friendly plastics through natural polymer blending. The environmentally friendly plastics through corn starch blending and thermoplastic chitosan have been synthesized [7]. The results show that the plastic film obtained has good thermal endurance, although the mechanical properties of polyblend are lower than the original chitosan. Futhermore, bioplastics can be obtained through blending of Polylactic acid and Poly-3-hydroxybutyrate [8]. The resulting polyblend show an increase in elongation at break. This characteristic is indispensable for plastic films, especially those used as food packaging.

Totally biodegradable plastics have been synthesized through blending of tapioca starch and chitosan with plasticizer from palm oil [9]. The results show that, in addition to being biodegradable, the polymer may be prepared in the form of films that can be used for food packaging. Other publication also reported the synthesis of poly (urethane-esters) of the lactone polyester prepolymer, and the resulting polymer is biodegradable [10]. Until now there has been limited publication about bioplastics from local raw materials of Aceh Province, such as pumpkin, taro, and others. Therefore, this research focuses on making bioplastic through blending of chitosan and starch from Aceh local raw material, that is from yellow pumpkin starch by using castor oil as natural plasticizer. The result of this research will contribute to the knowledge gap in the literature.

2. Experimental

2.1 Materials and equipment

The materials used in this study were yellow pumpkin starch, chitosan, castor oil, ethanol , sodium hydroxide pH 10, hydrochloride acid pH 5, acetic acid 5%, microbial Pseudomonas aeruginosa, nutrients agar (NA), Aquades, aluminum foil. The tools used in this research are glassware, hotplate, analytical scales, petri dish, tensile test equipment (IK Force Tester MCT-2150), needle, oven, spirit burner, laminar flow and autoclave.

2.2 Methods

2.2.1 Preparation of pumpkin yellow starch. Pumpkin fruit is firstly cleaned and separated from the hard skin and seeds. Then, it is washed using water to remove the fruit juice which is located in the middle of the pumpkin fruit. The next process is the pumpkin fruit is cut in small piece and adding enough water. Then it is blended until it becomes mush. The yellow pumpkin pulp that has been obtained is then filtered and allowed to stand for 50 minutes, to get the precipitate. After sterilizing for 50 minutes, the sediment still mixes with the water is filtered back so as to obtain wet starch deposition. The wet starch deposition is dried by an oven to obtain starch which is ready for use as a bioplastic feedstock.

2.2.2 Bioplastic synthesis. Bioplastic synthesis was performed by weighing chitosan and yellow Pumpkin starch with varying predetermined mass (Table 1). Then the chitosan is dissolved into 5% acetic acid while stirring with a magnetic stirrer. The same thing is also applied on yellow pumpkin starch until all the starch dissolves. Both solutions are mixed in a beaker and added with 15% of castor oil. After the sample begins to form the gel, the sample is poured into the mold and dried over the hotplate at 75 ° C until all the solvents evaporate and plastic films are obtained.

2.2.3 Tensile test. Bioplastic samples were cut to lengths of 4 cm and 1 cm wide. The measurement condition is a tensile velocity of 100 mm / min with a maximum load of 500 N. The sample is clamped

Table 1. Experimental Design on Bioplastic Synthesis

Code	Chitosan/starch (%)	Chitosan mass (g)	Starch mass (g)	Total mass (g)
CP46	40/60	0.56	0.84	1.40
CP55	50/50	0.70	0.70	1.40
CP64	60/40	0.84	0.56	1.40

on a tensile test apparatus, and runs the apparatus according to a predetermined condition until the sample is disconnected. Repeated procedure above for each sample is done 3 times tensile test. Recorded tensile test results are presented in the form of tables between tensile strength and extension. The magnitude of the breaking strength of a material depends on the amount of loading applied and the cross section of the material itself according to the following formula:

$$\sigma = \frac{F}{A} \quad (1)$$

where σ = strength (Mpa), F = load at break (N), A = the cross-sectional of the polymer material (mm^2). While Young Modulus determined using the following formula.

$$\text{Young Modulus (Mpa)} = \frac{\sigma}{\varepsilon} \quad (2)$$

where ε = elongation at break, ΔL = the difference between the length at break, and LO = the initial length.

2.2.4 Solvent uptake test. This test is based on the method performed by Pimpan, et al., 2001. Plastic is cut to the size of 1.0 cm x 1.0 cm. The plastic that has been cut is then weighed with an analytical balance sheet. The plastic is put into a 10 ml beaker filled with 5 ml solvent, then sterilized in room temperature. Every minute, the plastic is taken, the solvent on the plastic surface is wiped with a tissue, then weighed. The absorption capacity of the solvent is calculated using the following formula:

$$\text{Solvent uptake} = \frac{W - W_0}{W_0} \text{ x } 100\% \quad (3)$$

where: W_0 = weight of dry sample while W = weight of sample after immersion in the solvent

2.2.5 Biodegradation test. Bioplastic samples were cut to 1 x 1 cm size then sterilized using UV light in the laminar flow for 20 minutes. Sterilized bioplastic samples were incorporated into the biodegradation medium then incubated for 5, 10, 15 and 20 days

3. Result and Discussion

3.1 Bioplastic film

Bioplastic film obtained in this study is shown in Figure 1, which is a colored plastic film from light yellow to orange. The resulting of film color depended greatly on the composition of the pumpkin starch and chitosan. The color of bioplastic with a composition of at least 40% chitosan looks darker than the other two variations of bioplastic variation. This is influenced by the pigment base yellow pumpkin starch. The appearance of color pigment caused by starch used has not undergone bleacing and purification process.

Figure 1. Film bioplastic from pumpkin yellow starch and chitosan: (a) CP46, (b) CP55, (c)CP64

3.2 Solvent uptake.

Solvent uptake test is carried out to determine the resistance of a bioplastic sample to some type of solvent. The lyophobicity test is performed by weighing the bioplastic mass before and after dissolvedinto the solvent. The solvents used in this study were H_2O, C_2H_5OH, HCl, and NaOH. The observed results of the solvent absorption by bioplastics are shown in Table 2.

Table 2. Absorption capability bioplastic

code	composition	Quantity absorption(%)			
	chitosan/starch	H_2O	C_2H_5OH	HCl (pH=5)	NaOH (pH=10)
CP46	40/60	13.903	0.258	22.032	19.451
CP55	50/50	8.833	0.214	11.333	12.047
CP64	60/40	9.628	0.485	10.771	9.914

Based on the data in Table 2, the lowest absorption capacity of H_2O and CH_3COOH solvents was produced by bioplastic films at a 50/50 composition. While the lowest absorption capacity of HCl and NaOH solvents is produced by 60/40 bioplastic film. This is due to the basic nature of chitosan that is hydrophobic, so that in a balanced composition between starch and chitosan cause molecular interactions that can decrease the hydrophilic nature of starch [11]. In the HCl and NaOH solvents, the highest adsorption capacity of the solvent is produced by the film on bioplastic variations in which the more yellow squash starch content. It is caused by hydrophilic amylose so easily soluble in water, strong base solution, slightly soluble in dilute acid [12].

3.3 Tensile test

The tensile strength of the film is a specific strength obtained from the maximum pull before breaking/tearing. This measurement is to determine the magnitude of force required to reach the maximum pull point on every surface area of the film. The result of tensile strength test from bioplastic of pumpkin starch can be seen in Table 3.

Table 3. Tensile strength, elastisity, and young modulus of bioplastic film

No	chitosan/starch	Tensile strength (Mpa)	Elongation (%)	Young modulus (Mpa)
1	40/60	6.787 ±0.274	13.451 ±3.709	6.093
2	50/50	2.912 ±0.470	13.278 ±4.413	2.176
3	60/40	2.563 ±1.055	7.285 ±1.135	5.263

Based on the data in Table 3, the tensile strength of bioplastic samples is influenced by the variation of the composition of yellow pumpkin starch and chitosan, where the highest tensile strength is found in the composition of starch/chitosan 40/60 of 6.787 Mpa. While the lowest tensile strength is in the 60/40 composition of 2.563 Mpa. Based on the results of this tensile test, it showed that the composition of chitosan is very influential on the increase value of tensile strength bioplastik starch pumpkin produced. This is due to the greater chitosan composition that encourages the formation of hydrogen bonds with stronger starch molecules [13].

Young modulus is the ratio between the tensile strength at break of the extension. The lower the modulus of young value the more elastic the bioplastic samples are tested and vice versa if the higher the modulus of young value the less the elasticity value and the tensile strength value is very high [14]. Bioplastic samples with the best elasticity are 50/50 compositions of 2.076 MPa. This is triggered by the peak stress on the strain at break (mm), the lower the peak load and the value of the strain value at breaks are greater the more elastic the bioplastic is produced. The graph of tensile test results by using the tool IK Force Tester MCT-2150 can be seen in Figure 2.

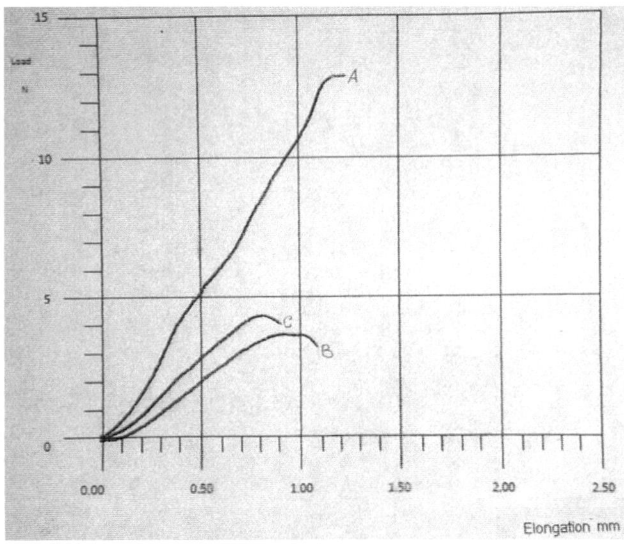

Figure 2. Tensile test graph, (A) CP46, (B) CP55 and (C) CP64

Based on Figure 2, the value of young modulus is greatly influenced by the chitosan content in the bioplastic film. The higher the chitosan content the modulus young value is also higher. The plastic film of the results of this study is better with the results achieved by publication on the manufacture of plastic films of corn starch and chitosan with a composition of 50/50, resulting from a modulus young value of 105 Mpa [15]

3.4 Biodegradation test

Biodegradation test is done by incubating bioplastic film in culture medium of pseudomonas aeruginosa. The film is incubated within a period of 5 days to 20 days. The surface of the biodegradation film at various time ranges is presented in figure 3. Based on figure 3, it is seen that the sample began to change after 5 days incubation in psedomonas aeruginosa media. After 10 days of incubation seen in the media there has been a change of color, especially films with more starch content. After a 30-day biodegradation process, the film has been totally biodegradable, where the film is difficult to separate from the media due to damage. Films with larger starch content are easier to biodegradable. This is due to the presence of glucosidic bonds in the amylose and amylopectin units. The existence of this bond causes the starch molecules more easily biodegraded through hydrolyzed mechanisms.

Figure 3. Biodegradation film: a = CP46 sample, b= CP55 sample, c = CP64, where the number of 0, 1, 2, 3, and 4 refers to before biodegradation, after 5 days biodegradation, after 10 days biodegradation, after 15 days biodegradation, and 20 days biodegradation

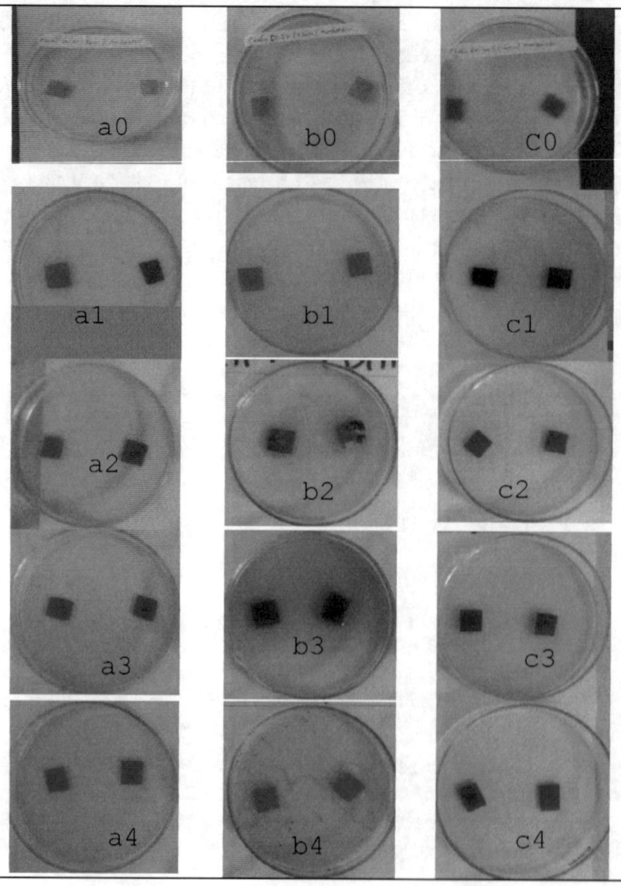

4. Conclusion

Bioplastics of yellow pumpkin and chitosan have been successfully synthesized by polblending pathway and its can be printed in the form of thin films. The resulting film has a biodegradable characteristic for 20 days incubation in the medium of psedumonas aeruginosa. The mechanical properties of bioplastics are highly determined by the composition of starch and chitosan, where the higher the chitosan content, the value of tensile strength and modulus of young increases. The adsorption capacity of water solvent, hydrochloric acid and sodium hydroxide is higher with the increase of starch content in plastic composition.

Reference

[1] Khalil HPSA, Tehrani MA, and Davoudpour Y 2013 *J Reinf. Plast. Compos.* **32** 330-356
[2] La Mantia FP and Morreale M 2011 *Compos Part A: Appl. Sci. Manuf* **42** 579-588
[3] Genovese L, Lotti N, Gazzano M, Siracusa V, Rosa M D, and Munari A 2016 *Polym. Deg. and Stab.* **132** 191-201
[4] Urtuvia V, Villegas P, Gonzalez M, and Seeger M 2014 *Macromolecules* **70** 2008-17
[5] Navarro-Baena I, Sessini V, Dominici F, Torre L, Kenny J M, and Peponi L 2016 *Polym. Deg. and Stab.* **132** 97-108
[6] Ali H E, and Abdel-Ghaffar A M 2017 *Rad. Phys. and Chem.* **130** 411-20
[7] Mendes J F, Paschoalin R T, Carmona V B, Neto A R S, Marques A C P, Marconcini J M, Mattoso L H C, Medeiros, E S, and Olivera J E 2016 *Carbohydrate polym.* **137** 452-58
[8] D'Amico D A, Montes, M L I, Manfredi L B, and Cyras V R 2016 *Polym. Test.* **27** 22-28

[9] Hasan M, Rusman, and Hanum L 2010 *J. Purifikasi* **11** 171-76
[10] Arcana I M, Bundjali B, Hasan M, Zulfikar M A, and Hariyawati K 2011 *Polym. Int.* **60** 1535-40
[11] Ghani S W A, Bakar A A, and Samsudin S A 2016 *J. Plast. Film and Sheeting* **32**(2) 140-62
[12] Peelman N, Ragaert P, De Maulenaer B, Adons D, Peeters R, Cardon L, van Impe F, and Devlieghere F 2013 *Trend in Food Sci. Techn.* **32** 128-41
[13] Akter N, Khan R A, Tuhin M O, Hagu M E, Nurnabi M, Parvin F, and Islam R 2014 *J. Therm. Comp. Mat.* **27**(7) 933-48
[14] Zuo Y, Gu J, Yang L, Qiau Z, and Zhang Y 2016 *J. Therm. Comp. Mat.* **29**(3) 397-409
[15] Bof, Julieta M, Bordagarauf V C, Locaso D E, and Garcia M A 2015 *J. Food Hydrocol.* **51** 281-94

IOP Conf. Series: Materials Science and Engineering **333** (2018) 012088 doi:10.1088/1757-899X/333/1/012088

The effect of recycled Natural Rubber Glove (rRG) Plasticizers to Deflection and Flexural Strength Properties of PP/MMt/rRG Smart Composites and Its Inflammability

N S Suharty[1], H Ismail[2], K Diharjo[3], D S Handayani[1], L N M Z Saputri[1], and N Ariesta[1]
[1]Chemistry Department, Faculty of Mathematics and Natural Sciences, Sebelas Maret University, Jl. Ir. Sutami 36A Surakarta, Indonesia
[2]School of Materials and Mineral Resources Engineering, Engineering Campus, University Sains Malaysia, 14300 Nibong Tebal, Pulau Pinang, Malaysia
[3]Faculty of Engineering, Sebelas Maret University, Surakarta, Indonesia

E-mail: nsuharty@yahoo.com

Abstract. Had been synthesized PP/rRG/MMt+ZB smart material composite in solution reactive processes with various rRG concentration. The addition of rRG plasticizers will improve the deflection properties and increase the filler capacity MMt loading to reach the optimum concentration. The addition of 3% rRG is capable of loading filler capacity MMt to 23% as the optimum condition. At the optimum conditions it can increase the deflection (Defl) and flexural strength (FS) up to 16% and 15% respectively compared to that of the composites without rRG. The rRG plasticizer serves as a bio-compatibilizer that can reduce surface tension of the mixture and leads to decrease the Defl., follow by the increase of loading filler capacity and well interaction finally can increase the FS properties. The increase of loading filler MMt up to 23% can also improve the inflammability of the composites. Time to Ignition (TTI) increase by 5% and Burning Rate (BR) decrease by 4.5% compared to that of the composites which is containing MMt 20% without rRG.

1. Introduction

The availability of metals in nature is increasingly depleted while the production of automotive that uses metal as its raw material production is increasing by 3.2% every year [1]. Therefore its need alternative material of metal substitute which have strong, light and flame resistance properties, that is composite. The use of 35% lightweight composite on automotive can save 12-20% fuel usage [2]. The composite is very suitable to be applied in automotive components such as upholstery, door panel, interior and car body [3]. Increased mechanical properties and flame resistance of a composite can be accomplished by the addition of clays such as montmorillonite (MMt) [4]. The type of MMt clay is abundant in Indonesia and can be used as a composite filler material. Lee made PP/WF/MMt composite with PP-g-MA coupling compound internally mixer producing composite reported to have high in tensile strength (TS) and melting temperature [5].

Elastic composite engineering is one of them with the addition of a composite-like plasticizer. Wang made the PP/EPDM/SGF composite to produce composites with high elasticity and mechanical

[1] To whom any correspondence should be addressed.

properties [6]. Xie and others make composite PP/styrene-butadienerubber (SBR)/SGF produce composites with high elasticity and mechanical properties [7]. Balakrishnan makes composite PP/epoxydized natural rubber (ENR)/OMMt produces composites with high elasticity and mechanical properties [8]. Ismail and Suryadiansyah made a composite PP/recycle natural rubber (rNR) reported to have high elasticity and mechanical properties compared to PP [9]. Riyajan make composite PS/recycle rubber glove (rRG)/compostable leaf cane produced with elasticity and high mechanical properties compared to starting materials [10]. The objective of the research is to study the effect of adding rRG plasticizer to rPP/DVB/PP-AA/MMt+ZB composites to deflection, flexural strength and its inflammability properties.

2. Experimental

2.1 Materials

Recycled polypropylene (rPP) was collected from cup recycled plastics packaging of mineral water in Surakarta, Indonesia. The rPP was washed and then chopped approximately in 5 mm x 2 mm. The montmorillonite (MMt) clay was obtained from Soka tile Kebumen Indonesia. The MMt prepared by crushed and sieved with a 200 mesh sieve and calcined at 800°C for an hour. The recycle natural Rubber Glove (rRG) was collected from Surakarta, Indonesia. Preparation of rRG by decrosslink processed using two roll mill and then crushed and sieve with a 40 mesh. All chemicals were pro-analysis without any purification. The acrylic acid (AA) purchased from Schuchai, divinyl benzene (DVB) purchased from Aldrich, xylene derived from PT. Brataco in Surakarta, Indonesia. While benzoyl peroxide (BPO) and zinc borate (ZB) purchased from E.Merck.

2.2 Composites manufacturing process

Fragments of rPP and a number of rRG are mixed in the shaker and then inserted in a chamber containing boiled xylene. The system is equipped with an oil bath heated at 135 °C., thermometer, mechanical stirrer, and N2 gas flow of 0.25 Barr/hr so that the rPP melts and then BPO initiator were added. Furthermore, PP-AA, DVB and MMt are inserted to the mixture compounds and the heating reactive process was carried out for 1 hour. The composite making composition is in accordance with the formula shown by Table 1.

Table 1. Formulation of rPP/DVB/PP-AA/rRG/MMt+ZB Composites

Formulas	Code	Composition (%)				
		rPP	rRG	DVB	PP-AA	MMt
rPP	MF0	100	0	0	0	0
rPP/DVB/PP-AA/MMt20%+ZB	MF1	65	0	0.1	15	20
rPP/DVB/PP-AA/0%rRG/MMt23+ZB	MF2	65	0	0.1	15	23
rPP/DVB/PPAA/1%rRG/MMt23+ZB	MF3	60	1	0.1	15	23
rPP/DVB/PP-AA/3%rRG/MMt23+ZB	MF4	60	3	0.1	15	23
rPP/DVB/PPAA/5%rRG/MMt23+ZB	MF5	60	5	0.1	15	23
rPP/DVB/PP-AA/7%rRG/MMt23+ZB	MF6	60	7	0.1	15	23

2.3 Testing of materials

The characterization of a composite functional group was done by the FTIR Prestige 21 Shimadzu. The testing of Flexural Strength (FS), Deflection, and Bending Elasticity (BE) was carried out using universal testing machine (UTM) model JTM-UTS510 by following ASTM D790 with a three point bending method. The testing of flame resilience of composites in accordance with ASTM D365 to determine the value of time to ignition (TTI) and burning rate (BR).

3. Results and Discussion

3.1 FTIR Analysis Fucntional Groups of Composites

The result of FTIR characterization on MMt was obtained Si-O at 1084cm^{-1}, Al-O-Si absorption at 686 cm^{-1}, Si-O-Si uptake at 480 cm^{-1}and –OH absorption at 3447 cm^{-1}. In the PP/DVB/PP-AA/rRG composites, the typical uptake of rPP is the >C-H - bond of methyne group at 1723 cm^{-1} and vibration stretching of carbonyl group (>C=O) at 1718 cm^{-1} indicates carbonyl acid of PP-AA [4,11,12,13]. In composites of PP/DVB/PP-AA/rRG/MMt+ZB, functional group of Al-O$_{str.}$ can be indicated as the absorption of Al-OH at 867 cm^{-1}, absorption of Si-O$_{str.}$ at 1088 cm^{-1}, and Si-O-Si uptake at 486 cm^{-1} indicates the presence of MMt with the main content of SiO_2 and Al_2O_3 [14,15]. The broad vibration stretching of O-H bond at 3460 cm^{-1} that indicate hydrogen bond of >Al-OH of MMt. This is supported by the previous research [14,16,17]. From the results of FTIR spectra analyzes mentioned above, the composites using raw materials of rPP, rRG, PP-g-AA, ZB and MMt have been clearly indicated.

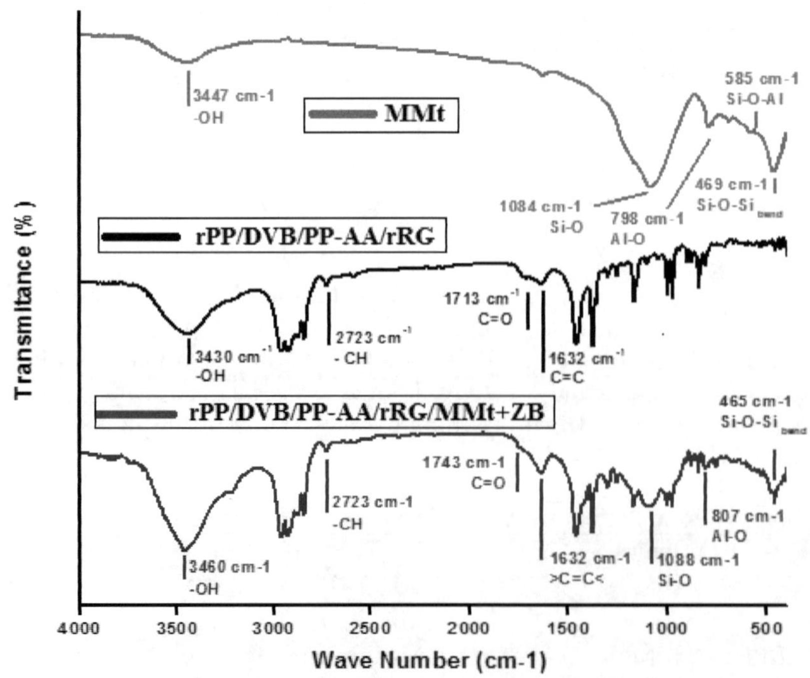

Figure 1. FTIR spectra of (a) MMt, (b) PP/DVB/PP-AA/rRG, (c) PP/DVB/PP-AA/rRG/MMt+ZB

3.2. Mechanical Properties of Composites

The mechanical properties of composites are presented in Figure 1-6. In Figure 2-3 show the effect of plasticizers rRG to strength properties of PP/DVB/PP-AA/rRG/MMt+ZB (MF2-MF6 containing 23% MMt with various concentration of rRG). The optimum FS of PP/DVB/PP-AA/MMt+ZB composites (MF1, without rRG plasticizer) have content 20% MMt was 39.32 MPa, however, when the contain of MMt added to 23% (MF2) the FS decreased to 5% compared to that of MF1. In the addition of 1% and 3% of rRG plasticizer (MF3 and MF4) containing 23% MMt can increase 8% and 15% of FS compared to MF1, respectively. This happens because of the addition of a rRG plasticizers on the composite will reduce the surface tension of the mixture, it causes the mixture to become softer, follows by the increase in elasticity and consequence the value of the deflection will increase. This softer material properties can receive a larger MMt filler capacity of up to 23% under optimum conditions. Increased the MMt filler automatically functions as an reinforcement thus increasing the FS value of composite [4,13,18]. While the addition of 5% and 7% of rRG (MF5 and MF6) containing

23% MMt decreased 11% and 19% of FS compared to MF1. The decrease in mechanical properties are due to the greater content of plasticizer material resulting in too soft material and finally lead to lower mechanical properties.

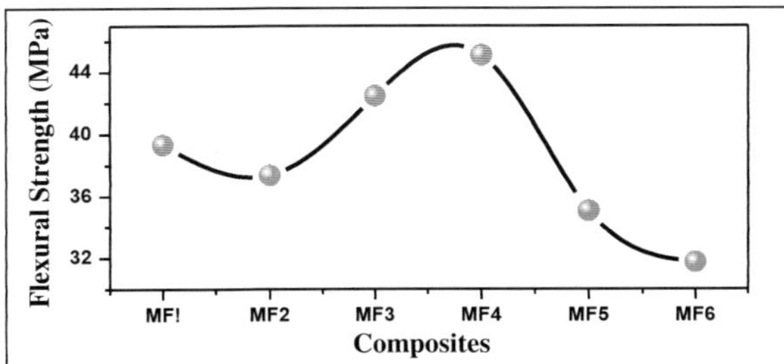

Figure 2. Flexural Strength (FS) of the composites (MPa)

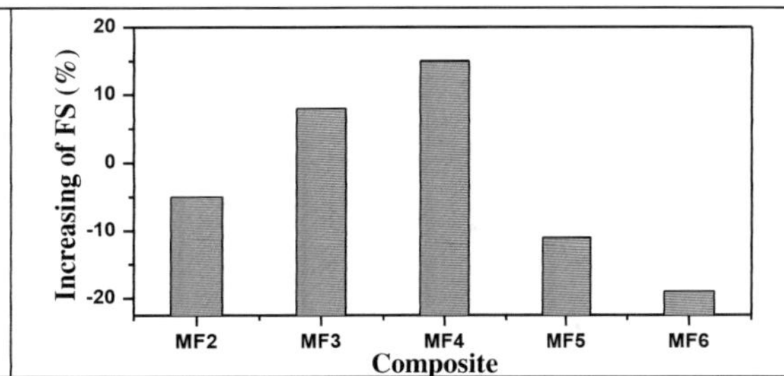

Figure 3. Comparison of increasing FS (%) to that of MF1

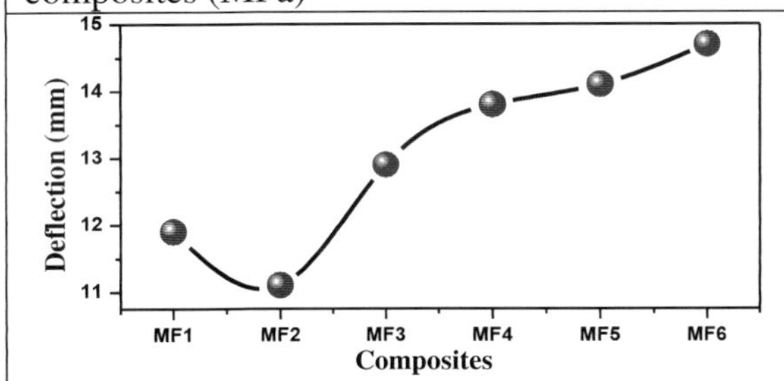

Figure 4. Deflection of the composites (mm)

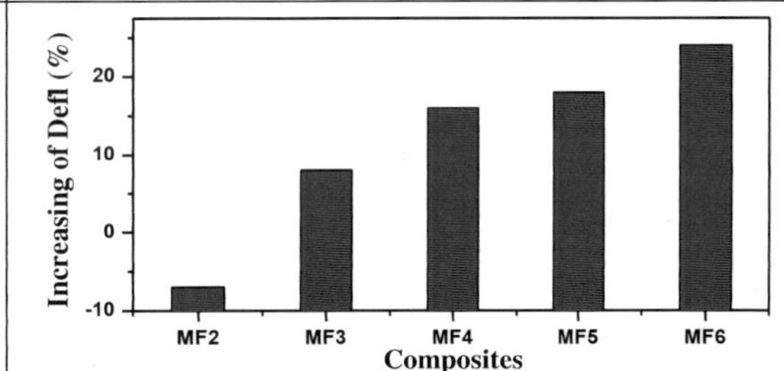

Figure 5. Comparison of increasing deflection (%) to that of MF1

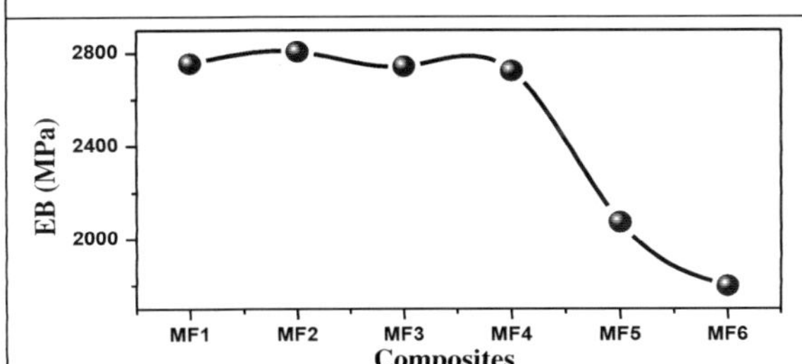

Figure 6. Bending Elasticity (BE) values of the composites (MPa)

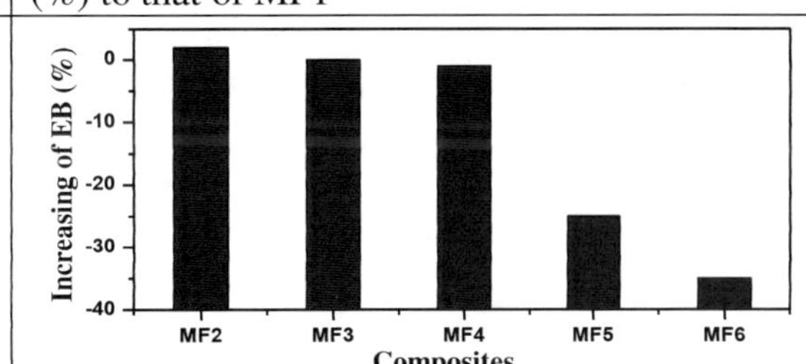

Figure 7. Comparison of increasing BE (%) to that of MF1

Figure 4 and 5 show the deflection (Defl) value of composites MF1-MF6. The Defl value of MF1 without rRG plasticizer and have content 20% MMt was 11.9 mm, however, when the MMt filler added to 23% of MMt (MF2), the Defl. value decreased 7% compared to that of MF1. This is because, the addition of filler above the optimum capacity conditions will result in interaction between raw material does not occur well, consequence in the brittle properties of the composite. This is supported by Othman which is reported that the higher filler content will decrease the value of strain (elasticity) which eventually decrease the deflection value [19]. The composites (MF2-MF6) which have content 1%, 3%, 5%, and 7% of rRG plasticizer respectively and contain 23% of MMt filler, have the Defl. value increase to 8%, 16%, 18%, 24% respectively compared to that of MF1. This increase is supported by the previous work, that the higher rubber will increase the value of the strain resulting in an increase in deflection [9,10]. Figure 6 display the bending elasticity (BE) and Figure 7 display its comparison to that of MF1. The stiffness properties of the material will decrease as elasticity increases. Likewise, what happens in Figure 6 and 4. The BE values of the composites in figure 6 shows a decreasing profile when Defl. values in Figure 4 increases. The BE composite value of MF1 without plasticizer is 2753.5 whereas MF2 is 2804.8 MPa. Its because the Defl value in MF2 is lower than MF1, due to the amount of filler MMt on HM2 is larger then MF1. Profile of composite EB value of MF2 to MF6, showing downward trend with respect to rRG plasticizer content go up (Defl value increase).

3.3 Flammability Properties

Figure 8-13 show the flammability of the rPP and filled composites (formulas MF1-MF6) measured by horizontal burning test following ASTM D635. The effect of addition of rRG plasticizer to time to ignition (TTI) and burning rate (BR) of rPP/DVB/PP-AA/rRG/MMt+ZB composites (MF2-MF6 which is containing 23% of MMt in various concentration of rRG) were showed in Figure 8-11.

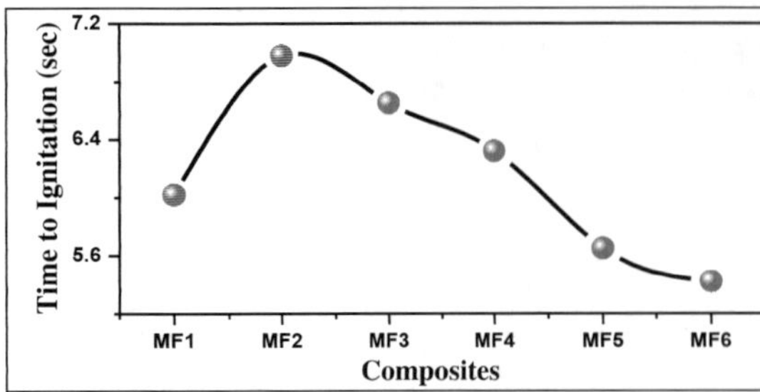

Figure 8. TTI of the composites sec)

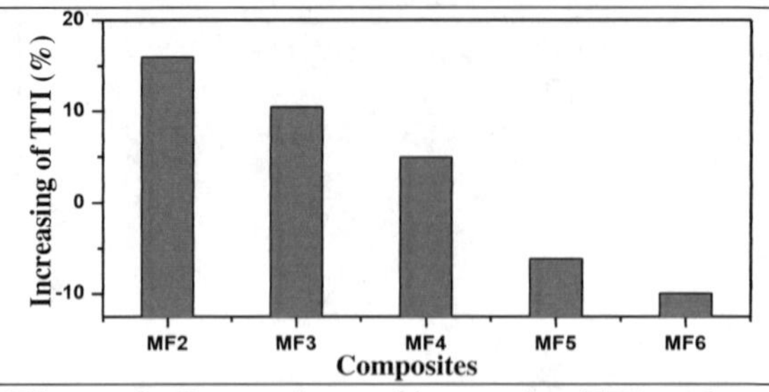

Figure 9. Comparison of increasing TTI to that of MF1 (%)

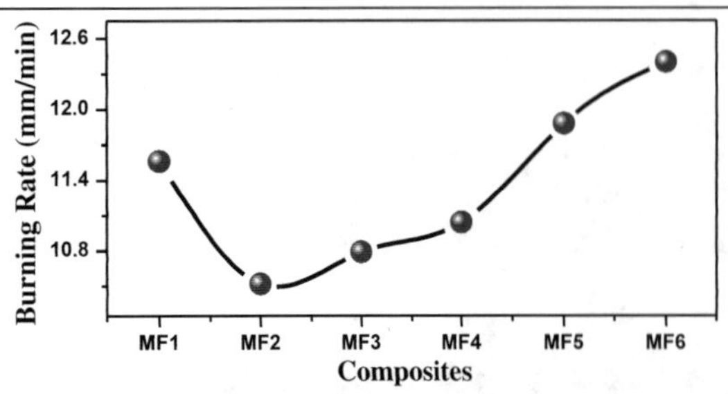

Figure 10. BR of the composites (mm/min)

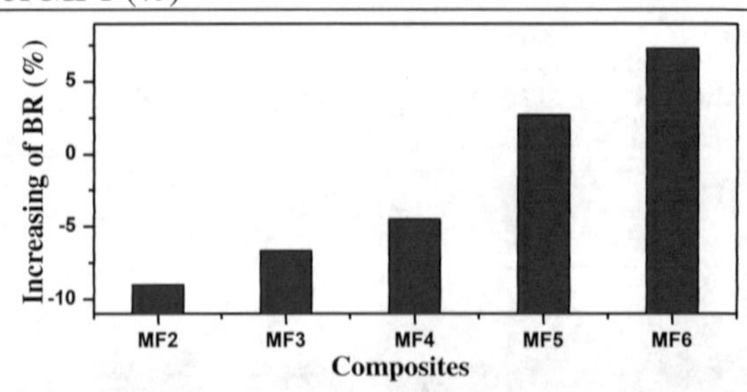

Figure 11. Comparison of increasing BR (%)

The TTI and BR of MF1 (without rRG plasticizer) were 6.02 s and 11.58 mm/minutes. When the filler become 23% of MMt (MF2), the TTI increased 15.95% while the BR decreased 9% to that MF1. Meanwhile the 1% and 3% concentration of rRG (MF3 and MF4) with 23% of MMt, the TTI increased 10.47% and 4.98%, while the BR decreased 6.68% and 4.5% compared to MF1. At 5% and 7% concentration of rRG (MF5 and MF6), the TTI decreased 6.15% and 9.97% while the BR increased 2.73% and 7.28% compared to MF1. MMt content and ZB in the composite on the combustion process undergo endothermic reactions and produce water vapor, charcoal and humid B_2O_3. This will lowering the temperature, limiting the intake of O_2 thus inhibiting the combustion process, this is supporting by the previous work [13,20]. The highest TTI and the lowest BR show that the material has high inflammability.

The heat release (HR) is the ability of a material to release heat after burning is extinguished at a certain period of time. Thus, the percentage of heat released during that period of time can be measured by using the equation HR (%)={(1-T_1/T_0)x100%, where, T_1 is the material's temperature after the period has elapsed, and T_0 is combustion temperature. In Figure 12-13 show the effect of rRG addition to that %HR of composites. The %HR of MF1 without rRG plasticizer, with 20% of MMt filler was 66.55%, however, when the filler became 23% the %HR increased 5.09% compare to

MF1. This is supported by Du and Othman who stated that the higher the clay concentration in the composite the flame retardant properties are increasing and the composite is not flammable [19,21].

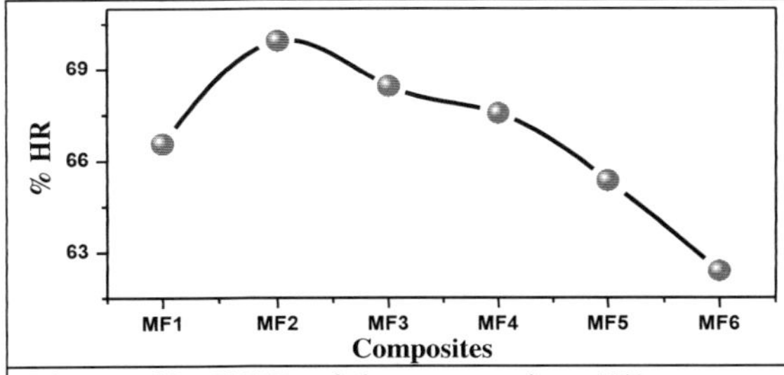

Figure 12. %HR of the composites (%)

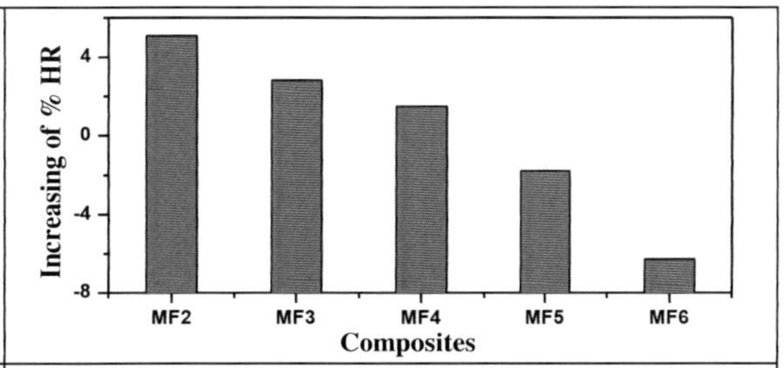

Figure 13. Comparison of Increasing %HR

Delhom reported that in the event of burning clay will form a charcoal that can cover the composite surface so as to reduce the contact between the composite with oxygen causing the burning process is inhibited [22]. In addition, the presence of clays in the polymer matrix causes an endothermic reaction resulting in reduced combustion temperatures and burning being impeded [23]. In addition of rRG plasticizer can increase the loading capacity resulting the high inflammability. The %HR of MF3 and MF4 can increase the HR% by 2.83% and 1.50% against MF1. However, composites MF5 and MF6 decreased %HR by 1.8% and 6.27% respectively compared to MF1. The decrease is due to the higher rRG concentration. This is in accordance with Balakrishnan who state the value of 10% mass degradation temperature that the higher the thermal endurance rubber content decreases [8]. This is also supporting by the previous work Patra and Suharty [24,25]. Material that has a high % heat release also indicates a high fuel resistance.

4. Conclusion

Addition of 3% rRG plasticizer of rPP/DVB/PP-AA/MMt+ZB is able to increase 15% filler capacity on composite. At optimum condition without plasticizer the filler capacity of MMt is only 20% but with addition of 3% pemplastis obtained optimum condition of filler loading capacity increased to 23%. The addition of 3% rRG pemplastis capable of increasing filler loading capacity, it will automatically improve the mechanical properties and combustion resistance of the composite. The mechanical properties of rPP/DVB/PP-AA/3% rRG/23% MMt+ZB new smart composite is as follows. The FS value is 45.08 MPa, deflection is 13.8 mm and and bending elasticity (BE) is 2722 MPa. The effect of addition 3% rRG plasticizer can increase the FS by 15%, increase the deflection by 16% and decrease the BE by 1% compared to that of MFI composite without rRG plasticizer. The inflammability properties value of new smart material is as follows. The TTI is 6.32 seconds, BR is 11.04 mm/mins, and HR is 67.55%. The effect of addition 3% rRG plastics can increase TTI to 4.98%, decrease BR to 4.50% and increase HR to 1.5% compared to that of MFI composites without rRG plasticizer.

Acknowledgment

The author would like to thank for financial support in this project through International Research Collaboration and Spesific Publication Grants to Directorate General of Higher Education Indonesia.

References

[1] Gordon R B, Bertram M, Graedel T E 2006 *PNAS* **103** 1209-1214
[2] Bandivadekar A, Bodek K, Cheah L, Evans C, Groode T, Heywood J, Kasseris E, Kromer M, Weiss M 2008 *On The Road in 2035: Reducing Transportation's Petroleum Consumption and GHG Emissions* (Report No. LFEE 2008-05 RP,Massachusetts Institute of Technology)
[3] Lyu M, Choi T G 2015 *Int. J. Precision Engineering and Manufacturing* **16** 213-220

IOP Conf. Series: Materials Science and Engineering **333** (2018) 012088 doi:10.1088/1757-899X/333/1/012088

[4] Suharty N S, Hanafi I, Fajar RW, Kuncoro D, Desi S H, Maulidan F, and Lupi L 2015 *Applied Mechanics and Materials* **736** 13-18
[5] Lee S Y, Kang I A, Doh G H, Kim W J, Yoon H G, Wu Q 2008 *eXPRESS Polymer Letters* **2** 78-87
[6] Wang W, Longxiang T, Baojun Q. 2003. *European Polymer Journal* **39** 2129–2134
[7] Xie H Q, Zhang S, Xie D 2005 *Journal of Applied Polymer Science* **96** 1414-1420
[8] Balakrishnan H, Attaran S A, Imran M, Hassan A, and Wahit M U 2012 *Journal of Thermoplastic Composite Materials* **27** 233-250
[9] Ismail H and Suryadiansyah 2002 *Polymer Testing* **21** 389-395
[10] Riyajan S A, Intharit I, Tangboriboonrat P 2012 *Industrial Crops and Products* **36** 376–382
[11] Mustafa I 2010 *Natural* **10** 7-13
[12] Benmesli S and Riahi F 2014 *Polymer Testing* **36** 54-61
[13] Suharty N S, Ismail H, Diharjo K, Nizam M, and Firdaus M 2014 *Advanced Materials Research* **950** 18-23
[14] Yuan P, Peter D S, Zongwen L, Malcolm E R G, James M H, Sarah J A, Camerom J K 2008 *Journal of Physical Chemistry C* **112** 15742-15751
[15] Ilic BR, Mitrovic A A, and Ljiljana L R 2010 *Hem. Ind* **64** 351-356
[16] Larena A, Urreaga J M, de la Orden M U 1991 *Journal of Colloid and Interface Science* **142** 111-115
[17] Petersen H N, Kusano Y, Bronsted P, and Almdal K 2013 *Risoe International Symposium on Materials Science. Proceedings* **34** 333-340
[18] Bendjaoahdou C and Bensaad S 2011 *Journal of Vinyl &Additive Technology* **17** 48-57
[19] Othman N, Ismail H, Mariatti M 2006. *Polymer Degradation and Stability* **91** 1761-1774
[20] Hassany Z Al, Genovese A, and Shanks R A 2010 *Journal of Polymer Letters* **4** 70-93
[21] Du Y, Zhang J, Xue Y, Lacy Jr T E, Toghiani H, Horstemeyer M F, Pittman Jr C U 2010 *Forest Products Journal* **60** 514-521
[22] Delhom C C, White-Ghoorahoo L S, and Pang S S 2010 *Composites: Part B* **48** 475-48
[23] Lee J H, Jung D, Hong C E, Rhee K Y, Advani S G 2005 *Composites Science and Technology* **65** 1996-2002
[24] Patra P K, Warner S B, Kim Y K, Qinguo Fan, Calvert P D and Adanur S 2005 *NTC Annual report*. No: M02-MD08 1-10
[25] Suharty N S, Almanar, Sudirman, Dihardjo K, Astasari N 2012 *Procedia Chemistry* **4** 282-287

IOP Conf. Series: Materials Science and Engineering **333** (2018) 012089 doi:10.1088/1757-899X/333/1/012089

Prediction of pavement remaining service life based on repetition of load and permanent deformation

R S Usman[1], **A Setyawan**[2], **M Suprapto**[2]
[1]Master Program of Civil Engineering, Universitas Sebelas Maret, Surakarta
[2]Civil Engineering Department, Universitas Sebelas Maret, Surakarta, Indonesia

E-mail: ramausman14@gmail.com

Abstract. One of the methods which was applied in the assessment of flexible pavement performance was mechanistic method assuming structures of road pavement to become multi-layer structure for flexible pavement, that the vehicle load working on the pavement layer under repetition with power failure worth 1 (one) unit which was assumed as evenly distributed static load, and therefore the pavement material would provide response in the form of stress, strain, and deflection.This is closely related in order to assess the structure of flexible pavement and to predict the remaining service life on the roads of Pulau Indah sta 0 + 000 to sta. 0 + 845 in Kota Kupang, Nusa Tenggara Timur. The performance appraisal indicator which was used was fatigue cracking happening bottom of the asphalt layer and permanent deformation (rutting) on the surface of subgrade.The strain estimate on the flexible pavement layer structure needs carefulness and high accuracy and therefore a software like KENPAVE which produces horizontal tensile strain of 8,802E-05 and vertical compressive strain of 2,642E-04 was used. By applying equation of The Asphalt Instituteit was obtained repetition of permit load when reaching fatigue cracking (Nf) was 16.071.516 ESAL and permanent deformation (rutting) was 14.703.867 ESAL and also it was predicted the remaining service life of pavement applied the equation of *AASTHO 1993* by considering Traffic Multiplier factor (TM 1.8, TM 1.9 and TM 2.0) obtained the remaining life service due to fatigue of 5.51% in the year of 13th (TM 1.8), 7.95% in the year of^{12th} (TM 1.9) and 3.11% (TM 2.0) in the year of 12th, also the remaining service life due to rutting of 4.69% in the year of 12^{th}(TM 1.8), 7.79% in the year of 11th (TM 1.9), and 2.94 in the year of 11th (TM 2.0).

1. Introduction

Road infrastructure performance service, from the stage of implementation to the planned service life keeps undergoing an increase of loading due to traffic volume, so that the road built must have reliable carrying capacity. Thus, it needs study and research to conduct about the behaviour of flexible pavement towards response of vehicle axle load or other factors. In flexible pavement structure, the load of vehicle under goes repitition has 1 (one) power failure which is assumed as evenly distributed static load, and therefore the material of pavement will provide response in the form of stress, strain, and deflection. [1]. This is closely related in order to predict the performance of pavement structure on the road of in the form of values of load repitition when reaching fatigue and rutting and to predict the remaining service life of the pavement.

Performance appraisal indicator which is applied often for flexible pavement is fatigue cracking occurred bottom of the asphalt layer and permanent deformation (rutting) on the surface of subgrade. The estimate of strain response in each pavement layer structure needs carefulness and high accuracy level and therefore a software named KENPAVE developed by Yang H. Huang P.E is required [2].

This research aims to study in detail in as certaining condition and remaining service life and therefore solution of handling improvement next day in particular related to behaviour of flexible pavement towards vehicle load.

2. Experimental

2.1 Flexible pavement response against traffic load.

Traffic load working on the flexible pavement is assumed as evenly distributed static load that the material of pavement will give response in the form of horizontal tensile strain (ε_t) bottom of the asphalt layer as an indicator of fatigue cracking and vertical compressive strain value (ε_c) on the surface of subgrade being the cause of permanent deformation (rutting) on the road pavement. According to the data of Pulau Indah road existing pavement, with the assistance of Kenpave software, it will ease the strain estimate applied in counting the repetition value of fatigue and rutting permit load. The Asphalt Institute recommended equation of flexible pavement fatigue cracking in order to estimate the number of load repetition as the following:

$$Nf = 0.0796(\varepsilon_t)^{-3.291}(E)^{-0.854} \quad (1)$$

with :

- Nf : the number of fatigue load repetition
- ε_t : horizontal tensile strain bottom of the asphalt layer
- E : Asphalt layer elastic modulus

And equation to count the number of rutting load repetition as the following:

$$Nd = 1.365 \times 10^{-9}(\varepsilon_c)^{-4.477} \quad (2)$$

with:

- Nd : the number of rutting load repetition
- ε_c : vertical compressive strain on the surface of subgrade

2.2 Prediction of traffic volume.

The analysis of traffic volume was based on the results of factual Avarage Daily Traffic (ADT) survey which was conducted by applying road pavement design manual of Highways year 2013 and therefore cumulative standard axle load ESAL (Equivalent Single Axle Load) and CESA (Cumulative Equivalent Single Axle Load) could be estimated using the following equation:

$$ESA = \Sigma\text{vehicle type ADT x VDF x DL} \quad (3)$$
$$CESA = \text{ESA x 365 x R} \quad (4)$$

with:

- *ESA* : Equivalent Single Axle
- *CESA* : Cumulative Equivalent Axle Load
- *R* : Vehicle Growth Factor
- *ADT* : Average Daily Traffic
- *VDF* : Vehicle Damage Factor
- *DL* : Lane Distribution

2.3 Analysis of remaining service life.

Analysis of remaining service life is a concept of pavement damage due to repetition load causing fatigue and permanent deformation (Rutting).AASTHO 1993 recommended equation to count remaining service life obtained from fatigue equation (Nf) and *rutting* (Nd).

$$RL = 100\left[1 - \left(\frac{\mathrm{Np}}{\mathrm{N1.5}}\right)\right] \quad (5)$$

with:

RL	:	Remaining Life (%)
NP	:	Total Traffic to date
$N_{1,5}$	:	Total Traffic to pavement failure

3. Results and Discussion

3.1 Flexible pavement response against traffic load.

In order to count flexible pavement response due to traffic load in the form of horizontal tensile strain and vertical compressive strain, it is required to have property material data from data material properties from existing pavement, will be analyzed by using KENPAVE software as presented in Table 1.

Table 1. Existing pavement, modulus typical value and poisson's ratio

Type of pavement	Thickness (mm)	Typical Modulus Mpa	Poisson's ratio
HRS-WC	30	800	0.40
HRS-Base	70	550	0.40
Base course class A	150	300	0.35
Sub base course class B	200	200	0.35
Selected embankment	300	100	0.35
Subgrade	~	40	0.35

The result of running output from Kenpave program and repetition causing fatigue and rutting damage is by in putting the value of horizontal tensile strain (ε_t) and vertical compressive strain (ε_c) in the equation 1 and 2 is presented in Table 2.

Table 2. Response of flexible pavement towards traffic load

No	Review point (cm)	Location	Response		Load repetition	
			horizontal tensile strain(ε_t)	vertical compressive strain(ε_c)	Nf	Nd
1	3,00	bottom of asphalt layer	8.802E-05		16.071.519	
2	75,00	surface of subgrade		2.642E-04		14.703.867

3.2 Traffic volume prediction.

Traffic volume data were obtained from secondary data of survey result year 2016 conducted by Department of Public Works Nusa Tenggara Timur Province on Pulau Indah roads in 2 directions as direction of Bundaran PU-Pulau Indah and reverse direction of Pulau Indah-Bundaran PU conducted for 6 days in busy-hours for 12 hours/day. From the Average Daily Traffic (ADT) data and therefore by using equation 3 and equation 4, it will be obtained ESA and CESA value as presented in Table 3 and Table 4.

Table 3. The result of ESA Value Estimation

Type of Vehicle	ADT	VDF	DL	ESA
Private Car	3164	0.0005	0.8	2
Truck 1.2, Mini Bus	105	0.8	0.8	67
Big bus, Tanker	1773	1.0	0.8	1385
Trailer, Truck 1.2.2	53	7.6	0.8	320
Total				**1774**

3.3 Analysis of

remaining service life

Analysis of remaining service life can be estimated by using equation of AASTHO 1993 when the condition reaches fatigue and rutting load repetition in standard CESA, and by assuming the condition of overloading based on TM factor times CESA for 20 years, the result of estimate presented in Table 5.

Table 4. The result of CESA Value Estimation

No	Year	R	CESA
ESA 1774			
1	2016	1.00	647,510
2	2017	2.00	1,295,149
3	2018	3.00	1,943,209
4	2019	4.00	2,591,594
5	2020	5.01	3,240,302
6	2021	6.01	3,889,335
7	2022	7.01	4,537,331
8	2023	8.01	5,186,558
9	2024	9.01	5,836,045
10	2025	10.02	6,485,793
11	2026	11.02	7,135,800
12	2027	12.03	7,786,067
13	2028	13.03	8,436,594
14	2029	14.04	9,087,381
15	2030	15.04	9,738,429
16	2031	16.05	10,389,737
17	2032	17.05	11,041,305
18	2033	18.06	11,693,134
19	2034	19.07	12,345,224
20	2035	20.08	12,997,574

Table 5. Standard Load Repetition and the effect of TM factor

No	Year	Load Repetition			
ESA	1774	Standard	TM 1.8	TM 1.9	TM 2.0
1	2016	647,510	1,165,343	1,230,084	1,294,825
2	2017	1,295,149	2,331,268	2,460,783	2,590,298
3	2018	1,943,209	3,497,777	3,692,098	3,886,419
4	2019	2,591,594	4,664,868	4,924,028	5,183,187
5	2020	3,240,302	5,832,544	6,156,574	6,480,604
6	2021	3,889,335	7,000,803	7,389,736	7,778,670
7	2022	4,537,331	8,167,195	8,620,928	9,074,661
8	2023	5,186,558	9,335,805	9,854,461	10,373,116
9	2024	5,836,045	10,504,882	11,088,486	11,672,091
10	2025	6,485,793	11,674,427	12,323,006	12,971,585
11	2026	7,135,800	12,844,439	13,558,019	14,271,599
12	2027	7,786,067	14,014,920	14,793,526	15,572,133
13	2028	8,436,594	15,185,868	16,029,528	16,873,187
14	2029	9,087,381	16,357,286	17,266,024	18,174,762
15	2030	9,738,429	17,529,171	18,503,014	19,476,857
16	2031	10,389,737	18,701,526	19,740,499	20,779,473
17	2032	11,041,305	19,874,349	20,978,480	22,082,610
18	2033	11,693,134	21,047,642	22,216,955	23,386,269
19	2034	12,345,224	22,221,404	23,455,926	24,690,448
20	2035	12,997,574	23,395,635	24,695,392	25,995,150

Analysis of the remaining service life was conducted towards the condition of fatigue and rutting which is presented in Table 6, Figure 1,and Figure 2.

Table 6. The analysis of the remaining service life of fatigue and rutting

No	Year	The remaining service life of fatigue *(%)*				The remaining service life of rutting *(%)*			
		Standard	TM 1.8	TM 1.9	TM 2.0	Standard	TM 1.8	TM 1.9	TM 2.0
1	2016	95.97	92.75	92.35	91.94	95.60	92.07	91.63	91.19
2	2017	91.94	85.49	84.69	83.88	91.19	84.15	83.26	82.38
3	2018	87.81	78.24	77.03	75.82	86.78	76.21	74.89	73.57
4	2019	83.87	70.97	69.36	67.75	82.37	68.27	66.51	64.75
5	2020	79.84	63.71	61.69	59.68	77.96	60.33	58.13	55.93
6	2021	75.80	56.44	54.02	51.60	73.55	52.39	49.74	47.10
7	2022	71.77	49.18	46.36	43.54	69.14	44.46	41.37	38.28
8	2023	67.73	41.91	38.68	35.46	64.73	36.51	32.98	29.45
9	2024	63.69	34.64	31.01	27.37	60.31	28.56	24.59	20.62
10	2025	59.63	27.36	23.32	19.29	55.89	20.60	16.19	11.78
11	2026	55.60	20.08	15.64	11.20	51.47	12.65	7.79	2.94
12	2027	51.55	12.80	7.95	3.11	47.05	4.69	*failure*	*failure*
13	2028	47.51	5.51	*failure*	*failure*	42.62	*failure*		
14	2029	43.46	*failure*			38.20			
15	2030	39.41				33.77			
16	2031	35.35				29.34			
17	2032	31.30				24.91			
18	2033	27.24				20.48			
19	2034	23.19				16.04			
20	2035	19.13				11.60			

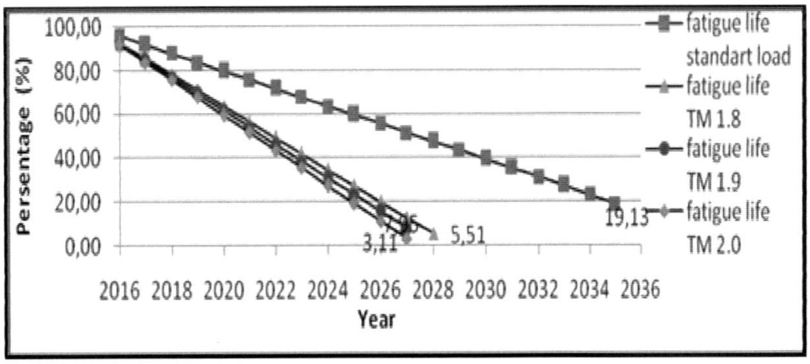

Figure 1. Analysis of the remaining service life of fatigue by TM factor

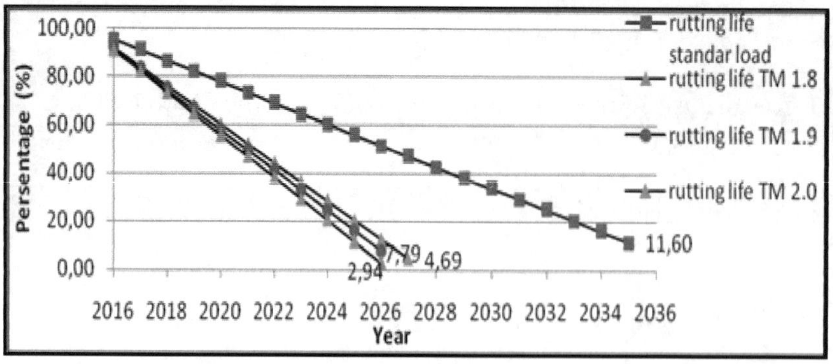

Figure 2. Analysis of the remaining service life due to rutting by TM factor

4. Conclusion

The result of running output obtained from Kenpave program in the form of horizontal tensile strain (ε_t) in the bottom of the asphalt layer was 8.802E-05 and vertical compressive strain (ε_c) on the surface of subgrade layer was 2.642E-04. Using the equation of fatigue and rutting obtained fromThe Asphalt Institute, it was obtained fatigue load repetition of 16.071.519 ESAL and *rutting* load repetition of 14.703.867 ESAL.

The result of CESA prediction estimate for 20 years with standard load of 8.16 Ton, it was obtained 12.997.574 ESAL value which was still in a safe condition under fatigue dan rutting load repetition value. If CESA prediction for 20 years times TM factor, it would result in load repetition of 23.395.635 ESAL (TM 1.8), 24.695.392 ESAL (TM 1.9) and 25.995.150 ESAL (TM 2.0) meaning that it went beyond fatigue dan rutting load repetition and therefore the pavement underwent failure condition.

The remaining fatigue service life in standard load in the end of year twentieth (2035) will remain 19.13 %, due to factor (TM 1.8) the remaining service life will remain 5.51 % in the 13th year (2028), due to factor (TM 1.9) the remaining service life will remain 7.95 % in the twelfth year (2027) and due to factor (TM 2.0) only remains 3.11 % in the twelfth year (2027). Meanwhile, the remaining rutting life service with standard load in the end of the twentieth year (2035) will remain 11.60 %, due to factor (TM 1.8) the service life only remains 4.69 % in the year twelfth (2027), due to factor (TM 1.9) the remaining service life only remains 7.79 % in the year eleventh (2026) and due to factor (TM 2.0) only remains 2.94 % in the eleventh year (2026) meaning that the damage due to rutting in the pavement will be reached first compared to the damage due to fatigue.

References

[1] Ahmed E, Abu E M B, 2012 *Ain. S. Eng. J.* **3** 367-374
[2] Huang H Y 2004 *Pavement Analysis and Design.* (New Jersey : University of Kentucky Prentice Hall)
[3] American Association of State Highway and Transportation Officials (AASHTO) 1993 *Guide for Design of Pavement Structures.*
[4] Citra K P 2014 *Prediksi Nilai kerusakan Perkerasan Lentur dengan metode Mekanistik–Empirik* (Studi Kasus: Rekonstruksi JL. Arteri Selatan Jogyakarta)
[5] Edovita S 2011*Civil. Eng. for.* **20** 1163-1174
[6] Ekwulo, Emmanuel O, and Dennins B E 2009 *Afric. J. Env. Sci. and Tech.* **3(12)** 412-421
[7] Salem H M A 2008 *Emi. J. Eng. Res.* **13** 21-28
[8] Kementrian Pekerjaan Umum Direktorat Jenderal Bina Marga 2013 *Manual Desain Perkerasan Jalan Nomor 02/M/BM/2013*
[9] Nainggolan J, Setyawan A, Budiarto A 2015 *Proc. Eng.* **125** 417-423
[10] Sukirman S, 1999 *Perkerasan Lentur Jalan Raya* (Bandung : Penerbit Nova)

IOP Conf. Series: Materials Science and Engineering **333** (2018) 012090 doi:10.1088/1757-899X/333/1/012090

Evaluation of Road Performance Based on International Roughness Index and Falling Weight Deflectometer

Hasanuddin[1], **A Setyawan**[2] **and B Yulianto**[1]
[1] Master program of Civil Engineering, Sebelas Maret University, Surakarta
[2] Roadmate Research Group Civil Engineering Department, Sebelas Maret University, Surakarta

E-mail: wonghasan80@gmail.com

Abstract. Assessment to the performance of road pavement is deemed necessary to improve the management quality of road maintenance and rehabilitation. This research to evaluate the road base on functional and structural and recommendations handling done. Assessing the pavement performance is conducted with functional and structural evaluation. Functional evaluation of pavement is based on the value of IRI (International Roughness Index) which among others is derived from reading NAASRA for analysis and recommended road handling. Meanwhile, structural evaluation of pavement is done by analyzing deflection value based on FWD (Falling Weight Deflectometer) data resulting in SN (Structural Number) value. The analysis will result in SN_{eff} (Structural Number Effective) and SN_f (Structural Number Future) value obtained from comparing SN_{eff} to SN_f value that leads to SCI (Structural Condition Index) value. SCI value implies the possible recommendation for handling pavement. The study done to Simpang Tuan-Batas Kota Jambi road segment was based on functional analysis. The study indicated that the road segment split into 12 segments in which segment 1, 3, 5, 7, 9, and 11 were of regular maintenance, segment 2, 4, 8, 10, 12 belonged to periodic maintenance, and segment 6 was of rehabilitation. The structural analysis resulted in 8 segments consisting of segment 1 and 2 recommended for regular maintenance, segment 3, 4, 5, and 7 for functional overlay, and 6 and 8 were of structural overlay.

1. Introduction

Evaluation of pavement performance is crucial for improved quality of road handling to materialize more efficient, reliable, and effective and accurately-targeted handling so as to facilitate sustainable monitoring and evaluation of road performance.

This study was aimed at investigating the functional and structural conditions of Simpang Tuan-Batas Kota Jambi road segment based on the International Roughness Index Method with NAASRA (SNI 03-3426-1994) and Non-destructive Method (NDT) with Falling Weight Deflectometer (FWD) and seeking for possible handling recommendations.

Functional evaluation comprises information concerning the characteristics of pavement that directly affect the safety and comfort of road users and road services. The Survey targeted skid resistance, surface texture, and road roughness and serviceability [4]. Simulated prediction of IRI value determines the plan for the functional road rehabilitation and maintenance [3].

International Roughness Index (IRI) serves as one of the parameters to determine the serviceability level of road segments that may affect the riding quality. Good roads should be strong, flat, watertight, durable and economical throughout the planned lifetime. Therefore, roads should be regularly/periodically monitored and evaluated for correct method of rehabilitation. Determining road roughness index can be done with NAASRA [7], a method that Bina Marga and AASHTO widely recommend.

Evaluation comprises information about the performance of pavement structure on traffic load and environmental condition. To this extent, surveys on road characteristics help obtain information on the pavement structure performance, pavement damage and road mechanical/structural properties [4]. Road structural capacity indicates the capability of road pavement to support the traffic load. It is commonly resulted from evaluating the mechanical properties of each pavement structure layer such as elastic modulus, fatigue properties, and deflection condition, measured by laboratory studies or on-site non-destructive (NDT) tests. One of the NDT methods is Falling Weight Deflectometer (FWD) [4]. The dynamic analysis done with Falling Weight deflectometer (FWD) test aims to predict the strength characteristics of flexible pavement and the contributing factors [4], one of which is overloaded vehicle axles that affect the service life of pavement [5].

2. Experimental

2.1. Functional Evaluation Analysis

Functional evaluation was done along with collecting NAASRA-based data of International Roughness Index (IRI) in a 100-meter range of halda setting. Segmentation was subsequently made based on uniformity of IRI value along 28,892 meters of road segment to facilitate grouping and deciding the recommended handling.

2.2. Traffic Volume Analysis

Traffic volume analysis was based on factual surveys. To support design, traffic volume was obtained from actual traffic surveys corresponding to the Traffic Enrollment Survey Manual by exercising Pd T-19-2004-B Manual or using equipment with the same approach to the results of previous traffic survey results, which incorporated traffic growth (R) factor based on historical growth data or correlation formulations with other valid growth factors, lane distribution factors and lane capacity referring to Ministry of Public Works Decree No.19/PRT/M/2011 and MKJI, Vehicle Damage Factor (VDF). The data were gained from static weighbridges and regional WIM data that the Directorate of Engineering issued, current Cumulative ESAL (Np), Cumulative ESAL at end of design life (N_f), calculated the remaining life that indicated gradual, repeated fatigue damage due to vehicle load by the percentage of traffic load in the pavement damage.

2.3. Structural Evaluation Analysis

Analysis of structural evaluation suggests that collecting data of deflection value derived from Falling Weight Deflectometer (FWD) shall consider parameters such as Resilient Modulus (M_R) of each pavement material both asphalt mix and subgrade with the stiffness value or resilient modulus. The study employed Subgrade Resilient Modulus (M_R), Effective Elastic Modulus (E_p) calculated on the basis of the deflection under the center plate load adjusted to the standard temperature of 68°F which was trial and error. The value of Structural Number Effective (SN_{eff}) was obtained based on the analysis of road segment deflection value obtained from FWD. Further, analysis of deflection value of segmented road was carried out adhering to the uniformity of deflection value. The analysis resulted in SN_{eff} value which refers to the value of pavement structural capacity currently suffering from structural deterioration. Structural Number Future (SN_f) refers to the value of pavement structural capacity for serving future/estimated traffic load (cumulative traffic load for the next 5 or 10 years).

IOP Conf. Series: Materials Science and Engineering **333** (2018) 012090 doi:10.1088/1757-899X/333/1/012090

2.4. Structural Condition Index (SCI) Calculation

The value of Structural Condition Index (SCI) serves as a parameter for determining road handling. SCI was obtained by comparing SN_{eff} to SN_f ratio resulting in classification value of segmented road handling. SCI value indicates the category of road handling in which SCI > 1 was categorized into maintenance type, SCI = 0.7-1 was categorized into functional overlay, SCI = 0.5-0.7 belonged to structural overlay, and SCI <0.5 was categorized into reconstruction.

3. Results and Discussion

3.1. Road Functional Evaluation Analysis

Evaluation of road pavement condition to Simpang Tuan-Batas Kota Jambi road segment split into 12 segments adhering to the uniformity of IRI value and handling recommendation which refers to the Minister of Public Works Regulation as shown in Figure 1 and Table 1 below.

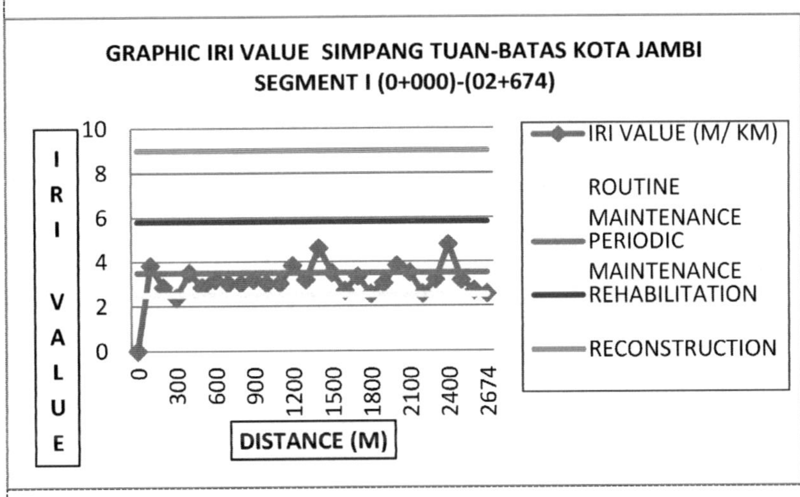

Figure 1. Typical of IRI Value.

Table 1. Recommended Handling Based on Range of IRI Value

Segment No.	Segment of STA	Length (m)	IRI	Condition	Recommendation
1	Km 0 + 000 up to 02 + 674	2,674	IRI ≥ 3.5	Good	Routine Maintenance
2	Km 02 + 674 up to 03 + 174	500	3.5 < IRI≤ 5.8	Fair	Periodic Maintenance
3	Km 03 + 174 up to 04 + 274	1,100	IRI ≥ 3.5	Good	Routine Maintenance
4	Km 04 + 274 up to 04 + 674	400	3.5 < IRI≤ 5.8	Fair	Periodic Maintenance
5	Km 04 + 674 up to 07 + 120	2,446	IRI ≥ 3.5	Good	Routine Maintenance
6	Km 07 + 120 up to 07 + 520	400	5.8 < IRI≤ 9.0	Damaged	Reconstruction
7	Km 07 + 520 up to 10 + 077	2,557	IRI ≥ 3.5	Good	Routine Maintenance
8	Km 10 + 077 up to 11 + 677	1,600	3.5 < IRI≤ 5.8	Fair	Periodic Maintenance
9	Km 11 + 677 up to 16 + 543	4,866	IRI ≥ 3.5	Good	Routine Maintenance
10	Km 16 + 543 up to 19 + 311	2,768	3.5 < IRI≤ 5.8	Fair	Periodic Maintenance
11	Km 19 + 311 up to 27 + 630	8,319	IRI ≥ 3.5	Good	Routine Maintenance
12	Km 27 + 630 up to 28 + 892	1,262	3.5 < IRI≤ 5.8	Fair	Periodic Maintenance

3.2. Traffic Volume Analysis

The analysis was based on secondary data of Routine Traffic Counting conducted by the National Road Planning and Implementation Agency of IV Jambi Region in May 2016. The analysis indicated that the growth value of traffic (R) at Simpang Tuan-Batas Kota Jambi road segment corresponded to five and ten years of design life, traffic growth factor (i) for Primary Arterial Road so that the percentage of traffic factor (i)=5%, Lane Distribution Factor by 80% for 2-way lane distribution, Vehicle Damage Factor (VDF) were used in reference to the Road Pavement Design Manual.

Current cumulative ESAL (Np or W18), cumulative ESAL at design life (Nf) and service life were derived from daily traffic volume (LHR). Np, N_f and service life can be seen in Table 2 and 3.

Table 2. Current cumulative ESAL (Np) or (W18)

Classification	Type of Vehicle	UNIT	LHR in 2016	VDF	DD	DL	R	DAYS IN A YEAR	W 18
2	PASSENGER CAR	VEHICLE	923	0.0005	0.5	0.8	1	365	67
3	COMBI VAN, MINIBUS (UTILITY 1)	VEHICLE	1,116	0.035	0.5	0.8	1	365	5,703
4	PICK UP, SHUTTLE CAR (UTILITY 2)	VEHICLE	758	0.035	0.5	0.8	1	365	3,873
5a	SMALL BUS	VEHICLE	132	0.3	0.5	0.8	1	365	5,782
5b	BIG BUS	VEHICLE	233	1	0.5	0.8	1	365	34,018
6a	2-AXLE LIGHT TRUCK	VEHICLE	590	0.8	0.5	0.8	1	365	68,912
6b	2-AXLE HEAVY TRUCK	VEHICLE	216	7.3	0.5	0.8	1	365	230,213
7a	3-AXLE HEAVY TRUCK	VEHICLE	221	28.9	0.5	0.8	1	365	932,487
7b	4-AXLE TOW TRUCK	VEHICLE	47	36.9	0.5	0.8	1	365	253,208
7c	SEMI TRAILER TRUCK	VEHICLE	258	13.6	0.5	0.8	1	365	512,285
		VEHICLE	**4,494**		**TOTAL (N_p) or W18**				**2,046,548**

Table 3. Cumulative ESAL at Design Life (Nf)

NO	YEAR	W18	R	ESAL ke-n	Umur layan
1	2015	2,046,548	1.0000	2,046,548	98.18454
2	2016	2,046,548	2.0005	4,094,119	94.55272
3	2017	2,046,548	3.0015	6,142,714	89.10363
4	2018	2,046,548	4.0030	8,192,333	81.83636
5	2019	2,046,548	5.0050	10,242,977	72.75000
6	2020	2,046,548	6.0075	12,294,647	61.84364
7	2021	2,046,548	7.0105	14,347,342	49.11637
8	2022	2,046,548	8.0140	16,401,064	34.56728
9	2023	2,046,548	9.0180	18,455,812	18.19546
10	2024	2,046,548	10.0225	20,511,588	-
			Nf	**20,511,588**	

3.3. Structural Evaluation

The analysis consists of traffic analysis, structural condition analysis, setting structural condition index (SCI), and recommendation for segmented road handling. Data comprising deflcection curve and pavement temperature were obtained from the Falling Weight Deflectometer, the tool that has a 300 mm diameter load plate, a 200 kg weight load and a 315 mm falling height. The distance between the deflectometers was set between 0, 200, 300, 450, 600, 00, 1200 and 1500 mm from the load center to fit the total pavement thickness which was 300-700 mm (normal). Data of deflection curve (FWD) were projected for structural analysis with AASHTO 1993 along with the data of traffic analysis and pavement thickness. Therefore, segmentation should be grouped based on the uniformity level of deflection value as seen in Table 4.

Table 4. Segmentation of Road Based on Uniformity of Deflection (D1)

Segment No.	Segment of STA	Length (m)	Standard Deviation (S)	Mean of Deflection Value (D1)	Uniformity Factor (Fk) %	Uniformity
1	Km 0 + 000 up to 03 + 602	3,602	32.11	157.11	20.4368	Fairly Good
2	Km 03 + 602 up to 06 + 411	2,839	40.05	171.26	23.3886	Fairly Good
3	Km 06 + 411 up to 08 + 004	1,593	46.61	178.20	26.1549	Fairly Good
4	Km 08 + 004 up to 11 + 440	3,436	56.10	200.83	27.9326	Fairly Good
5	Km 11 + 440 up to 15 + 408	3,968	41.69	179.64	23.2096	Fairly Good
6	Km 15 + 408 up to 19 + 641	4,233	93.61	341.01	27.4502	Fairly Good
7	Km 19 + 641 up to 23 + 750	4,109	66.04	226.52	29.1527	Fairly Good
8	Km 23 + 750 up to 27 + 777	4,027	52.20	192.90	27.0623	Fairly Good
	Total Road Length	**27,807**				

Subgrade Resilient Modulus (M_R) was calculated on the basis of per segment representative. M_R value shall meet the required distance of the farthest geophone sensor at R9=1500 mm away from the load center with greater or equal to 0.7 of stress basin radius on subgrade ($R \geq 0.7$ ae). M_R for each segment can be seen in table 5.

Tabel 5. Resilient Modulus (Mr) of Subgrade

SEGMENT No.	DISTANCE (m)	P Lbs	P Psi	C	r9 (mm)	r9 (inch)	d9 (x 0.001 mm)	d9 (inch)	M_R = C (0.24*P/d.r) (psi)
1	3,602	9565.89	87.22	0.33	1,500	59.06	45.08	0.00177	7,229
2	2,839	9568.35	87.24	0.33	1,500	59.06	38.52	0.00152	8,462
3	1,593	9534.77	86.94	0.33	1,500	59.06	46.81	0.00184	6,938
4	3,436	9472.49	86.37	0.33	1,500	59.06	51.65	0.00203	6,248
5	3,968	9296.92	84.77	0.33	1,500	59.06	44.43	0.00175	7,129
6	4,233	9370.20	85.44	0.33	1,500	59.06	54.89	0.00216	5,815
7	4,109	9352.01	85.27	0.33	1,500	59.06	40.35	0.00159	7,894
8	4,027	9322.09	85.00	0.33	1,500	59.06	37.54	0.00148	8,459

Table 6. Radius of stress basin on subgrade (Ae)

Segment No.	Distance (km)	A (inch)	D (inch)	Mr (psi)	Ep (psi)	R (inch)	Ae (inch)	0.7Ae	r ≥ 0.7Ae
1	3,602	5.,91	28.54	7228.91	85,328	59.06	65.26	45.68	OK
2	2,839	5.91	28.54	8461.58	76,050	59.06	59.638	41.75	OK
3	1,593	5.91	28.,54	6938.28	75,732	59.06	63.59	44.51	OK
4	3,436	5.91	28.54	6247.88	65,240	59.06	62.666	43.87	OK
5	3,968	5.91	28.54	7128.72	68,946	59.06	61.099	42.77	OK
6	4,233	5.91	28.54	5814.65	34,185	59.06	51.853	36.30	OK
7	4,109	5.91	28.54	7894.379	48,548	59.06	52.626	36.84	OK
8	4,027	5.91	28.54	8458,782	57,068	59.06	54.256	37.98	OK

Pavement Layer Modulus (E_P) was calculated as trial and error by iterating each segment as seen in table 10. The data of pavement layer thickness at Simpang Tuan-Batas Kota Jambi road segment, which were obtained from P2JN Work Unit of Jambi, can be seen in Table 7 below. To note, overlay was done to the road segment in 2015.

Table 7. Existing Pavement Layers Thickness(D)

Pavement Type	Thickness (mm)	Thickness (inch)
AC-WC	40	1.57
AC-BC	60	2.36
AC-BASE	75	2.95
LPA	300	11.81
Pavement Thickness	**475**	**28.54**

Table 8. Effective Elastic Modulus (Ep) of Pavement Layers

Seg No.	P (Lbs)	P (psi)	T (°C)	T (°F)	TAF	d1 (x 0.001 mm)	d1 (inch)	d1 x TAF (inch)	Ep (psi)
1	9565.89	87.22	34.92	94.86	0.83	221.33	0.0087	0.007232	85,328
2	9568.35	87.24	35.20	95.36	0.82	251.37	0.0099	0.008115	76,050
3	9534.77	86.94	37.40	99.32	0.76	271.42	0.0107	0.008121	75,732
4	9472.49	86.37	37.41	99.33	0.76	313.03	0.0123	0.009366	65,240
5	9296.2	84.77	33.40	92.12	0.84	263.02	0.0104	0.008698	68,946
6	9370.20	85.44	32.23	90.01	0.85	528.23	0.0208	0.017677	34,185
7	9352.01	85.27	29.85	85.73	0.88	358.59	0.0141	0.012423	48,548
8	9322.09	85.00	28.10	82.58	0.90	297.31	0.0117	0.010535	57,068

Structural Capacity Analysis was done by calculating SN_{eff} based on *Ep* value and asphalt-layered thickness with AASHTO 1993 for the analysis, based on reliability on Simpang Tuan-Batas Kota Jambi road segment referring to Manual for Planning Bending Pavement Thickness (Pt- 01-2002-B) that was classification of inter-city arterial roads with the value of 75% -95%. 95% of the value was selected so that value of standard deviation (ZR) was -1.645. Overall standard deviation (S0) represented local condition. Among the range of S_0 value, the highest value was 0.50. Surface index

(IP) used comprises final surface index (IPt), Initial Pavement Index (IP_0), Design serviceability loss ($\Delta PSI = IP_0 - IP_t$) and Subgrade Resilient Modulus (M_R)

3.4. Calculation of Structural Condition Index Value (SCI)

Structural Condition Index (SCI) implies compared Effective Structural Number (SN_{eff}) to Future Structural Number (SN_f). The comparison resulted in the classification value of segmented road handling as seen in Table 9.

Table 9. Comparison Value of SNeff to SNf

STA	Distance (km)	SNf 2020	SNeff	SCI	Recommendation
Km 00 + 000 up to 03 + 602	3,602	5.590	5.654544	1.012	Regular Maintenance
Km 03 + 602 up to 06 + 411	2,839	5.316	5.441688	1.024	Regular Maintenance
Km 06 + 411 up to 08 + 004	1,593	5.662	5.434092	0.960	Functional Overlay
Km 08 + 004 up to 11 + 440	3,436	5.850	5.,170574	0.884	Functional Overlay
Km 11 + 440 up to 15 + 408	3,968	5.614	5.266681	0.938	Functional Overlay
Km 15 + 408 up to 19 + 641	4,233	5.982	4.168504	0.697	Structural Overlay
Km 19 + 641 up to 23 + 750	4,109	5,436	4.685525	0.862	Functional Overlay
Km 23 + 750 up to 27 + 777	4,027	5,316	4.944987	0.930	Structural Overlay

4. Conclusion

The functional analysis to the segmented road indicated that the road segment split into 12 segments in which segment 1, 3, 5, 7, 9, and 11 were of regular maintenance, segment 2, 4, 8, 10, 12 belonged to periodic maintenance, and segment 6 was of rehabilitation.

Meanwhile, the structural analysis resulted in 8 segments consisting of segment 1 and 2 recommended for regular maintenance, segment 3, 4, 5, and 7 for functional overlay, and 6 and 8 were of structural overlay.

It could be concluded that the results of Functional Evaluation and Structural evaluation do not suggest the similar recommendation for pavements treatment, the combination of both functional and structural evaluation will result in the best recommendation for pavement rehabilitation, maintenance or reconstruction.

References

[1] Care F R A M, Subagio B S, Rahman H, Kusumawati A 2012 *Jurnal Teknik Sipil, Jurnal Teoritis dan Terapan Bidang Rekayasa Sipil ISSN* 0853-2982.
[2] Direktorat Jenderal Bina Marga 2002 *Pedoman Perencanaaan Tebal Perkerasan Lentur (Pt T-01-2002-B)* (Jakarta : Dirjen Bina Marga Departemen Pekerjaan Umum Republik Indonesia)
[3] Hermawan, Suprapto M, Setyawan A 2016 *IOP Conf. Series: Mat. Sci. Eng. 176* 012031
[4] Herry P 2012 *Penyusunan Program Pemeliharaan Jalan Berdasarkan Tinjauan Kondisi Struktural dan Fungsional Jalan* Tesis Institut Teknologi Bandung
[5] Setyawan A, Nainggolan J, Budiarto A 2015 *Proced. Eng.* **125** 417-423
[6] Nega A, Nikraz H, dan Al-Qadi I L 2016 *J. Traffic Trans. Eng.* **3(5)** 427-437
[7] Putra M Y M, Subagio B S, dan Hariadi E S *Jurnal Teoretis dan Terapan Bidang Rekayasa Sipil* ISSN 0853-2982

IOP Conf. Series: Materials Science and Engineering **333** (2018) 012091 doi:10.1088/1757-899X/333/1/012091

Pavement Condition Assessment Using IRI from Roadroid and Surface Distress Index Method on National Road in Sumenep Regency

T Arianto[1], M Suprapto[2] and Syafi'i[2]

[1]Student of Graduate Program of Civil Engineering, Universitas Sebelas Maret, Jl. Ir. Sutami 36A Kentingan Jebres Surakarta 57126, INDONESIA
[2]Lecture of Graduate Program of Civil Engineering, Universitas Sebelas Maret, Jl. Ir. Sutami 36A Kentingan Jebres Surakarta 57126, INDONESIA

E-mail: mamokuns@gmail.com

Abstract. The pavement condition will decrease due to the influence of traffic and environment, so that the maintenance effort is needed to maintain the road condition during the service period. In order to carry out road maintenance activities right on target, there needs to be a plan based on accurate pavement condition data. Road roughness is the most commonly used condition parameter in evaluating pavement conditions objectively because road roughness data is relatively easy to obtain, well correlated with vehicle operating costs and the most relevant parameter in road functional performance measurement. The Roadroid is an Android-based application that measures road roughness by using vibration sensors on a smartphone so it is possible to get an International Roughness Index (IRI) value as an indicator of pavement conditions more easily and efficiently. Besides based on road roughness, pavement condition evaluation can also be done visually by using Surface Distress Index (SDI) method that uses the total crack area parameters, average crack width, total number of potholes and the average depth of rutting. This study attempts to assess the condition of Jenderal Sudirman-Kalianget road by combining IRI Roadroid value and SDI value which will be used as the basis to determine the required road maintenance. This road segment is one of the national strategic road connecting the center of Sumenep regency with the Kalianget harbor. Based on IRI measurement and SDI calculation, the pavement condition of Jenderal Sudirman-Kalianget road can be described 4.2 kilometers (37.17%) were good and 2.3 kilometers (20.35%) were fair that need routine maintenance. While 2.1 kilometers (18.58%) were bad and 2.7 kilometers (23.89%) were poor that need periodical maintenance and reconstruction.

1. Introduction

Road is one of the important infrastructure in supporting the economic and social activities of the society. Sumenep regency, located at the eastern end of Madura island East Java, has potential in agriculture, fishery, industry and tourism. Seeing these conditions, it is necessary to properly manage the road network so that it is expected to improve the service and welfare of the society.

The pavement condition will decrease due to the influence of traffic and the environment, so that the maintenance effort is needed to maintain the road condition during its service period. In order to carry out maintenance activities right on target, it is necessary to have a plan based on accurate road condition data [1]. One of the road condition parameters that can be used to determine pavement conditions is road roughness that expressed in the International Roughness Index (IRI) values. The using of roadroid smartphone applications to obtain IRI value can be an alternative for road operators to monitor road conditions more easily, scalable, effective and efficient. Beside based on road roughness, pavement condition evaluation can also be done visually by using the Surface Distress Index (SDI) method.

This study attempts to assess the condition of the Jenderal Sudirman-Kalianget road by combining the IRI value of the Roadroid application with SDI values that will be used as the basis for determining the required road maintenance recommendations. The road is one of the strategic national roads connecting the center of Sumenep regency with the Kalianget harbor.

1.1. Pavement Condition Assessment based on IRI

Pavement condition evaluation should be done regularly to determine existing pavement performance and predict future pavement performance, whether the pavement has fulfilled its three basic functions, which provide efficient service, safety and has a structural capacity to support the traffic load and resistant to environmental impacts [2].

Road roughness is the most commonly used condition parameter in evaluating pavement conditions objectively. Road roughness data is relatively easy to obtain, well correlated with vehicle operating costs and the most relevant condition parameter in the road functional condition assessment in the long term [8].

The international roughness index was developed in 1986 which was the result of the International Road Roughness Experiment held in Brazil in 1980. Since then, IRI has become the standard for measuring road roughness received worldwide [4]. The IRI is expressed as a ratio of the accumulation of standard vertical vehicle movement (mm or inch) to the distance traveled by the vehicle during the measurement (m, km or miles) [10].

Generally, road roughness measuring method are divided into four classes. IE, direct measurement, indirect measurement, IRI estimation based on correlation equations such as response type road roughness measuring system (RTRRMS) and subjective judgments based on the opinion of the assessment group [9]. Measuring road roughness by using vibration sensors on a smartphone such as Roadroid basically includes the type of RTRRMS. Roadroid is an Android-based smart phone application developed by the Swedish National Road Administration can be an alternative to obtaining road roughness data that becomes an indicator of road conditions to grade 2 and grade 3 with easier and less costly [5].

The criteria of pavement conditions based on IRI values on asphalt road surface types are shown in Table 1 [6].

Table 1. The Criteria of Pavement Conditions based on IRI

IRI Value	Pavement Conditions
$IRI < 4$	Good
$4 \leq IRI \leq 8$	Fair
$8 \leq IRI \leq 12$	Bad
$IRI > 12$	Poor

1.2. Pavement Condition Assessment Using SDI Method

The SDI method is developed by the Directorate General of Bina Marga in 2011 in a national road condition survey manual that is intended for road managers within the Balai Besar/Balai Pelaksanaan Jalan Nasional to determine pavement conditions and as a basis for determining the type of road maintenance required [3]. In this method, the road condition evaluation is visually done by using the parameters of total crack area, average crack width, number of potholes and depth of vehicle rutting.

The criteria of road conditions based on the SDI value on the asphalt road surface type are shown in Table 2.

Table 2. The Criteria of Pavement Conditions based on SDI

SDI Value	Pavement Conditions
SDI < 50	Good
50 ≤ SDI ≤ 100	Fair
100 ≤SDI ≤ 150	Bad
SDI ≥ 150	Poor

1.3. Road Maintenance Based on IRI and SDI

Road maintenance activities can be both preventive and reactive in order to limit or delay further damage and repair damage to the road surface. Road maintenance activities consist of routine maintenance, periodic maintenance that can be followed by reconstruction/improvement [6].

The pavement condition is determined based on a combination of the IRI and SDI values obtained, are shown in Table 3 with the road maintenance recommendations are shown in Table 4 [3].

Table 3. The Pavement Conditions based on IRI and SDI Value

IRI Value (m/Km)	SDI Value			
	<50	50-100	100-150	>150
<4	Good	Fair	Bad	Poor
4-8	Fair	Fair	Bad	Poor
8-12	Bad	Bad	Bad	Poor
>12	Poor	Poor	Poor	Poor

Table 4. The Road Maintenance Recommendation Based on Pavement Condition

IRI Value (m/Km)	SDI Value			
	<50	50-100	100-150	>150
<4	Ruotine M.	Ruotine M.	Periodical M.	Reconstruction
4-8	Routine M.	Routine M.	Periodical M.	Reconstruction
8-12	Periodical M.	Periodical M.	Periodical M.	Reconstruction
>12	Reconstruction	Reconstruction	Reconstruction	Reconstruction

2. Experimental

The IRI value is obtained by using the Roadroid application on the smartphone mounted on the windshield of a survey vehicle and running it along the road segment that are observed. Prior to the measurement it is necessary to do Rodroid setting by determining the type of vehicle used, the distance measurement IRI, automatic image capture and minimum speed limits of the vehicle.

The observation of pavement conditions was conducted by using video recording along the observed roads. Further assessment is done based on the type and amount of damage (cracks, holes, rutting) that occur every 100 m of road length using the Bina Marga standard. The research was conducted on Jenderal Sudirman-Kalianget national road segment located in the Sumenep regency of East Java province. This road segment is divided into six segments; Jl. Sudirman (500 m), Jl. A. Yani (400 m), Jl. Urip Sumoharjo (1,600 m), Jl. Slamet Riyadi (1,000 m), Jl. Yos Sudarso (3,200 m) and Bts. Kota Sumenep-Kalianget (4,600 m).

3. Result and Discussion

3.1. Pavement Condition Assessment based on IRI

The results can be obtained from IRI measurements per 100 m using Roadroid applications such as longitude, latitude, distance, altitude, grade and IRI values. The examples of IRI survey results and road conditions based on IRI on the Jenderal Sudirman-Kalianget road at Jl. Slamet Riyadi segment (Sta. 2+500-3+500) is summarized in Table 5 and Figure 1.

Table 5. The IRI Survey Result of Slamet Riyadi Segment

No.	Latitude	Longitude	Distance (m)	Altitude (m)	Grade (%)	IRI (m/Km)	Remarks
1	-7.02053658	113.87677225	2,600	36.09	0.95	7.03	Fair
2	-7.02115698	113.87748518	2,700	41.52	5.43	8.18	Bad
3	-7.02177321	113.87806690	2,800	39.75	-1.77	5.20	Fair
4	-7.02245105	113.87868161	2,900	34.52	-5.23	6.47	Fair
5	-7.02314097	113.87916247	3,000	37.86	3.34	3.97	Good
6	-7.02390731	113.87961248	3,100	39.46	1.6	6.34	Fair
7	-7.02466185	113.88010418	3,200	36.94	-2.51	7.41	Fair
8	-7.02544359	113.88066806	3,300	35.12	-1.82	7.83	Fair
9	-7.02617907	113.88125786	3,400	31.98	-3.14	2.50	Good
10	-7.02693132	113.88171707	3,500	31.84	-0.14	5.88	Fair

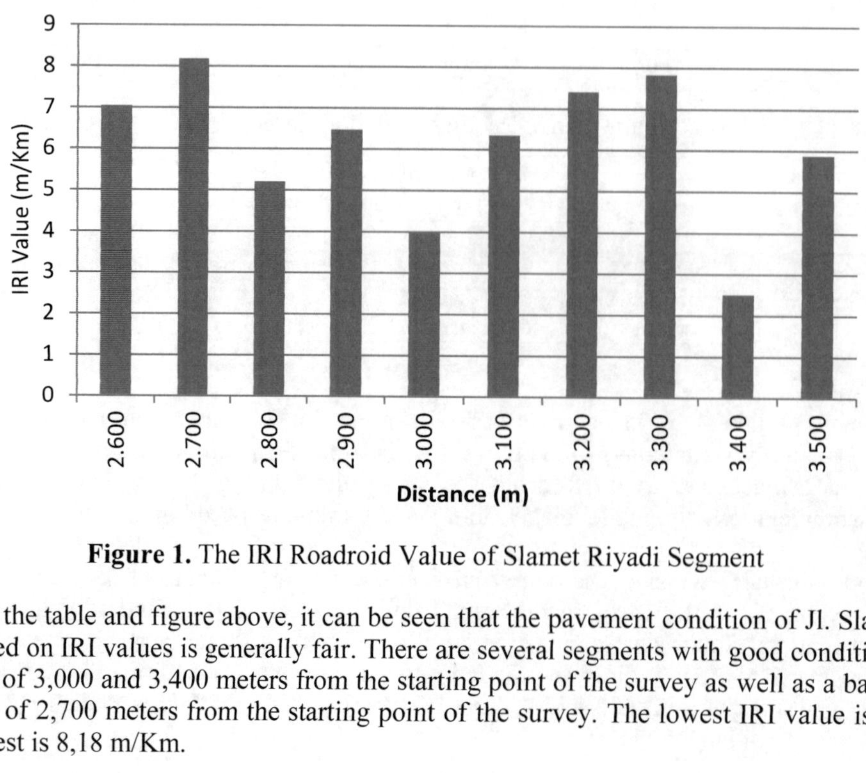

Figure 1. The IRI Roadroid Value of Slamet Riyadi Segment

Based on the table and figure above, it can be seen that the pavement condition of Jl. Slamet Riyadi segment based on IRI values is generally fair. There are several segments with good conditions such as at a distance of 3,000 and 3,400 meters from the starting point of the survey as well as a bad condition at a distance of 2,700 meters from the starting point of the survey. The lowest IRI value is 2,5 m/Km and the highest is 8,18 m/Km.

3.2. Pavement Condition Assessment Using SDI Method

From the results of pavement conditions observation, can be obtained amount for each type of damage that is cracked, potholes and ruts. Furthermore, the amount of damage data is used to calculate the SDI value and determine the conditions on a road segment. The example of SDI calculation result on Jenderal Sudirman-Kalianget road at Jl. Slamet Riyadi segment (Sta. 2 + 500-3 + 500) is shown in Table 6 and Figure 2.

Table 6. The SDI Survey Result of Slamet Riyadi Segment

No.	STA_0	STA_1	SDI Calculation								Remarks
			Percent of Cracks	Crack Width Classification	No. of Pothole /Km	Rutting Calculation	SDI_1	SDI_2	SDI_3	SDI	
1	2+500	2+600	22.50%	Wide	110	0	20	40	265	265	Poor
2	2+600	2+700	11.67%	Wide	70	0	20	40	265	265	Poor
3	2+700	2+800	10.00%	Wide	70	10	20	40	265	275	Poor
4	2+800	2+900	30.00%	Wide	60	0	20	40	265	265	Poor
5	2+900	3+000	14.17%	Wide	60	10	20	40	265	275	Poor
6	3+000	3+100	39.17%	Wide	100	10	40	80	305	315	Poor
7	3+100	3+200	17.50%	Wide	50	0	20	40	115	115	Bad
8	3+200	3+300	10.00%	Med	40	0	20	20	95	95	Fair
9	3+300	3+400	10.00%	Med	40	0	20	20	95	95	Fair
10	3+400	3+500	26.67%	Wide	110	0	20	40	265	265	Poor

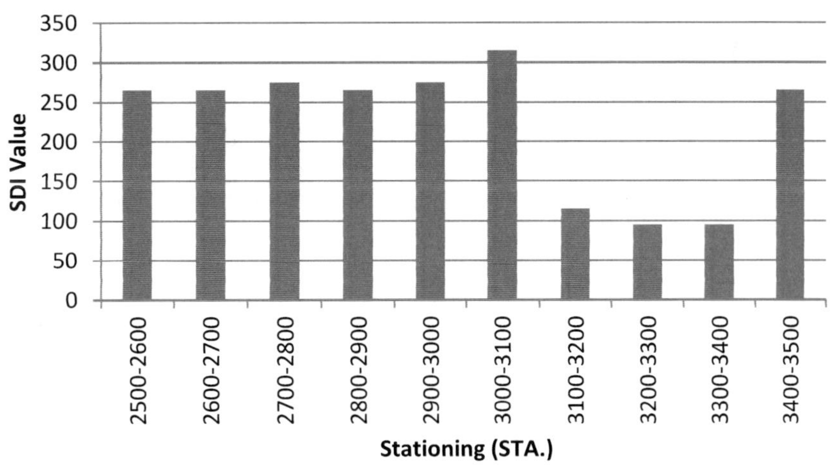

Figure 2. The SDI Value of Slamet Riyadi Segment

Based on the table and picture above, it can be seen that the condition of Jl. Slamet Riyadi segment based on SDI values is generally poor. This is due to the segment there is a lot of damage in the form of potholes that give a significant influence on the SDI value. There are several segment parts with bad conditions such as in STA. 3+100-3+200 and fair conditions at STA. 3+200-3+400. The lowest SDI value is 95 and the highest is 315.

3.3. Road Maintenance Recommendation Based on IRI and SDI

From the previous calculation results, IRI values and SDI values have been obtained, then combined with the results describing the pavement conditions per 100 m (Table 3.) and the type of road maintenance required (Table 4.). The examples of collaboration IRI values and SDI values to determine pavement conditions and road maintenance recommendations on Jenderal Sudirman-Kalianget road at Jl. Slamet Riyadi segment (STA. 2+500-3+500) is shown in Table 7.

Table 7. The Pavement Condition and Maintenance Recommendation of Slamet Riyadi Segment

No.	STA.	IRI Value	SDI Value	Pavement Condition	Maintenance Recommendation
1	2+500–2+600	7.03	265	Poor	Reconstruction
2	2+600–2+700	8.18	265	Poor	Reconstruction
3	2+700–2+800	5.20	275	Poor	Reconstruction
4	2+800–2+900	6.47	265	Poor	Reconstruction
5	2+900–3+000	3.97	275	Poor	Reconstruction
6	3+000–3+100	6.34	315	Poor	Reconstruction
7	3+100–3+200	7.41	115	Bad	Periodical M.
8	3+200–3+300	7.83	95	Fair	Routine M.
9	3+300–3+400	2.50	95	Fair	Routine M.
10	3+400–3+500	5.88	265	Poor	Reconstruction

From the table above it can be seen that the pavement condition of the Jl. Slamet Riyadi segment generally is poor, which requires reconstruction/improvement. The segment parts with fair and bad condition only need routine maintenance and periodical maintenance.

The using of SDI method in this research combined with the pavement condition assessment based on IRI as has been done by previous researchers [1], aims to determine the pavement condition and road maintenance recommendation more accurately and approriately. The IRI value as an indicator of the functional pavement condition based on the driving comfort level, while the SDI value is used to determine the level of pavement damage based on visual observation.

4. Conclusion

Roadroid applications can be used as an alternative to obtain IRI values efficiently that can be collaborated with SDI values to determine appropriate maintenance recommendations. Based on the IRI measurement using Roadroid combined with SDI calculation, the condition of the Jenderal Sudirman-Kalianget national road in Sumenep regency along 11.3 kilometers can be described 4.2 kilometers (37.17%) of roads were in good condition and 2.3 kilometers (20.35%) were in fair condition requiring routine maintenance. While the 2.1 kilometers (18.58%) road segment were bad and 2.7 kilometers (23.89%) were poor, requiring periodic maintenance and road improvements.

References

[1] Achmadi F, Suprapto M and Setyawan A 2016 *IOP Conf. Ser.: Mater. Sci. Eng.* **176** 012033
[2] Bennet CR, Chamorro A, Chen C, Solminihac H and Flintsch GW 2007 *Data Collection Technologies for Road Management* (Washington DC: East Asia Pasific Transport Unit) p 23-25
[3] Direktorat Jenderal Bina Marga 2011 *Indonesian Road Integrated Road Management System (IIRMS)* (Jakarta)
[4] Du Y, Liu C, Wu D and Jiang S 2014 *Math. Prob. in Eng.* **2014** 1-10
[5] Forslof L and Jones H 2014 *J. of Civ. Eng. and Arch.* **9** 485-496
[6] Kementerian Pekerjaan Umum 2011 *Peraturan Menteri Pekerjaan Umum No: 13/PRT/M/2011 tentang Tatacara Pemeliharaan dan Penilikan Jalan* (Jakarta) p 6-7
[7] Kementerian Pekerjaan Umum dan Perumahan Rakyat 2016 *Peraturan Menteri Pekerjaan Umum Perumahan Rakyat No: 33/PRT/M/2016 tentang Penyelenggaraan Dana Alokasi Khusus Bidang Infrastruktur* (Jakarta) p 8
[8] Robert JD and Martin TC 1999 *Recomendations for Monitoring Pavement Performance* (Vermont South: ARRB Transport Reserach) p 2

IOP Conf. Series: Materials Science and Engineering **333** (2018) 012091 doi:10.1088/1757-899X/333/1/012091

[9] Sayers MW, Gillespie TD and Paterson William DO 1986 *Guidelines for Conducting and Calibrating Road Roughness Measurement* (Washington DC: The World Bank) p 6-9
[10] Suherman 2008 *J. Tek. Sip.* **8** 206-214

IOP Conf. Series: Materials Science and Engineering **333** (2018) 012092 doi:10.1088/1757-899X/333/1/012092

Audit Technical of Kori Rubber Dam in the River of Keyang District of Ponorogo East Java Province

E Murnianto [1*], M Suprapto [2], C Ikhsan[2]
[1]Graduate students in maintenance and rehabilitation of infrastructure of Civil Engineering Sebelas Maret University (UNS), Jl. Ir. Sutami No. 36A Surakarta 57126, Indonesia
[2]Post graduate lecturer in maintenance and rehabilitation of infrastructure of Civil Engineering Sebelas Maret University (UNS), Jl. Ir. Sutami No. 36A Surakarta 57126, Indonesia

*E-mail: ekomurnianto22@gmail.com

Abstract. The development of science and technology for the utilization and protection of rivers has embodied various types of river infrastructure. Without proper maintenance, rapid river sediments undergo physical degradation and function. Problems that occur in Kori Rubber Dam, among others, the damage to the body of the rubber dam that is made of rubber, so that the function of flower deflection is not optimal. This happens because of limited operational and maintenance activities (OM). A technical audit is a process of identifying problems, analyzing, and evaluating ones conducted independently, objectively and professionally on the basis of examination, to assess the truth, accuracy, credibility, and reliability of information about a job. In this case an assessment of the Kori Rubber Dam, which is basically a benchmarking activity. Assessment of rubber dam components includes the physical conditions and functions that affect the weir. This research is expected to know the performance of Kori rubber Dam as a recommendation material in the implementation of OM Rubber Dam activities.

1. Introduction

The rivers in Ponorogo Regency are generally still a natural river. However, some water infrastructure has been built in several rivers, one of them in the Keyang River. Several weirs have been built in the Keyang River, one of which is the Kori Rubber Dam in Sawoo District to irrigate rice fields with a standard area of 676 ha covering 17 Villages 4 District.

Kori Rubber Dam was built in 1993 by the Territory of Bengawan Solo River Region. Problems that occur on the Rubber Clump, among others, the damage to the body of the rubber dam, so that the blowing deflatedam function is not optimal. This is still the result of OM activities.

Kori rubber battle is used for irrigation weir and flood control. The length of the lower rubber weir respectively is 15.11 m, the length of the upper rubber weir is 17.67 m with maximum height when developed is 2.5 m, the slope of the retaining wall is 1: 0.3

Because of less optimal each component of Kori Rubber Dam resulted in sedimentation in upstream weirs. This resulted in reduced service and service life of the Kori Rubber Dam.

IOP Conf. Series: Materials Science and Engineering **333** (2018) 012092 doi:10.1088/1757-899X/333/1/012092

In general, the current problems are: inadequate and not even provided funds from the Regional Revenue Budget (APBD) for OM Rubber Dam activities, lack of OM staff in the field (in terms of quality & quantity), the priority of OM activities is still low.

This research is based on Regulation of PUPR No. 12/PRT/M/2015 and No. 23/PRT/M/2015. Rubber Dam Performance can be known by the assessment of the physical condition and function of the rubber dam components so as to know the performance of the rubber dam. Some of the problem formulation in this research are as follows: 1. How is the physical condition of Kori Rubber Climb in Keyang River; 2. What is the condition of the Kori Rubber Garden function in the Keyang River; 3. How does the Kori Rubber Garden performance in the Keyang River.

2. Experimental

2.1 Research sites

This research was conducted in Bendung Kori located in UPTD Sambit, Ponorogo Regency, East Java Province. Located about 200 km southwest of the provincial capital, and about 800 km east of the capital of the State of Indonesia. Ponorogo Regency is located at 111 ° 7 'to 111 ° 52' East Longitude and 7 ° 49 'to 8 ° 20' South Latitude.

The area of Ponorogo Regency is directly adjacent to Magetan Regency, Ponorogo Regency and Nganjuk Regency in the north. Regency of Tulungagung in the east and Regency of Trenggalek in the east. To the south with Pacitan Regency. While in the west is bordered by Pacitan regency and Wonogiri regency (Central Java province).

The total area of Ponorogo regency which reached 1,371.78 km2 is divided into 21 districts consisting of 307 villages.

2.2 Data collection

What is needed in this study include from primary data and secondary data.

1. Primary data
   a. Conducting site visits to Kori weir for observation of the physical condition of the existing weir directly in the field with an inventory of all the components of damaged buildings and still functioning.
   b. Direct interview with field officer at UPT Sambit of DPU of Ponorogo Regency to know Operation and Maintenance at Kori Rubber Dam
2. Secondary Data
   a. Data and reports from previous studies;
   b. Watershed maps with watershed boundaries, river basin patterns / schemes, settlements and concentrations of economic activities, boundaries of work areas between interpreters, other areas: working area of UPTD, working area of BBWS / BWS, and district area;
   c. The Infrastructure Network Scheme of the River depicting the location of the flooded river infrastructure, and each equipped and the names of other complementary building structures with nomenclature;
   d. Post-construction drawing (as built drawing);
   e. Planning 5 (five) annual management of river infrastructure;
   f. Documents and other supporting data.

2.3 Research Steps

The steps of this study include:

a. Component Rating and Weighting
   Each major component is divided into several smaller components, each of which needs to be assessed for its condition. Each component will contribute the value of the condition to the overall physical condition of the rubber dam.

The contribution of each major component to the overall physical weir of rubber weighs is not the same. The weight of each component is prepared on the basis of the magnitude of the effect of each component on the assurance of irrigation water services.

The weight for each of the major components is a composite of each of its constituent components, and the weight distribution for both the main component and its constituent components.

The distribution of the assessment of the main components of the rubber dam is adjusted to the sub component of the existing building on the rubber dam. From sub component of existing building each weight multiply by weight of component of building then divided by main weight. Thus the weight for each of the smallest sub-components can be known. Distribution of components and weights in the complete rubber dam.

b. Method of Assessment of Physical Condition and Condition of Rubber Dam Function
Approach and test model of condition assessment system and function of component component (facet) in determining priority order of rehabilitation of AHP-based rubber dam, as shown in Figure 1. Assessment Weight of Rubber Bend Asset Component is approximated by AHP Method. This AHP method is a method used in decision making of double-weighted parameters (OECD, 2008 and Saaty, 1990).

2.4 Assessment of Physical Condition of Rubber Dam

Asset component condition assessment system carried out three approaches with defects that are adjusted in the maintenance program. Approach system assessment of the condition of the rubber dam components. Based on PUPR No.12/PRT/M/2015 assessment percentage damage asset to four criteria right.

$$\mathrm{NKBFi} = \sum_{i=1}^{i=n} \mathrm{W_k} \text{ x } \mathrm{NKFi_n} \qquad (1)$$

With:

NKBFi = Total Value Condition Rubber Dam
Wk = The weight factor of the Rubber Dam component
$\mathrm{NKFi_n}$ = value scale rubber dam conditions index component to the Rubber dam
n = Components to 1,2,3, ... etc

2.5 Assessment Condition of Function of Rubber Dam

Based on the Minister of Public Works Decree Number 23/PRT/M /2015 assess the percentage of functionality of the assets into four functioning criteria.

System assessment Functioning Component Rubber dam ability drain water to area. The value of the functioning of Rubber Dams come into one's vision Synthesis Assessment compound functioning Rubber Dam Value of functioning compound Rubber dam press a man as following:

$$\mathrm{NKBFu} = \sum_{i=1}^{i=n} \mathrm{W_f} \text{ x } \mathrm{NKFu_n} \qquad (2)$$

With:

NKBFu = Total Value Function Rubber Dam
Wf = The weight factor of the Rubber dam component component
$\mathrm{NKFu_n}$ = value index scale rubber dam component function Rubber dam
n = Components to 1,2,3, ... et

2.6 Performance Rating of Rubber Dam

To know the performance of a building, keep in mind the physical condition and condition of the function of the building. After both conditions are known, then the next performance of the building can be known that is the sum of physical condition and condition of the function. Furthermore, based on the performance value, it can be found the recommended activity recommendations

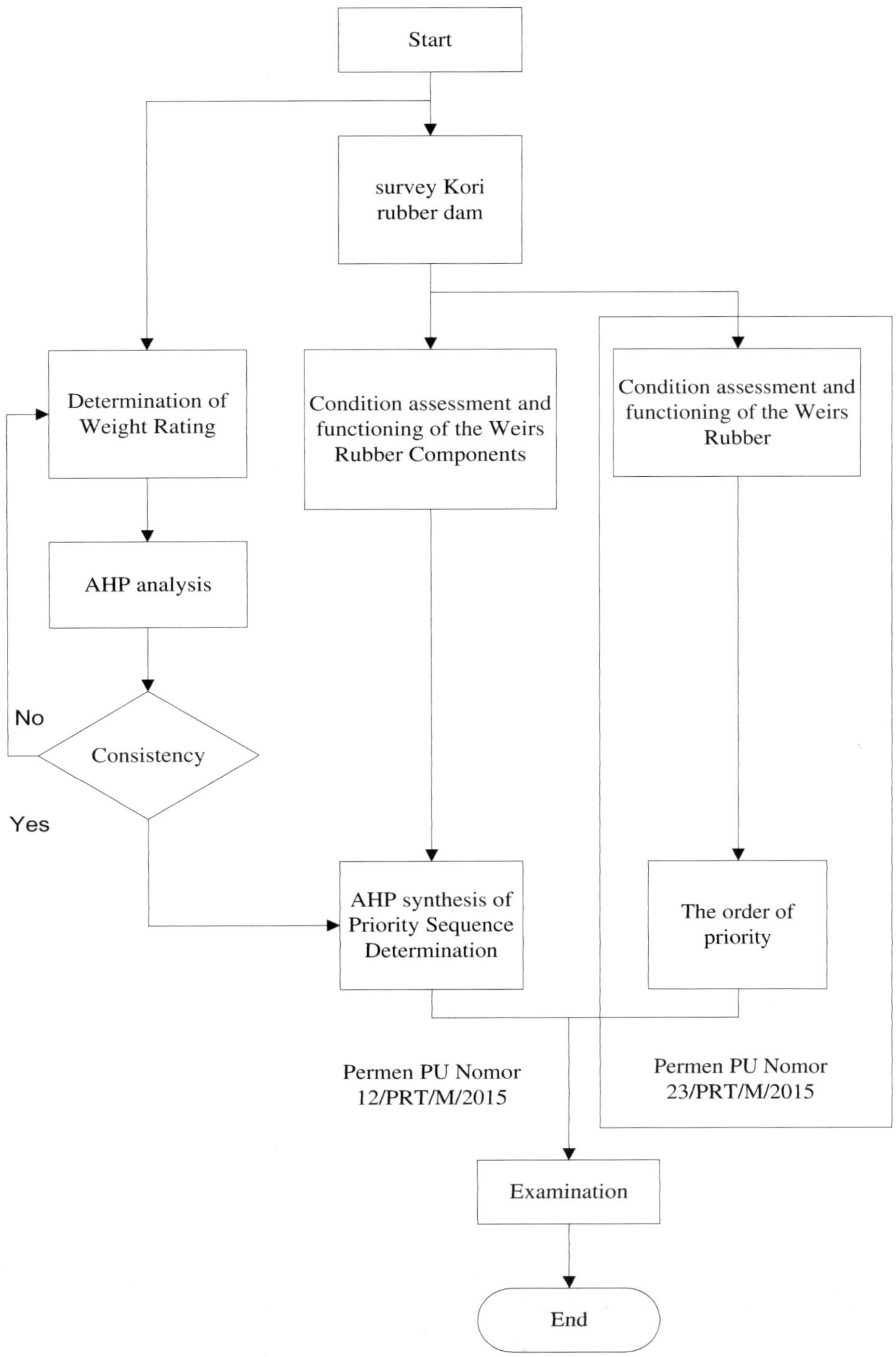

Figure 1. Approach Model of Condition Assessment System and Functioning of Rubber Frame Based on AHP-Based Rubber Component

IOP Conf. Series: Materials Science and Engineering **333** (2018) 012092 doi:10.1088/1757-899X/333/1/012092

3 Result and Discussion

The number of respondents is 46 (forty six) people. Targets of respondent questionnaire criteria of rubber weir weight are:

1. Practitioners in the field of water resources consisting of:
   a. Respondents from the Ponorogo District Public Works Office in charge;
   b. Respondent from Territory of Bengawan Solo River Region in charge;
   c. Respondents from HIPPA (Association of Water User Farmers).
2. Academic field of water resources consisting of the father of UNS Solo lecturer in charge.

Assessing the weight of each component of kori rubber dam is done by conducting questionnaires to practitioners and academics who know about the rubber dam component, is presented in Table 1.

Table 1. Calculation of the weight of the rubber dam

No	Sub component	Component				
		Building Retrieval	Building Drain	Wing	Body weir	Other buildings and surroundings
		31%	13%	12%	37%	7%
1	Doors Intake	10%				
2	Sediment / sludge	4%				
3	Debit Measurers	4%				
4	Exploitation Board	13%				
5	Door drain		10%			
6	Sediment / sludge		3%			
7	Wing			9%		
8	Koperan / Embankment Cover			3%		
9	Mercu / Rubber materials				14%	
10	Floor weir (downstream floor / Olakan)				9%	
11	Scale Board/Peil Schaal				4%	
12	Foundation and Pillar				10%	
13	Motor and Air Pump					2%
14	Installation Filling / Deflation					2%
15	Bridge Crossings					1%
16	Security Fence					1%
17	House Operation					1%

3.1 Assessment of Physical Condition of Kori Rubber Dam

From the observation in the field, obtained the physical condition of kori rubber dam. Recapitalization of the physical condition of the Kori rubber dam is presented in the table. 2

3.2 Assessment Condition of Function of Kori Rubber Dam

Based on survey results in Kori Rubber can be known the value of the condition of the Kori Rubber dam function. The recapitulation of the Kori Rubber Dam function status is presented in Table.3

3.3 Performance Rating of Kori Rubber Dam

Based on the result of questionnaire of respondent about weighting about physical condition of dam with condition of dam function got result as follows: physical condition weir 29% and condition of dam function 71%. This shows that performance bendung more affected by the condition of the weir function. From Table 2. and Table 3 we get physical condition value and condition condition of Kori Rubber Dam as follows: physical condition value of kori rubber dam 81.79%, this indicates that the

IOP Conf. Series: Materials Science and Engineering **333** (2018) 012092 doi:10.1088/1757-899X/333/1/012092

small risk and the condition of kori rubber dams are in good physical condition (40). Condition value of kori rubber dye function 68.12%, this indicates that the small risk and condition of kori rubber dam in good condition in function (40). From the analysis of rubber weir performance obtained a value of 40, which means the rubber dam has a moderate risk with sufficient performance so that corrective maintenance is required.

Table 2. Recapitulation of the physical condition of the Kori rubber dam

No.	Building	Physical Condition of Building				Value of Building Physical Condition
		Good	Damage Lightly	Damaged Medium	Heavy Damage	
		Condition Value	Condition Value	Condition Value	Conditio n Value	
		>90 – 100%	>80 – 90%	>60 – 79%	<60%	
		Damage level <10 %	Damage level 10–20 %	Damage level 21–40 %	Damage level >40%	
1	**Building Retrieval 31%**					**25.910%**
	Doors Intake 10%		80%			8.00%
	Sediment / sludge 4%		80%			3.20%
	Debit Measurers 4%			59%		2.36%
	Exploitation Board 13%	95%				12.35%
2	**Building Drain 13%**					**10.400%**
	Door drain 10%		80%			8.000%
	Sediment / sludge 3%		80%			2.400%
3	**Wings of 12%**					**9.750%**
	Wing 9%		80%			7.200%
	Koperan / Embankment Cover 3%		85%			2.550%
4	**Body weir 37%**					**29.750%**
	Mercu / Rubber materials 14%		80%			11.200%
	Floor weir (downstream floor/Olakan) 9%		85%			7.650%
	Scale Board/Peil Schaal 4%			55%		2.200%
	Foundation and Pillar 10%		87%			8.700%
5	**Other buildings and surroundings 7%**					**5.940%**
	Bridge Crossings 1%		88%			0.880%
	Motor and Air Pump 2%		89%			1.780%
	Installation Filling / Deflation 2%	92%				1.840%
	Security Fence 1%	91%				0.910%
	House Operation 1%			57%		0.570%
		Total Number				***81.79%***

Table. 3 Recapitulation of the condition of the Kori Rubber dam function the Kori Rubber Dam

No	Building	Building Function Condition				Value of Building Function Condition
		Excellent Functional Value <90%	Good Functional Value 70-90%	Medium Functional Value 55-69%	Bad Functional Value >55%	
1	**Building Retrieval 31%**					**24%**
	Doors Intake 10%		90%			9%
	Exploitation Board 13%		80%			3%
	Building Drain 13%		90%			12%
2	**Building Drain 13%**					**6%**
	Door drain 10%			60%		6%
3	**Wings of 12%**					**7%**
	Wing 9%			60%		5%
	Koperan / Embankment Cover 3%			56%		2%
4	**Mercu / Rubber materials 14%**					**25%**
	Mercu / Rubber materials 14%			56%		8%
	Floor weir (downstream floor / Olakan) 9%			60%		5%
	Scale Board/Peil Schaal 4%			60%		2%
	Foundation and Pillar 10%	95%				10%
5	**Other buildings and surroundings 7%**					**6%**
	Bridge Crossings 1%		85%			0.85%
	Motor and Air Pump 2%		85%			1.70%
	Installation Filling / Deflation 2%		85%			1.70%
	Security Fence 1.0%	95%				0.95%
	House Operation 1.0%		80%			0.80%
	Total Number					**68.12%**

4 Conclusion

Assessing the weight of each component of kori rubber dam is done by conducting questionnaires to Practitioners and academics who know about the rubber dam components. After the weights of each component of kori rubber dam can be assessed each component of kori rubber dam based on physical condition and condition of the function. The physical condition evaluation and the function of kori rubber dam is done by conducting survey or observation. After obtained the value of physical condition and condition of function Kori rubber dam will be obtained from the performance of kori rubber rubber. The performance of kori rubber dams is based on the sum of the physical condition and the value of the kori rubber dam condition.

Acknowledgment

This research is expected to know the performance of Kori rubber Dam as a recommendation material in the implementation of OM Rubber Dam activities.

References

[1] Ministry of Public Works and Housing 2015 *Regulation of the Minister of Public Works and Public Housing No. 06 / PRT / M / 2015*
[2] Ministry of Public Works and Housing 2015 *Regulation of the Minister of Public Works and Public Housing No. 12 / PRT / M / 2015*
[3] Ministry of Public Works and Housing 2015 *Regulation of the Minister of Public Works and Public Housing No. 23 / PRT / M / 2015*
[4] LGs Ponorogo, *Ponorogo* 2012 *RTRW Bylaw 01 2012*
[5] Ministry of Public Works 2013 *02 Planning Criteria - Main Building*
[6] PU Research Agency Ministry of Public Works 2005 *Operation and Maintenance Manual Rubber dam (Pd T-07-2005-A)*
[7] Nova S D 2016 *Journal of Civil Engineering* ISSN 2339-0271
[8] Dion K F 2015 *Journal of Civil Engineering* **4(4)** 546-56
[9] Djoko L 2002 *Role of Mathematical Models in Natural Disaster Mitigation* (Yogyakarta: Gajah Mada University)
[10] Imam S 2011 Teodoil Volta **12(2)** 1-11
[11] Joygopal J 2012 *Proc. Int. Conf. Struct. Civ. Eng. 2012*
[12] Bargas M F 2008 *Journal of Civil Engineering* **5(1)** ISSN 1411-9331
[13] XQ Zhang 2002 *J. Civ. Eng.* **29** 409-420
[14] Yousef P 2016 IJBR ISSN 0976-2612 **7(1)** 59 -166

IOP Conf. Series: Materials Science and Engineering **333** (2018) 012093 doi:10.1088/1757-899X/333/1/012093

The use of Crumb Rubber as Substitute of Fine Aggregate for Hot Asphalt Mixture using Polymer Modified Bitumen

A Setyawan[1], S K Nugroho[1], A M Irsyad[1], H F Mutaqo[1], P Ramadhan[1], A Sumarsono[1], F P Pramesti[1]

[1] Roadmate Research Group, Civil Engineering Department, Sebelas Maret University, Jl. Ir. Sutami 36 A Solo 57126 Indonesia

E-mail: cenase@yahoo.com

Abstract. The development of road pavement to fulfilled the need of modern life is not only focused on heavy duty road, but also a light duty road for the convenience of road users according to its function. For example the use of pavement on the jogging track, rail crossing, playground and so on. Due to the need of an alternative and the innovation of a comfortable pavement layer, but sufficiently strong in holding the load on the layer. The alternative innovation that can be used for the respective requirement is the utilization of waste old tires as substitute material in pavement construction. In this case the use of crumb rubber made from old tire rubber as an 100% fine aggregate substitute on the asphalt mixtures is investigated. To improve the strength and durability of the mixtures, the addition of polymer modified bitumen was incorporated. The two types of asphalt mixture selected in this study by using a continuous gradation of asphalt concrete and a gap gradation of hot roll asphalt. Testing to be implemented in this research is volumetric characteristics, Marshall characteristics, resistance to abrasion and impact and permeability. Replacement of fine aggregate with crumb rubber on asphalt concrete mixture with 60/70 penetration grade bitumen and polymer modified asphalt SBS E-55 in this research are expected to be an alternative in improving the quality of pavement and overcoming the environmental problems by reuse the waste materials.

1. Introduction

Crumb Rubber is one of the results of processing a waste old tires used in vulcanized industry. The number of used old tire rubber is increasing with increasing the number of vehicles. Therefore, tire processing is an important issue. The use of crumb rubber in the pavement industry is to use the crumb rubber as an aggregate replacement in asphalt [1] and concrete [2]. In Indonesia, there have been several studies on crumb rubber. As an aggregate, CR can be used as a substitute for stone dust for the filler of a mixture of the Hot Rolls Sheet - Abrasioning Coarse so that it has stability, flow, residual stability, durability, flexibility and abrasion tires better than conventional filler. Partial replacement of aggregates with used the crumb rubber can increase the resilience of asphalt mixture to water, thus reducing road damage [3]. This study has attempted to utilize crumb rubber as a substitute of 100% fine aggregate on the HRA and AC mixtures. The bitumen used in this study were 60/70 asphalt pen and SBS E-55 polymer modified asphalt. This study will review the results of asphalt mixtures test in volumetric and Marshall, permeability and resistance to disintegration properties. The product of this research is expected to be a new innovation in the field of pavement, and can be applied on special need of pavement especially for very flexible pavement.

2. Experimental

2.1. Mixture design

There are two types of aggregate gradation of Asphalt Concrete and Hot Rolled Asphalt, the illustration of each gradations are presented in Figure 1.

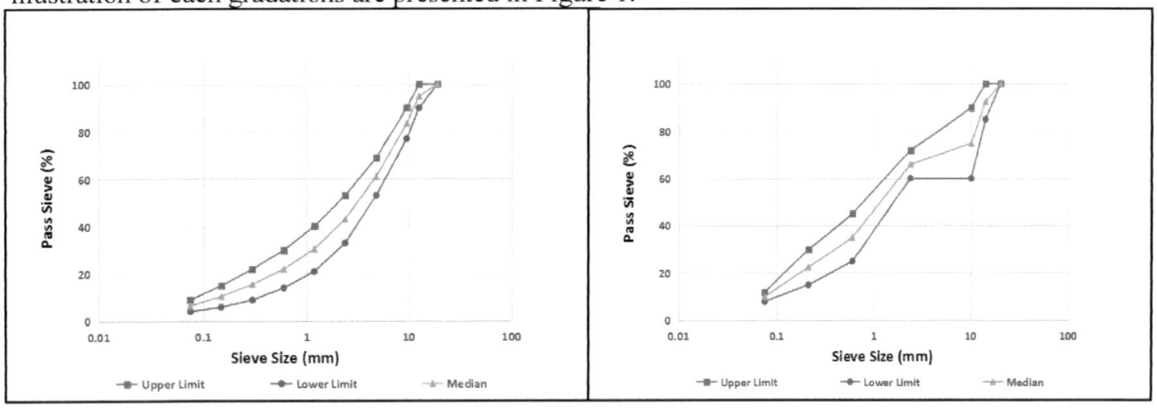

(a) (b)

Figure 1. Gradation for Asphalt Concrete (a) and Hot Rolled Asphalt (b)

Job mix design calculations are used for conversion from weight to volume to achieve the expected thickness of the test specimens. The aggregate were then weighted according to their proportion to acive total weight (aggregate + bitumen) of 900 gr. The example of the aggregate calculation as presented in Table 1.

Table 1. Aggregate calculation with average thickness of 6.32 cm with asphalt content of 5%

Sieve Number	% Pass	% Retain		Aggregate Volume		SG	Aggregate Weight		Information
		Cumulative	Per Sieve	Per Sieve	Cumulative		Per Sieve	Cumulative	
				Cm^3	Cm^3	Gr/cm^3	Gr	Gr	
1	2	3	4	5	6	7	8	9	10
3/4"	100	0	0	0	0.00	0	0.00	0.00	
1/2"	95	5	5	22.10	22.10	2.63	58.11	58.11	CA
3/8"	83.5	16.5	11.5	50.82	72.91	2.63	133.65	191.76	CA
# 4	61	39	22.5	99.43	172.34	2.63	261.49	453.26	CA
# 8	43	57	18	79.54	251.88	1.1	87.50	540.75	*CR*
# 16	30.5	69.5	12.5	55.24	307.12	1.1	60.76	601.52	*CR*
# 30	22	78	8.5	37.56	344.68	1.1	41.32	642.83	*CR*
# 50	15.5	84.5	6.5	28.72	373.41	1.1	31.60	674.43	*CR*
# 100	10.5	89.5	5	22.10	395.50	1.1	24.30	698.73	*CR*
# 200	6.5	93.5	4	17.68	413.18	1.1	19.44	718.18	*CR*
PAN	0.00	100.00	6.5	28.72	441.90	3.15	90.48	808.66	PC
Total			100	441.90			808.66		
Bitumen at %Volume			8.84	42.86	484.76	1.03	44.24	852.89	

3. Results and Discussion

3.1. Volumetric testing

Volumetric testing is intended to determine the porosity or VIM (Void in Mixture) of the specimen, which normally is required between 3-5% as reviewed by Daniel and Lachance [4]. Transportation Research Record: Journal of the Transportation Research Board, (1929), pp.28-36.. The guidance on assessing the performance testing for hot mix asphalt has been reported by Brown et al [5]. The porosity was calculated based on the specific gravity and density of the specimen. The comparison of density, specific gravity and porosity of four types of mixture can be seen in Figure 2. The use of crumb rubber as a substitute for fine aggregate in Asphalt Concrete with SBS E-55 yields the porosity value of 1.78%, while using conventional asphalt penetration 60/70 resulted in porosity of 1,957%. The porosity value of Hot Rolled Asphalt with SBS E-55 polymer modified bitumen gave the porosity value of 1,179%, while using asphalt Penetration 60/70 resulted in porosity value of 1,201%. All the porosity values are below the requirement of 3%. However, it is expected that the flexibility of crumb rubber's fine aggregates could accommodate the expansion of bitumen when the temperature increase, so it needs more experimentation to study this phenomenon. The advantages of using crumb rubber have been reported in Microscopic analysis of the interaction between crumb rubber and bitumen in asphalt mixtures using the dry process [6].

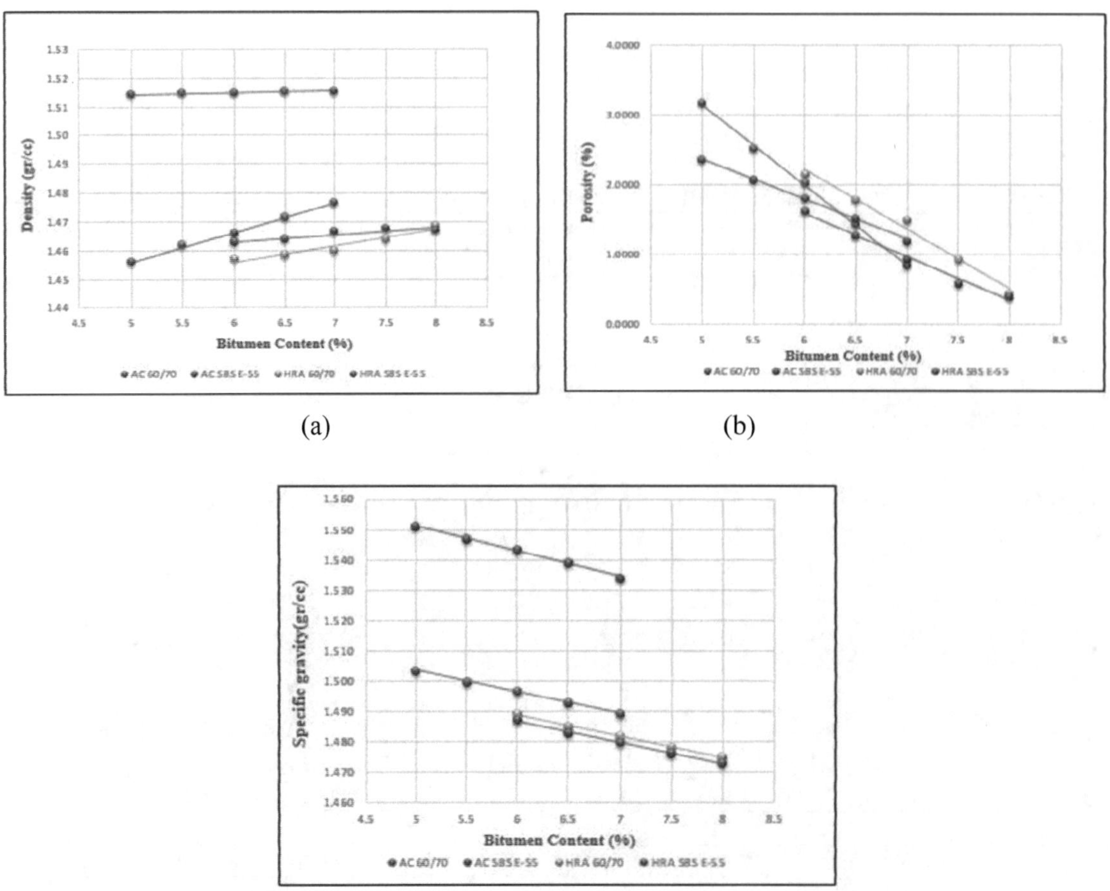

(a) (b)

(c)

Figure 2. Volumetric Properties of Rubberized Asphalt Mixture, (a) Density, (b) porosity and (c) Specific Gravity

IOP Conf. Series: Materials Science and Engineering **333** (2018) 012093 doi:10.1088/1757-899X/333/1/012093

3.2. Marshall performance

Marshall testing is an assessment procedure of the specimen to determine the optimum bitumen content (OBC) and by reviewing their stability, flow, and Marshall Quotient as also reviewed in the laboratory comparison study of the use of stone matrix asphalt in hot weather climates [7]. Comparison of Marshall characteristics of four types of mixture can be seen in Figure 3. The properties of Asphalt Concrete with SBS E-55 polymer modified bitumen gave the stability value of 130.23kg, flow 8.98mm and Marshall Quotient of 14.16 kg/mm, meanwhile with conventional asphalt penetration 60/70 gave the stability value of 115,49kg, flow 8,49mm and Marshall Quotient 13,44 kg. On Hot Rolled Asphalt SBS E-55Marshall Quotient 13,44 kg/mm, while with Asphalt Penetration 60/70 resulted in the stability value of 137,439 kg, flow 13,420 mm, Marshall Quotient 10,240 kg/mm. All the stability values are below the requirement of 800kg but having higher Marshall flow as expected to have more flexibility, the value of Marshall quotient were also indicated the similar pattern.

The result of bitumen content with crumb rubber as a substitute for fine aggregate was found as optimum asphalt content of 6.02% on Asphalt Concrete with SBS E-55 and 6.04% at Asphalt Concrete Penetration 60/70. Hot Rolled Asphalt SBS E-55 gave the optimum value of 6.86% and 7.19% on Hot Rolled Asphalt with asphalt penetration 60/70. The optimum bitumen contents of all rubberized specimens were quite similar to the conventional mixture, however, some researcher gave different conclusion on the effect of crumb rubber on the optimum bitumen content [8,9].

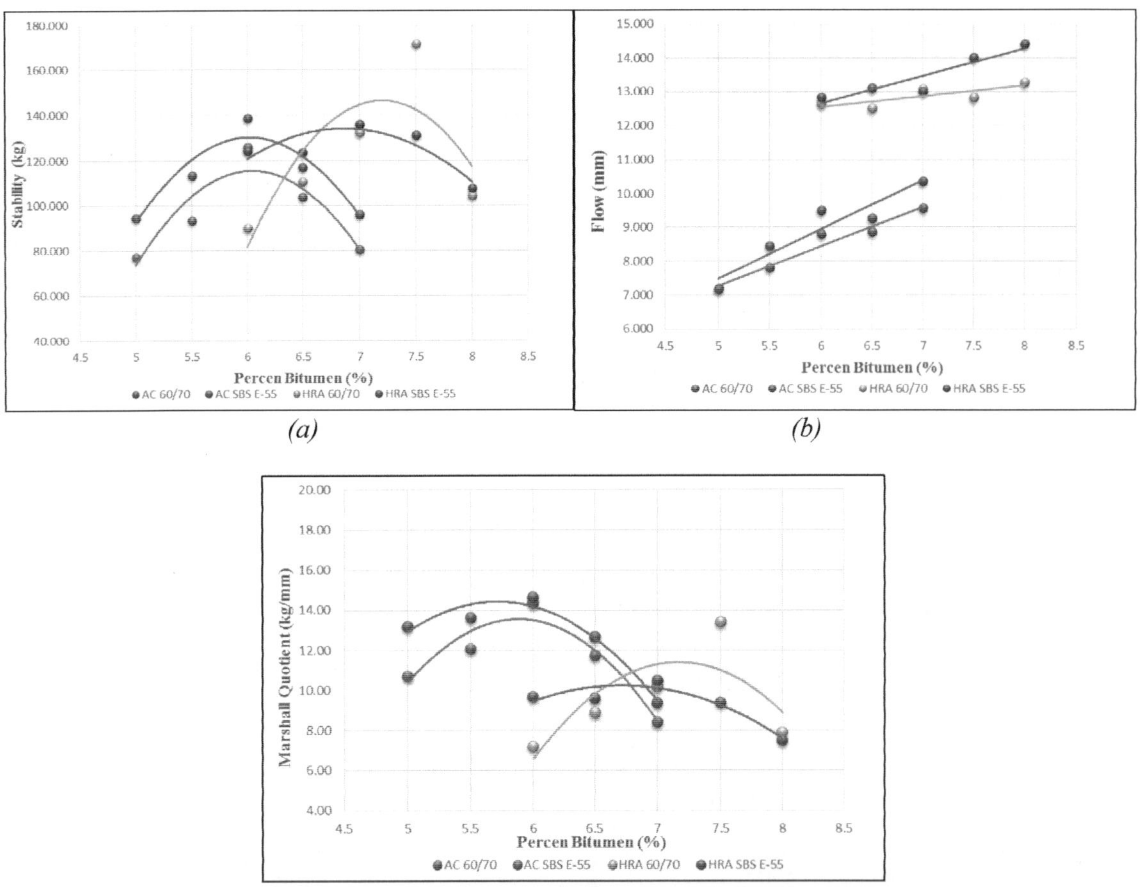

(a) (b)

(c)

Figure 3. Comparison of Marshall Quotient for four types of mixtures

3.3. Cantabro Test.

Testing Cantabro done to determine the weight loss of the test specimen after the abrasion and impact test using the Los Angeles machine for 300 rotations with the average speed of 30-33rpm without a steel ball. Previous research on assessing the resistance to abrasion and impact have been done by Chiu on the use of ground tire rubber in asphalt pavements, field trial and evaluation in Taiwan[10]. Comparison of the values of four mixed types can be seen in Figure 4. The properties of Asphalt Concrete with SBS E-55 polymer modified bitumen gave the lost of particles of 0.02%, meanwhile with 60/70 pen gave the value about 0.10%. On Hot Rolled Asphalt SBS E-55 gave the particle lost about 0.10%., meanwhile with asphalt Penetration 60/70 abrasion 0.98%. It could be seen that the resistance to abrasion and impact are extremely high, the use of polymer modified seems to strengthen the bond of aggregate particles on the mixture.

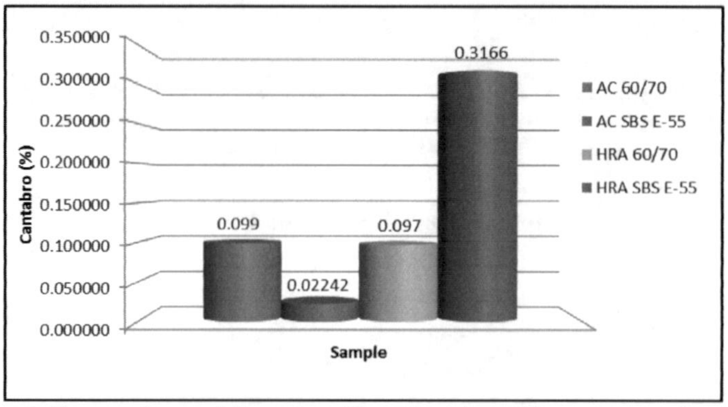

Figure 4. Comparison of Cantabro Test Results of four types of mixtures

3.4. Permeability test

The permeability test is performed to find out the permeability coefficient value. The permeability test has been carried out by Sharma and Goyal in the comparative study of performance of natural fibres and crumb rubber modified stone matrix asphalt mixtures[11]. The comparisons of the permeability values of the four mixture types are presented in Figure 5. The properties of Asphalt Concrete with SBS E-55 polymer modified bitumen gave the permeability index of of $3{,}24x10^{-4}$, while using 60/70 penetration gave the permeability index of $1{,}02x10^{-4}$. On Hot Rolled Asphalt with SBS E-55 polymer modified bitumen gave the permeability index of $1{,}90x10^{-4}$, meanwhile using asphalt penetration 60/70gave the permeability of $1{,}02x10^{-4}$. It could be concluded that the rubberized asphalt mixtures in this research are impermeable.

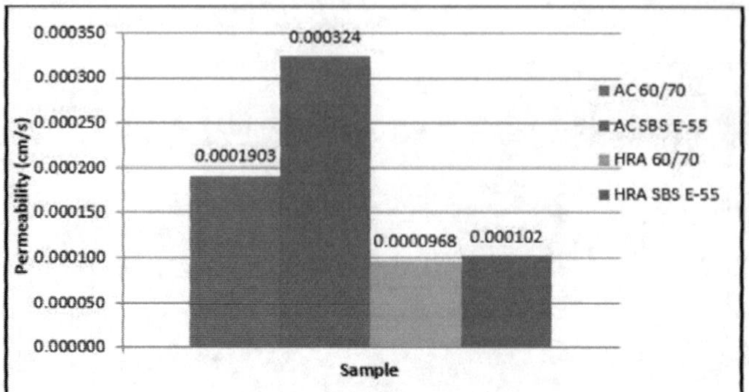

Figure 5. Comparison of Permeability Index of four types of mixtures

4. Conclusion

From the results of research and data analysis conducted can be obtained a conclusion as follows:

The addition of crumb rubber and polymer modified bitumen in Marshall properties of asphalt Concrete and Hot Rolled asphalt indicates the lower value of Marshall Stability, the higher the flow rate and the lower the Marshall quotient. The influences on the volumetric properties indicate the lower the density and specific gravity values, so that resulted in the low porosity or air voids.

This suggests that along with the substitution of fine aggregates with crumb rubber levels lead to the lower use of petroleum asphalt, so more friendly to the environment. With elastic and durable blending properties, the innovation of crumb rubber as a substitute for fine aggregates on asphalt mixes is quite feasible for pavement with special requirements as playgrounds, jogging track or rail crossing. The more resistance to disintegration and the impermeable layer are the advantages of rubber substitution to protect the underneath layer.

References

[1] Takal H B 1991 *Transp. Res. Rec.* **1339** 23
[2] Holmes N, Browne A, and Montague C 2014 *Constr. Build. Mater.* **73** 195-204
[3] Setyawan A, Febrianto N, and Sarwono D 2017 I*OP Conf. Series: Earth and Environmental Science* **75** 12-16
[4] Daniel J and Lachance A 2005 *Transp. Res. Rec: Journal of the Transportation Research Board* **1929** 28-36
[5] Brown E R, Kandhal P S, and Zhang J 2001 *National Center for Asphalt Technology Report* 01-05
[6] López-Moro FJ, Moro M C, Hernández-Olivares F, Witoszek-Schultz B, and Alonso-Fernández, M 2013 *Constr. Build. Mater.* **48** 691-699
[7] Asi I M 2006 *Constr. Build. Mater.* **20** 982-989
[8] Heitzman M 1992 *Transp. Res. Rec.* **1339**
[9] Cao W 2007 *Constr. Build. Mater.* **21(5)** 1011-1015
[10] Chiu C T 2008 *Resources, Conservation and Recycling* **52(3)** 522-532
[11] Sharma V and Goyal S 2006 *Can. J. Civil Eng.* **33(2)** 134-139

IOP Conf. Series: Materials Science and Engineering **333** (2018) 012094 doi:10.1088/1757-899X/333/1/012094

The influence of using accelerator addition on High strength self-compacting concrete (HSSCC) in case of enhancement early compressive strength and filling ability parameters

Wibowo[1] and Y Fadillah[2]

[1] *SmartConMart Research Group*, Civil Engineering Department, Universitas Sebelas Maret, Surakarta, Indonesia

[2] Civil Engineering Department, Universitas Sebelas Maret, Surakarta, Indonesia.

E-mail : wibowotsipil87@gmail.com

Abstract. Efficiency in a construction works is a very important thing. Concrete with ease of workmanship and rapid achievement of service strength will to determine the level of efficiency. In this research, we studied the optimization of accelerator usage in achieving performance on compressive strength of concrete in function of time. The addition of variation of 0.3% - 2.3% to the weight of cement gives a positive impact of the rapid achievement of hardened concrete, however the speed of increasing of concrete strength achievement in term of time influence present increasing value of filling ability parameter of self-compacting concrete. The right composition of accelerator aligned with range of the values standard of filling ability parameters of HSSCC will be an advantage guidance for producers in the ready-mix concrete industry.

1. Introduction

There are several types of concrete that are classified based on the value of compressive strength, i.e. normal concrete, high quality concrete, and ultra-high strength concrete. High quality concrete is a concrete with a compressive strength more than 40 MPa, while ultra-high strength concrete is concrete with a compressive strength above 80 MPa. High quality concrete defined as concrete having a minimum compressive strength of 6000 psi or about 41 MPa [1].

High quality concrete has a high durability and low cement water factor ratio compared to conventional concrete that has only a compressive strength of 30 MPa - 40 MPa. To obtain a compressive strength until 60 MPa or equivalent to 650 kg/cm2 can use the type of concrete self-compacting concrete. Self-compacting concrete is a very plastic and easy-flowing fresh concrete because its own weight can fill up throughout the mould. This concrete has the properties to solidify itself without the need for the aid of a vibrator commonly used in concrete compaction. SCC concrete is often called a flowing concrete because it has a very high slump value. Self-compacting concrete must have a homogeneous, cohesive, non-segregated, and no bleeding properties.

In order to achieve high workability Super plasticizer can used as an added ingredient that directly mixed into the mortar of concrete, and then to accelerate the hardening of concrete can use admixture materials. Added material (admixture) is a powder or liquid material, which added to the mixture of concrete during stirring in order to alter the nature of the mortar or its concrete.

In this research used accelerator as admixture. The use of the accelerator is very common for concrete casting in water, basalt waterproof work, seafront repair work in tidal areas, or if concrete structural elements are, need to be immediately burden by subsequent work in relation to strict execution time. This is because the accelerator main function is decrease duration of binding process and development of the initial strength of concrete. This material used to shorten cement-binding time to accelerate the achievement of concrete strength.

The use of this accelerator-added material should also consider the level of dissolved chloride ions in the required hard concrete, if exceeded would be at risk of corrosion in the iron or steel reinforcement. In the use of this accelerator should also pay attention to the time settings are faster and curing done should be as perfect as possible to achieve the desired higher initial strength.

2. Experimental

Objectives of the experimental program are planning accurately how to conduct research in laboratory. There are two type of approach for studying of SCC. Fresh concrete measured for observing self-compacting parameters. Hardened concrete gave the values of compressive strength that affected by accelerator addition [2].

2.1. Mix Proportion

In this study, repeated trial mixed design to obtain the proportion or composition of materials in accordance with the target research [3]. The composition of mortar mixture used is:

Table 1. Composition of self-compacting concrete

Material used	**Control material**	**Dosage of accelerator**		
		0,4%	**1,2%**	**2%**
Cement (kg)	15,47	15,41	15,26	15,106
Fine aggregates/sand (kg)	15,82	15,82	15,82	15,82
Coarse (kg)	17,65	17,65	17,65	17,65
Water (ml)	4180	4163	4120	4079
Super plasticizer (ml)	247,48	246,11	243,59	241,14
Accelerator (ml)	0	48,18	143,07	386,73

3. Result and Discussion

3.1. Tests of Parameters of Self-Compacting Concrete (SCC)

Fresh concrete testing performed to see mixed concentration as a basis for ease of work. Fresh concrete tests for SCC include testing of flow table, L-box, and V-funnel

3.1.1 L-Box Testing. In this research L-box, testing done to know the passing-ability of self-compacting concrete.

$$PA\ (\%) = \frac{H_1}{H_2} x100\% \qquad (1)$$

Where:

PA: The initial high difference divided by the concrete end of the flow (passing Ability)

H_1: Initial level of concrete flows (cm)

H_2: The final height of the concrete flows (cm)

IOP Conf. Series: Materials Science and Engineering **333** (2018) 012094 doi:10.1088/1757-899X/333/1/012094

Table 2. Passing ability of fresh concrete

No	Dosage of admixture		h1 (cm)	h2 (cm)	PL=h1/h2
	Super plasticizer	Accelerator			
1	1.70%	0%	8	8	1
2	1.70%	2%	6.5	8	0.8125
3	1.70%	1.20%	7.3	7.5	0.973333333
4	1.70%	0.40%	7.5	7.8	0.961538462

In fresh concrete with lower accelerator content, passing-ability or the ability of concrete to pass the spacing between the reinforcement is better because accelerator is not quite enough to reduce duration of setting time, and vice versa the higher accelerator content then passing-ability of concrete to be slightly faltered because the time setting decrease significantly.

3.1.2. Slump Flow Testing. Flow Table Testing conducted to determine the filling-ability of self-compacting concrete.

Table 3. Result of filling-ability testing

Dosage of admixture		d1	d2	Time (s)
Super plasticizer	*Accelerator*			
1.70%	0%	740	750	05.64
1.70%	2%	660	670	06.11
1.70%	1.20%	680	710	05.86
1.70%	0.40%	730	745	05.13

Filling ability of cement base material influenced by duration of setting time. In range accelerator addition between 0.4% until 2%, fresh concrete was satisfy with self-compacting concrete requirement properties.

3.1.3.V-funnel testing. V-funnel testing done to figure segregation resistance of self-compacting concrete. The test kit is made of a V-shaped steel plate and there is an opening valve at the bottom [4].

Table 4. Result of segregation resistance of self-compacting concrete

Dosage of admixture		Time (s)
Super plasticizer	*Accelerator*	
1.70%	0%	08.00
1.70%	2%	26.21
1.70%	1.20%	16.54
1.70%	0.40%	12.37

Result of the V-funnel test showed that accelerator influence performance of self-compacting concrete in term of segregation resistance, it could be understand that the duration of setting time decrease sharply when amount of accelerator raised.

The lower content of accelerator in cement paste show the best properties on segregation resistance or the ability of concrete to maintain the better homogeneity, and when the amount of accelerator raise, ability to maintain homogeneity of concrete material decrease clearly and then finally affected quality of concrete as a member of structure.

3.1.4 V-Vunnel Testing T 5 second. V-funnel T5 minute test was conduct to determine segregation resistance of self-compacting concrete in 5 second after mixing. Using same instrument as V-funnel test but adding time lag 5 minutes after mixing. The following table illustrate the result of V-funnel T 5 second.

Table 5. Result of V-funnel T5

Dosage of admixture		Time (s)
Super plasticizer	***Accelerator***	
1.70%	0%	12.21
1.70%	2%	65.33
1.70%	1.20%	56.43
1.70%	0.40%	52.65

V-funnel test T5 second is the most important result, because utilization of accelerator have to limit no more than 5 second time lag. Hence, all above result is out of range from minimal requirements SCC properties.

3.2. Compressive strength of self-compacting concrete

Compressive strength was measured use cylindrical specimen with diameter 15 cm and height 30 cm, where Compression Testing Machine (CTM) will crush this specimen at age 14 and 28 days. Content variations of accelerator used are 0.4%, 1.2%, and 2%. The following table show the detail various type of specimen used in this research.

Table 6. Specimen for compressive strength testing

Dosage of Super plasticizer	Dosage of Accelerator	Specimen code
1.7%	0%	BA A
1.7%	0.4%	SPACC A
1.7%	1.2%	SPACC B
1.7%	2%	SPACC C

The dosage of accelerator varied, because objective of testing is determining trend and optimum dosage that satisfy with the properties of SCC and giving an optimum of compressive stress. Nevertheless, dosage of plasticizer is constant in order for minimizing plasticizer influence towards compressive strength of concrete.

In this test, specimen classified to be two group: specimen tested on 14 days and 28 days after remolding and curing. Variations of mixed accelerator dosage added with super plasticizer presented in table form as follows:

Table 7. Compressive strength of self-compacting concrete

Compressive Strength (MPa)	Dosage of *Accelerator*			
	0%	0.4%	1.2%	2%
14 days	37.82	42.25	44.70	46.68
28 days	41.87	44.79	47.91	53.66

Selection of testing time on days 14 and 28 is intend to observe from the time aspect whether SCC behavior is shown in **Table 6** and **Table 7**, would be affected by result of the addition of accelerator and plasticizer.

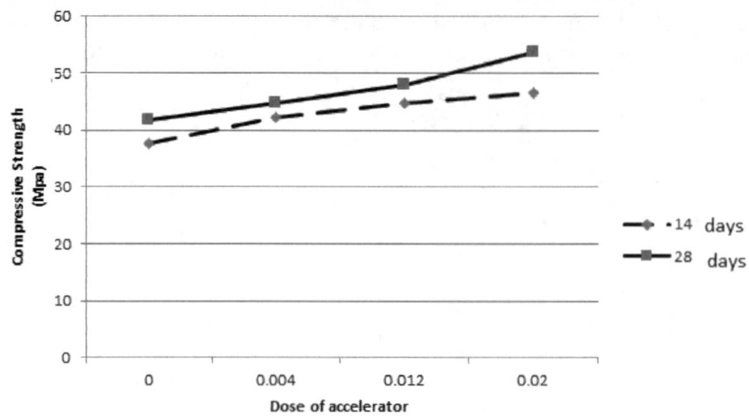

Figure 1. Compressive strength of SCC with addition accelerator and superplasticizer

Figure 1 show that the control specimen meet the requirement of high strength concrete. At the age of 14 days, specimen has value of compressive strength about 37.6 MPa, while at 28 days reached 41.87 MPa. Base on the previous observation and measurement, this specimen also meet the requirement of self-compacting concrete.

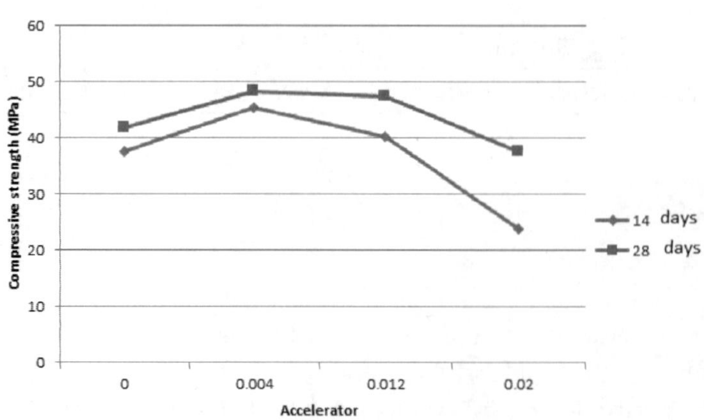

Figure 2. Compressive strength with only accelerator addition

If observation focused on accelerator dosage and how influence towards compressive strength of concrete, then it could be expressed that accelerator has significant role in order compressive strength of concrete attain as high as concrete without accelerator in shorter time.

In case of self-compacting concrete, accelerator addition is unfavorable. Rapid setting-time give consequences, self-compacting concrete have to cast fast. **Figure 1** also illustrate in term of quality or strength, accelerator give positive contribution in raising compressive strength even more than 2%, but **Figure 2** indicates that the effect of adding the accelerator alone without the addition of plasticizer will have the best contribution in the range of 0.4% to 1.2%. It complement the results of previous studies [7], [8] and [9].

4. Conclusion

The role of the accelerator in the concrete material is to accelerate the achievement of compressive strength before the usual time (28 days), but from this research got the fact that accelerator greatly affect the fulfillment of self-compacting concrete standard, so only doses ranged 0.4% which meet all SCC specifications.

HSSCC testing for passing-ability and filling-ability parameters meets EFNARC standard, while for segregation resistance parameter does not meet EFNARC standard unless V-funnel test is 0.4%

In terms of the effect of the quality of concrete, the addition of accelerator contributes significantly to raising the quality of concrete. However, due to the limited compliance of self-compacting concrete specs, the best-recommended use of accelerator be in the dose range of 0.4% - 1.2%.

Acknowledgement.

SmartConMat Research Group, Civil Engineering Universitas Sebelas Maret Surakarta, supports this research. Appreciation and thankful gratitude to all supported team of research group, which has provided many feedbacks, inputs and discussions for the refinement of the results of research studies of special topics High Strength Self-compacting Concrete (HSSCC)

Reference.

[1] *Nawy, Edward G 1998 Beton Bertulang : Suatu Pendekatan Dasar (Bandung : PT Rafika Aditama)*
[2] Sunarmasto, Kristiawan S, Putri N, & Basuki A 2013 *J. Teknik Sipil: Universitas Sebelas Maret.*
[3] *The European Federation of National Trade Associations (EFNARC) 2001 Specification and guidelines for sel-compacting concrete (Farnham : Association House)*
[4] Ozawa K, Sakata N, and Okamura H 1995 *JSCE Concrete Lib.* ***25*** *59-75*
[5] El Chabib H, Nehdi M 2006 *ACI Mat J* **103**(5)
[6] Tan H, Bauguo M A, Xianguo L, Jian S, Yang H 2014 *Wuhan University of Technology and Springer Verlag Berlin Heidelberg*
[7] Betancourt G H 1998 *J. Mat. Struct.* **21** 286-288
[8] S Chandra J, Bjornstro M 2002 J. *Cem. Conc. Res.* **32** 1605-1611
[9] Kadek A P 2015 *Perubahan Kuat Tekan Self Compacting Concrete yang Menggunakan Bahan Accelerator Concrete Admixture Type C yang disebabkan oleh Lingkungan Agresif ($MgSO_4$) di awal Perkerasan Beton* (Bali : Universitas Udayana)

IOP Conf. Series: Materials Science and Engineering **333** (2018) 012095 doi:10.1088/1757-899X/333/1/012095

Characteristic Asphalt Concrete Wearing Course (ACWC) Using Variation Lime Filler

R A Permana[1], F P Pramesti[1], A Setyawan[1]
[1] Civil Engineering, Universitas Sebelas Maret, Jl. Ir. Sutami 36A, Kentingan, Surakarta, 57126, Central Java, Surakarta, Indonesia

E-mail : asepperm05@yahoo.com

Abstract. This research use of lime filler Sukaraja expected add durability layers of concrete pavement is asphalt damage caused by the weather and load traffic.This study attempts to know how much value characteristic Marshall on a mixture of concrete asphalt using lime filler.This research uses experimental methods that is with a pilot to get results, thus will look filler utilization lime on construction concrete asphalt variation in filler levels 2 %, 3 %, 4 %.The results showed that the use of lime filler will affect characteristic a mixture of concrete asphalt.The more filler chalk used to increase the value of stability.On the cretaceous filler 2 % value of stability is 1067,04 kg.When lime filler levels added to the levels of filler 4 %, the value of stability increased to 1213,92 kg. The flexibility increased the number of filler as levels lime 2 % to 4 % suggests that are conducted more stiff mix.

1. Introduction

Asphalt Concrete is one type of layers construction for flexible pavement. A mixture of Asphalt Concrete consisted of Coarse aggregate, Fine aggregate, Filler, and by using asphalt as a belt. Filler commonly called filler material can be obtained from the breakdown of rocks naturally or artificially. Filler commonly used is the type of stone dust filler.

One of the filler used is lime extinguished (hydrated lime). Lime is one of the mineral much used by the industrial sector or construction and farm, among others to building materials, stabilization highway material, Lime for agriculture. Physically lime is a body white and fine. A starter lime is limestone. Limestone containing calcium carbonate ($CaCO_3$), the chemical or the properties of materials a base containing lime is different from one place to another. Even in one place is not necessarily same. Lime is economically since their prices cheap. Limestone used as filler in this research obtained from the results of villagers Sukapura mining in Sukaraja, West Java .

According to Peter E.Sebaaly (2006) [6], The advantages use of lime out as filler in concrete pavement in Hot Mix Asphalt include :

1. Improving the survival of concrete pavement is asphalt against exfoliation stripping the water.
2. Reduce or hinder oxidizing asphalt process.
3. Fix the properties of mechanical mixture as resistance against grooves (rutting) and exhaustion (fatique) on asphalt pavement.
4. Leveraging performance a mixture of asphalt concrete by raising a bond between the asphalt and aggregate.

Hence and attempted to do something research on the use of lime as filler on a mixture of Asphalt Concrete a surface coating especially a kind ACWC. To research will be emphasises the

extent to which the influence of variation lime filler levels characteristic mixture of Asphalt Concrete, which in turn can be seen characteristic mixed with respect to changes filler the cretaceous levels, and it is expected that the lime can be used as a alternative filler in highway construction.
This report aims to review the influence of the use of variation lime filler levels to characteristic a mixture of Asphalt Concrete (ACWC).

2. Experimental

The research method are arranged to give easy in the implementation of the research so that a walk more precise, effective, and efficient. Testing Course aggregate use of broken stone in a dry state. The Fine Aggregate, worn materials that sharp surface, and clean of dirtyes or other unwanted materials. The fine aggregate composed of sand clean, fine materials the breakdown of stone or combination of them in a dry state. Asphalt used in this study asphalt type Asphalt Penetration 60/70.

Before held mixing so gradations mix to be used should be determined first. Filler used is lime with variation lime 2 %, 3 %, and 4 % to the total aggregate. Estimates the asphalt in a mixture of the asphalt steady estimated by means of determined steady empirically with the equation (Pb) as follows:

$$Pb = 0{,}035\ (\%\ CA) + 0{,}045\ (\%\ FA) + 0{,}18\ (\%\ \textit{filler}) + K \quad (1)$$

Description :
CA = Percent aggregate suspended sieve no.8
FA = Aggregate escaped sieve no.8 and suspended sieve no.200
filler = percent aggregat at least 75 % escaped no.200
k = constant (0,5 - 1,0) to laston

The matrix sample of done systematically with recording all data resulting from series of testing from start to finish with still consider validation data collected. The determination of Optimum Asphalt Content determination of asphalt levels steady using charts by plotting the values characteristic of the results of the Marshall Testing.

3. Results and Discussion

3.1. *The Result Testing Materials*

Table 1. The Results Lime Filler

No.	Testing	Result Testing	Specification	Unit
1	Specific Gravity	1.75	-	gr/cc

Table 2. Characteristic of Marshall

	Lime	Asphalt Contents (%)							
	%	4.5	5.0	5.5	6.0	6.5	7.0	7.5	8.0
Density	2	2.177	2.223	2.272	2.290	2.304	2.308	2.314	2.310
	3	2.199	2.234	2.298	2.308	2.317	2.323	2.325	2.326
	4	2.214	2.267	2.321	2.331	2.330	2.331	2.340	2.340
VMA	2	20.568	19.289	17.912	17.648	17.525	17.762	17.924	18.455
	3	19.516	18.611	16.664	16.697	16.788	16.945	17.269	17.600
	4	18.703	17.143	15.582	15.617	16.061	16.413	16.458	16.861
VITM	2	13.049	10.655	8.070	6.733	5.537	4.781	3.931	3.507
	3	11.918	9.888	6.690	5.673	4.708	3.848	3.174	2.545
	4	11.051	8.284	5.499	4.469	3.932	3.288	2.280	1.683
VFWA	2	36.565	44.997	55.057	61.849	68.406	73.088	78.090	81.000
	3	38.941	46.872	59.880	66.053	71.967	77.290	81.619	85.551
	4	40.929	51.838	64.735	71.387	75.524	79.968	86.153	90.022
Stability	2	876.960	941.760	997.920	1067.040	1045.440	1041.120	920.160	773.280
	3	924.480	1062.720	1144.800	1200.960	1166.400	1054.080	924.480	799.200
	4	928.800	1067.040	1157.760	1213.920	1175.040	1062.720	933.120	812.160
Flow	2	3.600	3.733	3.800	3.873	3.930	3.997	4.190	4.430
	3	3.480	3.593	3.657	3.727	3.807	3.987	4.107	4.217
	4	3.273	3.313	3.400	3.637	3.700	3.817	3.923	4.130
Marshall Quotient	2	201.801	212.077	219.555	231.298	223.282	219.899	184.703	145.874
	3	221.665	248.622	265.299	271.897	255.884	223.158	189.982	159.256
	4	238.327	267.004	286.740	278.655	268.531	234.533	202.770	165.675

3.2. The relationship between Lime Filler Content of Hot Mix Asphalt

3.2.1. Density. With the additional Lime Filler content into a mixture of accompanied by increasing the Asphalt in a mixture indicated can raise the density of a mixture of Asphalt Concrete. The main cause to high density is mineral content delicate in this Lime Filler on a mixture of that makes density aggregate bituminous between with a high and cause of the nature of interlocking particles aggregate with increase asphalt.

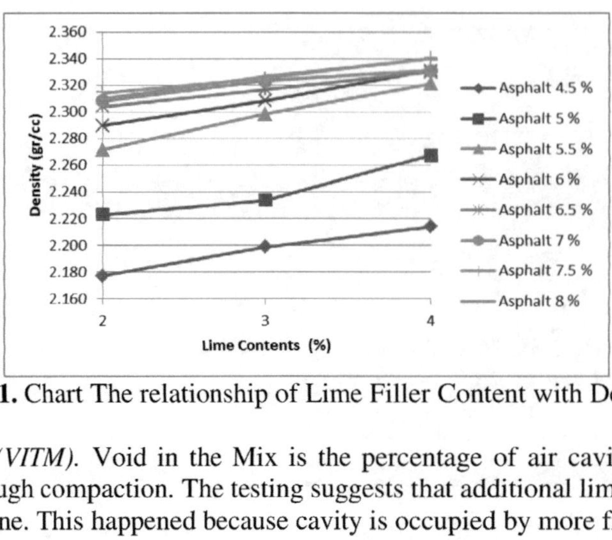

Figure 1. Chart The relationship of Lime Filler Content with Density

3.2.2. Void in The Mix (VITM). Void in the Mix is the percentage of air cavities in a mix of the aggregate and asphalt through compaction. The testing suggests that additional lime into the mix likely to cause value VITM decline. This happened because cavity is occupied by more filler.

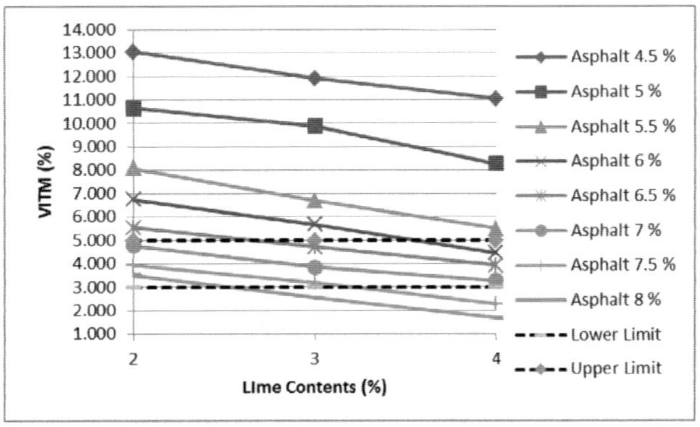

Figure 2. Chart The relationship of Lime Filler Content with VITM

3.2.3. Void in Mineral Aggregat (VMA). VMA is the percentage cavity is between the aggregate in a mixture of Asphalt Concrete expressed in (%) against the volume of a mixture of Asphalt Concrete. VMA tend to decrease far lime filler content. This happened because the lime filler content make space available to accommodate the volume of asphalt and volume air cavities required in a mixture of the less.

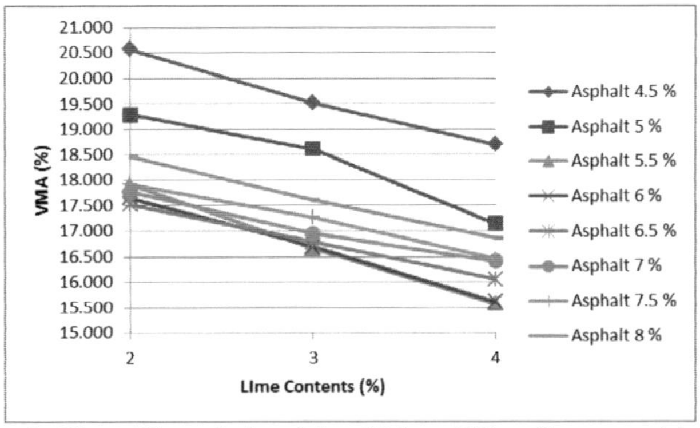

Figure 3. Chart The relationship of Lime Filler Content with VMA

3.2.4. Void Filled With Asphalt (VFWA). VFWA is cavity unfilled asphalt on a mixture of after being process density expressed in percent against the grain aggregate cavity (VMA). Factors that can affect VFWA include the asphalt, gradations aggregate, compaction energy, and heating asphalt. From the testing suggests that value VFWA tend to increase in line with the number of lime filler content. This hapened because lime filler content is absorb asphalt and fill a more cavity.

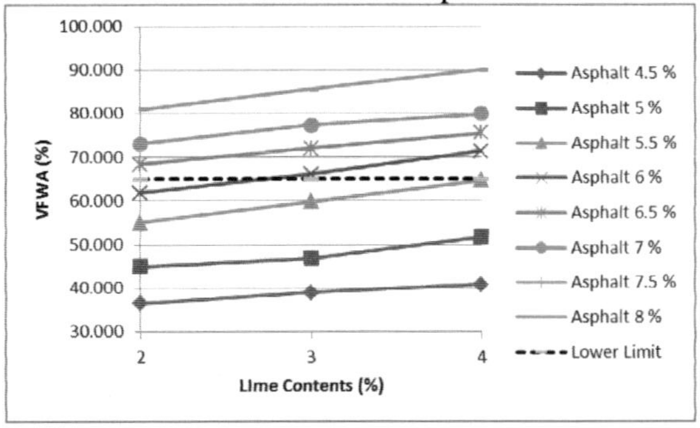

Figure 4. Chart The relationship of Lime Filler Content with VFWA

3.2.5. Stability. Stability is capability pavement layers receives the load traffic without changed a fixed form (permanent deformation) wavelike, rutting, and experienced bleeding. The value of stability influenced by asphalt cohesion, the asphalt content, internal friction, the nature of interlocking of aggregate particles, shape and texture surface and aggregate gradations. From the testing suggests that the stability obtained tend to increased by adding lime filler content.

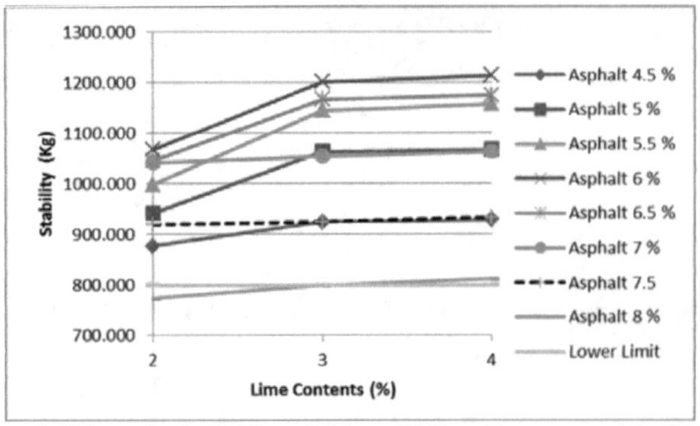

Figure 5. The relationship of Lime Filler Content with Stability

3.2.6. Flow. Flow been the large deformation that happened one another pavement due to the traffic. From the tests showed that the more lime filler content on a mixture of asphalt concrete this value flow the less. This happened because the lime filler content created a cocktail be meeting that deformation due to the reduced .

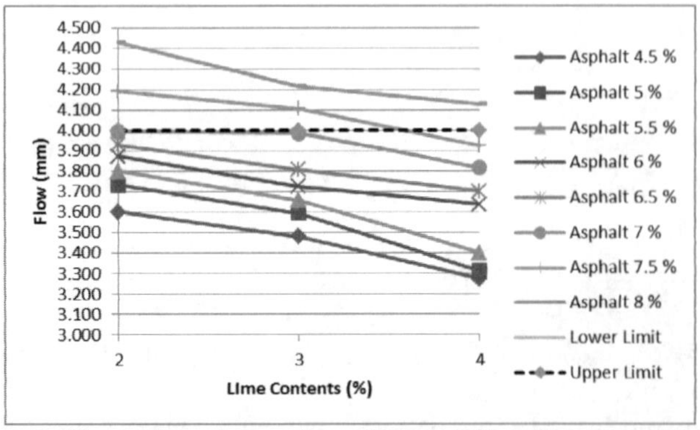

Figure 6. Chart The relationship of Lime Filler Content with Flow

3.2.7. Marshall Quotient (MQ). Marshall Quotient (MQ) is the result for between stability and flow. The value MQ is The mixed flexibility suggests that is the value MQ on a mixture of the more stiff the mix, similarly if the less value MQ so the malleability and plastisitas mix will be bigger. From the testing suggests that value mq tends to be up along with increasing lime filler content. This hapened because the lime filler content created a cocktail into a solid and rigid. It can be seen from the form of a curve open down so it can be additional lime filler content too high, MQ value could reduce, because it reduces cohesion between asphalt and aggregate.

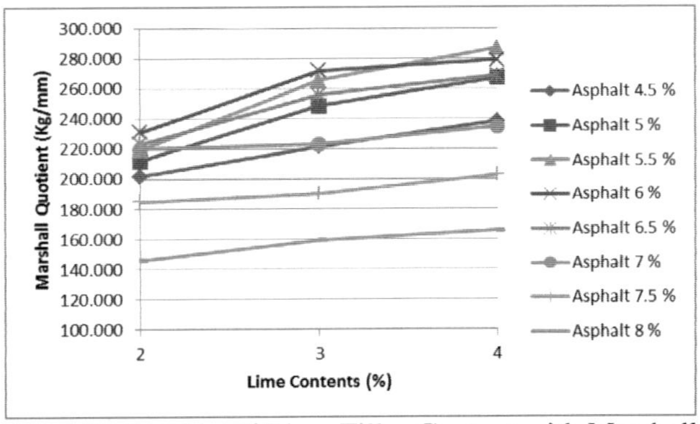

Figure 7. Chart The relationship of Lime Filler Content with Marshall Quotient (MQ)

4. Conclusion

Based on results and analysis characteristic a mixture of asphalt concrete ACWC using variation lime filler content indicated increasing a mixture of asphalt concrete ACWC expressed in characteristic as follows :

1. Stability in mixture of using lime filler lime tend to increased received an steady then fell. Stability highest levels of asphalt finish in 6 % with lime filler 4 %.
2. The value of mix fleksibilty expressed in the Marshall Quotient (MQ), suggests that value tend to increases the rising lime filler content into a mixture of asphalt concrete. The research obtained that mix will be more stiff worth MQ that tends to increases the rising variation lime filler into the mix.

References

[1] AASHTO 1993 *Guide For Design of Pavement Structure* Washington DC.
[2] Brown SF and Brunton 1984 *An Introduction to the Analytical Design of Bituminous Pavement, 2th Edition* University of Nottingham, England.
[3] Craus dkk 1981 *Durability of Bituminous Paving Mixture as Related to Filler Typaand Properties, Proceedings Association of Asphalt Paving Technologists Technical Sessions* San Diego California February 16,17, and 18, 1981, vol **50**.
[4] Departemen Pekerjaan Umum 2000 *Spesifikasi Campuran Beraspal Panas* Badan Penelitian dan Pengembangan PU Departemen Pekerjaan Umum Jakarta.
[5] Rohman A, Setyawan A., Syafii 2015 Pemilihan Material Overlay jalan yang tahan terhadap perubahan iklim *Tesis* Megister Teknik Sipil Universitas Sebelas Maret Surakarta.
[6] Sebaaly, Peter E 2006 *The Benefits of Hydrated Lime in Hot Mix Asphlat* The National Lime Association.
[7] Sukirman S 1999 *Perkerasan Lentur Jalan Raya* Penerbit Nova Bandung.
[8] Tahir A 2009 *Jurnal SMARTEK* **7(4)** Universitas Tadulaku Palu.

IOP Conf. Series: Materials Science and Engineering **333** (2018) 012096 doi:10.1088/1757-899X/333/1/012096

Infrastructure performance of irrigation canal to irrigation efficiency of irrigation area of Candi Limo in Mojokerto District

S Kisnanto[1], R R R Hadiani[1] and C Ikhsan[1]

[1]Master Program of Civil Engineering, Sebelas Maret University, Jl. Ir. Sutami 36A Surakarta

E-mail : kisnanto.satriyo@student.uns.ac.id

Abstract. Performance is a measure of infrastructure success in delivering the benefits corresponding it's design implementation. Debit efficiency is a comparison between outflow debit and inflow debit.Irrigation canal performance is part of the overall performance aspects of an irrigation area. The greater of the canal performance will be concluded that the canal is increasingly able to meet the planned benefits, need to be seen its comparison between the performance and debit efficiency of the canal. The existing problems in the field that the value of the performance of irrigation canals are not always comparable to the debit efficiency . This study was conducted to describe the rèlationship between the performance of the canal to the canal debit efficiency. The study was conducted at Candi Limo Irrigation Area in Mojokerto Disctrict under the authority of Pemerintahan Provinsi Jawa Timur. The primary canal and secondary canal are surveyed to obtain data. The physical condition of the primary and secondary canals into the material of this study also. Primary and secondary canal performance based on the physical condition in the field. Measurement inflow and outflow debit into the data for the calculation of the debit efficiency. The instrument used in this study such as the current meter for debit measurements in the field as a solution when there is a building measure in the field were damaged, also using the meter and the camera. Permen PU No.32 is used to determine the value of the performance of the canal, while the efficiency analysis to calculate a comparison value between outflow and inflow debit. The process of data running processing by performing the measurement and calculation of the performance of the canal, the canal debit efficiency value calculation, and display a graph of the relationship between the value of the performance with the debit efficiency in each canal. The expected results of this study that the performance value on the primary canal in the range of 0 to 100 % with debit efficiency value in the range of 0 to 100 %, while for the secondary canal 1 has a performance range between 0 to 100% with efficiency ranges between 0 to 100%, the performance of the secondary canals 2 ranges between 0 to 100% with efficiencies ranging from 0 to 100%, the secondary canal 3 performance ranges between 0 to 100% efficiency ranges between 0 to 100%, the secondary canal 4 performance ranges between 0 to 100% efficiency ranges between 0 to 100% and secondary canals 5 performance ranges between 0 to 100% efficiency ranges between 0 to 100%. For the tendency to expect from the performance and efficiency of the debit canal can have a proportional clockwise or counterclockwise, which amount can be random. The tendency to be graphed the relationship between performance and efficiency of the debit of each segment studied canal.

IOP Conf. Series: Materials Science and Engineering **333** (2018) 012096 doi:10.1088/1757-899X/333/1/012096

1. Introduction

Irrigation in Indonesia are mostly used for rice crops, and partly also for palawija, fruits and vegetables. Irrigation management continue to be improved both the operation and maintenance of irrigation networks, making it more advanced and more efficient in distributing water availability.

Water is one of the most valuable natural resources that has very diverse functions. Increasing population growth in Indonesia is also increasing the need for water for various purposes, such as agriculture, settlement, urban, industrial, fishery, energy, tourism, environment and others. This results in water becoming scarce at a certain time for some locations prone to water shortages [1]

Irrigated agriculture would seem to be a prime candidate for interventions aimed at reducing water demand and inducing water savings [2]. The conveyance efficiency is the efficiency of canal and conduit networks from the reservoir, river diversion, or pumping station to the offtakes of the distributary system [3]

Considering the importance of irrigation water efficiency in reducing water criterion, it is necessary to conduct research to know efficiency of irrigation water utilization and water availability index with pipe irrigation input so that available water can be utilized properly and efficiently. Water utilization in a good and efficient way can be one step in irrigation water management [4].

Seepage from canal cannot be controlled completely. Significant seepage losses do occur from a canal even if it is lined. This problem more aggravates in percolation soils. Thus, the conversion of canal lining to pipeline may lead to further reduction in seepage losses [5].

Performance assessment of irrigation canals carried out by water worker with the search, observe any damage, leakage and interference of wild plants that exist in the irrigation canals. Construction activities carried out so that the performance assessment of irrigation officers observation can be improved.

Conservation of water resources has become an international issue that always suggests the importance of saving water consumption, because many countries are experiencing a water crisis and drought. Indonesia's water availability are abundant enough, although its advanced management hasn't apply well. Conflicts among farmers occasionally occur during the dry season, because of scarcity of water and poor water management. The more efficient a canal that can reach farther distribution, so it is expected to reach areas with water scarcity problem.

A gradually decrease in the efficiency of irrigation canals were influenced by many factors, which can lead to water scarcity, particularly in the rice field far away from the source or the dam intake. Good maintenance for technical irrigation in addition to improving performance may also improve the efficiency of the irrigation canal as it has been planned accordingly.

This study discusses the relationship of the performance of the canal towards the discharge efficiency of irrigation canals, is expected to provide insights and another point in the maintenance and rehabilitation of irrigation infrastructure. The amount of the performance and efficiency of data derived from observations in the field.

1.1. Efficiency

The efficiency of the distribution (Conveyance efficiency) is the efficiency in the main canal of the primary and secondary of the weir to the tertiary tapping, and can be calculated by the

$$Ec = \frac{Wf}{Wr} \times 100\% \qquad (1)$$

Where:
Ec = Efficiency distribution
Wf = the amount of water in the canal
Wr = the amount of water taken from the river

1.2. Performance of channal

The performance of the irrigation area is considered good, when the realization of the field going according to plan has been created. Components of the performance of irrigation areas such as physical infrastructure, productivity, supporting facilities, personnel organization, documentation, institutional conditions (i.e *P3A*).

One element of the performance of the physical infrastructure is infrastructure performance of irrigation canal, which is a review of the research.

1.2.1. Search. Under the proposal sent by the interpreter damage on a regular basis, do a search to determine the level of damage to tissue in order to making the proposed maintenance work next year. Maintenance of an inventory of activities indispensable condition of irrigation network is data including data damage and its effect on the service area.

Based on the results of a survey conducted inventory identification of problems and maintenance needs in a participatory manner, and made a series of structured action plan with priorities and a description of the maintenance work. In determining criteria for the maintenance of physical damage was apparent from the irrigation network. In essence the maintenance of irrigation networks were delayed will result in more severe damage and require early rehabilitation.

1.2.2. Indicator. According Permen PU 32 2007 said about some success indicator maintenance activities. The performance assessment of irrigation canals can be written:

a) good condition, when the canal has a performance value greater than 90%.
b) slightly damaged condition, when the canal has the value performance of 80-90%.
c) broken and, when the canal has the value performance of 60-79%.
d) heavily damaged condition, when the canal has a value of less than 60% performance.

2. Experimental

This study took place in the Candilimo Irrigation Area. Candilimo Irrigation Area has an area of 1938 Ha services are in Mojokerto, East Java, Indonesia. This Irrigation area is the authority of the Government of East Java in the management of the Dinas PU Pengairan in East Java Province.[7]

Mojokerto save a lot of history, especially the buildings of the kingdom of Majapahit famous relic in Southeast Asia. Heritage sites that still exist as an Kolam Segaran, candi tikus and other temples as the cultural heritage. Residents in mojokerto fraction are farmers, but agricultural land in mojokerto holds the first rank in the appeal of other sectors with an area that is approximately 38% of total area 371.01 km^2 Mojokerto regency.

Candilimo Irrigation Area located around the coordinates 7^0 35 '52.35 "S and 112^0 26' 02.09" East.

This study concentrated on the performance and efficiency of discharge in each segment of primary and secondary canals in irrigated areas candilimo Mojokerto regency.

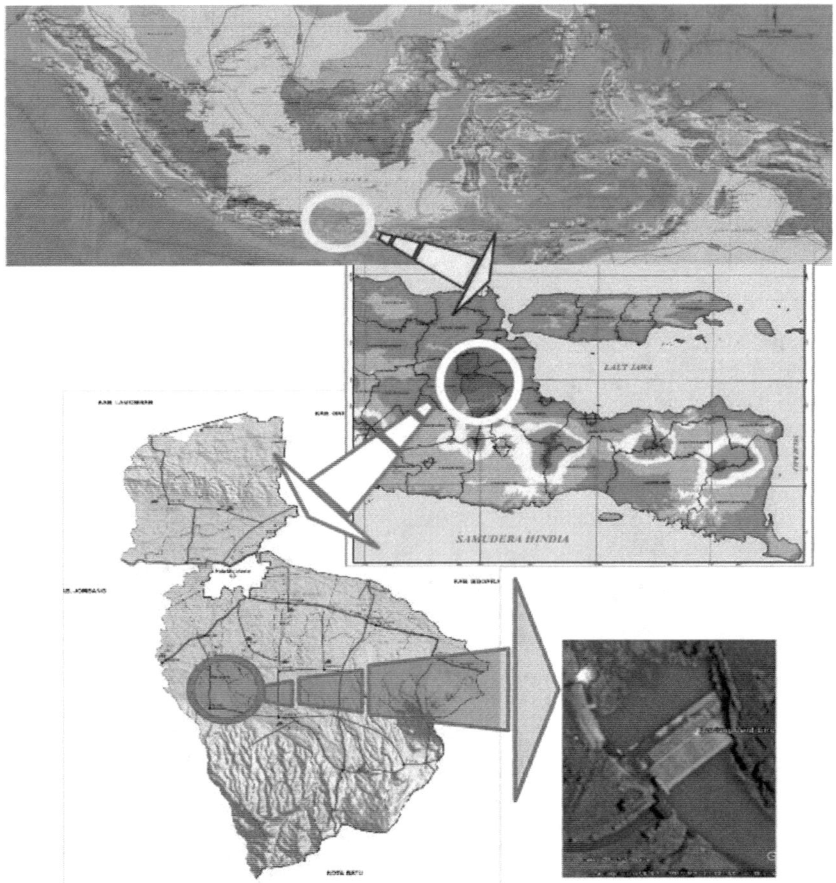

Figure 1. Map of research sites

3. Result and Discussion

Total length of primary canal in candilimo irrigation area were about 5.59 km length and total length of secondary canal were about 21.64 km length.

There are one intake and six tapper building (in some location tapper building was exist together with divider building in the same point) in the primary canal of candilimo irrigation area. Intake is a building that tapping the water from the river and flow it to the primary canal. Tapper building flow the water from primary canal to the paddy fields or tertiary canal.

B.CL is a symbol that writen in the building scheme mapping. The letter of *B* represent the word of building, and *CL* represent the words of candilimo primary canal. Every tapper building were represented by *B.CL* symbol letter. There are *B.CL.1*, *B.CL.2*, *B.CL.3*, *B.CL.4*, *B.CL.5*, and *B.CL.6* for every six tapper building.

There are six segment in the primary canal. First segment start from intake and end to *B.CL.1*, second segment start from *B.CL.1* and end to *B.CL.2*, and continued to sixth segment start from *B.CL.5* and end to *B.CL.6*. All of six segments in primary canal was detailed in the table 1.

Table 1. Primary canal 1 in the Candilimo irrigation area

No	Segment	Boundary		Length (km)
		Begin	End	
1	1 (primary canal)	Intake	*B.CL.1*	1,126
2	2 (primary canal)	*B.CL.1*	*B.CL.2*	0,2
3	3 (primary canal)	*B.CL.2*	*B.CL.3*	0,582
4	4 (primary canal)	*B.CL.3*	*B.CL.4*	2,175
5	5 (primary canal)	*B.CL.4*	*B.CL.5*	1,242
6	6 (primary canal)	*B.CL.5*	*B.CL.6*	0,265

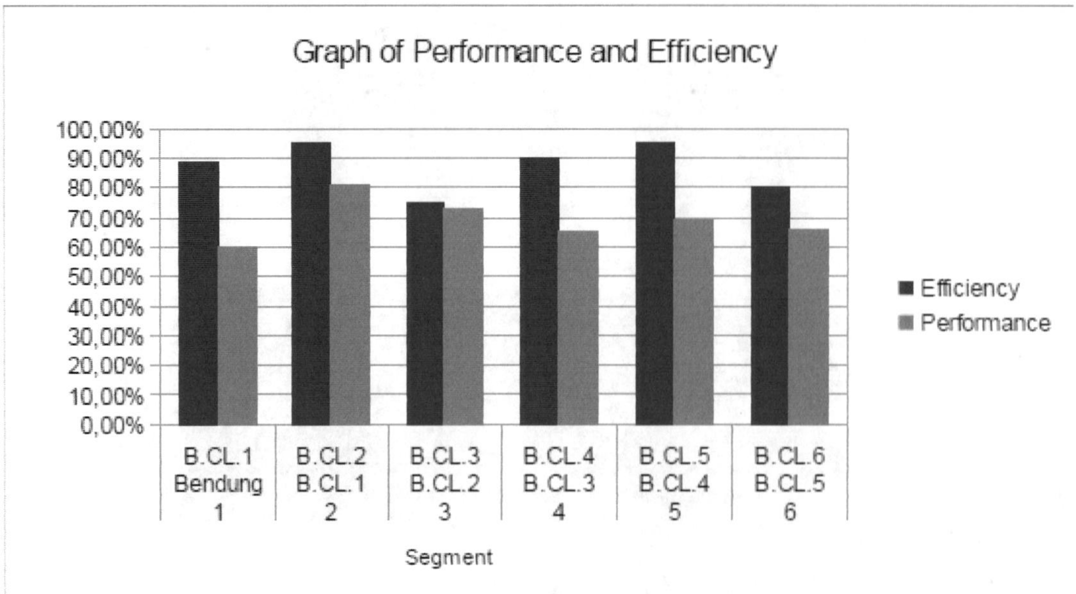

Figure 2. Relationship of performance and discharge efficiency in the canal segment in primary canal of Candilimo

Performance and efficiency are some of many charecteristic that exist in every canal segment. Performance and efficiency have its own value in all six of canal segments. Each of value of performance and efficiency in six of canal segments was ilustrated in a chart. The value of performance and efficiency was the expected value from search survey in the field. The chart was ilustrated in figure 2.

From chart in figure 2 ilustrated the value of performance and efficiency in the six segment of primary canal. The value is in percentage, it range between 0 % to 100 %. Segment 1 have performance value about 60 %, it is the lowest value of performance in all segment on primary canal. Segment 2 have performance value about 80 %, it is the highest value of performance on primary canal.

In this case, segment 1 have lowest performance, so it will have the first priority to maintained, instead segment 2 have the highest performance, it will have the last priority to maintained. In the aplication of maintenance work package, it take only performance for main indicator. It means that the canal with low performance will get the first priority.

There are thousand kilometers length of canal in east java province only that use performance for main indicator. In some location took 2 milions rupiah for maintain 150 meters length of canal, it is depend on the wide of canal and kind of degradation. It increase the performance only.

This chart will compare the performance and efficiency in each of segment, so the maintenance work package not only to increase the performance, but also to increase the efficiency in the canal. It will use efficiency and performance for indicator for succes in maintenance work package.

We can use the budget more efectively by increasing the efficiency in every canal. It will decrease the water conflict, that occur in the downstream of irrigation network. The conflict especially happen in the dry season.

4. Conslusion and Recomendation

Efficiency is the important indicator, beside performance especially in the canal segment maintenance. There are many building in irrigation network. measuring building is one of some building in irrigation network. Measuring buildingused for measure debit of water that flow on the measuring building. Measuring building was ussualy instaled after tapper building.

In the primary canal, measuring building was instaled after intake building that tapping the water from the river. After intake building, there are some tapper building that has current distance from each other, so from intake building. Tapper building flowing water from primary canal to tertiary canal.

In this case, point location is at the tapper building in the primary canal. There are measuring building that exist only after tapper building in the downstream of tapper building. So the measuring building lies in the tertiary canal closed to tapper building.

In this point location, there are no measuring building in the primary canal. It is needed to instal measuring building in the primary canal after tapper building or before tapper building. It is needed to know the debit that flows in the primary canal in this point location. So it is knowing the efficiency in this canal segment before it point location.

Then in every tapper building need to instal measuring building in the primary canal position after tapper building or before tapper building. So we hope that there are some advanced research about evaluation of instaling addition measuring building in the primary canal or secondary canal in the irrigation network.

References

[1] Bancin A A, Jayanti D S, and Ferijal T 2015 *Jurnal Rona Teknik Pertanian* **8(1)** 19 – 28

[2] Bekchanov M 2016 *J. Water Res. and Economic* **13** 30–45

[3] Bos M G, Nugteren J 1990 *On irrigation efficiencies* (Netherlands : International Institute for Land Reclamation and Improvement)

[4] Fajar A, Purwanto M Y J, and Tarigan S D 2016 *Jurnal Irigasi* **11(1)** 33 – 42

[5] Jadhav P B 2014 *Int. J. Eng. Innov. Res.* **3**(6) 820 – 826

[6] Kementerian PUPR 2013 *Standar Perencanaan Irigasi, Kriteria Perencanaan Bagian Saluran KP-03* (Jakarta : Direktorat Irigasi and Rawa, Direktorat Jenderal Sumber Daya Air)

[7] Kementerian PUPR 2015 *Kriteria dan Penetapan Status Daerah Irigasi* (Jakarta : Menteri Menteri Pekerjaan Umum and Perumahan Rakyat) Rakyat)
Rakyat)

IOP Conf. Series: Materials Science and Engineering **333** (2018) 012097 doi:10.1088/1757-899X/333/1/012097

The Effects of Substitution of The Natural Sand by Steel Slag in The Properties of Eco-Friendly Concrete with The 1:2:3 Ratio Mixing Method

A Rahmawati[1], I N Saputro[2]
[1,2]Department of Building Engineering Education, Faculty of Teacher Training and Education, Universitas Sebelas Maret, Indonesia

E-mail: anisrahmawati79@gmail.com

Abstract. This study was motivated by the need for the development of eco-friendly concrete, and the use of large quantities of steel slag as an industrial waste which is generated from the steel manufacturers. This eco-friendly concrete was developed with steel slag as a substitute for natural sand. Properties of concrete which used waste slag as the fine aggregate with the 1 cement: 2 sand : 3 coarse aggregate ratio mixing method were examined. That ratio was in volume. Then a part of natural sand replaced with steel slag sand in six variations percentages that were 0 %, 20 %, 40 %, 60 %, 80 % and 100 %. The compressive strength, tensile strength, and flexural strength of concrete specimens were determined after curing for 28 days. The research results demonstrate that waste steel slag can increase the performance of concrete. The optimal percentage substitution natural sand by steel slag sand reached of slag on the percentage of 20 % which reached strength ratios of steel slag concrete to the strength of conventional concrete with natural sandstone were 1.37 for compressive strength and 1.13 for flexural strength. While the tensile strength reached a higher ratio of concrete with steel slag sand to the concrete with natural sand on the 80% substitution of natural sand with steel slag sand.

1. Introduction

The development of eco-friendly and sustainable construction materials has gained major attention by the construction industry. One of the most significant activities stressed by the engineers and scientists related to construction industry especially in the concrete industry were aims to use raw materials with the possibility of improvement of cement characteristics and durability of concrete that will support the thought of green concept in concrete [1-2]. The three major objective behind green concept in concrete is to reduce greenhouse gas emission (carbon dioxide emission from cement industry, as one ton of cement manufacturing process, emits one ton of carbon dioxide), secondly to reduce the use of natural resources such as limestone, shale, clay, natural river sand, natural rocks that are being consume for the development of human mankind that are not given back to the earth, thirdly use of waste materials in concrete that also prevents the large area of land that is used for the storage of waste materials that results in the air, land and water pollution. This objective behind green concrete will result in the sustainable development without destruction natural resources [3].

Alternative materials can be used to substituted natural aggregates to support green concrete are industrial by-products that are easily available, require little or no pyro-processing and have inherent or latent cementitious properties. These materials are producing a greener concrete concerning resource and energy consumption [4]. Such industrial by-products are commonly called as supplementary cementitious materials or mineral additives [5]. One of the wastes generated from industries were looked upon as possible alternatives to be used in concrete production is steel slag, which is a waste product generated from the steel industry. Steel slag is the inevitable by-product which is 15~20% of the production of crude steel in steelmaking process [6] or equivalent with 130 – 200 kg of slag from one ton of steel production, depending on the composition of the steel and on the steel production process [7]. Slag often appears as granulated materials containing large clusters, coarse and very fine particles. A significant amount of steel slag was always disposed of as waste which results in the waste of resource, environmental pollution even ecological destruction. That's why the recycling of industrial waste slag is the core content of sustainable development. The recycling of steel slag will inevitably become an essential measure for the environment protection and therefore can lead to great social significance [8].Steel slag can be seen as a potential alternative to natural aggregates. The uses of steel slag as an alternative material help save a large share of natural resources and protect the environment [9] . This study used steel slag as a substitute material in concrete productions.

Although there are many studies that have been reported by researchers on the use of steel slag in concrete mixture [6,10-16], not much research has been carried out in Indonesia and other countries concerning the mechanical properties of steel slag concrete which use the most simple and easy concrete mix design that is 1:2:3 ratio mixing method. Therefore, to generate specific experimental data on properties of steel slag as the natural sand replacement in the concrete mixture, the present study has been performed.

In Indonesia, there are many small steel industries located in several areas, one of which is in Klaten, Central Java. In this central of steel casting industrial, there are more than 50 small-medium industries which concerned in steel casting industry. The average of production capacity is 1000 ton/week which produces about 100 m^3 steel slag. So far surrounding communities used steel slag as ground landfill, which can cause environmental pollution, especially water pollution. For this, this study conducted to study in using steel slag as substitute material of the natural sand in concrete with the 1:2:3 ratio mixing method. This method was chosen because it is a conventional method used by communities in around of this central of small steel industries. So the result from this study will be easier adopted by communities. Normal concrete used natural sand as fine aggregate, which usually comprises up to 30 percent of the total volume of concrete. Consequently, characteristics of sand significantly affect the performance of fresh and hardened concrete and have an impact on the cost-effectiveness of concrete [17].

The main objective of this research was to provide information about the effects of substitution of the natural sand by steel slag in the composition of concrete with the 1:2:3 Ratio Mixing Method to the mechanical properties of concrete. In this work, we have characterized the mechanical proprieties of concrete with steel slag sand regarding compressive strength, tensile strength, and flexural strength. These mechanical properties than compared to the performance of ordinary concrete.

2. Experimental

2.1. Materials

Portland Cement Composite (PCC) manufactured by HOLCIM was used. It's confirmed to Indonesian specification of SNI 7064-2004 [18] . Coarse aggregate was taken from Merapi mountain's quarry which crushed with the maximum size of 20 mm. Physical properties of coarse aggregate are presented in Table 1. The natural fine aggregate was taken from Merapi mountain's quarry, which selected by sieving on sieves size and selected the fine gradation as shown in figure 1. Physical properties of fine aggregate are listed in Table 2. Steel slag took from the steel industries in Klaten, Central Java, Indonesia. The large clusters and coarse steel slag than crushed with the maximum size of 4.8 mm as shown in figure 2. The granulometric curves which comparing of natural sand aggregate and steel slag sand resulting from the previous granulometric adjustment are shown in Figure 1, while the physical properties of steel slag are listed in table 3. The physical properties of all aggregates than compared with the standard from SNI 03-6861.1-2002 [19].

Table 1. The Physical Properties of Coarse Aggregate

Property	Point	Standard from SNI
Specific gravity	3.85	2.5 – 2.7
Abrasion	32.068	<50%
Fineness modulus	3.48	2.5-3.8

Table 2. The Physical Properties of Fine Aggregates.

Characteristic	Point	Standard from SNI
Water content	0.13%	1-3%
Specific gravity	2.58	2.5-2.7
Fineness modulus	3.44	2.5-3.8

Table 3. The Physical Properties of steel slag

Characteristic	Point	Standard from SNI
Water content	0.06%	1-3%
Specific gravity	2.83	2.5-2.7
Fineness modulus	4.7	2.5-3.8

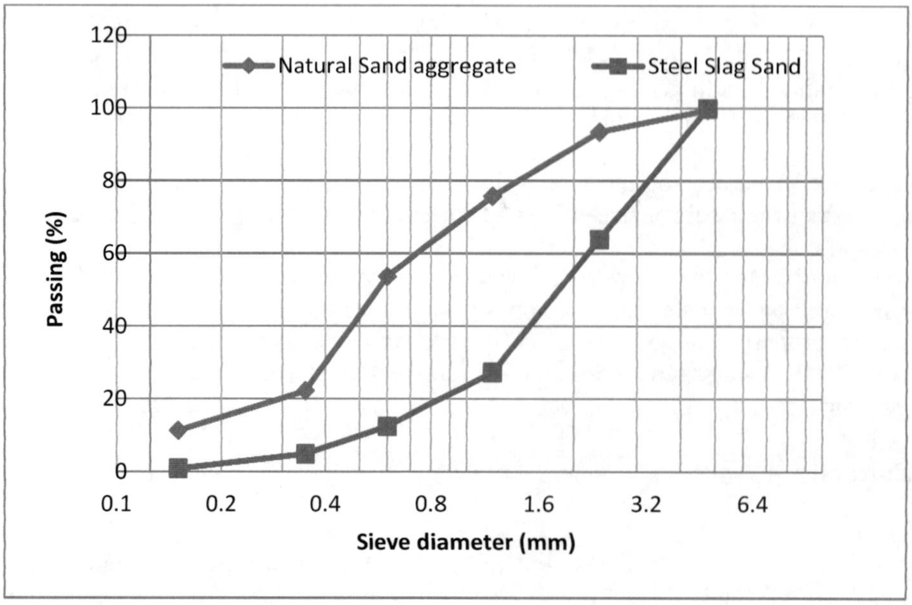

Figure 1. The gradation of aggregates

Figure 2. Steel Slag

2.2. Mixing proportions of concrete

Concrete mixing method using 1 cement: 2 fine aggregate:3 coarse aggregate ratio is widely used in the concrete made by the unprofessional one due to it's simple and easy way to be adapted. In this study, six concrete mixtures were designed. All concrete mixtures were prepared kept all cement (PC), sand, and coarse aggregate ratio constant that was 1pc: 2 sand:3 coarse aggregate. That ratio was in volume. Cement and coarse aggregate content were kept constant in all mixtures. All concrete mixtures used constant 0,8 water/cement ratio. The high water/cement ratio was obtained from the primary trial test of mixing concrete which produces a homogeneous mixture. Then the steel slag was substituted to the mixture with variations of concrete composition was listed in table 4. The concrete specimens prepared with mineral admixtures are summarized in Table 5.

Table 4. Variation of Composition of Steel slag in Concrete Mixture

Mix code	0	1	2	3	4	5
Cement	1	1	1	1	1	1
Natural sand	2	1.6	1.2	0.8	0.4	0
Coarse aggregate	3	3	3	3	3	3
Steel slag sand	0	0.4	0.8	1.2	1.6	2

Table 5. Mixing proportions of steel slag Concrete

Mixing code	Cement content (m^3/m^3 mixture)	Natural sand content (m^3/m^3 mixture)	Coarse aggregate content (m^3/m^3 mixture)	Steel slag sand content (m^3/m^3 mixture)
0	0.17	0.33	0.50	0.00
1	0.17	0.27	0.50	0.07
2	0.17	0.20	0.50	0.13
3	0.17	0.13	0.50	0.20
4	0.17	0.07	0.50	0.27
5	0.17	0.00	0.50	0.33

2.3. Castings of concrete specimens

Castings of cylinder concrete specimens with dimensions were 300 mm in height and 150 mm in diameter, were prepared to determine the compressive strength and tensile strength of steel slag sand concrete. While the specimens for the flexural test was the beam (150x150x600mm).

At first, steel slags, natural sands, and cement were mixed together in the dry state for about 3 minutes to obtain a uniform mixture. Next, the water was added to the blend slowly and mixed into a homogeneous mixture and then added the coarse aggregates and the residual water. The mixing was performed using a 350-litre mixer. The fresh concrete mixture was then poured into the molds and compacting manually using steel rod compactor.

After casting, the specimens were covered with wet burlap and left in the casting room at a temperature of (20±5)°C for a period of 24 h. The specimens were then de molded and submerged in water and measured at 28 d.

2.4. Testing

All concrete tested at 28 days age. Compressive strength was carried out on concrete cylinders using Compressive Testing Machine by following codes of SNI 1974-2011 [20] . The Tensile strength was determined using the standard of SNI 2491:2014 [21]. The flexural test using a simple beam with center-point loading was performed according to standard SNI 4154-2014 [22].

3. Results and Discussion

3.1. Compressive strength test results

The value of the considered compressive strength constitutes the average of the results from four specimens. The ratios of compressive strength used steel slag sand to the strength of conventional concrete with ordinary sandstone fine aggregate were illustrated in Figure 3. Results have shown that concrete with 20% substitution of fine aggregate by steel slag sand reached a highest compressive strength than others concrete mix.

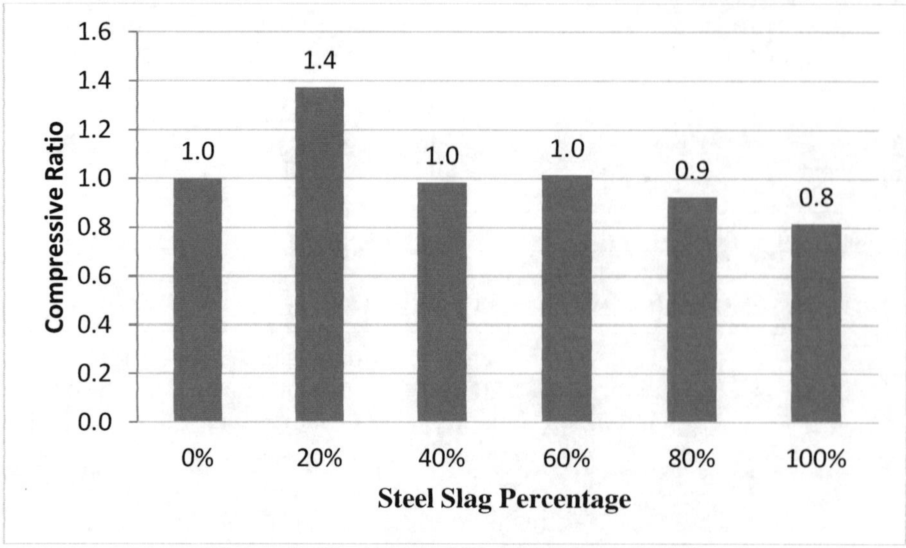

Figure 3. The Ratio of Compressive strength

3.2. Tensile Strength test results

The ratios of tensile strength used steel slag sand to the strength of conventional concrete with natural sandstone fine aggregate were illustrated in Figure 4. It was average from four specimens for each type of concrete studied here. While the results have shown that the largest recorded value of the tensile strength was that of concrete with 80% substitution of natural sand with steel slag sand, which reached 18% of tensile strength's increasing. But the overall difference of tensile strength's value from all concrete mixture in this study was not significant.

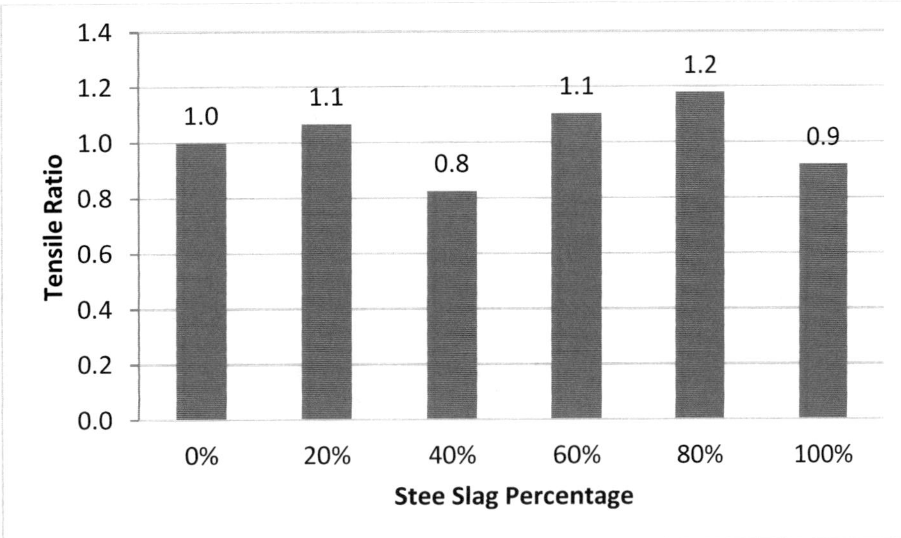

Figure 4. The Ratio of Tensile strength

3.3. Flexural strength test results

The ratio of flexural strength used steel slag sand to the strength of conventional concrete with natural sandstone fine aggregate which came from the average of five samples tested specimens for each type of concrete studied here shown in figure 5. From the figure noted that concrete with 20% substitution natural sand with steel slag sand reached higher flexural strength than the other concrete mix.

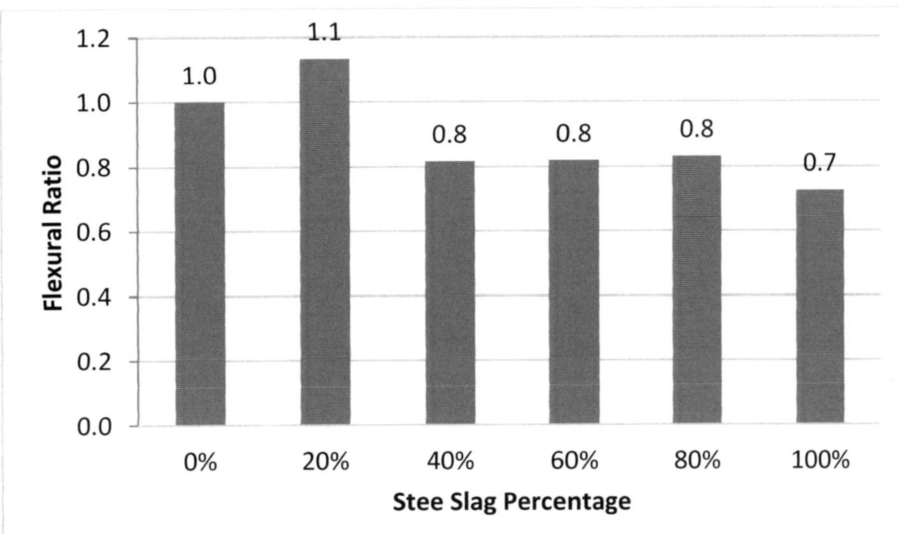

Figure 5. The Ratio of Flexural strength

The use of steel slag sand as 20% substitution of natural sand in concrete, increasing flexural strength by about 13%. Then more percentage of steel slag sand produced lower flexural strength.

IOP Conf. Series: Materials Science and Engineering **333** (2018) 012097 doi:10.1088/1757-899X/333/1/012097

3.4. Discussion
The test results indicated that concrete using steel slag as substitution material replacing a part of natural sand have a potential to produce better concrete performance than ordinary concrete. Theory said that steel slag concrete has lower strength at early ages. Therefore, steel slag can be used to control the hydration process so that concrete with high strength and good performance can be obtained [23]. According to recent studies, an increase in the compressive strength reported if steel slag used as a mineral admixture in the concrete mixture in term of EAF oxidizing slag aggregates [10-11], fine aggregate [12,13,14], partial replacement of sand [15], or fine and coarse aggregates [6,16]. Generally, using steel slag in concrete mix would improve the workability and compressive strength and does not give any deterioration as results of reaction with alkali in cement [21]. In this study, the highest compressive strength and flexural strength reached by 20% substitution natural sand by steel slag sand.

The increase in the compressive strength could be attributed to the high compactness of concrete achieved by adding sand slag in suitable proportions [24]. Steel slag sand used in this study have coarser gradation than the natural sand. Until the 20% substitution of steel slag, this coarser aggregate contributed to the higher compressive strength of the concrete. The improving of concrete performance also influenced by the shape of steel slug crushed sand. The shape and texture of steel slag crushed sand particles have a significsnt effect on the interlocking of paste and aggregate particles, providing good aggregate-binder bonding, which will lead to an improvement of strength of concrete [25,4]. After 20% substitution, then the higher percentage of steel slag sand produced lower compressive strength and flexural strength of concrete. The decrease in compressive and flexural strength with increasing steel slag sand after 20% substitution was believed to be due to the more existence of weak aggregate and paste interface due to the formation of the coating of calcite on steel slag surface which hinders the bonding between the paste and aggregates [26-27] . The thin coat of calcite cause steel slag's surface similar with glass surface which has weak bonds between the particles and the cement paste [28]. More steel slag sands mean more aggregates and paste interfaces exist in the concrete mixture. Furthermore, aggregate with good gradation and finenes modulus will be able to result in maximum density and minimum porosity, which will generate higher compressive strength in concrete [29]. In this study, natural sand has finer grain size than the steel slag sand.The decrease of concrete strength also can be explained by the poor quality of granulated slag as sand due to its high degree of absorption [16]. Steel slag have more pores than the natural aggregate, that will absorb more waters from concrete admixture which caused less water available for cement hydration process.

4. Conclusions
In this work the effect of using steel slag sand as substitution of natural sand in the concrete mixture was preliminary studied to investigate the possibility of using a compound mineral admixture of steel slag as an additive material for concrete production with 1:2:3 ratio mixing method. From the test results and calculated strengths of the steel slag concrete, the following conclusions are drawn:

1. Concrete with good performance could be produced using compound mineral admixture of steel slag with 1:2:3 ratio mixing method
2. Compressive strengths of steel slag concrete reached the highest compressive strength by mixture ratio 1 Pc: 1.6 Sand: 3 Coarse aggregate: 0.4 steel slag which get 1.37 of ratio of concrete steel slag sand compressive strength than concrete with natural sand.
3. Tensile strengths of steel slag concrete reached the highest tensile strength by mixture ratio 1 Pc: 0.4 Sand: 3 Coarse aggregate: 1.6 steel slag which get 1.18 of ratio of concrete steel slag sand tensile strength than concrete with natural sand.
4. Flexural strengths of steel slag concrete reached the highest flexural strength by mixture ratio 1 Pc: 1.6 Sand: 3 Coarse aggregate : 0.4 steel slag which get 1.13 of ratio of concrete steel slag sand flexural strength than concrete with natural sand.

In brief, it can be said that the steel slag sand could be an alternative for use in concrete applications using 1:2:3 ratio mixing method which has higher strength than concrete with natural sand in substitution till 20%. Finally, the use of steel slag as building materials should be considered as a viable solution in terms of environmental conservation, which satisfies society's current requirements.

References

[1] Muhammad I A K, Emmanuel R. 2007 *Proc. of Int. Conf. on Advances in Cement Based Materials and Appl. in Civil Infrastructure* 289–299
[2] Rai A, Prabakar J, Raju C B, Morchalle R K 2002 *Constr. and Building Materials* **16** 489–494
[3]Arun A 2015 Vit university http://www.researchgate.net/post/what_is_green_concrete retrieved at November 28, 2016
[4] Anastasiou E, Filikas K G, and Stefanidou M 2014 *Constr. and Build. Mater.* **50** 154–161
[5] Altun I A.and Yilmaz I. 2002 *Cement and Concrete res.* **32** 1247–1249
[6] Liu C, Zha K, and Chen D 2011 *Int. Conf. Adv. Eng. Proc. Eng.* **24** 412–416
[7] Subathra D B and K Gnanavel 2014 *Proc. of 12th Global Congress On Manufacturing And Management, Procedia Eng.* **97** 95 – 104
[8] Sun Shu-Shan, Zhu Gu I-Lin, Zhang Yu 2007 *China Waste Steel* **4**(2) 21–28
[9] K. Behfarnia, M. Shojaei and R. Mohebi 2015 *Asian J. of Civil Eng.* **16**(4) 505-513
[10] S-W Kim, Y-J Lee, and K-H. Kim 2012 *J. of Asian Arch. Build.Eng.* **11** 1 133–138
[11]Sekaran A, Palaniswamy M, and Balaraju S 2015 *The Sci. World J.*
[12]Singh G and Siddique R 2016 *Constr. and Build. Mater.* **128** 88–95.
[13] Ouda A S and Abdel-gawwad H A 2015 *Housing And Building National Research Center* 0–6
[14] Tarawneh S A, Gharaibeh E S, and Saraireh F M 2014 *American J. of Appl. Sci.* **11**(5) 700–706.
[15] Al-Rawi, R. S. 2001 *Using Local Steel Slag as a Fine Aggregate in Concrete* (Baghdad: College of Engineering)
[16] L Zeghichi 2006 *Asian J. of Civil Eng. (Building And Housing)* **7**(1) 27-35
[17] Gauhar S, Rafiqul A T and Syed M J 2016 *Int. Conf. on Sustainable Design, Eng. and Constr. Procedia Eng.* **145** 66 – 73
[18] SNI 7064:2014 2014 *Portland Cement Composite* (Jakarta: Indonesian National Standardization Agency)
[19] SNI 03-6861.1-2002 2002 *Specification Of Building Materials - Part A: Non Metals Building Materials* (Jakarta: Indonesian National Standardization Agency)
[20] SNI 1974-2011 2011 *Test method for compressive strength of concrete with cylindrical specimen cast* (Jakarta: Indonesian National Standardization Agency)
[21] SNI 2491:2014 2014 *Test Methode For Tensile Split Strength Of Concrete With Cylindrical Concrete Specimen (ASTM C496/C496M-04, IDT*) (Jakarta: Indonesian National Standardization Agency)
[22] SNI 4154-2014 2014 *Test Method For Flexural Strength Of Concrete Using A Simple Beam With Center-Point Loading* (Jakarta: Indonesian National Standardization Agency)
[23] Li Y-F, Yao Y, and Wang I 2009 *J. Cent. South Univ. Technol.* **16** 0768–0773
[24]Senani M, Ferhoune N, and Guettala A 2016 *Alexandria Eng. J.* 0–6
[25] Al-ameeri A S A and Al-baghdadi H M 2012 *Int. J. Art. Sci.* **5**(7) 187–196
[26] Palankar N, Shankar A U R, and Mithun B M 2015 *Int. J. Sustain. Built Environ.* **4**(2) 378–390
[27] Palankar N, Shankar A U R, and Mithun B M 2016 *J. Clean. Product.* **129** 437–448
[28] Guadalupe F, Manuel C J, Luis J, Paola A S, Mendivil-escalante J M, Gomez-soberon J M, Mendivil-escalante J M 2016 *Int. J. of Civil Eng.*
[29] Purwati A, As'ad S, Sunarmasto 2014 *E-Jurnal Matriks Teknik Sipil* **2** 2 58-63

IOP Conf. Series: Materials Science and Engineering **333** (2018) 012098 doi:10.1088/1757-899X/333/1/012098

Preparation and Mechanical Properties of Pressed Straw Concrete Brick

S Sumarni[1], W Wijanarko[2]
[1]Lecture Department of Civil Engineering Education, Faculty of Teacher Training And Educations, Sebelas Maret University, Jl. Ir. Sutami 36A Surakarta 57126 Indonesia
[2]Student Department of Civil Engineering Education, Faculty of Teacher Training And Educations, Sebelas Maret University, Jl. Ir. Sutami 36A Surakarta 57126 Indonesia

E-mail: marnis-ri@yahoo.com

Abstract. Rice straws have been widely used as wall filler material in China, Australia, and United States, by spinning them into hays with an approximate dimension of 40 cm of height, 40cm of thickness and 60 cm of width, using a machine. Then, the hays are placed into a wall frame until they fill it completely. After that, the wall frame is covered with wire mesh and plastered. In this research, rice straws are to be used as concrete brick fillers, by pressing the straws into hays and then putting them into the concrete brick mold along with mortar. The objective of this research is to investigate the mechanical properties of concrete brick, namely: compressive strength, specific gravity, and water absorption power. This research used experimental research method. It was conducted by using concrete bricks which had 400 cm of width, 200 cm of height, and 100 cm of thickness, made from rice straws, cement, sand, and water as the test sample. The straws were each made different by their volume. The mortars used in this research were made from cement, sand, and water, with the ratio of 1:7:0.5. The concrete bricks were made by pressing straws mixed with glue into hays, and then cut by determined variations of volume. The variations of hays volume were 0 m^3, 0.000625 m^3, 0.00075 m^3, 0.000875 m^3, 0.00125 m^3, 0.0015 m^3, 0.00175 m^3, 0.001875 m^3, 0.00225 m^3, and 0.002625 m^3. There were 3 samples for each volumes of hays. The result shows that the straw concrete bricks reached the maximum compressive strength of 1.92 MPa, specific gravity of 1,702 kg/m^3, and water absorption level of 3.9 %. Based on the provided measurements of products in the *Standar Nasional Indonesia* (Indonesian product standardization), the concrete bricks produced attained the prescribed standard quality.

1. Introduction

The research on wall made of rice straws, commonly called as straw bale, is necessary to be developed, since the use of straws as mixed material of lightweight concrete will create an eco-architecture building.

The facts should be noted that buildings are the largest energy user, starting from construction, construction material, building operation, and building maintenance to building demolition. If a life cycle analysis of a building is conducted, the building's impacts on the environment will be shown. The overall costs of unsustainable architecture are, therefore, greater than the sustainable one, so that following on the line of eco-architecture will save cost in longer term.[1-3]

This wall of straws is developed in countries such as China, United States, and Australia. The building wall is made by spinning rice straws into hays with an approximate length of 30-60 cm. Then, the hays are placed into a wall frame and then covered with wire mesh to strengthen it. It is then plastered with mortar.

The renewal in this research is the creation of a lighter block of hay material than ordinary brick. This brick Straw including material that the temperature is higher then suitable when used for cold areas and also for the insulation of the room soundproof. Because the weight of the structure is reduced, the workload will also be smaller so that the structure is expected to be safer and very suitable for housing in earthquake prone areas. Implementation on building materials especially for wall-shaped brick is more environmentally friendly than clay brick. Straw as a material brick is an environmentally friendly material so that this paper macth to construction materials conference topic.

Straws contain silica so that it is good to be used as mixed material for concrete. [4] They also weigh very light. If used as lightweight concrete mix in the form of concrete bricks, it will result in a significantly lighter concrete than other concretes in general. Because of its light weight, it will have some plus points, such as ease transportation and installation, cost reduction, heat resistance, and the ability to be used as isolation of soundproof rooms. [5-6]

The straws weigh lighter than most material. The earthquake load working in the building structure will be smaller Therefore, the structure will be safer from destruction, and it will be suitable for houses located in earthquake-prone areas.

The straw ashes used as substitute for sand in mortar mix will increase mechanical strength and affect water absorption. [7-8]

Based on *Standar Nasional Indonesia*, standard concrete must have the compressive strength between 2.5-10 MPa and water absorption level between 25-30%. [9]

Non-structural lightweight concrete has the compressive strength between 0.35 – 6.9 MPa and specific gravity under 1900 kg/m^3. [10]

In this research, rice straws were used as forming material of concrete bricks. Conblock (*concrete block*) is a building component made from mixture of Portland Pozzolana cement, sand, water, with or without any other additives. It is molded in such way that it meets the requirements and can be used as a material for wall installation. The objective of this research is to investigate the compressive strength, specific gravity, and water absorption level in pressed straw concrete brick.

2. Experimental

2.1 Materials

The materials used in this research were cement, sand, water, straws and glue. The cement used had to meet prescribed requirements in the specification of group A construction material, SK SNI S-04-1989-F. The cement used was Holcim cement. The sand used was taken from River Kaliworo. It met the prescribed requirements in the specification of group A construction material, SK SNI S-04-1989-F. The straws used were in dry condition. Their average length ranged from 350 mm to 400 mm. To glue the straws, Express wood glue was used.

2.2 Experimental procedure

2.2.1 Examination of material. The sand examination was done through the following steps: the examination of water level, specific gravity, unit weight, organic content, mud content, and sand gradation which had to meet the prescribed requirements of *Standar Nasional Indonesia*. The examination of straws was done through the following steps: drying the straws, cutting the straws according to the dimension, and examining the specific gravity.

2.2.2.The spinning of straws. The straw spinning process was conducted by using press machine, by firstly mixing the straws with glue and then pressing the straws into hays. The straw hay bale was cut according the volume variation. The produced hay bale is shown in Figure 1. The variation of hay bale sample is shown in Table 1.

Figure 1. Hay bale

Table 1. The variations of straw hay bale

No Variation	Volume of straw hays (m^3)	size of hay straw length*width*height (mm)	Quantity
1	0.000000	0	3
2	0.000625	250*50*50	3
3	0.000750	300*50*50	3
4	0.000875	350*50*50	3
5	0.001250	250*50*100	3
6	0.001500	300*50*100	3
7	0.001750	350*50*100	3
8	0.001875	250*50*150	3
9	0.002250	300*50*150	3
10	0.002625	350*50*150	3

2.2.3 *Concrete brick production.* The specimens were concrete bricks with the measurement of length, width, and height of 400 mm, 100 mm, and 200 mm. They were made by firstly making a mixture of concrete mould with the ratio of cement: sand:water of 1:7:0.5. The concrete bricks without straws were made by putting the concrete mould approximately 1/3 of mould volume, and then compact them. This way was then conducted for 3 times until the concrete mould volume was met. The concrete bricks with straw addition were made by putting the concrete mould to the block and then compacting them until the concrete filled the block by 50 mm (first compacting). After that, the hays were put and the mortar was added, and then they were compacted to the height of hays (second compacting). Finally, the mortar was added until filling the block and then the mixture was compacted until the full concrete brick volume (third compacting). The steps are shown in Figures 2-4.

Figure 2. First Compacting

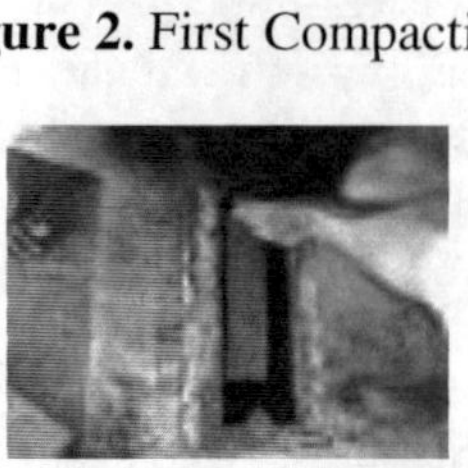

Figure 3. Second Compacting

Figure 4. Third Compacting

The produced samples were then stored in a location without much heat and humidity to preserve them.

2.2.4. The compressive strength test of concrete bricks. Concerning the compressive strength test of concrete bricks. Compressive strength of concrete bricks is the amount of load per area unit which cause the specimen to break if it is loaded with certain amount of compressive force produced by compressing machine. [11] The compressive strength test of concrete bricks was conducted 28 days after the production of the specimens, pursuant to the prescribed requirements of *Standar Nasional Indonesia* by using concrete compressive test apparatus as shown in Figure 5.

Figure 5. Compressive strength test

The compressive strength was calculated by using equation as follows:

$$\sigma = P/A \quad (1)$$

where σ= compressive strength, P= load, A= area of specimen

2.2.5.Specific gravity test. The concrete specific gravity is the comparison between the concrete weight and its volume. The test was conducted by weighing the weight of concrete bricks and calculating their volume. Specific gravity was calculated by using equation as follows:

$$Bj = \frac{W}{V} \quad (2)$$

where Bj = specific gravity, W = specimen weight, V = specimen volume.

2.2.6.Water absorption level test. Water absorption level of concrete bricks is the capability of concrete bricks to absorb water. The calculation was conducted by using the following equation.

$$\text{Absorption} = \frac{A-B}{B} \quad (3)$$

Where A : specimen weight after it is soaked in water for 24 hours, B: specimen weight after it is dried in the oven on 105°C.

3. Results and Discussion

The result of sand material test is shown in Table 2. Based on the examination result, the sand already met the requirements to be used as construction material. One of the requirements of sand, which is good to be used as construction material is having a specific gravity of 2.4-2.9.

Table 2. Properties of Sand

Properties	**Values**
Specific Gravity	2.39 gr/cm^3
Fineness Modulus	2.48
Zone	2
Absorption	1.85 %

Examination of fine modulus of sand is a way to investigate the value of fineness or coarseness of an aggregate. The fineness or coarseness of an aggregate can affect the workability of concrete mortar. If there is too many fine aggregates found in mortar, they will cause thin layer of fine aggregate to occur, and the cement will go up.

The grain fine modulus, besides serving as the measurement of grain fineness, can also be used to find the comparison value of weight between sand and pebble. The sand gradation classification can be obtained from the sand grain fine modulus graph. The result of the fine modulus test was 2.48. If it was put in the sand grain fine modulus graph, then the sand gradation of the specimen was included to area no. 2 (rather coarse), and by it, it was included in the sand grain fine modulus which could be used as construction material.
The data of the results of compressive strength test is shown in Figure 6.

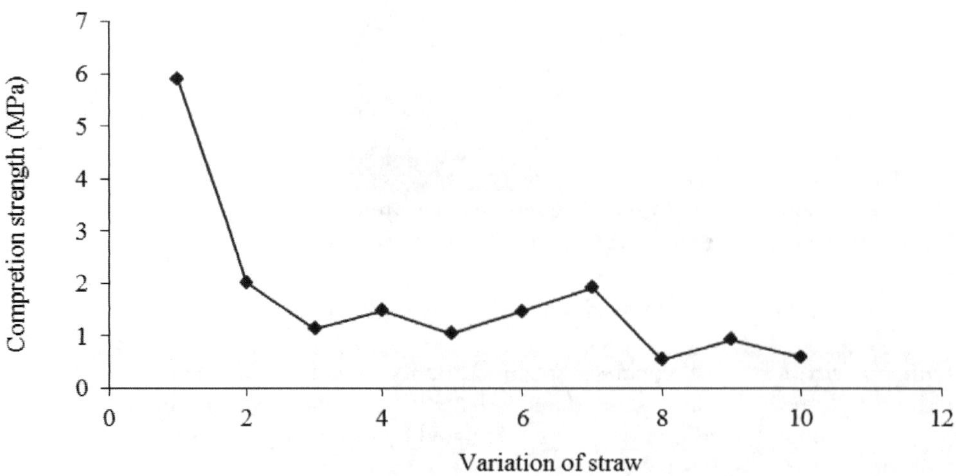

Figure 6. Result of compressive strength

Based on Figure 6, the compressive strength of concrete bricks without straws was 5.896 MPa, while straw variation no. 7 with the volume of 0.00175 m^3 had the maximum compressive strength of 1.92 MPa. It was an ideal variation for compressive strength value. The value itself met the requirements as non-structural lightweight concrete [10].

Data of the specific gravity test result of several concrete bricks variation is shown in Figure 7.

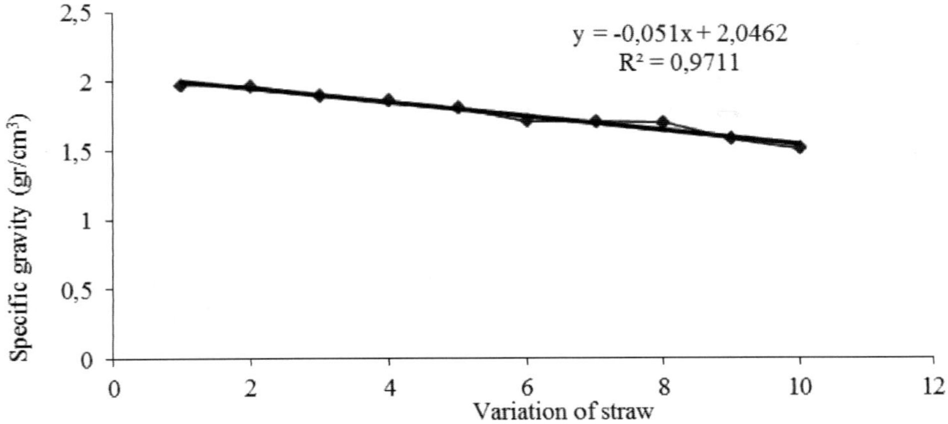

Figure 7. Brick variations versus specific gravity

The variation relation between the straw concrete bricks with specific gravity shows that the larger the volume of straws added to the concrete, the smaller the specific gravity would be. On variation of straw brick no. 7 the volume of 0.00175 m^3, the value of specific gravity was 1,702 kg/m^3. According to the standard of lightweight concrete, the specimen met the criteria as non-structural wall. [10]

The result of absorption test of concrete brick is shown in Figure 8.

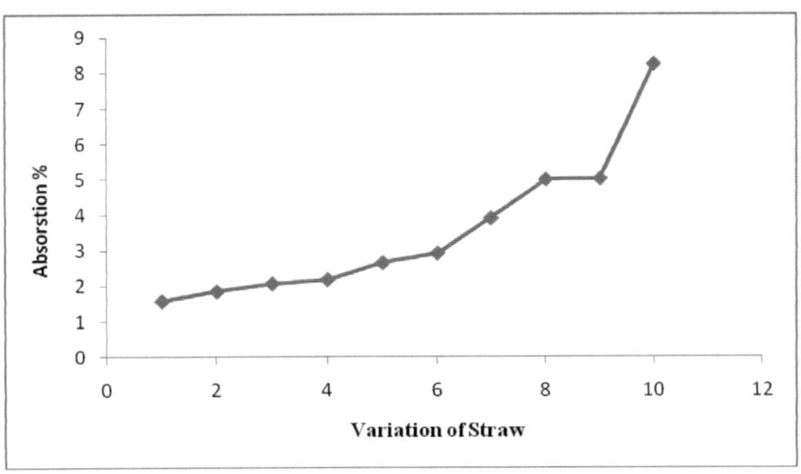

Figure 8. Variations of brick versus water absorption level.

From the result of experiment it is apparent that the larger the dimension of hay bale added, the larger the level of water absorption would be. The largest water absorption level was found in variation 10 with the absorption value of 8.23%. The water absorption level in variation 7 was valued 3.9%. The value met the requirements under maximum limit. [9]

4. Conclusions

The optimum value of concrete bricks with straw mixed material was found in the variation of straw brick no. 7 with the volume of 0.00175 m^3, with the value of compressive strength, specific gravity, and water absorption level of 1.92 MPa;1,702 kg/m^3; 3.9%. Based on the optimum value, the brick attained prescribed standards of quality.

References

[1] Igaz, Titusz, Krisztina Szirtesi, and Gyula Lakatos 2011 *Studia Universitatis Vasile Goldis Seria Stiintele Vietii (Life Sciences Series)* **1** **(21)** 127-132.

[2] Lolli V, V Panone, and A Benedetti 2015 *International multidisciplinary scientific geoconference* **2** 59-66.

[3] Carlos Cobreros, José Luis Reyes-Araiza, Alejandro Manzano-Ramírez, Rufino Nava, Mario Rodríguez, Mauricio Mondragón- Figueroa, Luis Miguel Apátiga, Eric M. Rivera-Muñoz 2015 *BioResources* **10(2)** 3757-3774

[4] Hasan Biricik, Fevziye Aköz, I.lhan Berktay, Ali N Tulgar 1999 *Cement and Concrete Research* **5** **(29)** 637–643.

[5] Soofia Tahira Elias-Ozkan and Francoise Summers 2013 *Journal of Green Building* **4** **(8)** 110-126.

[6] Adedeji, Adeola A, and Jimoh A Bello 2011 *Annals of the Faculty of Engineering Hunedoara* **3** **(9)** 255-262.

[7] Nabil M Al-Akhras, Bilal A Abu-Alfoul 2002 *Cement and Concrete Research* **6** **(32)** 859–863.

[8] B Larisa, and P L Clouston 2014 *ARPN Journal of Engineering and Applied Science* **9(9)** 1645-1652.

[9] Indonesian product standardization SNI 03-0349-1989

[10] Dobrowolski A J 1998 *Concrete Contruction Hand Book* Mc. Graw-Hill Companies Inc New York.

[11] Indonesian product standardization SNI M–14–1989–F

The Fogging Method with Variable of Nozle Diameter as the Mitigation Alternative for Spreading the Dust of Cement

J Purwanta[1], T Marnoto[1], P Setyono[2], A H Ramelan[2]
[1] Faculty of Mineral Technology, UPN Veteran Yogyakarta, Yogyakarta, Indonesia
[2] Faculty of Mathematics & Natural Science, Sebelas Maret University, Surakarta Indonesia

E-mail: jakapurwanta_tlupn@yahoo.co.id

Abstract. The cement plant impacts on the lives of people around the factory site, one of them on the air quality, especially dust. Cement plant has made various efforts to mitigate dust generated, but the reality on the ground is still a lot of dust flying around either of the cement factory chimneys and transportation. The purpose of this study was to find the optimum condition of nozle diameter from the cement dust catcher, for mitigation the dust spread to around the cement plant. This study uses research methods such as collecting secondary data which includes data intensity rainfall, the average long rains, wind speed and direction as well as data quality monitoring dust around PT. Semen Gresik (Persero) Tbk. Tuban plant. To determine the wind direction propensity models, use a soft Windrose file. To determine the impact on the spread of dust into the environment using secondary data monitoring air quality. Results of the study is that the mitigation of dust around the cement plant is influenced by natural factors, namely the tendency of wind direction, rainfall and rainy days, and the rate of dust emission from the chimney. I try for operate the cement dust catcher with variable of nozle diameter. Finally, I find the optimum condition of nozle diameter for cement dust catcher is 1.40 mm, with line equation is $y = 149.09.e^{1.6237.x}$ and error 5%. In that condition, nozle can make the fog with a good quality and it can catch the cement dust well.

1. Introduction

Similarly, the existence of the activities of PT. Semen Gresik (Persero) Tbk. Tuban plant also caused some impact, both positive and negative. According Fityatur [1] where the cement plant impact on the lives of people around the factory site as public revenue and public health. Some positive effects that create jobs; seeks opportunities in the community such as the establishment of food stalls, grocery stores, boarding houses/rented, and others, to support the fulfillment of the daily needs of employees; as well as rising incomes as a result of subsequent / follow-up of work opportunities and increasing the business opportunity. On the other hand, the activities of PT. Semen Gresik (Persero) Tbk. The Tuban plant had a negative impact in the form of air pollution, especially dust. The emergence of dust from the production process can be dispersed into the environment around the plant site. In effect, then the environment will be exposed to dust. Exposure to dust is inhaled dust particles public both outdoors and indoors. Exposure to this dust can interfere with the respiratory community outside the home [2, 3]. The purpose of this study was to determine the effect of the wind against the spread of dust and find alternative means of environmentally friendly dust catcher. Dust particulate characteristics including size, size distribution, shape density, adhesiveness, corrosive properties, reactivity and toxicity. One of the most important karakteriktik of dust particle suspensions is particle-size distribution of aerosols. Generally, particles with a diameter of less than 2.5 μm is considered fine and larger particles 2.5 μm is considered rude (Suhariyono et al [4]).

One is a cement manufacturing technology with Dry Process. Lower operating costs and greater production capacity required a major consideration using Dry Process Cement factory, but the negative impact is the emergence of dust which if not managed then it would be harmful to the environment [2]. According Dimitriou and Christidou [1], air pollution is one of the important environmental issues that contribute to the effects of high temperatures that affect public health, animal life, natural ecosystems and man-made environment. Air pollution is also responsible for climate change, greenhouse effect, acid rain, and others.

According Nurbiyantara [5], exposure via inhalation of dust particles in the respiratory tract will cause a variety of lung function impairment. The dust particles will accumulate in the respiratory tract according to the nature of chemistry, physics, and biology of the dust. According Naqpure [6] termed TSP dust are defined as PM (Particulate Material) with aerodynamic diameter not more than 30 μm.

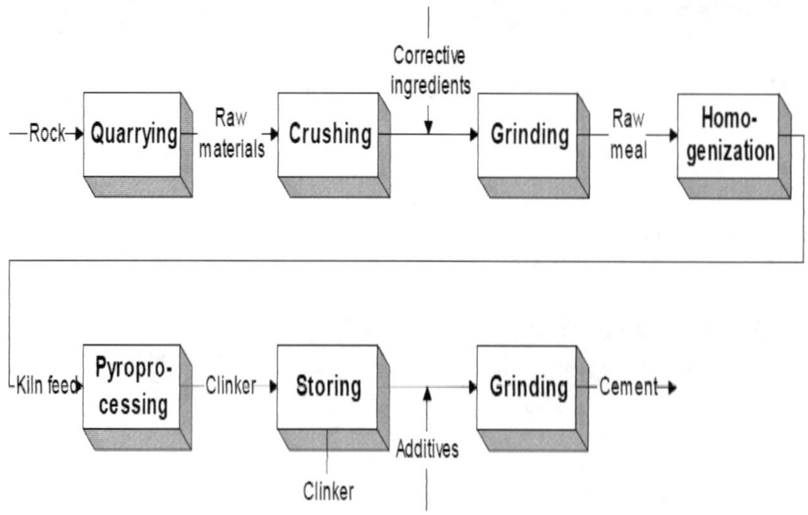

Figure 1. The process of making cement
(Source: Ibrahim et al [7])

Control of dust emissions in the cement industry generally use mechanical collectors/collectors mechanical/dust collector, electrostatic precipitators, fabric filters, or a combination of such equipment [8]. According Zimwara et al [9]), air pollution control technology that is Flexible Pulse Jet Filters, Electrostatic Pricipitators, and Wet scrubbers. In wet scrubbers, the gas stream is polluted taken and contact with the liquid that is sprayed with a liquid, and the liquid will catch dust.

2. Experimental

2.1. Determining the trends in wind direction by collecting secondary data rainfall intensity, long average rainfall, and wind speed and direction in Tuban last few years. The data was then entered into the file soft Wind Rose to know the trend of the wind direction.

2.2. Designing a dust catcher equipment fogging method.

3. Result and Discussion

3.1. Determining the trends in wind direction

Table 1. Data of wind speed and direction monitoring were done by. Semen Gresik (Persero) Tbk. Tuban Some village in the district of Kerek, Tuban, East Java Province (2016)

No	Measurement Locations (Village)	Measurement Time	Wind Speed (km/hour)	Wind Direction	No	Measurement Locations (Village)	Measurement Time	Wind Speed (km/hour)	Wind Direction
1	**Temandang**	07-08 March	0.90	East	13	Temandang	01-02 August	0.60	North
2	**Tlogowaru**	08-09 March	0.70	East	14	Tlogowaru	01-02 August	1.20	West
3	**Kasiman**	11-12 March	0.70	West	15	Kasiman	02-03 August	1.00	West
4	**Margomulyo**	14-15 March	1.40	East	16	Margomulyo	02-03 August	1.40	North
5	**Sumberarum**	15-16 March	1.30	East	17	Sumberarum	03-04 August	1.10	West
6	**Karanglo**	16-17 March	1.30	East	18	Karanglo	03-04 August	1.20	West
7	**Temandang**	09-10 May	0.90	West	19	Kasiman	20-21 Nov	1.30	East
8	**Tlogowaru**	09-10 May	0.50	South	20	Sumberarum	24-25 October	1.30	West
9	**Kasiman**	11-12 May	1.10	North	21	Karanglo	25-26 October	1.40	North
10	**Margomulyo**	11-12 May	0.90	North	22	Margomulyo	26-27 October	1.30	East
11	**Sumberarum**	12-13 May	2.00	South	23	Temandang	17-18 October	1.50	East
12	**Karanglo**	12-13 May	1.00	South	24	Tlogowaru	18-19 October	1.30	East

Source: The report of RKL-RPL PT. Semen Indonesia Semester I&II 2016 application, 2016

In the picture below it can be seen that the dominant wind speed in the area of 0.5 to 2.1 m/s and the average wind speed is 1.16 m/s. The tendency of the wind direction is east. This means that the spread of dust tends to the east so that the area on the east side impact sources will be potentially affected by the spread of dust more than the territory in the other direction.

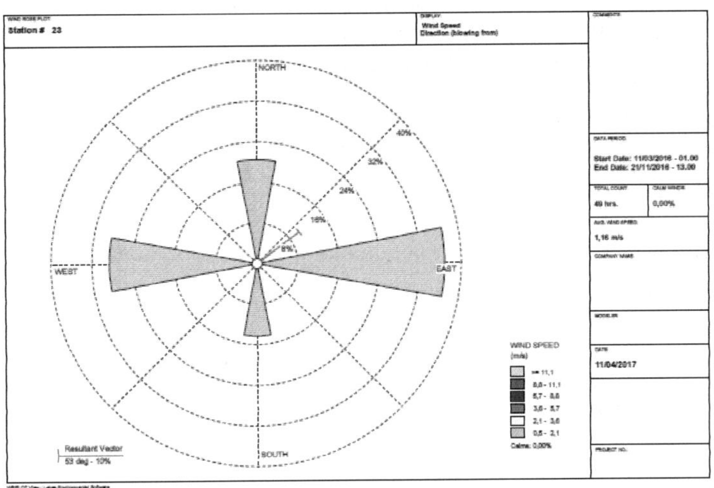

Figure 2. The tendency of the wind direction and wind speed on March-November 2016
(Source: data processing constituent, 2017)

3.2 Designing tools dust catcher with fogging method

The spread of dust can be overcome by using a dust catcher fogging method is environmentally friendly appliance because it uses the mist atomized water demand will be less than with other methods, Can catch cement particles whose size is very smal and Clots / floc capture results of this appliance can be recycled. To determine the concentration of dust particles, based on the data results of sampling and substituted into the following equation:

$$C = [(M_1 - M_0) / (t.v)]\ (\mu g/m^3) \qquad (1)$$

Information:

C : The concentration of dust particles measured ($\mu g/ m^3$)

M_1: Weight filter after dust measurements (pg)

IOP Conf. Series: Materials Science and Engineering **333** (2018) 012099 doi:10.1088/1757-899X/333/1/012099

M_0: Heavy dust filter before measurement (g)
t : Duration of measurement (hours)
v : velocity of volumetric air (m^3/ h), in this study v = 1,698 m^3/ h

Cement dust concentrations obtained in equation (1) to the model equations Canter conversion to obtain cement dust concentration at the time of sampling / sample for 24 hours so in accordance with Government Regulation No. 41 of 1999.

$$C_1 = C_2. [(t_2 / t_1) p] \quad (2)$$

Information:
C_1 : The concentration of cement dust with long average sampling instance t_1 (μg/ m^3)
C_2 : an average dust concentration at the sampling time instance t_2 (μg/ m^3)
t_1 : The duration of sampling sampel 1 (in this study = 24 hours)
t_2 : The duration of sampling sample 2 (h) = 0.5 hours
p : conversion factor value between 0.17 to 0.2.

The p-value in equation (2) obtained from the Government Regulation No. 41 of 1999 with C_1 = 150 μg/ m^3, t_1 = 1 day, C_2 = 50 g / m^3 and t_2 = 365 days in order to obtain the value of p = 0.186.
Factors temperature, humidity, and pressure greatly affects the air concentration including the concentration of dust. Therefore the weather conditions recorded and accounted for in this study either temperature, humidity, wind direction, wind speed, and the season. [7]. Equation (2) corrected for the effects of T, P, and RH normal to the following equation.

a. Normal wet air condition (≥ 50% RH): C n, f = CB. [T / TN] [PN / P] (3)
b. Normal dry air conditions (RH <50%): CN, tr = CN, f. [100 / (100-F)] (4)

Information:
C N, f : The concentration of dust in normal conditions (μg/m^3)
CB : The concentration of dust in the air in the normal condition at the measurement time: C_1μg/m^3.
C N, tr : concentration of dust in the dry air conditions (μg/ m^3).
T : temperature at the time of measurement (° K).
T N : air temperature in normal conditions (° K) = 25°C=298,15°K
P N : The air pressure in normal conditions (Pa) = 760 mmHg = 1Atm.
P : absolute air pressure at the time of measurement (Pa).
F : humidity at the time of measurement / TN] [PN / P]

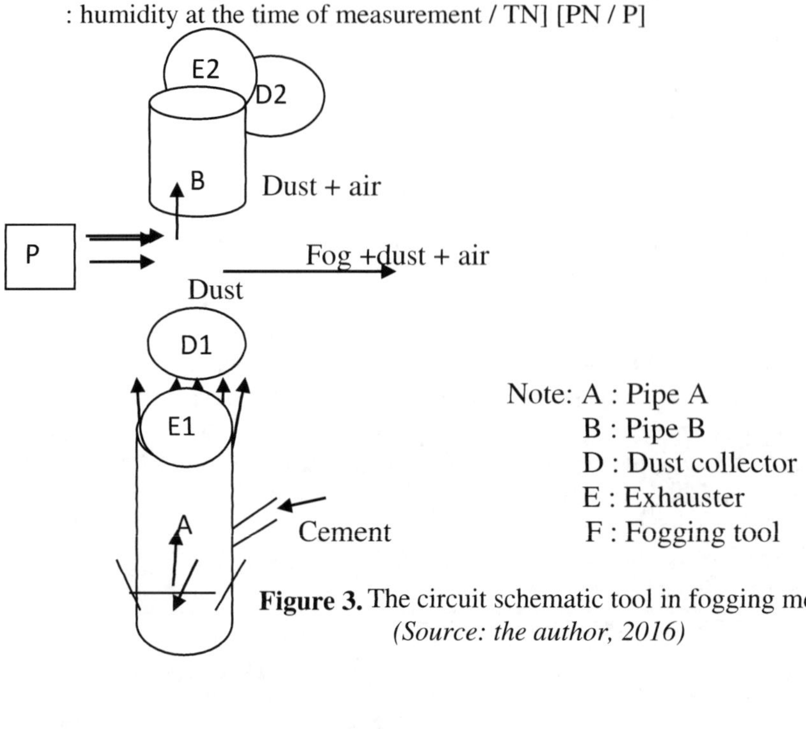

Figure 3. The circuit schematic tool in fogging method
(Source: the author, 2016)

The state variables are distance from sprayer to device (L)=2.5 m; device air velocity (Vu)= 430 m/minute; device water velocity (Vu)=110 ml/minute; stack high (Z)=100 cm. So the free variable is nozle diameter.

Table 2. Relationship of diameter nozzle (Dn) with concentration of cement dust was caught in fog (C)

No	Dn (mm)	Mass of cement dust was caught in fog (μg)	C (μg/m³) at time is 0,5 hour	C (μg/m³) at time is 24 hour	C (μg/m³) at RH=52%; T=29°C (=302°K); P= 0,99 atm
1	2.48	15,767.67	18,572.05	9,039.59	9,248.81
2	2.00	8,561.59	10,084.32	4,908.35	5,021.95
3	1.40	1,444.86	1,701.84	828.34	847.51
4	1.06	1,210.54	1.425.84	694.00	710.06
5	0.40	673.67	793.49	386.21	395.15

Source : primary data, 2017

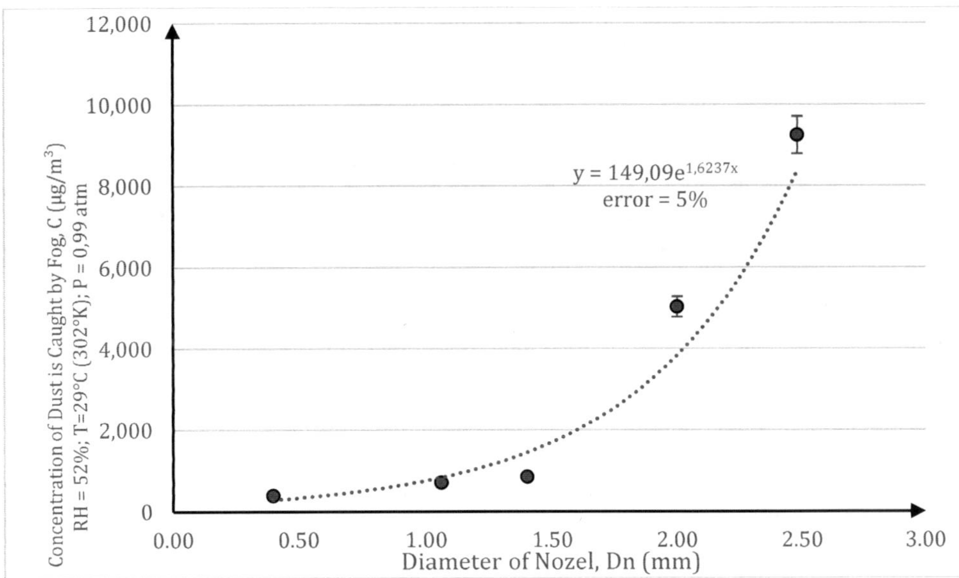

Figure 4. The relation of diameter of nozel (Dn) with concentration of cement dust is caught by fog (C)

The analisys for explaining relationship between nozle diameter vs. fog that produced:

a. 2.48 mm : The product still in the form of water so that most of the dust that passes through the chimney will be affected by water. Weight of dust on the measurement of the bottom is very large, while the measurement of dust on the top position will be small, so the weight difference is very large.

b. 2.00 mm : The product still in the form of water with a smaller grain / fine, so that the dust that passes through the chimney will be exposed to such fine water butitan. The measurement of the weight of the dust at the bottom is large, while the measurement of dust on the top is small that the weight difference is large.

c. 1.40 mm : The product shaped fog is perfectly formed, its position is fitting about the airflow that carries dust so as to capture the dust optimally.

d. 1.06 mm : The product tangible form of fog that is not perfect then it cannot bind the perfect dust so that the difference in weight obtained relatively little.

e. 0.40 mm : The product do not fog because the fog has not materialized then the capture power is almost non-existent. The measurement of the dust weight on the bottom and top is almost as large as the weight gain is very small.

From the analisys, the optimum variabel of nozle diameter is 1.40 mm.

4. Conclusion

1. The trend of wind direction is identical with the trend towards the spread of dust which is to the east with average wind speed is 1.16 m/s.
2. The optimum condition of nozle diameter for cement dust catcher is 1.40 mm, with line equation is $y = 149.09.e^{1.6237.x}$ and error= 5%. In that condition, nozle can make the fog with a good quality and it can catch the cement dust well.

Acknowledgments

I look forward to another opportunity to develop this research, namely by examining other variables to obtain optimal operating conditions for the dust catcher with the fogging method.

References

[1] Dimitriou A and Christidou V 2011*In Tech* **218**
[2] Duda W H 1983 *International Process Engineering in The Cement Industries Bauverlag GMBH Weisbaden and Berlin 2nd edition* (Berlin)
[3] Suhariyono, Sri Saeni M S and Bay A 2003 *P3TM Batan Yogyakarta.*
[4] Nurbiyantoro S 2010 *UNS Surakarta* 44-45
[5] Naqpure A S 2014 *Atmospheric Poll. Res. India* 373
[6] Kinsey J S 1987 *Cement Industry, Air and Energy Engineering Research Laboratory Research Triangle Park* **2**
[7] Ibrahim H G, Okasha A Y, Elatrash M S, and Al-Meshragi M A 2012 *J. Environ. Sci. Eng.* ISSN 1934-8932 620-628
[8] Thaib Y P 2014 *Universitas Sam Ratulangi Manado*
[9] Zimwara D, Mugwagwa L, and Chikowore T R 2012 *CIE4 2 Proceedings* 37-13

IOP Conf. Series: Materials Science and Engineering **333** (2018) 012100 doi:10.1088/1757-899X/333/1/012100

Analysis of fatigue on surface course using dissipated energy approach

Michael,[1] A Setyawan,[1] and F P Pramesti[1]
[1]Civil Engineering Department, Sebelas Maret University, Surakarta, Indonesia

E-mail: michaelsudianto@gmail.com

Abstract. As an important transportation infrastructure, pavement is subjected to repeated vehicle loads that may cause fatigue, which often leads to cracking. The point when this cracking initiates can be determined from the energy dissipated during the loading. This research investigates fatigue in Adi Soemarmo Airport mix-design using bitumen Pen 60/70 + EVA (Ethyl Vinyl Acetate) polymer. An Indirect Tensile Fatigue Test (ITFT) was conducted using stress-controlled loading mode to determine its fatigue life. The stress levels were 500, 600, and 700 kPa, while the loading frequency and the temperature were 10 Hz and 20 °C, respectively. The test exhibits strain levels for each loading cycle, which were used to determine the dissipated energy (DE). The result indicates that the DE increases when the number of loading cycles increases, due to progress of the strain levels. The values of DE are 7122.8, 8614.3, and 2654.9 J/m^3 for loading levels of 500, 600, and 700 kPa, respectively, whereas the failure points for stress levels of 500, 600, and 700 kPa are 8171, 5161, and 841 cycles, respectively. Thus, the longer the time until the pavement failure point is reached (fatigue life), the greater the amount of energy that is dissipated.

1. Introduction

A pavement is an extremely important infrastructure for a transportation system, because it is a facility for vehicles to move from one place to another. When vehicles run along the pavement, which forces the pavement vertically and dynamically, each course of the pavement deforms. As it is viscoelastic, it returns to the former condition. After this has happened repeatedly, there is indirect stress-strain in the longitudinal direction at the bottom of the surface course [1]. If it happens continuously, with insufficient repetitive vertical and force rest time, the surface course will become fatigue cracked. Fatigue cracking is connected to the stiffness and strength of the pavement. So, the decrease of the stiffness leads to weakening of the pavement, which can finally crack [2].

By identifying the stress-strain of the pavement, a fatigue analysis can be done by developing S-N (stress amplitude–life) curves, called classical fatigue analysis [3]. In this method, a decrease of the stiffness modulus is observed until the specimen totally collapses with a transversal crack in the middle of the specimen. However, the classical method analyses the life of the specimen using the lowered values of the stiffness modulus before its collapse. In this way, it is not possible to observe when the crack is initiated. So, a more specific analysis by observing the dissipated energy is required. Under repeated loading, the pavement dissipates the energy as a response to the force. Then, when the dissipated energy reaches its peak, the pavement has initially lost its stability, which means that micro-cracking is happening. Therefore, the latter method is able to identify the fatigue life earlier than the former method by observing the pavement's dissipated energy.

2. Experimental

2.1. Indirect Tensile Fatigue Test (ITFT)

According to the configuration, the fatigue test consists of a simple flexure test, a direct axial loading test, and a diameter loading test [4]. These can be executed using two modes of loading: the strain-controlled and stress-controlled modes. This research used the diameter loading test (Indirect Tensile Fatigue Test, ITFT) with the stress-controlled mode.

In the ITFT, a compressive force is applied to a cylindrical specimen. The force is then distributed in the vertical and horizontal directions of the specimen [5]. When the vertical stress compresses the specimen, horizontal stress (called indirect tensile stress) occurs in the middle of the specimen. Because of the tensile stress, the specimen is displaced and becomes bigger when the force keeps compressing it until it collapses. Illustrations of the ITFT are shown in Figures 1.

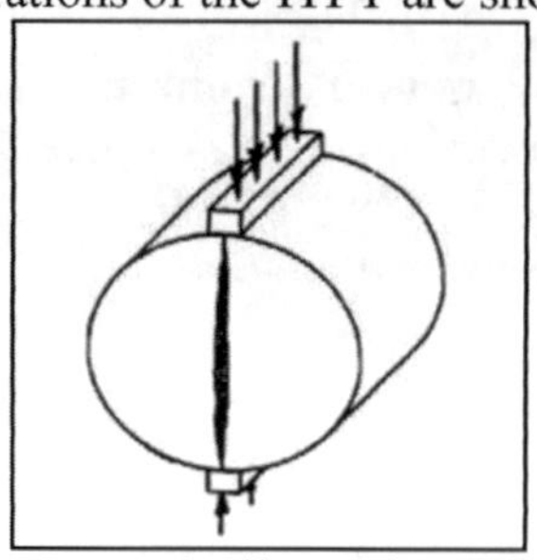

Figure 1. Illustration of ITFT when the specimen collapses.

2.2. Dissipated Energy Approach

During the performance of the test with repeated loading and relaxation, the energy of the specimen is dissipated, as it is viscoelastic at ambient temperatures. The viscoelastic materials make a different path on the time–load graph under loading and relaxation, forming an area. This phenomenon is called hysteresis. So, the dissipated energy is equal to the area of the graph [3].

The different path formed by loading and relaxation is caused by tensile stress-strain and phase-lag in each cycle. In stress-controlled mode, the stress is constant. However, the strain and phase lag get bigger during loading. This causes the dissipated energy to increase in line with the increase in the number of loading cycles. It was also reported by Francken and Clauwaert [6] that the dissipated energy rises in the stress-controlled mode but drops in the strain-controlled mode. So, the dissipated energy can be determined by Equation 1. After the energy dissipated in each cycle has been measured, the cumulative dissipated energy can be determined (2).

$$w_i = \pi \sigma_i \varepsilon_i sin\emptyset_i \qquad (1)$$

$$W = \pi \sum_{i=0}^{i=N} \sigma_i \varepsilon_i sin\emptyset_i \qquad (2)$$

where w_i is the energy dissipated in cycle i; σ_i is the *stress* in cycle i; ε_i is the *strain* in cycle i; $\emptyset_i$ is the *phase lag* in cycle i; and W is the cumulative dissipated energy.

Using the dissipated energy, the fatigue life is defined as the point at which the graph of the dissipated energy ratio (DER) versus the load cycle number deviates from a straight line [7]. So, DER and fatigue life can be determined by Equations 3 and 4, respectively.

$$R_\sigma = nE_i^* \qquad (3)$$

where R_σ is the DER in the stress-controlled fatigue test and E_i^* is the stiffness modulus.

Knowing the cumulative dissipated energy and DER, the relationship between them can be identified (4).

$$W = AN^Z \tag{4}$$

where A and Z are experimental coefficients [8].

2.3. Materials

The aggregate is taken from Kulonprogo quarry, Yogyakarta, and the gradation of the aggregate for the asphalt mixture using Adi Soemarmo runway's, which is shown in Table 1.

Table 1. Existing pavement gradation of Adi Soemarmo Airport

Sieve Opening Size		% Passing
Inch	**mm**	
1	25.4	100
3/4	19	100
1/2	12.7	84.54
3/8	9.5	79.13
No.4	4.76	76.57
10	2	65.30
40	0.42	26.47
80	0.177	11.97
200	0.074	5.41

(Source: Final Report of Inspection of the Quality and Quantity of Extension Construction of R/W, T/A, and Paved Shoulder Stage II)

However, the bitumen (Pen 60/70) of the mixture is modified using EVA (ethylene vinyl acetate) polymer. EVA is a polymer that can make the bitumen more plastic, so the strength improves. This polymer will improve the performance of the mixture, as confirmed by Indirect Tensile Strength (ITS) Test and Unconfined Compression Strength (UCS) Test. The use of 4% EVA can optimize the mixture strength as well as the temperature resistant [9].

3. Results and Discussion

From the Marshall test, the optimum bitumen content is determined, as shown in Table 2.

Table 2. Marshall test result using Bitumen Pen 60/70 and EVA 4%

Test Results	4.5%	5.0%	5.5%	6.0%	6.5%	7.0%	Specification
Stability (kg)	986.33	1017.81	1370.66	1375.52	1285.77	1156.13	975.51
Flow (mm)	3.17	3.20	4.55	4.00	3.47	2.90	2.5–4
Porosity (%)	10.137	8.909	6.636	8.116	4.389	2.409	2–5
Density (gr/cc)	2.211	2.225	2.264	2.213	2.286	2.317	
MQ (kg/mm)	316.337	323.669	287.254	345.114	373.070	404.818	

According to all of the properties of the Marshall test, the optimum bitumen content is 6.5%. So, it is used for the specimen's formula. Three specimens were used for the ITFT in Udayana University, Denpasar, and the results are listed in Table 3. The data in Table 3 were used for developing a model of fatigue and initial stiffness, which was done by Nugroho [10].

Table 3. ITFT result

No.	σ (kPa)	N_f (cycles)
1	500	10401
2	600	6600
3	700	1250

IOP Conf. Series: Materials Science and Engineering **333** (2018) 012100 doi:10.1088/1757-899X/333/1/012100

With the data, the cumulative dissipated energy is calculated, as shown in Table 4. The DER is determined using the stiffness modulus and then graphs of DER versus the number of cycles are built, as shown in Figures 2, 3, and 4.

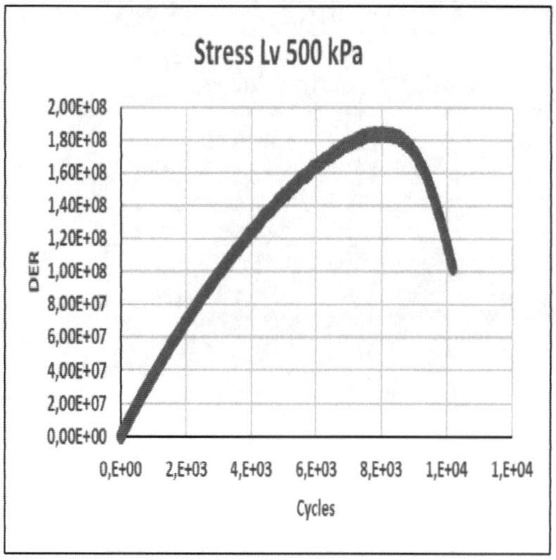

Figure 2. Relationship between DER and the number of load cycles under a stress level of 500 kPa

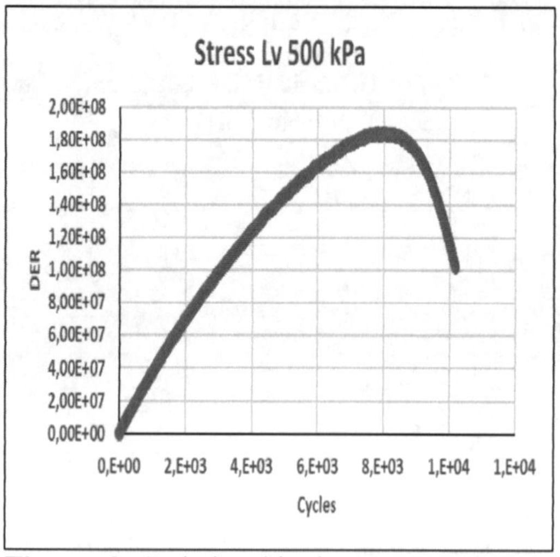

Figure 3. Relationship between DER and the number of load cycles under a stress level of 600 kPa

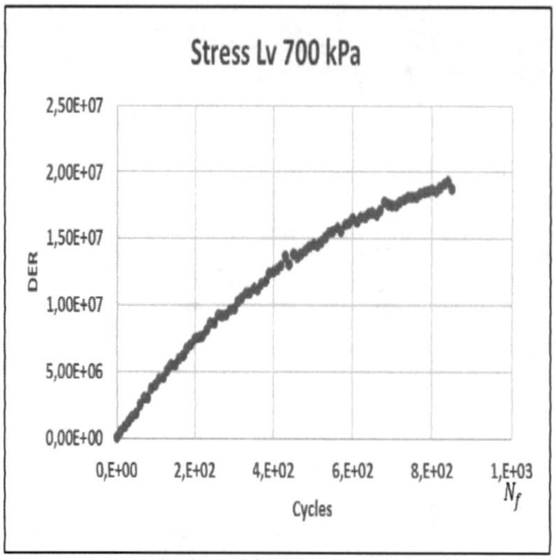

Figure 4. Relationship between DER and the number of load cycles under a stress level of 700 kPa

Table 4. Summary of N_R

No.	σ (kPa)	N_R (cycles)
1	500	8171
2	600	5161
3	700	841

If these results are compared with the results of the results of Nugroho's classical analysis [10], the fatigue life found by using the dissipated energy approach is lower than that found by the classical approach. The initiation of cracking of the specimen does not occur at the time of specimen collapse in the test. Cracking is initiated by micro-cracks, which are identified by observing the dissipated energy. Therefore, the dissipated energy analysis determines the fatigue life by using a more specific analysis that identifies the unseen cracks in the specimens.

Furthermore, the mathematic model of the relationship between N_R and W can be built as shown in Figure 8 using a graph of log N_R versus log W. From the graph, the trend line can be determined and the coefficients A and Z are calculated, resulting in values of 99.312 and 0.495, respectively. Finally, the model equation is completed (5).

$$W = 99.312\, N_R^{0.495} \tag{5}$$

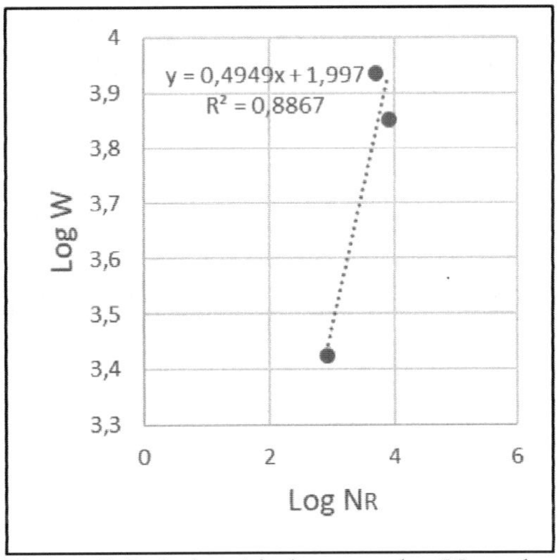

Figure 5. Relationship between log N_R and log W

4. Conclusion

After the analysis and discussion, it can be concluded from the research that the greater the number of loading cycles, the higher the strain that takes place in the specimens. The increase in strain leads to more dissipation of the energy of the specimens. However, the relationship between cumulative dissipated energy and stress level is inversely proportional, with a decrease of cumulative dissipated energy followed by an increase of the stress level. Furthermore, the fatigue life determined by using the dissipated energy approach (N_R) is directly and inversely proportional to the cumulative dissipated energy and stress level, respectively.

References

[1] Shahin M Y 2005 *Pavement Management for Airports, Roads, and Parking Lots* second edition (USA: Springer Science+Business Media, LLC)
[2] Pell P S 1965 *Fatigue of Bituminous Materials in Flexible Pavements* Proc. Institution of Civil Engineers **31**
[3] Li N 2013 *Asphalt Mixture Fatigue Testing: Influence of Test Type and Specimen Size* (Delft: Delft University of Technology)

[4] Tangella S R, Craus J, Deacon J A and Monismith C L 1990 *Summary Report on Fatigue Response of Asphalt Mixtures* SHRP Report TM-UCB-A-003A-89-3 for project A-003-A University of California Berkeley
[5] Kennedy T W and Anagnos J N 1983 *Procedures for the Static and Repeated-Load Indirect Tensile Tests* Center for Transportation Research University of Texas at Austin Research Record **183**-14
[6] Francken L and Clauwaert C 1987 *Characterization and Structural Assessment of Bound Materials for Flexible Road Structures* Proc. Sixth ICSDAP (Michigan: Ann Arbor) pp 130–144
[7] Pronk A C and Hopman P C 1991 *Energy Dissipation: The Leading Factor of Fatigue* Proc. Conference on the United States Strategic Highway Research Program pp 255–267
[8] Chomton G and Valayer P J 1972 *Applied Rheology of Asphalt Mixes – Practical Applications Proceedings* Third International Conference on the Structural Design of Asphalt Pavements London England pp 214–225
[9] Kurniawan R 2016 *Pengaruh Bitumen Modifikasi Ethylene Vinyl Acetate (EVA) Pada Asphalt Concrete Dan Thin Surfacing Hot Mix Asphalt Terhadap Pengujian Unconfined Compressive Strength (UCS) Dan Indirect Tensile Strength (ITS) [The Influence of Ethylene Vinyl Acetate (EVA) Modified Bitumen on Asphalt Concrete and Thin Surfacing Hot Mixed Asphalt to Unconfined Compressive Strength (UCS) Test and Indirect Tensile Strength (ITS) Test]* (Surakarta: e-Mateksi UNS) **4**-2 pp 586-590
[10]Nugroho A M 2016 *Analisis Fatigue pada Wearing Course dengan Metode Classical Fatigue [Analysis of Fatigue on Wearing Course Using Classical Fatigue Method]* (Surakarta: Sebelas Maret University)

IOP Conf. Series: Materials Science and Engineering **333** (2018) 012101 doi:10.1088/1757-899X/333/1/012101

Determination of Failure Point of Asphalt-Mixture Fatigue-Test Results Using the Flow Number Method

C E P Wulan,[1] **A Setyawan, and F P Pramesti**
[1] Road Mate Research Group, Civil Engineering Department, Sebelas Maret University, Jl. Ir. Sutami No. 36 A Surakarta 57126, Indonesia

E-mail: cahyaesther@gmail.com

Abstract. The failure point of the results of fatigue tests of asphalt mixtures performed in controlled stress mode is difficult to determine. However, several methods from empirical studies are available to solve this problem. The objectives of this study are to determine the fatigue failure point of the results of indirect tensile fatigue tests using the Flow Number Method and to determine the best Flow Number model for the asphalt mixtures tested. In order to achieve these goals, firstly the best asphalt mixture of three was selected based on their Marshall properties. Next, the Indirect Tensile Fatigue Test was performed on the chosen asphalt mixture. The stress-controlled fatigue tests were conducted at a temperature of 20°C and frequency of 10 Hz, with the application of three loads: 500, 600, and 700 kPa. The last step was the application of the Flow Number methods, namely the Three-Stages Model, FNest Model, Francken Model, and Stepwise Method, to the results of the fatigue tests to determine the failure point of the specimen. The chosen asphalt mixture is EVA (Ethyl Vinyl Acetate) polymer -modified asphalt mixture with 6.5% OBC (Optimum Bitumen Content). Furthermore, the result of this study shows that the failure points of the EVA-modified asphalt mixture under loads of 500, 600, and 700 kPa are 6621, 4841, and 611 for the Three-Stages Model; 4271, 3266, and 537 for the FNest Model; 3401, 2431, and 421 for the Francken Model, and 6901, 6841, and 1291 for the Stepwise Method, respectively. These different results show that the bigger the loading, the smaller the number of cycles to failure. However, the best FN results are shown by the Three-Stages Model and the Stepwise Method, which exhibit extreme increases after the constant development of accumulated strain.

1. INTRODUCTION

The fatigue test using the stress-controlled mode has a short failure period. This makes the failure point hard to determine. Several studies have been developed to define the failure point, one of which is called the Flow Number Method. The Flow Number Method is a part of the results of the NCHRP research project on asphalt performance. This research assumed that the Flow Number Method is better than other methods for determining the failure point. It was found that the Flow Number Method obtained a better correlation with the field performance [1], [2]. This study aims to determine the failure point of the results of fatigue tests on asphalt mixture using the Flow Number Method and to compare the most applicative theory of the Flow Number Method based on the graphical results of fatigue tests on asphalt mixtures for a particular asphalt mixture.

The failure point is determined just before the total collapse of the specimen. In stress-controlled mode, the flow number is defined as the cycle number at which the tertiary phase initiates. This study uses four Flow Number models, namely the Three-Stages Model, the FNest Model, the Francken Model, and the Stepwise Method. These models have the same correlation between the permanent strain (ε_p) and cycle number (N). The scientific approaches of the Flow Number models are explained below.

1.1 Three-Stages Model

The Three-Stages Model has three phases of failure, determined as the primary, secondary, and tertiary stages. Every phase has its own approach, shown in Eqs. 1–3 below [3].

Primary stage:

$$\varepsilon_p = aN^b;\ N < N_{PS} \quad (1.1)$$

Secondary stage:

$$\varepsilon_p = \varepsilon_{PS} + c(N - N_{PS})\ ;\ N_{PS} \leq N < N_{ST} \quad (1.2)$$

Tertiary stage:

$$\varepsilon_p = \varepsilon_{ST} + d(e^{f(N-N_{ST})} - 1);\ N \geq N_{ST} \quad (1.3)$$

N_{PS} is the load repetition number at the initiation of the secondary stage, N_{ST} is the load repetition at the initiation of the tertiary stage. ε_{PS} is the permanent strain at the initiation of the secondary stage, ε_{ST} is the permanent strain at the initiation of the tertiary stage, and a, b, c, d, and f are materials constants. In this model, N_{ST} is the flow number.

1.2 FNest Model

The FNest Model is shown in Eqs. 4 and 5 below [4].

$$\varepsilon_p = \frac{1}{\beta}\left[-\ln\left(1 - \frac{N}{\gamma}\right)\right]^{\frac{1}{\alpha}} \quad (1.4)$$

$$F_N = \gamma\left[1 - \exp\left(\frac{1}{\alpha} - 1\right)\right] \quad (1.5)$$

FNest Model was developed by Archilla and Diaz [5]. The α, β and γ parameters are calculated to develope relation of permanent strain and number of cycles. Furthermore, these parameters are used to determine FN.

1.3 Francken Model

The Francken Model is a composite model between the power part and the exponential component [6]. The Francken Model is defined in (1.6), where A, B, C, and D are the regression constants and (1.7) shows the second derivation [7]:

$$\varepsilon_p(N) = AN^B + C(e^{DN} - 1) \quad (1.6)$$

$$\frac{d^2\varepsilon_p}{dN^2} = (A \times B\,(B-1) \times N^{(B-2)}) + (C \times D^2 \times e^{DN}) \quad (1.7)$$

1.4 Stepwise Method

The Stepwise Method is a traditional method to find the minimum point on a correlation curve between strain rate and cycles by smoothing the measured permanent deformation. There are three simple steps as follows: Firstly, The measured permanent deformation is smoothed by reallocating the measured result under the assumption that the permanent strain will either remain the same or increase over the number of load cycles [5]. Second, the strain rate is calculated on the basis of the modified permanent deformation above mentioned. Equation (1.8) is used to calculate the strain rate [4]. Third, the flow number is defined as the minimum strain rate value by combining the curve of the permanent strain versus cycles and the curve of strain rate versus cycles.

Strain Rate = $\frac{\varepsilon_p}{N}$.. (1.8)

2. EXPERIMENTAL

This study used split aggregate with 19 mm (3/4 inch) as the maximum dimension and asphalt pen 60/70, EVA modified asphalt, and SBS (*Styrene Butadiene Styrene*) as the binder materials. A cylinder specimen with a thickness of ± 63.5 mm and diameter of ±100 mm was used in this research.

These kinds of mixtures were tested by the Marshall test to find their Optimum Bitumen Contents (OBCs). Also the Marshall test was used to find the Marshall characteristic of the asphalt mixtures in order to choose the best one. The best asphalt mixture will be tested by the fatigue test method. This research uses the Indirect Tensile Fatigue Test (ITFT) in the stress-controlled mode at a temperature of 20° C and frequency of 10 Hz. The ITFT method is based on BS EN-12697-24. The fatigue test used loads of 500, 600, and 700 kPa. The results of this test are resilient modulus, permanent strain, and deformation at related number of load repetitions (cycles). The next step is the determination of the failure point by the Flow Number method based on the fatigue test of the asphalt mixture.

3. RESULTS AND DISCUSSION

According to the results of Michael [8] and Nugroho [9], all of the asphalt mixtures showed OBCs as large as 6.5%. However, EVA-modified bitumen showed the best Marshall characteristic based on the stability, flow, porosity, density, and Marshall quotient. Therefore, the EVA-modified asphalt mixture was tested by the fatigue test. The results of the fatigue test were processed to determine the failure point using four Flow Number Method, namely the Three-stages Model, FNest Model, Francken Model, and Stepwise Method.

3.1 Three-Stages Model

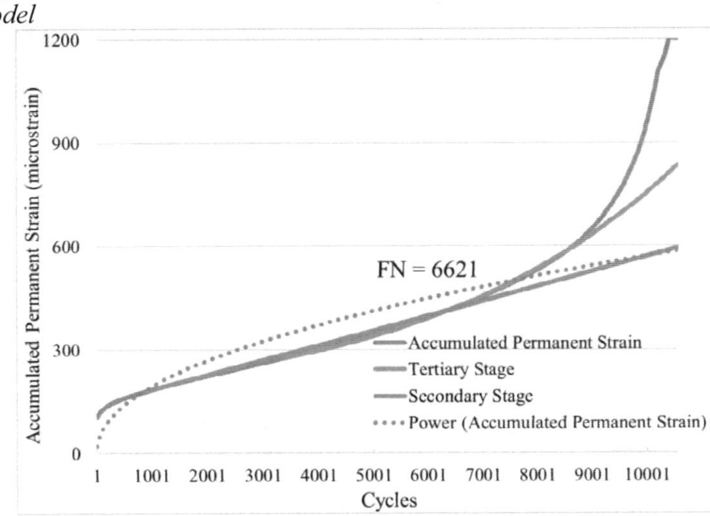

Figure 1. Diagram of Three-Stages Model P = 500 kPa

Table 1. Three-Stages Model Approach

P (kPa)	Primary Stage	Secondary Stage	Tertiary Stage
500	$\varepsilon_p = 22.184N^{0.7647}$	$\varepsilon_p = \varepsilon_{PS} + 4.29134E-06(N-761)$	$\varepsilon_p = \varepsilon_{ST} + 156.7793(e^{0.000212(N-6621)} - 1)$
600	$\varepsilon_p = 61.064N^{0.3546}$	$\varepsilon_p = \varepsilon_{PS} + 7.75801E-06(N-341)$	$\varepsilon_p = \varepsilon_{ST} + 152.1617(e^{0.000328(N-4841)} - 1)$
700	$\varepsilon_p = 145.43N^{0.3419}$	$\varepsilon_p = \varepsilon_{PS} + 4.06383E-05(N-611)$	$\varepsilon_p = \varepsilon_{ST} + 166.8984(e^{0.00204(N-611)} - 1)$

This research used Solver tools in Microsoft Excel to obtain the material constants. The result of N_{ST} under loads of 500, 600, and 700 kPa are 6621, 4841, and 611 cycles respectively.

3.2 FNest Model

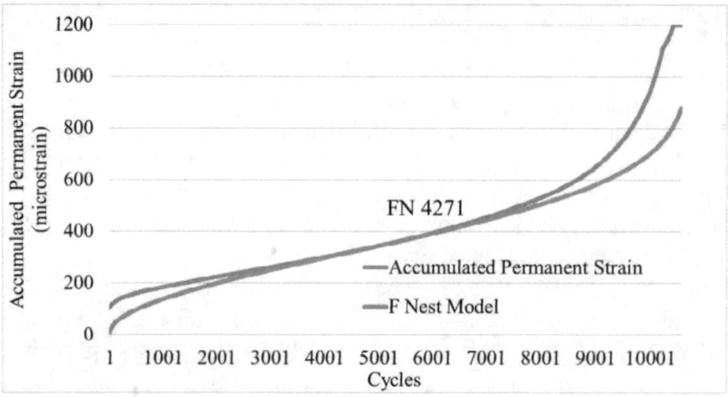

Figure 2. Diagram of FNest Model. P = 500 kPa

Table 2. Probability Distribution Parameters of FNest Model

No.	P (kPa)	A	B	ϒ	FN	Model
1	500	2.0387	0.00232	10700	4271	$\varepsilon_p = \frac{1}{0.00232}\left[-\ln\left(1-\frac{N}{10700}\right)\right]^{\frac{1}{2.0387}}$
2	600	3.4373	0.00208	6432.181	3267	$\varepsilon_p = \frac{1}{0.00208}\left[-\ln\left(1-\frac{N}{6432.181}\right)\right]^{\frac{1}{3.4373}}$
3	700	3.4257	0.00177	1058.899	537	$\varepsilon_p = \frac{1}{0.00177}\left[-\ln\left(1-\frac{N}{1058.899}\right)\right]^{\frac{1}{3.4257}}$

Table 2 shows the probability distribution obtained by the FNest Model approach to determine FN. In this model, FN is determined by (1.5) with the help of Solver tools to find the probability distribution parameter. Figure 2 shows the results of the permanent strain model based on (1.4). The failure points for the load, namely 500, 600, and 700 kPa, are 4271, 3267, and 537 cycles.

3.3 Francken Model

In the Francken model, FN is derived by the Francken Model approach. Based on the permanent strain approach, regression constants which have the same structure as (1.6) is defined. The first derivative of the permanent strain results in the strain slope. The minimum strain slope is FN in the Francken Model. The strain slope is the transition phase from the negative axis to the positive one in the second derivative of the Francken Model. So the failure points under loads of 500, 600, and 700 kPa are 3401, 2431, and 421 cycles respectively.

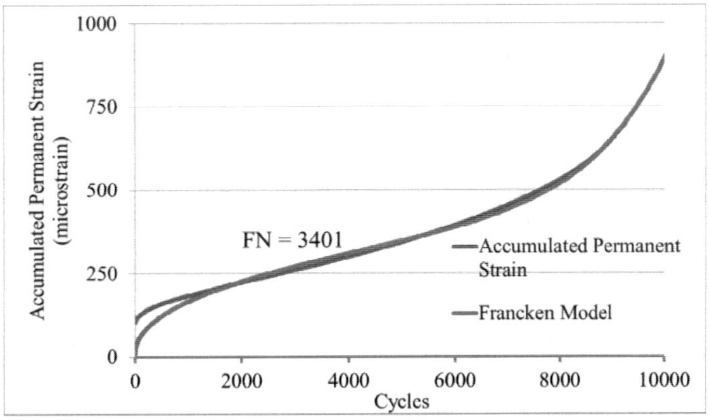

Figure 3. Diagram of Francken Model. P = 500 kPa

Table 3. Regression Constants of Francken Model

P (kPa)	A	B	C	D	FN	Model ε_p
500	8.367	0.433	1.23	0.0007	3401	$\varepsilon_p(N) = 8.367N^{0.433} + 1.23(e^{0.0007N} - 1)$
600	26.955	0.343	0.9	0.0011	2431	$\varepsilon_p(N) = 26.955N^{0.343} + 0.9(e^{0.0011N} - 1)$
700	78.546	0.296	1.16	0.0061	421	$\varepsilon_p(N) = 78.546N^{0.296} + 1.16(e^{0.0061N} - 1)$

3.4 Stepwise Method

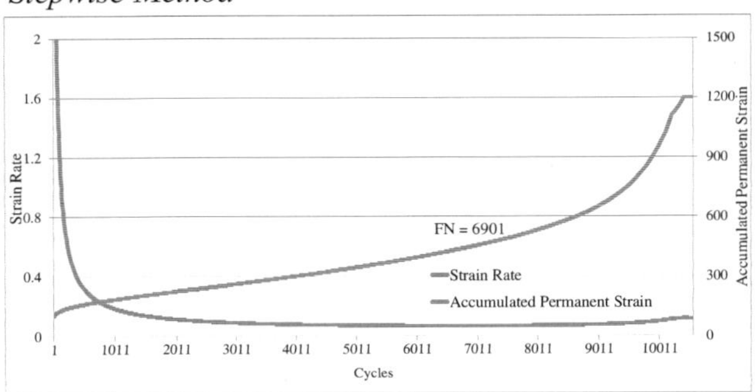

Figure 4. Diagram of Stepwise Method. P = 500 kPa

Table 4. FN calculation for Stepwise Method

P	Min. $\dot{\varepsilon}$	FN
500	0.6477	6901
600	0.1136	6841
700	0.68397	1291

The fatigue failure phase in the Stepwise Method are the same as in the Francken Model. In the primary phase, the strain rate decreases extremely from the beginning until N cycles of the test. In the secondary phase, the strain rate curve is relatively flat. Until the strain rate starts to increase again, this is the ending of the secondary phase and the beginning of the tertiary one. The tertiary phase is the crack initiation. FN is allocated at the minimum point of the strain rate. So the failure points for the load repetitions of 500, 600, and 700 kPa are 6901, 6841, and 1291 cycles.

Based on the model results, the bigger the loading, the smaller the number of cycles to failure. The results of each method are summarized in Table 5 and Figure 5. Moreover, each method has a different FN value. A high variability factor leads to different FN values [3]. Each model has a different failure criteria based on its equation. Each equation consists of different independent and dependent variables. These differences result in different values of FN.

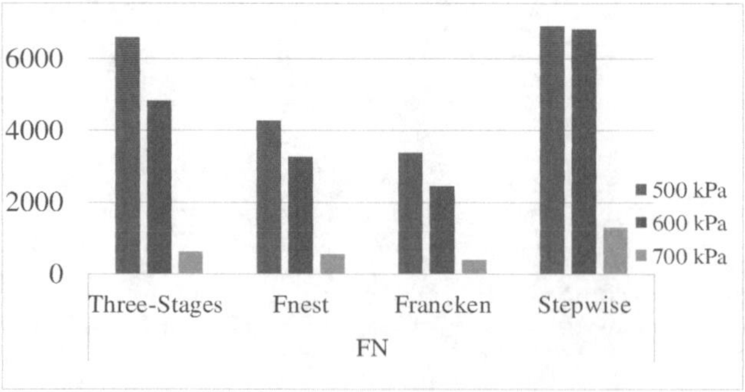

Figure 5. FN Comparation for Three-Stages Model, FNest Model, Francken Model, and Stepwise Method

Table 5. Summary of FN calculation

P (kPa)		500	600	700
FN	Three-Steps	6621	4841	611
	FNest	4271	3266	537
	Francken Model	3401	2431	421
	Stepwise Method	6901	6841	1291

4. CONCLUSION

This paper presents a summary of the determination of FN for EVA-modified asphalt mixture. The failure point of results of the fatigue tests of the EVA-modified asphalt mixture was determined using the Flow Number Method with four model approaches, namely the Three-Stages Model, FNest Model, Francken Model, and Stepwise Method. The failure points of the Three-Stages Method with load repetitions of 500, 600, and 700 kPa are 6621, 4841, 611. The failure points of the FNest Model with the same load repetitions as the Three-Stages Model are 4271, 3266, and 537. With the same load repetition, the failure points of the Francken Model are found to be 3401, 2431, and 421. Lastly, the failure points of the Stepwise Method are 6901, 6841, and 1291 for the same load repetition.

Based on the FN calculation, the Stepwise Method has the highest FN value. Moreover, based on the curve result, the FN values obtained by the Stepwise Method and the Three-Stage Model are allocated to the tertiary phase, when the strain rate increases extremely. Therefore, based on the FN calculation and the curve result, the Stepwise Method and the Three-Stage Model are fitter models for determining the failure point using the Flow Number Method for EVA-modified asphalt mixture at a temperature of 20°C and frequency of 10 Hz. Both failure points are allocated in a critical area but are still in the right place.

REFERENCES

[1] Witczak M, Kaloush K, Pellinen T, El-Basyouny M and Von Quintus H 2002 *NCHRP Rep.* **465**
[2] Zhou F and Scullion T 2003 *J. Transp. Res. Board* **1832**(1) 209–216
[3] Ameri M, Sheikhmotevali A H and Fasihpour A 2014 *Road Mater. Pavement Des.* **15** 182–206
[4] Goh S W and You Z 2009 *Const. Build. Mater.* **23** 3398-3405
[5] Archilla A R and Diaz L G 2011 *J. Transp. Res. Board* **2210** 1–8
[6] Biligiri K P, Kaloush K E, Mamlouk M S and Witczak M W 2007 *J. Transp. Res. Board* **2001**(1) 63–72
[7] Pramesti F P 2015 *Laboratory and Field Asphalt Fatigue Performance Matching Theory with Practice* (Delft: Delft University of Technology)
[8] Michael 2016 *Analisis Fatigue pada Surface Course dengan Pendekatan Dissipated Energy* (Surakarta: Universitas Sebelas Maret)

IOP Conf. Series: Materials Science and Engineering **333** (2018) 012101 doi:10.1088/1757-899X/333/1/012101

[9] Nugroho A M 2016 *Analisis Fatigue pada Wearing Course dengan Metode Classical Fatigue* (Surakarta: Universitas Sebelas Maret)

IOP Conf. Series: Materials Science and Engineering **333** (2018) 012102 doi:10.1088/1757-899X/333/1/012102

An Application of Multiplier Analysis in Analyzing the Role of Mining Sectors on Indonesian National Economy

S Subanti[1,2,5], A R Hakim[3,5], and I M Hakim[4]

[1]Department of Statistics, Faculty of Mathematics and Natural Science, Universitas Sebelas Maret
[2]Research Group for Applied Statistics, Universitas SebelasMaret
[3]Faculty of Economics and Business, Universitas Indonesia
[4]Department of Industrial Engineering, Faculty of Engineering, Universitas Indonesia
[5]Institute for Research and Social Services,Universitas Sebelas Maret

E-mail: sri_subanti@yahoo.co.id

Abstract. This purpose of the current study aims is to analyze the multiplier analysis on mining sector in Indonesia. The mining sectors defined by coal and metal; crude oil, natural gas, and geothermal; and other mining and quarrying. The multiplier analysis based from input output analysis, this divided by income multiplier and output multiplier. This results show that (1) Indonesian mining sectors ranked 6th with contribute amount of 6.81% on national total output; (2) Based on total gross value added, this sector contribute amount of 12.13% or ranked 4th; (3) The value from income multiplier is 0.7062 and the value from output multiplier is 1.2426.

1. Introduction

The development of Indonesian economy needs to pay attention to the development of every sector in the economy. The development of the economic sector could have an impact on increasing output although it does not necessarily reflect the equitable distribution of people's income and job opportunity [1]. In fact, every country can not always develop all economic sector. Thus, a country must choice the economic sectors that it's expected to increase income distribution and provide employment.

Early Indonesian economy, one of leading economic sector is mining sector besides agricultural sector. This sector defined as secondary sector or it's included in industrial sector. Based on Central Burearu of Statistics data, in 1960's, the contribution of this sector only 31.73%, then increasing until 46.78% in 2010 [2].Based from BPS data, in 2000, the contribution of mining sector compared with other economic sectors amount of 15.60%, it's decreasing until 13.13% in 2005; 10.46% in 2010; and last 2015's, it's not far from 9% [3]. The establishment of a solid structure in which the agricultural economy and mining quarrying sector, it make form the basis of an economy that produces products in both an efficient, an effective, and a modern manner, in which the manufacturing sector contains global competitiveness and becomes the engine of the Indonesia economy, and services become the glue of economic resilience [4].

Data from Indonesia"s Investment Co-ordinating Board (BKPM) suggest that foreign investment is concentrated in Java (60%) and Sumatra (21%), particularly in the Jakarta metropolitan area, Batam, Bintan and Karimun. Since 2004, FDI concentration has shifted from manufacturing toward the mining and quarrying and certain services sectors. Foreign investors tend to come from other Asian countries, with Japan, Korea, Malaysia and Singapore as important investors. However, other countries, such as the United States (particularly in the mining sector) and the United Kingdom are also important sources of FDI in Indonesia [5].

The economic sector'scan not be separated from community activities, either in the form of outward-oriented goods or services, in other words, it must be export oriented sector to other regions, otherwise the economic sector is not to be the imported sector's, in other words, all products (goods and services) came from other regions [6,7]. The purpose of this paper is to know the role of mining sector on Indonesian economy. The role of this sector is described by the contribution on national total output and gross value added, then the value from income and output multiplier. This paper focuses on the case of the Indonesian state and uses the input output analysis approach as an analytical instrument.

2. Experimental

2.1. Data

The data sources in this paper came from Indonesia's input output data in 2008. This data published by the Central Bureau of Statistics. The use of data is expected to capture the changes occurring along this periods as well as to know the impact of a priority sector on household income. The data are classified into 10 sectors. These are agriculture, other crops, forestry and hunting; mining & quarrying sector; manufacturing industry sector; electricity gas and water supply; construction; trade, hotel and restaurant; transportation sector; communications sector; financial, ownership, &business services; and services sector.

2.2. Input Output Method

The Input-Output Table, commonly called as Table I-O, was first introduced by W. Leontief in the 1930s. Table I-O is a table that provides information about transactions of goods and services that occur between production sectors within an economy. It's presented in the form of a matrix [8,9]. The data in table I-O shows trade relationships between sectors within a country's economy. Each line shows the number of sales from a sector. Since a sector does not sell its goods to an existing sector, it is common to find zeros in a row in table I-O. The columns in the I-O table record purchases made by a sector on goods and services produced by various sectors within the region. If the numbers within the columns of a sector are numerous zeros, this is because a sector does not always buy goods and services from all sectors in the country's economies [8,9,10]. The simplest form of the I-O table is as follows.

Table 1. Simple Form for Input Output Table

Seller Sector	Buyer Sector				Final Demand	TotalProduction
	1	2	...	n		
1	X_{11}	X_{12}	...	X_{1n}	f_1	X_1
..	...	...	...	...	...	...
N	X_{n1}	X_{n2}	...	X_{nn}	f_n	X_n
Value Added	V_1	V_2	...	V_n		
Import	M_1	M_2	...	M_n		
Total Input	X_1	X_2	...	X_n		

From table 1, we can made two balanced balance equations :

Row :

$$\sum_{j=1}^{n} x_{ij} + f_i = x_i ; \forall i = 1,2,3,\ldots,n \tag{1}$$

Column :

$$\sum_{i=1}^{n} x_{ij} + v_j + m_j = x_j ; \forall i = 1,2,3,\ldots,n \tag{2}$$

Where x_{ij} is the flow of goods and services from sector i to sector j; F_i is the total final consumption; V_j is value added; and M_j is import. The definition of a balanced balance is the amount of production equal to the number of inputs. The flow can be transformed into coefficients by assuming that the number of purchases is fixed for a total output level and there is no possibility of substitution between an input feed and another input material. From both equations we can constructed coefficient as follows:

$$a_{ij} = x_{ij} / x_j \quad (3)$$

or

$$x_{ij} = a_{ij} x_j \quad (4)$$

with substitute equation (4) to (1), we get :

$$\sum_{j=1}^{n} a_{ij} x_{ij} + f_i = x_i; \forall i = 1,2,3, \dots, n \quad (5)$$

From equation (2.5), we have basic relationship in IO Table as follows :

$$(I - A)^{-1} f = x \quad (6)$$

Notation $(I - A)^{-1} f = x$;it's named as Leontief's inverse matrix. This matrix provides important information on how increased production of a sector will lead to the development of other sectors. Because each sector has a different pattern, then the impact of a sector's production changes on the total production of other sectors will be different. The Leontief matrix summarizes the entire impact of a sector's production change on the total production of another sector into a coefficient. It's often known as multipliers.

Since the purpose of this paper is to calculate the impact of the mining sector on output, it is necessary to know the value of the output multiplier which is the amount of output created by the increased demand for sector j output by 1 unit. The output multipler calculation formula as follows:

$$W_j = \sum_{i=1}^{n} v_{ij} \Phi_i \quad (7)$$

Where W_j is the output multiplier sector j; v_{ij} is Leontief's inverse matrix; and Φ_i is is the output coefficient.

Second, the purpose of this paper is to calculate the impact of the mining sector on household income, it is necessary to know the value of the income multiplier which is the amount of income created by the increased demand for sector j output by 1 unit. The income multipler calculation formula as follows:

$$H_j = \sum_{i=1}^{n} v_{ij} \Pi_i \quad (8)$$

Where H_j is the income multiplier sector j; v_{ij} is Leontief's inverse matrix; and $\Pi_i$is is the income coefficient. The income coefficient can be calculated by dividing the total revenue of sector i by output from sector i.

3. Results And Discussions

Table 2. The Contribution of Economic Sector on Total Output

No	Sector	Output	Percentage
1	Manufacturing	3,882,504,357	36.87%
2	Trade, hotel, & restaurant	1,336,221,986	12.69%
3	Construction	1,243,975,535	11.81%
4	Agriculture	1,098,336,352	10.43%
5	Public administration	900,298,190	8.55%
6	Mining	717,482,885	6.81%
7	Finance	566,629,564	5.38%
8	Transportation	469,459,527	4.46%
9	Communication	190,642,096	1.81%
10	Electricity, gas, and water supply	124,490,705	1.18%

Based on table 2, it shows that the manufacturing industry sector has the highest value followed by the trade, hotel, and restaurant sector. While the sector with the lowest value is the electricity, gas, and water supply sector. The value of contribution on total output is 36.87% for the manufacturing industry, this value indicates that it have a strong ability to drive output growth. Not only this sector, but also other sector likes trade, hotel, & restaurant sector; construction sector; and agricultural sector can be a leading sector to drive output growth too. As we know that the output generated from these sectors is the intermediate output, then the output can be a raw material for industry or other economic sectors.

Table 3. The Economic Sector Contribution Based On Value Added

No	Sector	Value Added	Percentage
1	Agriculture	1,399,524,198	15.94%
2	Mining	1,065,408,831	12.13%
3	Manufacturing industry	2,468,458,179	28.11%
4	Electricity, gas, and water supply	60,499,186	0.69%
5	Construction	735,427,477	8.37%
6	Trade, hotel, and restaurant	1,166,246,423	13.28%
7	Transportation	296,926,731	3.38%
8	Communication	267,757,988	3.05%
9	Finance	684,665,646	7.80%
10	Public administration	637,498,651	7.26%

Next, from Table 3, the contribution of value added, it can be seen that the manufacturing industry sectors have the highest value, then followed by the agriculture sector. While the sector with the lowest value is the electricity, gas, and water supply sector. The value is 28.11% for the manufacturing sector and 15.94% for the agriculture sector. This value indicates that the sector has a strong ability to encourage the growth of the downstream and the upstream sector, because as each increase in one unit to the final demand will encourage an increase in output for this sector if it's used as an input. Then, if we show the total contribution of two biggest sectors, both manufacturing industry sector and agriculture sector have 45% on total value added. So, both sectors can be the priority economic sectors. The development of this sectors of the economy in the long run is expected to encourage the growth of other sectors, to reduce poverty, and it is expected to create sustainable growth in economy.

As we know that the mining sector has been one of the key sectors supporting Indonesian economic for many decades. Not only have a significant contribution to Indonesian Gross Domestic Product, but also the mining sector make a significant contribution to Indonesian exports, government revenue, employment, and the development of many remote regions. The development of many remote regions become exist because mining companies are located in remote regions where it gives significant employers. Eventhough this sector is still remains a significant contributor to the Indonesian economy, this contribution tend to decrease. If we compare with two period, in 2011, the contribution to GDP equal to 6.14%; then in 2016, the contribution to GDP equal to 4.23%. We have some reason for this condition, one of the reason is the fluctuations of commodity prices especially for mineral and coal. However, the contribution in 2016 is lower than 2011, the development of mining industries becomes more evenly distributed, because the spread of mining industries have captured until Papua Province, Central Sulawesi Province, Bangka Belitung Province, West Nusa Tenggara Province, and East Kalimantan Province [11].

According to report from Price Water House Coopers Indonesia, the mining sector contributes a higher proportion of Indonesian exports, particularly as mining products are generally priced in US dollars, however, the contribution of the sector to Indonesian exports has also fallen off in the last few years. The implementation of the ban on exports of unprocessed (or not-sufficiently processed) minerals in January 2014 and the introduction of a significant (and progressively increasing) export duty on mineral concentrates, have resulted in an ongoing decline in mineral production over recent years. The mining industry's contribution to total Indonesian export revenues during 2014-2016 was however consistent at 13% (down from 17% in 2013). The Government, however, hopes that the total contribution of the mineral sector will increase once mineral processing and refining facilities are in place, generating higher value products and when the relaxation of the export ban on low-grade nickel ores and washed bauxite takes effect in 2017.

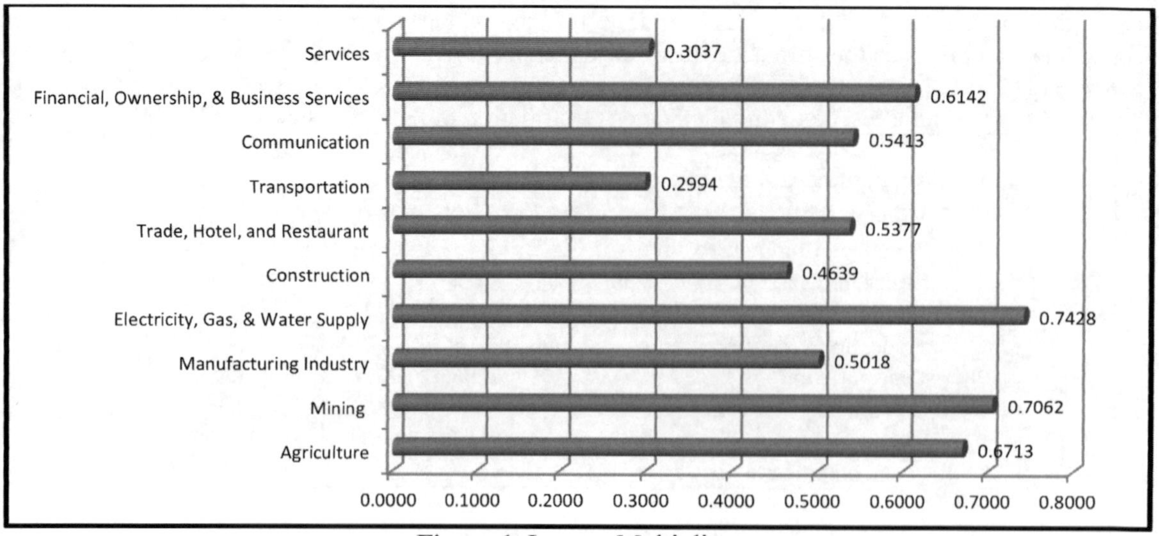

Figure 1. Income Multiplier

Based on Figure 1, the results of income multipliers show the value of mining sector is 0.7062;this value indicates that every increase of one billion rupiah output was generated from this sector, then the total income of the community will increase by 706million rupiah. Next, the multiplier value of manufacturing industry sector is 0.5018;this value indicates that every increase of one billion rupiah output was generated from this sector, then the total income of the community will increase by 501million rupiah. This finding suggests that it is possible that the high income multipliers generated from the mining sector come from the contribution of priority economic sectors. Where the priority economic sector are expected to create employment opportunities and increase revenues in the sector that this sector can be the foundation of an economy [1].

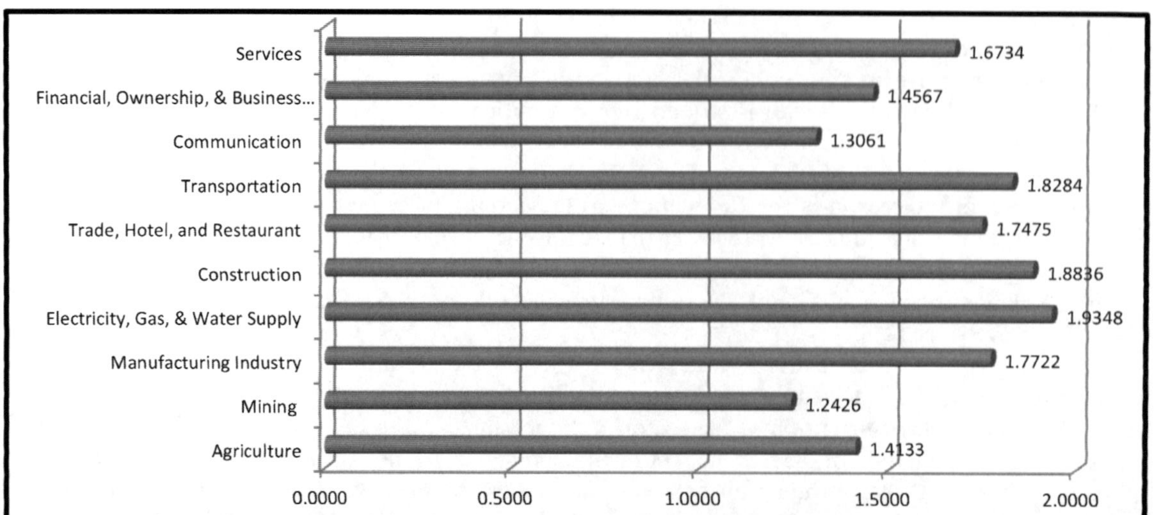

Figure 2.Output Multiplier

From Figure 2, the results of output multipliers show the value of mining sector is 1.2426; this value indicates that every increase of one billion rupiah output was generated from this sector, then the total output of this sector will increase by 1242 million rupiah. Next, the multiplier value of manufacturing industry sector is 1.7722; this value indicates that every increase of one billion rupiah output was generated from this sector, then the total output of this sector will increase by 1772 million rupiah. This findings cannot be separated by the role of labor, because it's become one of the importance production factor that it represents all of the people that are available to transform into goods or services. If the government wants to achieve this sector growth, the government should provide well educated and well trained labor to ensure that they can produce goods and services with a highly efficient and qualified.

From income and output multiplier, we found that the multiplier value of mining sector can be classified with high category. However, this sector have effect, its come from insite or outsite, such as degradation of water and air quality, flora and fauna habitats, erosion and sedimentation, potential of social impact, and others. One of important things is the environmental aspects include post mining land use, abandoned mining sites community development, and illegal mining. Government should have make many community development programs. The benefit of the program are it can be the solutions to alleviate poverty and it will be a key factor to make this industry more sustainable.

In line with above statement that our main economic objectives remain unchanged: to promote growth and enhance people's welfare supported by sustainable growth, resilient economic sector, inclusive economic development, and macroeconomic and financial stability. The Government remains committed to continuing structural reforms and accelerating infrastructure development to achieve these objectives. However, both domestic and international challenges still persist, which results adjustments on national development goals to be more realistic. In addition, the Government also put forward efforts to promote social welfare for low-level income society through jobs creation and social protection improvement .

On the other hand, we have the guidance for energy sector development. It came from The government's RPJMN for 2015–2019 reaffirms Indonesia's commitments to reducing its greenhouse gas emissions and strengthening its national energy security. The RPJMN states the government's primary energy-related priorities as (i) strengthening the availability of primary energy through oil production, supported by gas and coal production; (ii) increasing the domestic use of primary energy resources; and (iii) achieving a 96.6% electrification ratio. The RPJMN also establishes goals for biodiesel and bioethanol, and specifies that renewable energy should reach 10%–15% of the energy mix. Specifically, 7.5 GW of renewable energy generation should be added (geothermal, hydropower, and micro hydro), and initial trials of nuclear and marine-based energy should be implemented [5].

4. Conclusion

According to the analysis and findings, this study found that (1) Indonesian mining sectors ranked 6th with contribute amount of 6.81% on national total output; (2) Based on total gross value added, this sector contribute amount of 12.13% or ranked 4th; (3) The value from income multiplier is 0.7062 and the value from output multiplier is 1.2426.

The suggestion that can be given, (1) the Indonesian government should find the other alternative economic sector besides the mining sector and they must concern to develop this sector; (2) the Government of Indonesia can increase the export targets where not in the form of raw material but in the form of goods or services, because it can increase the value added and Indonesian economy generally; and (3) Government must be push the mining industries to make more community development program because this program could be one of the solutions to alleviate poverty. Currently it is regarded as being closely related to the environmental issue, and it constitutes a key factor in a successful sustainable mining development. It must be a special attention, since iron, steel, cement, ceramic, glass, and textile industries are all heavy energy consumers with potential for substantial improvements in energy utilization. Collectively, the industrial sector accounts for almost half of all energy consumption in Indonesia, but consumes energy inefficiently when compared with international benchmarks. Conversion efficiency at the power plant level has remained almost constant for 20 years.

IOP Conf. Series: Materials Science and Engineering **333** (2018) 012102 doi:10.1088/1757-899X/333/1/012102

References

[1] S Subanti and A R Hakim 2009 *Journal of Economy and Development Studies* **10 (1)**13 – 33.
[2] Jumadi 2015 The Structure Output Characteristics of Economic Sector and Regional Disparity in Indonesia, 1973 – 2005 *Master Thesis* Bogor Agricultural University.
[3] C B Statistics 2000-2015 *Gross Domestic Product 2000 – 2015* [In Bahasa Indonesia] Jakarta, Publisher of Central Bureau of Statistics.
[4] OECD 2012 *OECD Reviews of Regulatory Indonesia Market Opennes* (Paris, OECD) [Online : https://www.oecd.org/indonesia/Chap%204%20-%20market%20openness.pdf].
[5] ADB 2016 *Indonesia Energy Sector Asessment, Strategy, and Roadmap* Metro Manila Asia Development Bank [Online : https://www.adb.org/sites/default/files/institutional-document/189713/ino-energy-asr.pdf]
[6] B P Resosudarmo, D Hartono, A Tauhid, I L S Nina, Olivia and N Anang 2002 *Coastal & Ocean* **4 (3)** 17 – 28.
[7] C Virgowansyah and S Nazara 2003 Journal of Economy Policy **2(3).**
[8] C B Statistics 2009 *Input Output Table 2008* [In Bahasa Indonesia] Jakarta, Publisher of Central Bureau of Statistics.
[9] S Nazara 2005 *Input – Output Analysis 2nd Edition* [In Bahasa Indonesia] Jakarta, Publisher Institution of Faculty of Economics and Business Universitas Indonesia.
[10] R E Miller and P D Blair 1985 *Input – Output Analysis : Foundations and Extensions* New Jersey: Prentice Hall.
[11] PWC 2017 *Indonesia, Mining in Indonesia : Investment and Taxation Guide* Jakarta: PwC Indonesia [Online : https://www.pwc.com/id/en/energy-utilities-mining/assets/mining/mining-guide-2017-web.pdf]

IOP Conf. Series: Materials Science and Engineering **333** (2018) 012103 doi:10.1088/1757-899X/333/1/012103

Implementation of Augmented Reality Technology in Sangiran Museum with Vuforia

F A Purnomo[1], P I Santosa[2], R Hartanto[2], E H Pratisto[1] and A Purbayu[1]

[1]Informatics Engineering Department, Sebelas Maret University, Jl. Ir. Sutami 36A Surakarta 57126, Indonesia.
[2]Faculty of Engineering, Gadjah Mada University, Yogyakarta

E-mail: fendi_aji@mipa.uns.ac.id

Abstract. Archaeological object is an evidence of life on ancient relics which has a lifespan of millions years ago. The discovery of this ancient object by the Museum Sangiran then is preserved and protected from potential damage. This research will develop Augmented Reality application for the museum that display a virtual information from ancient object on display. The content includes information as text, audio, and animation of 3D model as a representation of the ancient object. This study emphasizes the 3D Markerless recognition process by using Vuforia Augmented Reality (AR) system so that visitor can access the exhibition objects through different viewpoints. Based on the test result, by registering image target with 25o angle interval, 3D markerless keypoint feature can be detected with different viewpoint. The device must meet minimal specifications of Dual Core 1.2 GHz processor, GPU Power VR SG5X, 8 MP auto focus camera and 1 GB of memory to run the application. The average success of the AR application detects object in museum exhibition to 3D Markerless with a single view by 40%, Markerless multiview by 86% (for angle 0° - 180°) and 100% (for angle 0° - 360°). Application detection distance is between 23 cm and up to 540 cm with the response time to detect 3D Markerless has 12 seconds in average.

1. Introduction

Augmented reality (AR) is a technology that combines visual objects or virtual world into the real world to see in real time [1]. Augmented Reality uses cameras to recognize the image of the marker on an ongoing basis, process and then generates a virtual interaction that appears in the real world to see both the display screen and the Head Mounted Display (HMD). The main problem in the AR system is the accuracy in the registration of 3D objects, which require the collaboration of virtual objects with the real environment. The results of the registration object is a target image that has specific keypoint feature. Keypoint feature that have been generated can be stored in a database and then used in a match with a known object by the camera in the AR system.

This research will test the ability of the AR in detecting Markerless which is 3D real objects in museum's showroom with visitor's camera angle position to the exhibition objects approach. Factors affect the Markerless detection process, including distance, angle, lighting, shadows and movement / motion. Device capability in implementing AR also to be considered that the recognition of 3D Markerless can be accessible and virtual information can be displayed next to the object of the exhibition. The purpose of this study is to design a mobile augmented reality application for the museum exhibition at the museum with objects as 3D Markerless and measure the ability of mobile devices to recognize 3D Markerless.

In the museum showrooms presentation according to Jing [2], the number of catalog that describe the object information for current exhibition has begun to lag. With the development of technology, visitor ables to choose their own show off objects that he/she want to know the information and the information can be generated virtual ly using visitor's smartphone. However, the interest of visitor to the Museum Sangiran still low, by 2015 only amounted to 314,000 visitors per year and 14% are foreign visitors. It is very unfortunate because Sangiran Museum is one of the best archaeological sites in the world. Research conducted by Ramírez [3] stated that smartphones are the right choice in AR application development for the museum. The device is lightweight making it easy to carry and can be easily manipulated. Research conducted by Ćuković [4] mentioned that the use of ID markers to deliver information in a book is inefficient because it requires additional space and not scalabe. Research conducted by Sari [5] has been successfully collaborate on the use of NFC (Near-Field Communication) and Augmented Reality technology with Vuforia as a guide to the visitor in the museum of Yogyakarta. In that study, the use of NFC as visitor's guide can only be enjoyed limited to smartphones with NFC feature. The accuracy of detected Markerless is 65% of 50 registered Markerless, this is due to Markerless' feature recognition still done at ideal angle that is at an angle of 90 degrees only.

2. Experimental

Research conducted was developed in stages as shown in Figure 1 which includes the tools installation, application development, local testing, real testing. Local testing aims to determine the maximum tilt angle of a detection device in 3D Markerless. Data obtained from local testing will be used in registering the target Markerless 3D image for each AR tilt in recognizing image targets range from 0° up to 360°. The stage of finding minimum angle deviation as shown in Figure 2.

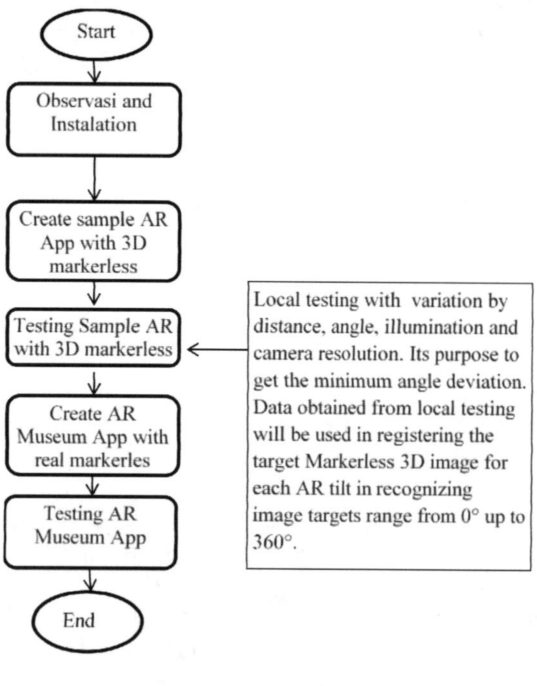

Figure 1. Research stage

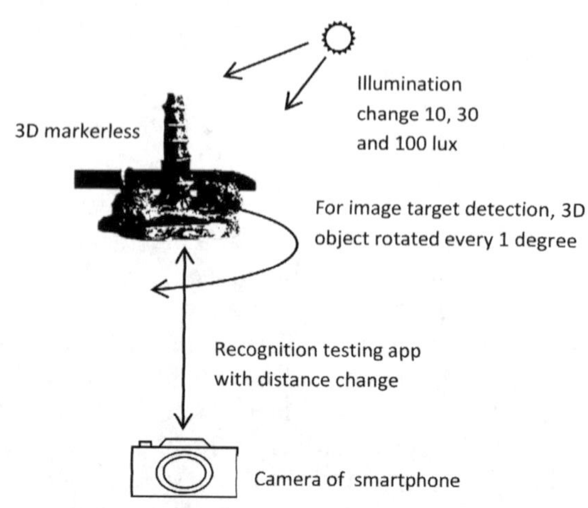

Figure 2. Determining minimum angle deviation for 3D markerless recognition

This study uses Markerless display in the Sangiran Museum. Markerless used in the form of 2D and 3D Markerless with the selection of 3D Markerless for a one-way visitor's perspective with 73 objects, selecting 3D Markerless with a visitor's perspective between 0° - 180° with 7 objects and selection of 3D Markerless with a visitor's perspective between 0° - 360° with 1 object. Each objects are given headlights lighting with intensity varying with the range between 8 Lux to 100 Lux.

Making an AR application for museum that use Vuforia using library which is imagetarget.unitypackage. The process of merging the information with the target image uses Unity. The AR museum applications is tested using various criteria in android smartphone hardware devices. Usefulness test of the application to the the visitor uses a questionnaire with five questions that measure aspects of interest in the application, ease of getting information aspect, the aspect of easy navigation, interactivity aspects of the application, and aspects of application innovation.

3. Result and Discussion

AR applications that have been implemented in Sangiran Museum using 3D Markerless aims to support the preservation of ancient objects on display because they are touchless, in addition the use of Markerless also maintains the initial settings that were neatly without compromising aesthetics of museum showroom space. Content virtual information presented in the form of text information, audio, video and 3D models that provide innovative delivery of information to the visitor. The results display the main menu AR application Sangiran Museum and the information is presented as shown in Figure 3 and Figure 4.

Figure 3. Main menu interface of AR Museum Sangiran

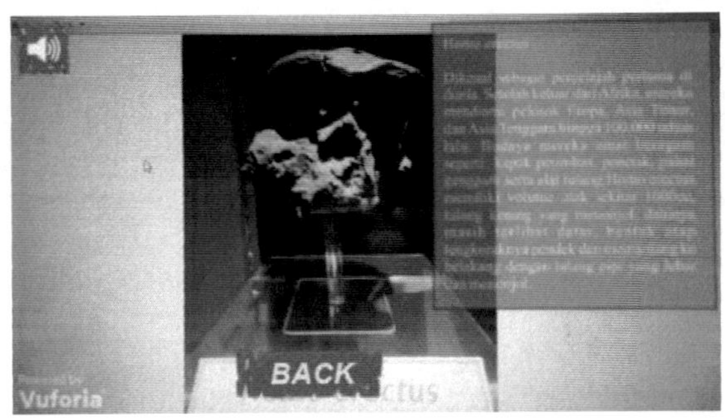

Figure 4. Displayed text information content when 3D markerless is detected by the camera

In detection testing, 3D Markerless keypoint began to be identified either from close range or from the furthest distance and based on tracking testing of the ability to maintain a virtual object that appears stating greater the angle of intersection, the smaller / narrower distance range for the Markerless recognition. However, for the greater the distance of provision of light intensity, the wider the range to maintain the appearance of the virtual object. From the graphical trend relationship between the distance, the angle and the influence of light intensity in the detection and tracking of the target image shown in Figure 5, it can be determined the minimum angle of 3D Markerless can be recognized at an average of 25°.

The successful accuracy of 3D Markerless recognition in the case of the use of images targeted for one-way visitor's perspective is low, this is because some of the weaknesses of the 3D Markerless method, the first weakness is very vulnerable to changes in the position of the object of the exhibition itself. Changes in position can be caused because the object of the exhibition is taken and not returned to the starting position.

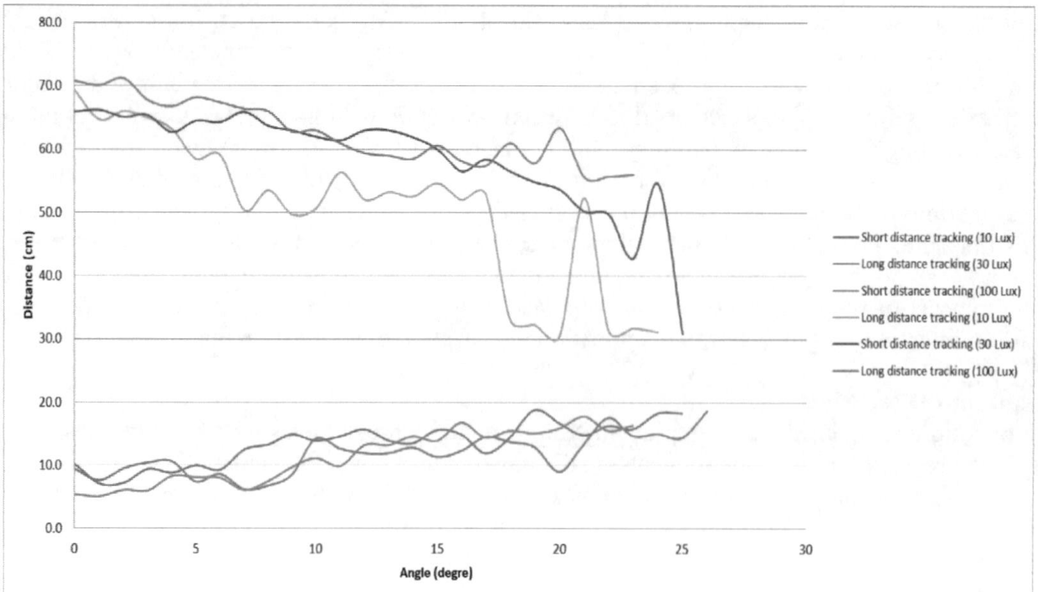

Figure 5. 3D markerless tracking result between detection distance and deviation angle

In early testing, the use of 3D Markerless can be detected when changing the angular position horizontally within the range of 0° up to 25°. The use of 3D Markerless at the initial assumption that the object of the exhibition placed in a glass static and not likely to not be touched or taken is proven wrong. Museum Sangiran Party purge of the exhibition objects, both objects show off, big and small objects showroom located inside a glass cabinet is also cleaned every Monday. As a result, a deep cleaning of the exhibition objects has chance to shift compare to initial position. This makes the target image that has been registered at the beginning of the exhibition objects has decreased the number of objects to show off a number of keypoint that 54.3% did not successfully recognized by AR app museum. In the image registration targets on Vuforia also have to be optimized feature on the exhibition objects that have a rating above the value of 3 or above, which means that the target image is easily recognizable by the AR system.

The second weakness, the use of 3D Markerless very vulnerable to changes in a great light. In a test conducted, the AR is given variation in light intensity changes in the value of 10 Lux, 30 Lux and 100 Lux. This means that the total amount of light incident on the object to show off with such intensity changes, the AR system can still recognize and bring up a virtual information. In the showroom, the lights were on exhibition objects will affect the 3D Markerless recognition if the initial condition when the registered image is different target with the current state of AR application testing museum. Showrooms in Sangiran Museum have been conditioned by both the influence of light from outside the room. The condition can be caused by changes in light intensity in the use of electric lighting showrooms museum. The use of electric lights when they are short-lived, the capacitor and resistor components will experience a reduction in the value and causing the light intensity is about the object of the exhibition will also be reduced. Reduction in the intensity of light that is too large will affect the image recognition system targeted by AR.

A third disadvantage, the use of 3D Markerless very influential in the shadow covering the object showrooms. In the Sangiran Museum there are many objects that is placed inside the glass so that the object is not marred by the exhibition visitors. Object placement in the glass cupboard in a certain distance will be seen the reflection of light and shadow of show off objects that affect detection of the object keypoint showroom.

As a first solution, an approach that can be done to address the disadvantages of the use of 3D Markerless in this method is the use of 3D objects Markerless is for an object that exactly has not changed position. Object placement on a permanent position makes the introduction of features more easily recognized by the AR system. In object placement showroom that can only be seen from one point of view only and likely to occur reflection of light by the glass, use Markerless can use name labels on objects such showrooms. This is because the use of 2D Markerless ranges reading corner by the AR system that is greater than the use of 3D Markerless are only in the range of viewpoints 0°-25° to the object show off. The second solution, in the case of the use of electric lights can be replaced with incandescent lamps so that light intensity on the object to show off can be maintained and does not affect too big to a target image on the object to show off.

While testing on 3D objects in the exhibition at the Museum Sangiran carried variation detection position between 0°-180° as shown in Figure 6. Position changed virtual information detection that appears also customized the appearance of perspective so that the information can be easily read by the visitor and equipped with audio dubbing containing a summary description of the object show.

AR application developed also displays 3D objects such as for example the appearance of information on the Ancient elephant bones as Markerless 3D model of Ancient Elephant in the form of 3D models shown include a rotating animation so that visitors can see an ancient elephant in 360°. Elephant virtual 3D model can be shown in Figure 6. Compared to the research conducted by Sari [5] this method is able to detect markerless between angles of 0 degrees to 360 degrees.

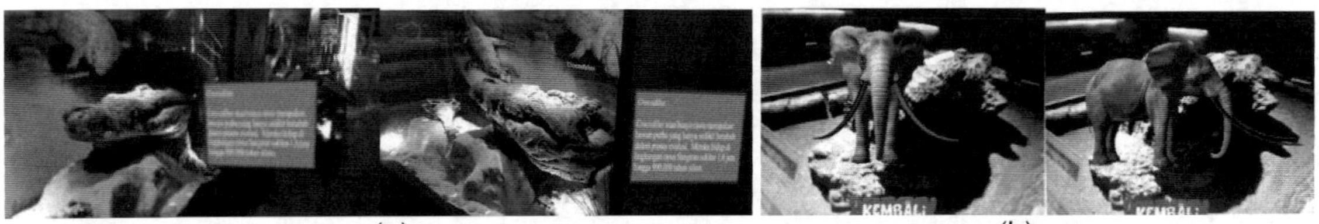

(a) (b)

Figure 6. Testing in Sangiran Museum showroom where object can be seen by the visitor from detection position between 0°-180° (a) and detection position between 360°(b)

In Figure 7 is test results of 81 objects exhibition at the Museum Sangiran give AR application museum in the detection of objects showroom for Markerless with a single view of 40%, Markerless with multiview of 86% (for angle 0° - 180°) and 100% (for angle 0° - 360°) with a detection distance of 23 cm up to 540 cm. The response time starts to detect 3D Markerless at average of 12 seconds. From the results of a survey of visitors to the museum, the analysis results in the form of validity and reliability generate Cronbach alpha of 0.97 so that all the questions declared valid and reliable. Respondents to the application of AR Museum Sangiran embodied in five aspects: the interest in applications worth agree-strongly agree 93%, aspects of the ease of getting information valuable agree-strongly agree 97%, aspects of ease of navigation is worth the agreed-strongly agree 100%, aspects of interactivity applications worth agree-strongly agree 94%, and aspects of application innovation worth agree-strongly agree 97%. According to visitor feedback, text is presented in a particular showroom object still looks small and has been offset by the provision of audio content in the delivery of information. In this study also tested the smartphone device hardware specifications variation. The test result indicates that AR museum can be run with the minimum criteria of a device with a hardware specification that is Dual Core 1.2 GHz processor, GPU Power VR SG5X, 8 MP auto focus camera, 1 GB of memory. The greater the performance, the faster its runs the application and detect the 3D Markerless.

IOP Conf. Series: Materials Science and Engineering **333** (2018) 012103 doi:10.1088/1757-899X/333/1/012103

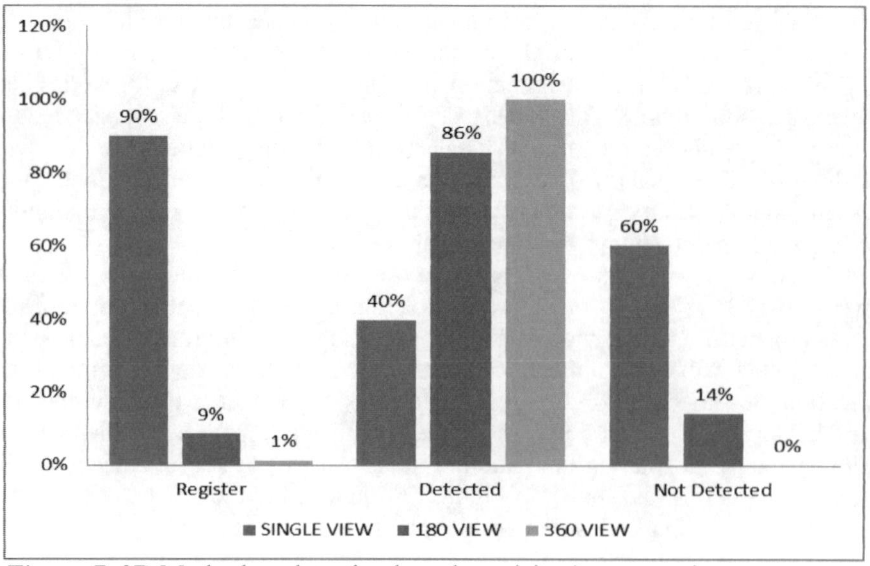

Figure 7. 3D Markerless detection based on visitor's perspective category at Museum Sangiran

4. Conclusions

Interactive application in the Archaeological Museum Sangiran using Android-based Augmented reality and Markerless is an object that is in the showroom Museum has been successfully created and run well. Vuforia ability to detect 3D Markerless has to minimum deviation angle of 25°, which is 0° tilt angle of up to 25°, 3D Markerless is recognizable while outside the those angle range the 3D Markerless is unrecognizable. Extra features of the target image every angle of the hose is capable of displaying virtual information from the different viewpoint horizontally.

Reference

[1] Azuma R and Azuma R 1997 *Presence Teleoperators Virtual Environ.* **6**(4) 355–385

[2] Jing C, Junwei G, and Yongtian W 2011 *Proc. Int. Commun. Conf. on Wireless Mobile and Computing* 262–265

[3] Ramirez M, Ramos E, Cruz O, Hernandez J, Perez-Cordoba E, and Garcia M 2013 *Proc. Int. Conf. Electron. Commun. Comput* 1–6

[4] Ćuković S, Gattullo M, Pankratz F and Devedžić G. 2015 *Proc. Int. Conf. Electric. Bio-med. Eng. Clean Energy and Green Computing* 24–31

[5] Sari I. P 2014 *Perancangan Sistem Pemandu Pengunjung Museum Berbasis Augmented Reality. In : Electrical Department* (Yogyakarta : Universitas Gadjah Mada)

IOP Conf. Series: Materials Science and Engineering **333** (2018) 012104 doi:10.1088/1757-899X/333/1/012104

Development of Applications about Hazards and Preventions of Drug Based On Android

Hartatik[1], F Febriyanto[1], and H Munawaroh[2]
[1]Informatics Engineering Department, Faculty of Mathematics and Natural Sciences, Sebelas Maret University, Surakarta, Indonesia
[2]Chemistry Department, Master Program Graduate School, Sebelas Maret University, Surakarta, Indonesia

E-mail: hartatik.uns@gmail.com

Abstract. The number of drug abuse was increase among the younger generation, it caused younger generation fall into drug abuse, and it will lead to physical and mental damage. The lack of knowledge of drugs danger is one of the most potential problems, so in this study we made an application about the types, dangers, and how to avoid its abusement. The application built using PHP programming language with codeiniter framework on admin part, while the parsing data between mobile application server using Javascript Object Notation (JSON). This application has been tested and 85% respondents stated that this application provides positive benefits especially for the socialization of drug abuse.

1. Introduction

Based on survey results from Badan Narkotika Nasional (BNN) on 2005, 13.710 student respondents showed 10 years drug abuse. Surveys from BNN reinforce the results of the study. Dr. Dadang Hawari in 1991 stated that 97% of the perpetrators were aged 17-24 years. From the results of the survey found that the trigger factors of drug abuse were teenagers ranging from family problems or broken home, that want to find a sensation, depression, and easily affected another person. In addition, the lack of drug abusing knowledge among adolescents showed the biggest factor[1].

Robby Robson stated that to adopt a technology in the world of education required two basic needs, that technology should make the teaching strategy more effective and the technology must be available and accessible [2]. For that we need a new innovation that can be used as a tool to facilitate the distribution of information about drug abusing prevention. In the previous research, not much has been developed about this application, in 2016, Eko Setyo Raharjo entitled "Development of Application of the dangers of smoking, alcohol, and drug-based android as a medium of learning in health education courses". The purpose of the study is as a medium of learning about the dangers of smoking, drinking, and drugs. In this application there are 3 menus of cigarettes, alcohol, drugs. The each menu contains submenu that explain about the impact caused by consuming the substance for health [3].

From this problem, it could be the reason the author makes an application that can help users in obtaining information about the types and categories of drug circulating in the community, drug abuse, the impact of drug use, and prevention and prevention of drug abuse. Accessible easily, the author makes this application based on android, where android smartphone users are very common and dominate the society today. With the application is expected to abuse drugs and drugs among

IOP Conf. Series: Materials Science and Engineering **333** (2018) 012104 doi:10.1088/1757-899X/333/1/012104

adolescents, and can add insight into the type of drug and its impact on health so that users are not interested and fall into drug abuse.

2. Experimental

The design of the application "NO DRUGS Application which provides information about knowledge of drug and drug abuse prevention" can be seen in Figure 1.

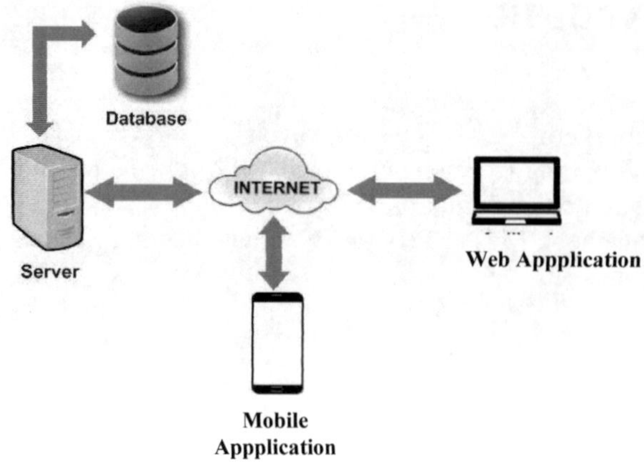

Figure 1. Application data flow

Web based application built for admin is built with PHP programming language with codeigniter framework, so it can communicate directly with the server through the network. As for mobile-based applications created using android studio tools used to create android interface applications, create android apps menu, and coding android applications. While parsing data between servers with mobile applications using Javascript Object Notation (JSON). The decomposition is done by a web service built with PHP via the HTTP protocol.

Preparation of system design [4] and relation between tables is needed in system development. From table that have been made, hence can made class diagram [5] which looks like picture 2 below:

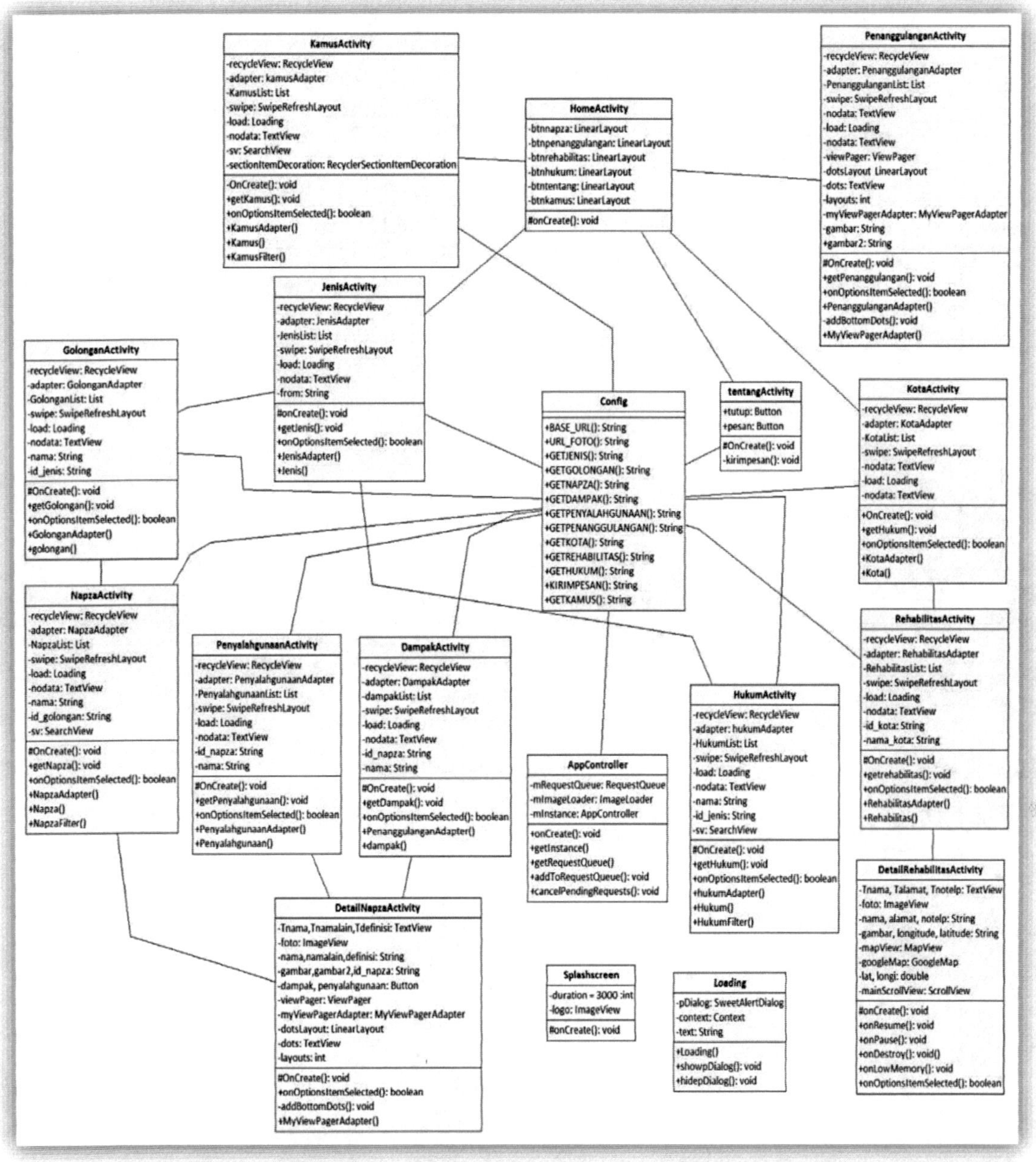

Figure 2. Class diagram NO Drug application

Then to see the usefulness of this application, taken respondents 30 respondents from the community guidance Interviews used as data collection techniques if researchers want to know things from the respondent more in depth and the number of respondents less or small (Sugiyono, 2010: 194) [6] .

3. Result and Discussion

In this application No Drugs, web applications are managed by the administrator. Application can be managed can be more than one admin that has been registered in charge of managing this web data. Admin can perform action in the form of data input, edit data, search data, and delete data (CRUD). This app comes with terms and descriptions of all types of drugs based on words published from BNN 2012 [7]. The development of this app using android studio, according to Felker (2011) Android Studio is an IDE from Google which was launched during the Google I / O event in May of 2013 and is an IDE. Alternative to the Eclipse IDE. In its official website it is said that Android Studio is the official IDE to develop android applications, which are IDEA-based intell [8]. Previous learning helps in the development of no drug applications among them study about knowledge of the dangers of drugs [9] and the influence of peer education on adolescent knowledge [10] and study about Effect of Community Health Education on Physical Activity Knowledge, Attitudes [11] .

The development of No Drug's assistant is expected to be helpful in the socialization of the dangers of drugs, so that the nation's children can avoid the moral damage caused by this illicit goods. This expert system is designed with web and android versions, so the bias makes it easy to operate. The version of the website has many feature of the login page, type, class, and abuse, as shown in Figure 3.

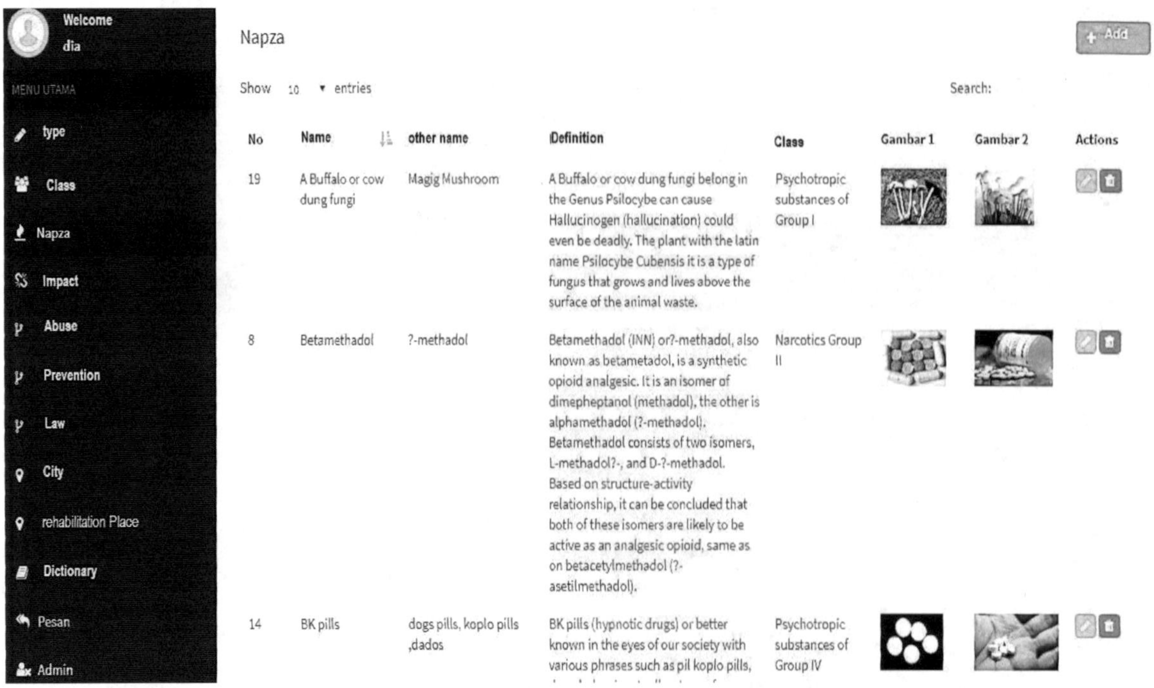

Figure 3. Interface of NO DRUG Application

Display on the countermeasures page displays a list of data countermeasures that have been stored in the database. On the menu there is a feature to set the number of rows to be displayed. In addition there is also a search facility to find the desired data. There is added button to add data, delete icon to delete data, and edit icon to display the edit data form. The countermeasure page can be shown in figure 4.

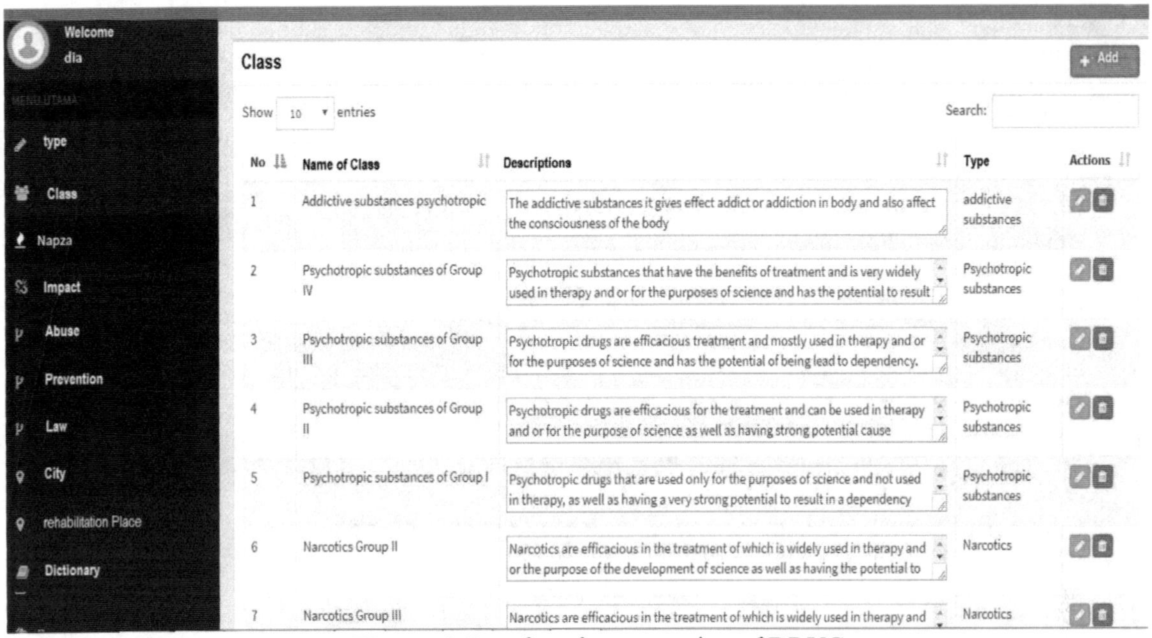

Figure 4. Interface for prevention of DRUG

Here is the view on the law page. On this page will display a list of legal data that has been stored in the database. On the menu there is a feature to set the number of rows to be displayed. In addition there is also a search facility to find the desired data. The button added to add data, delete icon to delete data, and edit icon to display edit data form.

In order to test the usefulness of this application, given the test to some respondents, in this case there are 30 respondents that and got the result 53% said it is very necessary to socialize the danger of drug misuse to the public. And 67% stated that no drugs applications can be useful in assisting in the socialization of hazards and drug abuse, and 47% stated that the application of NO DRUGS is sufficient to provide complete information on the dangers of drug abuse. More details can be seen in the following picture 5.

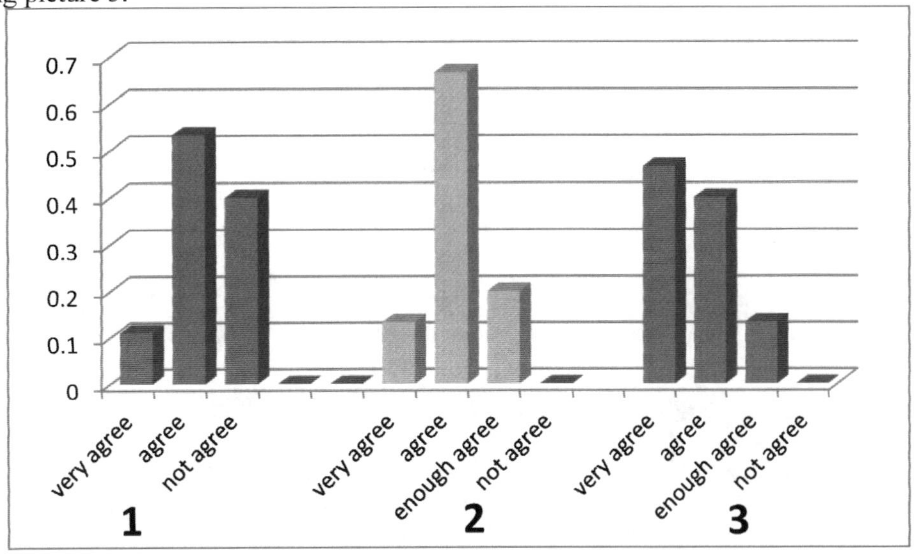

Figure 5. Result of observation NODRUG application

IOP Conf. Series: Materials Science and Engineering **333** (2018) 012104 doi:10.1088/1757-899X/333/1/012104

4. Conclusion

Applications "No Drugs" based on android has been running well on some android jellybean smartphone and needed for dissemination of drug abuse, so it can be one application that helps the field of education to increase knowledge of drug abuse.

References

[1] BNN 2012 *Hasil Penelitian BNN Tahun 2011* Jakarta: Pusat Penelitian dan Informasi BNN.
[2] Robby Robson 2003 *Mobile Learning and Handheld Devices in the Classroom.Eduworks Corporation.Corvallis* (Oregon:USA)
[3] Eko Setyo Raharjo 2016 *Skripsi* Universitas Negeri Yogyakarta
[4] Brady M, Loonam J 2010 *Exploring the use of entity relationship diagramming as a technique to support grounded theory inquiry* (Bingley : Emerald Group Publishing Bradford)
[5] Dennis A, Wixom BH, Tegarden D 2005 *Systems Analysis and Design with UML Version 2.0.* (New York: John Wiley & Sons Inc)
[6] Sugiyono 2010 *Metode Penelitian Pendidikan (Pendekatan Kualitatif, Kuantitatif dan R&D)* Bandung: Alfabeta.
[7] Kamus Narkoba 2016 *Badan Narkotika Nasional Republik Indonesia* Jakarta
[8] Felker D 2011 *Android Application Development for Dummies* (Indianapolis: Wiley Publishing Inc)
[9] Lisa J & Nengah 2013 *Narkoba, Psikotropika dan Gangguan Jiwa: Tinjauan Kesehatan dan Hukum* (Yogyakarta : Nuha Medika)
[10] Hulu F S 2005 *Tesis* Yogyakarta Program Pasca Sarjana Universitas Gajah Mada
[11] Young D R, Haskell W L, Taylor C B, Fortmann S P 2000 *Am. J. Epidemi.* **144(3)** 264-274

IOP Conf. Series: Materials Science and Engineering **333** (2018) 012105 doi:10.1088/1757-899X/333/1/012105

Naïve Bayes Approach for Expert System Design of Children Skin Identification Based on Android

Hartatik[1], A Purnomo[1], R Hartono[1], H Munawaroh[2]
[1] Informatics Engineering Department, Faculty of Mathematics and Natural Sciences, Sebelas Maret University, Surakarta, Indonesia
[2] Chemistry Department, Master Program Graduate School, Sebelas Maret University, Surakarta, Indonesia

E-mail: hartatik.uns@gmail.com

Abstract. The development of technology gives some benefits to each person that we can use it properly and correctly. Technology has helped humans in every way. Such as the excess task of an expert in providing information or answers to a problem. Thus problem that often occurs is skin disease that affecting on child. That because the skin of children still vulnerable to the environment. The application was developed using the naïve Bayes algorithm. Through this application, users can consult with a system like an expert to know the symptoms that occur to the child and find the correct treatment to solve the problems.

1. Introduction

Growth period becomes very important period for the development of the human body. In childhood, the body is still quite vulnerable to both food and the environment. The children's phase is an interesting phase between the other phases. At the stage of development of children, they begin to have a much greater curiosity. Know a lot and want a lot of things too. In this phase most of the human phase is very keen to be noticed (pampered). Rapid physical growth also occurs in childhood. For example, feet that can grow faster, to see the child's height is rapidly increasing. For that, the child's health should get more attention on the growth period, especially on the skin of children.

The development of information technology is currently very fast development and also plays an important role in various human activities [1]. Almost all areas utilize information technology to help complete human work. Similarly in the medical world, knowledge-based technology, facts and reasoning to help solve problems in various disciplines have been developed [2].

The technology is expected to assist people in solving problems encountered, such as about decision support system or expert system about skin disease in children. [3]. In previous research has been about the development of decision support system among others has been done by some researchers, including decision support system for news classification [4], decision support system for nutritional status [5]. In addition to the development of applications about decision support systems, some research that can help people in solving problems is the development of expert systems Prediction System for Heart Disease [6], expert systems for Prediction of Different Dermatological [7], Expert System for Diagnosis of Tropical Infectious Diseases [8], as well as an Expert System to Diagnose Corn Plant Diseases [9], and Expert System for Mineral Identification [10].

2. Experimental

In this study the authors use Naïve Bayes Classification algorithm the main focus in determining the classification of Throat Disease [11]. The meaning of the Naïve Bayes Classification algorithm is to use the probability approach to produce a classification of skin disease types.
Steps of work in this study can be seen in Figure 1:

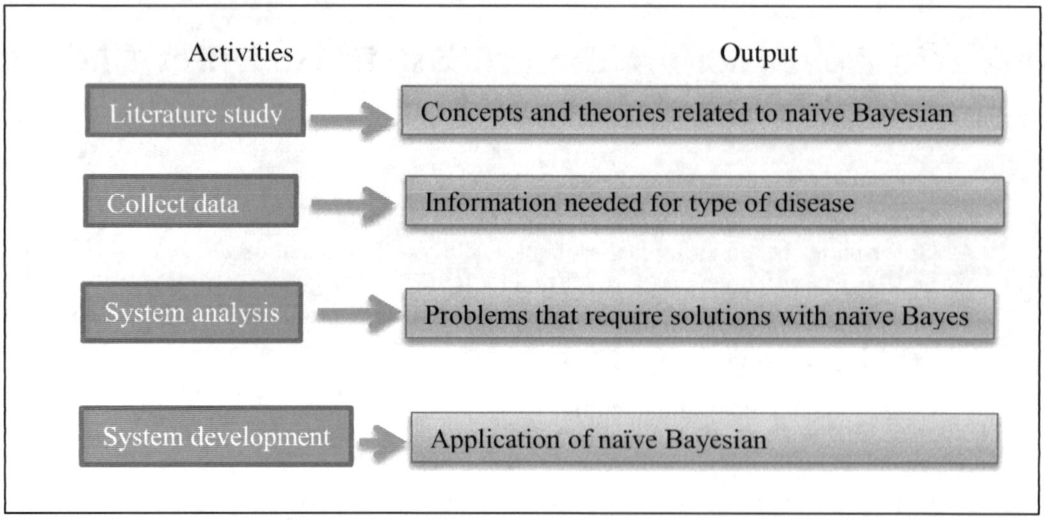

Figure 1. Research flow for detect naïve Bayes

To support good results, the object of research is children aged 1-12 years. Conducted interviews of children with complaints of skin diseases. Furthermore, based on the results of interviews formed categories of diseases and symptoms. This will be the basis for decision making. Sampling is done by Purposive sampling.

3. Result and Discussion

Naïve Bayes Classifier is a simple probability classifier based on Bayes theorem. Bayes's theorem is combined with "Naïve" which means every attributes / variables are independent (independent). Naïve Bayes Classifier can be trained efficiently in supervised learning.
In this research causes of skin diseases

1. Poor self-hygiene
2. Viruses
3. Bacteria
4. Allergic Reactions
5. Low immune system

Bayes method as a method for problem solving in decision support system to detect the type of skin disease, especially for children aged 1-12 years. In the process of forward chaining of the inference engine for the determination of appropriate healthy lifestyle categories, it is possible that the final diagnostic results have several categories selected. To select one of several selected categories requires management of uncertainty. This uncertainty will be calculated using the Bayes probability theorem, with equation 1.

$$P(H|X) = \frac{P(X|H)P(H)}{P(X)} \quad (1)$$

IOP Conf. Series: Materials Science and Engineering **333** (2018) 012105 doi:10.1088/1757-899X/333/1/012105

Where Variable X represents the class, while the variable X1 ... Xn represents the characteristics of the instructions needed to classify the types of skin diseases. Then the formula explains that the probability of entering a sample of certain characteristics in class H (Posterior) is the probability of occurrence of class H (before the entry of the sample, often called prior), multiplied by the probability of occurrence of sample characteristics in class H (also called likelihood) with the probability of occurrence of sample characteristics globally (also called evidence). Therefore, the above formula can also be written simply as follows:

$$Posterior = prior\ x\ likelihoodevidence \qquad (2)$$

Evidence values are always fixed for each class on a single sample. The value of the posterior will then be compared with the posterior values of the other classes to determine to which class a sample will be classified. A further description of the Bayes formula is made by elaborating (H | X1 , ..., Xn) using the multiplication rule as follows:

$$\begin{aligned}(H|X_1,...,X_n) &= (H)P(X_1,...,X_n\,|H)\\ &= (H)P(X_1|\,H)P(X_1,...,X_n\,|\,H,F_1)\\ &= P(H)P(X_1|\,H)P(X_2|\,H,X_1)P(X_3,...,X_n|\,H,X_1,X_2\\ &= (\,H)P(X_1|\,H)P(X_2|\,H,X_1)P(X_3|C,X_1,X_2)P(X_4,...,X_n|\,H,X_1,X_2,X_3)\\ &= P(H)P(X_1|\,H)P(X_2|\,H,X_1)P(X_3|\,H,X_1,X_2)...P(X_n|\,H,X_1,X_2,X_3,...,X_{n-1}) \qquad (3)\end{aligned}$$

It can be seen that the results of such descriptions lead to more and more complex factors that affect the probability value, which is almost impossible to analyze one by one. As a result, the calculation becomes difficult to do. Here is the very high (naive) assumption of independence, that each clue (X1, X2, ...,Xn) is independent of each other. With these assumptions, then apply a similarity as follows:

$$P(X_i|X_j)=P(X_i\cap X_j)P(X_j)=P(X_i)P(X_j)P(X_j)=P(X_i) \qquad (4)$$

For i $\neq$ j, so

$$(X_i|H,X_j)=(X_i|H) \qquad (5)$$

The above equation is a model of the Naive Bayes theorem which will then be used in the classification process. For classification with continuous data used Gauss Density formula:

$$(X_i = x_i|Y = y_j) = \frac{1}{\sqrt{2\pi\sigma_{ij}}} e^{\frac{-(x_i-\mu_{ij})^2}{\sigma 2ij}} \qquad (6)$$

Where:
P: Opportunities
Xi: Attribute to i
xi: The attribute value to i
Y: Classes searched
yi: Sub class Y is searched
μ: mean, states the mean of all attributes
σ: Standard deviation, declares a variant of all attributes.

Furthermore the naïve Bayes method is used to determine the type of disease, with the design as follows:

Table 1. Type of disease

Number	Code	Desease
1	D1	measles
2	D2	eczema
3	D3	water fleas
4	D4	Panu
5	D5	Boils

Based on the type of disease, then known the symptoms of each disease meaning. For example, measles, there are some symptoms as shown in figure 2.

Table 2. Symptoms of the disease

S11	Hot sores
S12	Limp
S13	blisters rupture and wound
S14	blister and yellow discharge
S15	Wet

Based on the symptoms of the disease it fits with all types of diseases, and will be searched for compatibility with the highest level of compatibility. Thus, a symptom with a high titled fit is expressed as a possible skin disease suffered, can be seen in Table 3

Table 3. Table decision

	D1	D2	D3	D4	D5
S1	√	√		√	
S2		√			√
S3			√		
S4				√	
S5					√
S11	√				
S12	√				
S13					
S14					
S15					
S21		√		√	
S22		√			
S23		√			

4. Conclusion

Simulations for the application of naïve Bayesian methods for the determination of further skin disease types to be developed in android applications, so that the community can use to detect the type of skin disease experienced.

References

[1] Jogiyanto H M 2005 *Analisis dan Desain Sistem Informasi: Pendekatan Terstruktur Teori dan Praktek Aplikasi Bisnis* (Yogyakarta: Andi Offset)
[2] Durkin J 1994 *Expert System Design and Development* (New Jersey: Prentice Hall International Inc)
[3] Efraim T 1995 *Decision Support Systems and Expert System* (New Jersey: Prentice Hall International Inc)
[4] Wibisono Y 2005 *Klasifikasi Berita Berbahasa Indonesia menggunakan Naïve Bayes Classifier, Seminar Nasional Matematika 2005* (Bandung: Universitas Pendidikan Indonesia)
[5] Nur Anggraeni Diana Rahmawati and Firli Irhamni 2012 *Jurnal Sarjana Teknik Informatika* **1(1)** 85 – 92
[6] Pattekari S A, Parveen A 2012 *Int. J. Adv. Comp. Math. Sci.* **3(3)** 290-294
[7] Manjusha K K, Sankaranarayanan K, Seena P 2014 *Int. J. Adv. Res. Comp. Sci. Software Eng.* **4(1)** 864-868
[8] Diema Hernyka Satyareni 2011 *Jurnal Teknologi* **1(2)** 61-67
[9] Munanda, Edi, Nanang P 2013 *Jurnal Litek* **10(2)** 113-117
[10] I O Folorunso, O C Abikoye, R G Jimoh and K S Raji 2012 *JETCIS* **3(2)** 205-210
[11] Ting S L, Ip W H, Tsang A H C 2011 *IJSEA* **5(3)** 37-46

IOP Conf. Series: Materials Science and Engineering **333** (2018) 012106 doi:10.1088/1757-899X/333/1/012106

DIJKSTRA METHODE FOR OPTIMALIZE RECOMMENDATION SYSTEM OF GARBAGE TRANSPORTATION TIME IN SURAKARTA CITY

Hartatik[1], A Purbayu[1] , L Triyono[1]
[1] Informatics Engineering Department, Faculty of Mathematics and Natural Sciences, Sebelas Maret University, Surakarta, Indonesia
[2] Informatics Department, Faculty of Mathematics and Natural Sciences, Politeknik Semarang, Indonesia

E-mail: hartatik.uns@gmail.com

Abstract. Major problem that often occurs in waste transportation in each region is the route of garbage transportation. Determination of this route should become a major concern because it affects fuel consumption and also the working time from the employee. Therefore, in this research we will develop an application to optimize with pigeonhole and dijsktra algorithm. Pigeonhole algorithm is used to determine which garbage trucks should be taken in a particular TPS. Time optimization is done by determining the shortest path that can be skipped for each garbage truck. Data generated from Pigeonhole then used to determine the shortest path by using Dijkstra algorithm.

1. Introduction

Environmental hygiene is a hot topic and is often a major issue. Stacking garbage especially in big cities is often a very disturbing problem. Increasing the amount of unbalanced waste by the environmentally friendly management will cause damage and environmental pollution [1]. The faster the garbage is transported by the janitor then the more cities that look beautiful, awake hygiene and health awake. Garbage that often accumulates in temporary dumps (TPS) one of them is caused by the lack of truck fleets and the right lane that truck passes. Garbage truck lanes should be tailored to the best paths that can be skipped and adjusted to the number of garbage tonnages in each different TPS. Incomplete waste management will trigger social problems, such as mass ammunition, clashes between residents, blocking of landfill facilities [2]. In addition there are some previous studies developed by Susanti et al. [3] on optimization of the transport of garbage vehicles. The problem of vehicle routing (VRP) is a combinatorial problem where its solution requires from existing algorithms and the development of new ideas [4]. The method of VRP settlement and its implementation is exemplified by Iskandar [5]. Lacomme et al. (2006) develop Memetic Algorithm (MA) for solving the capacitated arc routing problem [6].

The number of truck fleets, truck fleet lines, the location of TPS and the amount of waste tonnage is determined first. Then all truck fleets will be determined the best path from the early existence of truck fleet to TPS and from TPS to Final Disposal Site (TPA) or other TPS.

Initially from each truck fleet assigned to TPS, this election was performed using a pigeon hole algorithm. The method for determining the best path is to use the dijectstra algorithm for the selection of the shortest path on the path that the truck can pass. The purpose of moving trucks in TPS can be determined by the pigeon scheduling algorithm.

In this study it is assumed that the path has the same level of congestion and has the same speed. The time obtained is calculated based on trajectory length and time of the predetermined truck (time = distance / average velocity). The time required for preparation, transportation and waste reduction is also ignored. Start times for all trucks are also considered the same and determined at the beginning of the system. The number of trucks is determined early so that each trucking route can be determined in the system to be created so that the amount of time it takes to clean up the garbage at the TPS can be determined to a minimum.

2. Experimental

Waste transport planning uses the concept of route optimization [7] by applying the Dijkstra [8] algorithm. This method has been developed in multigraph [9]. Stages of work in this study are as follows in figure 1.

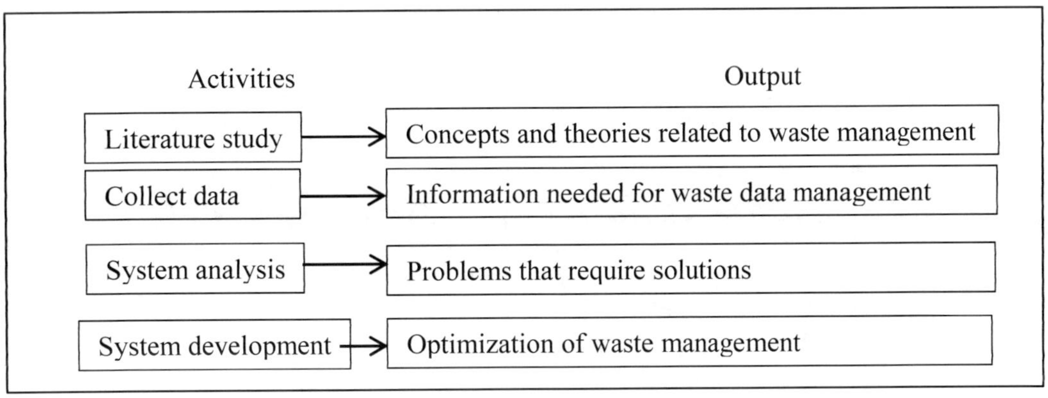

Figure 1. Research flow for optimize system

In accordance with the stages of the research in this study using two methods of pigeonhole method to determine the placement of the fleet at the TPS location of the destination algorithm dijkstra to determine the shortest path / trajectory of trucking. The optimization calculation by using dijkstra method is applied for rail optimization [10].

The same has also been done by Arinalhaq, Imran, and Fitria (2013) determination of Waste Transportation Vehicle Routes but Using the Nearest Neighbor Method [11].

The location of research observation is Surakarta area which has an area of about 44 Km^2, according to geographic location Surakarta City split and flowed by 3 (three) big river that is river of Bengawan Solo, Kali Jenes and Kali Pepe. Observations are devoted to the disposal site of the Surakarta region covering the districts of Banjarsari, jebres, Pasar kliwon, laweyan, serengan. And conducted data collection is by using the method of interviews, field surveys and also use data from the city cleanliness agency.

3. Result and Discussion

Solving waste management problems is an optimization problem where routes with high optimization values will be determined [12].

Implementation Database system is a database of applications built. In this program only used one table as storage of TPS data throughout Surakarta and longitude and latitude Google Maps. Tables are then called markers with detail structures as shown in Figures 2.

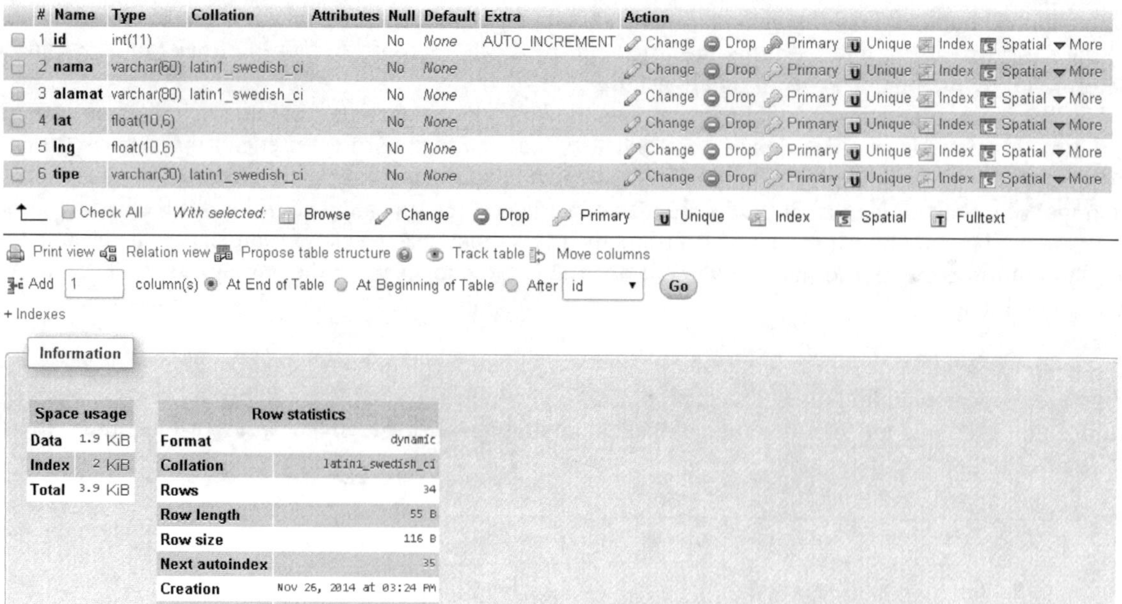

#	Name	Type	Collation	Attributes	Null	Default	Extra	Action
1	id	int(11)			No	None	AUTO_INCREMENT	Change Drop Primary Unique Index Spatial More
2	nama	varchar(60)	latin1_swedish_ci		No	None		Change Drop Primary Unique Index Spatial More
3	alamat	varchar(80)	latin1_swedish_ci		No	None		Change Drop Primary Unique Index Spatial More
4	lat	float(10,6)			No	None		Change Drop Primary Unique Index Spatial More
5	lng	float(10,6)			No	None		Change Drop Primary Unique Index Spatial More
6	tipe	varchar(30)	latin1_swedish_ci		No	None		Change Drop Primary Unique Index Spatial More

Check All With selected: Browse Change Drop Primary Unique Index Spatial Fulltext

Print view Relation view Propose table structure Track table Move columns

Add 1 column(s) At End of Table At Beginning of Table After id Go

+ Indexes

Information

Space usage	
Data	1.9 KiB
Index	2 KiB
Total	3.9 KiB

Row statistics	
Format	dynamic
Collation	latin1_swedish_ci
Rows	34
Row length	55 B
Row size	116 B
Next autoindex	35
Creation	Nov 26, 2014 at 03:24 PM

Figure 2. Table used for TPS data collection in Google Maps

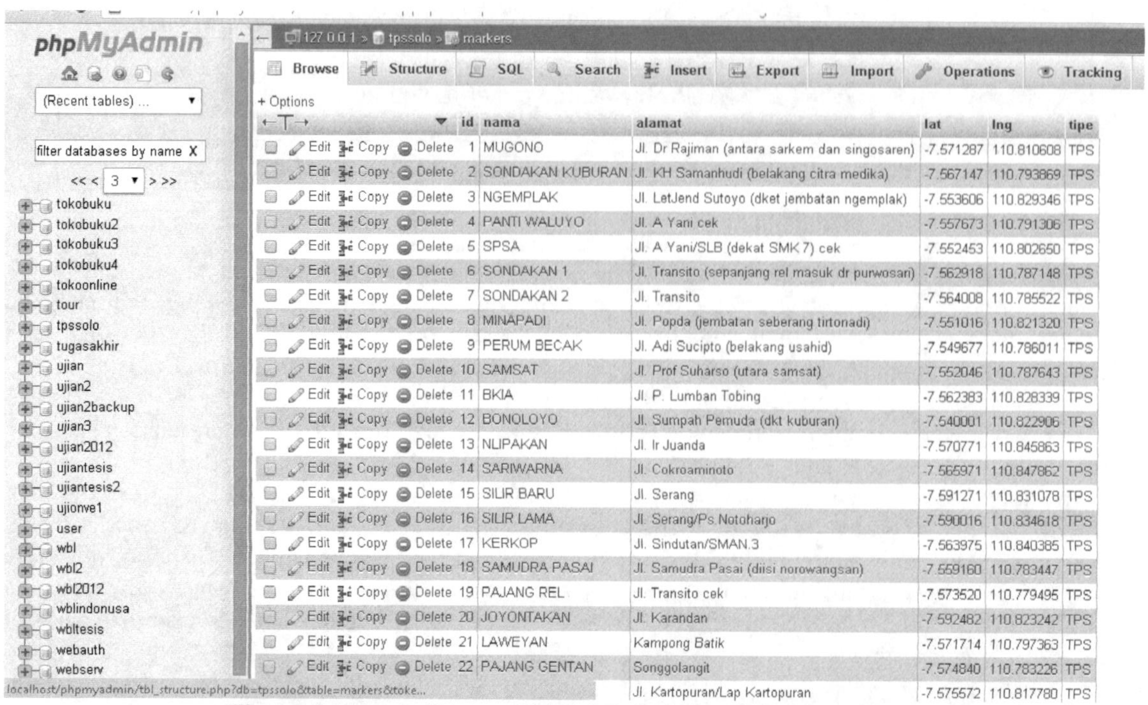

id	nama	alamat	lat	lng	tipe
1	MUGONO	Jl. Dr Rajiman (antara sarkem dan singosaren)	-7.571287	110.810608	TPS
2	SONDAKAN KUBURAN	Jl. KH Samanhudi (belakang citra medika)	-7.567147	110.793869	TPS
3	NGEMPLAK	Jl. LetJend Sutoyo (dket jembatan ngemplak)	-7.553606	110.829346	TPS
4	PANTI WALUYO	Jl. A Yani cek	-7.557673	110.791306	TPS
5	SPSA	Jl. A Yani/SLB (dekat SMK 7) cek	-7.552453	110.802650	TPS
6	SONDAKAN 1	Jl. Transito (sepanjang rel masuk dr purwosari)	-7.562918	110.787148	TPS
7	SONDAKAN 2	Jl. Transito	-7.564008	110.785522	TPS
8	MINAPADI	Jl. Popda (jembatan seberang tirtonadi)	-7.551016	110.821320	TPS
9	PERUM BECAK	Jl. Adi Sucipto (belakang usahid)	-7.549677	110.786011	TPS
10	SAMSAT	Jl. Prof Suharso (utara samsat)	-7.552046	110.787643	TPS
11	BKIA	Jl. P. Lumban Tobing	-7.562383	110.828339	TPS
12	BONOLOYO	Jl. Sumpah Pemuda (dkt kuburan)	-7.540001	110.822906	TPS
13	NLIPAKAN	Jl. Ir Juanda	-7.570771	110.845863	TPS
14	SARIWARNA	Jl. Cokroaminoto	-7.565971	110.847862	TPS
15	SILIR BARU	Jl. Serang	-7.591271	110.831078	TPS
16	SILIR LAMA	Jl. Serang/Ps.Notoharjo	-7.590016	110.834618	TPS
17	KERKOP	Jl. Sindutan/SMAN.3	-7.563975	110.840385	TPS
18	SAMUDRA PASAI	Jl. Samudra Pasai (diisi norowangsan)	-7.559160	110.783447	TPS
19	PAJANG REL	Jl. Transito cek	-7.573520	110.779495	TPS
20	JOYONTAKAN	Jl. Karandan	-7.592482	110.823242	TPS
21	LAWEYAN	Kampong Batik	-7.571714	110.797363	TPS
22	PAJANG GENTAN	Songgolangit	-7.574840	110.783226	TPS
		Jl. Kartopuran/Lap Kartopuran	-7.575572	110.817780	TPS

Figure 3. Sample Data on Table Markers for TPS data collection

Based on these data, then by using coordinates google Maps obtained the point Terminal, TPA and TPS-TPS in the city of Surakarta as in Figure 4.

IOP Conf. Series: Materials Science and Engineering **333** (2018) 012106 doi:10.1088/1757-899X/333/1/012106

Figure 4. Location of all polling stations in Surakarta.

In this application is divided into 4 transportation processes namely Transport 1 (P1), Transport 2 (P2), Transport 3 (P3), Transport 4 (P4). And obtained the results of them for the location Ngemplak. On Results in the program using the Google MAPS route a visualization of the path as will be drawn on one of the routes on the truck 15 (longest route) with route TERMINAL-TPS NGEMPLAK-TPS SAMSAT-TPA-TPS BONOLOYO-TPA-TPS BKIA-TPA with 45 km distance. The depiction of routes one by one as in the picture in figure 5.

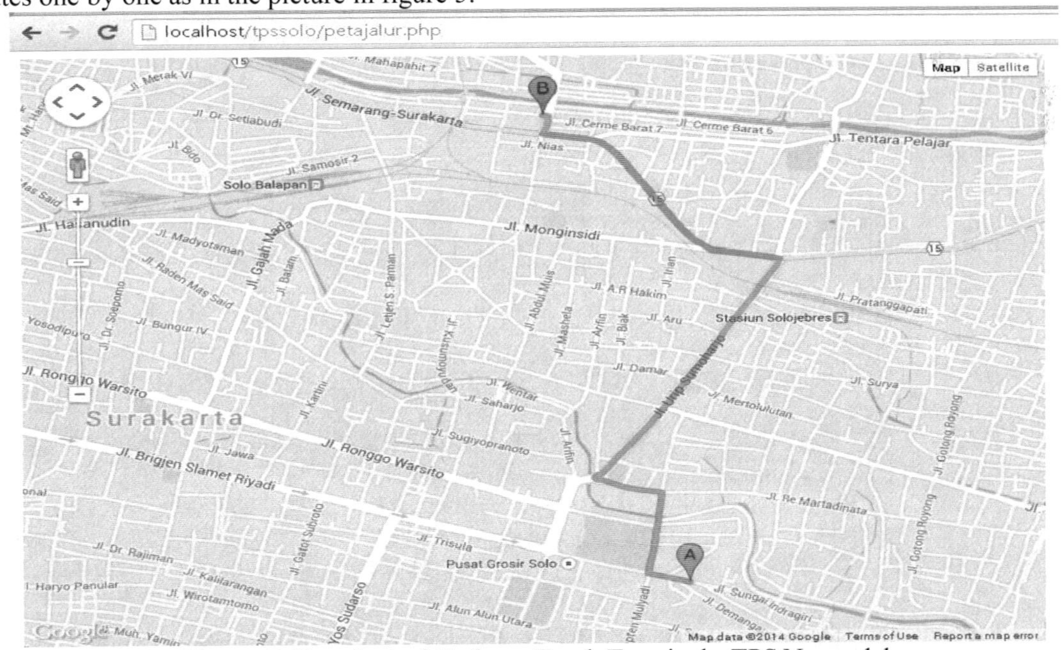

Figure 5. Terminal of Garbage Truck Terminal - TPS Ngemplak

Figure 6. TPS rute of ngemplak-TPA

Dan this process will be repeated for other areas. Based on the total area of Surakarta city, there are 23 optimum pathways, namely:

i. Rute-1 : TERMINAL-TPSBKIA-TPA-TPSSONDAKAN 1-TPA-TPSMOJOSONGO-TPA-TPSBKIA-TPA with mileage 39,9km
ii. Rute-2 : TERMINAL-TPSBONOLOYO-TPA-TPSSILIR LAMA-TPA-TPSDAWUNG-TPA with mileage 41 km
iii. Rute -3 :TERMINAL-TPSDAWUNG-TPA-TPSSAMPANGAN BARAT-TPA-TPSSAMSAT-TPA with mileage 42.6 km
iv. Rute -4 :TERMINAL-TPSJOYONTAKAN-TPA-TPSSAMBENG-TPA-TPSSONDAKAN 1-TPA with mileage 43.1 km
v. Rute -5 :TERMINAL-TPSJURUG-TPA-TPSSILIR BARU-TPA-TPSSARIWARNA-TPA with mileage 36.9 km
vi. Rute -6 :TERMINAL-TPSKARTOPURAN-TPA-TPSMUGEN LEPAS-TPA-TPSSILIR LAMA-TPA with mileage 41.6 km
vii. Rute -7 :TERMINAL-TPSKEDUNG TUNGKUL-TPA-TPSSONDAKAN KUBURAN-TPA-TPSBONOLOYO-TPA with mileage 35.5 km
viii. Rute -8 :TERMINAL-TPSKERKOP-TPA-TPSSAMUDRA PASAI-TPA-TPSBKIA-TPA with mileage 37.9 km
ix. Rute-9 :TERMINAL-TPSLAWEYAN-TPA-TPSJURUG-TPA-TPSBKIA-TPA with mileage 36.7 km
x. Rute-10 :TERMINAL-TPSMAKRO-TPA-TPSSARIWARNA-TPA-TPSSONDAKAN 2-TPA with mileage 44.3 km

IOP Conf. Series: Materials Science and Engineering **333** (2018) 012106 doi:10.1088/1757-899X/333/1/012106

xi. Rute-11 :TERMINAL-TPSMINAPADI-TPA-TPSDAWUNG-TPA-TPSSILIR BARU-TPA with mileage 41.6 km
xii. Rute-12 :TERMINAL-TPSMOJOSONGO-TPA-TPSSONDAKAN 2-TPA-TPSBKIA-TPA with mileage 38.3 km
xiii. Rute -13 :TERMINAL-TPSMUGEN LEPAS-TPA-TPSMUGONO-TPA-TPSMINAPADI-TPA with mileage 36.6 km
xiv. Rute -14 :TERMINAL-TPSMUGONO-TPA-TPSSAMPANGAN TIMUR-TPA-TPSSPSA-TPA with mileage 42.2 km
xv. Rute-15 :TERMINAL-TPSNGEMPLAK-TPA-TPSSAMSAT-TPA-TPSBONOLOYO-TPA-TPSBKIA-TPA with mileage 45 km
xvi. Rute-16 :TERMINAL-TPSNLIPAKAN-TPA-TPSSPSA-TPA-TPSSAMPANGAN BARAT-TPA with mileage 39.3 km
xvii. Rute -17 :TERMINAL-TPSNOROWANGSAN-TPA-TPSBKIA-TPA-TPSSAMPANGAN BARAT-TPA with mileage 40.2 km
xviii. Rute -18 :TERMINAL-TPSPAJANG GENTAN-TPA-TPSBONOLOYO-TPA-TPSBKIA-TPA with mileage 37.6 km
xix. Rute-19 :TERMINAL-TPSPAJANG REL-TPA-TPSKEDUNG TUNGKUL-TPA-TPSBONOLOYO-TPA with mileage 35.1 km
xx. Rute -20 :TERMINAL-TPSPANTI WALUYO-TPA-TPSMINAPADI-TPA-TPSJURUG-TPA with mileage 36.8 km
xxi. Rute -21 :TERMINAL-TPSPERUM BECAK-TPA-TPSNGEMPLAK-TPA-TPSBKIA-TPA with mileage 37.1 km
xxii. Rute-22 :TERMINAL-TPSSAMBENG-TPA-TPSNLIPAKAN-TPA-TPSSAMUDRA PASAI-TPA with mileage 44.4 km
xxiii. Rute-23 :TERMINAL-TPSSAMPANGAN BARAT-TPA-TPSSOLO SQUARE-TPA-TPSSILIR BARU-TPA with mileage 42.9 km.

Based on the result of application of Dijkstra method there are several route options which can be used as recommendation for garbage transportation in Surakarta city.

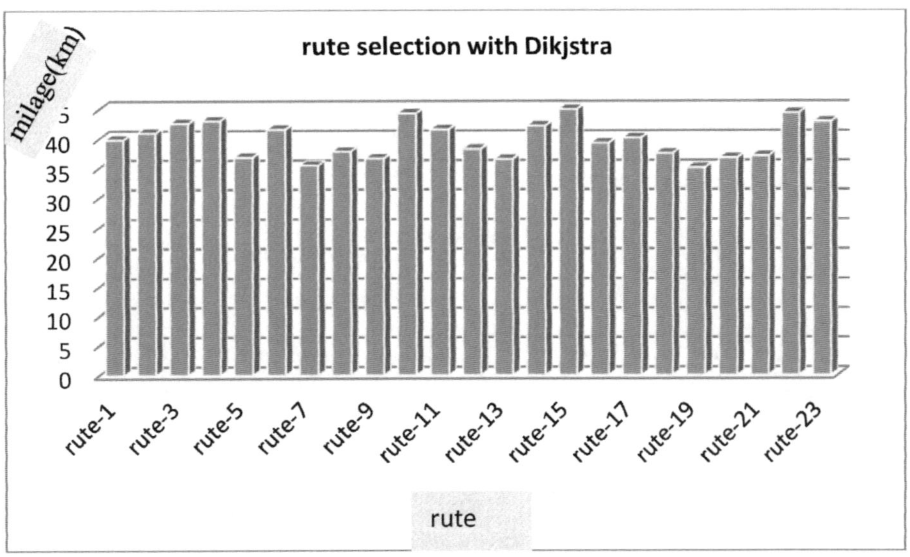

Figure 7. Rute selection with dikjstra

IOP Conf. Series: Materials Science and Engineering **333** (2018) 012106 doi:10.1088/1757-899X/333/1/012106

4. Conclusion

The simulation of pigeon hoel and dijkstra algorithm was used to optimize the garbage collection time in Surakarta City by using the distance between the TPS and the garbage tonnage at each TPS as the main parameter of determination. Generated precision of distance utilization which will optimize the time of garbage transportation and indirectly the target of city cleanliness can be achieved. From this research we get 23 optimal route for garbage transportation in Surakarta city.

References

[1] T Kustiah 2005 *Kajian Kebijakan Pengelolaan Sanitasi Berbasis Masyarakat*, Bandung, Pusat Penelitian & Pengembangan Permukiman, Balitbang Departemen Pekerjaan Umum

[2] Hadi S P 2004 *Sindrom Sampah* Kompas 7 Desember 2008 Jakarta

[3] Susanti E, Cahyono E S, and Dwi Purwani O 2016 *Jurnal Matematika* **6(2)** ISSN: 1693-1394

[4] Toth P and Vigo D 2002 *The Vehicle Routing Problem. Philadelphia* PA: SIAM

[5] Iskandar 2010 *Tesis* (Bogor: Institut Pertanian Bogor)

[6] Lacomme P, Prins C, and W Ramdane-Cherif 2001 *Applications of Evolutionary Computing* 473-483

[7] Rardin R 1998 *Optimization in Operation Research* (New Jersey: Prentice-Hall)

[8] Dijksta E W 1959 *Numerische Mathematic* **1(1)** 269-271

[9] Siddhartha Sankar Biswas, Bashir Alam and M N Doja 2013 *J. Comp. Sci.* **9 (3)** 377-382

[10] Pandey P, Dixxit S 2014 *IJRITCC* **2(3)** 435-440

[11] Arinalhaq F, Imran A and Fitria L 2013 *Jurnal Online Teknologi Nasional* **1(1)** 32-42

[12] Katzin 2008 *Route optimization application in waste collection system* (Lithuania: Vilniaus Gedimino technikos universities)

IOP Conf. Series: Materials Science and Engineering **333** (2018) 012107 doi:10.1088/1757-899X/333/1/012107

Effect of variation in active route timeout and my route timeout on the performance of AODV-ETX protocol in mobile adhoc network

A Purnomo[1], Widyawan [2], W Najib[2], R Hartono[1] and Hartatik[1]

[1] Department of Informatics Engineering, Faculty of Mathematics and Natural Sciences, Sebelas Maret University, Surakarta, Indonesia

[2] Department of Electrical Engineering And Information Technology, Faculty of Engineering, Gadjah Mada University, Yogyakarta , Indonesia

E-mail : agus.purnomo@staff.uns.ac.id

Abstract. Mobile adhoc network (MANET) consists of nodes that are independent. A node can communicate each other without the presence of network infrastructure. A node can act as a transmitter and receiver as well as a router. This research has been variation in active route timeout and my route timeout on the performance of AODV-ETX protocol in MANET. The AODV-ETX protocol is the AODV protocol that uses the ETX metric. Performance testing is done on the static node topology with 5 m x 5 m node grid model where the distance between nodes is 100 m and node topology that consists of 25 nodes moves randomly with a moving speed of 1.38 m/s in an area of 1500 m x 300 m. From the test result, on the static node, AODV protocol-ETX shows optimal performance at a value MRT and ART of 10 s and 15 s, but showed a stable performance in the value of MRT and ART ≥60 s, while in randomly moved node topology shows stable performance in the value of MRT and ART ≥80 s.

1. Introduction

Mobile adhoc network (MANET) consists of nodes that are independent. A node can communicate each other without the presence of network infrastructure. A node can act as a transmitter and receiver as well as a router. Wireless ad hoc network requires a routing protocol that is able to manage route reliably. AODV (Ad hoc On-Demand Distance Vector) is a routing protocol in an ad hoc network that uses hop count as a routing metric. AODV protocol has more minimal routing overhead compared to the proactive protocol type [1] and has a better energy efficiency compared to the proactive protocol type and other reactive protocol [2]. AODV combines the superior aspects of the two algorithms that is using the maintenance pattern of DSR and using hop per hop sequence number along with its update of DSDV [3]. AODV uses the routing table with one entry for every destination and will keep the timer-based state on each node. AODV has a type of unipath routing because it is only made with one path for one communication line [4]. AODV algorithm is inspired by Bellman-Ford algorithm such as DSDV. Performance evaluation of AODV protocol is needed to improve service quality (QoS) such as throughput , end to end delay, packet delivery ratio, overhead, drop packet etc. Performance of protocol is strongly influenced by the ability to maintain the route, the speed of route discovery, the choice of metric, the speed of route recovery. Several researchers have done analyzed of AODV protocol on routing metric by using different simulators. They have result AODV-ETX protocol with ETX as metric [5]. From the throughput measurements that the AODV-ETX protocol has better performance compared to the AODV protocol because ETX is more concerned with paths of higher delivery ratios [6]. The AODV-ETX protocol has the same characteristics as the AODV protocol. The route is formed by the route discovery process. The packet delivery RREQ (Route Request) is performed in the route discovery process by the source node to the neighboring node to form a reverse path and packet delivery RREP (Route Reply) by the destination node to the neighboring node towards the source node to form a forward path. If there is a damage in the link, the RRER (route error) packet will be delivered by a broadcast to all forwarder [7]. In AODV protocol or AODV-ETX there is a table of the reverse and forward lines that have a life time. The life time is declared in My Route Timeout (MRT) and Active Route Timeout (ART). When the destination node replies the RREQ packet by sending a RREP packet through the reverse path then the forward path will be formed. Forward Pathway is used to transmit data packet from the source node to the destination node. Forward pathway has an active time in which the AODV protocol is specified in the ART parameters [3]. From the discovery route, the ability to maintain a route of the AODV or AODV-ETX protocol is depends by MRT and ART. MRT standard value given in AODV protocol is 10 seconds [8]. We are interested this does not apply to the AODV-ETX protocol. This study focused on the increasing of performance of AODV-ETX protocol to maintain a valid route using MRT's and ART's values. The expected results was to obtain MRT's and ART's values to improve AODV-ETX protocol performance.

2. Experimental

This research has been variation in active route timeout and my route timeout on the performance of AODV-ETX protocol in MANET. This research has used software tools here were follows: NS2 (Network simulator), Gedit, setdest, NS2 Scenarios Generator 2 (NSG2), cbrgen, NS2 Trace Analyzer and AWK script. Variation in active route timeout and my route timeout was conducted by changing the My Route Timeout (MRT) and Active Route Timeout (ART) values as indicated on the Figure 1.

```
class AODV;
#define MY_ROUTE_TIMEOUT            10  //seconds
#define ACTIVE_ROUTE_TIMEOUT        10  //seconds
#define REV_ROUTE_LIFE              6   //seconds
#define BCAST_ID_SAVE               6   //seconds
```

Figure 1. A cut of aodv.h Script on NS2

Standard values of MRT and ART of the AODV protocol on NS2 was 10 seconds. In this study, MRT and ART values tuning would be conducted from 1second – 100 seconds. MRT value was made the same with ART value. Every time the ART and MRT values were changed, NS2 should be recompiled by using sudo make clean; sudo make all; sudo make install commands.

To determine the effect of variation values of ART and MRT dotesting of laboratory scale using NS2 simulator. There are 2 testing scenarios: test scenario 1 and scenario 2. In scenario 1, the topology of grid model with 25 nodes and the distance between nodes are 100 meters. Node-0 as the source node and node-24 act as the destination node. The source node sends a CBR packet with a size of 1000bytes /packet. Scenario 1 topology was shown in Figure 2. The simulation parameters in scenario 1 are shown in Table 1.

Table 1. Simulation Parameter set on NS2 of scenario 1

Parameter	Value
Channel type	Wireless channel
TX Power	0.28 Watt
TX dan RX Gain	1 dB
Antena position from the ground	1.5 m
Propagation model	shadowing
MAC layer	IEEE 802.11b
Routing protocol	AODV-ETX
Traffic type	UDP
Application type	CBR
Number of sender node and receiver node	1
Packet size	1000 byte
Packet delivery interval	0.005 seconds
Simulation time	0-300 seconds and 300 seconds
Number of nodes	25 (grid model)
Transmission range	250 m
Simulation location	Outdoor

Figure 2. Topologi grid

The scenario 2 made on randomly moving node condition usingtechnique of Random Way Point. There are 25 nodes that move randomly of simulation times at 300 second with speed at every nodes is 1,38 m/sec thetotal area of 1500 m x 300 m. This 1500 m x 300 m wide area refers to research conducted by Sung-Ju Lee and Mario Gerla [9]. The source node and destination node are randomly selected. The source node sends CBR packet with size at 1000 bytes/packet. The simulation parameters in scenario 2 are shown in Table 2.

Table 2. Simulation Parameter set on NS2 of scenario 2

Parameter	Value
Channel type	Wireless channel
TX Power	0.28 Watt
TX dan RX Gain	1 dB
Antena position from the ground	1.5 m
Propagation model	shadowing
MAC layer	IEEE 802.11b
Routing protocol	AODV-ETX
Traffic type	UDP
Application type	CBR
Number of sender node and receiver node	1

Packet size	1000 byte
Packet delivery interval	0.005 seconds
Simulation time	0-300 seconds and 300 seconds
Number of nodes	25
Transmission range	250 m
Moving Model	Random Way Point
Node speed	1.38 m/second
Probe interval AODV	1 second
Probe window AODV	10 seconds
Area	1500 m x 300 m
Simulation location	Outdoor

The performance parameter measured here were throughput, end to end delay and routing overhead and packet delivery ratio (PDR). Throughput was the total packet of data received per second by a recipient calculated in kbps [10]. End to end delay was the time required for a packet to be sent until the packet was received and gives acknowledgment to the sender of the packet [9]. Routing Overhead was the total routing package in proportion to the number of packets received by the receiver. This routing overhead can be obtained by calculating the number of routing packets divided by the number of packets received by the receiver [10]. The PDR was the ratio between the packet successfully received by the destination and the packet sent by the source [10].

3. Results and Discussion

Test of scenario 1 intended to see the effect of changes in the value of MRT and ART of the AODV-ETX protocol on wireless mesh network with fixed node condition. The node was not move shown weak change in the link quality. The results of the test in scenario 2 were shown in Figure 3.

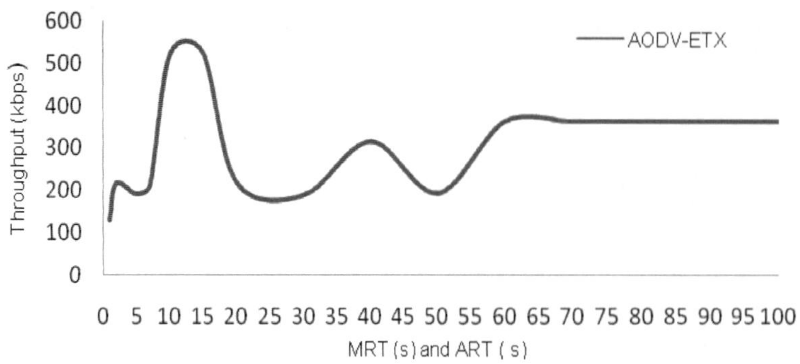

Figure 3. Association among MRT, ART and *throughput* on scenario 1

Based on the graph shown in Figure 3, the highest values of throughput AODV-ETX protocol was obtained the values of MRT and ART at 10 seconds and 15 seconds but the performance of AODV-ETX protocol was stable in the increase of MRT and ART at ≥60 seconds. The value MRT and ART at 10 seconds was the standard of a parameter value in the AODV-ETX protocol. This means that in the condition on the node does not move the protocol AODV-ETX will produce the most optimal throughput at the MRT and ART of 10 seconds.

The topology in scenario 2 was made with 25 nodes on moving randomly. To get different motion patterns then tested with different pause time. The small values of pause time reflected constant node motion and large of pause time reflected node motion to be silent. The change value of pause time will get different node motion pattern although use the same speed.

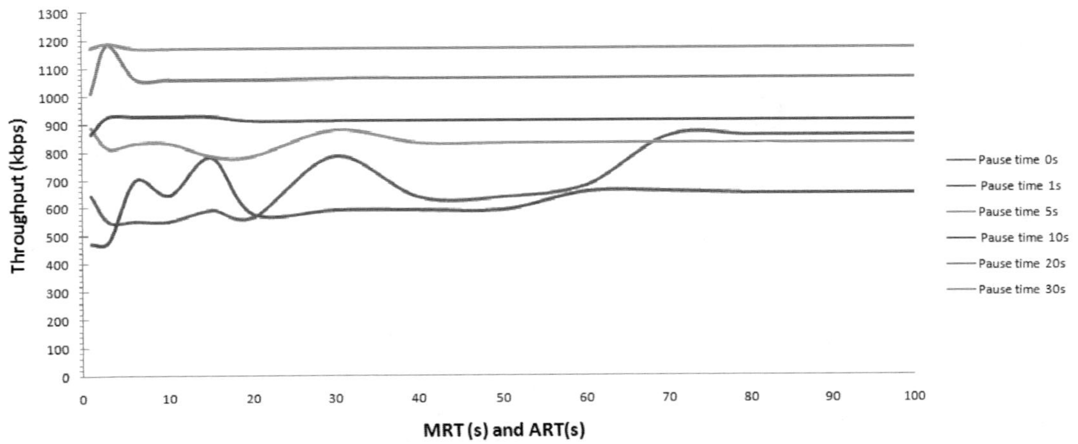

Figure 4. Association among MRT, ART and throughput on scenario 2

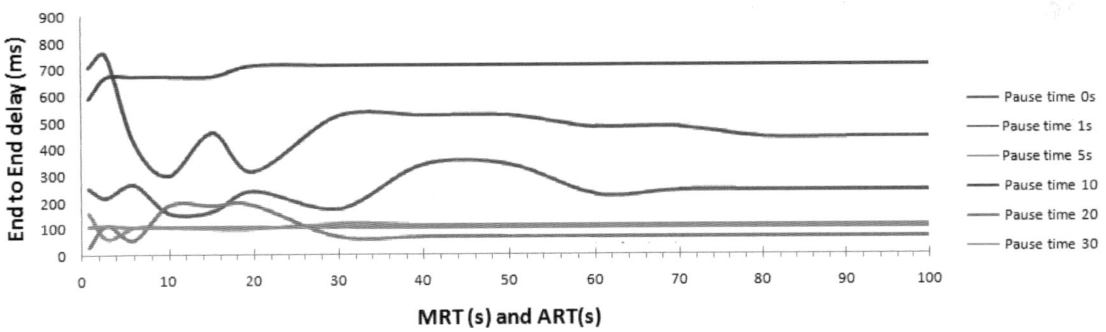

Figure 5. Association between MRT, ART and end to end delay on scenario 2

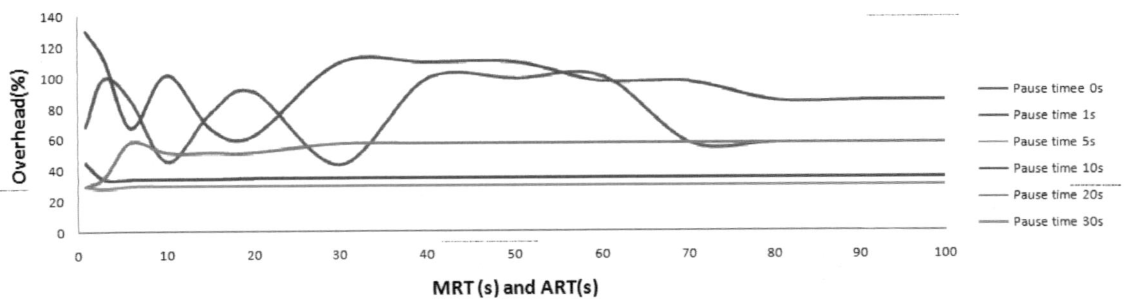

Figure 6. Association among MRT, ART and overhead on scenario 2

IOP Conf. Series: Materials Science and Engineering **333** (2018) 012107 doi:10.1088/1757-899X/333/1/012107

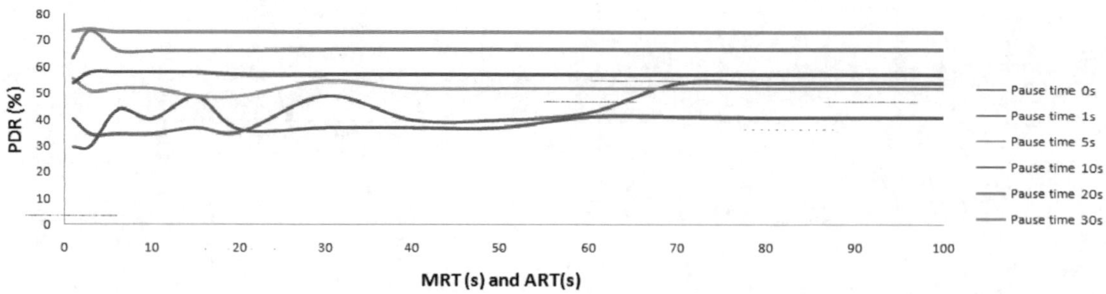

Figure 7. Association among MRT, ART and PDR on scenario 2

The data of scenario 2 shown that the variation of MRT and ART influences the AODV-ETX protocol. Figure 4 shows the value of throughput at pause time the 30s higher than at pause time 1s. This is because the pause time 30s mobility level of the node was lower than pause time 1s so the connection between nodes was much more stable. The large throughput was not necessarily smaller end to end delay, as shown in Figure 5 in pause time 10s has the largest end to end compared to the 1s and 5s pause time. The value of throughput, end to end delay, routing overhead, and PDR at different pause times in each change in MRT and ART values shows stable pattern when the value of MRT and ART was increased. The start values of MRT and ART showed stable pattern in each pause time are different but seen at MRT and ART of ≥80 seconds indicate a stable pattern.

4. Conclusion

This study have the following conclusions:

1. In the immobile node topology of AODV-ETX protocol showed optimal performance on the MRT and ART values of 10 seconds and 15 seconds, but showed a stable performance in the MRT and ART values of ≥60 seconds.
2. In the randomly mobile node topology showed stable performance on the MRT and ART values of ≥80 seconds.

As future work, we necessary to test the MRT tuning and ART protocols of AODV-ETX on the varied speed of the nodes, number of nodes, number of channels and the number of sender and receiver nodes to obtain the the most optimal MRT and ART values on any MANET network conditions.

References

[1] Richard C, Perkins C E, and Westphal C 2015 *Second Annual IEEE Communications Society Conf. on Sensor and Ad-Hoc Communications and Networks* (California:IEEE)

[2] Gouda B.S, Dass A.K, and Narayana 2013 K.L *IEEE Autom. Comput. Commun. Control Compress. Sens. Imac4s* (Kottayam: IEEE) 306 – 312.

[3] Nitnaware D. and Verma A. 2010 *IJCSC* **1** 67–71

[4] Mueller S, Tsang R P, Ghosal D 2004 *Springer* **2965** 209–234

[5] Hatti S, Kamakshi M.B 2013 *Springer* **131** 817-826

[6] Jonathan L J, Anne M P, Hilt B, and Boeglen H. 2012 *Int. Conf. On ICWCUCA* (Clermont Ferrand: IEEE) 1-5

[7] Patel D K, Shah S K, Thaker M P 2010 *Springer* **70** 276–280

[8] Dubey V and Dubey N 2014 *Proc. Int. Conf. On Computational Intelligence and Communication Network s* (Washington, DC : IEEE) 482 –485

[9] Lee S J and Gerla M. 2000 *WCNC* (Chicago:IEEE) 1311-1316

[10] Rohal P, Dahiya R, and Dahiya P 2013 *IJARET* **1** 54-58

IOP Conf. Series: Materials Science and Engineering **333** (2018) 012108 doi:10.1088/1757-899X/333/1/012108

Wireless Sensor Network Quality of Service Improvement on Flooding Attack Condition

R Hartono[1], Widyawan[2], S B Wibowo[2], A Purnomo[1], Hartatik[1]

[1] Department of Informatics Engineering, Faculty of Mathematics and Natural Sciences, Sebelas Maret University, Surakarta, Indonesia
[2] Department of Electrical Engineering and Information Technology, Gadjah Mada University, Yogyakarta, Indonesia

E-mail : rudi.hartono@staff.uns.ac.id

Abstract. There are two methods of building communication using wireless media. The first method is building a base infrastructure as an intermediary between users. Problems that arise on this type of network infrastructure is limited space to build any network physical infrastructure and also the cost factor. The second method is to build an ad hoc network between users who will communicate. On ad hoc network, each user must be willing to send data from source to destination for the occurrence of a communication. One of network protocol in Ad Hoc, Ad hoc on demand Distance Vector (AODV), has the smallest overhead value, easier to adapt to dynamic network and has small control message. One AODV protocol's drawback is route finding process' security for sending the data. In this research, AODV protocol is optimized by determining Expanding Ring Search (ERS) best value. Random topology is used with variation in the number of nodes: 25, 50, 75, 100, 125 and 150 with node's speed of 10m/s in the area of 1000m x 1000m on flooding network condition. Parameters measured are Throughput, Packet Delivery Ratio, Average Delay and Normalized Routing Load. From the test results of AODV protocol optimization with best value of Expanding Ring Search (ERS), throughput increased by 5.67%, packet delivery ratio increased by 5.73%, and as for Normalized Routing Load decreased by 4.66%. ERS optimal value for each node's condition depending on the number of nodes on the network.

1. Introduction

There are two ways to establish communication using wireless media. First is to build a base station (infrastructure) [1-2]. The problem that arise on the type of network infrastructure is limitation of space to build all of the physical infrastructure of the network and also the cost factor. Secondly, builds an ad hoc network between users who communicate each other. In an ad hoc network, each user must be willing to submit data from source to destination so that the communication occurs [2].
In the ad hoc wireless network is often called MANET (Mobile Ad hoc Network), network design is structured. Each node can communicate with each other in a certain transmission range. To be able to communicate with each node needs help another node, the node in MANET can act as terminals and routers [3-5].

MANET research concerned at the moment are the routing protocol, the security aspect, limited bandwidth and power consumption [6]. There are two routing protocols that have been set as the standard to the ad hoc network by the Internet Engineering Task Force (IETF), routing protocol that is both proactive and reactive [6][7][2].

Overall, MANET routing protocols has weak level of security [9]. There is also security weaknesses on AODV protocol that is in the process of finding a route for transmitting data. In AODV protocol, searching for a route is begin with sending a route request (RREQ) addressed to the destination node [10]. The problem that arise in AODV is the process of sending RREQ that is broadcast. It can be used by third parties to attack and immobilize MANET network.

2. Experimental

This study uses Network Simulator 2 (NS2) software as a tool to perform ad hoc network simulation to test the performance of AODV protocol that has been modified.
The research process to be conducted determine the optimal threshold value of TTL which provides a major influence on the value of the parameter packet delivery, average end-to-end delay, throughput, and normalized routing load. The flow of the research is shown in Figure 1.

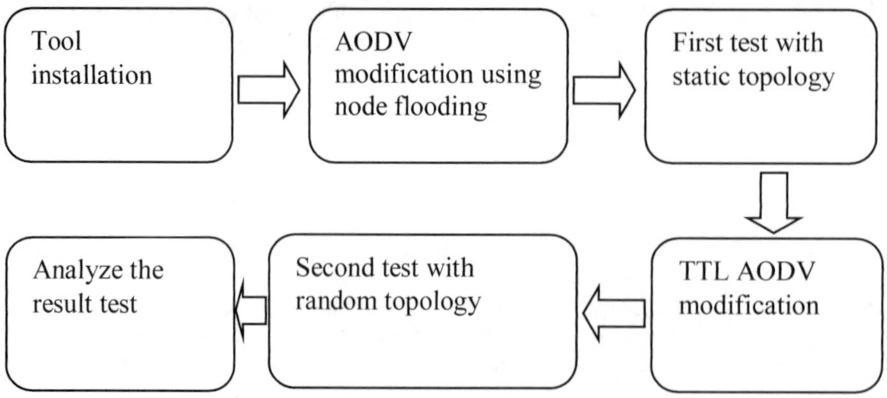

Figure 1. Research process

In routing search, a node will first perform RREQ delivery where RREQTTL delivery is highly influence so RREQ data packet does not become a packet flooding. AODV TTL influenced by three parameters, namely TTL_START, TTL_THRESHOLD, and TTL_INCREMENT as shown in Figure 2 :

```
/* Various constants used for the expanding ring search */
#define TTL_START     5
#define TTL_THRESHOLD 7
#define TTL_INCREMENT 2
```

Figure 2. ERS parameter

Default value of AODV of TTL_START with 5 hops, TTL_THRESHOLD with 7 hops and TTL_INCREMENT with 2 hops.

2.1. RREQ submission prior changes

Submission RREQ on AODV default first shipment has reached the five-layer or the network layer, it can be facilitate for attackers to immobilize MANET networks that have been built. The illustration shown in Figure 3.

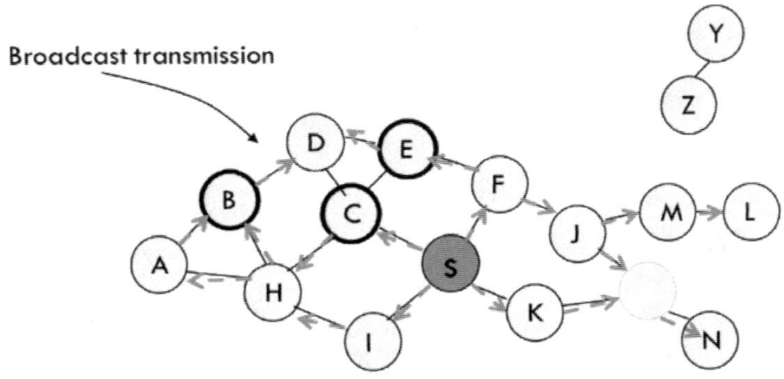

Figure 3. RREQ AODV default delivery process

Figure 3.3 shows that S node transmit RREQ. RREQ, as indicated by the red arrows, in one RREQ delivery will spread throughout the network. It is not effective when the destination node is in the area close by. The process is also used by third parties to immobilize network. Due to shipping RREQ RREQ once it has been flooded throughout the network.

2.2. Proposed RREQ submission

Submission of proposed TREQ is with the delivery layer by layer node, can be illustrated in Figure 4.

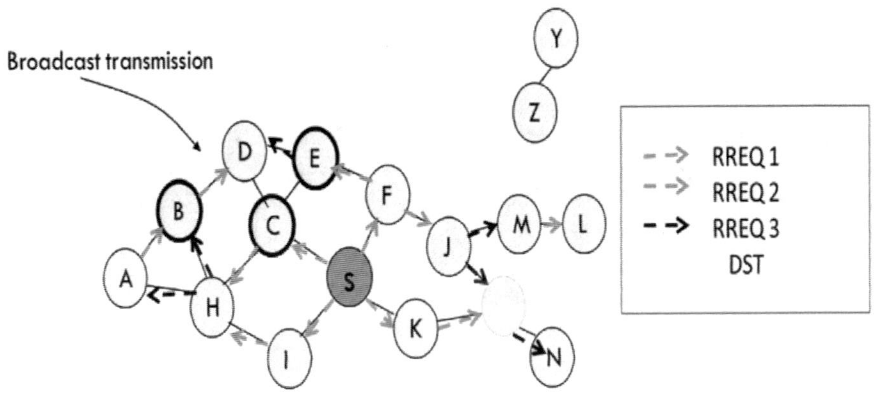

Figure 4. Proposed RREQ AODV delivery process As seen on Figure 3.4, RREQ is sent layer by layer to secure from flooding.

3. Results and Discussion

This chapter discuss the routing protocol AODV with changing to the ERS and also the condition of the network in case of flooding. The discussion includes trial details, details and implementation scenarios, test results and analysis.

3.1. Test scenario on static network

Parameter test to demonstrate the performance of a protocol in this test is throughput and average delay.

The test result indicates that the variation affect the results of the simulation time parameters of throughput and average delay of the data obtained. Value throughput and average delay is beginning to stabilize at simulation time of 500 seconds or above. This indicates that the next simulation is recommended using simulation time above 500 seconds. Next test is using parameter simulation time of 700 seconds.

At the time of 700 seconds with a variation simulation conducted ERS, ERS value with the value of the highest throughput parameter is the ERS ERS 2 and 6 with the same throughput on both the ERS is 258.87 Kbps, the value of ERS ERS 2 and 6 also have value parameter average delay is the lowest compare to the other ERS is 0.85 ms. With the value of detail ERS ERS 2 and 6 can be seen in Table 1.

Table. 1 Optimun ERS value from first test

ERS Parameter	*TTL start*	*TTL increment*	*TTL threshold*
ERS 4	1	1	6
ERS 6	1	2	16

3.2 *Optimum ERS test on mobile random network*

The next test is to get an optimum value of ERS on mobile random network with flooding network condition. The test will use wifi 802.11a standard.

For an optimum ERS testing on mobile random network, parameter used as a reference are Throughput, Packet Delivery Ratio, Delay and Average Normalized Routing Load. With ERS threshold values that varied from 1 to 12 and the value of the ERS AODV default.

Testing with 25 nodes in the network, mobile random and flood condition network with 4 node flooding, showed that the best value of ERS is ERS 3 with ERS TTL_start 3, TTL_increment by 1 and TTL_ threshold value is 3. Details of the results as follows:

1. Throughput parameter increases 7.49% compare to AODV ERS default.
2. Packet Delivery Ratio parameter increases 7.6% compare to AODV ERS default.
3. Average Delay paremeter decreases 4.17% compare to AODV ERS default.
4. Normalized Routing Load parameter increases 6.74% compare to AODV ERS default.

From the result of test, value obtained is ERS 3 as the optimal value compare to ERS default. With ERS value of 3, the next test will be the variation of the number of nodes.

3.3 *Optimum ERS testing on several condition*

From the previous test, the optimum value of ERS is ERS 3 with parameter of ERS TTL_start, TTL_increment by 1 and TTL_ threshold value of 3. From the ERS value, the test will conduct by varying the amount of network nodes compare to ERS value AODV default.

From the test results, with node variation will obtain Throughput and Packet Delivery Ratio parameter as seen on Figure 5.

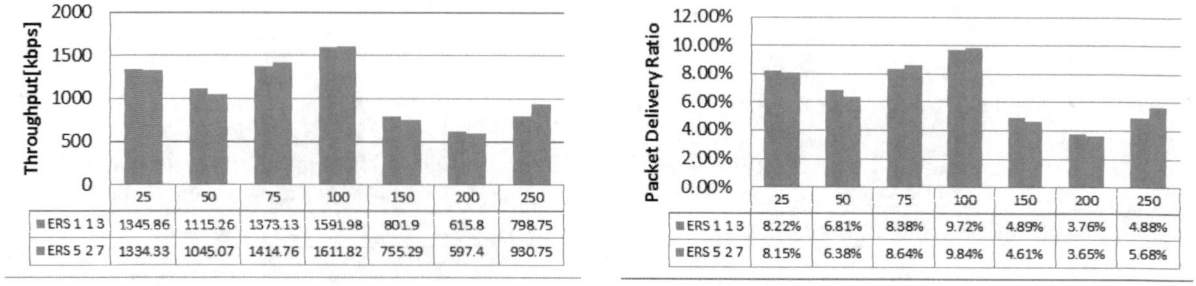

Figure 5. Throughput and Packet Delivery Ratio parameter test result

Test result with Average Delay and Normalized Routing Load parameter test can be seen on Figure 6.

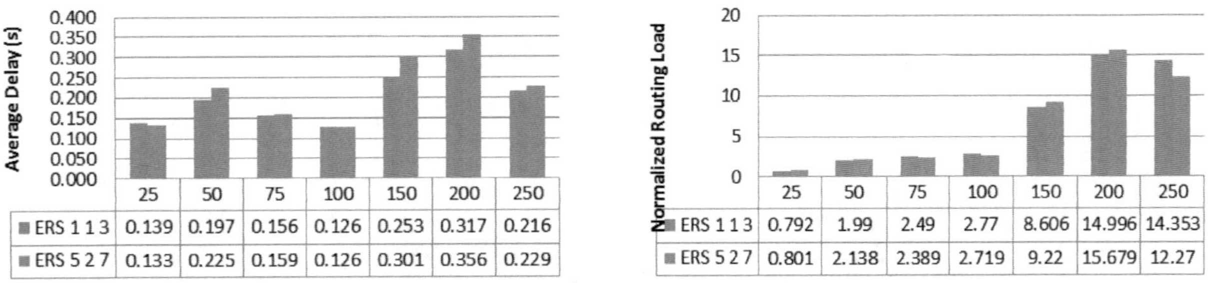

Figure 6. Average Delay and Normalized Routing Load parameter test result

T-Test results conducted on all data results indicate that the critical value of t> t Stat which means Ho is accepted and H1 is rejected. It can be concluded that changes on the ERS 3 and ERS default AODV does not provide changes to the parameter Throughput, Packet Delivery Ratio, Delay and Average Normalized Routing Load.

Testing with a variety of nodes on optimal ERS compare to ERS default AODV will result an optimum ERS can not applicable in general to the different number of nodes.

3.4 Optimum ERS searching on number of node variation

The test is then performed on searching optimum ERS with node variation. Node variations tested are 25, 50, 75, 100, 125 and 150. From the test results, an optimum throughput parameters, Packet Delivery Ratio and Normalized Routing Load shown in Table 2.

Table. 2 Optimum test result from Throughput, Packet Delivery Ratio and Normalized Routing Load parameter

Number of nodes	Throughput [kbps]	Packet Delivery Ratio	Normalized Routing	TTL Star	TTL Increment	TTL Threshold
25	1222,79	7,47%	0,846	1	1	3
50	1416,73	8,65%	1,586	1	1	3
75	928,72	5,67%	3,673	1	1	5
100	1143,02	6,97%	3,891	1	1	6
125	1252,27	7,64%	4,649	1	1	7
150	851,97	5,20%	8,042	1	1	8

From the results of the parameters in Table 2, a pattern of increasing the value of ERS on the number of nodes on the network shown on Figure 7.

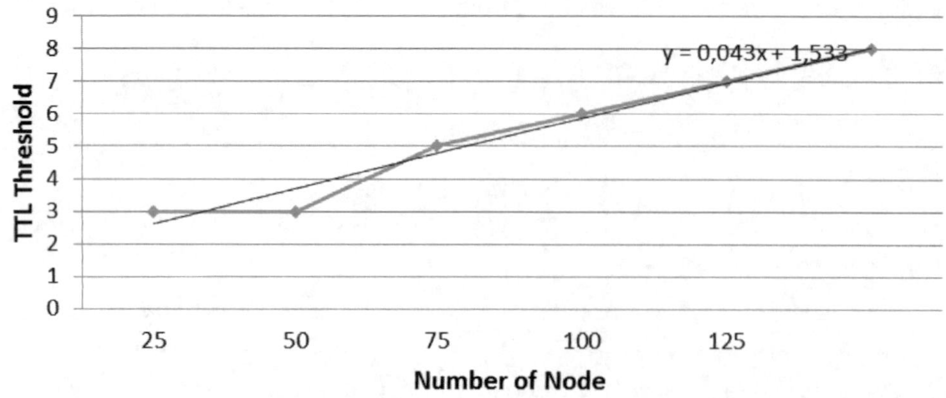

Figure 7. Relation between number of node with ERS parameter (TTL Threshold)

Relation between TTL Threshold and the number of nodes in the network will obtained an equation:

$$TTL_{Threshold} = (0{,}043 \text{ x number of node}) + 1{,}533 \quad (1)$$

4. Conclusions

From the research that has been done on ERS optimization on flooding attack condition, some conclusions can be drawn based on the research objectives and the results achieved:

By applying an optimum ERS (TTL threshold) in AODV protocol, the protocol has increased within the parameters of throughput, packet delivery ratio and Normalized Routing Load, compare to the AODV protocol default with one data sender and one data receiver. Throughput parameter increased by 5.67%, packet delivery ratio parameter increased by 5.73%, and Normalized Routing Load decreased by 4.66% with one data sender and one receiver data. An optimum ESR value in some number of node conditions with single data sender and single data recipient, relation between ERS on TTL threshold parameter with the number of nodes shown as follow equation:

$$TTL_{Threshold} = (0{,}043 \text{ x number of node}) + 1{,}533$$

References

[1] Gorantala k 2006 *Routing Protocols in Mobile Ad-hoc Networks* (Sweden: Ume˚a University)
[2] X. Zhang, T. Kunz, L. Li, and O. Yang 2010 *CNSR 2010 - Proc. 8th Annu. Conf. Commun. Networks Serv. Res.*, 1712–1715 (Canada: IEEE Publications)
[3] C Science and S Engineering 2012 *IJARCSSE* **2** 451–455
[4] B Krupa, A Talwar 2014 *Int. J. Innov. Res. Comput. Commun. Eng.* **2** 4541–4545
[5] I S Hendrawan 2010 *Jurnal Penelitian dan Pengembangan Telekomunikasi* **15** 22–29
[6] I Khalil and I Johannes 2012 *Proceedings of International Conference on Information Technology Department of Electrical Engineering* (Yogyakarta, Indonesia: IEEE Publications)
[7] R H Khan, K M Imtiaz-ud-Din, A A Faruq, A R M Kamal and A Mottalib 2008 *Electr. Comput. Eng.*

[8] B C K B C Kim, H S L H S Lee, and J S M J S Ma 2005 *VTC-2005-Fall. 2005 IEEE 62nd Veh. Technol. Conf.*
[9] P. Rathod, N. Mody, D. Gada, R. Gogri, and Z. Dedhia 2004 *Distributed Computing - IWDC 2004, 6th International Workshop*
[10] D. S. Karaulia and N. Bharot 2014 *Proc. - 2014 4th Int. Conf. Commun. Syst. Netw. Technol. CSNT 2014*

IOP Conf. Series: Materials Science and Engineering **333** (2018) 012109 doi:10.1088/1757-899X/333/1/012109

Sign Language Translator Application Using OpenCV

L Triyono[1], E H Pratisto[2], S A T Bawono[2], F A Purnomo[2], Y Yudhanto[2] and B Raharjo[2]

[1]Informatics Department, Politeknik Negeri Semarang
Jl. Prof. Sudharto, Tembalang, Semarang, Jawa Tengah 50275, INDONESIA
[2]Informatics Engineering Department, Sebelas Maret University
Jl. Ir. Sutami 36A Kentingan Jebres Surakarta 57126, INDONESIA

E-mail: sahirul@mipa.uns.ac.id

Abstract. This research focuses on the development of sign language translator application using OpenCV Android based, this application is based on the difference in color. The author also utilizes Support Machine Learning to predict the label. Results of the research showed that the coordinates of the fingertip search methods can be used to recognize a hand gesture to the conditions contained open arms while to figure gesture with the hand clenched using search methods Hu Moments value. Fingertip methods more resilient in gesture recognition with a higher success rate is 95% on the distance variation is 35 cm and 55 cm and variations of light intensity of approximately 90 lux and 100 lux and light green background plain condition compared with the Hu Moments method with the same parameters and the percentage of success of 40%. While the background of outdoor environment applications still can not be used with a success rate of only 6 managed and the rest failed.

1. Introduction

Sign language is mostly only dislocated by the disable and few people who understand the sign language that usually lives associated with disables such as families, activists, and teachers of *Sekolah Luar Biasa* (SLB). Sign language is divided into two namely natural gestures and formal cues[1]. The natural cue is a cue used by a deaf person (as opposed to body language), a manual (hand-handed) expression agreed between the user (conventionally), known to be limited in a particular group (esoteric), and a substitute for words. A formal gesture is a cue that is intentionally developed and has the same language structure as the spoken language of the community. Various forms of formal sign language developed, among other things, sign language called Sign English or also called Pidgin Sign English (PSE) which is a combination or mixture of original / natural sign language with English, standard sign language American Sign Language (ASL) to explaining the words and concepts[2].

1.1. American Sign Language

American Sign Language (ASL) is one of the most widely used sign language in the world with the method of fingerspelling as a representation of the alphabet on cues[3].

The development of sign language application has been developed by several researchers, including Nuriyanti[4] has made basic sign language introduction application based on Android. The developed application is limited to text recognition and can not translate gesture sign language. Ismawati[5] developed the application of communication aids tool with Deaf through webcam on computer. The

IOP Conf. Series: Materials Science and Engineering **333** (2018) 012109 doi:10.1088/1757-899X/333/1/012109

application will translate the movements of a finger that is demonstrated to be translated into an alphabet letter. This research will develop a sign language interpreter application on a mobile device to make it easier for everyone to communicate with the Deaf. However, some hand gestures (alphabets) such as K, M, Q, U, X, H, P, S, T have not been translatable because they have a more dominant range of other approaches to the HuMoments alphabet.

1.2. Fingerspelling

Fingerspelling describes the alphabet manually. Hand positions show each letter of the Latin alphabet. Fingerspelling is usually used as a complementary sign language. If there is no sign language for a single word, then fingerspelling is used. Fingerspelling is usually also used to mention names appropriately or when people are unsure of sign language for a particular word[6] .

2. Experimental

An overview of the application's design and implementation as seen at Figure 1.

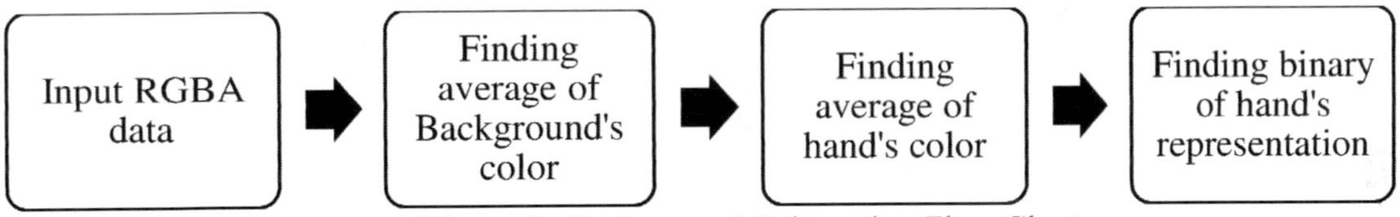

Figure 1. Background Subtraction Flow Chart

2.1. Finding background color

At the beginning, the application will detect the background color on 7 sections coordinates that are marked with 7 boxes as shown on Figure 2. The average color will be stored in a 7 x 3 matrix array. The application starts with two pre-sampling steps, which collects the background color and the user's hand using the 7 small squares displayed on the screen. This color data is used to calculate the threshold for obtaining binary imagery from RGBA data input. For simplicity, adaptive methods are not used to find the best threshold. Next, the user is asked to put his hand close to the screen to cover 7 boxes so that the app can get hand color data.

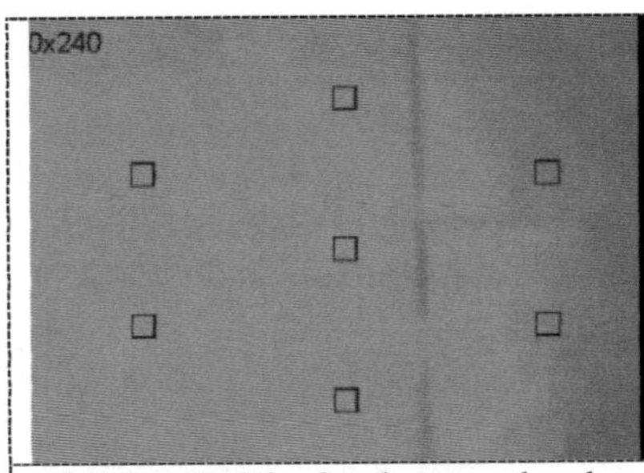

Figure 2. The background color sampling is stored in the array matrix

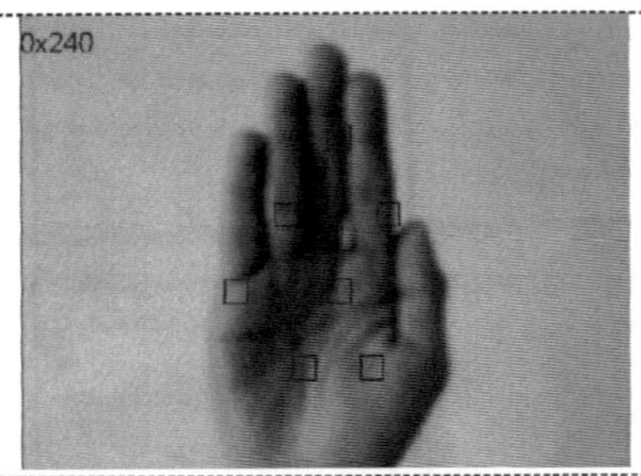

Figure 3. Hand color sampling that will be stored in array matrices

2.2. Finding hand color

The average color will be stored in a 7 x 3 matrix array like the image. After 7 color data for hand is obtained, 7 upper and lower borders for hand area are calculated, which can be represented as 2 dimensional array. The array's color boundaries are authorized by the author and can not be changed. Therefore, the actual segmentation performance depends on the choice of the array color limit. The original color of the frame ie RGBA will be converted to obtain the best color samples to detect the same hand in different light conditions.

2.3. Binary of the hand representation

After the limit of each 7 colors is obtained, the next 7 binary images of hand representation can be computed by performing on each binary image using the 'OR' logical operation. The same process is done in the background color to produce a binary image. The result of the binary image of the hand is then performed an operation logic 'AND' with binary image of the background. Furthermore, the blurring is done and the last process is by performing morphological operations (dilation and erosion) to remove noise (pixels that are not in want) to gain optimal result. After all process were done, the obtained representation of the hand as shown on Figure 4.

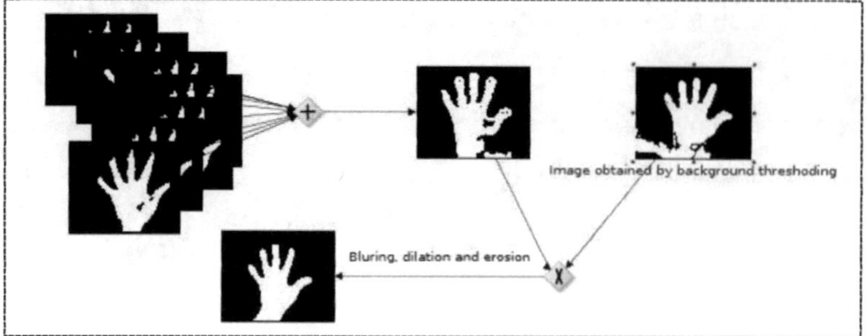

Figure 4. Hand representation in the form of binary logic operation
Source: http://eaglesky.github.io/images/posts/hand-gesture/thresholding-back.jpg

2.4. Finding contour and feature extraction

The process of Feature Extraction as seen at Figure 5. The result of hand representation on binary image will be used to search contours. Furthermore, the contour will be described next operation such as drawing bounding box, convex hull, and convexity defects. It will also be drawn a circle on the palm of the hand. This circle is useful for finding the midpoints that will be used as features along with the location of the finger to train and then in the prediction. Figure 6 shows an illustration of the results of the functionality found in OpenCV.

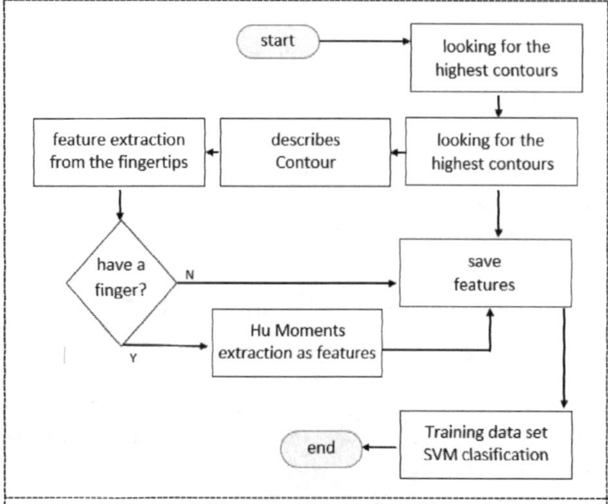

Figure 5. Feature Extraction Flowchart

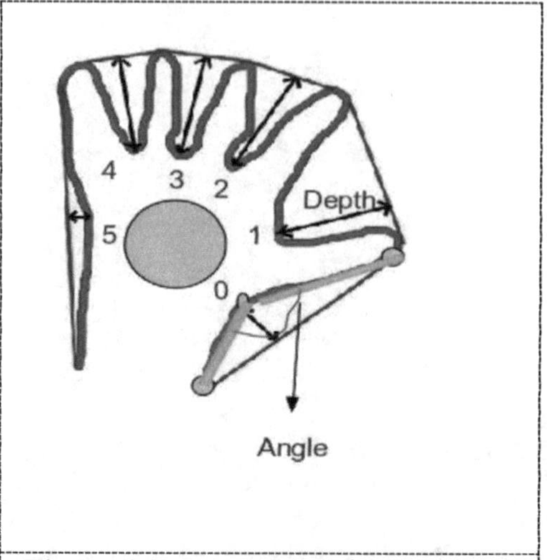

Figure 6. Features extraction on a hand
Source: http://eaglesky.github.io/images/posts/ hand-gesture/defects-elimination.jpg

Gray line represents contours, red line is convex hull. The location between the red lines of contours with convex hull is called convexity defects. The location of the fingertips is calculated using a defect point. Redundancy defect points will be eliminated by checking the depth and angle of the defect point. As illustrated in Figure 6, there will be 4 checks of conditions, ie Depth <= circleRadius * 0.7, Cosine (Angel) <= -0.7, close to the limit, the number of defects> 4. If there are unfulfilled consissions it will not be assume a hand or not processed.

The condition applies if a finger opens. However, when the hands are clenched or none of the fingers open then feature extraction is performed by another method by searching 7 Hu Moments values. The value of Hu Moments will be included in the SVM data model and subsequently used to predict the outcome.

2.4.1. *Data training.* After. The result of feature extraction ie fingertip coordinates or Hu Moments values will be entered on SVM data model with sequential label. The labels are sequential from 1 for A, 2 for B and so on. In doing the training then the sequence should not be reversed because it can be ascertained the prediction results will also be wrong.

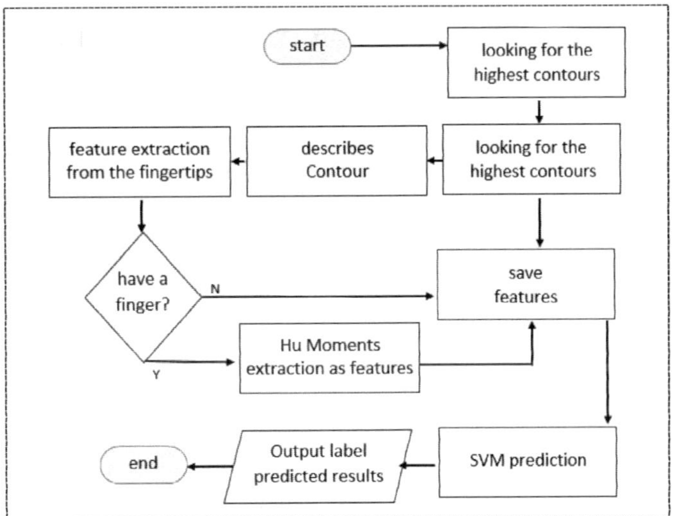

Figure 7. Test Mode Flowchart

The application test process as seen at Figure 7. The initial process in this section is the same as the previous explanation of contours search and feature extraction. Predicted results are only done in test mode. The result of feature extraction will be predicted by SVM based on the data set that has been stored during the training to produce sequence label then compared with alphabetical order and displayed on screen.

3. Result and Discussion

Data training is data generated from the training process and stored in the ReadMySign folder in the root of internal storage. This data will be used to predict the results of labels to display the results of hand gesture readings. Training data includes:

1. *Gesture image.* This data is an image of a saved gesture. Noteworthy of this data is that this image data is not used for application or programming purposes. Gesture image data is only a palm / fist with the naming of an Arabic number that is adjusted to the order of adding data. An image with the name 1 represents label 1 and in this study is considered the letter 'A' then label 2 as 'B' and so on.

Figure 8. Alphabet "A" with Hu Moments Features

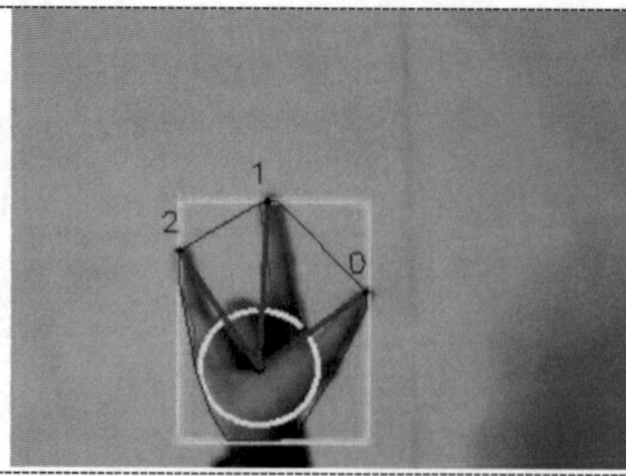

Figure 9. Alphabet " **I LOVE U** " with Fingers Features

2. *Features File.* This data is text that is named as train_data.txt which contains features of each gesture. Features for Hu Moments method there are 7 values while for the fingertip search method varies depending on the number of open fingers. Each gesture is labeled 1 to n and each gesture maintains the same 10 units of labels but contains different values but with the same amount for each gesture. Ten units of the label are taken from 10 frames during the training, each frame contains 1 unit of feature. The value of features with the fingertip search method numbered vary, according to the number of open fingers when there are 2 fingers open then there are 4 value features, if there are 3 fingers open then aka tone 6 value features. The value is obtained from the coordinates of each finger (x and y), each calculated first through the division with the radius of the circle in order to obtain the desired result.
3. *SVM Model File.* This file in the application is named 'model'. The model file is a generated file from the train_data.txt file performed by SVM. This file contains a training data model that SVM will use to predict the label.

Table 1. Test Condition

	Object distance from camera (cm)	Light intensity (lux)	Background color	Marker color
1	35	950 - 1050	light green	light green
2	35	105 - 115	light green	light green
3	55	105 - 115	light green	light green
4	35	2435 - 2521	outdoor	light green
5 (using gloves)	35	334 - 354	outdoor	light green

Figure 10. Gesture "A" on Test 1

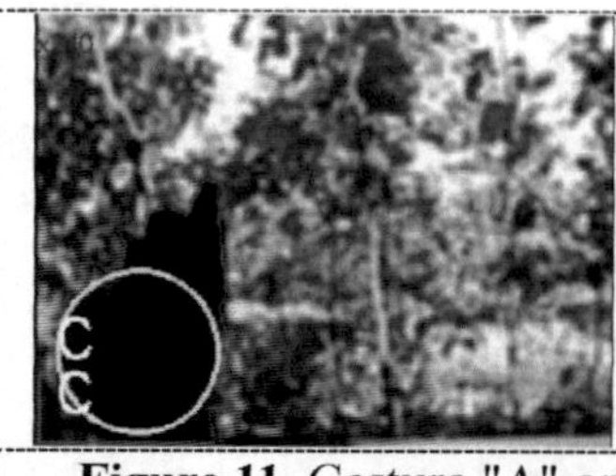

Figure 11. Gesture "A" on Test 4

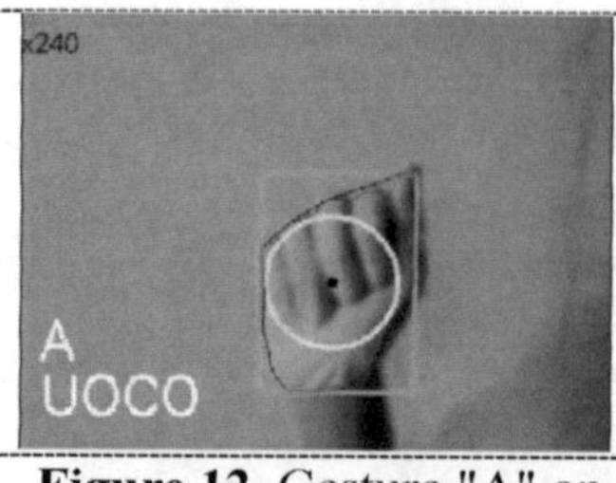

Figure 12. Gesture "A" on Test 5

Figure 13 shows that the search method of fingertip coordinates is better than the search for Hu Moments values. Almost all gestures that use the value of fingertip coordinates as features show 100% success except gesture letter H. This can happen because even in the training of the letter H using the coordinates of the fingertips but when the test is detected with Hu Moments and gesture the letter H is also very similar to gesture U with a little slope. Gestures of the letters E, M, N, and T have not been recognized because the three gestures are very similar to each other. Overall Hu Moments method is used for 11 gestures with 4 of them failing.

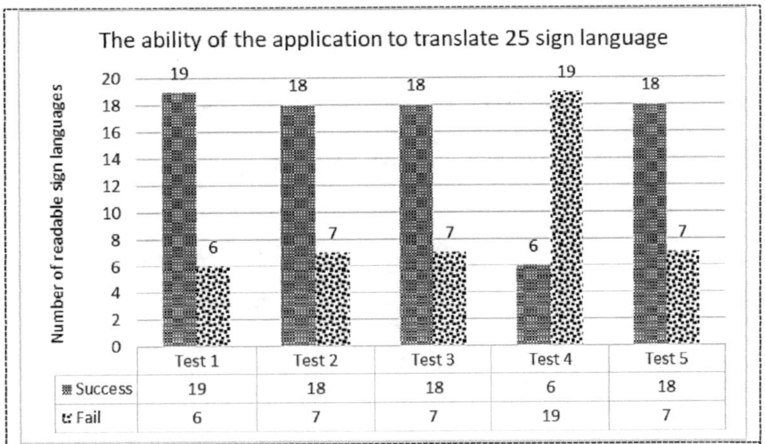

Figure 13. Test result

In the 4th experiment with a light intensity above 2000 lux and the background is a plantation environment, the application made poor hand gesture recognition and achieved little success. It happens because the background is too crowded with a color that could just resemble skin color, other than that the intensity of light that is too high coming from the front of the camera also gives a silhouette effect on the hand. In the experiments using hand shirts it appears that there is a difference that the application can read with both the gesture E and H. Compared to research by [5] this study can recognize the letters K, Q, U, X, H, P or 80% more efficient than the previously developed method of only 64%.

4. Conclusion

The method of color difference can be used to detect the color of the skin as a recognized object as a sign language. Some gestures still can not be translated because they have a high similarity to each other like M, N, T, and S; contour comparison motion with 7 Hu Moments values for hand sign language recognition has a small success rate of 50%. Applications can not translate gestures properly without a certain background. The solution for dynamic sign language on every word, further research can take advantage of Motion Templates on OpenCV. This method is very useful in detection motion in a video. The important feature of this technique is that motion can be detected even in small regions of a frame.

References

[1] McInnes J M and Treffry J A 1993 *Deaf-blind Infants and Children: A Developmental Guide* (Toronto : University of Toronto Press)
[2] Martin D S 2003 *Cognition, Education, and Deafness: Directions for Research and Instruction* (Washington : Gallaudet University Press)
[3] Williams P and Evans M 2013 *Social Work with People with Learning Difficulties* (California: SAGE Publications)
[4] Nuriyanti Y and Tresnawati D 2015 *Algoritma* **12** 1-9bvcgghfd
[5] Ismawati R D 2011 *Pembuatan Software Alat Bantu Komunikasi Penyandang Cacat Tuna Rungu-Tuna Wicara (Berbahasa Isyarat Tangan) Berbasis Webcam.* (EEPIS Repository: Politeknik Elektronika Negeri Surabaya)
[6] Smith M and Levack N 1996 *Teaching Students with Visual and Multiple Impairments: A Resource Guide* (Texas : Texas School for the Blind and Visually Impaired)

IOP Conf. Series: Materials Science and Engineering **333** (2018) 012110 doi:10.1088/1757-899X/333/1/012110

Latent Dirichlet Allocation (LDA) Model and kNN Algorithm to Classify Research Project Selection

M A Safi'ie[1,2], E Utami[1], H A Fatta[1]
[1] Department of Informatics Engineering, Postgraduate Programme, AMIKOM University, Yogyakarta
[2] Department of Informatics Engineering, Faculty of Mathematics and Natural Sciences, University of Sebelas Maret, Surakarta

E-mail: safiie99@mipa.uns.ac.id

Abstract. Universitas Sebelas Maret has a teaching staff more than 1500 people, and one of its tasks is to carry out research. In the other side, the funding support for research and service is limited, so there is need to be evaluated to determine the Research proposal submission and devotion on society (P2M). At the selection stage, research proposal documents are collected as unstructured data and the data stored is very large. To extract information contained in the documents therein required text mining technology. This technology applied to gain knowledge to the documents by automating the information extraction. In this articles we use *Latent Dirichlet Allocation* (LDA) to the documents as a model in feature extraction process, to get *terms* that represent its documents. Hereafter we use k-Nearest Neighbour (kNN) algorithm to classify the documents based on its terms.

1. Introduction

Proposal documents are collected at the selection are still in the category of unstructured data and very large, in order to extract the information contained its documents is required text mining technology. This technology can provide solutions such as processing, organizing/grouping and analyzing unstructured text in very large quantities. Text mining technology can be applied to these problems because the documets has characteristic unstructured text, the volume of data is stored very large. This technology can be done to gain knowledge of text by automating the extraction of information. The techniques commonly used in text mining technology are summarizing text, information extraction, text categorization, text groupings, text tracking, and etc. The proposed of this research is to get the performance of LDA model and kNN algorithm to classify proposal documents.

1.1 Text mining

Text Mining also known as *Intelligent Text Analysis*, *Text Data Mining* or *Knowledge Discovery in Text* (KDT), its generally refers to interesting extracting processes and non-trivial information and knowledge of unstructured text. In the Data Mining area the tools is designed to handle structured data, but in the Text Mining area, its can work with unstructured or semi-structured data sets. Text Mining is a new invention by computer from unknown information, automatically information extracted from different written sources[1]. The key is linking together to form new facts or new hypotheses to be explored further with conventional way experiments.

Text Mining is similar to Data Mining, which is about Knowledge Discovery, so in the process of its knowledge discovery has stages as in Data Mining as described in Figure 1. The stages are document collection, pre-processing, data transformation, pattern matching and evaluation. Text Mining also uses the same technology as Natural Language Processing (NLP) such as information extraction, topic tracking, summarization, categorization, clustering, concept linkage, information visualization and question answering.

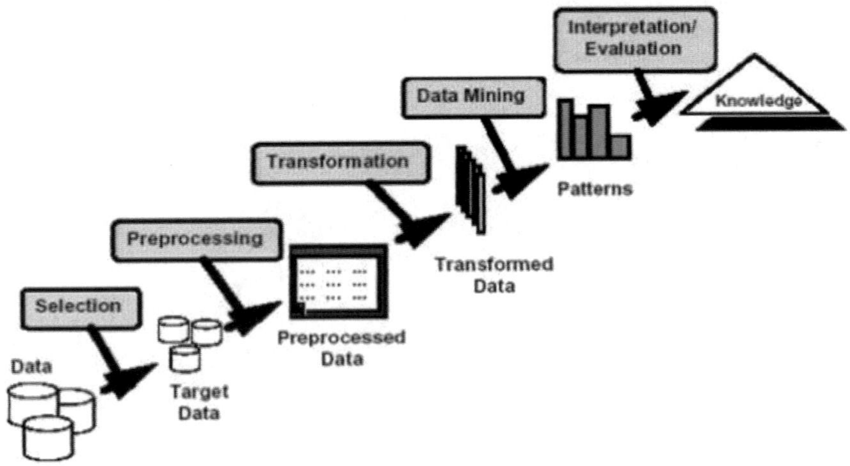

Figure 1. Knowledge discovery process[2]

1.2 Porter Stemmer Algorithm

Stemming is a process that provides the mapping of variations in word morphology into its basic word (stem)[3]. This process is also called merging, based on the assumption that terms with the same stem usually have the same meaning. Stemming Porter algorithm is a combination of words proposed by Martin Porter. The algorithm is based on the idea that English words are mostly made up of a combination of compact imbalances.

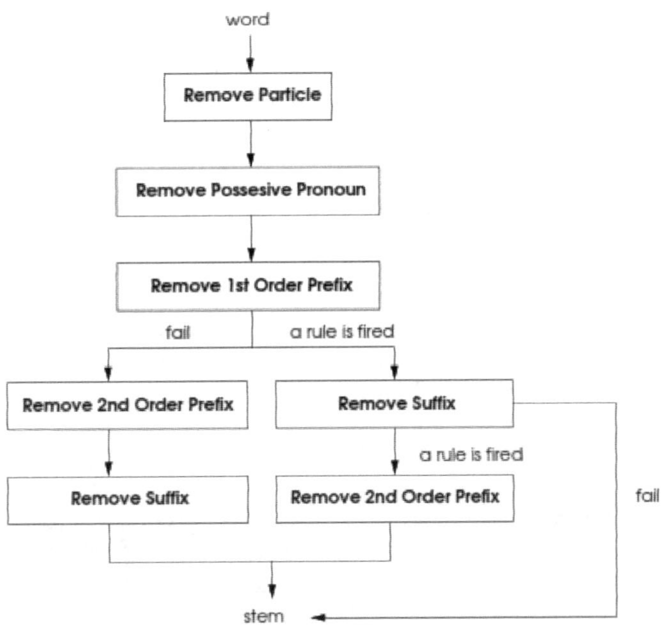

Figure 2. Stemmer Porter algorithm design[3]

1.3 Latent Dirichlet Allocation (LDA) Model

Latent Dirichlet Allocation is a probabilistic, generative model for searching for latent semantic topics in large data text collections[4]. Each topic found is characterized by a certain distribution of the words of the document. Each document is characterized as a random mix of topics that indicate proportion at a time in the document for each topic. The random mix of these topics is a short description, not only expressing the semantic content of a document but also a principle approach to describing the document quantitatively. The key in the LDA model is a premise that words contain strong semantic information from document text.

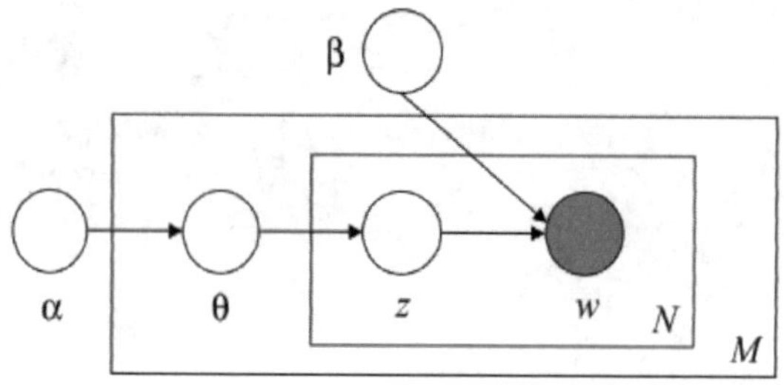

Figure 3. Graphical representation of LDA[4]

1.4 K-Nearest Neighbour (kNN) Algorithm

The kNN algorithm is done by searching for the k group of objects in the closest training data (similar) to the object in the new data/data testing. The kNN algorithm is a method for classifying objects based on learning data closest to the object. To define the distance between two points ie the point in the training data (x) and the point in the test data (y) then use the Euclidean formula, as shown in the equation:

$$D(x,y) = \sqrt{\sum_{k-1}^{n} (x_k - y_k)^2} \quad (1)$$

The kNN algorithm is done by searching for the k group of objects in the closest training data (similar) to the object in the new data / data testing. The kNN algorithm is a method for classifying objects based on learning data closest to the object. To define the distance between two points ie the point in the training data (x) and the point in the test data (y) then use the Euclidean formula, as shown in the equation:With D is the distance between the points in the training data x and the data testing point y to be classified, where x = x1, x2, ..., xi and y = y1, y2, ..., yi.

2. Experimental

The steps of this research, are as follows:

a. Data collection: manually collecting P2M document, the data collected from the system and has format * .pdf.
b. Convert pdf to text: the collected document data then parsed and converted to rawtext. In addition on converting documents, its also stores the attribute of the document into the database.
c. Text Cleanup: is the process of removing punctuation, symbols or character-character contained in the text and its not required to the next process.
d. Tokenizing: clean text data then changed to lowercase letters which subsequently breaks the text into its constituent words.
e. Stopwords Removal: the stage to remove words that do not represent text, stopwords are usually additional words. The removal of these words to obtain terms representing the document in other other words is called filtering.
f. Stemming: is the stage to change the words to its roots word, by removing affix (affix) either in the form of prefix, suffix, confix and particle.

g. Information Extraction: terms that represent the text document then using the LDA model is searched for probability, the document has the highest probability of representing a particular theme / topic.

The documents will be experimented with several test scenarios described in table 1 below:

Table 1. Testing scenario

No	Training Data	Testing Data	Description
1	100%	100%	The overall data will be data learning and testing
2	75%	25%	75% data will be used as data learning and 25% data will be used as data testing
3	50%	50%	50% data will be used as data learning and the remaining 50% used as data testing
4	25%	75%	25% data will be used as data learning and the remaining 50% is used as data testing.

3. Result and Discussion

The results of research testing scenario are shown in Table 2, total documents were used are 20 samples with 5 categories of topics. Based on the table 2, the highest values of classification were obtained at 20 sample documents but the value decreases with the lower number of data sample.

Table 2. Testing result

No	Data set	Result (%)
1	20 Dokumen	85,00
2	15 Dokumen	73,33
3	10 Dokumen	70,00
4	5 Dokumen	60,00

4. Conclusion

This study have following conclusions:

1. The high value of accuracy is obtained by the increasing use of documents as training data and the accuracy will decrease if less data training.
2. Higher accuracy is achieved when the number of documents topics in the LDA model are closer to the actual number of documents topics.

References

[1] Gupta V, Science L C, and Lehal G S 2009 *A Survey of Text Mining Techniques and Applications* **1(1)** 60–76
[2] Agarwal C, Charu 2015 *Data Mining: The Text Book* (New York : Springer International Publishing)
[3] Tala Fadillah Z 2003 *A Study of Stemming Effect on Information Retrieval* in Bahasa Indonesia
[4] Hu D J 2009 *Latent Dirichlet Allocation for Text, Images, and Music* (San Diego: University of California)

IOP Conf. Series: Materials Science and Engineering **333** (2018) 012111 doi:10.1088/1757-899X/333/1/012111

Priority Scale of Drainage Rehabilitation of Cilacap City

Jatmiko Rudiono[1]

[1] Magister Maintenance and Rehabilitation Infrastructure, Sebelas Maret University Surakarta, Jl. Ir. Sutami 36 A Solo 57126 Indonesia

E-mail: Jatmikorudiono@student.uns.ac.id

Abstract. Characteristics of physical condition of Cilacap City is relatively flat and low to sea level (approximately 6 m above sea level). In the event of a relatively heavy rainfall resulting in inundation at several locations. The problem of inundation is a serious problem if there is in a dense residential area or occurs in publicly-used infrastructure, such as roads and settlements. These problems require improved management of which include how to plan a sustainable urban drainage system and environmentally friendly. The development of Cilacap City is increasing rapidly, this causes drainage system based on the Drainage Masterplan Cilacap made in 2006 has not been able to accommodate rain water, so, it is necessary to evaluate the drainage masterplan for subsequent rehabilitation. Priority scale rehabilitation of the drainage sections as a guideline is an urgent need of rehabilitation in the next time period.

1. Introduction

Flood is a natural disaster that often occurs in Indonesia, especially in the rainy season. This incident hit almost all cities in Indonesia and repeated every year, but this problem has not been resolved, even tends to increase, both frequency, extent, depth and duration [1,2].

Therefore every development of the city should be followed by the evaluation and improvement of the drainage system as a whole, not only on the development site, but also the surrounding areas affected [3]. For example, the development of a settlement area in upstream of a drainage system, then drainage planning is not only done in the settlement area, but the downstream drainage system should also be evaluated and redesigned if necessary.

The drainage development masterplan needs to be revised to suit the development of the city. Since the last five years the growth of housing construction, office buildings, and shophouses is quite rapid, so a lot of open land that turned into buildings [4,5,6]. Complete drainage system improvement is necessary but drainage improvement at the same time is not possible. Therefore, priority scale is needed to determine which drainage sections need to be prioritized for rehabilitation [7].

2. Experimental

Analithycal Hierarchy Process (AHP) is a method used in the decision-making process of complex issues such as planning problems, alternative determination, prioritization, policy selection, resource allocation, needs determination, forecasting performance planning needs, optimization and conflict resolution. Saaty establishes a quantitative scale of 1 (one) to 9 (nine) to assess the comparative importance of an element to another. [8].

IOP Conf. Series: Materials Science and Engineering **333** (2018) 012111 doi:10.1088/1757-899X/333/1/012111

According to Wignyosukarto, the use of Analytical Hierarchy Process (AHP) method in urban drainage system has the following strengths [9] :

a. Structuring the problem systematically.
b. Designed to use ratios and intuition to choose the best alternative, in the event of flooding in urban / an area. The best alternative is the one that has the least disadvantage, and has the greatest advantage.
c. Match the factors that make decisions gradually from the general to the specific

The reason for choosing the AHP method, according to Marimin, is that AHP has many advantages in explaining the decision making process :

a. Determination of the most dominant criteria greatly affects the outcome.
b. The end result can be graphically depicted, so it is easy to be understood by all parties involved in decision making.
c. A complex decision process can be decomposed into smaller decisions. [10] .

3. Results and Discussion

Cilacap City area is relatively flat field typology and there are many river. The relatively flat surface of land makes the drainage slope limited, so it is very possible that the flow becomes substandard. While the proximity of areas of the city with a coastline allows the back water (reverse flow) on the rivers and drainage canals during high tides. Generally Cilacap City drainage system is shown in Figure 1.Cilacap City Drainage System outline served with drainage that flow by gravity, divided into 17 Main Drainage (Major and Sub Maj Drain) as shown in Table 1. In preparing the proposed priority criteria used are Condition Values, Debit Plan and Rehabilitation Costs.

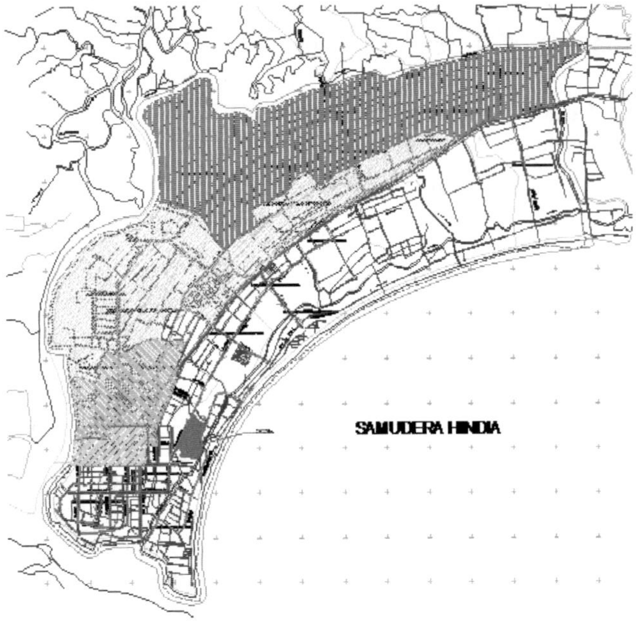

Figure 1. Drainage System of Cilacap City.

Table 1. Main Drainage (Major and Sub-Major Drain)

Number	The name of river segment
1	Kali Sabuk
2	Kali Yasa
3	Kali Menganti
4	Kali Irigasi
5	Kali Ciglagah
6	Kali Tanjung
7	Kali Karangwaru
8	Kali Cinyemeh
9	Kali Sentul
10	Kali Kodok
11	Kali Gubed
12	Kali Donan
13	Kali Watu
14	Kali Beji
15	Kali Sendangsari
16	Kali Cidapur
17	Kali Sendang

3.1. Hierarchical structure

The diagram in Figure 2 presents the decision to select the priority of rehabilitation of the drainage network, while the criteria for making the decision are the value of condition, debit plan, rehabilitation cost.

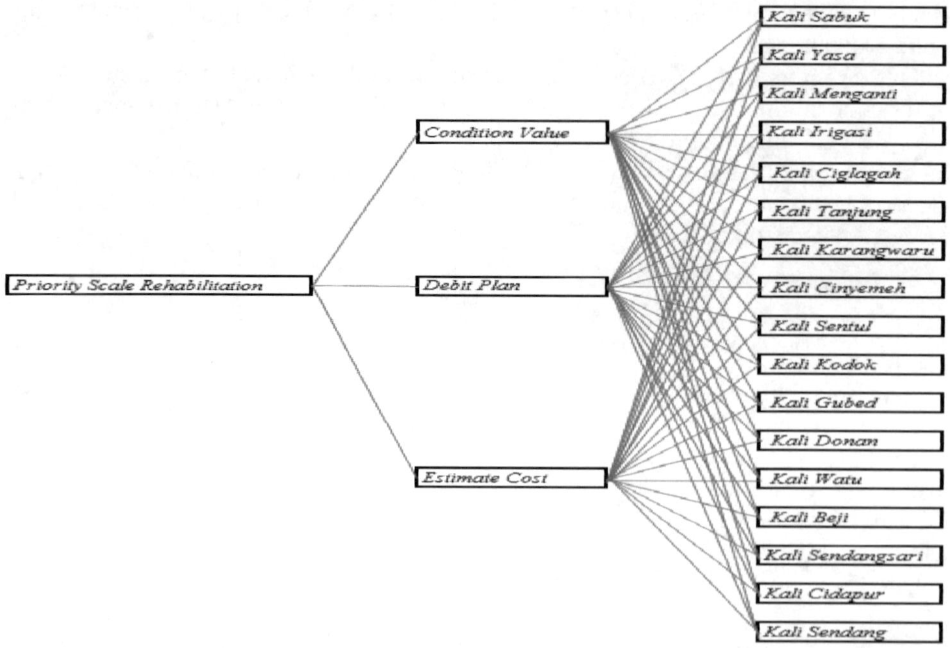

Figure 2. Hierarchy Structure Diagram

The alternative in making the decision is the main drainage section Kali Sabuk, Kali Yasa, Kali Menganti, Kali Irigasi, Kali Ciglagah, Kali Tanjung, Kali Karangwaru, Kali Cinyemeh, Kali Sentul, Kali Kodok, Kali Gubed, Kali Donan, Kali Watu, Kali Beji, Kali Sendangsari, Kali Cidapur dan Kali Sendang.

3.2. Assessment of criteria by filling in comparative data between criteria

Table 2 shows the comparison between criteria

Table 2. Comparison Between Criteria

	Condition value	Debit plan	Rehabilitation costs
Condition value	1	7	9
Debit plan	1/7	1	5
Rehabilitation costs	1/9	1/5	1

In the comparison of criteria can be explained as follows: Condition value is very important than the debit plan; Absolute Condition Value is more important than the cost of rehabilitation; Debit Plan is more important than the cost of rehabilitation

3.3. The weight data of each criteria on each alternative (river segment).

The obtained data are in the form of value/ weight of the degree of importance between the criteria and sub criteria that affect the decision making of the priority of drainage maintenance and rehabilitation. The weight result according to the Condition Values in Table 3., The weighted results according to Plan Debit in Table 5., and Weight Results According to Rehabilitation Cost in Table 7.

Table 3. The Assessment of Alternative Weight of Condition Value

River	Condition Value (%)	Weight assesment	River	Condition Value (%)	Weight assesment
Kali Sabuk	62	7	Kali Kodok	77	8
Kali Yasa	75	8	Kali Gubed	74	8
Kali Menganti	53	6	Kali Donan	68	7
Kali Irigasi	67	7	Kali Watu	67	7
Kali Ciglagah	54	6	Kali Beji	56	6
Kali Tanjung	51	6	Kali Sendangsari	66	7
Kali Karangwaru	55	6	Kali Cidapur	57	6
Kali Cinyemeh	66	7	Kali Sendang	62	7
Kali Sentul	68	7			

Table 4. Weight scale of condition value

Condition Value (%)	Weight Scale	Condition Value (%)	Weight Scale
0-10	1	51-60	6
11-20	2	61-70	7
21-30	3	71-80	8
31-40	4	81-90	9
41-50	5	91-100	10

Table 5. The Assessment of Alternative Weight of Debit Plan

River	Debit Plan (m3/s)	Weight assesment	River	Debit Plan (m3/s)	Weight assesment
Kali Sabuk	44	9	Kali Kodok	37	8
Kali Yasa	44	9	Kali Gubed	41	9
Kali Menganti	43	9	Kali Donan	40	9
Kali Irigasi	39	8	Kali Watu	33	7
Kali Ciglagah	35	8	Kali Beji	31	7
Kali Tanjung	34	7	Kali Sendangsari	36	8
Kali Karangwaru	32	7	Kali Cidapur	39	8
Kali Cinyemeh	35	7	Kali Sendang	42	9
Kali Sentul	38	8			

Table 6. Weight scale of Debit Plan

Debit Plan (m3/s)	Weight Scale	Debit Plan (m3/s)	Weight Scale
0-5	1	26-30	6
6-10	2	31-35	7
11-15	3	36-40	8
16-20	4	41-45	9
21-25	5	46-50	10

Table 7. The Assessment of Alternative Weight of Estimate Cost

River	Estimate Cost in Billion	Weight assesment	River	Estimate Cost in Billion	Weight assesment
Kali Sabuk	88	2	Kali Kodok	95	1
Kali Yasa	98	1	Kali Gubed	92	1
Kali Menganti	68	3	Kali Donan	82	2
Kali Irigasi	49	6	Kali Watu	41	6
Kali Ciglagah	45	6	Kali Beji	56	5
Kali Tanjung	57	5	Kali Sendangsari	59	5
Kali Karangwaru	55	5	Kali Cidapur	57	5
Kali Cinyemeh	53	5	Kali Sendang	45	6
Kali Sentul	88	2			

Table 8. Weight scale of Estimate Cost

Estimate cost in billion	Weight Scale	Estimate cost in billion	Weight Scale
0-10	10	51-60	5
11-20	9	61-70	4
21-30	8	71-80	3
31-40	7	81-90	2
41-50	6	91-100	1

Priority Scale with AHP method indicates that the lowest priority scale value is 5.667 in Sentul and Kali Kodok, meaning First Priority of rehabilitation of Cilacap City drainage system is done in Sentul and Kali Kodok, second priority in Kali Sabuk, Kali Yasa, Kali Menganti, Kali Tanjung, Kali Karangwaru, Kali Gubed, Kali Donan, Kali Beji. Third Priority in Cinyemeh River. Fourth Priority in Ciglagah River, Kali Watu, Sendangsari River. Fifth Priority in Kali Irrigasi. The Sixth Priority in Kali Sendang. Shown in Table 9.

Table 9. Results of Priority Scale Determination with AHP Method

No.	River	Weight Criteria			Result	Priority Scale
		Condition Value	Debit Plan	Rehabilitation Cost		
1	Kali Sabuk	7	9	2	6.000	2
2	Kali Yasa	8	9	1	6.000	2
3	Kali Menganti	6	9	3	6.000	2
4	Kali Irigasi	7	8	6	7.000	5
5	Kali Ciglagah	6	8	6	6.667	4
6	Kali Tanjung	6	7	5	6.000	2
7	Kali Karangwaru	6	7	5	6.000	2
8	Kali Cinyemeh	7	7	5	6.333	3
9	Kali Sentul	7	8	2	5.667	1

10	Kali Kodok	8	8	1	5.667	1
11	Kali Gubed	8	9	1	6.000	2
12	Kali Donan	7	9	2	6.000	2
13	Kali Watu	7	7	6	6.667	4
14	Kali Beji	6	7	5	6.000	2
15	Kali Sendangsari	7	8	5	6.667	4
16	Kali Cidapur	6	8	5	6.333	3
17	Kali Sendang	7	9	6	7.333	6

4. Conclusion

Based on the survey and analysis of drainage conditions in Cilacap overall is good, although drainage rehabilitation must be done in several places to cope with flooding. This can be seen from the percentage of drainage condition weight, that is Sentul 68% and 77% Kali Kodok. Rehabilitation of damaged drainage network should be implemented gradually based on priority scale sequence with priority order, first priority of rehabilitation is done in Sentul and Kodok. This is based on the high weight of Condition Values, Debit Plans and Rehabilitation Costs.

References

[1] Kleidorfer M 2013 *J. Procedia Engineerin. Elsevier.*
[2] Tingsanchali T 2011. *J. Procedia Engineering. Elsevier.*
[3] Suripin 2004 *Sistem Drainase Perkotaan yang Berkelanjutan* (Jogyakarta: Andi)
[4] Kodoatie R 2003 *Manajemen dan Rekayasa Infrastruktur* (Jogyakarta: Pustaka Pelajar)
[5] BAPPEDA Kab.Cilacap 2012 *Master Plan dan DED Sistem Drainase Kota Cilacap*
[6] Dinas Cipta Karya dan Tata Ruang Kab.Cilacap 2015 *KAK Kajian Penanganan Banjir Kota Cilacap*
[7] Chow J f 2013 *J. Procedia Engineering. Elsevier.*
[8] Saaty T L 1980 *The Analytic Hierarchy Process* (New York: McGraw-Hill)
[9] Wignyosukarto B 2001 *Malang J. App. Himpunan Ahli Teknik Hidraulik Indonesia (HATHI) Kongres VII dan PIT VIII*
[10] Marimin 2004 *Pengambilan Keputusan Kriteria Majemuk* (Jakarta: Grasindo)